DISCARD

NUMERICAL ANALYSIS AND COMPUTATION THEORY AND PRACTICE

E. K. BLUM
University of Southern California

ADDISON-WESLEY PUBLISHING COMPANY
Reading, Massachusetts • Menlo Park, California • London • Don Mills, Ontario

This book is in the
ADDISON-WESLEY SERIES IN MATHEMATICS

Richard S. Varga
Consulting Editor

Copyright © 1972 by Addison-Wesley Publishing Company, Inc. Philippines copyright 1972 by Addison-Wesley Publishing Company, Inc.

All rights reserved. No part of this publication may be reproduced, stored in a retrieval system, or transmitted, in any form or by any means, electronic, mechanical, photocopying, recording, or otherwise, without the prior written permission of the publisher. Printed in the United States of America. Published simultaneously in Canada. Library of Congress Catalog Card No. 79–150574.

PREFACE

Numerical analysis, in essence, is a branch of mathematics which deals with the numerical—and therefore constructive—solution of problems formulated and studied in other branches of mathematics. To illustrate, consider the (linear) algebraic problem of solving the equation $Ax = b$, where A is a real $m \times n$ matrix and b is an m-dimensional vector. In linear algebra, the problem is disposed of by proving the theorem that a solution exists if and only if the rank of the augmented matrix equals the rank of A. Similarly, the matrix eigenvalue problem $Ax = \lambda x$ is treated in theorems concerning diagonalizable matrices and culminating in the Jordan normal form theorem. Nowhere in the theory is there found any clue as to how one should proceed to solve for x and λ numerically when specific numerical instances of A and b are given. The specification and analysis of such procedures is the province of numerical analysis.

Another example is afforded by the theory of polynomial approximation of a real function f of a real variable x. In classical analysis, the Weierstrass theorem disposes of the problem for continuous f. The more recent theory of Chebyshev approximation sheds further light. This theory is not concerned, in general, with the numerical determination of the best approximating polynomial, although mathematicians have paid some attention to constructive existence proofs, which then suggest computational algorithms. In the main, however, the subject matter of numerical analysis is the analysis of algorithms for the numerical determination of a low-degree polynomial approximation to a specific function f.

A final example is the theory of differential equations, ordinary and partial. Much of the theory is concerned with qualitative properties of solutions of differential equations. Existence theorems are often nonconstructive, and the constructive ones, such as the Picard theorem, generally do not lead to practical numerical algorithms. Once again, this is where numerical analysis begins.

In each of these examples, there is no sharp boundary line separating numerical considerations from the rest of the theory. The numerical analyst—if he is to understand the power and limitations of existing numerical methods, and especially if he is to be able to devise new methods—must be familiar with, if not expert in, many of the non-numerical parts of these theories. One of the aims of this book is to present background material in the theories underlying each of several numerical methods in branches of mathematics with which numerical analysis has made its principal contacts. This is one meaning of the word "theory" in the title. At the same time, numerical analysis has a theoretical aspect of its own. This pertains mainly to the analysis of error. When we consider a numerical method, we shall analyze the error in the numerical solution and prove theorems concerning convergence and error bounds. A general basis for such analyses can be found in modern functional analysis. The first three chapters provide an introduction to that subject, as well as a summary of the main concepts and results of real analysis and algebra which concern us most.

Functional analysis permits us to deal with measures of error in a unified manner. Thus many aspects of one-dimensional and n-dimensional problems are subsumed in a single theory, as for example in Newton's method for solving $f(x) = 0$. By using a functional analysis approach, we treat the scalar case and the vector case $f = (f_1, \ldots, f_n)$ and $x = (x_1, \ldots, x_n)$ in a single theory. But more important, functional analysis—true to its name—allows us to consider the infinite-dimensional case, as for example, in solving the differential equation $x' - f(x) = 0$, where the solution is a function $x(t)$, or the "control problem" $x' - f(x, u) = 0$, where both x and u are functions in some infinite-dimensional space. Modern numerical analysis is more and more concerned with such problems. Of course, a numerical method must succeed only by reducing these infinite-dimensional continuous problems to a sequence of discrete and ultimately finite problems. Nevertheless, or perhaps even more so because of this requirement, the methods of functional analysis provide a convenient and unifying approach to the numerical solution of such problems. These considerations are closely related to one of the fundamental, and perhaps most fascinating, aspects of numerical analysis: Numerical analysis provides a bridge between the mathematics based on the real continuum and discrete mathematics, that is, between mathematics based on the real number system and mathematics based on the integers.

Classical mathematical analysis—and most of its modern abstract extensions—rests ultimately on the real numbers. The idea of the real number continuum (i.e., the real line) has played a dominant role in much of pure and applied mathematics. For example, the ordinary and partial differential equations which arise in the physical sciences are based on the real continuum. Its influence extends beyond the boundaries of analysis to other branches of mathematics, such as geometry, topology, and algebra. Yet the intuitive basis of the continuum has always been open to question, whereas the intuitive basis of the integers has, by and large, been free of doubt. This led nineteenth-century mathematicians (Cauchy, Weierstrass, Dedekind, and Cantor) to attempt to "arithmetize" the real number system by establishing its foundation in the integers. At the turn of the century, the achievements of these mathematicians in this endeavor were generally regarded as entirely successful. In 1900, Poincaré made his oft-quoted assertion, "Today there remains in analysis only integers and finite or infinite systems of integers." The equally famous remark of Kronecker, "God made the integers; all the rest is the work of man," was made in 1886. The supposedly rigorous foundation of the real continuum required a "theory of sets," and this was developed by Cantor. (A rational number is an equivalence class of pairs of integers. A real number is a "cut" separating two sets of rationals.)

Unfortunately, at about the time the continuum was placed on a supposedly firm foundation, certain paradoxes were discovered in the theory of sets which cast doubt on the entire development. Although these paradoxes have been eliminated in the modern theory of sets, one cannot yet say that others will not arise. Furthermore, there are other objections which, although of a logical nature, are also of practical importance. In a certain strict technical sense, the continuum is not a constructively defined set. In fact, only a denumerable set of real numbers is computable by a machine (e.g., by a Turing machine). Also, many of the theorems of real analysis are proved in a nonconstructive manner, e.g., the existence of a point at which a continuous function on a closed bounded interval assumes its least upper bound is proved by methods which do not tell us how to construct the point. This has led to a recent redevelopment of analysis by constructive techniques.

However, analysis based on the continuum (nonconstructive) continues to thrive. In this book, we accept the classical real continuum and the nonconstructive theory of real and complex functions and anchor one end of the numerical analysis bridge to it. Strictly speaking, the other end should be firmly fixed in the integers and the theory of recursive functions, since the final results of numerical analysis should be effective (in the Turing sense) algorithms for digital computers. The practice and theory of numerical analysis is concerned — in large part — with bridging the gap between nonconstructive real analysis and constructive discrete mathematics. Fortunately, we need not bridge the entire gap to provide a useful body of knowledge. We do not have space to enter into a discussion of the technical questions concerning constructivity in mathematics. See references [0.1], [0.2], [0.3], [0.4], for example. (Numbers enclosed in brackets allude to the references at the end of each chapter.)

For purposes of numerical analysis, our position in this regard is essentially that of most non-intuitionistic analysts; namely, as we have said, we accept the nonconstructive real number continuum and the results of real and complex analysis based on it. Then, for the purposes of numerical analysis, we are obliged to translate these results into computationally meaningful ones, and this means, ultimately, into practical computer algorithms. Yet we shall not always pursue a subject to this ultimate end. We shall begin a subject by considering its nonconstructive continuum setting and proceed to obtain "constructive" results, usually in the form of convergent sequences still within the real continuum framework, analyzing "rates of convergence" and "stability with respect to perturbations." In many cases, we shall proceed further and obtain an algorithmic form of a "constructive" method. This will involve finite (integer) representations of all quantities. The connection with the nonconstructive continuum will evidence itself in the form of "rounding errors." This is best exemplified by matrix computations (Chapter 4). However, our study of rounding errors is carried out within the framework of the real continuum (nonconstructive). Thus we have not based this end of our subject firmly on the integers.

When we consider the practical side of our subject, we journey into a realm where art and ingenuity must play a role. It is impossible to cover all the practical material that one might wish. Indeed, in our opinion, there is no substitute for experience as a teacher. Therefore, we have included a number of exercises requiring actual computation. We pay particular attention to procedures suitable for modern digital computers and we interpret theoretical results in terms of computable quantities wherever possible. We also indicate where theory ends and the art and practice (ingenuity and common sense) of numerical computation begins. In the final analysis, theory and practice are forever entwined one with the other. Each serves the other and contributes to its growth and development. This is true in all the sciences, indeed in life itself. If it seems truer in numerical analysis than in some other parts of mathematics, this is only because the interaction is more pronounced in this age of the computer and abstract mathematics.

The modern electronic computer—distinguished by its stored-program feature—is perhaps two decades old. Along with the recent development of the computer as a physical machine, there has grown a body of knowledge concerning the computer as a conceptual or ideal machine and concerning computation: that which the machine does, either in real life or conceptually. This body of knowledge has come to be known as computer science. Many of the roots of computer science lie deep in mathematics.

In this volume we consider numerical computation such as one finds in the physical sciences and engineering, and which is usually referred to as "scientific computing" as opposed to, say, "business data processing." The somewhat diffuse subject matter of scientific computing is collected under the traditional heading of "numerical analysis." As we have stated above, numerical analysis has a theoretical side and a practical side. It is our contention that these are two facets of a single indissoluble entity, and this idea has guided the planning of this book. It is our hope that mathematicians interested in numerical analysis will agree with this view. Computer scientists, on the other hand, may question the inclusion of some of the theoretical mathematics. However, we contend that students of computer science will find this inclusion even more useful than mathematics students. The latter will have been exposed to the fundamentals of mathematical analysis and algebra often enough in their education to make the presentation here more of a review than anything else. Students of computer science will probably have had less exposure to the basic mathematics on which numerical computation rests. Therefore it seems even more worthwhile for them to study this mathematics directly in connection with numerical computation.

This point of view presents a formidable problem for the writer of a book. Obviously, one cannot include all the mathematics related to numerical computation. One must rely on the previous education and experience of the student. The graduate student in computer science, mathematics, science or engineering, for whom this book is intended, has generally had a good undergraduate course in mathematical analysis and linear algebra. On this assumption, we have, in the first two chapters, collected in a compressed form the basic principles of analysis and algebra which are most frequently used in numerical analysis. This material can be quickly reviewed by those students who are already well acquainted with it. Others may have to proceed more slowly, using the references to supplement the text. Chapter 3 may contain material that is new to some. It is intended as a brief introduction to the ideas and topics in functional analysis which provide the unifying thread running through our approach to numerical analysis. The student is advised to spend somewhat more time on this chapter, but not to dwell on it too long. As he proceeds through the later chapters on specific topics in numerical analysis, he will be referred back to Chapter 3, at which time he will be better able to digest its contents. Thus Chapters 1, 2, and 3 may be regarded as the pier anchoring the numerical analysis "bridge" to classical mathematics founded on the real continuum. The remaining chapters constitute a partial span both to practical computation and to the theory of computability.

Chapters 1 through 6 can be used as the basis of the first semester of a two-semester course in numerical analysis, with Chapters 7 through 10 providing the subject matter for the second semester. Each chapter contains, in addition to the usual subject matter of numerical analysis under the chapter heading, background material drawn from the related mathematical discipline. The time spent on the background material may be varied at the option of the student and instructor. In a course of lectures, it may be assigned as home reading for the most part, thereby enabling the instructor to cover more topics in numerical analysis, while yet providing the able student with a deeper understanding of the relation of the parts to the whole. The exercises serve the same purpose, as well as providing practical experience in numerical computation. Naturally, it is not possible to pursue each topic to its fullest in a book designed as a basic introduction to the entire subject. In a computer science or applied mathematics curriculum, this course would be followed by special elective courses in such subjects as the numerical solution of partial differential equations, the numerical solution of ordinary differential equations, the numerical solution of ordinary differential equations with emphasis on boundary-value problems and control problems, and a course in the numerical solution of mathematical programming problems, to mention but a few. Chapters 11 and 12 can serve as a core for the last two.

A preliminary version of this book, in the form of lecture notes, has been used by the author and his colleagues in a two-semester course at the University of Southern California. The course has been attended by graduate students in both mathematics and computer science. To provide the practical experience which cannot be imparted to a student by classroom lectures, we supplement the course by informal laboratory work. In the first semester, each student is required to write, check out, and run computer programs for the following problems and methods: Gaussian elimination with pivoting and iteration on the residuals (Exercises

4.11–4.13); Newton's method for systems of n equations in n unknowns (Exercise 5.10); and computation of matrix eigenvalues and eigenvectors by the Jacobi and Householder methods (Exercise 6.11). There is no substitute for this kind of experience. It is advisable that the programming be done in FORTRAN, ALGOL or PL/I (or the equivalent) to avoid undue concentration on fine programming details. For those computer science majors whose main interest is in the nonnumerical side of computer science, this may well be their only exposure to the problems peculiar to scientific programming, e.g., to questions of rounding errors, iteration, convergence, stability, etc. For mathematics majors interested primarily in the theory, the required laboratory work should illuminate the theory and enliven it. In the second semester, a similar laboratory regime accompanies the lectures and reading. For example, the following problems are programmed: approximation of e^x by the best approximating polynomial to be determined by the single-exchange algorithm (Exercise 7.16); Chebyshev quadrature (Exercise 9.12); and solution of initial-value problems for systems of ordinary differential equations by the Adams-Moulton and Runge-Kutta methods (Exercise 10.14). Each of these problems and methods is discussed in the text, both from a theoretical and practical standpoint. The practical details are frequently collected under the heading "Computational Aspects," to separate them clearly from what can be established with mathematical rigor. However, we have avoided such devices as the inclusion of specific computer programs, since, first of all, we assign this programming task to the student to force him to exercise his ability to think his way from the theory of a method to its realization as a computer program. Second, our aim is not to present a catalog of algorithms. (We do, however, cite references for various algorithms to inform the student of the existence of such sources.) Rather, our aim is to present many of the basic ideas, techniques, and results of scientific computation as a foundation which should serve the computer scientist or numerical analyst throughout his career.

As must always be the case in an undertaking of this kind, I am indebted to many for their assistance. I have tried to acknowledge this implicitly in the references listed at the end of each chapter. However, these lists are certainly not complete. To Richard Varga, who read the entire manuscript with great care, I express my special thanks. And I do acknowledge the skillful typing of Mrs. Reva Bennett and Mrs. Elsie Walker. Finally, I wish to thank the editorial staff of Addison-Wesley for their efforts on my behalf.

E.K.B.

Los Angeles, California
December 1971

REFERENCES

0.1. R. L. GOODSTEIN, *Recursive Analysis*. Amsterdam: North-Holland Publishing Co. (1961)

0.2. E. BISHOP, *Foundations of Constructive Analysis*. New York: McGraw-Hill (1967)

0.3. W. MILLER, "Recursive Function Theory and Numerical Analysis," *J. Computer and System Sci.* **4**, 5, October 1970, pp. 465–472

0.4. J. CLEAVE, "The Primitive Recursive Analysis of Ordinary Differential Equations and the Complexity of Their Solutions," *J. Computer and System Sci.* **3**, 4, November 1969, pp. 447–456

To Lori, Debbie, Beth, and Amy

CONTENTS

Chapter 1 Basic Concepts of Analysis
- 1.1 Sets and functions — 1
- 1.2 Topologies — 4
- 1.3 Metric spaces — 8
- 1.4 Compactness — 11
- 1.5 Real functionals — 15
- 1.6 Measurability — 16
- 1.7 Connectedness — 18
- 1.8 Notation — 18

Chapter 2 Linear and Multilinear Algebra
- 2.1 Vector spaces and linear operators — 22
- 2.2 Examples of infinite-dimensional spaces — 36
- 2.3 Linear functionals and multilinear mappings — 38

Chapter 3 Topological Vector Spaces
- 3.0 Introduction and guide to Chapter 3 — 48
- 3.1 Topologies and norms — 48
- 3.2 Normed vector spaces — 51
- 3.3 Pre-Hilbert (inner product) spaces — 55
- 3.4 Convexity in topological vector spaces — 62
- 3.5 Metric properties of linear operators on normed spaces — 62
  - The conjugate of a normed space — 69
  - The adjoint of a continuous linear operator — 71
- 3.6 The spectrum of a linear operator; eigenvalues — 73
- 3.7 Convergence of operators — 84

Chapter 4 Numerical Solution of Linear Algebraic Systems
- 4.1 Introduction — 102
- 4.2 Gaussian elimination with real numbers — 104
  - Gaussian elimination with finite-precision numbers
- 4.3 Estimates of error in Gaussian elimination — 113
- 4.4 Iterative methods — 123
  - Iteration on the residuals — 123
  - Jacobi iteration — 129
  - Diagonal dominance and irreducible matrices — 136
  - The self-adjoint case — 143
- 4.5 Relaxation methods, acceleration of convergence — 145
- 4.6 Solution of singular linear systems; the pseudo inverse — 153

Contents

Chapter 5 Solution of Nonlinear Equations
- 5.1 Fixed points and contractive mappings ... 164
  - Computational aspects (rounding errors, rates of convergence) ... 170
  - Stability of successive approximations ... 172
- 5.2 Newton's method—the one-dimensional case ... 174
- 5.3 Differential calculus of vector functions ... 178
  - Higher order-derivatives and differentials ... 189
  - Partial differentiation ... 194
- 5.4 Newton's method for vector functions ... 195
- 5.5 Modifications of Newton's method ... 199
- 5.6 Zeros of polynomials in one variable ... 208
  - Sturm sequences ... 209
  - Bernoulli's method ... 210
  - Bairstow's method ... 211
  - Error analysis ... 212
- 5.7 Acceleration of convergence of sequences ... 213
- 5.8 Rates of convergence ... 215

Chapter 6 The Computation of Matrix Eigenvalues and Eigenvectors
- 6.1 Introduction ... 230
- 6.2 Error estimates ... 231
- 6.3 The power method ... 238
- 6.4 The Jacobi method (for a symmetric matrix) ... 240
  - Rate of convergence ... 243
  - Error analysis ... 244
- 6.5 Householder's method ... 245
- 6.6 Givens' method (eigenvalues of a symmetric tridiagonal matrix) ... 247
- 6.7 Eigenvectors of a symmetric tridiagonal matrix ... 250
- 6.8 Eigenvalues of an arbitrary matrix by reduction to upper Hessenberg form ... 250
- 6.9 The QR Algorithm (for reduction of a matrix to triangular form by similarity transformations) ... 251
- 6.10 External properties of eigenvalues ... 256
- 6.11 Singular values and the pseudo inverse ... 257

Chapter 7 Approximation Theory
- 7.1 Introduction ... 265
- 7.2 The Weierstrass theorem (1885) ... 266
- 7.3 Closest approximation in normed and seminormed spaces ... 269
- 7.4 Closest least-squares approximation in pre-Hilbert spaces ... 276
- 7.5 Polynomial approximation in the uniform norm ... 280
- 7.6 Generalized polynomial approximation in the uniform norm ... 290
- 7.7 Least squares approximation of functions of a real variable; orthogonal families of polynomials ... 293
- 7.8 Orthogonal polynomials as eigenfunctions ... 309
- 7.9 Rational approximation ... 311
- 7.10 Multivariate approximation ... 319
- 7.11 Mathematical programming techniques ... 321

Chapter 8 Interpolation
8.1	The general interpolation problem	335
8.2	Other forms of the interpolating polynomial	339
8.3	The error in polynomial interpolation	345
8.4	Numerical differentiation	353
8.5	Spline functions	355
8.6	Inverse interpolation	360
8.7	Trigonometric interpolation	363
8.8	Curve-fitting and smoothing of data	365

Chapter 9 Numerical Integration
9.1	Introduction	380
9.2	Linear quadrature functionals	381
9.3	Composite quadrature functionals	388
9.4	Gaussian quadrature	392
	Gaussian quadrature with assigned nodes	395
9.5	Integration over infinite intervals, singular integrals	403
9.6	Integration of periodic functions	406
9.7	Extrapolation to the limit; Romberg integration	408
9.8	Quadrature functionals with minimum remainder	411
9.9	Multiple integrals	416

Chapter 10 Numerical Solution of Ordinary Differential Equations
10.1	Introduction; differential equations and difference equations	431
10.2	Numerical solution by numerical quadrature	439
10.3	Error analysis and convergence of quadrature methods	444
10.4	Rounding errors in quadrature methods	447
10.5	One-step methods	450
10.6	Stability of numerical solutions by difference methods	464
10.7	Stiff equations and A-stability	475

Chapter 11 Boundary-Value Problems for Differential Equations
11.1	Existence and uniqueness of exact solutions	489
11.2	Solution of boundary-value problems by iteration on the initial values	502
11.3	Difference methods	507
11.4	Linear boundary-value problems, eigenvalue problems and variational methods	510
11.5	The Bubnov-Galerkin method	512
11.6	Boundary-value problems for partial differential equations	513

Chapter 12 Extremum Problems
12.1	Introduction	520
12.2	Extremum problems with equality constraints	523
12.3	Gradient methods for extremum problems without constraints	534
12.4	Gradient methods for extremum problems with equality constraints	546
12.5	The eigenvalue problem for a self-adjoint operator	552
12.6	Computational solution of the optimal control problem of Mayer by gradient techniques	559

12.7	The Rayleigh-Ritz method	565
12.8	Linear programming	571
12.9	Inequality constraints, mathematical nonlinear programming	577
12.10	Penalty function techniques (sequential unconstrained methods)	582

Appendix I   Proof of convergence of Galerkin's method   590

Appendix II   The Hahn-Banach theorem   592

Appendix III   The Stone-Weierstrass theorem   594

Appendix IV   The implicit-function theorem   596

Appendix V   Results on the spectrum of linear operators   599

Appendix VI   Proof of fundamental theorem of linear programming   602

Index   605

CHAPTER 1

BASIC CONCEPTS OF ANALYSIS

1.1 SETS AND FUNCTIONS

We shall assume the reader is acquainted with intuitive set theory. Intuitively, a *set* is a collection of elements.

We shall be concerned with various kinds of sets and especially with sets of real numbers, sets of vectors, and sets of real functions. If X is any set, we write $x \in X$ to signify that x is a *member* or *element* or *point* of the set X. Likewise, $x \notin X$ means that x is not a member of X. The set of all $x \in X$ having a property P is denoted by $\{x \in X : P\}$ or simply $\{x : P\}$ when X is known; e.g., for x a real number, $\{x : x \geq 0\}$ denotes the set of nonnegative real numbers. If there are no elements having property P, then the set $\{x : P\}$ is empty. The empty set is denoted by $\varnothing$. If every element of a set X is contained in a set Y, then X is a *subset* of Y and we denote this by $X \subset Y$ or $Y \supset X$. We define $X = Y$ if and only if $X \subset Y$ and $Y \subset X$. If $X \subset Y$ and $X \neq Y$, then X is called a *proper subset* of Y and X is *properly contained* in Y. Thus, the inclusion symbol $\subset$ need not denote proper inclusion. Note that we usually use capital letters to denote sets and lower case letters for elements of sets. To display the members of a set we sometimes write $X = \{x, y, z, \ldots\}$.

Let A be a set. The set of all subsets of A is denoted by $P(A)$. Thus, $X \in P(A)$ if and only if $X \subset A$. A subset $\mathscr{F} \subset P(A)$ is called a family of sets. Let I be another set such that to each $i \in I$ there corresponds precisely one set E_i of the family $\mathscr{F}$ and every set in $\mathscr{F}$ is some E_i, that is, $\mathscr{F} = \{E_i : i \in I\}$. We say that $\mathscr{F}$ is a family of sets indexed by I and write $\mathscr{F} = \{E_i\}_{i \in I}$. For example, if I is the set of natural numbers $\{1, 2, 3, \ldots\}$ and E_i is the straight line in the xy-plane having slope one and y-intercept equal to i, then $\{E_i\}_{i \in I}$ is the family of lines given by the equations $y = x + i$, $i = 1, 2, 3, \ldots$.

If A and B are sets, then the *union* $A \cup B$ is the set consisting of all elements which are in A or in B (or both). The intersection $A \cap B$ is the set of all elements which are in both A and B. Likewise, if $\{E_i\}_{i \in I}$ is a family of sets, then $\bigcup_{i \in I} E_i$ is the set of all x such that $x \in E_i$ for some $i \in I$. Similarly, $\bigcap_{i \in I} E_i$ is the set of all x such that $x \in E_i$ for all $i \in I$. If $A \cap B = \varnothing$, A and B are said to be *disjoint*. The *difference* of two sets is defined as $A - B = \{x : x \in A \text{ and } x \notin B\}$. $A - B$ is also called the *complement* of B with respect to A. The Cartesian product $A_1 \times A_2 \times \cdots \times A_n$ of the sets $A_1, \ldots, A_n$ is the set of ordered n-tuples $(a_1, \ldots, a_n)$ where $a_i \in A_i$ for $i = 1, \ldots, n$.

The real line or real number system is denoted by R and we write

$$R^n = R \times \cdots \times R,$$

where there are n factors in the Cartesian product. We shall sometimes consider the extended real number system (or extended real line) R_∞ consisting of R together with two additional elements, $-\infty$ and ∞, where these elements have their usual ordering and meaning. For $-\infty \leq a \leq b \leq \infty$, we define the *closed interval* $[a, b]$ by $[a, b] = \{x : a \leq x \leq b\}$. Similarly, we have the *open interval*

$$(a, b) = \{x : a < x < b\}$$

and the *half-open* intervals $[a, b) = \{x : a \leq x < b\}$, $(a, b] = \{x : a < x \leq b\}$. Thus, $[-\infty, \infty]$ is precisely the extended real line. Note that (a, ∞) and $(-\infty, a)$ are (infinite) open intervals in R.

A set X is said to be *partially ordered* if there is a binary relation, $a \leq b$ (or $b \geq a$) defined for some pairs $(a, b) \in X \times X$ such that the following three properties hold: $a \leq a$ (reflexive); $a \leq b$ and $b \leq a$ implies $a = b$ (antisymmetric); $a \leq b$ and $b \leq c$ implies $a \leq c$ (transitive). The extended real line is partially ordered by the usual relation of less than or equal, where, of course, we also have $-\infty \leq x \leq \infty$ for all $x \in R$.

Let $E \subset X$, where X is partially ordered. By an *upper bound* of E, we mean an element $u \in X$ such that $x \leq u$ for all $x \in E$. Similarly, a *lower bound* of E is an element $l \in X$ such that $l \leq x$ for all $x \in E$. The *supremum* (or *least upper bound*) of E is an upper bound u^* of E such that $u^* \leq u$ for all upper bounds u of E. We denote the supremum by $\sup E$. The *infimum* (or *greatest lower bound*) is defined similarly and is denoted by $\inf E$. If E has an upper (lower) bound, it is said to be *bounded from above (below)*. If E is bounded from above and below, then E is called a *bounded* set. It is a property of the real numbers that every nonempty subset E of R which is bounded from above has a supremum. Likewise, a set E which is bounded from below has an infimum. When considering subsets E of the real line, it is convenient to extend the meaning of $\sup E$ to the case where E is not bounded from above. In that case we define $\sup E = \infty$, considering E now to be a subset of the extended real line. Similarly, $\inf E = -\infty$ if and only if E is not bounded from below. Under this convention, any subset $E \subset [-\infty, \infty]$ has both a supremum and an infimum. If $\sup E \neq \infty$ we shall often write $\sup E < \infty$, and similarly $\inf E > -\infty$ shall mean that $\inf E \neq -\infty$.

A set X is *totally ordered* if it is partially ordered and if for any two of elements a, b either $a \leq b$ or $b \leq a$.

Note that the definitions of open interval and closed interval apply to an arbitrary partially ordered set X. (The relation $a < b$ means that $a \leq b$ and $a \neq b$.) The set $\{x \mid x > a\}$ is called the *open interval* (a, ∞).

We state *Zorn's lemma* for future reference: If X is a partially ordered set (nonempty) such that every totally ordered subset of X has an upper bound in X, then X has a *maximal* element (i.e., an element m such that $m \leq x$ implies $x = m$.)

A *mapping* (or *transformation*) f of a set X into a set Y is a correspondence which associates with each $x \in X$ a unique $y \in Y$. As usual, we write $y = f(x)$. f is also called a *function on X to Y* and we denote this by

$$f : X \longrightarrow Y.$$

To indicate that $y = f(x)$ we shall also write $f: x \longmapsto y$. X is called the *domain* of f and is denoted by $D(f)$. The set $y \in Y$ such that $f(x) = y$ for some $x \in X$ is called the *range* of f and denoted by $R(f)$. Thus, $R(f) \subset Y$, but $R(f)$ need not be equal to Y. If $R(f) = Y$, then f is called an *onto* mapping (or a *surjective* mapping). For example, let $X = A_1 \times \cdots \times A_n$. The mapping $\pi_i : (a_1, \ldots, a_n) \longmapsto a_i$ is called the *ith projection* of the Cartesian product on A_i, $i = 1, \ldots, n$. $\pi_i : A_1 \times \cdots \times A_n \longrightarrow A_i$ is surjective. For any $A \subset X$, $f(A)$ is the set $\{f(x) : x \in A\}$ and is called the *image* of A under the mapping f. Thus, $R(f) = f(X)$. If $B \subset Y$, then $f^{-1}(B)$ is the set $\{x : f(x) \in B\}$ and is called the *inverse image* (or *pre-image*) of B under f. If $B = \{y\}$ is a singleton set, then we write $f^{-1}(y)$ for the inverse image. If $f^{-1}(y)$ consists of at most one point for every $y \in Y$, then f is a *one-to-one* (1 : 1) (or *injective*) mapping. Thus, f is injective if and only if $a \neq b$ implies $f(a) \neq f(b)$. A *bijective* mapping is one which is both injective and surjective (i.e., one-to-one and onto). If f is a bijective map of X onto Y, then for each $y \in Y$ there is exactly one x such that $f(x) = y$. This defines a mapping $f^{-1} : Y \longrightarrow X$, called the *inverse* of f, according to the equation $f^{-1}(y) = x$. Thus, $f^{-1}(f(x)) = x$ for all $x \in X$ and $f(f^{-1}(y)) = y$ for all $y \in Y$.

Occasionally, we shall wish to consider the situation in which f is a mapping from a subset of X into Y, i.e., $f(x)$ need not be defined for all $x \in X$. In this case, we shall call f a *partial mapping* from X to Y and write $f: X \longrightarrow Y$.

If $f: X \longrightarrow Y$ and $g: Y \longrightarrow Z$, then the *composite* map $g \cdot f$ is defined in the usual way by $g \cdot f(x) = g(f(x)) = g(y) \in Z$. Thus, $g \cdot f : X \longrightarrow Z$. The *identity* mapping on X is the mapping I_x defined by $I_x(x) = x$ for all $x \in X$. If f^{-1} exists, we may write (1) $f^{-1} \cdot f = I_x$ and (2) $f \cdot f^{-1} = I_y$, where I_y is the identity transformation on Y. These two conditions are also sufficient for f to be bijective, that is, the mapping $f: X \longrightarrow Y$ is bijective if and only if there exists a mapping $f^{-1} : Y \longrightarrow X$ such that (1) and (2) hold. To prove sufficiency we simply note that for $x, x' \in X$, $f(x) = f(x')$ implies $x = f^{-1}(f(x)) = f^{-1}(f(x')) = x'$. Thus, (1) implies that f is one-to-one. To show it is onto, we use (2). For any $y \in Y$ we have $f(f^{-1}(y)) = y$, so that $x = f^{-1}(y) \in X$ is mapped onto the element y.

A set X is *finite* if there is a bijection of X onto a set of natural numbers $\{1, \ldots, n\}$ for some natural number n. The set X is *denumerable* (or *countable*) if there is a bijection of X onto the set of all natural numbers. Two sets have the same *cardinality* if there is a bijection of one onto the other.

The function f is called an *extension* of g and g is called a *restriction* of f if f and g are functions such that $D(g) \subset D(f)$ and $f(x) = g(x)$ for all $x \in D(g)$.

If $f: X \longrightarrow [-\infty, \infty]$, and $E \subset X$, we shall write $\sup_{x \in E} f(x)$ instead of $\sup f(E)$.

A *sequence* in a set X is a mapping f from the nonnegative integers $N = \{0, 1, 2, \ldots,\}$ into X. As is customary, we write $x_i = f(i)$, $i = 0, 1, 2, \ldots$, and denote the sequence by $(x_i : i \in N)$ or simply by (x_i). If E_1 is an infinite subset of N, the restriction of f to E_1 is called a *subsequence* of (x_i) and is denoted by $(x_{i_j} \mid i_j \in E_1)$ or simply by (x_{i_j}). (One may regard x_{i_j} as the result of composing maps $f(i(j))$, where $i(j) = i_j$ is a 1:1 map of the natural numbers onto E_1.)

1.2 TOPOLOGIES

We shall be interested in continuous functions of various kinds. When we say that a function $f: X \longrightarrow Y$ is continuous at a point, we are making an assertion concerning its behavior in a *neighborhood* of the point, that is, *near* the point. To make the notions of neighborhood and nearness mathematically precise, we introduce the concept of a *topology*. A topology in a set X is given by specifying a family of subsets called the *open* subsets of X. For example, a subset E of the real line R is said to be *open* if it is empty or if for every $x \in E$ there is an open interval containing x and contained in E. In other words, *an open set in R is a set which is the union of open intervals*. It is easy to verify that the following properties hold for the family τ of open sets of R:

O_1: The empty set $\varnothing$ and the set X are in τ.
O_2: If $A_i \in \tau$ for $i = 1, \ldots, n$, then $A_1 \cap A_2 \cap \cdots \cap A_n \in \tau$.
O_3: If $\{A_i\}_{i \in I}$ is an arbitrary family of sets in τ, then $\bigcup_{i \in I} A_i \in \tau$.

Thus, finite intersections of open sets are open and every union (finite or not) of open sets is open. When we speak of the line R, we shall assume the open sets are defined as above. We observe that to establish O_1, O_2 and O_3 only the total ordering property of R is used. Hence, these properties hold for an arbitrary totally ordered set. (It suffices to prove that the intersection of two open intervals is an open interval.) A set $E \subset R_\infty$ is said to be *open* if it is a union of open intervals and half-open infinite intervals of the form $[-\infty, b), (a, \infty]$.

For an arbitrary set X, the above properties may be used as the definition of a topology in X.

Definition 1.2.1 A family τ of subsets of a set X is said to be a *topology in X* if τ satisfies properties O_1, O_2, and O_3 above. The sets in τ are called the *open* sets of the *topological space* (X, τ).

A given set X may have many topologies, for example:

1. the set of all subsets of X is clearly a topology (called the *discrete* topology);
2. $\tau = \{\varnothing, X\}$ consisting of just two subsets is called the *coarse* topology.

The useful topologies will usually lie somewhere between these two. For example, if X is totally ordered, we have seen that we can define a subset to be open if it is the union of open intervals. This is called the *order* topology. When the topology being used is evident from the context we shall omit mention of τ and call X itself a topological space. If τ_1 and τ_2 are two topologies in X, τ_1 is said to be *stronger* (or *finer*) than τ_2, and τ_2 is *weaker* (or *courser*) than τ_1 if $\tau_2 \subset \tau_1$. A subset A of a topological space X becomes a topological space if we define the open sets of A to be all the intersections with A of the open sets of X. This topology is called the *relative* topology of A as a subset of X.

Definition 1.2.2 A subset E of a topological space X is called *closed* if its complement $X - E$ is open.

By the familiar relations of set theory (the complement of a union of sets is the intersection of the complements, etc.), we find that finite unions of closed sets are closed and infinite intersections of closed sets are closed. Also, $\emptyset$ and X are closed. In the discrete topology, every subset of X is both open and closed. If X is totally ordered, then every closed interval $[a, b]$ is a closed set in the order topology, since the complement of $[a, b]$ is $(-\infty, a) \cup (b, \infty)$. Note that some sets are neither open nor closed; e.g., $(a, b]$ in R. However, for any subset E of a topological space X, there is a smallest closed set containing E, namely, the intersection of all closed sets which contain E. This intersection is called the *closure* of E and is often denoted by $\bar{E}$.

Definition 1.2.3 Let x be a point in a topological space X. A subset $E \subset X$ is called a *neighborhood* of x if E contains an open set containing x.

For example, in R, E is a neighborhood of x if E contains an open interval containing x. An open set E is a neighborhood of each $x \in E$. Conversely, if E is a neighborhood of each $x \in E$, then E is open. To see this, note that every $x \in E$ is contained in some open set $E_x \subset E$ and $E = \bigcup E_x$ is open. Thus, the open sets are known if the neighborhoods of every x are given.

In the definition of *neighborhood*, we have sacrificed the intuitive connotation of *closeness*. However, we recover some of this connotation by considering families of neighborhoods as in the next two definitions.

Definition 1.2.4 Let E be a subset of a topological space X. A point $x \in X$ is called an *accumulation point* of E if every neighborhood of x contains at least one point of E different from x.

Note that x need not be in E to be an accumulation point. If $E \subset R$, every neighborhood of an accumulation point x of E must contain infinitely many points of E, since otherwise we could find an open interval containing x and only finitely many points of E. But then there would be a smaller interval which contains x and no points of E different from x, contradicting the definition of accumulation point.

A point x is called an *isolated point* of a subset E if $x \in E$ and x is not an accumulation point of E. Thus, there is a neighborhood of x containing no points of E other than x.

If $\tau_2 \subset \tau_1$ are two topologies of X, then if x is an accumulation point of $E \subset X$ in the stronger τ_1-topology, it is also an accumulation point in the τ_2-topology. A set E is closed if and only if it contains all its accumulation points. (If E is closed, its complement $X - E$ is open, hence is a neighborhood of every point in the complement. This neighborhood contains no points of E. Hence, $x \in X - E$ cannot be an accumulation point of E. Conversely, if $X - E$ contains no accumulation point of E, every $x \in X - E$ has some neighborhood contained in $X - E$, so that $X - E$ is open. Hence, E is closed.) It follows from this that the closure of E is the union of E and the set of its accumulation points.

The union of all open sets contained in a subset $E \subset X$ is an open set called the *interior* of E. Thus, a point $x \in E$ is an *interior point* of E if x is contained in an open set contained in E. The *boundary* of E is the set of points x such that every neighborhood of x has a nonempty intersection with E and with $X - E$. The boundary of any set E is a closed set. A subset $E \subset X$ is *dense* (or *everywhere dense*) in X if the closure of E is X. (The rational numbers are everywhere dense in the real line R.) The subset E is *nondense* (or *nowhere dense*) if its closure has no interior points. (The integers are nondense in R, with the open interval topology.)

Having introduced a topology in a set X, we can define the important concept of the *limit of a sequence*.

Definition 1.2.5 Let (x_i) be a sequence of points in a topological space X. A point $x \in X$ is called a *limit* of the sequence if for every neighborhood U of x there exists an integer n such that $x_i \in U$ for all $i \geq n$. The sequence also is said to be *convergent* and to *converge* to x and we write "$\lim_{i \to \infty} x_i = x$" or "$x_i \to x$ as $i \to \infty$," or simply "$x_i \to x$."

If $\tau_2 \subset \tau_1$ are two topologies of X, then $x_i \to x$ in the stronger τ_1-topology implies that $x_i \to x$ in the τ_2-topology. For a sequence to have a unique limit, it is sufficient that the space X be *separated* (or *Hausdorff*), by which we mean that any two distinct points have disjoint neighborhoods. *All the spaces we shall consider are assumed to be separated unless we state otherwise.* For example, the line R is separated with its usual topology. However, if the coarse topology is used, it is not separated and every point is a limit point of every sequence.

To generalize the notion of limit to include such familiar situations as $\lim_{x \to 0} (x + 1) = 1$ and $\lim_{x \to 0} \sin x/x = 1$, we introduce the concept of a *filter base* on X. A *filter base* $\mathscr{B}$ on an arbitrary set X is a family of nonempty subsets of X such that for every $B_1, B_2 \in \mathscr{B}$ there exists $B_3 \in \mathscr{B}$ such that $B_3 \subset B_1 \cap B_2$. (Observe that this implies that any finite subfamily of B has nonempty intersection.) For example, if X is the nonnegative integers the subsets of the form $\{n, n+1, \ldots,\}$ form a filter base. If X is any topological space, by a *neighborhood base at* $x \in X$ we mean a family $\mathscr{N}$ of neighborhoods of x such that every neighborhood of x contains an element of $\mathscr{N}$. A family of neighborhoods is a *neighborhood base of* X if it is a neighborhood base at x for every $x \in X$. For example, the open sets containing x constitute a neighborhood base at x. If $X = R$, the family $\mathscr{F}_x$ of open intervals $(x - 1/n, x + 1/n)$ form a neighborhood base at x and $\bigcup_{x \in X} \mathscr{F}_x$ is a base of X. It is clear that a neighborhood base at x is also a filter base. In particular, the set of all neighborhoods of x is a neighborhood base and therefore a filter base. (See Exercise 1.10.)

Definition 1.2.6 Let $f: X \longrightarrow Y$ be a mapping of a set X into a topological space Y. Let $\mathscr{B}$ be a filter base on X. We say that $f(x)$ *converges to a point* $b \in Y$ along $\mathscr{B}$ (or $\lim_{\mathscr{B}} f(x) = b$) if for every neighborhood V of b there exists a $B \in \mathscr{B}$ such that $f(B) \subset V$.

Again, if Y is separated, $\lim_{\mathscr{B}} f(x)$ is unique.

This definition obviously agrees with Definition 1.2.5.

We shall be interested in the special case where f is defined on some nonempty subset E of a topological space X. Let a be a point in the closure of E. Then the family of subsets of the form $E \cap U$, where U is an arbitrary neighborhood of a, constitutes a filter base $\mathscr{B}_a$. If $\lim_{\mathscr{B}_a} f(x) = b$, we say that $f(x)$ *has limit b as x tends to a in E* and write briefly $\lim_{x \to a} f(x) = b$. For example, if $X = R$ and $E = (a, \infty)$, then $\lim_{\mathscr{B}_a} f(x)$ is just the usual limit from the right and is written as $f(a+)$ or $\lim_{x \to a+} f(x)$. Similarly, with $E = (-\infty, a)$ we obtain the limit from the left $f(a-)$ or $\lim_{x \to a-} f(x)$. If $\lim_{x \to a+} f(x) = \lim_{x \to a-} f(x)$, then $\lim_{x \to a} f(x) = b$, for any open set E containing a.

Definition 1.2.7 A mapping $f : X \longrightarrow Y$, where X and Y are topological spaces, is *continuous at a point* $x \in X$ if for every neighborhood U_y of $y = f(x)$ in Y there exists a neighborhood U_x of x such that $f(U_x) \subset U_y$. The mapping f is *continuous on X* if it is continuous at every point of X.

Remark If $\tau_2 \subset \tau_1$ are two topologies of X and $f : X \longrightarrow Y$ is continuous at x in the τ_2-topology, then it is continuous at x in the τ_1-topology. This is an immediate consequence of Definition 1.2.7.

We have noted that in a topological space X, the family of all neighborhoods of a point $a \in X$ is a filter base $\mathscr{B}_a$. It follows immediately from Definitions 1.2.6 and 1.2.7 that f is continuous at a point $a \in X$ if and only if $f(x)$ converges to $f(a)$ along $\mathscr{B}_a$; i.e., $\lim_{\mathscr{B}_a} f(x) = f(a)$.

As an immediate consequence of Definition 1.2.7 we have that $f : X \longrightarrow Y$ is continuous on X if and only if the pre-image $f^{-1}(V)$ of every open set $V \subset Y$ is an open set of X. We shall be concerned with the special case $f : R \longrightarrow R$. It is clear that Definition 1.2.7 reduces to the usual ε, δ definition for real functions. Our remarks on filters show that a real function f is continuous at an interior point a of its domain if and only if $f(a+) = f(a) = f(a-)$. If f is defined on a closed interval $[a, b]$, then f is continuous at a if and only if $f(a+) = f(a)$. (We use the relative topology in $[a, b]$.)

We shall also be concerned with the case $f : R^n \longrightarrow R$. To introduce a topology into a Cartesian product $X = X_1 \times X_2 \times \cdots \times X_n$ of n topological spaces X_i ($i = 1, \ldots, n$), we shall define the open sets of X to be unions of products of open sets in the X_i. Thus, if $O = A_1 \times A_2 \times \cdots \times A_n$, where each A_i is an open set of X_i ($i = 1, \ldots, n$), then O is taken to be an open set in X and any union of such sets is also taken to be an open set in X. Axioms O_1, O_2, O_3 are readily verified. With this *product topology*, X is called the *topological product* of the X_i. In R, an open set is a union of open intervals. If we define an *open interval in* R^n to be a product $E_1 \times E_2 \times \cdots \times E_n$, where each E_i is an open interval in R, then it is easy to see that an open set in the topological product R^n is either empty or is a union of open intervals in R^n. Thus, $f : R^n \longrightarrow R$ is continuous at a point $x = (x_1, \ldots, x_n)$ in R^n if for every neighborhood V of $f(x)$ in R there is an open interval U in R^n containing x and such that $f(U) \subset V$. Henceforth, unless otherwise specified, R^n shall denote the topological product. (As we shall see, other common methods of introducing a topology into the Cartesian product R^n lead to the product topology.)

Let $f : R^2 \longrightarrow R$ be continuous in R^2. The restriction of f to every subspace of

R^2 is also continuous. In particular, if we consider the line $\{a\} \times R$, where a is an arbitrary fixed point of R, this is a subspace and the restriction $y \longmapsto f(a, y)$ is a continuous map of R into R. Similarly, $x \longmapsto f(x, b)$ is continuous. It is well known that the converse is false, as is shown by the example $f(x, y) = xy/(x^2 + y^2)$ for $(x, y) \neq (0, 0)$ and $f(0, 0) = 0$. (Both $f(0, y) = 0$ and $f(x, 0) = 0$ are continuous at $(0, 0)$ but $f(x, y)$ is not.)

We shall frequently consider the case $f: X \longrightarrow R^n$, where X is a topological space (e.g., $X = R^m$). The function f gives rise to n component functions f_i ($i = 1, \ldots, n$), where $f_i: X \longrightarrow R$ is such that $f_i(x)$ is the ith component of $f(x)$ in R^n. In order that f be continuous (using the product topology in R^n) at a point x it is necessary and sufficient that each component f_i be continuous at x. The necessity follows easily from the result that if $f: X \longrightarrow Y$ and $g: Y \longrightarrow Z$ are continuous mappings, then the composite mapping gf is continuous. (We take for g the projection mapping $\pi_i: (y_1, \ldots, y_n) \longrightarrow y_i$, which is easily shown to be continuous, and note that $f_i = \pi_i f$.)

Let $\mathscr{B}$ be a filter base on X and $f: X \longrightarrow R^n$. Let $y = (y_1, \ldots, y_n) \in R^n$. Then it is simple to verify that $\lim_{\mathscr{B}} f(x) = y$ if and only if $\lim_{\mathscr{B}} f_i(x) = y_i$ for all i. In particular, if (x_j) is a sequence of points in R^n, then $\lim_{j \to \infty} x_j = x^*$ if and only if $\lim_{j \to \infty} (x_j)_i \longrightarrow x_i^*$ ($i = 1, \ldots, n$).

A bijection $f: X \longrightarrow Y$ of a topological space X onto a topological space Y is called a *homeomorphism* if f and f^{-1} are continuous mappings. We say that X and Y are homeomorphic. Homeomorphic spaces have the same topological properties, since there is a bijective mapping of their respective topologies.

1.3 METRIC SPACES

Of particular importance are topological spaces in which the topology is induced by a *metric* or *distance function*.

Definition 1.3.1 A *metric space* is a set X in which a function $d: X \times X \longrightarrow R$ is defined such that for all $x, y, z \in X$

M_1: $d(x, y) \geq 0$ and $d(x, y) = 0$ if and only if $x = y$;
M_2: $d(x, y) = d(y, x)$;
M_3: $d(x, z) \leq d(x, y) + d(y, z)$ (triangle inequality).

The function $d(x, y)$ is called the *distance* between x and y; d is also called a *metric*.

For $x_0 \in X$ and r a nonnegative real number or $r = \infty$, the *open* (*closed*) *ball* with center x_0 and radius r is the set $B(x_0, r)$ ($\bar{B}(x_0, r)$) of points x such that $d(x_0, x) < r$ ($\leq r$). The set $S(x_0, r)$ of points x such that $d(x_0, x) = r$ is called the *sphere* of radius r with center x_0.

A subset of a metric space is *bounded* if it is contained in some ball of finite radius. A mapping $f: X \longrightarrow Y$ of a set X into a metric space Y is *bounded* on X if the image

$f(X)$ is a bounded subset of Y. The diameter of a subset E is the number $\operatorname{diam}(E) = \sup_{x,y \in E} \{d(x, y)\}$. Let τ be the family of all subsets which are unions of open balls. It is easy to verify that τ is a topology in X. τ is the *topology induced by the metric*. (We must show that if B_1 and B_2 are open balls and $x \in B_1 \cap B_2$, then there exists $r > 0$ such that $B(x, r) \subset B_1 \cap B_2$. Then properties O_1, O_2, O_3 follow directly. See Exercise 1.6.) Thus an *open set in a metric space* is a union of open balls. A point x is an accumulation point of a subset E in a metric space if and only if for every $\varepsilon > 0$ there exists a $y \in E$, $y \neq x$, such that $d(x, y) < \varepsilon$. This follows immediately from the definitions of accumulation point (Definition 1.2.4) and open set in a metric space.

In R, we may define $d(x, y) = |x - y|$. This is the usual distance between points. In R^n, if we define the distance between $x = (x_1, \ldots, x_n)$ and $y = (y_1, \ldots, y_n)$ by $d(x, y) = (\sum_{i=1}^{n} (x_i - y_i)^2)^{1/2}$, we obtain the usual Euclidean distance. Another distance function is given by $d(x, y) = \max_{1 \leq i \leq n} \{|x_i - y_i|\}$. Properties M_1, M_2 and M_3 are easily verified for both of these functions. Thus, R^n is a metric space with either of these distance functions. Furthermore, the topology induced by each of these metrics is identical to the product topology induced by the order topology of R. Hence, both metrics give rise to the same family of open sets and to the same family of convergent sequences. We shall explore this further when we consider linear spaces.

Given any set X, we can introduce the *discrete metric* defined by $d(x, y) = 1$ if $x \neq y$. Properties M_1, M_2, M_3 hold and for any x, $B(x, 1) = \{x\}$. Thus, the topology induced by this metric is the discrete topology.

If x and y are distinct points of a metric space, the open balls $B(x, r)$ and $B(y, r)$ where $r < d(x, y)/2$ are disjoint neighborhoods of x and y. Hence, a metric space is separated and every sequence has at most one limit point. We note that for every sequence (r_n) of positive numbers such that $\lim_{n \to \infty} r_n = 0$, the balls $B(x, r_n)$ form a neighborhood base at the point x. From Definition 1.2.5 we then see that $\lim_{i \to \infty} x_i = x$ if and only if $\lim_{i \to \infty} d(x_i, x) = 0$.

If $f: X \longrightarrow Y$, where X is a metric space and Y is a topological space, then $\lim_{x \to a} f(x) = b$ if and only if $\lim_{n \to \infty} f(x_n) = b$ for every sequence (x_n) which converges to a. (The necessity is obvious, since $x \to a$ includes the special case $x_n \to a$. This holds for any topological space X. Now, suppose $\lim_{n \to \infty} f(x_n) = b$ for every sequence such that $\lim_{n \to \infty} x_n = a$. If $\lim_{x \to a} f(x) \neq b$, then there exists a neighborhood V of b in Y such that in every ball $B(a, 1/n)$ there exists a point x_n with $f(x_n) \notin V$ ($n = 1, 2, \ldots$). The sequence (x_n) converges to a, but $f(x_n)$ does not converge to b.) It follows from this result and Definition 1.2.7 that f is continuous at the point a in a metric space if and only if $\lim_{n \to \infty} f(x_n) = f(a)$ for every sequence (x_n) which converges to a. This, of course, is well known for the case $f: R \longrightarrow R$. For $f: X \longrightarrow Y$, where X and Y are metric spaces, we also have the following generalization of the ε, δ-definition: f is continuous at $a \in X$ if for every $\varepsilon > 0$ there exists $\delta > 0$ such that $d(a, x) < \delta$ implies $d(f(x), f(a)) < \varepsilon$.

If $\{E_i : i = 1, \ldots, n\}$ is a family of n metric spaces with metrics $d^{(i)}$, the product $X = \Pi_{i=1}^{n} X_i$ can be metrized in many ways. If $x = (x_1, \ldots, x_n)$ and $y = (y_1, \ldots, y_n)$

are two points in X, then it is easy to verify that the following are metrics on X:

$$d_2(x,y) = \left(\sum_{i=1}^{n} [d^{(i)}(x_i, y_i)]^2\right)^{1/2},$$

$$d_\infty(x, y) = \sup_i \{d^{(i)}(x_i, y_i)\},$$

$$d_1(x, y) = \sum_{i=1}^{n} d^{(i)}(x_i, y_i).$$

It is clear that $d_\infty(x, y) \leq d_2(x, y) \leq d_1(x, y) \leq nd_\infty(x, y)$. Thus, if $E \subset X$ is a union of open balls in one of these metrics, it is also a union of open balls in the other two metrics. Therefore, all three metrics induce the same topology in X. To see that this is the product topology, we choose the metric d_∞. The open ball $B_\infty(x, r)$ in X with this metric is just the product of the open balls $B_\infty(x_i, r)$ in X_i. Hence, the balls $B_\infty(x, r)$, $r > 0$, are a neighborhood base of x in the product topology as well as in the d_∞-topology.

Note that the discrete metric does not induce the same topology, in general. Thus, not all metric topologies in R^n are identical.

Any metric $d : X \times X \longrightarrow R$ is a continuous function on the topological product $X \times X$. Indeed, we have for any $x, y, a, b \in X$,

$$d(x, y) - d(a, b) = d(x, y) - d(x, b) + d(x, b) - d(a, b),$$

$$-d(b, y) \leq d(x, y) - d(x, b) \leq d(b, y),$$

$$-d(x, a) \leq d(x, b) - d(a, b) \leq d(x, a),$$

the latter two by the triangle inequality. Hence, for all (x, y) in the neighborhood $B(a, \varepsilon/2) \times B(b, \varepsilon/2)$ we have $|d(x, y) - d(a, b)| < \varepsilon$.

The *distance* $d(E, F)$, *between two subsets* E *and* F of a metric space is defined as the infimum of $d(x, y)$ for $x \in E$, $y \in F$. Thus, if $F = \{y\}$,

$$d(y, E) = \inf_{x \in E} \{d(y, x)\}.$$

Let $f : X \longrightarrow Y$ be a mapping of a metric space X (with metric d_X) into a metric space Y (with metric d_Y).

Definition 1.3.2 The mapping f is said to be *uniformly continuous* on X if for any $\varepsilon > 0$ there exists $\delta > 0$ such that $d_Y(f(x), f(x')) < \varepsilon$ whenever $d_X(x, x') < \delta$.

Definition 1.3.3 Let ω be a monotone nondecreasing mapping of the extended nonnegative real line into itself and such that $\omega(0) = 0$ and $\lim_{u \to 0+} \omega(u) = 0$. ω is called a *modulus of continuity* of f if for all $x', x'' \in X$

$$d_Y(f(x'), f(x'')) \leq \omega(d_X(x', x'')).$$

We see at once that if f has a modulus of continuity ω, then f is uniformly continuous since $\lim_{u \to 0+} \omega(u) = 0$.

Conversely, if f is uniformly continuous, then we define

$$\omega(u) = \sup_{d_X(x,x') \leq u} \{d_Y(f(x), f(x'))\},$$

and obtain a modulus of continuity for f. (Note that $\omega(0+) = 0$ follows from the uniform continuity, since $d_Y < \varepsilon$ whenever $u \leq \delta$.)

In case $\omega(u) = Lu$, we obtain the *Lipschitz condition* defined as follows.

Definition 1.3.4 Let $L > 0$. f is said to *satisfy a Lipschitz condition* with constant L if for all $x, y \in X$

$$d_Y(f(x), f(y)) \leq L d_X(x, y).$$

If $L < 1$, f is called a *contractive* mapping. More generally, if

$$d_Y(f(x), f(y)) \leq L(d_X(x, y))^\alpha$$

for some $0 < \alpha \leq 1$, then f satisfies a *Lipschitz condition* (sometimes *Hölder*) of order α.

We have seen that two different metrics can induce the same topology on X. Two such metrics give rise to the same class of continuous mappings. For two metrics to give rise to the same uniformly continuous functions, they must be *equivalent* in the sense of the next definition.

Definition 1.3.5 Let d and d' be two metrics on a set X. d and d' are called *equivalent* metrics if the identity map of the metric space (X, d) onto the metric space (X, d') and its inverse map are both uniformly continuous.

If d is equivalent to d', then for all x, y, $d(x, y) < \delta$ implies $d'(x, y) < \varepsilon$ and conversely. Thus, $d(x, y) \to 0$ implies $d'(x, y) \to 0$ and conversely, so that equivalent metrics give rise to the same topology. The metrics d_2, d_1 and d_∞ are equivalent. Referring to Definition 1.3.2, we see that for two equivalent metrics on X, a function which is uniformly continuous in one metric is uniformly continuous in the other. (Two metrics can induce the same topology without being equivalent; e.g., $|x - y|$ and $|1/x - 1/y|$ both define metrics on the nonnegative real line R_+ and they both induce the order topology. However, $|x - y| \to 0$ does not imply $|1/x - 1/y| \to 0$ (e.g., take $x = 1/n$ and $y = 1/n^2$.)

Definition 1.3.6 A topological space is *separable* if it contains a countable everywhere dense subset.

If X has a countable neighborhood base, then it is separable. (We choose one point from each neighborhood of the base to obtain an everywhere dense subset.)

A topological space is said to be *metrizable* if there is a metric which induces the topology. If X has a countable neighborhood base and is such that for any two disjoint closed subsets F_1 and F_2 there exist disjoint open sets O_1 and O_2 with $F_1 \subset O_1$ and $F_2 \subset O_2$, then X is metrizable.

1.4 COMPACTNESS

A family of sets $\{E_i\}_{i \in I}$ is called a *covering* of a set E if $E \subset \bigcup_{i \in I} E_i$. If the E_i are open sets of a topological space, the covering is said to be an *open covering*.

Definition 1.4.1 A subset E of a topological space X is called *compact* if every open covering of E (by open sets of X) contains a finite subfamily which also covers E. (If $E = X$, then the space X is *compact*.)

A compact set E is closed, i.e., it must contain all its accumulation points. For suppose $x' \notin E$. Since X is assumed to be a separated space, for each $x \in E$ there exists a pair of disjoint open sets O_x containing x, and O'_x containing x'. $\{O_x\}$ is an open covering of E, hence has a finite subcovering $\{O_{x_1}, \ldots, O_{x_n}\}$. Consider $O' = \bigcap_{i=1}^{n} O'_{x_i}$, where the O'_{x_i} are the sets paired with the O_{x_i}. The set is open and contains x' and $O' \cap E = \varnothing$. Hence, x' is not an accumulation point of E.

The Heine-Borel theorem for the real line shows that *every closed bounded interval $[a, b]$ is compact*. Not every closed subset of R is compact. However, *every closed subset F of a compact subset E of a topological space X is compact*. To prove this, we let $\{G_i\}$ be an open covering of F. Since F is closed, the complement $F^c = X - F$ is open in X. But $\{G_i, F^c\}$ is an open covering of E and therefore has a finite subcovering $\{G_1, \ldots, G_n, F^c\}$ and $\{G_1, \ldots, G_n\}$ must cover F. It follows from these results that *the compact subsets of R are precisely the closed bounded sets*, for if $F \subset R$ is closed and bounded, it is contained in some closed interval $[a, b]$. Since $E = [a, b]$ is compact and $F \subset E$ is closed, F is compact. Conversely, if F is compact, it must be closed. Also, F has a finite subcovering of bounded open intervals, hence is bounded.

A useful property of compact spaces is obtained by taking the set-theoretic dual of Definition 1.4.1. Let X be compact. If $\{F_i\}_{i \in I}$ is a family of closed sets such that the intersection of any finite number of the F_i is nonempty, then $\bigcap_{i \in I} F_i$ is nonempty. If $\bigcap_{i \in I} F_i = \varnothing$, then taking complements, we find that $\bigcup_{i \in I} (X - F_i) = X$. Thus, the family of open sets $\{X - F_i\}$ covers X and there must be a finite subcovering, say $\{X - F_1, \ldots, X - F_n\}$. Since $\bigcup_{j=1}^{n} (X - F_j) = X$, again taking complements we find that $\bigcap_{j=1}^{n} F_j = \varnothing$, contradicting the hypothesis on finite intersections.

Theorem 1.4.1 Let $f: X \longrightarrow Y$ be a continuous mapping of the compact space X into the space Y. The image $f(X)$ is a compact subset of Y.

Proof. Let $\{G_i\}$ be an open covering of $f(X)$. $\{f^{-1}(G_i)\}$ is an open covering of X. Since X is compact, there is a finite subcovering $\{f^{-1}(G_1), \ldots, f^{-1}(G_n)\}$ of X. But $G_i = f(f^{-1}(G_i))$. Hence, $\{G_1, \ldots, G_n\}$ covers $f(X)$. ∎

Corollary Let $f: X \longrightarrow R$ be a continuous mapping of a compact space X into the real line. Then f is bounded and attains its supremum and infimum on X.

Proof. Since $f(X)$ is a compact subset of R it is closed and bounded. Hence, $f(X)$ must contain its infimum l and its supremum u, since they are accumulation points or isolated points of $f(X)$. This means that there exists $x, y \in X$ such that $f(x) = l$ and $f(y) = u$. ∎

If $E \subset R^n$ is compact, it must be closed. But the image $\pi_i(E)$ under the ith projection of R^n is the continuous image of a compact set, hence is a compact subset of R. Thus, $\pi_i(E)$ is a bounded set for $i = 1, \ldots, n$, which implies that E itself is bounded. This shows that *every compact subset of R^n is closed and bounded*. To prove the converse, we use Tychonov's theorem, which asserts that the topological product $X =$

$\Pi \, X_i$ of a family of compact spaces is a compact space. If $E \subset R^n$ is a closed bounded set, it is a closed subset of a product X of closed bounded intervals $[a_i, b_i]$ ($i = 1, \ldots, n$). Since $[a_i, b_i]$ is compact, it follows from Tychonov's theorem that X is compact. Since E is a closed subset of X, E is also compact.

Remark 1 The *Bolzano–Weierstrass theorem* for the real line asserts that every infinite subset of a closed bounded interval $[a, b]$ has at least one accumulation point in $[a, b]$. This theorem and its proof apply to an arbitrary compact space. Let us say that a space X has the *Bolzano–Weierstrass property* if every infinite subset of X has at least one accumulation point. *If X is compact, it has the Bolzano–Weierstrass property.* To prove this, suppose E is a subset of X with no accumulation point. Then for any $x \in X$, there is an open set O_x containing x and no point of E other than x. $\{O_x\}$ is a covering of X and has a finite subcovering $\{O_{x_1}, \ldots, O_{x_n}\}$. Since $E \cap O_{x_i} = \{x_i\}$ or the empty set, E consists of at most the points $x_1, \ldots, x_n$. Hence, E is finite. Taking the contrapositive, we obtain the Bolzano–Weierstrass property.

Remark 2 *In a metric space X, the Bolzano–Weierstrass property implies that every sequence has a convergent subsequence.* Suppose (x_n) is a sequence of points in X. If the set of distinct points $E = \{x_n \in X\}$ is finite, then there is obviously a convergent subsequence. If E is an infinite point set, it has an accumulation point x^*. Each open ball $B(x^*, 1/j), j = 1, 2, \ldots$, contains a point $x_{n_j} \in E$, where $n_j \geq n_{j-1}$. The sequence (x_{n_j}) is convergent to x^* and is a subsequence of (x_n). If every sequence in a space X has a convergent subsequence, X is said to be *sequentially compact*. Thus, for a metric space, we have shown that "compact" implies "Bolzano–Weierstrass" implies "sequentially compact."

A metric space X is said to be *totally bounded* if for every $r > 0$ there is a finite set of open balls of radius r which covers X. *If X is sequentially compact, it is totally bounded.* For if X is not totally bounded, then we can find some $r > 0$ such that no finite set of balls of radius r covers X. Choose $x_1 \in X$; then $B(x_1, r)$ doesn't cover X. Hence, there exists $x_2 \notin B(x_1, r)$. $\{B(x_1, r), B(x_2, r)\}$ doesn't cover X. Hence, there exists x_3 such that $d(x_1, x_3) > r$ and $d(x_2, x_3) > r$. Proceeding in this way, we obtain an infinite sequence $\{x_n\}$ of distinct points such that $d(x_n, x_m) > r$ for each pair of points. This sequence can have no convergent subsequence. Thus, *for a metric space, "sequentially compact" implies "totally bounded."* (Loosely speaking, a totally bounded set can be uniformly approximated to arbitrary accuracy ε by a finite set.) A compact metric space X is separable. To obtain a countable everywhere dense set we proceed as follows. For every integer n the open balls $B(x, 1/n), x \in X$, cover X. Hence, a finite family $\mathcal{B}_n$ also covers X. The union of the $\mathcal{B}_n$ also covers X and is a countable family $\mathcal{B}$. Choose one x_i in each ball of $\mathcal{B}$. The set $E = \{x_i\}$ is everywhere dense, for if $x \in X$ and U is any open set containing x, there exist balls in $\mathcal{B}$ containing x and having arbitrarily small radius. Hence, one of these is contained in U, showing that x is in the closure of E.

A separable metric space has a countable base.

Definition 1.4.2 A sequence (x_n) of points in a metric space X is called a *Cauchy* sequence if for every $\varepsilon > 0$ there exists an integer N such that $d(x_q, x_m) < \varepsilon$ whenever $q, m \geq N$. A metric space is said to be *complete* if every Cauchy sequence is convergent (i.e., has a limit in X).

A bijection f of two metric spaces is called an *isometry* if it preserves distances; i.e., $d(x, x') = d(f(x), f(x'))$ and the spaces are said to be *isometric*. If X is a metric space, there is a complete metric space $\hat{X}$ which contains a dense subset which is isometric to X. Thus, an incomplete metric space can be regarded as a dense subset of a complete metric space. The space $\hat{X}$ is called the *completion* of X. To construct it, we follow the Cantor procedure of constructing the reals from the incomplete space of rational numbers. ($\hat{X}$ consists of equivalence classes of Cauchy sequences, where $x = (x_n)$ and $y = (y_n)$ are equivalent if $d(x_n, y_n) \to 0$. We define $d(x, y) = \lim_{n \to \infty} d(x_n, y_n)$ for two equivalence classes x and y. For details see [2.2].) If $E \subset X$, the completion $\hat{E}$ of E can be identified with a subset of $\hat{X}$.

A sequentially compact metric space X is complete. Let (x_n) be a Cauchy sequence in X. There exists a convergent subsequence $\{x_{n_i}\}$ converging to a point $x^* \in X$. Since $d(x_i, x^*) \leq d(x_i, x_{n_i}) + d(x_{n_i}, x^*)$ and both terms on the right tend to zero as $i \to \infty$, the sequence (x_n) converges to x^*. Hence, X is complete.

We have shown that a *sequentially compact metric space is totally bounded and complete. We shall prove that the converse is also true.* Let (x_n) be a sequence in a totally bounded complete metric space X. For every $\varepsilon > 0$, there is a ball $B(x(\varepsilon), \varepsilon/2)$ which contains an infinite number of the x_n (since X is covered by a finite number of balls of radius $\varepsilon/2$). Setting $\varepsilon = 1$, we obtain a subsequence $(x_n^{(1)})$ such that $d(x_p^{(1)}, x_m^{(1)}) \leq d(x_p^{(1)}, x(1)) + d(x(1), x_m^{(1)}) < 1$. Setting $\varepsilon = \frac{1}{2}$, we repeat the process and obtain a subsequence $(x_n^{(2)})$ of $(x_n^{(1)})$ all of whose points lie in some sphere of radius $\frac{1}{2}$. Continuing for $\varepsilon = (\frac{1}{2})^j$, $j = 1, 2, \ldots$, we get a countable family of subsequences $(x_n^{(j)})$ each lying in a sphere of radius $(\frac{1}{2})^j$. Selecting the diagonal sequence $(x_n^{(n)})$, we obtain a convergent subsequence.

To complete the circle of implications, we now show that "*sequential compactness*" implies "*compactness*" *for a metric space X*. If X is sequentially compact, it is totally bounded and complete. Hence, for each radius $r = 1/n$, $n = 1, 2, \ldots$, there exists a finite family of balls of radius $1/n$ covering X. The totality of these balls is evidently a countable family $\mathscr{B}$ such that for any open set O and any $x \in O$ there exists $B \in \mathscr{B}$ such that $x \in B \subset O$. Let $\mathcal{O} = \{O_i\}$ be an open covering of X. Let $\mathscr{B}'$ be the subfamily of all $B \in \mathscr{B}$ such that $B \subset O_i$ for some $O_i \in \mathcal{O}$. Since there exists a B such that $x \in B \subset O_i$ for every O_i and $x \in O_i$, $\mathscr{B}'$ is a covering of X. Now, choose for each $B \in \mathscr{B}'$ just one $O_i \in \mathcal{O}$ containing B. These O_i form a countable subfamily $\mathscr{A}'$ which covers X. Index these open sets in $\mathscr{A}'$ as $O_1, O_2, \ldots$, and consider the unions $U_n = \bigcup_{j=1}^n O_j$, $n = 1, 2, \ldots$. If for all n, $U_n \neq X$, choosing $x_n \in X - U_n$ yields a sequence (x_n) having a subsequence (x_{n_i}) converging to some point $x^* \in X$. But $x^* \in O_q$ for some $O_q \in \mathscr{A}'$, since $\mathscr{A}'$ is a covering. Hence, for infinitely many values of n the points x_n are in a ball $B(x^*, \varepsilon) \subset O_q$. But then these points cannot be in $X - U_q$, which contradicts the choice of the x_n for $n > q$. Hence, $U_n = X$ for some n and $\{O_1, \ldots, O_n\}$ is the desired finite subcovering.

The preceding results also show that *in a complete metric space a subset is compact if and only if it is closed and totally bounded*. In R^n, if a set is bounded, it is totally bounded. Hence, we have another proof that a closed bounded set in R^n is compact, and conversely.

A *continuous mapping* $f: X \longrightarrow Y$, where X is a compact metric space and Y is any metric space, *is uniformly continuous* (Definition 1.3.2). For $\varepsilon > 0$, and $x \in X$ choose δ_x such that $d_Y(f(x), f(y)) < \varepsilon/2$ for all $y \in B(x, \delta_x)$. Let $B_x = B(x, \delta_x/2)$. The family $\{B_x\}$ covers X. There is a finite subcovering $\{B_{x_1}, \ldots, B_{x_n}\}$. Choose $\delta = \min_{1 \le i \le n} \{\delta_{x_i}\}$. Then for two points x, y with $d(x, y) < \delta/2$, one of them, say x, is in some B_{x_j}. Hence, $y \in B(x_j, \delta_{x_j})$ and $d_Y(f(x), f(y)) \le d_Y(f(x), f(x_j)) + d_Y(f(x_j), f(y)) < \varepsilon/2 + \varepsilon/2$.

Definition 1.4.3 A space X is *locally compact* if each point in X has a compact neighborhood.

Every compact space is obviously locally compact. R^n is locally compact. The subspace of rational points on the real line is not locally compact, hence not compact. (Take a sequence of rationals in any neighborhood converging to a real point.)

If X is a topological space, we can construct a space Y consisting of X and an additional point $y \notin X$ such that Y is compact and the relative topology of X as a subset of Y coincides with the topology of X. The space Y is called the *one-point compactification* of X. In fact, we define a set in Y to be open if either it is an open subset of X or it is the union of $\{y\}$ with an open set $O \subset X$ such that $X - O$ is compact. This is clearly a topology in Y with the specified relative topology in X. To show compactness, let $\{O_i\}$ be an open covering of Y. Some O_i must be of the form $O \cup \{y\}$, where $X - O$ is a compact subset of X. The set $X - O$ is covered by the family $\{O_i \cap X\}$ of open sets in X. Hence, a finite subfamily, say $\{O_1 \cap X, \ldots, O_n \cap X\}$, covers $X - O$. But then $\{O_1, \ldots, O_n, O \cup \{y\}\}$ is a finite subcovering of Y.

The extended real line R_∞ is a two-point compactification of R, since the mapping $x \longmapsto \tan x$, with $\tan \pi/2 = \infty$ and $\tan(-\pi/2) = -\infty$, is a continuous mapping of the compact set $[-\pi/2, \pi/2]$ onto R_∞.

1.5 REAL FUNCTIONALS

A function $f: X \longmapsto R_\infty$ defined on a nonempty set X and having values in the extended real line will be called a *real functional*. For any subset $E \subset X$ the supremum (infimum) of the image $f(E)$ is denoted by $\sup_{x \in E} f(x)$ ($\inf_{x \in E} f(x)$). Note that $\sup_{x \in E} f(x)$ always exists. We say that f is *bounded from above* (*below*) on E if $\sup_{x \in E} f(x) < \infty$ ($\inf_{x \in E} f(x) > -\infty$). Clearly, f is bounded if and only if it is bounded from above and below.

Let (x_n) be a sequence of points in X. The sets $B_i = \{x_n : n \ge i\}$ form a filter base $\mathscr{B}$ on X. Let $\overline{f(B_i)}$ be the closure of $f(B_i)$ in R_∞. The intersection of any finite number of the sets $\overline{f(B_i)}$ is nonempty. Since R_∞ is compact, $\bigcap_{i=1}^{\infty} \overline{f(B_i)}$ is nonempty. Therefore, the supremum (infimum) of $\bigcap_{i=1}^{\infty} \overline{f(B_i)}$ exists and is called the *limit superior* (*limit inferior*) of f along $\mathscr{B}$. We denote it by $\lim_{i \to \infty} \sup f(x_i)$ ($\lim_{i \to \infty} \inf (fx_i)$). These remarks apply to any filter base $\mathscr{B}$ on X. For example, if X is the real line and $E = (a, b)$, a filter base is given by the sets $O \cap E$ where O is any neighborhood of a. In this case we write $\lim_{x \to a+} \sup f(x)$. In general, for an arbitrary filter base $\mathscr{B}$ we write $\lim_{\mathscr{B}} \sup f$. Note that $\lim_{\mathscr{B}} \inf f = -\lim_{\mathscr{B}} \sup (-f)$.

Now, let f be a real functional defined on a topological space X.

Definition 1.5.1 A function $f: X \longrightarrow R_\infty$ is *lower* (*upper*) *semicontinuous* at the point $x_0 \in X$ if for every $\alpha < f(x_0)$ ($\alpha > f(x_0)$) there exists a neighborhood U of x_0 such that $f(x) > \alpha$ ($f(x) < \alpha$) for all $x \in U$. The function f is *lower semicontinuous* on X if the above holds at every $x_0 \in X$.

For example, the function $f: R \longrightarrow R$ defined to be 0 at rational points and 1 at irrational points is lower semicontinuous at each rational point and upper semicontinuous at each irrational point. Lower semicontinuity at x_0 implies that there are no jumps in $f(x)$ from below $f(x_0)$.

It is evident that f is continuous at x_0 if and only if it is both upper and lower semicontinuous. Also, f is lower semicontinuous if and only if $f(x_0) = \lim_{x \to x_0} \inf f(x)$. It is upper semicontinuous if and only if $f(x_0) = \lim_{x \to x_0} \sup f(x)$.

If X is a metric space, then we can consider only sequences $x_n \to x_0$. Since x_0 may not be one of the x_n, we have f is lower semicontinuous if $f(x_0) \leq \lim_{n \to \infty} \inf f(x_n)$ for all $x_n \to x_0$. Also, f is lower semicontinuous on X if and only if for $r \in R_\infty$ the set of points x such that $r < f(x)$ is an open set.

Lower semicontinuity of f is equivalent to the upper semicontinuity of $-f$.

The following theorem is related to Theorem 1.4.1 and its corollary.

Theorem 1.5.1 Let $f: X \longrightarrow R_\infty$ be a lower semicontinuous functional on a compact topological space X. There exists at least one point $x_0 \in X$ such that $f(x_0) = \inf_{x \in X} f(x)$. If $f: X \longrightarrow (-\infty, \infty]$, then f is bounded from below.

Proof. Let $l = \inf_{x \in X} f(x)$. For all $r > l$, the sets $E_r = \{x : f(x) \leq r\}$ are closed. Clearly, $E_r \subset E_s$ for $r < s$. Hence, the intersection of any finite number of these sets is nonempty. Since X is compact $\bigcap_{r > l} E_r$ is nonempty. For $x_0 \in \bigcap E_r$ we have $f(x_0) \leq r$ for all $r > l$. This implies $f(x_0) \leq l = \inf f(x) \leq f(x_0)$; hence $f(x_0) = l$. If f maps X into $(-\infty, \infty]$, we cannot have $\inf f(x) = f(x_0) = -\infty$. Hence, f is bounded from below. ∎

1.6 MEASURABILITY

A family of subsets $\mathscr{M}$ of a set X is called a *σ-algebra* in X if
  i) $X \in \mathscr{M}$,
  ii) if $E \in \mathscr{M}$, then $X - E \in \mathscr{M}$, and
  iii) if $E_i \in \mathscr{M}$, $i = 1, 2, \ldots$, then $\bigcup_{i=1}^{\infty} E_i \in \mathscr{M}$. In short, a σ-algebra of subsets of X is a family of subsets containing X and closed under complementation and countable union. (It follows that $\mathscr{M}$ is closed under countable intersection.) $(X, \mathscr{M})$, or simply X, is called a *measurable space* and the sets of $\mathscr{M}$ are the *measurable sets* in X. A function $f: X \longrightarrow Y$ from a measur-

able space X into a topological space Y is called *measurable* if $f^{-1}(U)$ is a measurable set for every open set $U \subset Y$.

If $\mathscr{F}$ is any family of subsets, there is a smallest σ-algebra containing $\mathscr{F}$, namely, the intersection of all σ-algebras which contain $\mathscr{F}$. If X is a topological space and $\mathscr{F}$ the family of all open sets, the smallest σ-algebra $\mathscr{B}$ containing $\mathscr{F}$ is called the family of *Borel sets* of X. By the remark following Definition 1.2.7, every continuous mapping of X is Borel measurable.

If $f : X \longrightarrow Y$ is measurable and $\varphi : Y \longrightarrow Z$ is continuous, then $\varphi f : X \longrightarrow Z$ is measurable. For suppose V is open in Z. Then $\varphi^{-1}(V)$ is open in Y and $f^{-1}(\varphi^{-1}(V))$ is measurable. The latter set is the pre-image of V under the mapping φf.

If f and g are measurable functions on X to the real line R, then the mapping ψ given by $x \longmapsto (f(x), g(x))$ is a measurable function on X to R^2. For suppose V is any open rectangle which is the Cartesian product of two open intervals U_1 and U_2. Then $\psi^{-1}(V) = f^{-1}(U_1) \cap g^{-1}(U_2)$ is the intersection of two measurable sets, hence measurable. Since every open set O is a countable union of such rectangles V_i and $\psi^{-1}(O) = \psi^{-1}(\bigcup_{i=1}^{\infty} V_i) = \bigcup_{i=1}^{\infty} \psi^{-1}(V_i)$, we see that ψ is measurable.

It follows from the preceding two paragraphs that if f and g are real measurable functions, then so is $f + g$. We simply observe that $f + g$ is obtained by the composition $x \longmapsto (f(x), g(x)) \to f(x) + g(x)$ of the measurable mapping ψ and the continuous mapping $\varphi(y, z) = y + z$. Similarly, $f \cdot g$ is seen to be measurable and therefore $\lambda \cdot f(x)$ for any constant $\lambda \in R$.

A *measure* is a function μ defined on a σ-algebra into the interval $[0, \infty]$ and such that $\mu(\bigcup_{i=1}^{\infty} E_i) = \sum_{i=1}^{\infty} \mu(E_i)$ for any countable family of sets $E_i \in \mathscr{M}$ which are pairwise disjoint. The triple $(X, \mathscr{M}, \mu)$, or simply X, is called a *measure space*. (The *Lebesgue measure* in R coincides with length for intervals and the Lebesgue-measurable sets include all Borel sets.)

A measurable function φ on a measure space X to the interval $[0, \infty)$ is called *simple* if its range is a finite set of (nonnegative) values $\alpha_1, \ldots, \alpha_n$. Let $E_i = \varphi^{-1}(\alpha_i)$. The integral of φ with respect to the measure μ on a set $E \in \mathscr{M}$ is the number $\int_E \varphi \, d\mu = \sum_{i=1}^{n} \alpha_i \mu(E_i \cap E)$. (Here, we follow the convention that $0 \cdot \infty = 0$.) If $f : X \longrightarrow [0, \infty]$ is measurable, we define the *Lebesgue integral* of f over E with respect to μ to be the number $\int_E f \, d\mu = \sup_{\varphi \leq f} \int_E \varphi \, d\mu$, where the supremum is with respect to the set of all simple functions φ such that $\varphi(x) \leq f(x)$ for all $x \in X$. Note that $\int_E f \, d\mu$ may equal ∞. For measurable $f : X \longrightarrow [-\infty, \infty]$, we define $f^+(x) = \max\{f(x), 0\}$ and $f^-(x) = -\min\{f(x), 0\}$. Clearly, $f = f^+ - f^-$ and $|f| = f^+ + f^-$. We define $\int_E f \, d\mu = \int_E f^+ \, d\mu - \int_E f^- \, d\mu$ provided at least one of the integrals is finite. If $\int_X |f| \, d\mu < \infty$, we say that f is *Lebesgue integrable* or *summable* on X.

Definition 1.6.1 If f is a measurable function on X, define for $1 \leq p < \infty$

$$\|f\|_p = \left(\int_X |f|^p \, d\mu\right)^{1/p}.$$

The set of all f such that $\|f\|_p < \infty$ is denoted by $L^p(\mu)$. If μ is the Lebesgue measure on R^n, we write $L^p(R^n)$.

Definition 1.6.2 Let $f : X \longrightarrow [0, \infty]$ be measurable. Let E be the set of real numbers α such that $\mu(f^{-1}((\alpha, \infty])) = 0$. If $E = \varnothing$, define $\lambda = \infty$. If $E \neq \varnothing$, define $\lambda = \inf E$. λ is called the *essential supremum* of f.

If f is a real measurable function on X, the essential supremum of $|f|$ is denoted by $\|f\|_\infty$. If $\|f\|_\infty < \infty$, f is said to be *essentially bounded* and the set of such f is denoted by $L^\infty(\mu)$.

If $\|f\|_\infty < \infty$, then $|f(x)| \leq \|f\|_\infty$ almost everywhere (i.e., except on a set of measure zero). Any $\alpha > \|f\|_\infty$ is an upper bound of $|f(x)|$ almost everywhere, since $|f(x)| > \alpha$ only for x in a set of measure zero. Thus, $\|f\|_\infty$ is the least such upper bound and is also called the *essential* least upper bound. It is also the largest of the numbers β such that for $\varepsilon > 0$, $\mu(f^{-1}(\beta - \varepsilon, \infty]) > 0$.

1.7 CONNECTEDNESS

Definition 1.7.1 A topological space X is said to be connected if it is not the union of two nonempty disjoint open sets. A subset $E \subset X$ is said to be *connected* if it is a connected topological space in the relative topology.

Thus, $E \subset X$ is connected if there do not exist two disjoint open sets O_1, O_2 such that $E = (E \cap O_1) \cup (E \cap O_2)$, where $E \cap O_1$ and $E \cap O_2$ are both nonempty. For example, the set which is the union of the open intervals $(0, 1)$ and $(2, 3)$ is not connected. Neither is $[0, 1] \cup [2, 3]$ connected.

It is easy to see that the continuous image of a connected set is connected. Suppose $f: X \longrightarrow Y$ is a continuous mapping of the topological space X onto the topological space Y. If $Y = f(X)$ is not connected, there exist two disjoint nonempty open sets O_1, O_2 such that $Y = O_1 \cup O_2$. But this implies that $X = f^{-1}(O_1) \cup f^{-1}(O_2)$. Now, $f^{-1}(O_1)$ and $f^{-1}(O_2)$ are open, nonempty and they must be disjoint, since O_1 and O_2 are. Hence, X is not connected.

There is another concept of connectedness which is useful. Let X be a topological space. By a *Jordan arc* C in X we mean the image of a continuous mapping of the unit interval $[0, 1]$ into X. Thus, $C = \{f(t) : 0 \le t \le 1\}$, where $f: [0, 1] \longrightarrow X$ is continuous. The point $f(0)$ is called the *initial point* of the arc and $f(1)$ is the *end point*. A set $E \subset X$ is said to be *arcwise connected* if for any two points x, y in E there exists a Jordan arc contained in E and having x as initial point and y as end point. (If $\{f(t)\}$ is the arc, then $g(t) = f(1 - t)$ defines an arc having y as initial point and x as end point.) If E is arcwise connected, it is connected. For suppose $E = O_1 \cup O_2$, where O_1 and O_2 are two nonempty (relatively) open sets. Let $x \in O_1$ and $y \in O_2$. There exists an arc $C = \{f(t)\}$ such that $C \subset E$ and $x = f(0)$, $y = f(1)$. Furthermore, C is connected, since it is the continuous image of the connected set, $[0, 1]$. (See Exercise 1.7.) Therefore, $O_1 \cap C$ and $O_2 \cap C$ cannot be disjoint. Hence, $O_1 \cap O_2 \ne \varnothing$, which shows that E is connected. A connected set need not be arcwise connected. (See Exercise 1.7d.)

Let $\mathscr{F}$ be the family of connected subsets of X. Then $\mathscr{F}$ is a partially ordered set under the inclusion relation. Suppose $\mathscr{E}$ is a totally ordered subfamily of $\mathscr{F}$. Let $E = \bigcup_{E_i \in \mathscr{E}} E_i$. We shall show that $E \in \mathscr{F}$. Suppose $E = O \cup O'$, where O, O' are relatively open disjoint sets. Consider any E_i. Since E_i is connected, we must have $E_i \subset O$ or $E_i \subset O'$. Say $E_i \subset O$. Consider any other subset E_j in $\mathscr{E}$. If $E_j \subset E_i$, then $E_j \subset O$. If $E_i \subset E_j$, then $E_j \cap O \ne \varnothing$. Since $E_j = (E_j \cap O) \cup (E_j \cap O')$ and E_j is connected, $E_j \cap O'$ must be empty. Hence, $E_j \subset O$ for every E_j in $\mathscr{E}$, and therefore, $E \subset O$, which proves that E cannot be the union of two disjoint *nonempty* open sets. Thus, $E \in \mathscr{F}$. By Zorn's lemma, $\mathscr{F}$ must contain a maximal open connected set (i.e., a connected set not contained in any other connected set). A maximal open connected set is called a *component* of X. Two distinct components must be disjoint (Exercise 1.7e). Hence, X is the union of its components (Exercise 1.7f). This applies to an arbitrary subset $E \subset X$.

1.8 NOTATION

Let f and g be real functions of the real variable x. If $\lim_{x \to 0} f(x)/g(x) = 0$, we sometimes indicate this by writing $f(x) = o(g(x))$. If $|f(x)| < K|g(x)|$ for some constant K and all $|x|$ sufficiently small, we write $f(x) = O(g(x))$. We shall also use the $O(g(x))$ notation when $x \to a$, where $-\infty \le a \le \infty$, to indicate that $|f(x)|/|g(x)| \le K$ for x near a.

EXERCISES

1.1 a) In R^n an open interval O is the product of n open intervals (a_i, b_i) of R. Show that O is a union of open balls in the d_∞-metric.

b) Define $d_1(x, y) = \sum_{i=1}^n |\xi_i - \eta_i|$ for $x, y \in R^n$. Show that d_1 is a metric and it induces the product topology. (Prove that every open interval O in R^n is a union of open balls in the d_1 metric. Conversely, every open ball is the union of open intervals.)

1.2 Show that $\sin(1/x)$ is continuous and bounded on the open interval $(0, 1)$ but is not uniformly continuous (hence has no modulus of continuity).

1.3 a) Let $(f_n(x))$ be a sequence of real continuous functions defined on a subset X of a topological space. Suppose that $(f_n(x))$ is uniformly convergent on X; i.e., there exists $f(x)$ such that $\|f(x) - f_n(x)\|_\infty \to 0$ as $n \to \infty$ where $\|f\|_\infty = \sup_{x \in X} |f(x)|$. Prove that $f(x)$ is continuous on X. [*Hint:*

$$|f(x') - f(x)| \leq |f(x') - f_n(x')| + |f_n(x') - f_n(x)|$$
$$+ |f_n(x) - f(x)| \leq 2\|f - f_n\|_\infty + |f_n(x') - f_n(x)|.$$

Choose n sufficiently large so that $\|f - f_n\|_\infty < \varepsilon/3$ and then prescribe a neighborhood of x such that $|f_n(x') - f_n(x)| < \varepsilon/3$ for all x' in this neighborhood.]

b) Show that a necessary and sufficient condition that a sequence of real functionals (f_n) be uniformly convergent is that for any $\varepsilon > 0$ there exists N such that

$$\|f_p(x) - f_q(x)\|_\infty < \varepsilon \qquad (*)$$

for any $p, q > N$. [*Hint:* The sequence $(f_n(x))$ is a Cauchy sequence for each x. By completeness there exists a number $f(x)$ such that $f_n(x) \to f(x)$. To show $\|f_p(x) - f(x)\| < \varepsilon$ for $p > N$, let $q \to \infty$ in inequality (*) above. If $q, p > N$, then $\|f_p(x) - f_q(x)\| < \varepsilon$ for all x. For any x, $f_q(x)$ can be made arbitrarily close to $f(x)$ by taking q large enough. Hence, $\|f_p(x) - f(x)\| < \varepsilon$ for all x.]

c) Let (f_n) be a sequence of functions having values in a complete metric space. Show that (b) holds if (*) is replaced by

$$d(f_p(x), f_q(x)) < \varepsilon.$$

(Note that the common domain X of all f_n may be an arbitrary set.)

1.4 *Ascoli–Arzela Theorem.* A family $\mathscr{F}$ of real functions defined on a subset E of a metric space is called *equicontinuous* on E if for $\varepsilon > 0$ there exists $\delta > 0$ such that for all $x, y \in E$ with $d(x, y) < \delta$ and all $f \in \mathscr{F}$ $|f(x) - f(y)| < \varepsilon$. Prove that any sequence (f_n), $f_n \in \mathscr{F}$, where $\mathscr{F}$ is equicontinuous, which is pointwise bounded on a compact set E (i.e., $\{|f_n(x)|\}$ is bounded for each $x \in E$) contains a uniformly convergent subsequence. Further, (f_n) is uniformly bounded on E. [*Hint:* See [1.1] p. 144 for example.]

1.5 Let E be a subset of a metric space X with distance function d. The number diam $E = \sup_{x, y \in E} \{d(x, y)\}$ is called the *diameter* of E.

a) Let (x_n) be a sequence of points in X. Define $E_m = \{x_n : n > m\}$. Show that (x_n) is a Cauchy sequence if and only if $\lim_{m \to \infty} (\text{diam } E_m) = 0$.

b) Let $\bar{E}$ be the closure of E. Prove that diam E = diam $\bar{E}$.

c) Let (E_n) be a sequence of nonempty compact subsets of X such that $E_{n+1} \subset E_n$, $n \geq 1$. Prove that $\lim_{n \to \infty}$ diam $E_n = 0$ implies that $\bigcap_1^\infty E_n$ contains precisely one point.

1.6 a) Verify that if X is a metric space with metric d and $B_1 = B(x_1, r_1)$, $B_2 = B(x_2, r_2)$ are two open balls with $x \in B_1 \cap B_2$, then there exist r such that $B(x, r) \subset B_1 \cap B_2$. This shows that $B_1 \cap B_2$ is a union of open balls. [*Hint:* Let $\rho_1 = r_1 - d(x, x_1) > 0$, $\rho_2 = r_2 - d(x, x_2) > 0$. Take $0 < r < \min(\rho_1, \rho_2)$. By the triangle inequality (M_3 of Definition 1.3.1, $d(y, x_1) \le d(y, x) + d(x, x_1) \le r + d(x, x_1) < r_1$ for any $y \in B(x, r)$. Similarly, $d(y, x_2) < r_2$.]

b) Prove from (a) by induction that $\bigcap_{i=1}^n B_i$ is a union of open balls for any finite family $\{B_i\}$ of open balls with nonempty intersection. Hence, verify property O_2 of open sets (Definition 1.2.1) for the family of subsets which are unions of open balls. [*Hint:* $\bigcup_{i \in I} B_i \cap \bigcup_{j \in J} B_j = \bigcup_{\substack{i \in I \\ j \in J}} (B_i \cap B_j)$.]

Also verify O_3, ($\bigcup_i E_i = \bigcup_j B_{ji}) = \bigcup B_{ji}$.) Note that property M_1 of the metric was not used to prove that unions of open balls form a topology.

1.7 Prove that if $E \subset X$ is everywhere dense in X, then every point $x \in X$ is either an accumulation point of E or an isolated point of E.

1.8 a) Prove that the unit closed interval $[0, 1]$ is a connected set. [*Hint:* Suppose $[0, 1] = O \cup O'$, where O and O' are relatively open sets (i.e., intersections of open sets in R with $[0, 1]$). Let $1 \in O$ say. Then $(a, 1] \subset O$ for some $a < 1$. Let $l = \inf\{a : (a, 1] \subset O\}$. If $l \ge 0$, then $l \in O'$ and $[l, b) \subset O'$ for some $b > l$, which implies $O \cap O' \neq \emptyset$. If $l < 0$, then O' is empty.]

b) Prove that a subset $E \subset R$ is connected if and only if for any $a, b \in E$, $a < b$, the interval $[a, b] \subset E$.

c) Prove that in R a set is connected if and only if it is arcwise connected.

d) Show that a set in R^2 may be connected and not arcwise connected. [*Hint:* Consider the set E' consisting of the interval $[-1, 1]$ on the y-axis and the points of the graph of $y = \sin(1/x)$, $0 < x < 1$.]

e) Prove that if E, F are connected subsets such that $E \cap F \neq \emptyset$, then $E \cup F$ is connected. [*Hint:* Let $E \cup F \subset O \cup O'$, where $O \cap O' = \emptyset$, $E \subset O$ implies $F \subset O$.]

f) Prove that every connected set E is contained in a maximal connected set. [*Hint:* Use Zorn's lemma applied to the family of connected sets which contain E.] In particular, every singleton set $\{x\}$ is contained in a component.

1.9 The cardinality of the set of natural numbers is denoted by $\aleph_0$. Show that the set of rational numbers has cardinality $\aleph_0$. Show that the set of real numbers in the interval $[0, 1]$ has cardinality greater than $\aleph_0$. (For two sets X, Y we define card $X \ge$ card Y if there is a subset of X which has the same cardinality as Y: card $X >$ card Y if card $X \ge$ card Y and card $X \neq$ card Y.)

1.10 An alternative method of specifying a topology on a set X is to prescribe the family of neighborhoods of each point $x \in X$. Then an *open set* O can be defined as one which is a *neighborhood of each point in* O. The neighborhoods of a point x can be defined by using the notion of a *filter*. Given a filter base (Section 1.2) $\mathscr{B}$, the *filter generated by* $\mathscr{B}$ is the family of all subsets E such that there exists $B \in \mathscr{B}$ contained in E. Thus, a filter is a family of subsets generated by some filter base. A *filter of neighborhoods* of a point x is a filter $\mathscr{B}_x$ such that every $E \in \mathscr{B}_x$ contains x. A base of $\mathscr{B}_x$ is called a neighborhood base of x. Suppose a filter of neighborhoods has been given for each $x \in X$. Suppose further that for any $x \in X$ and $E \in \mathscr{B}_x$ there exists $F \in \mathscr{B}_x$ such that for any $y \in F$, $E \in \mathscr{B}_y$.

Show that the family $\bigcup_{x \in X} \mathscr{B}_x$ can be used as the neighborhoods of a topology τ; i.e., verify that the above defined open sets O satisfy the properties O_1, O_2, O_3. Since each $\mathscr{B}_x$ is determined by a neighborhood base at x, a *neighborhood base of* X (i.e., the union of the bases at each x) determines a topology.

REFERENCES

1.1. G. CHOQUET, *Topology*, Academic Press, New York (1966)
1.2. J. DUGUNDJI, *Topology*, Allyn and Bacon, Boston (1966)
1.3. W. RUDIN, *Principles of Mathematical Analysis*, McGraw-Hill, New York (1964)

CHAPTER 2

LINEAR AND MULTILINEAR ALGEBRA

2.1 VECTOR SPACES AND LINEAR OPERATORS

We assume that the reader has had at least an introductory course in finite-dimensional linear algebra covering vector spaces, linear operators and matrices, and linear algebraic equations. This chapter is intended as a very brief review of the main algebraic concepts and theorems of finite-dimensional vector spaces. At the same time, it serves as a convenient and natural place to introduce algebraic concepts for the infinite-dimensional case. The exercises elaborate on some of the theory presented in the text.

Definition 2.1.1 An $m \times n$ *system of linear algebraic equations* is a set of equations of the form,

$$
\begin{aligned}
a_{11}x_1 + a_{12}x_2 + \cdots + a_{1n}x_n &= b_1 \\
a_{21}x_1 + a_{22}x_2 + \cdots + a_{2n}x_n &= b_2 \\
&\vdots \\
a_{m1}x_1 + a_{m2}x_2 + \cdots + a_{mn}x_n &= b_m,
\end{aligned}
\tag{2.1.1}
$$

where the *coefficients*, a_{ij} ($i = 1, \ldots, m, j = 1, \ldots, n$), are known complex numbers, as are the b_i, $i = 1, \ldots, m$. The x_j, $j = 1, \ldots, n$, are variables or *unknowns* whose complex values are to be determined so as to satisfy system (2.1.1). If $m = n$, (2.1.1) is called an *nth-order linear algebraic system*.

Definition 2.1.2 The coefficients form a rectangular array called a *matrix* and denoted by A or (a_{ij}). Thus,

$$
A = (a_{ij}) = \begin{pmatrix} a_{11} & a_{12} & \cdots & a_{1n} \\ a_{21} & a_{22} & \cdots & a_{2n} \\ \vdots & \vdots & & \vdots \\ a_{m1} & a_{m2} & \cdots & a_{mn} \end{pmatrix}
\tag{2.1.2}
$$

(More generally, an $m \times n$ *matrix* (a_{ij}) is a function of two variables i and j which range over the sets $\{1, \ldots, m\}$ and $\{1, \ldots, n\}$ respectively. In this chapter, we specify that the values of the function are complex numbers.)

The variable i denotes the row of the matrix and is called the *row index*. Similarly, j is the *column index*. The *element* or entry of the matrix in row i, column j is denoted by a_{ij}. Since there are m rows and n columns, A is an $m \times n$ matrix. If $m = n$, A is a *square matrix of order n*. Two *matrices, A and B*, are equal if and only if $a_{ij} = b_{ij}$ for all i and j. (This implies that they have the same number of rows and columns.)

Definition 2.1.3 An *n*-tuple of complex numbers $(x_1, \ldots, x_n)$ is a $1 \times n$ matrix and is called an *n-dimensional row vector*. Similarly, the $n \times 1$ matrix,

$$\begin{pmatrix} x_1 \\ x_2 \\ \vdots \\ x_n \end{pmatrix}$$

is called an *n-dimensional column vector*. We shall denote the above row vector by x^T and the corresponding column vector by x. The x_i are called the *coordinates* of the vector. If the x_i are all real numbers, we shall refer to x as a *real column vector*.

The columns of an $m \times n$ matrix A are m-dimensional column vectors which we shall denote by $A_1, A_2, \ldots, A_n$. We shall sometimes write $A = (A_1, A_2, \ldots, A_n)$.

Definition 2.1.4 If the column vector, b, of system (2.1.1) is adjoined to the coefficient matrix, A, as column $n + 1$, the resulting matrix

$$(A, b) = \begin{pmatrix} a_{11} & a_{12} & \cdots & a_{1n} & b_1 \\ a_{21} & a_{22} & \cdots & a_{2n} & b_2 \\ \vdots & & & & \vdots \\ a_{m1} & a_{m2} & \cdots & a_{mn} & b_m \end{pmatrix}$$

is called the *augmented matrix* of the system.

Definition 2.1.5 The *transpose* of the $m \times n$ matrix A in (2.1.2) is the $n \times m$ matrix, A^T, whose columns are the corresponding rows of A. Thus, if we let $a_{ij}^T = a_{ji}$,

$$A^T = (a_{ij}^T) = \begin{pmatrix} a_{11} & a_{21} & \cdots & a_{m1} \\ a_{12} & a_{22} & \cdots & a_{m2} \\ \vdots & \vdots & & \vdots \\ a_{1n} & a_{2n} & \cdots & a_{mn} \end{pmatrix}$$

This justifies our use of the notation x^T for the row vector corresponding to the column vector x.

Obviously, $(A^T)^T = A$.

In calculations which involve matrices, certain operations occur in a natural way. An operation called *matrix multiplication* is defined as follows.

Definition 2.1.6 Let A be an $r \times n$ matrix with elements a_{ij} and let B be an $n \times s$ matrix with elements b_{ij}. The $r \times s$ matrix, C, with elements c_{ij} defined by the relation,

$$c_{ij} = \sum_{k=1}^{n} a_{ik} b_{kj}, \quad i = 1, \ldots, r; \; j = 1, \ldots, s, \quad (2.1.3)$$

is called the *product* matrix of A by B. We denote the product by AB and write $C = AB$. A is a *left factor* of C and B is a *right factor*.

Multiplication of matrices is usually *not* commutative; i.e., $AB \neq BA$. In fact, BA may not be defined, since Definition 2.1.6 requires that the number of columns in the left factor must equal the number of rows in the right factor. For square matrices of the same order, AB and BA are both defined but generally are not equal. (See Exercise 2.3.)

As an immediate consequence of (2.1.3) we have $(AB)^T = B^T A^T$ whenever AB is defined.

Using matrix multiplication, we can write system (2.1.1) as the matrix equation,

$$Ax = b. \tag{2.1.4}$$

In this equation, Ax is the product of the $m \times n$ matrix A by the $n \times 1$ matrix,

$$x = \begin{pmatrix} x_1 \\ \vdots \\ x_n \end{pmatrix}.$$

The result is the $m \times 1$ matrix

$$b = \begin{pmatrix} b_1 \\ \vdots \\ b_m \end{pmatrix}.$$

x is a "vector variable" ranging over the set of n-dimensional column vectors.

Since $(Ax)^T = x^T A^T$, (2.1.1) can also be written using row vectors as $x^T A^T = b^T$.

An operation of *matrix addition* is defined as follows:

Definition 2.1.7 Let $A = (a_{ij})$ and $B = (b_{ij})$ be two $m \times n$ matrices. The $m \times n$ matrix $C = (c_{ij})$ defined by the relation

$$c_{ij} = a_{ij} + b_{ij}, \quad i = 1, \ldots, m; \quad j = 1, \ldots, n, \tag{2.1.5}$$

is called the *sum of A and B*. We write $C = A + B$.

If $x^T = (x_1, \ldots, x_n)$ and $y^T = (y_1, \ldots, y_n)$ are two n-dimensional row vectors, then their *vector sum* is defined by (2.1.5) to be $x^T + y^T = (x_1 + y_1, \ldots, x_n + y_n)$. Since $(A + B)^T = A^T + B^T$, this also defines addition of column vectors.

Finally, an operation called *scalar multiplication* can be defined between a complex number and a matrix. Complex (or real) numbers will be referred to as *scalars*. More generally, the elements of any field K are called *scalars*. In general, unless otherwise specified, we assume that K has characteristic zero (e.g., the rationals, reals, or complex fields.) However, most of the results hold for finite fields of characteristic p, where p is a prime; e.g., $p = 2$ gives the field $\{0, 1\}$.

Definition 2.1.8 Let $A = (a_{ij})$ be an $m \times n$ matrix and let λ be a scalar. The matrix B, having elements b_{ij} defined by the relation

$$b_{ij} = \lambda a_{ij}, \quad i = 1, \ldots, m; \quad j = 1, \ldots, n \tag{2.1.6}$$

is called the *scalar product of λ and A*. We write $B = \lambda A = A \lambda$.

Now let $A^{(n)}$ be the set of all square matrices of order n. In $A^{(n)}$, the matrix operations given in Definitions 2.1.6, 2.1.7, 2.1.8 possess certain familiar algebraic properties. We shall state these properties for an abstract set $\mathscr{A}$, that is, without referring to the particular set $A^{(n)}$. We shall continue to use the symbol $+$ to denote any operation having properties to be specified below, realizing that $+$ may have different interpretations in specific sets.

Property α_1. Let $\mathscr{A}$ be a set. A binary operation $+$ (called *addition*) is defined such that for any A, B, C in the set $\mathscr{A}$,

α_{10}: $A + B$ is in the set $\mathscr{A}$ (closure under addition),
α_{11}: $A + B = B + A$,
α_{12}: $(A + B) + C = A + (B + C)$.

There is an element, 0, in $\mathscr{A}$ such that for all $A \in \mathscr{A}$,

α_{13}: $A + 0 = A$.

For every A in $\mathscr{A}$ there is an element B such that

α_{14}: $A + B = 0$.

(α_{10}, α_{11}, α_{12}, α_{13}, α_{14} are just the axioms for an Abelian group. Thus, $\mathscr{A}$ is an Abelian group under $+$.)

Property α_2. For every pair of elements A and B in $\mathscr{A}$, an operation of *multiplication*, AB, is defined such that

α_{20}: AB is in $\mathscr{A}$,
α_{21}: $A(BC) = (AB)C$,
α_{22}: $A(B + C) = AB + AC$.

Property α_3. If A is in $\mathscr{A}$ and λ is a scalar, then an operation of scalar multiplication, λA, is defined such that λA is in $\mathscr{A}$ and if λ, μ, ν, are scalars and A, B are elements of $\mathscr{A}$, then

α_{31}: $\lambda(\mu A + \nu B) = (\lambda\mu)A + (\lambda\nu)B$,
α_{32}: $(\lambda A)(\mu B) = (\lambda\mu)AB$,
α_{33}: $(\mu + \lambda)A = \mu A + \lambda A$.
α_{34}: $\lambda A = A\lambda$, $1A = A$.

The above three properties can easily be shown to hold for the set $A^{(n)}$.

Definition 2.1.9 If properties α_1, α_2, α_3 hold in a set $\mathscr{A}$, the set $\mathscr{A}$ is called a *linear algebra over the complex field*. $A^{(n)}$ is called the *matrix algebra of order n over the complex field*.

Let V^n be the set of all n-dimensional column vectors. Let $^T V^n$ be the set of all n-dimensional row vectors. One can abstract the following properties of both V^n and $^T V^n$ to obtain a set V called a *linear vector space*. The elements in V are called *vectors*.

Property ω_1. V is an Abelian group under addition. (See Property α_1.)

Property ω_2. For x and y in V and λ, μ, ν scalars,

ω_{20}: $\lambda x = x\lambda$ is in V (closure under scalar multiplication),
ω_{21}: $\lambda(\mu x + \nu y) = (\lambda\mu)x + (\lambda\nu)y$,
ω_{22}: $1x = x$,
ω_{23}: $(\lambda + \mu)x = \lambda x + \mu x$.

In some V, we also have the additional property:

Property ω_3. There exists a set of n vectors, $\{u_1, u_2, \ldots, u_n\}$ in V called a (finite) *basis* of V, such that any vector x can be expressed as a unique linear combination of the u_i; that is,

$$x = x_1 u_1 + x_2 u_2 + \cdots + x_n u_n, \tag{2.1.7}$$

where the x_i are uniquely determined scalars. The x_i are the *coordinates* of x with respect to the basis.

Properties ω_1 and ω_2 can be derived easily for V^n and $^T V^n$.

Definition 2.1.10 A set V satisfying ω_1 and ω_2 is called a *linear vector space over the complex field*. If the scalars are restricted to be real numbers, then the set is a *real linear vector space*. If the scalars are in some arbitrary field K, V is a *linear vector space over K*. (We frequently call V a *linear space* or a *vector space*.) If ω_3 is also satisfied, the vector space is said to be *finite-dimensional* and *of dimension n*. Otherwise, it is said to be *infinite-dimensional*.

A subset of a vector space V which is closed under vector addition and scalar multiplication is itself a vector space and is called a *subspace* of V.

Remark We see that any n-dimensional vector space is *isomorphic* to V^n, since every abstract vector x can be represented uniquely as a column vector; i.e., as the column vector having coordinates $x_1, \ldots, x_n$ given by (2.1.7) with respect to some basis.

An *isomorphism* of two algebraic systems $\mathscr{A}$ and $\mathscr{A}'$ is a bijection $\varphi : \mathscr{A} \longrightarrow \mathscr{A}'$ which "preserves the corresponding operations" in $\mathscr{A}$ and $\mathscr{A}'$. Thus, if $\mathscr{A}$ and $\mathscr{A}'$ are linear algebras, we require that

$$\varphi(x + y) = \varphi(x) + \varphi(y), \qquad \varphi(\lambda x) = \lambda \varphi(x) \quad \text{and} \quad \varphi(xy) = \varphi(x)\varphi(y).$$

For vector spaces, φ must satisfy the first two conditions. A *homomorphism* is any mapping $\varphi : \mathscr{A} \longrightarrow \mathscr{A}'$ which preserves the operations in $\mathscr{A}$ and $\mathscr{A}'$; it need not be injective or surjective. For vector spaces, a homomorphism is usually called a *linear operator* (see Definition 2.1.12 below). If V is an n-dimensional vector space and $\{u_1, \ldots, u_n\}$ is a basis, then the mapping $x \longmapsto (x_1, \ldots, x_n)$, where $x \in V$ and $x = \sum_{i=1}^n x_i u_i$, is clearly an isomorphism of V and $^T V^n$.

To establish ω_3 for V^n, we take u_i to be the column vector in which all components are 0 except the ith component, which is 1.

These remarks apply to $^T V^n$ as well, with the obvious replacement of columns by rows. Thus, $u_1^T = (1, 0, 0, \ldots, 0)$, $u_2^T = (0, 1, 0, \ldots, 0), \ldots, u_n^T = (0, \ldots, 0, 1)$ is a basis for $^T V^n$. We call this basis the *canonical basis*.

Definition 2.1.11 A finite set of vectors, $\{u_1, \ldots, u_k\}$ is *linearly dependent* if there exist scalars, $\lambda_1, \ldots, \lambda_k$, such that not all λ_i are zero, and $\sum_{i=1}^{k} \lambda_i u_i = \lambda_1 u_1 + \lambda_2 u_2 + \cdots + \lambda_k u_k = 0$. If $u_1, \ldots, u_k$ are not linearly dependent, then they are *linearly independent*.

Let the set of vectors, $\{u_i : i = 1, \ldots, n\}$, be a basis. Then the u_i are linearly independent, for suppose $\sum_{i=1}^{n} \lambda_i u_i = 0$. This expresses the zero vector as a linear combination of the u_i. By definition ω_3 of a basis, the λ_i are uniquely determined. Hence, $\lambda_i = 0$ for $i = 1, \ldots, n$, which shows that the u_i are linearly independent.

Theorem 2.1.1 Let V be an n-dimensional vector space. Let $\{v_i : i = 1, \ldots, r\}$ be a set of nonzero vectors in V. If $r > n$, then the v_i are linearly dependent.

Proof. Let $\{u_1, \ldots, u_n\}$ be a basis in V. Then $v_i = \sum_{k=1}^{n} \lambda_{ki} u_k$, $i = 1, \ldots, r$. Consider v_1. Some λ_{k1} is not zero, say $\lambda_{k_1 1}$. Then we can divide by $\lambda_{k_1 1}$ and express u_{k_1} as a linear combination of v_1 and the remaining u_i; that is,

$$u_{k_1} = \frac{1}{\lambda_{k_1 1}} v_1 - \sum_{k=1}^{n} {}' v_{k1} u_k, \qquad v_{k1} = \lambda_{k1}/\lambda_{k_1 1},$$

where the $\sum'$ denotes summation with $k \neq k_1$. Replacing u_{k_1} by the right-hand side, each of $v_2, \ldots, v_r$ can be expressed as a linear combination of the remaining u_i and v_1. Now consider v_2.

$$v_2 = \mu v_1 + \sum_{k=1}^{n} {}' \mu_k u_k.$$

If all $\mu_k = 0$, then $\mu \neq 0$ and v_1 and v_2 are linearly dependent, hence so is $\{v_i\}$ and the proof is complete. If some $\mu_{k_2} \neq 0$, we can express u_{k_2} as a linear combination of v_1, v_2, and the remaining u_i. By replacing u_{k_2} in the expressions for $v_3, \ldots, v_r$, these v's are expressed as linear combinations of v_1, v_2, and the remaining u_i. Suppose $v_1, \ldots, v_n$ are linearly independent. Then we can replace all the u_i by the v_i, $i = 1, \ldots, n$. This implies that

$$v_{n+1} = \sum_{k=1}^{n} \alpha_k v_k,$$

which shows that $v_1, \ldots, v_{n+1}$ are linearly dependent, hence also $\{v_1, \ldots, v_r\}$. ∎

If V is a linear vector space (over any field of scalars), a subset $U \subset V$ is said to *generate* V if every $x \in V$ can be expressed as a linear combination of elements of U, i.e., there exist finitely many $u_1, \ldots, u_n \in U$ and scalars $\lambda_1, \ldots, \lambda_n$ such that $x = \sum_{i=1}^{n} \lambda_i u_i$. An infinite subset $U \subset V$ is said to be *linearly independent* if every finite subset of U is linearly independent (Definition 2.1.11). A subset $B \subset V$ is called a (*algebraic* or *Hamel*) *basis* if it is linearly independent and generates V. The linear independence implies the uniqueness of linear combinations; i.e., $x = \sum_{i=1}^{n} \lambda_i u_i =$

$\sum_{i=1}^{n} \lambda'_i u_i$ implies $\sum_{i=1}^{n} (\lambda_i - \lambda'_i) u_i = 0$ and therefore $\lambda_i - \lambda'_i = 0$ for all i. Thus this definition of basis coincides with the previous definition (see property ω_3 above) when B is finite.

Theorem 2.1.2 Let $U \subset V$ be a set of generators of V (e.g., $U = V$). Any linearly independent set $L \subset U$ can be extended to a basis B.

Proof. We use Zorn's lemma. Let $\mathscr{S}$ be the family of all linearly independent subsets which contain L. (Since $L \in \mathscr{S}$, $\mathscr{S}$ is not empty.) $\mathscr{S}$ is partially ordered by the set inclusion relation $\subset$. Let $\mathscr{T}$ be a totally ordered subset of $\mathscr{S}$. The set $\bar{T} = \bigcup_{T \in \mathscr{T}} T$ contains L and is linearly independent (since any finite subset $\{u_1, \ldots, u_n\}$ is contained in some T because $\mathscr{T}$ is totally ordered). Thus, $\bar{T} \in \mathscr{S}$ and is an upper bound of $\mathscr{T}$. By Zorn's lemma, $\mathscr{S}$ has a maximal set B; i.e., a linearly independent set B containing L and not properly contained in any other linearly independent set. To see that B generates V note that for any $y \in V$ there exist finitely many $u_i \in U$ such that $y = \sum \lambda_i u_i$. If $u_i \notin B$, then $B \cup \{u_i\}$ is linearly dependent and we have $u_i = \sum \alpha_j b_j$ for finitely many $b_j \in B$. Thus, y is a linear combination of elements of B, which proves that B is a basis. ∎

Incidentally, we have shown that any basis is a maximal linearly independent set and conversely. We have also shown that any vector space has a basis. (Take $L = \{v\}$, where v is any nonzero vector.)

Theorem 2.1.3 Any two bases of a vector space V over a field K have the same cardinality.

Proof

CASE 1. Suppose V has a finite basis B of $n > 1$ elements, that is, V is finite dimensional. By Theorem 2.1.1, any linearly independent set in V has at most n elements. Since any other basis B' is a linearly independent set, the number of elements m in B' is less than or equal to n. By symmetry, we also have $n \leq m$. Hence, $m = n$.

CASE 2. Suppose V has an infinite basis B. Let B' be another basis. By case 1, B' must be an infinite set. For each $y \in B'$ there is a finite subset $S_y \subset B$ such that y is a linear combination of the vectors in S_y. The set $S = \bigcup_{y \in B'} S_y$ generates B' and therefore V. Since $S \subset B$, it is linearly independent. Hence, $S = B$, since otherwise some element of B is a linear combination of elements of S, contradicting the linear independence of B. Since $B = \bigcup_{y \in B'} S_y$ and each S_y is finite while B' is infinite, we conclude that card $B \leq$ card B'. By symmetry, card $B' \leq$ card B. Hence, card $B =$ card B'. ∎

Corollary. If $\{u_1, \ldots, u_n\}$ is a linearly independent set in V^n, then this set is a basis.

This justifies the definition (Definition 2.1.10) of the dimension of a vector space.

Let A be an $m \times n$ matrix. Let x and y be n-dimensional column vectors. Then Ax and Ay are m-dimensional column vectors. By properties α_{22}, α_{32} we have for any scalars λ, μ,

Property τ: $A(\lambda x + \mu y) = \lambda A x + \mu A y$.

The product Ax defines a mapping of x onto the vector Ax and the matrix A is called a *linear matrix operator* or *linear matrix transformation* which maps V^n into V^m. The set V^n is the *domain* of A and the set of vectors $\{Ax : x \in V^n\}$ is the *range* of A. (Corresponding statements hold for $x^T A^T$ and $^T V^n$.)

Definition 2.1.12 Let $A : V \longrightarrow W$ be a mapping of a vector space V into a vector space W (over the same field). Let Ax denote $A(x)$. If property τ holds, A is called a *linear operator on V into W*.

Any linear operator, T, on V^n to V^m can be represented as an $m \times n$ matrix as follows. Let $\{u_j\}$ be a basis in V^n. Let $\{v_i\}$ be a basis in V^m. Then since $Tu_j \in V^m$, we have $Tu_j = a_{1j}v_1 + a_{2j}v_2 + \cdots + a_{mj}v_m$, that is,

$$Tu_j = \sum_{i=1}^{m} a_{ij} v_i; \qquad j = 1, \ldots, n. \tag{2.1.8}$$

The $m \times n$ matrix $A = (a_{ij})$ *represents the operator T relative to the bases* $\{u_j\}$ and $\{v_i\}$. The columns of A correspond to the vectors Tu_j relative to the basis $\{v_i\}$.

Conversely, any $m \times n$ matrix $A = (a_{ij})$ defines a linear matrix operator on V^n to V^m. Choose as the basis $\{u_j\}$ in V^n the *canonical basis* of column vectors

$$e_j = \begin{pmatrix} 0 \\ \vdots \\ 0 \\ 1 \\ 0 \\ \vdots \\ 0 \end{pmatrix} \quad j\text{th row}, \quad j = 1, \ldots, n, \tag{2.1.9}$$

and similarly for v_i in V^m, $i = 1, \ldots, m$. Clearly,

$$Ae_j = \sum_{k=1}^{m} a_{kj} v_k \tag{2.1.10}$$

defines a linear operator on V^n to V^m and the matrix which represents this linear operator by (2.1.8) relative to the canonical bases in (2.1.9) is again the matrix A.

If we had used row vectors with bases $\{e_j^T\}$ and $\{v_i^T\}$, similar results would hold for $u_j^T A^T$.

Definition 2.1.13 Let T_1 and T_2 be linear operators on V into W. The *sum*, $T_1 + T_2$, is a linear operator on V into W defined by $(T_1 + T_2)x = T_1 x + T_2 x$. The *scalar product* λT is a linear operator defined by $\lambda T(x) = \lambda(T(x))$.

With these definitions, the set $L(V; W)$ of all linear operators on V to W forms a vector space.

If T_1 is a linear operator on V to W and T_2 is a linear operator on W to Y, then for any x in V, the composite mapping $T_2(T_1 x)$ is defined and is a linear operator on

V to Y. This leads to the natural definition of the product of two such linear operators, T_1 and T_2 as the composite mapping.

Definition 2.1.14 $(T_2 T_1)x = T_2(T_1 x)$

The set of all linear operators on a vector space V into itself forms a linear algebra, $L(V; V)$, according to Definitions 2.1.9, 2.1.12–2.1.14. The correspondence (2.1.8) with $v_i = u_i$ establishes an isomorphism between the *operator algebra* $T^{(n)} = L(V^n; V^n)$ and the matrix algebra $A^{(n)}$. (See Exercise 2.4.)

The identity transformation, I, is defined by $Ix = x$ for all x. For *any* basis $\{u_i\}$, $Iu_i = u_i$. Hence, the corresponding *identity matrix* is

$$I = \begin{pmatrix} 1 & 0 & \cdots & 0 \\ 0 & 1 & & 0 \\ \vdots & \vdots & & \vdots \\ 0 & 0 & \cdots & 1 \end{pmatrix} = (\delta_{ij})$$

where $\delta_{ij} = 0$ for $i \neq j$ and $\delta_{ii} = 1$; i.e., δ_{ij} is the *Kronecker* delta. Clearly, $IT = TI = T$ for all T in $T^{(n)}$ and $IA = AI = A$ for all A in $A^{(n)}$.

Remark Let $T: V \longrightarrow W$ be a linear operator. Since T is a mapping, the general remarks in Section 1.1 apply. In particular, by the *inverse* of T we mean a mapping T^{-1} such that $T^{-1}T = I_V$ and $TT^{-1} = I_W$, where I_V and I_W are the identity (linear) operators on V and W respectively. Thus, T has an inverse if and only if it is a bijection. In this case, T^{-1} is also a linear operator, since for any $w_1, w_2 \in W$ there exist $v_1, v_2 \in W$ such that $Tv_1 = w_1$, $Tv_2 = w_2$, $T^{-1}w_1 = v_1$, $T^{-1}w_2 = v_2$ and $T(v_1 + v_2) = Tv_1 + Tv_2 = w_1 + w_2$. Hence, $T^{-1}(w_1 + w_2) = v_1 + v_2 = T^{-1}w_1 + T^{-1}w_2$. Similarly, since $T(\lambda v_1) = \lambda Tv_1 = \lambda w_1$, we have $T^{-1}(\lambda w_1) = \lambda v_1 = \lambda T^{-1}(w_1)$. Thus, T^{-1} satisfies property τ.

Definition 2.1.15 If a linear operator T has an inverse, we say that T is *nonsingular*. Otherwise, T is said to be *singular*. A matrix A is *nonsingular* if the corresponding operator is nonsingular.

If A is an $n \times n$ square matrix, the corresponding operator T is a linear operator on V^n into V^n. By the isomorphism of $T^{(n)}$ and $A^{(n)}$, the existence of T^{-1} implies the existence of a matrix A^{-1} such that $AA^{-1} = A^{-1}A = I$, where I is the $n \times n$ identity matrix.

Definition 2.1.16 Let T be a linear operator on V to W. The set of $x \in V$ such that $Tx = 0$ is called the *null space* of T. When finite, the dimension of the null space is called the *nullity* of T.

The null space N_T of T is a subspace of V, since for $x, y \in N_T$, $T(\alpha x + \beta y) = \alpha Tx + \beta Ty = 0$. Likewise, $R(T)$, the range of T, is a subspace of W, since for $w_1, w_2 \in R(T)$ there exist v_1, v_2 in V such that $T(\alpha v_1 + \beta v_2) = \alpha Tv_1 + \beta Tv_2 = \alpha w_1 + \beta w_2$.

At this point, we recall that $T(0) = 0$ for any linear operator (since $T(x) =$

2.1 Vector Spaces and Linear Operators 31

$T(0 + x) = T(0) + Tx$). Also, $T(-x) = -Tx$, where $-x = (-1)x$. We remark that a linear operator is also called a *K-homomorphism* where K is the underlying field of scalars. The null space is the kernel of the homomorphism. Thus two vector spaces V, W are *isomorphic* if there exists a bijective linear operator $T: V \longrightarrow W$.

Remark If T is a nonsingular linear operator, the nullity of T must be zero. Indeed, $Tx = 0$ implies that $x = (T^{-1}T)x = T^{-1}(Tx) = T^{-1}(0) = 0$. If the nullity of T is zero, T must be injective, since $Tx = Ty$ implies $T(x - y) = 0$ and therefore $x = y$. However, T need not be surjective, as for example is the case with the matrix

$$\begin{pmatrix} 1 & 0 \\ 0 & 1 \\ 0 & 0 \end{pmatrix}$$

which defines an injective linear operator on $V^{(2)}$ into $V^{(3)}$. Thus, zero nullity is not a sufficient condition for the existence of an inverse of an arbitrary linear operator T. However, if $T: V^{(n)} \longrightarrow V^{(n)}$, this condition is sufficient, as is proved in Theorem 2.1.5 below. Further, since all n-dimensional spaces are isomorphic, we see that a linear operator $T: V^n \longrightarrow V^m$ cannot have an inverse if $n \neq m$.

The range of $T: V \longrightarrow W$ is a subspace of W. When finite, the dimension of the range is called the *rank* of T. These terms also apply to a matrix, A, considered as a linear operator by (2.1.10). The following theorem relates the rank of A to its columns, A_j.

Theorem 2.1.4 For any $m \times n$ matrix, A, the rank of A is the maximum number of linearly independent columns of A.

Proof. Let $A = (a_{ij})$ be regarded as a linear operator, A, from V^n to V^m. Then for the bases given in Eq. (2.1.9) we have $Au_j = \sum_{k=1}^{m} a_{kj}v_k$, or $Au_j = A_j$. The vectors Au_j span R, the range of A, i.e., any vector, y, in R is equal to a linear combination of the Au_j. In fact, since y is in the range of A, $y = Ax$ for some $x = \sum_{j=1}^{n} x_j u_j$ and we have $y = Ax = \sum_{j=1}^{n} x_j Au_j$ as the desired linear combination. Select from Au_1, $Au_2, \ldots, Au_n$, in that order, the largest linearly independent subset. This subset will still span the range, R. Hence, it is a basis of R and must contain precisely r vectors, where r is the dimension of R. Denote these basis vectors by $Au_{j_1}, Au_{j_2}, \ldots, Au_{j_r}$. Then columns $A_{j_1}, A_{j_2}, \ldots, A_{j_r}$ of matrix A are linearly independent column vectors and the remaining column vectors are linearly dependent on these. (By Theorem 2.1.3, any other maximal linearly independent set has r columns.) ∎

Theorem 2.1.5 A necessary and sufficient condition that a linear operator T in $T^{(n)}$ be nonsingular is that its nullity be zero.

Proof. Necessity. See the above remark after Definition 2.1.16.
Sufficiency. Let $\{u_i\}$ be a basis for V^n and let $v_i = Tu_i$. The v_i are linearly independent, since by hypothesis $Tx = 0$ implies $x = 0$ and therefore $0 = \sum \lambda_i v_i =$

$\sum \lambda_i T u_i = T(\sum \lambda_i u_i)$ implies $\sum \lambda_i u_i = 0$. This in turn implies that all $\lambda_i = 0$. By the corollary to Theorem 2.1.3, the set $\{v_i\}$ is a basis for V^n. We define a linear operator U by defining it on the v_i such that $Uv_i = u_i$. Let x be any vector in V^n. $x = \sum x_i u_i$ and $UTx = U \sum x_i Tu_i = \sum x_i Uv_i = \sum x_i u_i = x$. Thus, $UT = I$. Similarly, $TU = I$. Hence, $U = T^{-1}$. ∎

Corollary 1 If $m \geq n$ and T is a linear operator on V^n to V^m and if the nullity of T is zero, then its rank is n.

Proof. The above sufficiency proof carries over to show that the set $\{v_i : i = 1, \ldots, n\}$ in V^m is a basis for the range of T. ∎

Corollary 2. If $m < n$, and T is a linear operator on V^n to V^m, the nullity of T must be at least $n - m$.

Proof. The set $\{v_i : i = 1, \ldots, n\}$ of the theorem must be linearly dependent in V^m by Theorem 2.1.1. In fact, there are r linearly independent v_i, where r is the rank of T. Clearly, $r \leq m$. Reindex so that $v_1, \ldots, v_r$ are linearly independent and $v_j = \sum_{i=1}^{r} \lambda_{ij} v_i$, $j = r + 1, \ldots, n$. Then $0 = v_j - \sum_{i=1}^{r} \lambda_{ij} v_i = T(u_j - \sum_{i=1}^{r} \lambda_{ij} u_i)$. Since the u_i are linearly independent, the vectors $(u_j - \sum_{i=1}^{r} \lambda_{ij} u_i)$, $j = r + 1, \ldots, n$, are nonzero and linearly independent and they are in the null space of T. Thus the nullity of T is $\geq (n - r) \geq n - m$. ∎

Theorem 2.1.6 Let A be a linear matrix operator on V^n to V^m. Let b be a nonzero vector in V^m in Eq. (2.1.4). If the rank of the augmented matrix (A, b) is not greater than the rank of A, then Eq. (2.1.4) has $v + 1$ linearly independent solutions, where $v \geq 0$ is the nullity of A. The converse is also true. Finally, if $b = 0$, (2.1.4) has v linearly independent solutions.

Proof

CASE 1. $b \neq 0$. By Theorem 2.1.4, the column vector b must be linearly dependent on the columns of A. By (2.1.8), b must be in the range of the matrix operator A; i.e., b is linearly dependent on the column vectors $\{Au_i\}$ which span the range. Hence, there exists a vector x such that $Ax = b$, which means that the corresponding matrix equation (2.1.4) is satisfied by the column vector corresponding to x relative to the basis $\{u_i\}$.

If $v \geq 1$, let $\{z_1, \ldots, z_v\}$ be a basis for the null space of A. Since $Az_i = 0$, then $A(x + z_i) = b$ and the column vectors $x + z_i$, $i = 1, \ldots, v$ are solutions of (2.1.4). The set $x, z_1, \ldots, z_v$ is linearly independent since x is not in the null space. If x' is any other solution of (2.1.4), then $A(x' - x) = b - b = 0$ so that $x' - x$ is in the null space. Hence, $x' = x + \sum \lambda_i z_i$, which shows that x' is linearly dependent on x and the z_i. If $v = 0$, then $x' = x$.

CASE 2. $b = 0$. The solutions of (2.1.4) constitute the null space. Hence, there are v linearly independent solutions. ∎

Remark 1 If the nullity of A is greater than zero and a solution exists, then system

(2.1.4) has infinitely many solution vectors. (This is not true for a finite field of scalars.) The theorem shows that they are all linear combinations of some linearly independent set of solution vectors.

Remark 2 If A is $n \times n$ and $b \neq 0$, then a unique solution of (2.1.4) exists if and only if A is nonsingular.

Theorem 2.1.7 Let T be a linear operator on V^n to V^n. Suppose $\{u_j\}$ is a basis and let A be the $n \times n$ matrix which represents T relative to $\{u_j\}$; i.e., $A = (a_{ij})$ where

$$Tu_i = \sum_{k=1}^{n} a_{ki} u_k, \quad 1 \leq i \leq n.$$

Suppose $\{v_j\}$ is another basis in $V^{(n)}$ and let $B = (b_{ij})$, where

$$v_j = \sum_{i=1}^{n} b_{ij} u_i, \quad 1 \leq j \leq n.$$

Then the matrix $B^{-1}AB$ represents T relative to the basis $\{v_j\}$.

Proof.

$$Tv_j = \sum_{i=1}^{n} b_{ij} Tu_i = \sum_{i=1}^{n} b_{ij} \sum_{k=1}^{n} a_{ki} u_k = \sum_{k=1}^{n} \left(\sum_{i=1}^{n} a_{ki} b_{ij} \right) u_k.$$

Let $C = (c_{ij})$ represent T relative to $\{v_j\}$, so that

$$Tv_j = \sum_{i=1}^{n} c_{ij} v_i = \sum_{i=1}^{n} c_{ij} \left(\sum_{k=1}^{n} b_{ki} u_k \right) = \sum_{k=1}^{n} \left(\sum_{i=1}^{n} b_{ki} c_{ij} \right) u_k.$$

Equating coefficients of u_k, we obtain

$$\sum_{i=1}^{n} a_{ki} b_{ij} = \sum_{i=1}^{n} b_{ki} c_{ij},$$

which shows that $AB = BC$ or $C = B^{-1}AB$. Note that B^{-1} exists because the column vectors of B are linearly independent, i.e., they correspond to the vectors $\{v_j\}$ which form a basis. This completes the proof. ∎

Remark A vector space V is said to be the *sum* of two subspaces V_1, V_2 if for every $v \in V$, there exists $v_1 \in V_1$, $v_2 \in V_2$ such that $v = v_1 + v_2$. In this case, we write $V = V_1 + V_2$. If the vectors v_1 and v_2 are determined uniquely by v, then we say V is a *direct sum* of V_1 and V_2 and write $V = V_1 \oplus V_2$. A necessary and sufficient condition that a sum $V_1 + V_2$ be direct is $V_1 \cap V_2 = \{0\}$. (The proof is immediate.) (See Definition 2.2.1.) If T is a linear transformation of V into V, such that $T(V_1) \subset V_1$, then we say T leaves V_1 invariant. The restriction of T to V_1 is a linear operator T_1 on V_1.

Suppose $V = V_1 \oplus V_2$ and T leaves V_1 and V_2 invariant. Let $T_1 = T \mid V_1$ and $T_2 = T \mid V_2$ be the respective restrictions of T. Then for any $v = v_1 + v_2$,

$$Tv = Tv_1 + Tv_2 = T_1 v_1 + T_2 v_2.$$

Thus, T is decomposed into two operators T_1, T_2 acting on direct summands of V

and we say that T is the *direct sum* of T_1 and T_2, writing $T = T_1 \oplus T_2$. In the case $V = V^n$, we have $V_1 = V^m$ and $V_2 = V^p$, where $m + p = n$. Choosing a basis $\{u_1, \ldots, u_m\}$ in V^m and a basis $\{v_1, \ldots, v_p\}$ in V^p, it is clear that $\{u_i\} \cup \{v_i\}$ is a basis in V^n and the matrix A of T with respect to this basis is of the form

$$m \left\{ \begin{array}{c|c} A_1 & 0 \\ \hline 0 & A_2 \end{array} \right.$$
$$p \left\{ \phantom{\begin{array}{c|c} A_1 & 0 \\ \hline 0 & A_2 \end{array}} \right.$$

We say that A is the direct sum of A_1 and A_2, denoting this by $A = A_1 \oplus A_2$. (Compare this with Exercise 2.11c.)

Definition 2.1.17 Let A be an $n \times n$ matrix. The *determinant of A*, (denoted by det A) is a scalar defined as follows:

$$\det A = \sum \varepsilon_{i_1 i_2 \ldots i_n} a_{1 i_1} a_{2 i_2} \ldots a_{n i_n},$$

where the summation is over all $n!$ permutations, $(i_1, i_2, \ldots, i_n)$, of the sequence $(1, 2, \ldots, n)$ and $\varepsilon_{i_1 i_2 \ldots i_n} = +1$ if the permutation consists of an even number of transpositions and -1 if it consists of an odd number of transpositions.

Definition 2.1.18 Let I be the $n \times n$ identity matrix. An *elementary row (column) matrix* is one obtained from I by any one of the following *elementary row (column) operations*:

Type (1). Interchange any two row (column) vectors;

Type (2). add a scalar multiple of a row (column) vector to another row (column) vector;

Type (3). multiply a row (column) vector by a nonzero scalar α.

Remark 1 Operations (1) and (2) do not change the absolute value of the determinant of a matrix. Operation (3) multiplies the determinant by α.

Remark 2 If E is an elementary row (or column) matrix, and A is any matrix, then EA (AE) is the matrix obtained from A by performing the corresponding row (column) operation on A.

Remark 3 $\det EA = \det AE = \alpha \det A$ for any elementary matrix E, where $\alpha = 1$ if E is of type 2, $\alpha = -1$ if E is of type 1 and α is the scalar in type 3 above.

Remark 4 An elementary row (column) matrix is nonsingular, since the inverse is the elementary matrix obtained by applying the same column (row) operation to I.

Definition 2.1.19 Two matrices A and B are called *equivalent* if there exist two nonsingular matrices P and Q such that $A = PBQ$. The matrices A and B are called *similar* if $A = PBP^{-1}$.

Definition 2.1.20 $A = (a_{ij})$ is a *diagonal matrix* if $a_{ij} = 0$ for $i \neq j$. It is an *upper* *triangular* matrix if $a_{ij} = 0$ for $i > j$ and a *lower triangular* matrix if $a_{ij} = 0$ for $i < j$. (A is $n \times n$.)

Theorem 2.1.8 For any $n \times n$ matrix A there is a nonsingular matrix S such that $U = SA$ is upper triangular.

Proof. The matrix S is the product of elementary row matrices obtained by following the familiar *Gaussian elimination* procedure.

Consider column 1 of matrix A. We may assume $a_{11} \neq 0$. (If $a_{11} = 0$, and some $a_{i1} \neq 0$, interchange rows i and 1. This is accomplished by multiplying A by an appropriate elementary row matrix, E_1. Otherwise, skip col. 1.) Let E_{i1} be the elementary row matrix obtained from I by adding $(-(a_{i1}/a_{11}))$ times row 1 to row i. Then $A_1 = E_{n1} \cdots E_{21}A$ has zero's in the first column except for a_{11}. This process is repeated for the second column of matrix A_1, starting with the element in the second row. Continuing in this way, we obtain $U = E_{n,n-1} \cdots E_{21}A$, wherein all elements below the diagonal are zero. ■

Corollary 1 For any $n \times n$ matrix A, there is a nonsingular matrix B such that $L = AB$ is lower triangular.

Proof. Let $U = E_r \cdots E_1 A^T$. Then $L = U^T = A(E_1^T \cdots E_r^T)$ is lower triangular. The matrix $B = E_1^T \cdots E_r^T$ is nonsingular. ■

Corollary 2 For any $n \times n$ matrix, there is a nonsingular matrix S such that $L = SA$ is lower triangular.

Proof. As in Theorem 2.1.8, but eliminating x_n first in equations $n - 1, \ldots, 1$, then x_{n-1}, etc. ■

Corollary 3 Any $n \times n$ matrix A is equivalent to a diagonal matrix D such that $\det A = \pm \det D$.

Proof. Using S and B of the theorem and Corollary 1, we have $D = SAB$; i.e., first reduce to upper triangular form, obtaining nonzero elements $u_{11}, \ldots, u_{rr}, r \leq n$, and the remainder of zero's. Then reduce to lower triangular form.

Since S and B consist of products of elementary matrices of types 1 and 2 (Definition 2.1.18), and since $\det EA = \det AE = \pm \det A$ for any such elementary matrix E, $\det D = \det (SAB) = \pm \det A$. ■

Corollary 4 $\det A = \det A^T$.

Proof. $D = SAB$ implies $D = B^T A^T S^T$.

Theorem 2.1.9 $\det A \neq 0$ if and only if A is nonsingular.

Proof. By Corollary 3 of Theorem 2.1.8, $\det A = \pm \det D$. But $\det D = d_{11} d_{22} \cdots d_{nn}$. Further, a diagonal matrix is obviously nonsingular if and only if $d_{ii} \neq 0$ for all i. Hence $\det D \neq 0$ if and only if D is nonsingular. Since $D = SAT$, we have $A = S^{-1} D T^{-1}$ and $A^{-1} = TD^{-1}S$ if D^{-1} exists. Thus, A is nonsingular if and only if D is nonsingular. Hence, $\det A = \pm \det D \neq 0$ if and only if A is nonsingular. ■

Corollary If A is nonsingular, then there exists a sequence of elementary row matrices of type (1) (Definition 2.1.18), $E_1, \ldots, E_r$ such that the matrix $E_1 \ldots E_r A$ has only nonzero elements on the diagonal.

Proof. Using Definition 2.1.17 and Corollary 4 of Theorem 2.1.8, we can write

$$\det A = \sum \pm a_{i_1 1} a_{i_2 2} \cdots a_{i_n n},$$

where $\sigma = (i_1, i_2, \ldots, i_n)$ is a permutation of $(1, 2, \ldots, n)$. Since $\det A \neq 0$, at least one term in this sum is nonzero, say $a_{i_1 1} a_{i_2 2} \cdots a_{i_n n} \neq 0$. Using row interchanges to carry out the row permutation indicated, we can permute row i_1 and row 1 to obtain $E_r A$, which has $a_{i_1 1}$ in the (1, 1)-position. We next permute row i_2 and row 2 of $E_r A$ to obtain $E_{r-1} E_r A$ which has $a_{i_2 2}$ in the (2, 2)-position. Continuing in this way, we obtain the desired result. ■

Theorem 2.1.10 $\det (AB) = \det (A) \cdot \det (B)$.

Proof. By Corollary 3 of Theorem 2.1.8, $A = S_1 D_1 T_1$ and $B = S_2 D_2 T_2$ where D_1 and D_2 are diagonal and S_1, T_1, S_2, T_2 are products of elementary matrices. Hence, $AB = S_1 D_1 T_1 S_2 D_2 T_2$ and $\det AB = \det S_1 \det D_1 \det (T_1 S_2 D_2) \det T_2 = \pm \det D_1 \det D_2 = \det A \det B$. This follows from the fact that $\det (DC) = (d_{11} d_{22} \cdots d_{nn}) \det C$ for any diagonal matrix D and $\det SC = \det S \det C$ for any elementary row matrix S. ■

For later reference, we recall that by a *leading principal minor* of a square matrix $A = (a_{ij})$ we mean the determinant of an $r \times r$ matrix (a_{ij}) $1 \leq i, j \leq r$, for $r \leq n$.

2.2 EXAMPLES OF INFINITE-DIMENSIONAL SPACES

We have seen that any vector space of finite dimension n is isomorphic to the space V^n of n-dimensional column vectors. Thus, all vector spaces of a given finite dimension are isomorphic. The situation is rather different for infinite-dimensional vector spaces. Let us consider some examples of infinite-dimensional vector spaces.

Example 1 The set, s, of all real sequences (ξ_n) becomes a vector space if we define vector addition and scalar multiplication in the obvious way. Let $x = (\xi_n)$ and $y = (\eta_n)$ be two vectors in s. We define $x + y = (\xi_n + \eta_n)$ and for any real scalar λ we define $\lambda x = (\lambda \xi_n)$. Properties ω_1 and ω_2 (page 26) are readily verified.

Consider the set of vectors $u_1 = (1, 0, 0, \ldots), u_2 = (0, 1, 0, \ldots), u_3 = (0, 0, 1, 0, \ldots), \ldots$, and, in general, consider u_j to be the sequence having 1 as its jth term and 0 everywhere else, $j = 1, 2, \ldots$. The set $\{u_j\}$ is linearly independent. Hence, s is infinite-dimensional. Note that $\{u_j\}$ does not generate s and therefore is not a basis.

The *subspace generated by a set of vectors E* in a vector space V is defined to be the intersection of all the subspaces of V which contain E. (The intersection of a family of subspaces is obviously a subspace.) The subspace generated by E coincides with the set $V[E]$ of all linear combinations of vectors in E. The set $V[E]$ is clearly a subspace containing E and since any subspace is closed under (finite) linear combinations, $V[E]$ is contained in any subspace which contains E. In the space s of infinite sequences, the set of vectors $\{u_j\}$, defined in the preceding paragraph, generates a subspace s' consisting of all sequences (ξ_n) such that all but a finite number of the terms ξ_n are zero.

Definition 2.2.1 Let V be a vector space and $\{V_i\}_{i \in I}$ a family of subspaces of V. If every $x \in V$ can be expressed uniquely as a sum $\sum_{i \in I} x_i$, where $x_i \in V_i$ and all but a finite number of the x_i are 0, then V is said to be the *direct sum* of the V_i and we write $V = \sum'_{i \in I} V_i$.

In the space s, let $V_i = \{\lambda u_i : \lambda \in R\}$ where u_i is the vector having a 1 as its ith component and all other components 0. V_i is a one-dimensional subspace of s. The subspace s' is the direct sum of the V_i, since for every sequence $\{\xi_n\}$ in s', all but a finite number of the ξ_n are zero. Hence, such a sequence can be expressed uniquely as a sum $\sum \xi_n u_n$, where all but a finite number of the ξ_n are zero.

If $\{V_i\}_{i \in I}$ is a family of subspaces of a vector space V, the subspace generated by $\bigcup_{i \in I} V_i$ is called the *sum* of the V_i and is denoted by $\sum_{i \in I} V_i$. The sum $\sum_{i \in I} V_i$ is a direct sum if and only if every $x \in \sum_{i \in I} V_i$ can be expressed uniquely as a finite sum $\sum_{i \in I} x_i$. A necessary and sufficient condition that a sum $\sum_{i \in I} V_i$ of subspaces be direct is that $V_j \cap \sum_{i \neq j} V_i = \{0\}$ for all $j \in I$. The space V^n can be expressed as a direct sum in many ways: for example, as $\sum_{i=1}^{n} V_i$, where V_i is the one-dimensional subspace generated by the vector having 1 as its ith component and 0 for all other components.

Remark If X and Y are vector spaces, we can define a new vector space V consisting of pairs (x, y) of vectors $x \in X$, $y \in Y$ with an addition operation defined by $(x_1, y_1) + (x_2, y_2) = (x_1 + x_2, y_1 + y_2)$ and scalar multiplication $\lambda(x, y) = (\lambda x, \lambda y)$. With proper conventions, we can write $V = X \oplus Y$ (Exercise 2.11).

Example 2 The set m of bounded sequences is a subspace of s. The vectors u_j defined above are contained in m. Hence, m is infinite dimensional. The set c of convergent sequences is a subspace of m which also contains the u_j.

Example 3 For any real $p \geq 1$, let l_p be the set of all sequences (ξ_n) such that $(\sum_{n=1}^{\infty} |\xi_n|^p)^{1/p} < \infty$. For $p > 1$, the Minkowski inequality (see Exercises 3.9 and 3.10) shows that l_p is a subspace of m. For $p = 1$, we have $\sum |\xi_n + \eta_n| \leq \sum |\xi_n| + \sum |\eta_n| < \infty$.

Example 4 The set of real functions (Section 1.5) $x = x(t)$ defined for t in a set E is a vector space under the operations of "pointwise addition" $(x + y)(t) = x(t) + y(t)$ and $(\lambda x)(t) = \lambda x(t)$. If E is infinite, the space is infinite dimensional.

Example 5 Let X, Y be topological spaces. The set of continuous functions on X to Y is a vector space under pointwise addition and scalar multiplication as in example 4; i.e., if f and g are continuous functions, then $f(x) + g(x)$ is continuous. In particular, if $[a, b]$ is a closed interval of R, the set of continuous real-valued functions on $[a, b]$ is a vector space which is denoted by $C[a, b]$. $C[a, b]$ includes all polynomials in x, $x \in [a, b]$, since it includes the polynomial functions $1, x, x^2, x^3$, ..., and these form an infinite linearly independent set. (See Theorem 7.3, Corollary.)

Example 6 Using the Minkowski inequality for integrals, we prove that $f, g \in L^p(\mu)$, $1 < p < \infty$, implies $f + g$ is in $L^p(\mu)$ (Definition 1.6.1). Also, $\lambda f \in L^p(\mu)$. For $p = 1$, and $p = \infty$ the result follows from the triangle inequality $|f + g| \leq |f| + |g|$. Hence, $L^p(\mu)$, $1 \leq p \leq \infty$, is a vector space. (Strictly speaking, the vectors in $L^p(\mu)$ are equivalence classes of functions which differ only on a set of measure zero. We agree to regard $f = g$ if $f(x) = g(x)$ *almost everywhere*, i.e., except on a set of measure zero.) For $L^p(R^n)$, it is easy to exhibit an infinite linearly independent set of functions.

2.3 LINEAR FUNCTIONALS AND MULTILINEAR MAPPINGS

Let V be a real vector space. The set of all linear operators (Definition 2.1.13) $L(V; R)$, where R is the real line taken as a one-dimensional vector space, is a vector space called the *algebraic dual* or *algebraic conjugate* space of V. We shall denote it by V^* and call its elements *linear functionals*. (If V is a vector space over the complex field C, then $L(V; C)$ is the dual of V.) For any $x \in V$ and $x^* \in V^*$ we denote the value $x^*(x)$ by $\langle x, x^* \rangle$.

The dual of V^* is again a vector space and is denoted by V^{**}. Holding x fixed and allowing x^* to vary over V^*, we obtain from $\langle x, x^* \rangle$ a linear functional on V^*. Thus, $V \subset V^{**}$. It can be shown that $V = V^{**}$ if and only if V is finite dimensional [2.2]. (Compare this with the conjugate space in Section 3.5.)

Example 1 Suppose $V = V^n$ (real). We shall show that $V^* = V^n$. Let $x^T = (\xi_1, \ldots, \xi_n)$ and $y^T = (\eta_1, \ldots, \eta_n)$. Holding y fixed and allowing x to vary over V^n, we obtain a linear functional y^* by defining

$$\langle x, y^* \rangle = x^T y = \sum_{j=1}^n \xi_j \eta_j.$$

Hence, $V^n \subset V^*$. Conversely, if $y^* \in V^*$, let $y = \sum_{j=1}^n \eta_j e_j$, where $\eta_j = \langle e_j, y^* \rangle$ and $\{e_j\}$ is the canonical basis in V^n. Since $x = \sum_1^n \xi_j e_j$, the linearity of y^* yields

$$\langle x, y^* \rangle = \sum_{j=1}^n \xi_j \langle e_j, y^* \rangle = \sum_{j=1}^n \xi_j \eta_j = x^T y.$$

Thus, we may identify y^* with y, and $V^* \subset V^n$. In fact, the mapping $y^* \longmapsto y$ is an isomorphism. (See Exercise 2.9.)

If the scalar field is the complex numbers, we define $\langle x, y^* \rangle = x^T \bar{y}$, where $\bar{y}^T = (\bar{\eta}_1, \ldots, \bar{\eta}_n)$ is the complex conjugate of y. As above, we can show that $V^{(n)*} = V^{(n)}$.

The mapping f defined by $(x, x^*) \mapsto \langle x, x^* \rangle$ is an example of a bilinear mapping

of $V \times V^*$ into R, since it is linear in each variable, that is,
$$f(x + y, x^*) = f(x, x^*) + f(y, x^*),$$
$$f(x, x^* + y^*) = f(x, x^*) + f(x, y^*),$$
$$f(\alpha x, x^*) = \alpha f(x, x^*) = f(x, \alpha x^*).$$
We extend this concept to the general case as follows:

Definition 2.3.1 Let $V_1, \ldots, V_n, V$ be vector spaces over the same field. A mapping $f : V_1 \times \cdots \times V_n \longrightarrow V$ is called *multilinear* if for each $1 \leq i \leq n$ the mapping
$$x_i \mapsto f(x_1, \ldots, x_{i-1}, x_i, x_{i+1}, \ldots, x_n)$$
defined by f and the arbitrary fixed elements $x_j \in V_j, j \neq i$, is a linear operator on V_i into V. The set of such multilinear mappings is denoted by $L^n(V_1, \ldots, V_n; V)$. (For $n = 2$, f is called a *bilinear mapping*.)

The set $L^n(V_1, \ldots, V_n; V)$ is a vector space under the operations of pointwise addition and scalar multiplication just as for the case $L(V_1; V)$; that is, if f, $g \in L^n(V_1, \ldots, V_n; V)$, so is $f + g$ and λf. A multilinear map $f \in L^n(V_1, \ldots, V_n; V)$ can also be regarded as (or associated with) a mapping $f_1 \in L(V_1; L^{n-1}(V_2, \ldots, V_n; V))$ by defining $f_1(x_1) = f_{x_1}$ where $f_{x_1}(x_2, \ldots, x_n) = f(x_1, \ldots, x_n)$.

By induction, we obtain a linear mapping
$$\bar{f} \in L(V_1; L(V_2; L \ldots ; L(V_{n-1}; L(V_n; V) \ldots).$$
The association $f \to \bar{f}$ is an isomorphism. We illustrate the proof of this in the case $n = 2$.

Theorem 2.3.1 Let V_1, V_2, V be vector spaces over the same field. Then $L(V_1; L(V_2; V))$ is isomorphic to $L^2(V_1, V_2; V)$.

Proof. Let $f \in L^2(V_1, V_2; V)$. For $x \in V_1$ and $y \in V_2$ we define $f_x : V_2 \longrightarrow V$ by $f_x(y) = f(x, y)$. The mapping f_x is obviously linear in y so that $f_x \in L(V_2; V)$. The mapping $\bar{f} : x \mapsto f_x$ is a linear mapping of V_1 into $L(V_2; V)$. The function $\psi : f \mapsto \bar{f}$ is the desired isomorphism of $L^2(V_1, V_2; V)$ onto $L(V_1; (V_2; V))$. To see that ψ is a homomorphism, suppose $\psi : g \mapsto \bar{g}$. Then
$$\overline{f + g} : x \mapsto (f + g)_x$$
and
$$(f + g)_x(y) = (f + g)(x, y) = f(x, y) + g(x, y) = f_x(y) + g_x(y)$$
for all $y \in V_2$. Hence,
$$(f + g)_x = f_x + g_x$$
and
$$\overline{f + g} : x \mapsto f_x + g_x = \bar{f}(x) + \bar{g}(x).$$
Thus, $\bar{f} + \bar{g} = \overline{f + g}$. Similarly, we can show that $\lambda \bar{f} = \overline{\lambda f}$.

Now, let $\bar{f} \in L(V_1; L(V_2; V))$. Define $f: V_1 \times V_2$ by $f(x, y) = \bar{f}(x)(y)$. The function $\varphi: \bar{f} \longmapsto f$ is the inverse of ψ. Thus ψ is an isomorphism. ∎

Remark Since $\bar{f}$ is a linear operator, we shall follow our usual convention and write $\bar{f}(x) = \bar{f}x$. Thus, $f(x, y) = \bar{f}xy$. We now agree to identify f and $\bar{f}$, in view of the isomorphism. Thus, we shall write

$$f(x_1, \ldots, x_n) = fx_1 x_2 \cdots x_n,$$

where $fx_1 = \bar{f}x_1 = \bar{f}(x_1), fx_1 x_2 = \bar{f}(x_1)(x_2)$, etc.

A multilinear mapping f is called *symmetric* if $V_1 = V_2 = \cdots = V_n = W$ and for all n-tuples $(x_1, \ldots, x_n)$, $x_i \in W$ ($1 \leq i \leq n$), $fx_1 x_2 \cdots x_n = fx_{i_1} \cdots x_{i_n}$, where $i_1, \ldots, i_n$ is an arbitrary permutation of the indices $1, \ldots, n$. For $n = 2$, we have simply $fx_1 x_2 = fx_2 x_1$ for all $x_1, x_2 \in W$.

Consider two vector spaces V, W and their algebraic conjugates V^*, W^*. Suppose $T \in L(V; W)$. Consider $w = Tv$, where $v \in V$ is arbitrary. For arbitrary $w^* \in W^*$, $\langle Tv, w^* \rangle$ defines a linear mapping $V \longrightarrow R$. Hence, there exists $v^* \in V^*$ such that $\langle v, v^* \rangle = \langle Tv, w^* \rangle$. Define the operator $T^*: W^* \longrightarrow V^*$ by $T^*: w^* \longmapsto v^*$. Then we can write

$$\langle Tv, w^* \rangle = \langle v, T^*w^* \rangle \qquad \text{all } v \in V, w^* \in W^* \tag{1}$$

as the defining equation of T^*, which is called the (*algebraic*) *adjoint* of T. It is easy to prove that T^* is a linear operator, i.e., $T^* \in L(W^*; V^*)$.

Example 2 If $V = V^n$ and $W = V^m$ are finite-dimensional real vector spaces, we have seen that $V^* = V^n$ and $W^* = W^m$. If A is an $m \times n$ matrix representing T, then, since we can take $w^* = w \in V^m$, we may write (1) above as

$$\langle Av, w \rangle = \langle v, A^*w \rangle \qquad \text{all } v \in V^n, w \in V^m.$$

Letting $A = (a_{ij})$, $v^T = (v_1, \ldots, v_n)$ and $w^T = (w_1, \ldots, w_m)$, we have $(Av)^T = (\sum_{j=1}^n a_{1j}v_j, \ldots, \sum_{j=1}^n a_{mj}v_j)$. Then

$$\langle Av, w \rangle = (Av)^T w = \sum_{i=1}^m w_i \sum_{j=1}^n a_{ij}v_j = \sum_{i=1}^m \sum_{j=1}^n a_{ij} w_i v_j$$
$$= \sum_{j=1}^n v_j \sum_{i=1}^m a_{ij} w_i = v^T(A^T w) = \langle v, A^T w \rangle.$$

Since this holds for all v, $A^T w = A^* w$ for all w; i.e., $A^* = A^T$. Thus, the adjoint of a real matrix is simply its transpose. (If A is complex, then $A^* = \bar{A}^T$, the transpose of the complex conjugate matrix $(\bar{a}_{ij})$, since we represent w^* by the vector $\bar{w}$.)

If $A \in L(V; W)$ and $B \in L(W; X)$, then for $v \in V$, $x^* \in X^*$,

$$\langle BAv, x^* \rangle = \langle Av, B^*x^* \rangle = \langle v, A^*B^*x^* \rangle,$$

showing that $(BA)^* = A^*B^*$. It follows that if A^{-1} exists in $L(W; V)$, then

$$I_v^* = (A^{-1}A)^* = A^*A^{-1*} \quad \text{and} \quad (AA^{-1})^* = A^{-1*}A^* = I_w^*.$$

Since $I^* = I$ for any identity map, we see that $(A^*)^{-1} = A^{-1*}$.

REFERENCES

2.1. S. MacLane and G. Birkhoff, *Algebra*, Macmillan, New York (1967)
2.2. A. E. Taylor, *Introduction to Functional Analysis*, Wiley, New York (1958)
2.3. S. Lang, *Algebra*, Addison-Wesley, Reading, Mass. (1965)
2.4. P. C. Shields, *Linear Algebra*, Addison-Wesley, Reading, Mass. (1964)
2.5. R. Penrose, "A generalized inverse for matrices," *Proc. Cambridge Phil. Soc.* **51**, 406–418 (1955)
2.6. C. A. Rohde, "Some results on generalized inverses," *SIAM Review* **8**, No. 2, 201–205 (April 1966)
2.7. A. Ben-Israel and A. Charnes, "Contributions to the theory of generalized inverses," *J. SIAM* **11**, No. 3, 667–699 (Sept. 1963)
2.8. R. P. Tewarson, "On some representations of generalized inverses," *SIAM Review* **11**, No. 2, 272–275 (April 1969)
2.9. T. N. E. Greville, "The pseudo-inverse of a rectangular or singular matrix and its applications to the solution of systems of linear equations," *SIAM Review* **1**, 38–43 (1959)
2.10. T. N. E. Greville, "Some applications of the pseudoinverse of a matrix," *SIAM Review* **2**, No. 1, 15–22 (Jan. 1960)
2.11. B. Noble, *Applied Linear Algebra*, Prentice-Hall, Englewood Cliffs, N.J. (1969)

EXERCISES

2.1 Establish that the set of all nth-order square matrices form a linear algebra under Definitions 2.1.6, 2.1.7 and 2.1.8; i.e., show that the properties α_1, α_2, and α_3 hold. Show that α_{10}–α_{14} holds for all $m \times n$ matrices. Show that $A(BC) = (AB)C$ whenever all required products are defined.

2.2 Establish that the set of all n-dimensional column vectors forms an n-dimensional vector space as defined by properties ω_1, ω_2, ω_3.

2.3 Let $A = \begin{pmatrix} 1 & 0 \\ 1 & 1 \end{pmatrix}$ and $B = \begin{pmatrix} 1 & 1 \\ 0 & 1 \end{pmatrix}$. Show that $AB \neq BA$.

2.4 Show that the algebra of linear operators $T^{(n)}$ is isomorphic to the linear algebra of matrices $A^{(n)}$, by the correspondence (2.1.8); i.e., show that this correspondence establishes a one-one mapping, ϕ, from T^n onto A^n such that $\phi(\lambda T) = \lambda \phi(T)$,

$$\phi(T_1 + T_2) = \phi(T_2) + \phi(T_2) \quad \text{and} \quad \phi(T_1 T_2) = \phi(T_1)\phi(T_2).$$

2.5 Show that the null space of a linear operator T is a vector space.

2.6 If L_1 and L_2 are lower triangular matrices, then $L_1 L_2$, $L_2 L_1$, and $L_1 + L_2$ are likewise lower triangular. If U_1 and U_2 are upper triangular, so are $U_1 U_2$, $U_2 U_1$, $U_1 + U_2$. Prove these assertions.

2.7 If L is an $n \times n$ lower triangular matrix having zero elements on the diagonal, then $L^n = 0$. Prove this. (A matrix A such that $A^n = 0$ for some n is called *nilpotent*.)

2.8 Show that the unit step functions f_n given by

$$f_n(x) = \begin{cases} 1 & n \leq x \leq n+1 \\ 0 & \text{otherwise} \end{cases}$$

are an infinite linearly independent set in $L^p(-\infty, \infty)$, $p \geq 1$.

2.9 Let $V = V^n$.

a) Show that $V^{**} = V^n$; i.e., show that the mapping $y^{**} \mapsto y$ given in Section 2.3 (Example 1) is an isomorphism. [*Hint:* The sum of two functionals is defined by $\langle y_1^* + y_2^*, y \rangle = \langle y_1^*, y \rangle + \langle y_2^*, y \rangle$. Also, y^* is uniquely determined by its values $\eta_j = \langle e_j, y^* \rangle$ on the basis $\{e_j\}$.]

b) Let $y_1^*, \ldots, y_n^* \in V^*$. Prove that the set $\{y_i^*\}$ is linearly independent if and only if $\langle y, y_i^* \rangle = 0, 1 \le i \le n$ implies $y = 0$. [*Hint:* Since $V^{**} = V^n$, we may write $y_i^* = (\eta_{i1}, \ldots, \eta_{in})$. Then for $y = (\eta_1, \ldots, \eta_n)$, $\langle y, y_i^* \rangle = \sum_{j=1}^n \eta_j \eta_{ij}$. Consider the matrix (η_{ij}).]

c) Let $\{F_1, \ldots, F_n\}$ be a linearly independent set of functionals in V^*. Show that there exists a linearly independent set $v_1, \ldots, v_n \in V$ which is *biorthonormal* to the F_i, that is, $F_i(v_j) = \delta_{ij}$. Also, show that for any $x \in V$, $x = \sum_{i=1}^n F_i(x) v_i$. [*Hint:* let $\{x_1, \ldots, x_n\}$ be a basis in V. Prove that det $(F_i(x_j)) \ne 0$, since otherwise there exist scalars $a_1, \ldots, a_n$ not all zero such that

$$F_i\left(\sum_1^n a_j x_j\right) = 0, 1 \le i \le n,$$

and therefore $\sum_1^n a_j x_j = 0$ by part (b). Then set $v_j = \sum_{i=1}^n a_{ij} x_i$ and determine the a_{ij} by requiring $F_i v_j = \delta_{ij}, 1 \le i \le n$.]

d) An alternative scheme for obtaining a biorthonormal set is possible if $F_i \in V^*$ and $x_i \in V$ are such that the principal minors of the matrix $(F_i(x_j))$ are all nonzero. In this scheme, we define $F_1' = a_{11} F_1$, $F_2' = a_{21} F_1 + a_{22} F_2$, etc., and $x_1' = x_1$, $x_2' = b_{21} x_1 + x_2$, $x_3' = b_{31} x_1 + b_{32} x_2 + x_3$, etc., and impose the biorthonormality conditions $F_i'(x_j') = \delta_{ij}$. Prove that the a_{ij} and b_{ij} are uniquely determined and $a_{ii} \ne 0$. Let $A = (a_{ij})$, $B = (b_{ij})$, and $C = (F_i(x_j))$. Show that $C = A^{-1}(B^T)^{-1}$ is the LU factorization of C.

e) Let $V = V^n$ be the n-dimensional vector space over the field of complex numbers. Show that $V^* = V^n$; i.e., that for any $y^* \in V^*$ there exists $y \in V$ such that $\langle x, y^* \rangle = x^T \bar{y}$, where $\bar{y}$ is the complex conjugate of y. [*Hint:* $\bar{y} = \sum_{j=1}^n \bar{\eta}_j e_j$, where $\eta_j = \langle e_j, y^* \rangle$. See Exercise 3.3(a). Note that we could also represent y^* by the vector $y = \sum_{j=1}^n \eta_j e_j$, but then $y^*(x) \ne x^T \bar{y}$ as required in the complex inner product.]

2.10 Let A be an $m \times n$ matrix.

a) Prove that A is equivalent to a matrix D of the form

$$\left[\begin{array}{c|c} I_r & 0 \\ \hline 0 & 0 \end{array}\right]$$

where I_r is the $r \times r$ identity, $r \le \min(m, n)$. [*Hint:* Use elementary row and column matrices as in Theorem 2.1.8 and its corollaries.]

b) Let $B = PAQ$, where P and Q are nonsingular matrices. Prove that A and B have the same rank, r. [*Hint:* Q is $n \times n$ and has rank n, P is $m \times m$ and has rank m, and B is $m \times n$. The domain of A is the range of Q, hence is of dimension n. The domain of P is the range of A, hence is of dimension $r \le n$. Since P is nonsingular, its range (which is also the range of B) has dimension r]

c) Prove that the row rank of A (i.e., the maximum number of linearly independent rows) equals its (column) rank. [*Hint:* $A = PDQ$, where D is as given in part (a). Rank $A =$ rank D by part (b). $A^T = Q^T D^T P^T$.]

2.11 a) Let X and Y be vector spaces. Define the vector space V as in the Remark following Definition 2.2.1. Identify X with the (isomorphic) subspace of V consisting of all pairs $(x, 0)$, $x \in X$. Similarly, identify Y with the subspace $(0, y)$. Prove that $V = X \oplus Y$. (This is sometimes called the *external* direct sum of X and Y.)

b) Extend (a) to any family $\{V_i\}$ of vector spaces.

c) Let $T: X \oplus Y \longrightarrow Z$ be a linear operator mapping a direct sum into the vector space Z. Show that there exist two linear operators T_x, T_y such that $T(x, y) = T_x x + T_y y$. [*Hint:* Define $T_x x = T(x, 0)$ and $T_y y = T(0, y)$. Note that T need not be a direct sum of T_x and T_y.]

2.12 Let A be an $n \times n$ real matrix.

a) Show that there exists a polynomial $p(\lambda)$ with real coefficients such that $p(A) = 0$. [*Hint:* $I, A, A^2, \ldots, A^m$ cannot be linearly independent for $m \geq n^2$.]

b) The polynomial $m(\lambda)$ of minimum degree such that $m(A) = 0$ is called the minimum polynomial of A. Prove: If $f(A) = 0$ for some polynomial $f(\lambda)$, then $m(\lambda)$ divides $f(\lambda)$. [*Hint:* $f(\lambda) = q(\lambda)m(\lambda) + r(\lambda)$, where degree $r(\lambda) <$ degree $m(\lambda)$.]

c) Let $A = (a_{ij})$. By a k-rowed minor of A we mean a matrix

$$A_{(i_1,\ldots,i_k;\,j_1,\ldots,j_k)} = \begin{bmatrix} a_{i_1 j_1} & \cdots & a_{i_1 j_k} \\ a_{i_2 j_1} & & a_{i_2 j_k} \\ \vdots & & \vdots \\ a_{i_k j_1} & & a_{i_k j_k} \end{bmatrix}$$

If $j_1 = i_1, j_2 = i_2, \ldots, j_k = i_k$, then the minor is called a *principal minor* of A. Now, let $A_{ij} = \pm \det A_{(1,\ldots,i-1,i+1,\ldots,n;\,1,\ldots,j-1,j+1,\ldots,n)}$, where the sign is $+$ or $-$ according as the permutation

$$\begin{pmatrix} i, 1, \ldots, i-1, i+1, \ldots, n \\ j, 1, \ldots, j-1, j+1, \ldots, n \end{pmatrix}$$

is even or odd. A_{ij} is called the cofactor of a_{ij}. (It is obtained by deleting the ith row and jth column.) Prove that $\delta_{ij} \det A = \sum_{k=1}^{n} a_{ik} A_{jk}$ for any $1 \leq i, j \leq n$. (δ_{ij} is Kronecker's delta.) The matrix $A^{\mathrm{adj}} = (A_{ij})^T$ is the *classical adjoint* of A. Prove $A A^{\mathrm{adj}} = \det(A) I$.

d) Show how the above can be used to evaluate the determinant of a tridiagonal matrix A (i.e., a matrix having zero's everywhere except possibly on the diagonal a_{ii}, the superdiagonal $a_{i,i+1}$ and the subdiagonal $a_{i+1,i}$). [*Hint:* Obtain a recursion by expanding by minors of the last column.]

e) State and prove Cramer's rule for the solution of $Ax = b$ using determinants.

2.13 A *semigroup* S is a set with a binary operation (usually called *multiplication*) which is associative; i.e., for x, y in S, the *product* xy is an element of S and $(xy)z = x(yz)$ for any x, y, z in S.

a) Show that the set $A^{(n)}$ of all $n \times n$ real matrices is a semigroup under the operation of matrix multiplication.

b) An element I in a semigroup S is called an *identity* element if $IA = AI = A$ for all

A in S. Prove that an identity element must be unique; i.e., there is at most one identity.

c) Let S be a semigroup with an identity element I. An element $A \in S$ is called *invertible* (also *nonsingular*) if there exists an element A^{-1} such that $A^{-1}A = AA^{-1} = I$. A^{-1} is called the *inverse* of A. Prove that the inverse is unique. [*Hint*: Suppose B is also an inverse of A. Then $B = BI = B(AA^{-1}) = (BA)A^{-1} = IA^{-1} = A^{-1}$.]

d) $A \in S$ is called *regular* if there exists $C \in S$ such that $ACA = A$. Clearly, if A is invertible, then it is regular. Show, by example, that the converse is false. *Hint:* In $A^{(2)}$, take

$$A = \begin{bmatrix} 2 & 0 \\ 0 & 0 \end{bmatrix}, \quad C = \begin{bmatrix} \frac{1}{2} & 0 \\ 0 & 0 \end{bmatrix}.$$

The element C is called a *generalized inverse* of A.

e) Two elements $A, B \in S$ are called *reflexive generalized inverses* of each other if

$$ABA = A \quad \text{and} \quad BAB = B.$$

Prove that A is regular if and only if it has at least one reflexive generalized inverse. [*Hint:* If $ACA = A$, take $B = CAC$. This also shows the nonuniqueness of the generalized inverse.]

f) Two matrices A and $A^\dagger$ are called *pseudoinverses* of each other if they are generalized inverses of each other and

$$(AA^\dagger)^* = AA^\dagger \quad \text{and} \quad (A^\dagger A)^* = A^\dagger A.$$

Show that a pseudoinverse is unique. (See [2.5], [2.6].)

g) Show that $AA^gA = A$ implies that $P = A^gA$ is an *idempotent*; i.e., $P^2 = P$. Also show that $(I - A^gA)y$ is in the null space of A for any y. Prove that $A^gAA^g = A^g$ implies that the intersection of the range of A^g and the null space of A is $\{0\}$. This implies that $\{(I - A^gA)y\}$ is the null space of A. [*Hint:* $Az = 0$ and $z = A^gy$ implies $A^gAz = A^gAA^gy = A^gy = z$.] Note that, in general, A is a rectangular $m \times n$ matrix and $A^\dagger$ must then be $n \times m$. Thus, the triple products $AA^\dagger A$ and $A^\dagger AA^\dagger$ are defined. The definitions in parts (b) and (c) of *regular* and *generalized inverse* make sense in a *partial semigroup*—i.e., a set S with a partial binary operation of multiplication defined only for certain pairs $A, B \in S$—and when $(AB)C$ and $A(BC)$ are defined, then they are equal. The set of all rectangular (real) matrices is a partial semigroup under the usual matrix multiplication operation.

h) Let A^g be a generalized inverse of the $m \times n$ matrix A. Let u be an n-dimensional vector and b an m-dimensional vector such that $Au = b$. Prove that the system $Ax = b$ has A^gb as a solution; i.e., if b is in the range of A, then A^gb is a solution. Prove that the general solution of $Ax = b$ is

$$x = A^gb + (I - A^gA)y, \tag{1}$$

where y is an arbitrary n-dimensional vector. [*Hint:* See [2.5] or note that

$$b = Au = AA^gAu = A(A^gb) \quad \text{and} \quad A(I - A^gA)y = (A - AA^gA)y = 0.]$$

i) Let A be an $m \times n$ matrix of rank n. Show that the pseudoinverse of A is given by

$$A^\dagger = (A^*A)^{-1}A^* \tag{2}$$

[*Hint*: A^*A is an $n \times n$ matrix of rank n.

$$A(A^*A)^{-1}A^*A = A. \quad (AA^\dagger)^* = (A(A^*A)^{-1}A^*)^* = A(A^*A)^{*-1}A^* = AA^\dagger.]$$

Indeed, we see that $A^\dagger$ is a left inverse, since

$$A^\dagger A = I_n. \tag{3}$$

In this case $x = A^\dagger b = (A^*A)^{-1}A^*b$ is the unique solution of $Ax = b$. Hence,

$$A^*Ax = A^*b. \tag{4}$$

These equations (4) are called the *normal* equations. (See Section 7.4.1.)

j) Let A be an $m \times n$ matrix of rank $r > 0$. Prove that $A = BC$, where B is $m \times r$ and C is $r \times n$ and both B and C are of rank r. [*Hint*: Take B to be a matrix consisting of r linearly independent columns of A. Then $B : V^r \longrightarrow V^m$ is onto. Hence, $BC_i = A_i$ has a solution for each column A_i of A.] Now prove that the pseudoinverse of A is given by

$$A^\dagger = C^*(CC^*)^{-1}(B^*B)^{-1}B^*. \tag{5}$$

Verify that

$$(A^\dagger)^* = (A^*)^\dagger. \tag{6}$$

k) Prove that $P = AA^\dagger$ has the following two properties which characterize a *projection operator*: $P^2 = P$ and $P^* = P$. [*Hint*: Use Eq. (5) in part (j) above.] The set of vectors x such that $Px = x$ is a subspace M_p associated with the projection operator P. Show that for $P = AA^\dagger$ the subspace $M_p = R(A)$, the range of A. [*Hint*: $AA^\dagger(Ay) = (AA^\dagger A)y = Ay$. Thus, $R(A) \subset M_p$. Conversely, $(AA^\dagger)x = x$ implies $x = A(A^\dagger x) \in R(A)$.]

By the symmetry of the relations in parts (e) and (f) it follows that $A^\dagger A$ is a projection P' and $M_{p'} = R(A^\dagger)$. Use these results to establish (1) in part (h).

2.14 Let $V = V_1 + V_2$ be a sum of two subspaces V_1 and V_2. (See Remark following Theorem 2.1.7.) If V is finite dimensional, prove that dim V = dim V_1 + dim V_2 − dim $(V_1 \cap V_2)$. [*Hint*: Let $\{u_1, \ldots, u_r\}$ be a basis for the subspace $V_1 \cap V_2$. (If $V_1 \cap V_2 = \{0\}$, the basis is empty.) By Theorem 2.1.2, we can extend this to a basis for V_1, $\{u_1, \ldots, u_r, u_{r+1}, \ldots, u_m\}$, and to another basis for V_2, $\{u_1, \ldots, u_r, v_{r+1}, \ldots, v_n\}$. Show that $\{u_1, \ldots, u_m, v_{r+1}, \ldots, v_n\}$ is a basis for V.]

2.15 Bilinear mappings which have scalar values are called *bilinear forms*; e.g., if V_1, V_2 are real vector spaces, then a bilinear mapping $f : V_1 \times V_2 \longrightarrow R$ is a (*real*) *bilinear form*. (The real line R is a one-dimensional space.) If $V_1 = V_2 = V$, then a symmetric bilinear mapping $f : V \times V \longrightarrow R$ gives rise to a mapping $g : V \longrightarrow R$ defined by $g(x) = f(x, x)$. The mapping g is called a *quadratic form* on V.

a) Let $T \in L(V; W)$. For arbitrary $v \in V$ and $w \in W^*$, let $f(v, w^*) = \langle Tv, w^* \rangle$. Verify that $f : V \times W^* \longrightarrow R$ is a bilinear form and if $V = V^n$ and $W = V^m$, then T is represented by an $m \times n$ matrix A and $f(v, w) = w^T Av$.

b) Conversely, if $f : V^n \times V^m \longrightarrow R$ is a bilinear form, then there exists an $m \times n$ matrix A such that $fvw = w^T Av$, $v \in V^n$, $w \in V^m$. Prove this. [*Hint*: Choose bases $\{v_i\}$ in V^n, $\{w_j\}$ in V^m and define $a_{ij} = fv_i w_j$. Take $A = (a_{ij})$.] Show further that f is symmetric if and only if A is symmetric.

c) An alternative definition of a *quadratic form* g on a real vector space V is the following: g is a mapping of V into R such that $g(-v) = g(v)$ for all v and the mapping f given by

$$2f(u, v) = g(u + v) - g(u) - g(v) \qquad (1)$$

is a bilinear form $f : V \times V \longrightarrow R$. The mapping f is said to be obtained by *polarizing* g. Verify that f is symmetric. Prove that $g(2u) = 4g(u)$, $g(0) = 0$, and therefore, that $f(u, u) = g(u)$. For any bilinear form $F : V \times V \longrightarrow R$, $F(u, u)$ is a quadratic form $g(u)$ with $2f(u, v) = F(u, v) + F(v, u)$. Verify this. [*Hint:* $g(u + v + w) - g(u + v) - g(u + w) - g(v + w) + g(u) + g(v) + g(w) = 0$ by the additivity of $f(u, v + w)$. Then set $v = u$ and $w = -u$ to get the first result and $u = v = w = 0$ to get the second.] Thus, from a quadratic form arises a symmetric bilinear form. Conversely, let $g(u) = f(u, u)$, where f is a symmetric bilinear form. Show that (1) above is satisfied and $g(u) = g(-u)$. This shows that g is a quadratic form according to the alternative definition.

2.16 A subset C of a real vector space V is called a *cone* if $\alpha C \subset C$ for all $\alpha > 0$; i.e., for any vector $v \in C$, every positive multiple $\alpha v \in C$ also.

a) A cone C which is a convex set (section 3.4) is called a *convex cone*. Show that a cone $C \subset V$ is a convex cone if and only if $C + C \subset C$. ($C + C = \{x + y : x, y \in C\}$.)

b) Let C be a convex cone containing 0. Define the relation $x \geq y$ to hold for $x, y \in V$ if and only if $x - y \in C$. Show that $\geq$ is reflexive and transitive. (A reflexive transitive relation is called an *order relation*.) Show that this relation is *compatible with the linear structure of* V (i.e., $x \geq y$ implies $x + z \geq y + z$ and $\alpha x \geq \alpha y$ for all $z \in V$, $\alpha > 0$). V is then called a *partially ordered vector space*.

c) Let $\geq$ be an order relation compatible with the linear structure of V. Show that the set of nonnegative vectors $x \geq 0$ is a convex cone. E.g. in V^n, define $x \geq y$ to mean $x_i \geq y_i$, $1 \leq i \leq n$ and verify compatibility.

d) Let P be the vector space of polynomials in one variable with real coefficients. (See Chapter 7.) Let C be the subset of polynomials having positive leading coefficient. Prove that C is a convex cone such that $C \cap (-C) = \{0\}$ and $C \cup (-C) = P$.

2.17 a) Let V be a vector space and $W \subset V$ a subspace. Regarding V as an additive group, we may form the factor group V/W consisting of equivalence classes, where x and y are in the same equivalence class if and only if $x - y \in W$. Denote this equivalence by $x \equiv y \mod W$ or simply, $x \equiv y$. Show that $x \equiv y$ implies $\lambda x \equiv \lambda y$ for any scalar λ. Hence, for any equivalence class $\hat{x}$ we can define $\lambda \hat{x} = \widehat{\lambda x}$ where x is any element in the class $\hat{x}$. Show that this scalar multiplication and the usual addition of equivalence classes make V/W a vector space. This is called the *factor space* or *quotient space* determined by W. The mapping $\varphi : V \longrightarrow V/W$ which maps every $x \in V$ into the equivalence class $\hat{x}$ containing x is called the *natural* (or *canonical*) *homomorphism* of V onto V/W. Verify that φ is a linear operator on V to V/W.

b) The dimension of V/W is called the *codimension* of W. Suppose the codimension is $n < \infty$ and $\{\bar{u}_1, \ldots, \bar{u}_n\}$ is a basis of V/W. Show that any set $\{u_1, \ldots, u_n\}$ such that $\varphi(u_i) = \hat{u}_i (1 \leq i \leq n)$ is linearly independent in V. (We say that $\{\bar{u}_i\}$ has been *lifted* into the set $\{u_i\}$.) Let U be the subspace spanned by $\{u_i\}$. Prove that $V = W \oplus U$, the direct sum.

c) If W has codimension $n = 1$, then W is called a *hyperplane*. (If dim $V = 3$, then a hyperplane is simply a plane passing through the origin.) Show that a hyperplane is a maximal proper subspace.

d) A topology on V induces a topology on V/W, called the *quotient topology*, as follows: a subset $\hat{E} \subset V/W$ is a neighborhood of zero in W if and only if the pre-image E of $\hat{E}$ under the canonical mapping φ is a neighborhood of zero in V. Verify that φ is continuous with respect to this topology.

CHAPTER 3

TOPOLOGICAL VECTOR SPACES

3.0 INTRODUCTION AND GUIDE TO CHAPTER 3

In this chapter, we fuse together the topological concepts of Chapter 1 and the algebraic ideas of Chapter 2 to form the important notion of a topological vector space. Almost all the sets which we shall consider in the sequel are topological vector spaces. Indeed, they are metric vector spaces and, most often, normed vector spaces, although seminorms play a significant role too. Since the wide range of subjects covered may be somewhat bewildering to the reader who has had no previous exposure to functional analysis, we offer the following suggestions as a guide to the study of this chapter.

Since one of our basic aims is to provide a unified treatment of finite-dimensional and infinite-dimensional cases wherever possible, we follow the abstract approach—dealing in general concepts which have been abstracted from both cases. At the same time, another basic aim is to point up the differences between the finite-dimensional and infinite-dimensional cases as they influence computational methods. Therefore, after each abstract concept is introduced, we illustrate it with a finite-dimensional case and, if feasible, with an infinite case. However, on a first reading, the student may ignore or read cursorily the infinite-dimensional material. He will still be adequately prepared for Chapters 4 and 6 and much of 5 if he proceeds roughly as follows.

Ignore (or read lightly) the Examples 1 to 8. Similarly, one may skip Exercises 3.26 through 3.33 and 3.42 through 3.45. Central definitions and theorems are

Definitions 3.1.2, 3.1.3, 3.3.1, 3.3.2, 3.3.4, 3.5.1, matrix norms induced by l_1, l_2, l_∞-norms, 3.5.2, adjoint of an operator, self-adjoint, Hermitian, symmetric matrix, 3.7.1 (eigenvalues), 3.6.2.

Theorems 3.2.1 through 3.2.4, 3.3.1, 3.3.3, 3.5.1 through 3.5.6, 3.6.3 through 3.6.9.

Before proceeding with Chapter 7, Chapter 3 should be read again, this time without skipping the material pertaining to the infinite-dimensional case.

3.1 TOPOLOGIES AND NORMS

In analysis, many of the sets of interest are at the same time topological spaces and vector spaces. Furthermore, the algebraic operations are continuous in the topology.

Definition 3.1.1 Let V be a real (or complex) vector space and also a topological space. It is called a *topological vector space* if the mappings $(x, y) \longmapsto x + y$ and $(\lambda, x) \longmapsto \lambda x$ are continuous as mappings of the topological products $V \times V \longrightarrow V$ and $R \times V \longrightarrow V$ (or $C \times V \longrightarrow V$) respectively. (For brevity, we say that the operations $+$ and scalar multiplication are *continuous operations* in V.)

Two topological vector spaces V and W are said to be *topologically isomorphic* if there exists a bijective linear mapping $T : V \longrightarrow W$ which is also a homeomorphism.

The most familiar examples of topological vector spaces are those in which the topology is given by a metric. We can introduce a metric into each of the vector spaces of Examples 1 to 6 (Section 2.2) as follows:

Example 1 (*contd.*) *s*. Let $x = \{\xi_n\}$ and $y = \{\eta_n\}$ be arbitrary sequences in s. We define

$$d_s(x, y) = \sum_{n=1}^{\infty} \frac{1}{2^n} \frac{|\xi_n - \eta_n|}{1 + |\xi_n - \eta_n|}.$$

We see immediately that $d(x, y) = d(y, x)$, $d(x, y) \geq 0$ and $d(x, y) = 0$ if and only if $x = y$. To establish the triangle inequality, we use the inequality

$$\frac{a+b}{1+a+b} \leq \frac{a}{1+a} + \frac{b}{1+b},$$

which holds for any two real numbers $a, b \geq 0$, and also the fact that $t/(1 + t)$ is monotone increasing for $t \geq 0$.

A metric $d(x, y)$ on a vector space V is said to be *translation invariant* if $d(x + h, y + h) = d(x, y)$ for every $h \in V$. If d is translation invariant, then $x \in B(x_0, r)$ if and only if $x = x_0 + h$, where $d(h, 0) < r$. This follows from $d(x_0 + h, x_0) = d(h, 0)$, since any $x \in B(x_0, r)$ can be written as $x = x_0 + h$. We also have for any $x \in V, d(x, 0) = d(0, -x) = d(-x, 0)$. This implies

$$d(x + y, 0) = d(x, -y) \leq d(x, 0) + d(0, -y) = d(x, 0) + d(y, 0).$$

Also $d(x, y) = d(x - y, 0)$.

If d is translation invariant, the operation $x + y$ is continuous. Let $z = x + y$ and let $B(z, r)$ be any open ball with center z. Choose $B(x, r/2)$ and $B(y, r/2)$ as neighborhoods of x, y respectively. Then for any $x + \delta x \in B(x, r/2)$ and $y + \delta y \in B(y, r/2)$ we have

$$d(x + \delta x + y + \delta y, x + y) = d(\delta x + \delta y, 0) \leq d(\delta x, 0) + d(\delta y, 0) < r.$$

Hence, $(x + \delta x) + (y + \delta y) \in B(z, r)$, proving continuity. All metrics of interest to us here are translation invariant. The metric d_s of Example 1 is translation invariant.

If in addition to being translation invariant, a metric d has the properties

S(i) $d(\lambda_n x, 0) \to 0$ as $\lambda_n \to 0$,
S(ii) $d(\lambda x_n, 0) \to 0$ as $x_n \to 0$ and
S(iii) $d(\lambda_n x_n, 0) \to 0$ as $\lambda_n \to 0$ and $x_n \to 0$,

then scalar multiplication is a continuous operation. This results from

$$d(\lambda_n x_n, \lambda x) = d(\lambda_n x_n - \lambda x, 0)$$
$$\leq d((\lambda_n - \lambda)x, 0) + d(\lambda(x_n - x), 0) + d((\lambda_n - \lambda)(x_n - x), 0).$$

Properties S(i), S(ii), S(iii) hold for the metric d_s. Hence, s is a metric vector space.

Example 2 *(contd.)* m. If $x = (\xi_n)$ and $y = (\eta_n)$ are two bounded sequences in m we define $d(x, y) = \sup_n \{|\xi_n - \eta_n|\}$. Axioms M_1 and M_2 (Definition 1.3.1) are immediate. For $z = (\zeta_n) \in m$, we have $|\xi_n - \zeta_n| \leq |\xi_n - \eta_n| + |\eta_n - \zeta_n| \leq \sup_n \{|\xi_n - \eta_n|\} + \sup_n \{|\eta_n - \zeta_n|\}$. Consequently, $d(x, z) = \sup \{|\xi_n - \zeta_n|\} \leq d(x, y) + d(y, z)$. We shall denote this metric by $\|x - y\|_\infty$. It is clear that $\|x - y\|_\infty$ is translation invariant, and has properties S(i), S(ii), S(iii) in the preceding paragraph. In fact, $d(\lambda x, 0) = \|\lambda x\|_\infty = |\lambda| \|x\|_\infty$, and $d(x, 0) = 0$ implies $x = 0$. This makes $d(x, 0)$ a *norm*, according to the following definition.

Definition 3.1.2 A *seminorm* on a vector space V is a real-valued function $p(x)$ having the properties for any $x, y \in V$ and scalar λ:

N_1: $p(x + y) \leq p(x) + p(y)$,
N_2: $p(\lambda x) = |\lambda| p(x)$.

If in addition to N_1, N_2, $p(x)$ has the property

N_3: $p(x) = 0$ implies $x = 0$,

then it is called a *norm* and V is called a *normed vector space*. We denote a norm by $\|x\|$. If $\|x\| = 1$, we call x a *unit* vector.

Remark A seminorm has nonnegative values. $p(0) = 0$ by N_2 and $0 = p(x - x) \leq p(x) + p(-x) = 2p(x)$, so that $0 \leq p(x)$.

The space m has norm $\|x\|_\infty = \sup_n \{|\xi_n|\}$.

A normed vector space becomes a metric vector space if we define the metric by $d(x, y) = \|x - y\|$. We call this the *metric induced by the norm*. It is obviously a translation-invariant metric. We have $d(x, 0) = \|x\|$. Since $d(x, 0)$ is continuous in x, so is $\|x\|$. Thus, $x_n \to 0$ implies $\|x_n\| \to \|0\| = 0$. Properties S(i), S(ii), S(iii) above certainly hold for $d(x, 0) = \|x\|$. Hence, scalar multiplication is a continuous operation in a normed vector space.

Example 3 *(contd.)* l_p for $p \geq 1$ is a normed vector space with norm $\|x\|_p$ given by

$$\|x\|_p = \left(\sum_{i=1}^{\infty} |\xi_n|^p \right)^{1/p}.$$

For $p > 1$, the Minkowski inequality yields property N_1, while N_2 and N_3 are immediate. (See Exercise 3.10.)

Definition 3.1.3 In the finite-dimensional spaces ${}^T V^n (V^n)$, we can define for any row (column) vector $x = (\xi_1, \ldots, \xi_n)$ the norms,

$$\|x\|_p = \left(\sum_{i=1}^n |\xi_i|^p\right)^{1/p}, \qquad \|x\|_\infty = \sup_{1 \le i \le n} \{|\xi_i|\}.$$

We call these the l_p-*norms*, $1 \le p \le \infty$. (We assume that ξ_i are not scalars in a finite field. Otherwise, $\|x\|_p$ is a seminorm (Exercise 3.2).)

Remark Recall that R^n denotes the topological product of R taken n times. An open interval in R^n is a product of open intervals of R; e.g., in R^2 an open interval is an open rectangle. For each pair of points $x = (\xi_1, \ldots, \xi_n)$, $y = (\eta_1, \ldots, \eta_n)$ in R^n we may define a distance

$$d_1(x, y) = \|x - y\|_1 = \sum_{i=1}^n |\xi_i - \eta_i|,$$

and call the topology induced by the metric d_1 the l_1-*topology in R^n*. (R^n has no algebraic structure in these considerations.) It is easy to see that the l_1-topology is identical with the product topology. (If $\mathcal{O}$ is an open interval in R^n and $x \in \mathcal{O}$, then there is an l_1-ball $B(x, r)$ contained in $\mathcal{O}$ for r sufficiently small. Conversely, every open l_1-ball contains some open interval in R^n. See Exercise 1 of Chapter 1.)

Example 5 (*contd.*) In $C[a, b]$ a norm for $x = x(t)$ is given by $\|x\|_\infty = \sup_{a \le t \le b} \{|x(t)|\}$. We observe that convergence in this norm is equivalent to uniform convergence. For if $(x_n = x_n(t))$ is a sequence of functions and $x_n \to x$, then $\|x_n - x\|_\infty \to 0$. Hence, $\sup_t |x_n(t) - x(t)| \to 0$, which implies that $x_n(t)$ converges to $x(t)$ uniformly in t. Hence, we call this the *uniform* or *sup* norm.

Example 6 (*contd.*) Extending the results for l_p to L^p, we see that $\|f\|_p$ given in Definition 1.6.1 is a norm in $L^p(\mu)$.

Example 7 The set $L^\infty(\mu)$ of essentially bounded measurable functions becomes a normed vector space with norm $\|f\|_\infty$ given in Definition 1.6.2 and addition defined in the usual way as the pointwise sum of two functions.

Example 8 Let $C^k(X)$ be the set of all real (or complex) functions defined on a subset $X \subset R^n$ and having continuous partial derivatives of order $\le k$ (of order $< \infty$ for $k = \infty$). Let Y be any compact subset of X (relative topology) and let $0 \le m \le k$. A seminorm p_{Ym} is defined by

$$p_{Ym}(f) = \sup_{\substack{\sum_{r=1}^n j_r \le m \\ x \in Y}} \left\{\left|\frac{\partial^{j_1 + j_2 + \cdots + j_n} f(x)}{\partial x_1^{j_1} \partial x_2^{j_1} \ldots \partial x_n^{j_n}}\right|\right\}, \qquad x = (x_1, \ldots, x_n).$$

3.2 NORMED VECTOR SPACES

In Definition 1.3.5, we defined the concept of equivalent metrics. We shall say that two norms on a vector space V are *equivalent* if their induced metrics are equivalent. Thus, two norms p_1 and p_2 are equivalent if and only if $p_1(x - y) \to 0$ implies $p_2(x - y) \to 0$ and conversely.

52 Topological Vector Spaces

If X and Y are normed vector spaces, we may consider continuous mappings $f: X \to Y$ in terms of the topologies defined by the norms. In particular, if $f \in L(X; Y)$, we may consider continuous linear mappings. A linear mapping $f: X \longrightarrow Y$ is said to be *bounded* if there exists a positive constant $L > 0$ such that $\|f(x)\| \le L\|x\|$ for all $x \in X$. Compare with the concept of bounded given after Definition 1.3.1.

Theorem 3.2.1 Let $T: X \longrightarrow Y$ be a linear operator, where X and Y are normed vector spaces. The operator T is continuous in X if and only if it is bounded, that is, there exists a constant $L > 0$ such that

$$\|T(x)\| \le L\|x\|$$

(and therefore T is Lipschitz continuous).

Proof. The sufficiency is immediate from $\|T(x + h) - T(x)\| = \|Th\| \le L\|h\|$. Conversely, if T is continuous at 0, there is an open ball $B = B(0, r)$ in X such that $T(B_r) \subset B(0, 1)$, the open unit ball of Y. Since $T(\alpha x) = \alpha T(x)$, we have $\|T(\alpha x)\| = |\alpha| \|Tx\|$ and, therefore, $T(B(0, \alpha r)) \subset B(0, \alpha)$. Since α is arbitrary, T maps a bounded set into a bounded set. Taking $\alpha = (1 + \varepsilon)/r$, we obtain $T(B(0, 1 + \varepsilon)) \subset B(0, (1 + \varepsilon)/r)$, $\varepsilon > 0$. In particular, $\|T\bar{x}\| < (1 + \varepsilon)/r$ for every vector $\bar{x}$ with $\|\bar{x}\| = 1$. But any $x \in X$ can be written as $x = \|x\|\bar{x}$, where $\|\bar{x}\| = 1$. Hence, $\|Tx\| = \|x\| \|T\bar{x}\| \le L\|x\|$, where $L = (1 + \varepsilon)/r$. ∎

Theorem 3.2.2 Two norms p_1 and p_2 on a vector space V are equivalent if and only if there exist positive constants L_1, L_2 such that $p_2(x) \le L_1 p_1(x)$ and $p_1(x) \le L_2 p_2(x)$.

Proof. Let X_i be the normed vector space V with norm p_i, $i = 1, 2$. Let $I_{12}: X_1 \longrightarrow X_2$ and $I_{21}: X_2 \longrightarrow X_1$ be the identity mappings of V onto itself. Clearly, I_{12} and I_{21} are both continuous mappings if and only if the norms are equivalent. The result then follows from Theorem 3.2.1. (An alternative direct proof of the sufficiency can be given. Suppose $x_n \to x$ in the metric topology induced by p_1. This means that $p_1(x - x_n) \to 0$. Since $p_2(x - x_n) \le L_1 p_1(x - x_n)$, we have $p_2(x - x_n) \to 0$.) ∎

Corollary Two norms are equivalent if and only if they induce the same topology.

Remark If p_1 is equivalent to p_2 and p_2 is equivalent to p_3, then p_1 is equivalent to p_3, that is, the relation $\equiv$ is transitive. It is obviously reflexive and symmetric.

Theorem 3.2.3 Let V be a real (or complex) finite-dimensional vector space. Any two norms on V are equivalent.

Proof. In view of the preceding remark on transitivity, it suffices to prove that an arbitrary norm $\|x\|$ is equivalent to the l_1-norm $\|x\|_1$.

Let $\{u_1, \ldots, u_n\}$, $n \ge 1$, be a basis of V. For any $x \in V$ there is a unique representation, $x = \xi_1 u_1 + \cdots + \xi_n u_n$, which establishes an isomorphism $\varphi: x \longmapsto (\xi_1, \ldots, \xi_n)$ between V and V^n. (See the remark following Definition 2.1.10.)

We define $\|x\|_1 = \sum_{i=1}^n |\xi_i|$, which is the l_1-norm in V^n. This makes φ a homeomorphism (using l_1-topologies). Now, letting $M = \max_{1 \le i \le n} \{\|u_i\|\}$, we have

$\|x\| \leq \sum_{i=1}^{n} |\xi_i| \|u_i\| \leq M \|x\|_1$. To complete the proof, we must show the existence of a positive constant m such that $\|x\|_1 \leq m\|x\|$ or $\|x\| \geq K\|x\|_1$, $K = 1/m$. It suffices to show that $\|x\| \geq K > 0$ for all x on the unit l_1-sphere ($\|x\|_1 = 1$), since any x can be written as $x = \|x\|_1 \bar{x}$, where $\|\bar{x}\|_1 = 1$.

Now, the set of vectors in R^n such that $\sum_{i=1}^{n} |\xi_i| = 1$, being a closed bounded subset of R^n, is compact. The unit l_1-sphere in V is a homeomorphic image of this set by the mapping φ. Hence, the unit l_1-sphere in V is compact. But $\|x\|$ is a function on this unit sphere to the real line R. This function is continuous in the l_1-topology because

$$\big|\|x\| - \|y\|\big| \leq \|x - y\| \leq \sum_{i=1}^{n} |\xi_i - \eta_i| \|u_i\| \leq M \|x - y\|_1.$$

Hence, by Theorem 1.4.1, Corollary, $\|x\|$ is bounded from below and attains its infimum $K \geq 0$ on the unit l_1-sphere. Since $\|0\|_1 = 0$, 0 is not shown on the unit l_1-sphere. Hence, we must have $K > 0$; i.e., $\|x\| \geq K > 0$ as was to be shown. ∎

Corollary 1 Any norm on a finite-dimensional vector space is continuous in the topology defined by any other norm.

Corollary 2 Any finite-dimensional normed vector space V is complete. (V may be real or complex.)

Proof. V^n with the l_1-norm is complete, since $\|x_q - x_m\|_1 \to 0$ implies $\sum_{i=1}^{n} |\xi_i^{(q)} - \xi_i^{(m)}| \to 0$. By the completeness of R, there exists ξ_i, $1 \leq i \leq n$, such that $\xi_i^{(q)} \to \xi_i$ as $q \to \infty$. $x = (\xi_1, \ldots, \xi_n)$ is the l_1-limit of the sequence $\{x_q\}$. By the equivalence of norms, V is complete in any norm. ∎

Corollary 3 Any finite-dimensional subspace V of a normed space is closed.

Proof. V is homeomorphic to R^n with the l_1-norm. ∎

Corollary 4 Every closed bounded set X of a finite-dimensional normed space V is compact.

Proof. X is closed bounded in the given norm if and only if it is closed bounded in the l_1-norm. Thus, X is the homeomorphic image of a compact set in R^n. ∎

As a converse of Corollary 4, we have the following theorem.

Theorem 3.2.4 If V is a normed vector space in which the unit sphere $S = \{x : \|x\| = 1\}$ is compact, then V is finite dimensional.

Proof. Let $X \subset V$ be a closed proper subset which is also a subspace. For any $0 < \varepsilon < 1$, there is a unit vector $x_\varepsilon \in V$ whose distance from X is greater than ε. To prove this, choose $x_1 \in V - X$. Since X is closed, $d = d(x_1, X) > 0$. Since $d/\varepsilon > d$, there exists $x_0 \in X$ such that $\|x_0 - x_1\| = d/\varepsilon$. Let $\alpha = 1/\|x_1 - x_0\|$ and $x_\varepsilon = \alpha(x_1 - x_0)$. For any $x \in X$, $y = \|x_1 - x_0\| x + x_0$ is in X because X is a subspace. Hence, $\|x - x_\varepsilon\| = \|x - \alpha x_1 + \alpha x_0\| = \alpha \|y - x_1\| \geq \alpha d = \varepsilon$.

Now, cover S by open balls $B(x, \frac{1}{2})$, $x \in S$. Since S is compact, there is a finite subcovering, $\{B(x_i, \frac{1}{2}), i = 1, \ldots, n\}$. Let X be the finite-dimensional subspace generated by $\{x_1, \ldots, x_n\}$. By Corollary 3 above, X is closed. If X is a proper subset of V, there is a unit vector $x_{1/2} \in S$ whose distance from X is greater than $\frac{1}{2}$. Hence, $x_{1/2}$ cannot be in any $B(x_i, \frac{1}{2})$. This contradicts the fact that these balls cover S. Hence, $X = V$ and V is finite dimensional. ∎

If a normed vector space is complete (Definition 1.4.2) it is called a *Banach space*. That R^n is complete follows from the completeness of the real number system. The spaces in Examples 1 to 7 are complete. (For proofs, see Liusternik and Sobolev.) As an example of a space which is not complete we cite the subspace P of $C[0, 1]$ generated by the polynomials. By the Weierstrass theorem (Section 7.2), a continuous function f which is not a polynomial is the uniform limit of a sequence of polynomials p_n. The sequence (p_n) is a Cauchy sequence in P which has no limit in P.

If V is not complete, it has a completion (Definition 1.4.2 ff.) $\hat{V}$ such that V is dense in $\hat{V}$. Then $\hat{V}$ is a metric space. To make it a normed vector space we define addition of two equivalence classes x and y in terms of representative sequences, that is, $(x_n) + (y_n) = (x_n + y_n)$. Likewise, $\alpha(x_n) = (\alpha x_n)$. The norm is given by $\|x\| = \lim_{n \to \infty} \|x_n\|$. It is easy to verify that $\hat{V}$ is a vector space and that the metric defined by the norm is the metric in $\hat{V}$ given by the completion process.

Theorem 3.2.5 If V is a locally compact normed vector space, then V is finite dimensional.

Proof. There is a compact neighborhood U of 0. Since the closed balls $\{\bar{B}(0, r) : r > 0\}$ form a neighborhood base of 0, one of them, $\bar{B}(0, r_0)$, is contained in U and therefore is compact. Since $\bar{B}(0, 1) = (1/r_0) \bar{B}(0, r_0)$, $\bar{B}(0, 1)$ is compact. Hence, the unit sphere is compact, which by Theorem 3.2.4 implies that V is finite dimensional. ∎

Remark Let V be a topological vector space. For any set $E \subset V$, the *subspace generated* (or *spanned*) *by* E is the intersection of all subspaces containing E. We denote it by $V[E]$. It consists of all finite linear combinations of elements of E. The *closed subspace* generated by E is the intersection of all closed subspaces which contain E. It consists of $V[E]$ and all its accumulation points. (The closure of a subspace is again a subspace by continuity of the algebraic operations.)

In a normed vector space V, we may consider infinite series of elements $x_i \in V$. As with scalars, we define the partial sums $s_n = \sum_{i=1}^{n} x_i$, $n = 1, 2, \ldots$. We write $\sum_{i=1}^{\infty} x_i = \lim_{n \to \infty} s_n = s$ if the limit exists and we say that the *series converges* to the element $s \in V$. If the real series, $\sum_{i=1}^{\infty} \|x_i\|$ converges, then we say that the series $\sum_{i=1}^{\infty} x_i$ *converges absolutely*. If a series converges absolutely, then it may not converge. Indeed, we have only $\|s_q - s_p\| = \|\sum_{i=p+1}^{q} x_i\| \leq \sum_{i=p+1}^{q} \|x_i\| \to 0$ as $p, q \to \infty$, so that the sequence s_n is a Cauchy sequence. However, if V is complete, then $\lim_{n \to \infty} s_n$ exists in V, that is, the series converges.

We have considered the concept of a Hamel basis in a vector space. (See remark following Theorem 2.1.1.) In a topological vector space V, we may consider "infinite linear combinations" of vectors. Thus, if $\{u_i : i = 1, 2, \ldots\}$ is an infinite set of vectors in V, we may consider all convergent series of the form $\sum_{i=1}^{\infty} \alpha_i u_i = \lim_{n\to\infty} \sum_{i=1}^{n} \alpha_i u_i$. If every element $v \in V$ can be expressed uniquely as a series $v = \sum_{1}^{\infty} \alpha_i u_i$, then $\{u_i\}$ is again called a *basis* (sometimes a *Schauder basis*) of V. If $\{u_i\}$ is a (Schauder) basis, then the closed subspace $\bar{V}[E]$ generated by the set $E = \{u_i\}$ is the entire space V.

3.3 PRE-HILBERT (INNER-PRODUCT) SPACES

The real inner product of two real n-dimensional column vectors $x, y \in V^n$ is the real scalar which is the element of the 1×1 matrix $x^T y$. If $\xi_i, \eta_i, 1 \le i \le n$ are the coordinates of x and y respectively, we may write

$$x^T y = \sum_{i=1}^{n} \xi_i \eta_i.$$

We see immediately that $\sqrt{x^T x} = \|x\|_2$. This suggests that if an abstract vector space has an inner product defined on it, then the inner product induces a norm. Abstracting from the properties of $x^T y$, we arrive at the next definition.

Definition 3.3.1 Let V be a real vector space. V is called a *pre-Hilbert space* if there is a real-valued function defined on $V \times V$ which assigns to every pair $(x, y) \in V \times V$ a real value $\langle x, y \rangle$ called the *inner product* of x and y and such that

i$_1$) $\langle x, y \rangle = \langle y, x \rangle$
i$_2$) $\langle \lambda x, y \rangle = \lambda \langle x, y \rangle$
i$_3$) $\langle x + y, z \rangle = \langle x, z \rangle + \langle y, z \rangle$
i$_4$) $\langle x, x \rangle \ge 0$ and $\langle x, x \rangle = 0$ implies $x = 0$.

It follows directly from i$_1$, i$_2$, i$_3$ that $\langle x, y \rangle$ is a bilinear mapping of $V \times V$ to R. (For the complex-valued case, see Exercise 3.3.) (Many of our results are developed for the complex case.) Holding y fixed makes $\langle x, y \rangle$ a linear functional of x. Hence, our notation is consistent with the linear functional notation of Section 2.3. From the properties of matrix multiplication and addition, it follows that $x^T y$ satisfies i$_1$–i$_4$. Therefore, $x^T y$ defines an inner product in any finite-dimensional real vector space. (See Exercises 2.9(e), 3.3(a) for the complex case.) A pre-Hilbert space is also called an *inner-product space*.

Definition 3.3.2 The *inner-product norm* in a pre-Hilbert space is defined by

$$\|x\| = \sqrt{\langle x, x \rangle}.$$

We shall prove that $\|x\|$ is indeed a norm. Properties N$_2$ and N$_3$ (Definition 3.1.2) are immediate consequences of i$_1$, i$_2$ and i$_4$. To verify N$_1$, we prove Corollary 2 below.

Theorem 3.3.1 In any pre-Hilbert space, the following inequality (called the *Schwarz inequality*) holds with respect to the inner-product norm:

$$|\langle x, y \rangle| \le \|x\| \|y\|. \tag{1}$$

56 Topological Vector Spaces

Proof. For any real scalar α, we obtain (using i_1–i_3),

$$0 \leq \|x + \alpha y\|^2 = \langle x + \alpha y, x + \alpha y \rangle$$
$$= \|x\|^2 + 2\alpha \langle x, y \rangle + \alpha^2 \|y\|^2. \tag{2}$$

If $y \neq 0$, set $\alpha = -\langle x, y \rangle / \|y\|^2$ and obtain

$$0 \leq \|x\|^2 - \frac{2|\langle x, y \rangle|^2}{\|y\|^2} + \|y\|^2 \frac{(\langle x, y \rangle)^2}{\|y\|^4} \tag{3}$$

and (1) follows. If $y = 0$, (1) becomes $0 \leq 0$. ∎

Corollary 1 Let $x \neq 0$ and $y \neq 0$. Then $|\langle x, y \rangle| = \|x\| \|y\|$ if and only if $y = \beta x$ for some scalar $\beta \neq 0$.

Proof. Referring to the proof of the theorem, we see that equality in (1) implies that there is a nonzero real root, α, which makes (2) an equality. Hence, $\|x + \alpha y\| = 0$. This implies $y = (-1/\alpha)x$. ∎

Corollary 2 The inner-product norm of Definition 3.3.2 satisfies the triangle inequality; $\|x + y\| \leq \|x\| + \|y\|$.

Proof.

$$\|x + y\|^2 = \langle x + y, x + y \rangle = \langle x, x \rangle + \langle y, x \rangle + \langle x, y \rangle + \langle y, y \rangle$$
$$\leq \|x\|^2 + 2|\langle x, y \rangle| + \|y\|^2 \leq \|x\|^2 + 2\|x\| \|y\| + \|y\|^2$$
$$= (\|x\| + \|y\|)^2. \blacksquare \tag{4}$$

Remark $\|x + y\|^2 + \|x - y\|^2 = 2(\|x\|^2 + \|y\|^2)$.

Definition 3.3.3 A norm is said to be *strict* if $\|x + y\| = \|x\| + \|y\|$ implies that $y = \beta x$ for some scalar $\beta \geq 0$. The vector space is then said to be *strictly normed*.

Theorem 3.3.2 The norm induced by an inner product is a strict norm.

Proof. Suppose $\|x + y\| = \|x\| + \|y\|$. Referring to (4) in the proof of Corollary 2 above, we see that this implies equality throughout (4). Hence, $|\langle x, y \rangle| = \|x\| \|y\|$ and by Corollary 1, $y = \beta x$. Then

$$|1 + \beta| \|x\| = \|x + \beta x\| = \|x + y\| = \|x\| + \|y\| = \|x\| + |\beta| \|x\|,$$

whence $|1 + \beta| = 1 + |\beta|$. This implies $\beta \geq 0$. ∎

Remark It is clear that the l_∞-norm is not strict. ($\sup_i |\xi_i + \eta_i| = \sup_i |\xi_i| + \sup_i |\eta_i|$ does not imply $\xi_i = \beta \eta_i$, all i.)

Remark The inner product $\langle x, y \rangle$ is a continuous function on $V \times V$ (in the inner-product norm topology). (See Exercise 3.3b.)

Definition 3.3.4 Two vectors x, y in a pre-Hilbert space V are said to be *orthogonal* if $\langle x, y \rangle = 0$. A set of vectors $S \subset V$ is called an *orthonormal set* if $\langle x, y \rangle = 0$ for $x, y \in S$, $x \neq y$ and $\langle x, x \rangle = 1$ for all $x \in S$.

3.3 Pre-Hilbert (Inner-Product) Spaces

Theorem 3.3.3 Any orthonormal set S is linearly independent.

Proof. Let $\{e_1, \ldots, e_r\} \subset S$ and suppose $\lambda_1 e_1 + \cdots + \lambda_r e_r = 0$. Since $\langle e_j, 0 \rangle = 0$ for any $j = 1, \ldots, r$, we have $\lambda_1 \langle e_j, e_1 \rangle + \cdots + \lambda_r \langle e_j, e_r \rangle = 0$. But $\langle e_j, e_i \rangle = 0$ for $i \neq j$. Hence, $\lambda_j \langle e_j, e_j \rangle = 0$, which implies $\lambda_j = 0$ since $\langle e_j, e_j \rangle = 1$. ∎

Theorem 3.3.4 Any n-dimensional vector space V has an orthonormal basis.

Proof. Let $\{u_i : i = 1, \ldots, n\}$ be a basis. We shall construct an orthonormal basis $\{e_1, \ldots, e_n\}$ from the u_i by the *Gram-Schmidt* orthonormalization procedure. This is a recursive procedure defined as follows.

$$v_1 = u_1,$$

$$e_1 = \left(\frac{1}{\|v_1\|_2}\right) v_1,$$

$$\left.\begin{aligned} v_{i+1} &= u_{i+1} - \sum_{j=1}^{i} (e_j, u_{i+1}) e_j, \\ e_{i+1} &= \left(\frac{1}{\|v_{i+1}\|_2}\right) v_{i+1}. \end{aligned}\right\} i = 1, \ldots, n-1.$$

To prove $\{e_i\}$ is an orthonormal set, we use an induction argument. Certainly $\{e_1\}$ is orthonormal since $\|e_1\|_2 = (1/\|v_1\|_2)\|v_1\|_2 = 1$. Now suppose $\{e_1, \ldots, e_k\}$ is an orthonormal set; i.e., $\langle e_i, e_j \rangle = \delta_{ij}$ for $i \leq k$ and $j \leq k$. Then for any e_i with $i \leq k$, we have

$$\langle e_i, v_{k+1} \rangle = \langle e_i, u_{k+1} \rangle - \sum_{j=1}^{k} \langle e_j, u_{k+1} \rangle \langle e_i, e_j \rangle.$$

Since $\langle e_i, e_j \rangle = \delta_{ij}$ by the induction hypothesis, all terms in the summation are zero except for $j = i$. This gives

$$\langle e_i, v_{k+1} \rangle = \langle e_i, u_{k+1} \rangle - \langle e_i, u_{k+1} \rangle \langle e_i, e_i \rangle = 0.$$

(Note that $v_{k+1} \neq 0$ by the linear independence of the vectors $u_1, \ldots, u_{k+1}$.) Hence, v_{k+1} is orthogonal to each e_i for $i = 1, \ldots, k$. Since $e_{k+1} = (1/\|v_{k+1}\|_2)v_{k+1}$ is just a scalar multiple of v_{k+1}, it is also orthogonal to all e_i, $i \leq k$. This completes the induction.

Since $\{e_1, \ldots, e_n\}$ is orthonormal, it is linearly independent. Therefore, it is a basis for V. ∎

Corollary If $\{u_i\} \subset V$ is a denumerably infinite linearly independent set, the Gram-Schmidt orthonormalization process produces a denumerable orthonormal set which generates the same subspace as $\{u_i\}$.

The preceding three theorems apply to any finite-dimensional vector space with real inner product. In particular, they apply to V^n with the inner product defined as

$x^T y$. One orthonormal basis for V^n is the canonical basis $\{e_j\}$ defined by Eq. (2.1.9).

The coordinates ξ_i of any vector x in a finite-dimensional vector space V relative to an orthonormal basis $\{e_i\}$ are given by $\xi_i = \langle x, e_i \rangle$, $i = 1, \ldots, n$. (To see this, write $x = \sum_1^n \xi_k e_k$ and calculate the inner product using the orthonormality conditions.) Also, we have for $y = \sum_1^n \eta_i e_i$, $\langle x, y \rangle = \sum_1^n \xi_i \bar\eta_i$. In particular, $\|x\|^2 = \langle x, x \rangle = \sum_1^n |\xi_i|^2 = \sum_1^n |\langle x, e_i \rangle|^2$. Thus, the inner product norm of x in V is equal to the l_2-norm of $(\xi_1, \ldots, \xi_n)$ in $^T V^n$, where the ξ_i are the coordinates of x relative to an orthonormal basis. For infinite-dimensional pre-Hilbert spaces, we do not always have equality, as we shall now see.

Theorem 3.3.5 Let V be a pre-Hilbert space and let $E \subset V$ be an orthonormal set. For $x \in V$ and $e_j \in E$, $1 \leq j \leq n$ (n an arbitrary integer),

$$\sum_{j=1}^n |\langle x, e_j \rangle|^2 \leq \|x\|^2. \tag{1}$$

The subset $E_x \subset E$ of elements which are not orthogonal to x is either finite or denumerably infinite, so that we have *Bessel's inequality*,

$$\sum_{e \in E} |\langle x, e \rangle|^2 \leq \|x\|^2. \tag{2}$$

The closed subspace $\overline{V}[E]$ generated by E consists of all vectors $\sum_{e_j \in E} \langle x, e_j \rangle e_j$ provided that V is complete.

Proof. Let $\xi_i = \langle x, e_i \rangle$. Then

$$0 \leq \langle x - \sum_1^n \xi_j e_j, \ x - \sum_1^n \xi_j e_j \rangle = \|x\|^2 - \sum_1^n |\xi_j|^2,$$

establishing (1) above.

(1) implies that there cannot be more than $n^2 \|x\|^2$ elements $e \in E$ such that $|\langle x, e \rangle| \geq 1/n$; i.e., for each integer n, there are at most a finite number of elements e satisfying $|\langle x, e \rangle| \geq 1/n$. Since $\langle x, e \rangle \neq 0$ implies $|\langle x, e \rangle| \geq 1/n$ for some n, the set of $e \in E$ not orthogonal to x is a denumerable union of finite or empty sets and therefore is denumerable or finite. Thus the sum in (2) above is either a finite sum or an infinite series and converges as a result of (1).

By (2) and the completeness of V the series $\sum_{e_j \in E} \langle x, e_j \rangle e_j$ is convergent and is therefore in $\overline{V}[E]$. Furthermore, the series remains convergent under any rearrangement because of (2). Suppose (v_j) and (u_j) are two sequential arrangements of the $\{e_j\}$. Let

$$x_1 = \sum_{j=1}^\infty \langle x, u_j \rangle u_j \quad \text{and} \quad x_2 = \sum_{j=1}^\infty \langle x, v_j \rangle v_j.$$

Then $\langle x_1, u_j \rangle = \langle x, u_j \rangle$ by continuity of the inner product. Likewise, $\langle x_2, v_j \rangle = \langle x, v_j \rangle$. For $u_j = v_{i(j)}$, we have

$$\langle x_1 - x_2, u_j \rangle = \langle x, u_j \rangle - \langle x_2, v_{i(j)} \rangle = \langle x, u_j \rangle - \langle x, v_{i(j)} \rangle = 0.$$

Hence,

$$\|x_1 - x_2\|^2 = \langle x_1 - x_2, \sum \langle x, u_j \rangle u_j - \sum \langle x, v_j \rangle v_j \rangle = 0.$$

Thus $x_E = \sum_{e_j \in E_x} \langle x, e_j \rangle e_j$ is a uniquely defined vector. For any e_k,

$$\langle x - x_E, e_k \rangle = \langle x, e_k \rangle - \sum \langle x, e_j \rangle \langle e_j, e_k \rangle = \langle x, e_k \rangle - \langle x, e_k \rangle = 0.$$

Thus, $x - x_E$ is orthogonal to every element of $\overline{V}[E]$. If $x \in \overline{V}[E]$, so is $x - x_E$ and therefore is orthogonal to itself, that is, $x - x_E = 0$. ∎

Example 3 (*contd.*) In l_2, we define for the sequences $x = (\xi_i)$ and $y = (\eta_i)$, $\langle x, y \rangle = \sum_1^\infty \xi_i \bar{\eta}_i$. Convergence of the series is absolute, since

$$\left(\sum_1^n |\xi_i \bar{\eta}_i| \right)^2 \leq \sum_1^n |\xi_i|^2 \sum_1^n |\eta_i|^2 \leq \|x\|_2^2 \|y\|_2^2,$$

by the Schwarz inequality. The set of sequences e_k having a 1 as the kth element and all others 0 is an orthonormal set.

Theorem 3.3.6 If V is a separable pre-Hilbert space (in the topology of the inner product norm), then every orthonormal set E is either finite or denumerable.

Proof. For $x, y \in E$, $\|x - y\|^2 = \|x\|^2 - \langle x, y \rangle - \langle y, x \rangle + \|y\|^2 = 2$. Let S be a denumerable set everywhere dense in V. There exists $x_p, x_q \in S$ such that $\|x - x_p\| < \sqrt{2}/3$ and $\|y - x_q\| < \sqrt{2}/3$. Then

$$\sqrt{2} = \|x - y\| \leq \|x - x_p\| + \|x_p - x_q\| + \|x_q - y\| < 2\sqrt{2}/3 + \|x_p - x_q\|$$

so that $\|x_p - x_q\| > \sqrt{2}/3$. Hence, $x_p \neq x_q$ and we have established a one-to-one correspondence between E and a subset of S, which proves the theorem. ∎

Definition 3.3.5 An orthonormal set is called *complete* if it is not a proper subset of an orthonormal set.

Let $x \in V$, $x \neq 0$. The singleton set $\{x/\|x\|\}$ is orthonormal so that every space has an orthonormal set. Let $E \subset V$ be orthonormal. Let $\mathscr{F}$ be the class of all orthonormal sets which contain E, ordered by the inclusion relation. Zorn's lemma (Chapter 1) yields a maximal element in $\mathscr{F}$, which is a complete orthonormal set. Hence, every pre-Hilbert space contains a complete orthonormal set.

Theorem 3.3.7 If V is an infinite-dimensional separable pre-Hilbert space, then V contains a denumerable complete orthonormal set E.

Proof. Let S be a countable everywhere dense subset of V. Select a subset $\{y_n\} \subset S$ by choosing $y_1 \neq 0$ and $y_{k+1} \in S$ such that y_{k+1} is not in the subspace generated by $\{y_1, \ldots, y_k\}$. The subsets S and $\{y_n\}$ generate the same closed subspace, namely V. Applying the Gram-Schmidt process to $\{y_n\}$, we obtain an orthonormal set which also generates the closed subspace V. The set $\{y_n\}$ must be complete, for suppose $x \in V$ is such that $\langle x, y_n \rangle = 0$ for all y_n. Since $\{y_n\}$ generates V as its closed subspace, $\langle x, y \rangle = 0$ for all $y \in V$. In particular, $\langle x, x \rangle = 0$, which implies $x = 0$. ∎

Corollary If an orthonormal set E generates V as its closed subspace, then E is complete.

We shall be interested in pre-Hilbert spaces which are complete in the inner product norm (i.e., Cauchy sequences converge).

Definition 3.3.6 A complete pre-Hilbert space is called a *Hilbert* space.

In a Hilbert space, the converse of the above corollary holds.

Theorem 3.3.8 Let V be a Hilbert space. If E is a complete orthonormal set in V, then the closed linear subspace generated by E is V.

Proof. Suppose $x \in V - \bar{V}[E]$. Let $E_x = \{e_1, e_2, \ldots,\} \subset E$ be the finite or denumerable subset of E such that $\langle e_i, x \rangle \neq 0$. (Theorem 3.3.5.) Let $y_n = \sum_{i=1}^n \langle x, e_i \rangle e_i$. By Bessel's inequality, $\{y_n\}$ is a Cauchy sequence. Since V is complete, the sequence has a limit $y \in V$. Let $z = y - x$. The vector z is orthogonal to all $e \in E$, since

$$\langle z, e_j \rangle = \langle y, e_j \rangle - \langle x, e_j \rangle = \lim_{n \to \infty} \langle y_n, e_j \rangle - \langle x, e_j \rangle = 0$$

for all $e_j \in E_x$. For $e \notin E_x$, $\langle y_n, e \rangle = \langle x, e \rangle = 0$. Hence, $\langle y, e \rangle = 0$ and $\langle z, e \rangle = 0$. Normalizing z we obtain an orthonormal set $E \cup \{z\}$, contradicting the maximality of E. Hence, $V = \bar{V}[E]$. ∎

Theorem 3.3.9 Let E be an orthonormal set in a pre-Hilbert space V. If all $x \in V$ satisfy the *Parseval identity*,

$$\|x\|^2 = \sum_{e \in E} |\langle x, e \rangle|^2$$

then E is complete. If E is a complete orthonormal set and V is a Hilbert space, then Parseval's identity holds.

Proof. If E is not complete, there is an $x \neq 0$ such that $\langle x, e \rangle = 0$ for all $e \in E$. Parseval's identity does not hold for this x, proving the first part of the theorem.

If V is complete and E is a complete orthonormal set, then, by Theorem 3.3.8, $V = \bar{V}[E]$. Let $x \in V$. By Theorem 3.3.5, the set of $e \in E$ which are not orthogonal to x is finite or denumerable. Call this set $\{e_i\}$. By completeness of V and Bessel's inequality,

$$y = \lim_{n \to \infty} \sum_{i=1}^n \langle x, e_i \rangle e_i \text{ exists.}$$

But

$$\langle x - y, e_j \rangle = \langle x, e_j \rangle - \lim_{n \to \infty} \sum_{i=1}^n \langle x, e_i \rangle \langle e_i, e_j \rangle$$

$$= \langle x, e_j \rangle - \langle x, e_j \rangle = 0.$$

For $e \notin \{e_i\}$, we again have $\langle x - y, e \rangle = 0$. Since E is complete, $x - y = 0$. We write $x = \sum_i \langle x, e_i \rangle e_i$. Then $\|x\|^2 = \langle \sum \langle x, e_i \rangle e_i, \sum \langle x, e_j \rangle e_j \rangle$ and the result follows. ∎

Corollary. In a Hilbert space, a denumerable complete orthonormal set is a basis.

Example 3 (*contd.*) The orthonormal set $\{e_k\}$ in l_2 is complete. Let $x = (\xi_i) \in l_2$. Consider the vectors $x_n = \sum_1^n \xi_i e_i$, which are in the subspace generated by $\{e_k\}$. Then $\|x - x_n\|_2^2 = \sum_{i=n+1}^\infty |\xi_i|^2 \to 0$ as $n \to \infty$. Hence, x is in the closed linear subspace generated by $\{e_k\}$. By Theorem 3.3.7, corollary, $\{e_k\}$ is complete.

Theorem 3.3.10 If V is a separable infinite-dimensional Hilbert space, then V is isomorphic to l_2. If $x \in V$ corresponds to the sequence (ξ_n) under this isomorphism, then $\|x\|^2 = \sum_{n=1}^\infty |\xi_n|^2$. Thus, the isomorphism is also an isometry.

Proof. By Theorem 3.3.7, V contains a denumerable complete orthonormal set $\{e_i\}$. For $x \in V$, let $\xi_i = \langle x, e_i \rangle$. By Parseval's identity, $\|x\|^2 = \sum_1^\infty |\xi_i|^2$, so that $(\xi_n) \in l_2$. Conversely, if $(\xi_n) \in l_2$, then $\sum_{i=1}^\infty \xi_i e_i = x \in V$, and $\xi_i = \langle x, e_i \rangle$. ∎

Referring to Definition 2.3.1 and the properties of an inner product, we see that if T is a linear operator in a pre-Hilbert space V, the inner product $\langle x, Ty \rangle$ defines a bilinear mapping from $V \times V$ to the reals (or complex numbers). $\langle x, Ty \rangle$ is a bilinear form (Exercise 2.15) in x and y. If $x = y$, $\langle x, Tx \rangle$ is a quadratic form. If V is n-dimensional and $A = (a_{ij})$ is the matrix which represents T relative to an orthonormal basis $\{e_i\}$, then $x = \sum_{i=1}^{n} \xi_i e_i$, $y = \sum_{j=1}^{n} \eta_j e_j$ and $\langle x, Ay \rangle = \sum_{i,j=1}^{n} a_{ij} \xi_i \eta_j$ is a bilinear form in ξ_i, η_j which represents $\langle x, Ty \rangle$ relative to the basis $\{e_i\}$. Treating $x^T = (\xi_1, \ldots, \xi_n)$ and $y^T = (\eta_1, \ldots, \eta_n)$ as vectors in $^T V^n$ and taking $\langle x, y \rangle = x^T y$, we have $\langle x, Ay \rangle = x^T A y$. Taking transposes (see Example 2, Section 2.3) we find $y^T A^T x = (x^T A y)^T = x^T A y$, since $x^T A y$ is a scalar.

If $M \subset V$ is a closed subspace, the *orthogonal complement* $M^\perp$ of M is the set of vectors $x \in V$ such that $\langle x, m \rangle = 0$ for all $m \in M$. The set $M^\perp$ is obviously a closed subspace and $M \cap M^\perp = \{0\}$.

Theorem 3.3.11 Let M be a closed subspace of a Hilbert space V. Then for any $x \in V$ there exist unique vectors $m \in M$ and $m^\perp \in M^\perp$ such that $x = m + m^\perp$. The mapping $P : x \longmapsto m$ is a bounded linear operator and

$$P = P^2, \quad \langle Px, y \rangle = \langle x, Py \rangle.$$

Proof. We may assume $M \neq V$. If $x \in M$, we have the trivial decomposition $x = x + 0$. If $x \notin M$, then $d = \inf_{m' \in M} \|x - m'\| > 0$. Let (m_i) be a sequence in M such that $\|x - m_i\| \to d$. Then using the remark following Theorem 3.3.1,

$$\begin{aligned}
\|m_p - m_q\|^2 &= \|(x - m_q) - (x - m_p)\|^2 \\
&= 2(\|x - m_q\|^2 + \|x - m_p\|^2) - \|2x - m_q - m_p\|^2 \\
&= 2(\|x - m_q\|^2 + \|x - m_p\|^2) - 4\|x - (m_p + m_q)/2\|^2 \\
&\leq 2(\|x - m_q\|^2 + \|x - m_p\|^2) - 4d^2 \to 0.
\end{aligned}$$

$((m_p + m_q)/2 \in M$ since M is a subspace. Hence, $\|x - (m_p + m_q)/2\| \geq d$.) Since V is complete, there exists $m = \lim m_i$. $m \in M$, since M is closed and by continuity $\|x - m\| = d$.

Let $m^\perp = x - m$. For any $m' \in M$ and real scalar λ, $(m - \lambda m') \in M$. Hence (assuming a complex inner product)

$$\begin{aligned}
d^2 \leq \|x - m + \lambda m'\|^2 &= \langle m^\perp + \lambda m', m^\perp + \lambda m' \rangle \\
&= \|m^\perp\|^2 + \lambda \langle m^\perp, m' \rangle + \lambda \langle m', m^\perp \rangle + \lambda^2 \|m'\|^2, \\
0 &\leq 2\lambda \, \mathrm{Re} \, \langle m^\perp, m' \rangle + \lambda^2 \|m'\|^2.
\end{aligned}$$

This implies $\mathrm{Re} \, \langle m^\perp, m' \rangle = 0$. Replacing m' by im', we find $\mathrm{Im} \, \langle m^\perp, m' \rangle = 0$. Thus, $m^\perp \in M^\perp$.

Since $\|x\|^2 = \|m\|^2 + \|m^\perp\|^2 \geq \|Px\|^2$, we have $\|P\| \leq 1$. The linearity and other properties follow directly from the uniqueness of the decomposition $x = m + m^\perp$, which follows from the orthogonality. ∎

Corollary If M is a subspace not dense in V, then there exists $m^\perp \neq 0$ such that $\langle m^\perp, m \rangle = 0$ all $m \in M$.

Proof. The closure $\overline{M} \neq V$. For $y \notin \overline{M}$, $y \neq 0$, $y = m + m^\perp$, $m^\perp \neq 0$, where $m^\perp \in \overline{M}^\perp$ and $m \in \overline{M}$. ∎

The operator P of the theorem is called the *projection operator* determined by the closed subspace M and $m = Px$ is called the *orthogonal projection* of x on M.

Referring to the Remark preceding Definition 2.1.17, we may write $V = M \oplus M^\perp$.

Further properties of pre-Hilbert spaces are given in Section 7.4. Also, see Exercises 3.29, 3.32, 3.33.

Remark A study of the proof of the preceding theorem reveals that V may be only a pre-Hilbert space provided that M is a complete subspace of V (e.g., M is finite dimensional). In these circumstances, we still have $V = M \oplus M^\perp$.

3.4 CONVEXITY IN TOPOLOGICAL VECTOR SPACES

Let V be a topological vector space. If E, F are subsets of V, then $E + F = \{x + y : x \in E, y \in F\}$ and $\lambda E = \{\lambda x : x \in E\}$. If E consists of a single vector v, we write $v + F$ instead of $\{v\} + F$. If E is open, then $v + E$ is open. For if $y = v + x$, $x \in E$, there is a neighborhood N_x of x contained in E. Since $x = y + (-v)$ and $+$ is continuous, there exist neighborhoods N_y and N_{-v} such that $N_y + N_{-v} \subset N_x$. Hence, $-v + N_y \subset N_x \subset E$ and so $N_y \subset v + E$, proving $v + E$ is open. Similarly, λE can be shown to be open. It follows that every neighborhood N_x of a vector x is of the form $x + U$, where U is a neighborhood of the vector 0. Further, if $\{U_j\}$ is a neighborhood base at 0, then $\{x + U_j\}$ is a neighborhood base at x.

A set $E \subset V$ is said to be *balanced* if $\lambda E \subset E$ for all $|\lambda| \leq 1$. If $x, y \in V$, the *line segment* joining x and y is the set $[x, y] = \{\lambda x + (1 - \lambda)y : 0 \leq \lambda \leq 1\}$. The subset with $0 < \lambda < 1$ is the *open line segment*. If E is balanced and $x \in E$, then evidently $[x, -x] \subset E$. A subset E is called *convex* if $x, y \in E$ implies that the line segment $[x, y] \subset E$. We say that subset E *absorbs* subset F if there is a $\lambda_0 > 0$ such that $F \subset \lambda E$ for all $|\lambda| \geq |\lambda_0|$ (or $\alpha F \subset E$ for $|\alpha| \leq 1/|\lambda_0|$). If E absorbs $\{x\}$ for every $x \in V$, then E is called *absorbing*.

Since any intersection of balanced sets is balanced, any set $E \subset V$ is contained in a unique smallest balanced set E_B called the *balanced hull* of E. In fact, $E_B = \{\lambda x : x \in E, |\lambda| \leq 1\}$. Likewise, if $E \neq 0$, the *convex hull* E_C of E is the intersection of all convex sets containing E and consists of all finite linear combinations $\sum \lambda_j x_j$, $x_j \in E$, $\lambda_j \geq 0$, $\sum \lambda_j = 1$.

3.5 METRIC PROPERTIES OF LINEAR OPERATORS ON NORMED SPACES

Suppose X and Y are normed vector spaces.

A linear operator $T : X \longrightarrow Y$ is continuous if and only if $\|Tx\| \leq L \|x\|$ for some constant $L > 0$. (Theorem 3.2.1.) As we have seen, the set $L(X; Y)$ of linear operators is a vector space. The subset $L_c(X; Y)$ of continuous operators is a subspace, for if $T_1, T_2 \in L_c(X; Y)$, then $\|(T_1 + T_2)x\| = \|T_1 x + T_2 x\| \leq \|T_1 x\| +$

3.5 Metric Properties of Linear Operators on Normed Spaces

$\|T_2 x\| \leq (L_1 + L_2)\|x\|$. Also, $\|\lambda T x\| \leq |\lambda| L \|x\|$. The subset $L_c(X, Y)$ can be normed as follows.

Definition 3.5.1 Let $T \in L_c(X, Y)$. The *norm* of the operator T is the real number $\|T\|$ given by any of the following equivalent formulas:

$$\|T\| = \sup_{x \neq 0} \frac{\|Tx\|}{\|x\|}, \qquad \|T\| = \sup_{\|x\|=1} \|Tx\|, \qquad \|T\| = \sup_{\|x\| \leq 1} \|Tx\|.$$

We see that $\|T\| = \inf \{L : \|Tx\| \leq L \|x\|\}$. To show that $\|T\|$ is a norm on the space $L_c(X, Y)$, we note that $\|(T_1 + T_2)x\| \leq (\|T_1\| + \|T_2\|)\|x\|$ and therefore $\|T_1 + T_2\| \leq \|T_1\| + \|T_2\|$, establishing property N_1. Also, $\|T\| = 0$ implies $Tx = 0$ for all x. Hence, $T = 0$, the zero operator. Property N_2 is easily verified.

Remark If Y is a Banach space, then $L_c(X; Y)$ is also a Banach space. Suppose $\{T_n\}$ is a Cauchy sequence in $L_c(X; Y)$. Then for any $x \in X$,

$$\|T_n x - T_m x\| = \|(T_n - T_m)x\| \leq \|T_n - T_m\| \|x\|.$$

Since $\|T_n - T_m\| \to 0$ as $n, m \to \infty$, $\{T_n x\}$ is a Cauchy sequence in Y. Hence, there exists $y \in Y$ such that $T_n x \to y$. Define $Tx = y$. Then T is linear, since $T(x_1 + x_2) = \lim T_n(x_1 + x_2) = \lim T_n x_1 + \lim T_n x_2 = Tx_1 + Tx_2$. Similarly, $T(\lambda x) = \lambda Tx$. To prove that $\|T_n - T\| \to 0$ as $n \to \infty$, let $\varepsilon > 0$ and choose n_0 such that $\|T_n - T_m\| < \varepsilon$ for $n, m > n_0$. Then $\|T_n x - T_m x\| < \varepsilon$ for all x with $\|x\| = 1$. Allowing $m \to \infty$, we obtain $\|T_n x - Tx\| < \varepsilon$. Then $\|T_n - T\| = \sup_{\|x\|=1} \|(T_n - T)x\| < \varepsilon$. Hence $T_n \to T$ in the operator norm.

By the remark following Definition 2.1.14, the inverse of a linear operator is itself a linear operator.

Lemma 3.5.1 Let $T \in L(X; Y)$ be surjective (onto), where X and Y are normed vector spaces. Then T is nonsingular and $T^{-1} \in L_c(Y; X)$ if and only if there exists $m > 0$ such that $\|Tx\| \geq m \|x\|$ for all $x \in X$ (i.e., $\|Tx\| \geq m > 0$ for all x on the unit sphere).

Proof. Suppose m exists. Then $Tx = 0$ implies $x = 0$. Hence, T is one–one onto and T^{-1} exists. But $T^{-1}y = x$ implies $Tx = y$. Therefore, $\|T^{-1}y\| = \|x\| \leq (1/m) \|y\|$ and T^{-1} is continuous. Conversely if T^{-1} is continuous, $\|x\| = \|T^{-1}y\| \leq \|T^{-1}\| \|y\| = \|T^{-1}\| \|Tx\|$ and $m = 1/\|T^{-1}\|$. ∎

Theorem 3.5.0 (Banach). Let X and Y be Banach spaces. If $T \in L_c(X; Y)$ is bijective, then T^{-1} is a linear continuous operator.

Proof. See [3.1], p. 75, where this theorem is a consequence of the *open mapping theorem*, which asserts that any surjective linear continuous operator on a Banach space X onto a Banach space Y maps an open set of X onto an open set of Y. ∎

Example Let T be a continuous linear operator on a separable Hilbert space V into itself. Suppose $Te_n = (1/n) e_n$, $n = 1, 2, \ldots$, for $\{e_n\}$ a complete orthonormal basis.

Since $\|Te_n\| \to 0$ as $n \to \infty$, $\|Tx\|$ is not bounded away from zero on the unit sphere. Therefore, T cannot have a continuous inverse. We have for $x \in V$, $x = \sum_{n=1}^{\infty} \xi_n e_n$, where $\sum_{n=1}^{\infty} |\xi_n|^2 < \infty$, and

$$Tx = \sum_{n=1}^{\infty} \xi_n Te_n = \sum_{n=1}^{\infty} \frac{\xi_n}{n} e_n.$$

T is injective, since $Tx = Tx'$ implies $\sum_{n=1}^{\infty} [(\xi_n - \xi_n')/n] e_n = 0$, which implies $\xi_n = \xi_n'$, $n = 1, 2, \ldots$, so that $x = x'$. The operator T is not surjective, since the vector $\sum_{n=1}^{\infty} (1/n) e_n \in V$ is not in the range of T, $R(T)$. The range is a subspace of V. We may consider the inverse $T^{-1} \in L(R(T); V)$, since T^{-1} is linear. However, since $T^{-1} e_n = n e_n$, T^{-1} is not bounded.

Example An example of a linear operator on an infinite-dimensional space is the integral operator

$$Tx = y(t) = \int_a^b K(t, s) x(s) \, ds,$$

where $x = x(t) \in C[a, b]$ and $K(t, s)$ is a real function continuous in the square $a \leq t \leq b, a \leq s \leq b$. It follows from the linearity of the integral that $T \in L(C[a, b]; C[a, b])$. Since $K(t, s)$ is continuous on a closed bounded set, it is bounded; i.e., $|K(t, s)| \leq M$. This implies

$$\|Tx\|_\infty = \max_{a \leq t \leq b} |y(t)| \leq \max_t \int_a^b |K(t, s)| \, |x(s)| \, ds$$

$$\leq M \|x\|_\infty (b - a).$$

Hence, $\|T\| \leq M(b - a)$. In fact $\|T\| = \max_t \int_a^b |K(t, s)| \, ds$. (See Exercise 3.4.2.)

The special case $L_c(X; X)$ is important. Since we may form the composite map $(T_1 T_2) x = T_1(T_2 x)$, $L_c(X; X)$ is a linear algebra (see Definition 2.1.14) and we have the following additional property of the norm:

N_4: $\|T_1 T_2\| \leq \|T_1\| \, \|T_2\|$.

This follows directly from $\|T_1 T_2 x\| \leq \|T_1\| \, \|T_2 x\| \leq \|T_1\| \, \|T_2\| \, \|x\|$.

A linear algebra in which there is defined a norm satisfying N_1–N_4 is called a *normed algebra* (or *normed ring*). If it is complete with respect to the norm topology, it is called a *Banach algebra*. If a normed algebra $\mathscr{A}$ has a multiplicative unit element, e, we require that $\|e\| = 1$. The *inverse* of an element $a \in \mathscr{A}$ is the element a^{-1} such that $aa^{-1} = a^{-1}a = e$. If a^{-1} exists, a is called *nonsingular*.

Example $L_c(X; X)$ is a normed algebra and the identity mapping $I : X \longrightarrow X$ is a unit element. Since $Ix = x$, we obviously have $\|I\| = 1$. We shall denote the normed algebra $L_c(X; X)$ by $\mathscr{A}_x$. If X is a Banach space, then $\mathscr{A}_x$ is a Banach algebra (see Remark preceding Lemma 3.5.1). By Theorem 3.5.0, if $T \in \mathscr{A}_x$ and T^{-1} exists in $L(X, X)$ then $T^{-1} \in \mathscr{A}_x$.

3.5 Metric Properties of Linear Operators on Normed Spaces

We observe that multiplication is a continuous operation in a normed algebra, since
$$\|(a + h)(b + k) - ab\| \le \|h\|\,\|b\| + \|a\|\,\|k\| + \|h\|\,\|k\|.$$

Theorem 3.5.1 Let $\mathscr{A}$ be a Banach algebra with unit element e. If $a \in \mathscr{A}$ is such that $\|a\| < 1$, then $e - a$ has an inverse and
$$\frac{1}{1 + \|a\|} \le \|(e - a)^{-1}\| \le \frac{1}{1 - \|a\|}.$$

Proof. Since $\|a^n\| \le \|a\|^n$ by property N_4, the series
$$\sum_{n=0}^{\infty} a^n = e + a + a^2 + \cdots$$
is absolutely convergent; i.e., $\|e\| + \|a\| + \|a^2\| + \cdots$ converges. Thus, $\sum_{n=0}^{\infty} a^n = b$ is an element of $\mathscr{A}$ and
$$(e - a)b = \lim_{n \to \infty} (e - a)(e + a + \cdots + a^n)$$
$$= \lim_{n \to \infty} (e - a^{n+1}) = e,$$
since $\lim_{n \to \infty} a^{n+1} = 0$. Hence, $b = (e - a)^{-1}$ and
$$\|b\| \le \sum_{n=0}^{\infty} \|a\|^n = \frac{1}{1 - \|a\|}.$$

To obtain the lower bound, we use the fact that $1 = \|e\| = \|(e - a)(e - a)^{-1}\| \le \|e - a\|\,\|(e - a)^{-1}\| \le (\|e\| + \|a\|)\|(e - a)^{-1}\|$. ∎

Corollary 1 Under the conditions of the theorem, $(e + a)^{-1} = \sum_{n=0}^{\infty} (-a)^n$ exists and the norm satisfies the same inequalities.

Corollary 2 If $\|e - b\| \le k < 1$, then b^{-1} exists and
$$\|b^{-1}\| \le \frac{1}{1 - \|e - b\|} \le \frac{1}{1 - k}.$$

Proof. We write $b = e - (e - b)$. Taking $a = e - b$, the theorem applies. ∎

Theorem 3.5.2 Let $\mathscr{A}$ be a Banach algebra with unit element. If $x \in \mathscr{A}$ has an inverse x^{-1} and $\Delta x \in \mathscr{A}$ is such that $\|\Delta x\| < 1/\|x^{-1}\|$, then $(x + \Delta x)^{-1}$ exists. ∎

Proof. We write $(x + \Delta x) = x(e + x^{-1}\Delta x)$. Since $\|x^{-1}\Delta x\| \le \|x^{-1}\|\,\|\Delta x\| < 1$, we see by Theorem 3.5.1 that $(e + x^{-1}\Delta x)^{-1}$ exists. Hence
$$(x + \Delta x)^{-1} = (e + x^{-1}\Delta x)^{-1} x^{-1}. \blacksquare$$

Corollary 1 Under the hypotheses of the theorem,
$$\|(x + \Delta x)^{-1} - x^{-1}\| < \frac{\|x^{-1}\|^2\,\|\Delta x\|}{1 - \|x^{-1}\|\,\|\Delta x\|}.$$

Proof. We write $(x + \Delta x)^{-1} - x^{-1} = (e + x^{-1}\Delta x)^{-1}x^{-1} - x^{-1}$. Hence,
$$\|(x + \Delta x)^{-1} - x^{-1}\| \le \|(e + x^{-1}\Delta x)^{-1} - e\|\, \|x^{-1}\|.$$
$$\|(x + \Delta x)^{-1} - x^{-1}\| \le \left\|\sum_1^\infty (-x^{-1}\Delta x)^n\right\| \|x^{-1}\| \le \frac{\|x^{-1}\Delta x\|}{1 - \|x^{-1}\Delta x\|}\|x^{-1}\|. \quad (1)$$

Since $\|\Delta x\|\, \|x^{-1}\| < 1$, the result follows. ∎

Corollary 2 If $\|\Delta x\| < 1$, then $(e + \Delta x)^{-1}$ exists and
$$\|e - (e + \Delta x)^{-1}\| \le \frac{\|\Delta x\|}{1 - \|\Delta x\|}.$$

Proof. Equation (1) of Corollary 1 applies with $x = e$. ∎

Corollary 3 Let $xy - e = r$. If $\|r\| < 1$ and x has a left inverse, then x and y are nonsingular and
$$\|x^{-1}\| \le \frac{\|y\|}{1 - \|r\|}, \qquad \|y^{-1}\| \le \frac{\|x\|}{1 - \|r\|}, \qquad \|y - x^{-1}\| \le \frac{\|y\|\, \|r\|}{1 - \|r\|}.$$

Proof. Since $xy = e + r$ and $\|r\| < 1$, xy has an inverse by Theorem 3.5.1. Then the element $x_R^{-1} = y(xy)^{-1}$ is a right inverse of x. Since x has a left inverse, x_L^{-1}, we have $x_L^{-1} = x_L^{-1}(xx_R^{-1}) = (x_L^{-1}x)x_R^{-1} = x_R^{-1}$. Thus, x^{-1} exists. Then $y^{-1} = (x^{-1}(xy))^{-1} = (xy)^{-1}x$. $\|y^{-1}\| \le \|(e + r)^{-1}\|\, \|x\| \le \|x\|/(1 - \|r\|)$.

Writing $x = (e + r)y^{-1}$, we have $x^{-1} = y(e + r)^{-1}$, and the first inequality follows from Theorem 3.5.1.

Since $y - x^{-1} = y(e - (e + r)^{-1})$, the third inequality follows from Corollary 2 above. ∎

Corollary 4 Let X, Y be Banach spaces. Let $T, S \in L_c(X; Y)$ and suppose S^{-1} exists; i.e., $S^{-1}S = I_x$ and $SS^{-1} = I_y$. If $\|S^{-1}T - I_x\| \le k < 1$, then T^{-1} exists and $\|T^{-1}\| \le \|S^{-1}\|/(1 - k)$. Further, $\|T^{-1} - S^{-1}\| \le \|S\|\, \|S^{-1}\|^2 k/(1 - k)$.

Proof. By Theorem 3.5.0, $S^{-1} \in L_c(Y; X)$. Hence, $S^{-1}T \in \mathscr{A}_x = L_c(X, X)$. By Theorem 3.5.1 (Corollary 2), $(S^{-1}T)^{-1}$ exists in $\mathscr{A}_x$. Hence
$$U = (S^{-1}T)^{-1}S^{-1} \in L_c(Y; X) \qquad \text{and} \qquad U^{-1} = S(S^{-1}T) = T.$$
Therefore, $T^{-1} = U$ and
$$\|T^{-1}\| \le \|S^{-1}\|\, \|(S^{-1}T)^{-1}\| \le \frac{\|S^{-1}\|}{1 - k}.$$
Since
$$T^{-1} - S^{-1} = T^{-1}S(I - S^{-1}T)S^{-1},$$
the second inequality follows. ∎

The matrix algebra $A^{(n)}$ (Definition 2.1.9) can be made into a Banach algebra in many ways. For example, we would consider $A^{(n)}$ as a vector space of dimension

3.5 Metric Properties of Linear Operators on Normed Spaces

n^2 and use the l_2-norm. Let $A = (a_{ij})$ have row vectors A_i^T. Let $B = (b_{ij})$ have column vectors B_j. The element c_{ij} of the product $C = AB$ is given by $c_{ij} = A_i^T B_j$. Hence, by the Schwarz inequality,

$$\sum_{i,j=1}^{n} |c_{ij}|^2 = \sum_{i,j=1}^{n} |A_i^T B_j|^2$$

$$\leq \sum_{i,j=1}^{n} \|A_i\|_2^2 \|B_j\|_2^2 \leq \sum_{i=1}^{n} \|A_i\|_2^2 \sum_{j=1}^{n} \|B_j\|_2^2,$$

showing that $\|AB\|_2 \leq \|A\|_2 \|B\|_2$. This calculation holds if A is $n \times n$ and B is $n \times 1$. Therefore, $\|Ax\|_2 \leq \|A\|_2 \|x\|_2$. If we define

$$\|A\|_S = \sup_{\|x\|_2 = 1} \|Ax\|_2,$$

we see that $\|A\|_S \leq \|A\|_2$.

We call $\|A\|_S$ the *spectral norm* of A. It is the *norm induced by the l_2-norm* in $V^{(n)}$. Since $\|A\|_S$ is an operator norm, it satisfies N_4. Other operator norms for A are induced by each of the l_p-norms in $V^{(n)}$. We shall calculate the operator norms of A induced by the l_1-, l_2- and l_∞-norms in $V^{(n)}$. Let $x^T = (\xi_1, \ldots, \xi_n)$ and $A = (a_{ij})$.

1. *The l_∞-norm.* If $\|x\|_\infty = 1$, then $\max_{1 \leq i \leq n} |\xi_i| = 1$ and

$$\|Ax\|_\infty = \max_{1 \leq i \leq n} \left\{ \sum_{j=1}^{n} |a_{ij} \xi_j| \right\} \leq \max_{1 \leq i \leq n} \left\{ \sum_{j=1}^{n} |a_{ij}| \right\}.$$

Denoting the induced operator norm by $\|A\|_\infty$, we have

$$\|A\|_\infty \leq \max_{1 \leq i \leq n} \sum_{j=1}^{n} |a_{ij}|.$$

Suppose the maximum absolute row sum occurs for $i = K$. For x having coordinates $\xi_j = \bar{a}_{Kj}/|a_{Kj}|$ if $a_{Kj} \neq 0$ and $\xi_j = 0$ if $a_{Kj} = 0$, we obtain

$$\|Ax\|_\infty = \sum_{j=1}^{n} |a_{Kj}|.$$

Hence,

$$\|A\|_\infty = \max_{1 \leq i \leq n} \left\{ \sum_{j=1}^{n} |a_{ij}| \right\}.$$

the *maximum row sum*.

2. *The l_1-norm.* Denote the induced operator norm by $\|A\|_1$. Then $\|x\|_1 = 1$ implies $\sum_{i=1}^{n} |\xi_i| = 1$ and

$$\|Ax\|_1 = \sum_{i=1}^{n} \left| \sum_{j=1}^{n} a_{ij} \xi_j \right| \leq \sum_{i=1}^{n} \sum_{j=1}^{n} |a_{ij}| |\xi_j| = \sum_{j=1}^{n} |\xi_j| \sum_{i=1}^{n} |a_{ij}|$$

$$\leq \max_{1 \leq j \leq n} \left\{ \sum_{i=1}^{n} |a_{ij}| \right\} \sum_{j=1}^{n} |\xi_j| = \max_{1 \leq j \leq n} \sum_{i=1}^{n} |a_{ij}|.$$

If the maximum column sum occurs for $j = K$, then taking $\xi_K = 1$ and $\xi_i = 0$ for $i \neq K$, we obtain $\|Ax\|_1 = \sum_{i=1}^{n} |a_{iK}|$. Hence, $\|A\|_1 = \max_{1 \leq j \leq n} \sum_{i=1}^{n} |a_{ij}|$, the *maximum column sum*.

The l_2-*norm*. Let $\mu = \sup_{\|x\|_2 = \|y\|_2 = 1} |\langle x, Ay \rangle|$, so that for any x, y

$$|\langle x, Ay \rangle| \leq \mu \|x\|_2 \|y\|_2.$$

By the Schwarz inequality,

$$|\langle x, Ay \rangle| \leq \|x\|_2 \|Ay\|_2 \leq \|A\|_S \|x\|_2 \|y\|_2.$$

Thus, $\mu \leq \|A\|_S$. Also

$$\|Ay\|_2^2 = |\langle Ay, Ay \rangle| \leq \mu \|Ay\|_2 \|y\|_2,$$

or $\|Ay\|_2 \leq \mu \|y\|_2$, which implies $\|A\|_S \leq \mu$. Hence, $\|A\|_S = \mu$.

Note that

$$\|A^T\|_S = \sup_{\|x\|_2 = \|y\|_2 = 1} |x^T A^T y| = \sup_{\|x\|_2 = \|y\|_2 = 1} |y^T A x| = \|A\|_S.$$

Also, $\|A^T A\|_S = \|A\|_S^2$, since

$$\|A^T A\|_S \leq \|A^T\|_S \|A\|_S = \|A\|_S^2$$

and

$$\|Ax\|_2^2 = |\langle Ax, Ax \rangle| = |\langle x, A^T A x \rangle| \leq \|A^T A\|_S \|x\|_2^2.$$

Hence, $\|A\|_S \leq (\|A^T A\|_S)^{1/2}$. See Theorems 3.6.5, 3.6.6 for further results.

Note that since $\|I\|_2 = \sqrt{n}$, the l_2-norm of a matrix cannot be an operator norm induced by a vector norm.

Since $V^{(n)}$ is complete in any l_p-norm, $1 \leq p \leq \infty$, the corresponding normed algebra $A^{(n)}$ is also complete.

Definition 3.5.2 Let $A \in L_c(X; X)$ have a continuous inverse (i.e., $A^{-1} \in L_c(X; X)$), where X is a Banach space. The number $\|A\| \|A^{-1}\|$ is called the *condition number* of A.

Note that $\|A\| \|A^{-1}\| \geq \|AA^{-1}\| = \|I\| = 1$.

Theorem 3.5.3 Let X be a Banach space and suppose $A, A^{-1} \in L_c(X; X)$. Let $\delta A \in L_c(X, X)$ be such that $\|\delta A\| < 1/\|A^{-1}\|$. If $x \in X$ and $x + \delta x \in X$ are such that

$$Ax = b, \tag{1}$$

and

$$(A + \delta A)(x + \delta x) = b, \tag{2}$$

and $\gamma = \|A\| \|A^{-1}\|$ is the condition number of A, then

$$\frac{\|\delta x\|}{\|x\|} \leq \frac{\gamma}{1 - \gamma \|\delta A\|/\|A\|} \frac{\|\delta A\|}{\|A\|}. \tag{3.5.1}$$

Proof. Since $\|\delta A\| < 1/\|A^{-1}\|$, it follows from Theorem 3.5.2 that $(A + \delta A)^{-1}$ exists in the Banach algebra $L_c(X; X)$. From eqs. (1) and (2), we obtain

$$(A + \delta A) \delta x + Ax + (\delta A)x = b,$$
$$(A + \delta A) \delta x = -(\delta A)x,$$
$$\delta x = -(A + \delta A)^{-1}(\delta A)x.$$
$$\|\delta x\| \leq \|(A + \delta A)^{-1}\| \|\delta A\| \|x\|. \qquad (3)$$

Now, $(A + \delta A)^{-1} = (A(I + A^{-1} \delta A))^{-1} = (I + A^{-1} \delta A)^{-1} A^{-1}$, and by Theorem 3.5.1 (Corollary 1),

$$\|(I + A^{-1} \delta A)^{-1}\| \leq \frac{1}{1 - \|A^{-1} \delta A\|} \leq \frac{1}{1 - \|A^{-1}\| \|\delta A\|};$$

(the second inequality because $\|A^{-1} \delta A\| \leq \|A^{-1}\| \|\delta A\|$). Applying this to eq. (3) above, we get

$$\|\delta x\| \leq \frac{\|A^{-1}\|}{1 - \|A^{-1}\| \|\delta A\|} \|\delta A\| \|x\|,$$

and (3.5.1) follows.

3.5a THE CONJUGATE OF A NORMED SPACE

Let X and Y be normed vector spaces. We have seen that $L_c (X; Y)$, the set of continuous linear operators on X to Y, is a normed vector space. If $Y = R$, the operators are real functionals. Hence, $L_c(X; R) \subset X^*$. We shall call $L_c(X; R)$ the *conjugate* of X and denote it by X'.

Example Let $X = V^{(n)}$ with the l_2-norm. We have seen that $X^* = V^{(n)}$. Hence, for any $y^* \in X^*$ there exists $y \in V^{(n)}$ such that $y^*(x) = \langle x, y \rangle$, where $\langle x, y \rangle$ is the inner product. By the Schwarz inequality, $|y^*(x)| = |\langle x, y \rangle| \leq \|x\|_2 \|y\|_2$. Hence, $\|y^*\| \leq \|y\|_2$. Taking $x = y$, $|y^*(y)| = \|y\|_2^2$, showing that $\|y^*\| = \|y\|_2$. Therefore, every linear functional on $V^{(n)}$ is continuous in the l_2-norm, and $(V^{(n)})' = (V^{(n)})^* = V^{(n)}$ with l_2-norm.

Example 3 (*contd.*) l_p $(1 \leq p < \infty)$. We shall show that $(l_p)' = l_q$, where q is such that $1/p + 1/q = 1$ for $p > 1$ and $q = \infty$ for $p = 1$.

Proof

CASE 1. $(1 < p < \infty)$. As before, let e_K be the sequence having a 1 as its Kth term and all other terms 0. Let $y' \in (l_p)'$ and $\alpha_K = y'(e_K)$. For any $x \in l_p$, $x = (\xi_K)$, we have

$$x = \lim_{n \to \infty} \sum_1^n \xi_K e_K = \sum_1^\infty \xi_K e_K.$$

Since y' is continuous

$$y'(x) = \lim_{n\to\infty} \sum_1^n \xi_K y'(e_K) = \sum_1^\infty \xi_K \alpha_K. \tag{1}$$

Now, let $x_n = \sum_{K=1}^n \eta_K e_K$, where

$$\eta_K = \begin{cases} |\alpha_K|^{q-1} \bar{\alpha}_K/|\alpha_K| & \text{if } \alpha_K \neq 0, \\ 0 & \text{if } \alpha_K = 0. \end{cases}$$

Then

$$\alpha_K \eta_K = |\alpha_K|^q = |\eta_K|^p$$

and

$$\|x_n\|_p = \left(\sum_{K=1}^n |\alpha_K|^q\right)^{1/p},$$

$$y'(x_n) = \sum_{K=1}^n |\alpha_K|^q.$$

Since $|y'(x_n)| \leq \|y'\| \|x_n\|_p$,

$$\sum_{K=1}^n |\alpha_K|^q \leq \|y'\| \left(\sum_1^n |\alpha_K|^q\right)^{1/p},$$

$$\left(\sum_{K=1}^n |\alpha_K|^q\right)^{1/q} \leq \|y'\|.$$

Hence, $y = \sum_1^\infty \alpha_K e_K$ is in l_q and $\|y\|_q \leq \|y'\|$. Conversely, if $y = \sum_1^\infty \alpha_K e_K \in l_q$, we can define $y' \in (l_p)'$ by $y'(x) = \sum_1^\infty \xi_K \alpha_K$, since by Hölder's inequality,

$$|y'(x)| = \left|\sum_1^\infty \alpha_K \xi_K\right| \leq \left(\sum_1^\infty |\alpha_K|^q\right)^{1/q} \left(\sum_1^\infty |\xi_K|^p\right)^{1/p} = \|y\|_q \|x\|_p.$$

Thus, $\|y'\| \leq \|y\|_q$. Since $y'(e_K) = \alpha_K$, it follows from the preceding calculations that $\|y\|_q \leq \|y'\|$. Hence, $\|y\|_q = \|y'\|$.

CASE 2. $p = 1$. Define $\eta_K = \bar{\alpha}_K/|\alpha_K|$ if $\alpha_K \neq 0$ and $\eta_K = 0$ if $\alpha_K = 0$. Let $x_n = \eta_n e_n$. Then $\|x_n\|_1 = 1$ and $y'(x_n) = \alpha_n \eta_n = |\alpha_n|$. Thus, $|\alpha_n| \leq \|y'\| \|x_n\|_1 \leq \|y'\|$, which shows that $y = (\alpha_n) \in l_\infty$ and $\|y\|_1 \leq \|y'\|$. As before, we can show $\|y'\| \leq \|y\|_1$. Thus, $(l_1)' = l_\infty$. ∎

Remark We have written $(l_p)' = l_q$ to indicate that there is a bijection of $(l_p)'$ and l_q which is both an isomorphism and an isometry; i.e., any continuous linear functional on l_p can be represented in the form (1) above, where $(\alpha_n) \in l_q$, and conversely.

It follows that $(l_p)'' = l_q' = l_p$ for $1 < p < \infty$. A normed vector space V such that $V'' = V$ is called *reflexive*. The spaces l_p, $1 < p < \infty$, are reflexive. However, it can be shown that $(l_\infty)' \neq l_1$. (See [3.2], pp. 193 ff.) Since $l_1'' = l_\infty'$, we see that l_1 is not reflexive.

3.5b THE ADJOINT OF A CONTINUOUS LINEAR OPERATOR

If $T \in L_c(X; Y)$, where X and Y are normed vector spaces, then the adjoint T^* is continuous on the subspace $Y' \subset Y^*$. In proof, recall (Section 2.3) that the algebraic adjoint T^* is in $L(Y^*, X^*)$. We consider the restriction of T^* to Y', denoting the restriction again by T^*. The defining equation

$$\langle Tx, y' \rangle = \langle x, T^*y' \rangle. \qquad x \in X, \quad y' \in Y',$$

shows that $x' = T^*y' \in X'$ since

$$|\langle x, x' \rangle| = |\langle x, T^*y' \rangle| = |\langle Tx, y' \rangle| \le \|T\| \|x\| \|y'\| = K \|x\|,$$

where $K = \|T\| \|y'\|$ and $\|y'\|$ is the bound of y' as a linear operator. This also shows that $\|T^*y'\| \le \|T\| \|y'\|$. Hence, $T^* \in L_c(Y', X')$. (In fact, it can be shown that $\|T^*\| = \|T\|$. This requires the Hahn–Banach theorem. See Appendix II.)

In a Hilbert space X, the adjoint operator T^* is an operator on V itself, since $X = X'$, as the next theorem, called the *Riesz representation theorem*, shows.

Theorem 3.5.4 Let $x' \in X'$ be a bounded linear functional on a Hilbert space X. There exists a unique $x \in X$ such that for all $y \in X$,

$$x'(y) = \langle x, y \rangle$$

and $\|x'\| = \|x\|$. The mapping $x' \longmapsto x$ is an isomorphism and an isometry, so that we may write $X = X'$.

Proof. Let N be the null space of x'; i.e., $N = \{m : x'(m) = 0\}$. N is a closed subspace, since x' is continuous and linear. If $N = X$, we take $x = 0$. If $N \ne X$, then by Theorem 3.3.11 we have $X = N + N^\perp$. For $y_0 \notin N$, there exist $y_1 \in N$ and $y_2 \in N^\perp$, $y_2 \ne 0$, such that $y_0 = y_1 + y_2$. Then $x'(y_0) = x'(y_1) + x'(y_2) = x'(y_2)$. Since $x'(y_2) = a \ne 0$, the element $y_3 = (1/a)y_2 \in N^\perp$ is such that $x'(y_3) = 1$. Take $x = y_3/\|y_3\|^2$. For any $y \in X$, if $x'(y) = b$, then

$$x'(y - by_3) = x'(y) - b = 0.$$

Hence,

$$y_4 = y - by_3 \in N \qquad \text{and} \qquad \langle x, y \rangle = \langle x, y_4 \rangle + b\langle x, y_3 \rangle = 0 + b = x'(y).$$

The uniqueness is immediate.

Since $|\langle x, y \rangle| \le \|y\| \|x\|$ (by Schwarz inequality), $\|x'\| \le \|x\|$. But $\|x'\| \ge |x'(x/\|x\|)| = |\langle x, x \rangle|/\|x\| = \|x\|$. This completes the proof. ∎

Remark If X and Y are Hilbert spaces, then the equation defining the adjoint T^* of $T \in L_c(X; Y)$ can be written as

$$\langle Tx, y \rangle = \langle x, T^*y \rangle, \qquad x \in X, \quad y \in Y.$$

Taking $y = Tx$, we see that $\|Tx\|^2 \le \|x\| \|T^*\| \|Tx\|$, or $\|T\| \le \|T^*\|$. Together with the result $\|T^*\| \le \|T\|$ given above, this proves $\|T\| = \|T^*\|$. Note that $T^* \in L_c(Y; X)$ and therefore we can consider $T^*T \in L_c(X; X)$ and $TT^* \in L_c(Y; Y)$.

Theorem 3.5.5 If T and T^* are as in the above Remark, then $\|T^*T\| = \|T\|^2 = \|TT^*\|$. Also, $\|T\| = \|T^*\|$.

Proof. $\|T^*Tx\| \le \|T^*\| \|Tx\| \le \|T^*\| \|T\| \|x\| = \|T\|^2 \|x\|$. Hence, $\|T^*T\| \le \|T\|^2$. Also, $\|Tx\|^2 = \langle Tx, Tx \rangle = \langle x, T^*Tx \rangle \le \|x\|^2 \|T^*T\|$, which shows that $\|T\|^2 \le \|T^*T\|$. Reversing the roles of T^* and T, we prove $\|TT^*\| = \|T\|^2$.

If $X = V^{(n)}$ and $Y = V^{(m)}$ in the preceding discussion, then we must use the l_2-norm and the induced operator norm; i.e., we have

$$\|T\|_S = \|T^*\|_S$$
$$\|TT^*\|_S = \|T\|_S^2 = \|T^*T\|_S. \quad \blacksquare$$

If $T \in L_c(X; X)$, where X is a Hilbert space, then $T^* \in L_c(X; X)$. If $T = T^*$, then T is said to be *self-adjoint*. If $TT^* = T^*T$, T is called a *normal* operator.

Remark Again, suppose $T \in L_c(X; Y)$, where X and Y are Hilbert spaces. Since $T^* \in L_c(Y; X)$, we also have $T^{**} \in L_c(X; Y)$ and $\langle Tx, y \rangle = \langle x, T^*y \rangle = \langle T^{**}x, y \rangle$ for all $x, y \in X$. Hence, $T^{**} = T$. It follows that for any $T \in L_c(X; Y)$ that T^*T is self-adjoint, since $(T^*T)^* = T^*T^{**} = T^*T$.

Example If $X = V^n$ and $T \in L_c(X; X)$ is self-adjoint, then $A = \bar{A}^T$, where A is a matrix representing T. Thus, we may say that a matrix is self-adjoint if it is equal to its conjugate transpose.

A self-adjoint matrix is said to be *Hermitian*. In the real case, we have simply $A = A^T$ and the matrix A is then called a *symmetric* matrix and we have

$$\langle Ax, y \rangle = \langle x, Ay \rangle, \quad x, y \in X.$$

Let X be a Hilbert space. An operator $U \in L(X; X)$ such that $UU^* = U^*U = I$ is called a *unitary* operator. A unitary operator is normal and $U^{-1} = U^*$. Since $\langle Ux, Uy \rangle = \langle x, U^*Uy \rangle = \langle x, y \rangle$, it follows that $\|Ux\|^2 = \|x\|^2$, or $\|U\| = 1$. Also, if $\{e_i\}$ is an orthonormal set, then $\langle Ue_i, Ue_j \rangle = \langle e_i, e_j \rangle = \delta_{ij}$. Hence, $\{Ue_i\}$ is also orthonormal.

Obviously there are many possible orthonormal bases in $V^{(n)}$. Any two such bases are related by a linear operator which transforms one into the other. A linear operator T on $V^{(n)}$ to $V^{(n)}$ which transforms every orthonormal basis into an orthonormal basis is called an *orthogonal operator* in the real case and a *unitary operator* in the complex case.

Theorem 3.5.6 Let T be an orthogonal operator on an n-dimensional vector space V. The matrix, A, corresponding to T relative to any orthonormal basis has column vectors $A_1, \ldots, A_n$ which form an orthonormal set; i.e., $A_i^T A_j = \delta_{ij}$. (Such a matrix is called *orthogonal*. In the complex case, if $A_i^* A_j = \delta_{ij}$, A is called *unitary*.)

Proof. Let $\{e_1, \ldots, e_n\}$ be an orthonormal basis of V. Since T is orthogonal, the set $\{Te_j\}$ is also orthonormal. Let $Te_j = \sum_{k=1}^{n} a_{kj}e_k$. The matrix A corresponding to T relative to the basis $\{e_i\}$ is then (a_{ij}). The columns of A are

$$A_j = \begin{pmatrix} a_{1j} \\ a_{2j} \\ \vdots \\ a_{nj} \end{pmatrix}, \quad j = 1, \ldots, n.$$

We have

$$\delta_{ij} = \langle Te_i, Te_j \rangle = \langle \sum_{k=1}^{n} a_{ki}e_k, \sum_{k=1}^{n} a_{kj}e_k \rangle = \sum_{k=1}^{n} a_{ki}a_{kj} = A_i^{\mathrm{T}} A_j. \qquad (1)$$

This completes the proof. ∎

Corollary 1 A unitary $n \times n$ matrix defines a unitary operator on $V^{(n)}$.

Corollary 2 A linear operator T on a finite-dimensional vector space is unitary if it transforms a particular orthonormal basis into an orthonormal basis.

Proof. By Eq. (1), T is represented by a unitary matrix. ∎

Corollary 3 The matrix A is unitary if and only if $\bar{A}^{\mathrm{T}} A = A\bar{A}^{\mathrm{T}} = I$.

The matrix representation of a linear operator T in $T^{(n)}$ depends on the basis.

3.6 THE SPECTRUM OF A LINEAR OPERATOR; EIGENVALUES

We have defined the inverse of a mapping $f: X \longrightarrow Y$ to exist if f is bijective. When considering a linear operator $T: X \longrightarrow Y$, it is convenient to generalize the notion of inverse. Since the range $R(T)$ of T is a subspace Y_1 of Y, we may always regard $T \in L(X; Y_1)$ as surjective. The operator T is injective if and only if it has an inverse $T^{-1} \in L(Y_1; X)$. In this case, we shall again say simply that T^{-1} exists. Having allowed the domain $R(T)$ of T^{-1} to be a subspace of Y, it is natural to allow the domain of T to be a subspace X_1 of X. Thus, $T \in L(X_1; Y_1)$ and T^{-1} exists if and only if T is injective. In this case, $T^{-1} \in L(Y_1; X_1)$; i.e., the inverse of a linear operator is linear. If X and Y are Banach spaces and X_1 is a dense subspace of X and $T \in L_c(X_1; Y)$, then there is a continuous extension of T which has X as its domain. To see this, note that for $x \in X$ there is a sequence (x_n), $x_n \in X_1$, such that $x_n \to x$. Now, since T is bounded, $\|Tx_n - Tx_m\| \le \|T\| \|x_n - x_m\| \to 0$ as $n, m \to \infty$. Since Y is complete, there exists $y \in Y$ such that $Tx_n \to y$. Define $Tx = y$. (This is well defined, since for any other sequence $x'_n \to x$ and $Tx'_n \to y'$, we must have $\|y - y'\| \le \|y - Tx_n\| + \|T\| \|x_n - x'_m\| + \|y' - Tx'_m\| < \varepsilon$, for n, m sufficiently large.) It is apparent that now $T \in L_c(X; Y)$.

Remark 1 The operator T is called *closed* if the *graph* $\{(x, Tx) : x \in D(T)\}$ of T is a closed subset of $X \times Y$ (in the product topology). Suppose $T \in L_c(D(T); Y)$, where $D(T) \subset X$ and X, Y are Banach spaces. If $D(T)$ is dense in X and T is closed, then $D(T) = X$. For any $x \in X$, choose a sequence $x_n \to x$, $x_n \in D(T)$ as above. Since T is bounded, $Tx_n \to y \in Y$ and $(x_n, Tx_n) \to (x, y)$ in $X \times Y$. Since T is closed, (x, y) is in its graph; i.e., $Tx = y$. Thus, $D(T) = X$. Any operator $T \in L_c(X; Y)$ is closed, since $(x_n, y_n) \to (x, y)$ implies that $x_n \to x$, $y_n = Tx_n \to Tx$ by continuity, and therefore that $y = Tx$. The *closed graph theorem* asserts that a closed $T \in L(X; Y)$ must be continuous. (See Exercise 3.43.)

If T is closed and T^{-1} exists, then T^{-1} is closed. (Its graph is $\{(Tx, x)\}$.) If $T \in L(X; Y)$ where X and Y are pre-Hilbert spaces, then T^* is closed. (Exercise 3.46.)

In this section, we shall be concerned with a linear operator $T: D(T) \longrightarrow X$, where X is a normed vector space and the domain $D(T)$ and range $R(T)$ of T are both subspaces of X. Given T and any scalar λ, the operator

$$T_\lambda = \lambda I - T$$

is linear and has domain $D(T)$. We classify λ as follows.

Definition 3.6.1 Let T_λ be as in the preceding paragraph. The set $\rho(T)$ of scalars λ such that the range $R(T_\lambda)$ of T_λ is dense in X and a continuous inverse T_λ^{-1} exists (in $L_c R(T_\lambda), D(T))$ is called the *resolvent set* of T. The complement of $\rho(T)$ is called the *spectrum* $\sigma(T)$ of T.

The spectrum $\sigma(T)$ is the union of the following three disjoint sets

i) the *point spectrum* $P_\sigma = \{\lambda \in \sigma(T) : T_\lambda \text{ has no inverse}\}$; if λ is in the point spectrum, it is called an *eigenvalue* of T.

ii) the *continuous spectrum* $C_\sigma = \{\lambda : T_\lambda^{-1} \text{ exists but is not continuous and } R(T_\lambda) \text{ is dense in } X\}$;

iii) the *residual spectrum* $R_\sigma = \{\lambda : T_\lambda^{-1} \text{ exists but } R(T_\lambda) \text{ is not dense in } X\}$.

(If T is closed, then so is T_λ. If T_λ^{-1} exists, it is also closed. In this case, if X is complete, then $\lambda \in \rho(T)$ implies that T_λ^{-1} is defined on all of X. See Remark above.)

We observe that λ is in the point spectrum if and only if there exists $x \neq 0$ such that $T_\lambda x = 0$, or

$$Tx = \lambda x.$$

Such a vector x is called an *eigenvector of T belonging to the eigenvalue λ*. The zero vector and the set of all eigenvectors of T belonging to λ form a subspace.

Example If $X = V^{(n)}$ and $T \in L(X; X)$ is represented by an $n \times n$ matrix A, then λ is an eigenvalue of T if and only if

$$\det(\lambda I - A) = 0.$$

This is called the *characteristic equation* of A and $\det(\lambda I - A)$ is called the *characteristic polynomial* of A. Since $(\lambda I - A) \in A^{(n)}$ is an $n \times n$ matrix, $(\lambda I - A)^{-1}$ exists if and only if $\lambda I - A$ is bijective. Thus, in the finite-dimensional case, either $\lambda \in \rho(A)$ or $\lambda \in P_\sigma(A) = \sigma(A)$. The scalar $\lambda \in \sigma(A)$ if and only if λ is a root of the characteristic equation. Since $\det(\lambda I - A)$ is an nth-degree polynomial in λ, A has at least one (complex) eigenvalue and at most n distinct eigenvalues. If λ is a multiple root of det $(\lambda I - A)$ of multiplicity r, we shall call r the *multiplicity* of λ.

If A is a real matrix, then its characteristic polynomial has real coefficients. If λ is an eigenvalue (and therefore a root of the characteristic equation), then $\bar{\lambda}$ is also a root and therefore an eigenvalue. If u is an eigenvector belonging to λ, we have $Au = \lambda u$ and $\overline{Au} = \bar{\lambda}\bar{u}$ or since $A = \bar{A}$, $A\bar{u} = \bar{\lambda}\bar{u}$. Therefore, $\bar{u}$ is an eigenvector belonging to $\bar{\lambda}$. Note that if $Im(\lambda) \neq 0$, then u cannot be a real vector.

Since det $(\lambda I - A) = $ det $(\lambda I - A^T)$, A and its transpose A^T have the same spectrum. More generally, if $\lambda \in \sigma(A)$, then $\bar{\lambda} \in \sigma(A^*)$ and conversely. Regarding A as a linear operator on a Hilbert space, note that $Ax = \lambda x$ implies

$$\langle x, A^*y \rangle = \langle Ax, y \rangle = \langle \lambda x, y \rangle = \langle x, \bar{\lambda}y \rangle.$$

Hence, $\langle x, A^*y - \bar{\lambda}y \rangle = 0$ for all vectors y. If $(A^* - \bar{\lambda}I)y \neq 0$ for all $y \in V^n$, then $A^* - \bar{\lambda}I$ is nonsingular and hence $(A^* - \bar{\lambda}I)y_o = x$ for some y_o. This would imply $\langle x, x \rangle = 0$, which contradicts $x \neq 0$. Therefore, $A^*y = \bar{\lambda}y$ for some $y \neq 0$ and $\bar{\lambda} \in \sigma(A^*)$. Since A is a matrix, taking complex conjugates, we have $y^*A = \lambda y^*$. The vector y is sometimes called a *left* eigenvector of A (i.e., it is an eigenvector of A^* belonging to $\bar{\lambda}$).

Note that $y^*x = 0$, where y is eigenvector of A^* belonging to λ_1 and x is an eigenvector of A belonging to $\lambda_2 \neq \lambda_1$ (Exercise 3.24b).

Example Let X be the space of bounded real functions $f(x)$, $-\infty < a \leq x \leq b < \infty$. Define $Tf(x) = xf(x)$. Clearly, $T \in L_c(X; X)$. If $a \leq \lambda \leq b$, then $T_\lambda f(x) = \lambda f(x) - xf(x) = (\lambda - x)f(x) = g(x)$. Hence, $g(\lambda) = 0$. This is a necessary condition for g to be in the range of T_λ. Also, if $f_1(\lambda) \neq f_2(\lambda)$ and $f_1(x) = f_2(x)$ for $x \neq \lambda$, then $T_\lambda f_1 = T_\lambda f_2$ and $f_1 \neq f_2$. Hence, T_λ^{-1} does not exist and $\lambda \in \sigma(T)$. The corresponding eigenvectors are functions f such that $f(x) = 0$ for $x \neq \lambda$ and $f(\lambda) \neq 0$. If $\lambda \notin [a, b]$, then $\lambda \in \rho(T)$.

Example In $L^2[a, b]$, we regard two functions f, g as equal if $f(x) = g(x)$ except on a set of measure zero. As in the previous example, let $Tf(x) = xf(x)$, $-\infty < a \leq x \leq b < \infty$. Then $T_\lambda f(x) = (\lambda - x)f(x) = 0$ implies that $f(x) = 0$ except possibly for $x = \lambda$. But then $f = 0$. Thus, T_λ^{-1} exists and $\lambda \in [a, b]$ is not an eigenvalue. However, $T_\lambda f = (\lambda - x)f(x) = g(x)$ implies $f(x) = g(x)/(\lambda - x)$. Therefore, if $g(x) = $ constant $\neq 0$ in $[a, b]$, then since $K^2/(\lambda - x)^2$ is not integrable on $[a, b]$, $g(x) = K$ is not in the range of T_λ. Hence, the range is not dense in $L^2[a, b]$. Thus, $\lambda \notin \rho(T)$. In fact, λ is in the residual spectrum.

Example Let X be a complex Hilbert space and T a self-adjoint operator on X. Then the spectrum of T is real. We shall prove that $Im(\lambda) \neq 0$ implies $\lambda \in \rho(T)$. For $\lambda \in \rho(T)$, T_λ^{-1} is continuous and $\|T_\lambda^{-1}\| \leq 1/|Im(\lambda)|$. In proof, note that $\langle Tx, x \rangle = \langle x, Tx \rangle = \overline{\langle Tx, x \rangle}$. So $\langle Tx, x \rangle$ is real and therefore, for any complex scalar λ,

$$Im \langle (\lambda I - T)x, x \rangle = Im(\lambda) \|x\|^2.$$

By the Schwarz inequality,
$$|\text{Im}(\lambda)|\,\|x\|^2 \le |\langle(\lambda I - T)x, x\rangle| \le \|(\lambda I - T)x\|\,\|x\|.$$
Hence, $\|(\lambda I - T)x\| \ge |\text{Im}(\lambda)|\,\|x\|$. By Lemma 3.5.1, $T_\lambda^{-1} \in L_c(R(T); X)$ if $\text{Im}(\lambda) \ne 0$. In this case, $R(T_\lambda)$ is dense in X. Otherwise, there exists $y \ne 0$ such that $\langle T_\lambda x, y\rangle = 0$, all $x \in X$. (Theorem 3.3.11, Corollary.) But $\langle T_\lambda x, y\rangle = \langle x, (\lambda I - T)y\rangle$ since T is self-adjoint. This implies $(\lambda I - T)y = 0$ or $Ty = \lambda y$, which contradicts the fact that $\langle Ty, y\rangle$ is real. Therefore, if $\text{Im}(\lambda) \ne 0$, $\lambda \in \rho(T)$.

In particular, a Hermitian $n \times n$ matrix A has only real eigenvalues. In this case, the quadratic form $\langle Ax, x\rangle$ is real, as we have just seen. If $\langle Ax, x\rangle > 0$ for all nonzero x, then the Hermitian matrix A is said to be *positive definite*. If $\langle Ax, x\rangle \ge 0$ for all x, A is called *positive semidefinite*. Letting $A = (a_{ij})$ and $x^T = (\xi_1, \ldots, \xi_n)$, we have, since $a_{ij} = \overline{a_{ji}}$,

$$\langle Ax, x\rangle = \bar{x}^T A x = \sum_{i,j=1}^n a_{ij}\xi_i\bar{\xi}_j$$
$$= \sum_{i=1}^n a_{ii}|\xi_i|^2 + 2\,\text{Re}\sum_{1 \le i \le j \le n} a_{ij}\xi_i\bar{\xi}_j.$$

If A is positive semidefinite (definite), then $a_{ii} \ge 0$ (>0), $1 \le i \le n$. To see this, take $x = e_i = (0, \ldots, 0, i, 0, \ldots, 0)$. Then $\langle Ax, x\rangle = a_{ii}$.

This necessary condition is not sufficient; e.g.,

$$A = \begin{pmatrix} 1 & 2 \\ 2 & 1 \end{pmatrix}$$

is not positive definite. $\langle Ax, x\rangle = x_1^2 + x_2^2 + 4x_1x_2$. For $y = (1, -1)$, $\langle Ay, y\rangle = -2$. A necessary and sufficient condition is:

Lemma 3.6.1 A symmetric matrix A is positive definite if and only if all principal minors are positive.

Proof. By induction on the order n.

Theorem 3.6.1 Let X be a normed vector space and $T \in L_c(X; X)$. If $|\lambda| > \|T\|$, then $T_\lambda = (\lambda I - T)$ is injective and $(\lambda I - T)^{-1}$ exists as an operator in $L_c(R(T_\lambda); X)$. Further, for $y \in R(T_\lambda)$,

$$(\lambda I - T)^{-1}y = \sum_1^\infty \lambda^{-n}T^{n-1}y. \tag{3.6.0}$$

If X is a Banach space, $|\lambda| > \|T\|$ implies $\lambda \in \rho(T)$ and

$$(\lambda I - T)^{-1} = \sum_{n=1}^\infty \lambda^{-n}T^{n-1}, \tag{3.6.1}$$

where convergence is in the Banach algebra $L_c(X; X)$ i.e., $(\lambda I - T)^{-1} \in L_c(X; X)$.

3.6 The Spectrum of a Linear Operator; Eigenvalues   77

Proof. If $|\lambda| > \|T\|$, then $\|\lambda x - Tx\| \geq |\lambda|\|x\| - \|Tx\| \geq (|\lambda| - \|T\|)\|x\| > m\|x\|$. By Lemma 3.5.1, $(\lambda I - T)^{-1} \in L_c(R(T_\lambda); X)$. If $(\lambda I - T)x = y$, then $Tx = \lambda x - y$ and

$$x = \lambda^{-1}y + \lambda^{-1}Tx$$
$$= \lambda^{-1}y + \lambda^{-1}T(\lambda^{-1}y + \lambda^{-1}Tx)$$
$$= \lambda^{-1}y + \lambda^{-2}Ty + \cdots + \lambda^{-n}T^n x$$

and since $\|\lambda^{-n}T^{n-1}y\| \leq \|T\|^{n-1}/|\lambda|^n \|y\| < K^n \|y\|\|T\|$, where $0 < K < 1$, the first series converges absolutely. Since $\lambda^{-n}T^n x \to 0$, it converges to x.

Now, if X is complete, $L_c(X; X)$ is the Banach algebra $\mathscr{A}_x$ and Theorem 3.5.1 applies. Since $\lambda I - T = \lambda(I - \lambda^{-1}T)$ and $\|\lambda^{-1}T\| = |\lambda^{-1}|\|T\| < 1, (I - \lambda^{-1}T)^{-1}$ exists in $\mathscr{A}_x$. Hence,

$$T_\lambda^{-1} = (\lambda I - T)^{-1} = \lambda^{-1}(I - \lambda^{-1}T)^{-1}$$

exists and

$$T_\lambda^{-1} = \lambda^{-1}\sum_{n=0}^{\infty}(\lambda^{-1}T)^n = \sum_{n=1}^{\infty}\lambda^{-n}T^{n-1}.$$

Since $T_\lambda^{-1} \in \mathscr{A}_x$, the domain of T_λ^{-1} is X. ∎

Corollary Suppose X is a Banach space and $T \in L_c(X; X)$. If λ is in the spectrum of T, then

$$|\lambda| \leq \|T\|. \tag{3.6.2}$$

Remark 2 If A is an $n \times n$ matrix and $\|A\|$ is any norm induced by a vector norm in $V^{(n)}$, then $|\lambda| \leq \|A\|$ for any eigenvalue λ of A.

Definition 3.6.2 Let $T \in L_c(X; X)$, where X is a Banach space. If $\sigma(T)$ is nonempty, then

$$r_\sigma(T) = \sup_{\lambda \in \sigma(T)} |\lambda|$$

is called the *spectral radius* of T.

Remark 3 In general, T_λ^{-1} may exist as a continuous operator with domain $D(T_\lambda^{-1})$ a proper subset of X; i.e., $T_\lambda^{-1} \notin \mathscr{A}_x$. However, if X is a Banach space and T is closed, then $\lambda \in \rho(T)$ implies $T_\lambda^{-1} \in \mathscr{A}_x$. To show this, take an arbitrary $y \in X$. Since $D(T_\lambda^{-1})$ is dense in X, there exist $y_n \in D(T_\lambda^{-1})$ such that $y_n \to y$. Writing $x_n = T_\lambda^{-1}y_n$, we have $\|T_\lambda x_n\| = \|y_n\| \geq c\|x_n\|$, $c > 0$, by Lemma 3.5.1. Therefore, $\|x_p - x_q\| \leq (1/c)\|y_p - y_q\| \to 0$ as $p, q \to \infty$. Let $x = \lim x_n$; (completeness of X). Since T is closed and $(x_n, y_n) \to (x, y)$, $T_\lambda x = y$, or $y \in D(T_\lambda^{-1})$. In particular, by Remark 1 of this section, any $T \in L_c(X; X)$ is closed and $T_\lambda^{-1} \in \mathscr{A}_x$ for $\lambda \in \rho(T)$. We may regard $R(\lambda, T) = (\lambda I - T)^{-1}$ as a function defined on a subset $\{\lambda I : \lambda \in \rho(T)\}$ of the Banach algebra $\mathscr{A}_x$ to $\mathscr{A}_x$. The function $R(\lambda, T)$ is called the *resolvent* of T. It is easy to see that $R(\lambda, T)$ is a holomorphic function of $\lambda = \lambda I$, $\lambda \in \rho(T)$; i.e. R

can be expanded in a unique convergent power series around λ. Indeed, $R(\lambda, T)$ has a Laurent series. One such series is given by equation (3.6.1). This series converges at least for $|\lambda| > \|T\|$. In the next two theorems, we determine its radius of convergence.

Theorem 3.6.2 Suppose $T \in L_c(X; X)$, where X is a Banach space. Then $\lim_{n \to \infty} \|T^n\|^{1/n}$ exists and

$$r_\sigma(T) = \lim_{n \to \infty} \|T^n\|^{1/n}. \qquad (3.6.3)$$

The series in (3.6.1) converges absolutely for $|\lambda| > \lim_{n \to \infty} \|T^n\|^{1/n}$.

Proof. Let $m = \inf_{n \geq 1} \|T^n\|^{1/n}$. Then $\lim_{n \to \infty} \sup \|T^n\|^{1/n} \geq m$. For $\varepsilon > 0$, there exists a k such that $\|T^k\|^{1/k} \leq m + \varepsilon$. For arbitrary n, write $n = qk + r$, $0 \leq r < k$, so that $\|T^n\|^{1/n} = \|T^{qk}T^r\|^{1/n} \leq \|T^k\|^{q/n}\|T\|^{r/n} \leq (m + \varepsilon)^{qk/n}\|T\|^{r/n}$. But $qk/n \to 1$ and $r/n \to 0$ as $n \to \infty$. Therefore, $\lim_{n \to \infty} \sup \|T^n\|^{1/n} \leq m$, which shows that

$$\lim_{n \to \infty} \sup \|T^n\|^{1/n} = \inf_{n \geq 1} \|T^n\|^{1/n} \leq \lim_{n \to \infty} \inf \|T^n\|^{1/n}$$

and so $\lim_{n \to \infty} \|T^n\|^{1/n}$ exists. Now take $|\lambda| > \lim_{n \to \infty} \|T^n\|^{1/n}$. The series in Eq. (3.6.1) converges absolutely. Indeed, for $\varepsilon > 0$,

$$|\lambda^n| \geq (\lim_n \|T^n\|^{1/n} + \varepsilon)^n = b^n,$$

say, and

$$\|T^n\| \leq (\lim \|T^n\|^{1/n} + \varepsilon/2)^n = a^n,$$

say, for n sufficiently large. Hence, $\|\lambda^{-n}T^n\| \leq (a/b)^n$, where $|a/b| < 1$, which proves that (3.6.1) converges absolutely. This implies $T_\lambda^{-1} \in \mathscr{A}_x$ and $\lambda \in \rho(T)$, which establishes that $r_\sigma(T) \leq \lim_{n \to \infty} \|T^n\|^{1/n}$. (Also see Exercise 3.39.)

Suppose $|\lambda| > r_\sigma(T)$, so that $\lambda \in \rho(T)$. Since $R(\lambda, T)$ is holomorphic for all $|\lambda| > r_\sigma(T)$, by standard results on complex functions, it must have a Laurent series in this region which coincides with (3.6.1) for $|\lambda| > \lim \|T^n\|^{1/n}$. Therefore, (3.6.1) converges absolutely for $|\lambda| = r_\sigma(T) + \varepsilon$, $\varepsilon > 0$. This implies $\|T^n\| \leq (\varepsilon + r_\sigma(T))^n$ for n large. Hence, $\lim \|T^n\|^{1/n} \leq r_\sigma$. This completes the proof. ■

In particular, (3.6.3) holds for an $n \times n$ matrix operator in $A^{(n)}$. In studying iterative methods of solving matrix equations such as $Ax = b$, one encounters arbitrary powers of A. For example, we have seen that if $\|A\| < 1$, then

$$(I - A)^{-1} = \sum_{m=0}^{\infty} A^m,$$

where the series converges in the norm of the Banach algebra $A^{(n)}$. A necessary condition that the above series converge is that $\|A^m\| \to 0$, where 0 is the zero matrix. By the continuity of the norm, this implies

$$\lim_{m \to \infty} A^m = 0. \qquad (3.6.4)$$

An operator satisfying condition (3.6.4) will be called *quasinilpotent*. (A *nilpotent* operator T in $L(X; X)$ is one such that $T^m = 0$ for some integer m.) In the case of a matrix $A = (a_{ij})$, (3.6.4) implies that $a_{ij}^{(m)} \to 0$, where $A^m = (a_{ij}^{(m)})$; i.e., A^m converges to 0 elementwise. This follows from the fact that $\|A^m\| \to 0$ implies $|a_{ij}^{(m)}| \le \|A^m\|_\infty \to 0$ by the equivalence of any two norms in the finite-dimensional space $A^{(n)}$. Conversely, if $a_{ij}^{(m)} \to 0$ as $m \to \infty$ for $1 \le i, j \le n$, then

$$\|A^m\|_\infty = \max_{1 \le i \le n} \sum_{k=1}^{n} |a_{ik}^{(m)}| \to 0.$$

Hence, $\|A^m\| \to 0$ for any operator norm of A.

Further, by Theorem 3.6.2,

$$\lim_{m \to \infty} \|A^m\|^{1/m} = r_\sigma(A).$$

Writing

$$\|A^m\| = (r_\sigma(A) + \varepsilon_m)^m, \qquad \text{where } \varepsilon_m \to 0,$$

we see that A is quasinilpotent if and only if $r_\sigma(A) < 1$. We state this as corollary of Theorem 3.6.2.

Corollary 1 Let $T \in L_c(X; X)$, where X is a Banach space. Then $\lim_{m \to 0} \|T^m\| = 0$ if and only if $r_\sigma(T) < 1$.

Corollary 2 If $T \in L_c(X; X)$, X a Banach space, then a necessary and sufficient condition that $\sum_{m=0}^{\infty} T^m$ converge is that $r_\sigma(T) < 1$.

Proof. If the series converges, T^m is quasinilpotent, which implies $r_\sigma(T) < 1$. Conversely, $r_\sigma(T) < 1$ implies that the $1 \in \rho(T)$. Therefore, $(I - T)^{-1}$ exists in $\mathscr{A}_x$. But

$$(I - T)(I + T + \cdots + T^m) = I - T^{m+1},$$

or

$$\sum_{j=0}^{m} T^j = (I - T)^{-1}(I - T^{m+1}).$$

Since $T^{m+1} \to 0$, the series converges to $(I - T)^{-1}$. (See Exercise 3.40.) ∎

A linear operator T having its domain and range in a pre-Hilbert space X is called *symmetric* if $\langle Tx, y \rangle = \langle x, Ty \rangle$ for all $x, y \in D(T)$. For example, a self-adjoint operator in a Hilbert space is symmetric. A symmetric matrix A defines a symmetric operator. Note that $\overline{\langle Tx, x \rangle} = \langle x, Tx \rangle = \langle Tx, x \rangle$, so that $\langle Tx, x \rangle$ is real for symmetric T. If X is a Hilbert space and T is a bounded symmetric operator with $D(T) = X$, then T is self-adjoint. (See the Remark preceding Theorem 3.5.5 above.)

Theorem 3.6.3 If T is a symmetric operator, and λ is an eigenvalue of T, then λ is real and

$$\inf_{\|x\|=1} \langle Tx, x \rangle \le \lambda \le \sup_{\|x\|=1} \langle Tx, x \rangle. \tag{1}$$

Eigenvectors belonging to distinct eigenvalues are orthogonal. If $D(T)$ is dense in

X, then
$$\lambda_{\min} = \inf_{\|x\|=1} \langle Tx, x\rangle \quad \text{and} \quad \lambda_{\max} = \sup_{\|x\|=1} \langle Tx, x\rangle$$
are eigenvalues of T whenever these bounds are finite and attained for some $x \in X$.

Proof. Suppose $Tx = \lambda x$, $\|x\| = 1$. Then $\langle Tx, x\rangle = \langle \lambda x, x\rangle = \lambda$. Since $\langle Tx, x\rangle$ is real, so is λ and the inequality (1) is satisfied. If $Tx_1 = \lambda_1 x_1$ and $Tx_2 = \lambda_2 x_2$, where $\lambda_1 \neq \lambda_2$, then $\lambda_1 \langle x_1, x_2\rangle = \langle Tx_1, x_2\rangle = \langle x_1, Tx_2\rangle = \lambda_2 \langle x_1, x_2\rangle$, or $(\lambda_1 - \lambda_2)\langle x_1, x_2\rangle = 0$. This implies $\langle x_1, x_2\rangle = 0$.

Suppose $\lambda_{\max} = \sup_{\|x\|=1} \langle Tx, x\rangle < \infty$. Let $T_{\max} = T_{\lambda_{\max}} = \lambda_{\max} I - T$, and let y be such that $\lambda_{\max} = \langle Ty, y\rangle$, $\|y\| = 1$. Then $T_{\max}$ is symmetric and $\langle T_{\max} y, y\rangle = 0$. Also,
$$\inf_{\|x\|=1} \langle T_{\max} x, x\rangle = \inf_{\|x\|=1}(\lambda_{\max} - \langle Tx, x\rangle) = 0.$$
Then
$$0 \le \langle T_{\max}(y + \alpha x), y + \alpha x\rangle = \langle T_{\max} y, y\rangle + \alpha \langle T_{\max} x, y\rangle$$
$$+ \bar\alpha \langle T_{\max} y, x\rangle + |\alpha|^2 \langle T_{\max} x, x\rangle.$$
Substituting $\alpha = \theta \langle T_{\max} y, x\rangle$, where θ is real, and using $\bar\alpha = \theta \langle x, T_{\max} y\rangle = \theta \langle T_{\max} x, y\rangle$, we get
$$0 \le \theta |\langle x, T_{\max} y\rangle|^2 (2 + \theta \langle T_{\max} x, x\rangle).$$
For θ negative and $|\theta|$ sufficiently small, the right-hand side of the inequality would be negative unless $\langle x, T_{\max} y\rangle = 0$. Since $x \in D(T)$ is arbitrary and $D(T)$ is dense in X, $T_{\max} y = 0$, or $Ty = \lambda_{\max} y$. Thus, $\lambda_{\max}$ is an eigenvalue.

Similarly, it can be shown that if $\lambda_{\min} > -\infty$ is an attained value, then it is an eigenvalue. ∎

Theorem 3.6.4 Suppose T is symmetric. Then T is continuous if and only if $|\langle Tx, x\rangle|$ is bounded and
$$\|T\| = \sup_{\|x\|=1} |\langle Tx, x\rangle|.$$

Proof. In Section 3.5, we showed that
$$\|A\|_s = \sup_{\|x\|=\|y\|=1} |\langle x, Ay\rangle|$$
for a matrix A. The same calculation establishes that T is continuous if and only if
$$\sup_{\|x\|=\|y\|=1} |\langle x, Ty\rangle| = \mu < \infty.$$
Since $\sup_{\|x\|=1} |\langle Tx, x\rangle| \le \mu$ it suffices to prove $\mu \le \sup_{\|x\|=1} |\langle Tx, x\rangle| = \nu$.
$$\operatorname{Re}\langle Tx, y\rangle = \tfrac{1}{2}(\langle Tx, y\rangle + \langle Ty, x\rangle)$$
$$= \left\langle T\frac{(x+y)}{2}, \frac{x+y}{2}\right\rangle - \left\langle T\frac{(x-y)}{2}, \frac{x-y}{2}\right\rangle,$$
$$|\operatorname{Re}\langle Tx, y\rangle| \le \frac{\nu}{4}(\|x+y\|^2 + \|x-y\|^2) = \frac{\nu}{2}(\|x\|^2 + \|y\|^2).$$

Let $\|x\| = \|y\| = 1$ and $\alpha = |\langle Tx, y\rangle|/\langle Tx, y\rangle$. Then
$$|\langle Tx, y\rangle| = \langle T\alpha x, y\rangle = |\text{Re}\,\langle T(\alpha x), y\rangle| \leq v,$$
and therefore, $\mu \leq v$. ∎

Theorem 3.6.5 Let A be a Hermitian $n \times n$ matrix. Let $\|A\|_S$ be the spectral norm (induced by the l_2-norm). Then
$$\|A\|_S = r_\sigma(A).$$

Proof. The operator is symmetric on V^n, regarded as a pre-Hilbert space. By Theorem 3.6.4, $\|A\|_S = \sup_{\|x\|=1} |\langle Ax, x\rangle|$. Since A is a continuous operator, $|\langle Ax, x\rangle|$ is a continuous function on the unit sphere $\|x\|_2 = 1$, which is a compact subset of V^n. Hence,
$$\lambda_{\max} = \sup_{\|x\|_2 = 1} |\langle Ax, x\rangle| < \infty,$$
and the value $\lambda_{\max}$ is attained for some y on the unit sphere. By Theorem 3.6.3, $\lambda_{\max}$ is the largest eigenvalue of A. ∎

Theorem 3.6.6 Let A be an arbitrary complex $n \times n$ matrix. Then
$$\|A\|_S = \sqrt{r_\sigma(A^*A)}.$$

Proof. By Theorem 3.5.5, regarding A as an operator on the Hilbert space V^n (with l_2-norm), we have $\|A\|_S^2 = \|A^*A\|_S$. Since A^*A is Hermitian, $\|A^*A\| = r_\sigma(A^*A)$. ∎

The positive square roots of the eigenvalues of AA^* are called the *singular values* of A. The preceding theorem shows that for the largest singular value, $\sqrt{\lambda_{\max}(AA^*)} \geq |\lambda_i|$, where λ_i is any eigenvalue of A. We also have $|\lambda_i| \geq \sqrt{\lambda_{\min}(AA^*)}$, since
$$\lambda_{\min}(AA^*) = \min_{\|x\|=1} \langle AA^*x, x\rangle = \min \|A^*x\|^2 \leq |\lambda_i|^2.$$

Let $T \in L_c(X; X)$. Then T^n is defined by $T^0 = I$ and $T^n x = T(T^{n-1}x)$ all $x \in X$. If $f(\lambda) = \alpha_n \lambda^n + \cdots + \alpha_0$ is a polynomial in λ with complex coefficients, then we define the operator $f(T) = \alpha_n T^n + \cdots + \alpha_0 I$. Clearly, $f(T) \in \mathscr{A}_x$. In fact, $f(T)$ is in the subalgebra $\mathscr{A}[T]$ of $\mathscr{A}_x$ generated by T. Since $\mathscr{A}[T]$ is commutative, we see that if $f(\lambda) = g(\lambda)h(\lambda)$, where g and h are polynomials, then $f(T) = g(T)h(T)$.

Theorem 3.6.7 Let $T \in L_c(X; X)$, where X is a complex Banach space. Suppose $f(\lambda)$ is a polynomial. Then μ is in the spectrum of $f(T)$ if and only if $\mu = f(\lambda)$ for some $\lambda \in \sigma(T)$; i.e., $\sigma(f(T)) = f(\sigma(T))$.

Proof. We may take $\alpha_n = 1$ and $n \geq 1$. Let μ be arbitrary but fixed and let $\xi_1, \ldots, \xi_n$ be the zero's of $f(\lambda) - \mu$. Then
$$f(T) - \mu I = (T - \xi_1 I) \cdots (T - \xi_n I). \tag{1}$$
If $\mu \in \sigma(f(T))$, then some $\xi_K \in \sigma(T)$, since otherwise, $(T - \xi_i)^{-1}$ exists in $L_c(X; X)$

for all $i = 1, \ldots, n$, and

$$(f(T) - \mu I)^{-1} = (T - \xi_n I)^{-1} \cdots (T - \xi_1 I)^{-1} \in L_c(X; X).$$

Since $f(\xi_K) = \mu$, $\sigma(f(T)) \subset f(\sigma(T))$.

Conversely, suppose some $\xi_K \in \sigma(T)$. If $(T - \xi_K)^{-1}$ exists, by Theorem 3.5.0 the range of $(T - \xi_K)$ cannot be all of X. Hence, the range of $f(T) - \mu I$ cannot be all of X by (1) above. Therefore, $\mu \in \sigma(f(T))$. If $(T - \xi_K)$ has no inverse, since the factors in (1) commute, we see that $f(T) - \mu I$ has no inverse. ∎

Theorem 3.6.8 If A and B are similar matrices, then they have the same eigenvalues.

Proof. Since $A = C^{-1}BC$, we have

$$A - \lambda I = C^{-1}BC - \lambda I = C^{-1}(B - \lambda I)C,$$

$\det(A - \lambda I) = \det C^{-1} \det(B - \lambda I) \det C = \det(B - \lambda I)$. ∎

Theorem 3.6.9 Let A be a symmetric $n \times n$ matrix. There exists an orthogonal matrix R such that $RAR^{-1} = D$, where D is diagonal.

Proof. Consider the real quadratic form, $x^T A x$. On the unit sphere, $\|x\|_2 = 1$, $x^T A x$ is a continuous function of the coordinates $x_1, \ldots, x_n$; hence it is bounded and assumes its least upper bound for some vector u_1. Choose a new orthonormal basis $\{e'_i\}$ such that $e'_1 = u_1$. The coordinates of u_1 with respect to this new basis are $(1, 0, \ldots, 0)$.

Let R_1 be the orthogonal matrix defined by $R_1 e'_i = e_i$. Let $x' = R_1 x$. Then $x = R_1^{-1} x' = R_1^T x'$ and $x^T = (x')^T R_1$. Hence, $x^T A x = (x')^T R_1 A R_1^{-1} x'$. Let $x' = \sum x'_i e'_i$ and $R_1 A R_1^{-1} = A_1 = (a_{ij}^{(1)})$. Clearly, A_1 is symmetric. Hence,

$$x^T A x = a_{11}^{(1)} x_1'^2 + \cdots + a_{nn}^{(1)} x_n'^2 + 2 \sum_{i \neq j} a_{ij}^{(1)} x'_i x'_j. \tag{1}$$

Since $x^T A x$ has a maximum at $u_1 = e'_1$, its partial derivatives must all be zero at u_1. Thus, for example, differentiating (1) above with respect to x'_2 we obtain

$$\frac{\partial(x^T A x)}{\partial x'_2} = 2a_{22}^{(1)} x'_2 + 2a_{12}^{(1)} x'_1 + \cdots + 2a_{n2}^{(1)} x'_n.$$

Since $x'_1 = 1$ and $x'_i = 0$ for $i \neq 1$ at the point u_1, we must have $a_{12}^{(1)} = 0$. Repeating this procedure for $x'_3, \ldots, x'_n$, we get $a_{1j}^{(1)} = 0$ for $j = 2, \ldots, n$. Thus

$$x^T A x = a_{11}^{(1)} x_1'^2 + 2 \sum_{2 \leq i, j \leq n} a_{ij}^{(1)} x'_i x'_j. \tag{2}$$

Also, we see that

$$a_{11}^{(1)} = u_1^T A u_1 = \max_{|x|_2 = 1}(x^T A x). \tag{3}$$

From $AR_1^T = R_1^T A_1$ and the fact that $a_{j1}^{(1)} = 0$ for $j \geq 2$, it follows that the first column of R_1^T is a unit eigenvector belonging to the eigenvalue $a_{11}^{(1)}$.

Now consider the $(n - 1)$-dimensional subspace, V^{n-1}, spanned by $e'_2, \ldots, e'_n$

and the $(n-1) \times (n-1)$ matrix $\bar{A}_1 = (a_{ij}^{(1)})$, $i, j = 2, \ldots, n$. Applying the above procedure to the quadratic form $\sum_{2 \leq i, j \leq n} a_{ij}^{(1)} x'_i x'_j$, we obtain an orthogonal transformation, $\bar{R}_2$, on V^{n-1} such that $\bar{R}_2 \bar{A}_1 \bar{R}_2^{-1}$ has zero elements in the first row and column except on the diagonal. We extend $\bar{R}_2$ to an orthogonal transformation R_2 on V^n by defining $R_2 e'_1 = e'_1$ and $R_2 e'_i = \bar{R}_2 e'_i$, $i \geq 2$. The matrix of R_2 relative to the e'_i-basis is of the form,

$$R_2 = \begin{pmatrix} 1 & 0 & \cdots & 0 \\ 0 & & & \\ \vdots & & \bar{R}_2 & \\ 0 & & & \end{pmatrix}$$

where $\bar{R}_2$ denotes the $(n-1) \times (n-1)$ matrix of $\bar{R}_2$. Hence, the matrix $R_2 A_1 R_2^{-1} = A_2$ has zero elements in the first and second rows and columns except on the diagonal.

Continuing in this way, we obtain $n-1$ orthogonal transformations $R_1, R_2, \ldots, R_{n-1}$ such that $x^T A x = y^T D y$, where $y = R_{n-1} R_{n-2} \cdots R_2 R_1 x$, and $y^T D y$ has the form $\sum_{i=1}^{n} d_{ii} y_i^2$. Since $R = R_{n-1} R_{n-2} \cdots R_2 R_1$ is the product of orthogonal transformations, R itself is orthogonal. The matrix $D = RAR^{-1}$ is the diagonal matrix of the theorem. ∎

Corollary 1 The diagonal elements of $D = RAR^{-1}$ are the eigenvalues of A.

Proof. By Theorem 3.6.8. ∎

Corollary 2 The columns of R^T are the eigenvectors of A.

Proof. $AR^T = R^T D$, since $R^T = R^{-1}$. If R_i are the columns of R^T, this equation implies $AR_i = d_{ii} R_i$, and in fact $R_i^T A R_i = d_{ii} R_i^T R_i = d_{ii}$. ∎

The following theorem generalizes Theorem 3.6.5.

Theorem 3.6.10 If X is a Hilbert space and $T \in L_c(X, X)$ is a normal operator, then $r_\sigma(T) = \|T\|$. Furthermore, there exists $\lambda \in \sigma(T)$ such that $|\lambda| = \|T\|$.

Proof. First, observe that

$$\|Tx\|^2 = \langle Tx, Tx \rangle = \langle *Tx, x \rangle$$

and

$$\|T^*x\|^2 = \langle T^*x, T^*x \rangle = \langle T^{**}T^*x, x \rangle = \langle TT^*x, x \rangle.$$

Since $T^*T = TT^*$, it follows that $\|Tx\| = \|T^*x\|$. (The converse is also true.)
Now, this result implies $\|T^2 x\| = \|T^*Tx\|$, whence $\|T^2\| = \|T^*T\| = \|T\|^2$, by Theorem 3.5.5. Clearly, T^n is normal, since $(T^n)^* = (T^*)^n$. Hence, for $n = 2^k$, $k = 1, 2, \ldots$, $\|T^n\| = \|T\|^n$. Then by Theorem 3.6.2,

$$r_\sigma(T) = \lim_{k \to \infty} \|T^{2^k}\|^{1/2^k} = \|T\|.$$

From Corollary 4 of Theorem 3.5.2, it follows that the resolvent set of any linear operator T is an open subset of the complex plane. Hence, the spectrum $\sigma(T)$ is closed. If T is bounded, so is $\sigma(T)$. Therefore, $\sigma(T)$ is compact. But $|\lambda|$ is a continuous function of λ on $\sigma(T)$. Hence, there exists $\lambda_{max} \in \sigma(T)$ with $|\lambda_{max}| = \sup\{|\lambda| : \lambda \in \sigma(T)\} = r_\sigma(T)$. By the previous paragraph, $|\lambda_{max}| = \|T\|$ when T is normal, which completes the proof. ∎

If $T \in L_c(X; X)$, where X is a Hilbert space, the set $W(T) = \{\langle Tx, x\rangle : \|x\| = 1\}$ is called the *numerical range* of T (also *field of values* of T). The next theorem establishes an important connection between the numerical range and the spectrum.

Theorem 3.6.11 Let $T \in L_c(X; X)$, where X is a Hilbert space. Let $\overline{W(T)}$ be the closure of the numerical range. Then $\sigma(T) \subset \overline{W(T)}$.

Proof. We prove that $\lambda \notin \overline{W(T)}$ implies $\lambda \in \rho(T)$; i.e., $(\lambda I - T)^{-1}$ exists, is bounded and has a domain dense in X. Let $d > 0$ be the distance from λ to $\overline{W(T)}$. For any x with $\|x\| = 1$,

$$|\langle Tx - \lambda x, x\rangle| = |\langle Tx, x\rangle - \lambda| \geq d.$$

Applying the Schwarz inequality, we obtain $\|(\lambda I - T)x\| \geq d$ for x of unit norm, so that $\|(\lambda I - T)x\| \geq d\|x\|$ for all $x \in X$. By Lemma 3.5.1 (with $Y = R(\lambda I - T)$, the range of $\lambda I - T$), we conclude that $(\lambda I - T)^{-1}$ exists and is bounded. Hence, if we can show that $R(\lambda I - T)$ is dense in X, it will follow (by Definition 3.6.1) that $\lambda \in \rho(T)$. Suppose that $R(\lambda I - T)$ is not dense in X. Then its closure $\overline{R} \neq X$, and therefore $R^\perp$ contains a vector y of unit norm. For all $x \in X$, $\langle(\overline{\lambda}I - T^*)y, x\rangle = \langle y, (\lambda I - T)x\rangle = 0$. Hence, $T^*y = \overline{\lambda}y$ and $\langle Ty, y\rangle = \langle y, T^*y\rangle = \lambda$, so that $\lambda \in W(T)$. This is a contradiction. Therefore, $\overline{R} = X$ and $\lambda \in \rho(T)$. ∎

3.7 CONVERGENCE OF OPERATORS

The space $L_c(X; Y)$ of linear operators on a normed vector space X to a normed space Y is itself a normed space with the norm given by Definition 3.5.1. When we speak of convergence of a sequence of operators $T_n \in L_c(X; Y)$ to an operator $T \in L_c(X; Y)$, we mean that $\|T_n - T\| \to 0$ as $n \to \infty$. There is another kind of operator convergence which arises in applications. (See Section 9.3 for example.) If (T_n) is a sequence of operators in $L_c(X; Y)$ such that $T_n x$ converges for every x in X, then (T_n) is said to be *strongly* (or *pointwise*) *convergent*. Letting $y = \lim_{n\to\infty} T_n x$, we see that this defines a linear operator, $Tx = y$. The operator T is called the *strong limit* of (T_n). To distinguish strong convergence from convergence in the operator norm, the latter is sometimes called *uniform* convergence. Obviously, a uniformly convergent sequence (T_n) is strongly convergent, since $\|T_n - T\| \to 0$ implies that $\|T_n x - Tx\| \leq \|T_n - T\|\|x\| \to 0$. The converse is not always true. In Section 9.3, we shall encounter sequences of bounded linear functionals which converge strongly but not uniformly. However, if (T_n) converges strongly to T and X is a Banach space, then we can derive an important property of (T_n) and we can prove that T is bounded.

Theorem 3.7.1 (Banach). Let $T_n \in L_c(X; Y)$, $n = 1, 2, \ldots$, where X is a Banach space and Y a normed vector space. If (T_n) is strongly convergent, then the sequence of norms $(\|T_n\|)$ is bounded.

Proof. If we can show that there exists a closed ball $B = B(x_0, \varepsilon)$ and an integer N such that for all $n > N$, $\|T_n x\| \le K$ for all $x \in B$, the result will follow, since for any $x \in X$, $x_0 + (\varepsilon/\|x\|)x$ is in B. Hence,

$$\frac{\varepsilon}{\|x\|}\|T_n x\| - \|T_n x_0\| \le \left\| T_n x_0 + \frac{\varepsilon}{\|x\|} T_n x \right\| \le K,$$

and

$$\|T_n x\| \le (\|T_n x_0\| + K)\|x\|/\varepsilon.$$

By the strong convergence, $\{\|T_n x_0\|\}$ is bounded, say by K_1. Therefore, $\|T_n\| \le (K_1 + K)/\varepsilon$ for $n > N$.

To prove the existence of B, we shall prove that if no such B exists, then (T_n) is not strongly convergent. Suppose for any closed ball B_0 there exists an integer n_1 and $x_1 \in B_0$ such that $\|T_{n_1} x_1\| > 1$. By the continuity of T_{n_1}, there is a closed ball $B_1 = B_1(x_1; \varepsilon_1)$, $B_1 \subset B_0$, such that $\|T_{n_1} x\| > 1$ for all $x \in B_1$. For B_1 there must be an integer $n_2 > n_1$ and $x_2 \in B_1$ such that $\|T_{n_2} x_2\| > 2$. (Otherwise, we take $B = B_1$ and $N = n_1$.). As before, there exists $B_2 \subset B_1$, $B_2 = B_2(x_2; \varepsilon_2)$ such that $\|T_{n_2} x\| > 2$ for all $x \in B_2$. Continuing in this way, we obtain a subsequence (n_k) such that $\|T_{n_k} x\| > k$ for all $x \in B_k$ and $B_k \subset B_{k-1}$. The ε_k can be chosen so that $\varepsilon_k \to 0$. Then the sequence (x_k) is a Cauchy sequence. By the completeness of X, $\lim_{k \to \infty} x_k = x'$ exists. By the construction, $x' \in B_k$ for all k. Hence, $\|T_{n_k} x'\| \to \infty$ as $k \to \infty$, and $T_n x'$ does not converge. Therefore, (T_n) is not strongly convergent. ∎

Corollary If (T_n) converges strongly to T in the space $L_c(X; Y)$ where X is a Banach space, then T is a bounded operator.

Proof. By the theorem, $\|T_n x\| \le \|T_n\| \|x\| \le K \|x\|$ for some constant K. Since $T_n x \to Tx$, we must have $\|Tx\| \le K \|x\|$.

Remark Various generalizations of Theorem 3.7.1 are possible. For example, it is sufficient to require only that $\sup_n \|T_n x\| < \infty$ for all $x \in X$ instead of strong convergence to obtain the boundedness of $\{\|T_n\|\}$. (See [3.7], for example.)

If $\{\|T_n\|\}$ is bounded, we say that the set of operators $\{T_n\}$ is *uniformly bounded*. Theorem 3.7.1 is called the *uniform boundedness principle*.

Theorem 3.7.2 Let X be a normed space and let $E \subset X$ be a set everywhere dense in X. Let Y be a Banach space. If (T_n) is a uniformly bounded sequence of linear operators on X to Y which is strongly convergent on E, then (T_n) is strongly convergent on X.

Proof. There exists a constant K such that $\|T_n\| < K$, by the uniform boundedness. For any $f \in X$ and $\varepsilon > 0$, there exists $p \in E$ such that $\|f - p\| < \varepsilon/K$. Then for all m, n sufficiently large,

$$\|T_m f - T_n f\| \le \|T_m f - T_m p\| + \|T_m p - T_n p\| + \|T_n p - T_n f\|$$
$$< K\varepsilon/K + \varepsilon + K\varepsilon/K = 3\varepsilon.$$

Hence, $(T_n f)$ is a Cauchy sequence and, by the completeness of Y, it converges. Therefore, (T_n) is strongly convergent. ∎

Convergence implies the existence of a topology. The topology for strong convergence is called the *strong topology* of $L(X; Y)$ and is given in Exercise 3.44.

REFERENCES

3.1. K. YOSIDA, *Functional Analysis*, Springer-Verlag, New York (1966)
3.2. A. E. TAYLOR, *Introduction to Functional Analysis*, Wiley, New York (1958)
3.3. L. LIUSTERNIK and W. SOBOLEV, *Elemente der Funktionalanalysis*, Akademie-Verlag, Berlin (1955)
3.4. A. WILANSKY, *Functional Analysis*, Blaisdell, New York (1964)
3.5. H. SCHAEFER, *Topological Vector Spaces*, Macmillan, New York (1966)
3.6. L. V. KANTOROVICH and G. P. AKILOV, *Functional Analysis in Normed Spaces*, Pergamon (Macmillan), New York (1964
3.7. E. R. LORCH, *Spectral Theory*, Oxford University Press, New York (1962)
3.8. F. TREVES, *Topological Vector Spaces, Distributions and Kernels*, Academic Press, New York (1967)
3.9. N. DUNFORD and J. SCHWARTZ, *Linear Operators*, Part I, Interscience, New York (1958)
3.10. P. HALMOS, *A Hilbert Space Problem Book*, Van Nostrand, Princeton, N.J. (1965)
3.11. G. BACHMAN and L. NARICI, *Functional Analysis*, Academic Press, New York (1966)

EXERCISES

3.1 Prove the Schwarz inequality,

$$\left| \sum_{i=1}^{n} x_i y_i \right| \leq \left(\sum_{1}^{n} |x_i|^2 \right)^{1/2} \left(\sum_{1}^{n} |y_i|^2 \right)^{1/2}.$$

(See Exercise 3.9.)

3.2 Show that the l_1-, l_2-, and l_∞- norms as defined in Definition 3.1.3 possess properties N_1–N_3. Given that the scalar field is the field of characteristic 2 show that $\sum_1^n x_i = 0$ does not imply all $x_i = 0$. Writing $x = x - y + y$, prove that

$$|\,\|x\| - \|y\|\,| \leq \|x - y\|.$$

3.3 a) The *complex inner product* of two n-dimensional column vectors (over C) x and y is defined to be the complex scalar

$$x^T \bar{y} = \sum_{i=1}^{n} \xi_i \bar{\eta}_i,$$

where $\bar{\eta}_i$ is the complex conjugate of η_i. Property i_1 of the inner product (Definition 3.3.1) becomes

$$i_{c_1} \langle x, y \rangle = \overline{\langle y, x \rangle}.$$

Verify the other properties for $x^T \bar{y}$ and show that $\langle x, \lambda y \rangle = \bar{\lambda} \langle x, y \rangle$. (A vector space with a complex inner product is also called a *unitary space*.)

b) Prove the Schwarz inequality for a complex inner product. [*Hint:* Use the proof in Theorem 3.3.1, paying attention to complex conjugates.] Show that the continuity of the inner product (in the norm topology) follows from the Schwarz inequality.

c) Show that any bilinear form $\sum_{i=1}^{n} a_i \xi_i \eta_i$ such that $a_i > 0$, $1 \leq i \leq n$ defines an inner product $\langle x, y \rangle$ in V^n. (Verify i_1 to i_4 of Definition 3.3.1.) Show that the norm induced by this inner product is $\|x\|^2 = \sum_{i=1}^{n} a_i \xi_i^2$.

d) If $A = (a_{ij})$ is a positive definite symmetric matrix, let

$$\langle x, y \rangle_A = \sum_{1 \leq i,j \leq n} a_{ij} \xi_i \eta_j$$

for any $x^T = (\xi_1, \ldots, \xi_n)$ and $y^T = (\eta_1, \ldots, \eta_n)$. Show that $\langle x, y \rangle_A$ is an inner product in V^n. Show that any real inner product $\langle x, y \rangle$ in V^n must have this form. [*Hint:* Choose a basis $\{v_1, \ldots, v_n\}$ and let $a_{ij} = \langle v_i, v_j \rangle$.]

e) Show that an inner product in a pre-Hilbert space is completely determined by the values of the quadratic form $\langle x, x \rangle$. [*Hint:* In the real case,

$$\langle x, y \rangle = \tfrac{1}{4} \langle x+y, x+y \rangle - \tfrac{1}{4} \langle x-y, x-y \rangle.$$

In the complex case, add the terms $i/4 \langle x+iy, x+iy \rangle - i/4 \langle x-iy, x-iy \rangle$. Note that property i_{C1} of Exercise 3.3(a) is not required if we assume $\langle x, \lambda y \rangle = \bar{\lambda} \langle x, y \rangle$.]

f) Let A be a linear operator in $L(X; X)$, where X is a complex Hilbert space. If $\langle Ax, x \rangle$ is real for all $x \in X$, then $A = A^*$. [*Hint:* Define $\langle x, y \rangle_A = \langle Ax, y \rangle$ and $\langle x, y \rangle_{A^*} = \langle A^*x, y \rangle = \langle x, Ay \rangle$. Since $\langle x, x \rangle_A = \langle x, x \rangle_{A^*}$ for all x, we must have $\langle x, y \rangle_A = \langle x, y \rangle_{A^*}$ for all x, y by part (e).]

g) Let X be a real Hilbert space and T a linear operator on X to X. Show that $g(x) = \langle Tx, x \rangle$ is a real quadratic form according to the alternative definition in Exercise 2.15. Show that the associated symmetric bilinear form is $\tfrac{1}{2} \langle (T + T^*)x, y \rangle$

h) Let X be a real Hilbert space and $f : X \times X \longrightarrow R$ a continuous bilinear form. Show that there exists a bounded linear operator $T \in L_c(X; X)$ such that $fxy = \langle Tx, y \rangle$. If f is symmetric, then T is symmetric. [*Hint:* For x fixed, $f(x, y)$ defines a continuous linear functional f_x on X; i.e., $f_x y = fxy$. By Theorem 3.5.4, there exists $x' \in X$ such that $fxy = \langle x', y \rangle$. The mapping $x \mapsto x'$ is linear.]

i) A complex inner-product space X (see part (a)) is sometimes called a *unitary space*. A *Hermitian bilinear form* on X is a mapping $f : X \times X \longrightarrow C$ (the complex field) having the following two properties:

$$f(ax + by, v) = af(x, v) + bf(y, v), \qquad f(u, v) = \overline{f(v, u)}.$$

Show that if X is a Hilbert space and f is a continuous Hermitian bilinear form, then there is a self-adjoint transformation T such that $fxy = \langle Tx, y \rangle$. Conversely, for any such transformation T, $\langle Tx, y \rangle$ is a continuous Hermitian bilinear form.

3.4 A complex $n \times n$ matrix $A = (A_1, \ldots, A_n)$ such that $\bar{A}_i^T A_j = \delta_{ij}$ is called a *unitary matrix*. Show that a unitary matrix A defines a unitary operator. Let A^* denote the conjugate transpose of A (i.e., $A^* = (a_{ij}^*)$ where $a_{ij}^* = \bar{a}_{ji}$.) Show that $A^*A = AA^* = I$.

3.5 Show that $\|Ax\|_2 = \|x\|_2$ for any unitary matrix operator A.

3.6 a) Show that there exists an $x \in V^n$ with $\|x\| = 1$ such that $\|Tx\| = \|T\|$, if T is in $L(V^n; V^n)$. [*Hint:* The unit sphere in R^n is compact.]

b) Let $T \in L_c(X; Y)$. Prove that the image of the unit sphere consists of points y which satisfy an equation of the form $\langle By, y \rangle = 1$, when T^{-1} exists and X, Y are Hilbert spaces. Show that B is a self-adjoint positive definite operator. [*Hint:* Let $\|x\| = 1$ and $y = Tx$. Then $\langle T^{-1}y, T^{-1}y \rangle = 1$.]

c) Let T be as in part b. Prove that $\|T^{-1}\| = 1/\inf_{\|x\|=1} \|Tx\|$.
Hint:
$$\|T^{-1}\| = \sup_{y \in Y} \frac{\|T^{-1}y\|}{\|y\|} = \sup_{x \in X} \frac{\|x\|}{\|Tx\|}.$$

This gives a "geometric" interpretation of condition number.

3.7 Show that
$$\|A\|_s \leq \|A\|_2 \leq n \|A\|_s.$$

Hint:
$$\|A\|_2^2 \leq n^2 \max_{i,j} |a_{ij}|^2, \quad \leq n^2 (\sup_{\|x\|=\|y\|=1} \langle Ax, y \rangle)^2.$$

3.8 Show that $\|x\|_\infty = \lim_{p \to \infty} \|x\|_p$.

3.9 For $p > 1$ and q such that $(1/p) + (1/q) = 1$, prove the *Hölder inequality*,
$$|x^T y| \leq \|x\|_p \|y\|_q,$$
that is,
$$|\sum x_i y_i| \leq (\sum |x_i|^p)^{1/p} (\sum |y_i|^q)^{1/q}.$$

[*Hint:* Show that for $x \geq 0$, $0 < m < 1$,
$$x^m - 1 \leq m(x - 1).$$
Then let $m = 1/p$ and $x = a/b$ and show that
$$a^{1/p} b^{1/q} \leq (1/p)a + (1/q)b.$$
Let
$$a_i = |x_i|^p / \sum |x_i|^p, \quad b_i = |y_i|^q / \sum |y_i|^q$$
and show
$$\sum a_i^{1/p} b_i^{1/q} \leq 1.\ ]$$

Generalize this for $x = x(t)$ and $y = y(t)$, where $x \in L^p(\mu)$ and $y \in L^q(\mu)$.

[*Hint:* Set
$$a(t) = |x(t)|^p / \int |x(t)|^p \, dt$$
and
$$b(t) = |y(t)|^q / \int |y(t)|^q \, dt.\ ]$$

3.10 Using Hölder's inequality of Exercise 3.9, prove the Minkowski inequality,
$$(\sum |x_i + y_i|^p)^{1/p} \leq (\sum |x_i|^p)^{1/p} + (\sum |y_i|^p)^{1/p}.$$

[Hint:
$$\sum |x_i + y_i|^p \le \sum |x_i| \, |x_i + y_i|^{p-1} + \sum |y_i| \, |x_i + y_i|^{p-1},$$
$$\sum |x_i| \, |x_i + y_i|^{p-1} \le (\sum |x_i|^p)^{1/p} (\sum |x_i + y_i|^p)^{1/q}$$
by the Hölder inequality, where $q = p/(p-1)$.] Generalize to the case of functions $x(t)$, $y(t)$ as in Exercise 3.9.

3.11 Verify that the matrix
$$A = \begin{pmatrix} \cos\theta & \sin\theta \\ -\sin\theta & \cos\theta \end{pmatrix}$$
is orthogonal. Let $x' = Ax$, where $x^T = (x_1, x_2)$ and $x'^T = (x_1', x_2')$. Show that $\|x\|_2 = \|Ax\|_2$. Show that $A^T = A^{-1}$.

Let
$$A = \begin{pmatrix} 1 & -\frac{1}{5} \\ -\frac{1}{5} & 1 \end{pmatrix}.$$

For $x^T = (x_1, x_2)$, form $x^T A x$. Obtain an orthogonal matrix R such that $R^{-1}AR$ is the diagonal matrix
$$\begin{pmatrix} \frac{6}{5} & 0 \\ 0 & \frac{4}{5} \end{pmatrix}.$$

3.12 Let λ be an eigenvalue of an arbitrary square matrix A. Show that $|\lambda| \le \|A\|$.

3.13 For any square matrix A, show that $I + A$ is nonsingular if $\|A\| < 1$ by using the fact that $0 = (I + A)x = x + Ax$ implies that -1 is an eigenvalue unless $x = 0$.

3.14 Let $p(\lambda) = \sum_{j=0}^{k} c_j \lambda^j$ be a polynomial in λ. Let A be an $n \times n$ matrix. The matrix $p(A)$ is defined by
$$p(A) = \sum_{j=0}^{k} c_j A^j.$$

Prove that if λ is an eigenvalue of A, then $p(\lambda)$ is an eigenvalue of $p(A)$. Also prove that if A^{-1} exists, then $1/\lambda$ is an eigenvalue of A^{-1}.

3.15 a) If $\lambda_1 \ne \lambda_2$ are two distinct eigenvalues of an $n \times n$ matrix A, and x_1 and x_2 are two eigenvectors belonging to λ_1 and λ_2 respectively, then x_1 and x_2 are linearly independent. Prove this assertion and then extend it by induction to any number λ_1, $\lambda_2, \ldots, \lambda_k$ of distinct eigenvalues of A and their eigenvectors $x_1, \ldots, x_k$. If $k = n$, the eigenvectors form a basis for V^n. *Hint:* If $a_1 x_1 + a_2 x_2 = 0$, then
$$a_1 \lambda_1 x_1 + a_2 \lambda_2 x_2 = 0$$
$$a_1 \lambda_1 x_1 + a_2 \lambda_1 x_2 = 0$$
$$a_2(\lambda_2 - \lambda_1) x_2 = 0 \Rightarrow a_2 = 0 \Rightarrow a_1 = 0.$$

b) Since $\det(\lambda_i - A) = 0$, λ_i may be a multiple root of the characteristic equation. Let r_i be its *multiplicity*. Prove that *if A is symmetric*, then $\lambda_i I - A$ has nullity equal to r_i; i.e., there are r_i linearly independent eigenvectors belonging to λ_i. [*Hint:* Let $x_1^{(i)}$ be a unit eigenvector belonging to λ_i. Choose an orthonormal basis $x_1^{(i)}, e_2^{(i)}, \ldots, e_n^{(i)}$, using $x_1^{(i)}$ as the first vector in the basis. Let R be the orthogonal operator taking the natural basis into this new one. $R^T AR$ is the matrix representing A relative to this new basis and it has the form

$$R^T AR = \begin{pmatrix} \lambda_i & 0 & \cdots & 0 \\ 0 & & & \\ \vdots & & A_1 & \\ 0 & & & \end{pmatrix}$$

where A_1 is a symmetric $(n-1) \times (n-1)$ matrix. Clearly, $\det(\lambda I - R^T AR) = (\lambda - \lambda_i) \det(\lambda I - A_1)$. But $\det(\lambda I - R^T AR) = (\lambda - \lambda_i)^{r_i} q(\lambda)$. Thus, $\det(\lambda_i I - A_1) = 0$ if and only if $r_i > 1$. If $r_i > 1$, consider A_1 as a linear operator on the subspace spanned by $e_2^{(i)}, \ldots, e_n^{(i)}$. There is an eigenvector of A_1, $x_2^{(i)}$, belonging to λ_i. We see that $x_2^{(i)}$ is also an eigenvector of A belonging to λ_i. It is orthogonal to $x_1^{(i)}$. Continuing in this way, the result follows.]

Observe that this yields an alternative proof that any symmetric matrix is orthogonally similar to a diagonal matrix.

3.16 Let A be an $n \times n$ matrix. The *trace* of A, denoted by tr (A), is defined by

$$\text{tr}(A) = \sum_{i=1}^{n} a_{ii}.$$

Show that tr $(A^T A) = \|A\|_2^2$. Also show tr $(A) = \sum_1^n \lambda_i$, and $\det A = \pi_{i=1}^n \lambda_i$, where the λ_i are the eigenvalues of A. (Refer to the coefficients of the characteristic polynomial and the fact that the λ_i are the roots of $\det(\lambda I - A)$.)

Show that if $B = U^{-1} AU$, then tr $(B) =$ tr (A). Further, if U is a unitary matrix, then tr $(B^*B) = \text{tr}(A^*A)$, that is,

$$\|B^*B\|_2^2 = \|A^*A\|_2^2 \quad \text{or} \quad \sum_{i,j=1}^{n} |b_{ij}|^2 = \sum_{i,j=1}^{n} |a_{ij}|^2.$$

3.17 Prove: If A is symmetric, then

$$\|A\|_2 = \left(\sum_{i=1}^{n} \lambda_i^2\right)^{1/2}.$$

[*Hint:* $A^T A = A^2$, and use Exercises 3.14 and 3.16.]

3.18 Prove the *Cayley–Hamilton* theorem that $\theta(A) = 0$ where $\theta(\lambda) = \det(\lambda I - A)$ is the characteristic polynomial of A. [*Hint:* Use Exercise 2.12c to obtain $B_\lambda(\lambda I - A) = \theta(\lambda)I$. Do not substitute A for λ. See [2.1].]

3.19 If λ is an eigenvalue of the $n \times n$ matrix A, then any vector x such that $(\lambda I - A)^r x = 0$ and $(\lambda I - A)^{r-1} x \neq 0$, $(r \geq 1)$ is called a *principal vector of grade* r, belonging to λ. Suppose $\theta(\lambda) = \det(\lambda I - A) = (\lambda - \lambda_1)^{m_1} \cdots (\lambda - \lambda_k)^{m_k}$, $k \leq n$. Then the polynomials $q_i(\lambda) = \theta(\lambda)/(\lambda - \lambda_i)^{m_i}$, $i = 1, \ldots, k$ have 1 as their greatest common divisor. Hence, there exist polynomials $p_i(\lambda)$ such that

$$\sum_{i=1}^{k} p_i(\lambda) q_i(\lambda) = 1.$$

Since $(\lambda_i I - A)^{m_i} q_i(A) = \theta(A) = 0$, the $q_i(A)x$ which are nonzero are principal vectors. Now show that any vector, x, is a sum of principal vectors, $x = \sum v_i$, where the v_i

belong to distinct λ_i. (For any principal vector x of λ_i of grade r,

$$x = f(A)q_i(A)x + g(A)(\lambda_i I - A)^r x = f(A)q_i(A)x.$$

Hence, $(\lambda_i I - A)^{m_i}x = f(A)\theta(A)x = 0$. This implies that $m_i \geq r$.)

3.20 To generalize Exercise 3.15 to an arbitrary $n \times n$ matrix, prove that A is similar to a matrix of the form (called the *Jordan Canonical form*)

$$J = \begin{pmatrix} J_1 & & & & 0 \\ & J_2 & & & \\ & & J_3 & & \\ & & & \ddots & \\ 0 & & & & J_p \end{pmatrix}$$

where each J_i is a *Jordan matrix* of the form

$$J_i = \begin{pmatrix} \lambda_i & 1 & 0 & \cdots & 0 \\ & & & & \vdots \\ 0 & \lambda_i & & \ddots & 0 \\ \vdots & & \ddots & \ddots & 1 \\ 0 & \cdots & 0 & \cdots & \lambda_i \end{pmatrix}$$

where λ_i is an eigenvalue of A. [*Hint:* Let λ_i, $i = 1, \ldots, k$ be the distinct eigenvalues of A, $k \leq n$, and let

$$\det(\lambda I - A) = (\lambda - \lambda_1)^{m_1}(\lambda - \lambda_2)^{m_2}\cdots(\lambda - \lambda_k)^{m_k}.$$

The integer m_i is the *multiplicity* of λ_i. Consider λ_1. Its principal vectors (Exercise 3.19) are of maximum grade $r = m_1$. Let W_j be the subspace of vectors of grade $\leq j$ ($1 \leq j \leq r$). Choose a set of linearly independent vectors x_{1i} of grade 1 spanning W_1. Extend this to a basis for W_2 by adding vectors of grade 2, x_{2i}. Continue in this way, finally adjoining vectors x_{ri} of grade r to form a basis for W_r. Now, replace this basis X_1 by another basis as follows.

Let $u_r = x_{r1}$, $u_{r-i} = (\lambda_1 I - A)^i x_{r1}$, $1 \leq i \leq r - 1$. Since u_{r-i} is of grade $r - i$, the u's are linearly independent. Also, $Au_{i+1} = \lambda_1 u_{i+1} + u_i$, $1 \leq i \leq r - 1$, and $Au_1 = \lambda_1 u_1$. Continue this procedure with each x_{ri}, adjoining the new u's for each i. The resulting set is linearly independent, for suppose

$$\sum a_{ij}(\lambda_1 I - A)^i x_{rj} = 0.$$

Then

$$\sum a_{0j}x_{rj} + \sum_{i \geq 1} a_{ij}(\lambda_1 I - A)^i x_{rj} = 0.$$

The second sum is in W_{r-1}, and hence is a linear combination of $x_{1i}, \ldots, x_{r-1,i}, \ldots$. By the linear independence of X_1, all $a_{0j} = 0$. The remaining sum is of the form $(\lambda_1 I - A)z$, where $z = \sum a_{1j}x_{rj} + \sum_{i \geq 2} a_{ij}(\lambda_1 I - A)^{i-1} x_{rj}$. Since z is in W_1, it is a linear combination of the x_{1i}. The second sum in z is a vector in W_{r-1}, and hence is a linear combination of $x_{1i}, \ldots, x_{r-1,i}, \ldots$. As before, this implies all $a_{1j} = 0$. Continuing, we find that all $a_{ij} = 0$.

92   Topological Vector Spaces

Next, consider all vectors $x_{r-1,i}$ of grade $r-1$ which are not dependent on the u's already found. Apply a similar procedure to them, forming $v-1$ vectors $x_{v-1,i}$, $(\lambda_1 I - A)x_{v-1,i}, \ldots, (\lambda_1 I - A)^{v-2}x_{v-1,i}$, etc. Complete the process for the entire set X_1. Then repeat it for the set X_i of principal vectors of λ_i, $i = 2, \ldots, k$. The total set of u's is linearly independent, spans the set of principal vectors and by Exercise 3.19 the entire space V^n. Hence, it is a basis. The matrix A is similar to a matrix, J, with respect to the u-basis. Since $Au_{i+1} = \lambda_1 u_{i+1} + u_i$, with similar equations for $\lambda_2, \ldots, \lambda_k$, the result follows.

3.21 Show that

$$J = \begin{pmatrix} \lambda_i & 1 \\ 0 & \lambda_i \end{pmatrix}$$

has the eigenvalue λ_i with multiplicity 2 and that $(1, 0)^T$ is the only eigenvector of norm 1. More generally, show that any Jordan matrix J_i of order n (see 3.20) has a single eigenvalue λ_i of multiplicity n and one linearly independent eigenvector u where $u^T = (1, 0, \ldots, 0)$. (Show that $\lambda_i I - J_i$ has rank $n - 1$.) Also show that J_i^T has only one linearly independent eigenvector v, where $v^T = (0, \ldots, 0, 1)$. Thus, $v^T u = 0$.

3.22 a) Let $A = B^{-1}JB$, where J is the Jordan Canonical form of A, as in 3.20. Then the polynomials $\det(J_i - \lambda I)$ are called the *elementary divisors* of A. Prove that
i) A has only linear elementary divisors if and only if A is *diagonalizable* (i.e., similar to a diagonal matrix) (although its eigenvalues may have multiplicities >1); thus, the fact that A is diagonalizable implies its eigenvectors are a basis.

ii) if A has n distinct eigenvalues, then it has linear elementary divisors.

b) A matrix A is called *normal* if $AA^* = A^*A$. Prove that a matrix A is normal if and only if there exists a unitary matrix U such that $U^{-1}AU$ is a diagonal matrix. (We could say A is *unitarily diagonalizable*. See [2.1].)

3.23 Let J be a Jordan matrix of order r. Let I be the $r \times r$ identity matrix and U the $r \times r$ matrix having 1's on the *super diagonal*, i.e., $u_{12} = u_{23} = \cdots u_{r-1,r} = 1$ and $u_{ij} = 0$ everywhere else. Show that $J = \lambda I + U$. Show that $U^2 = V$ has zeros everywhere except on the diagonal beginning at the first row, third column (i.e., $v_{13} = v_{24} = \cdots = v_{r-2,r} = 1$). In general, U^k has 1's on the diagonal beginning at the first row and $(k + 1)$st column for any $k \leq r - 1$. Also $U^k = 0$ for $k \geq r$.

Show that

$$J^m = \sum_{k=0}^{m} \binom{m}{k} \lambda^{m-k} U^k.$$

Since

$$\lim_{m \to \infty} \binom{m}{k} \lambda^{m-k} = 0$$

if and only if $|\lambda| < 1$, this shows that $\lim_{m \to \infty} J^k = 0$ if and only if $|\lambda| < 1$.

$$J^m = \begin{pmatrix} \lambda^m & \binom{m}{1}\lambda^{m-1} & \cdots & \binom{m}{r-1}\lambda^{m-r+1} & & \\ 0 & \lambda^m & & \vdots & & \\ \vdots & & \ddots & & \binom{m}{1}\lambda^{m-1} & \\ 0 & \cdots & & 0 & \lambda^m & \end{pmatrix}$$

3.24 a) Let A be Hermitian $n \times n$. Show that $A = R + iS$, where R is symmetric and S is skew symmetric (i.e., $S^T = -S$). Let λ be an eigenvalue of A and $u = x + iy$ an eigenvector belonging to λ. Show that

$$\begin{pmatrix} R & -S \\ S & R \end{pmatrix} \begin{pmatrix} x \\ y \end{pmatrix} = \lambda \begin{pmatrix} x \\ y \end{pmatrix}.$$

(Thus, the complex problem $Au = \lambda u$ is reduced to a real problem of twice the order, $2n$.)

Further show that if λ is a simple eigenvalue of A, then it is an eigenvalue of multiplicity two of the $2n \times 2n$ matrix.

b) Prove that $y^*x = 0$ if y is an eigenvector of A^* belonging to $\bar{\lambda}_1$ and x is an eigenvector of A belonging to $\lambda_2 \neq \lambda_1$. [Hint: $\lambda_1 y^*x = \lambda_1 \langle x, y \rangle = \langle x, A^*y \rangle = \langle Ax, y \rangle = \lambda_2 \langle x, y \rangle = \lambda_2 y^*x$.]

3.25 a) Let

$$A = \begin{pmatrix} a_{11} & a_{12} \\ a_{21} & a_{22} \end{pmatrix}$$

with $a_{12} = a_{21}$. Let U be the rotation matrix

$$\begin{pmatrix} \cos\theta & \sin\theta \\ -\sin\theta & \cos\theta \end{pmatrix},$$

with $\tan 2\theta = 2a_{12}/(a_{22} - a_{11})$, $-\pi/4 \leq \theta \leq \pi/4$. Verify that $B = U^*AU$ is a diagonal matrix. Use Exercise 3.16 to prove that $b_{11}^2 + b_{22}^2 = a_{11}^2 + 2a_{12}^2 + a_{22}^2$. Verify that $U^* = U^{-1}$.

b) Let $A = (a_{ij})$ be an $n \times n$ real symmetric matrix. Let $U = (u_{ij})$ be a *two-dimensional rotation matrix* defined by

$$u_{rr} = u_{ss} = \cos\theta, \quad u_{rs} = -u_{sr} = \sin\theta, \quad u_{ii} = 1, i \neq r, s$$

and $u_{ij} = 0$ otherwise.

Show that $U^* = U^{-1}$. Let $B = UAU^*$. Verify that for $i \neq r, s$ and $j \neq r, s$, $b_{ij} = a_{ij}$, whereas in the r and s rows and columns, we have

$$\begin{aligned} b_{ri} &= b_{ir} = a_{ir}\cos\theta + a_{is}\sin\theta, \\ b_{si} &= b_{is} = -a_{ir}\sin\theta + a_{is}\cos\theta, \end{aligned} \bigg\} i \neq r, s$$

$$b_{rr} = a_{rr}\cos^2\theta + 2a_{rs}\cos\theta\sin\theta + a_{ss}\sin^2\theta,$$
$$b_{ss} = a_{rr}\sin^2\theta - 2a_{rs}\cos\theta\sin\theta + a_{ss}\cos^2\theta,$$
$$b_{sr} = b_{rs} = (a_{ss} - a_{rr})\sin\theta\cos\theta + a_{rs}(\cos^2\theta - \sin^2\theta).$$

[*Hint:* Show that the multiplication by U affects only the r and s rows and the multiplication by U^* affects only the r and s columns.]

Show that a rotation through angle θ given by
$$\tan 2\theta = 2a_{rs}/(a_{rr} - a_{ss}), \qquad 0 \leq |\theta| \leq \pi/4,$$
makes $b_{rs} = 0$.

c) Referring to part (b), show that for $i \neq r, s$,
$$b_{ir}^2 + b_{is}^2 = a_{ir}^2 + a_{is}^2$$
and
$$\sum_{i \neq r, s} (b_{ir}^2 + b_{is}^2) = \sum_{i \neq r, s} (a_{ir}^2 + a_{is}^2).$$
Similarly,
$$\sum_{i \neq r, s} (b_{ri}^2 + b_{si}^2) = \sum_{i \neq r, s} (a_{ri}^2 + a_{si}^2).$$
Show that, if $b_{rs} = b_{sr} = 0$,
$$\sum_{i \neq j} b_{ij}^2 = \sum_{i \neq j} a_{ij}^2 - 2a_{rs}^2.$$
(Recall that $a_{ij} = b_{ij}$ if neither index i, j is r or s.)

3.26 a) Prove Jensen's inequality. If $0 < p < q$, then
$$\left(\sum |a_i|^q\right)^{1/q} \leq \left(\sum |a_i|^p\right)^{1/p} = \|a\|_p.$$
[*Hint:* Take $\|a\|_p = 1$. Then $|a_i| \leq 1$ and $|a_i|^q \leq |a_i|^p$. Then use $\|\lambda a\|_p = |\lambda| \, \|a\|_p$.]
(See G. Hardy, J. Littlewood, G. Polya, *Inequalities*. Cambridge U. Press, 1952.)

b) Show that $l_p \subset l_q$ for $1 \leq p < q < \infty$.

c) Let $[a, b]$ be a bounded real interval. Show that $L^q[a, b] \subset L^p[a, b]$ for $1 \leq p < q < \infty$. [*Hint:* Let $E = \{x : |f(x)| \geq 1\}$. Then
$$\int_E |f|^p \, dx \leq \int_E |f|^q \, dx. \;]$$

3.27 a) Prove that for $1 \leq p < \infty$ the space l_p is separable. [*Hint:* The set of all sequences $r = (\rho_n)$ with ρ_n rational and all but a finite number of the ρ_n zero is a denumerable everywhere dense set in l_p, since
$$\sum_1^\infty |\xi_n - \rho_n|^p = \sum_1^N |\xi_n - \rho_n|^p + \sum_{N+1}^\infty |\xi_n|^p < N\left(\frac{\varepsilon^p}{2N}\right) + \frac{\varepsilon^p}{2}. \;]$$

b) Prove that l_∞ is not separable. [*Hint:* Suppose $x_n = (\xi_j^{(n)})$, $n = 1, 2, \ldots$, is a denumerable set in l_∞. Then define $x = (\xi_j)$ by a diagonal procedure so that $\xi_j = \xi_j^{(j)} + 1$ if $|\xi_j^{(j)}| \leq 1$ and $\xi_j = 0$ if $|\xi_j^{(j)}| > 1$. Clearly $\|x - x_n\|_\infty \geq 1$. Hence, x_n cannot be everywhere dense.]

3.28 Prove that the spaces L^p, $p > 1$, are strictly normed (Definition 3.3.3). Prove that L_1 and C are not strictly normed. [*Hint:* Let $f, g \in C[0, 1]$ be two linearly independent functions such that $\|f\|_\infty = \|g\|_\infty$ and both assume their maximum value at the same point $x \in [0, 1]$. Then $\|f + g\|_\infty = \|f\|_\infty + \|g\|_\infty$ but $f \neq \alpha g$. In $L[-1, 1]$, take $f(x) = 1$ for $x \geq 0$ and $f(x) = 0$ for $x < 0$ and $g(x) = 1 - f(x)$. Then $\|f + g\|_1 = 2$ and $\|f\|_1 = \|g\|_1 = 1$. Also see Exercise 7.7.]

3.29 a) Prove that the norm induced by an inner product satisfies the *parallelogram law*
$$\|x + y\|^2 + \|x - y\|^2 = 2\|x\|^2 + 2\|y\|^2.$$
[*Hint:* Show that $\|x \pm y\|^2 = \|x\|^2 \pm 2\text{Re} \langle x, y\rangle + \|y\|^2.$]

b) Prove that, conversely, if a norm satisfies the parallelogram law, then
$$\langle x, y\rangle = \tfrac{1}{4}(\|x + y\| - \|x - y\|^2)$$
defines a (real) inner product such that $\langle x, x\rangle = \|x\|^2$. Thus, the space is a pre-Hilbert space. [*Hint:*
$$\langle x, z\rangle + \langle y, z\rangle = \tfrac{1}{4}(\|x + z\|^2 - \|x - z\|^2 + \|y + z\|^2 - \|y - z\|^2)$$
$$= \tfrac{1}{2}\left(\left\|\frac{x+y}{2} + z\right\|^2 - \left\|\frac{x+y}{2} - z\right\|^2\right) = 2\left\langle \frac{x+y}{2}, z\right\rangle.$$
If we take $y = 0$, then $\langle x, z\rangle = 2\langle x/2, z\rangle$ all x, z. Hence $\langle x, z\rangle + \langle y, z\rangle = \langle x + y, z\rangle$. Hence, $\langle mx, y\rangle = m\langle x, y\rangle$ and $\langle x, y\rangle = n\langle (1/n)x, y\rangle$, giving $\langle (m/n)x, y\rangle = (m/n)\langle x, y\rangle$ for any rational number m/n. Since $\langle \alpha x, y\rangle$ is continuous in α, $\langle \alpha x, y\rangle = \alpha\langle x, y\rangle$ for all real α.]

Note. In the complex case $\langle x, y\rangle_c = \langle x, y\rangle + i\langle x, iy\rangle$, $i = \sqrt{-1}$, is a complex inner product with $\langle x, x\rangle_c = \|x\|^2$.

c) Let H be a pre-Hilbert space. The angle θ between two vectors $x, y \in H$ is defined by
$$\theta = \arccos \langle \bar{x}, \bar{y}\rangle, \qquad 0 \leq \theta \leq \pi$$
where $\bar{x} = x/\|x\|$ and $\bar{y} = y/\|y\|$ are unit vectors in the inner-product norm. Show that the orthogonal projection (Theorem 3.3.11) of x on the one-dimensional subspace spanned by y is the vector $m = (\|x\| \cos \theta)\bar{y}$. [*Hint:* Verify that $\langle x - m, \bar{y}\rangle = 0$.] Verify that $\|m\|^2 + \|x - m\|^2 = \|x\|^2$ (Pythagorean theorem) and that $\sin^2 \theta = \|x - m\|^2/\|x\|^2$.

d) Let M be a closed subspace of a Hilbert space H, and $M^\perp$ the orthogonal complement (see Theorem 3.3.11). Verify that $(M^\perp)^\perp = M$. For an arbitrary set $E \subset H$, we define E to be the set of all vectors x such that $\langle x, e\rangle = 0$ for all $e \in E$. Show that E is a closed subspace and $E \subset (E^\perp)^\perp$, but the inclusion is proper if E is not a closed subspace.

3.30 Let S be an open connected set in R^n. Let $C_0^K(S)$ be the space of real functions which have continuous partials on S of all orders up to and including k, and which have *compact support* (i.e., the closure of the set $x \in S : f(x) \neq 0$ is compact). Show that C_0^K is a pre-Hilbert space with inner product
$$\langle f, g\rangle = \sum_{j \leq k} \int_S D^j f(x)\overline{D^j g(x)}\, dx,$$
where
$$D^j f(x) = \frac{\partial^j f(x_1, \ldots, x_n)}{\partial x_i^{j_1} \ldots \partial x_n^{j_n}}, \qquad \sum_{r=1}^n j_r = j.$$

3.31 Let $C[a, b]$ be the space of continuous functions on the closed bounded interval $[a, b]$.

96   Topological Vector Spaces

Define for $f \in C[a, b]$,
$$\|f\| = \max_{1 \leq i \leq n} |f(x_i)|,$$
where $x_i \in [a, b]$. Show that this is a seminorm.

3.32 Let $\rho(x)$ be a positive integrable function on $[a, b]$, $\int_a^b \rho(x)\,dx < \infty$. Let $L_2^{(\rho)}[a, b]$ be the set of functions f such that
$$\int_a^b |f(x)|^2 \rho(x)\,dx < \infty.$$

a) Show that $L_2^{(\rho)}[a, b]$ is a normed vector space. (We identify functions which differ on a set of measure zero. Thus, $\rho(x)$ may actually be zero on a set of measure zero. In particular, it may have a finite number of zeros.)

b) Define $\langle f, g \rangle = \int_a^b f(x) g(x) \rho(x)\,dx$ and show that $L_2^{(\rho)}$ is a pre-Hilbert space with this definition of inner product.

c) Prove that $L_2^{(\rho)}$ is complete. (See [3.3], or [7.1], p. 38.)

d) Prove that $L^2[0, 1]$ is separable (see [7.1], p. 34) and therefore isomorphic and isometric to l_2.

e) If $\int_a^b \rho(x)\,dx < \infty$, show that $L_2^{(\rho)}$ is separable and complete, hence is isomorphic to l_2.

3.33 Prove that
$$\left\{ \frac{1}{\sqrt{2\pi}}, \frac{1}{\sqrt{\pi}} \cos k\theta, \frac{1}{\sqrt{\pi}} \sin k\theta, (k = 1, 2, \ldots) \right\}$$
is a complete orthonormal set in the space $L^2[0, 2\pi]$. [*Hint*: The second Weierstrass theorem of Section 7.2 shows that the subspace generated by these functions is dense in the subspace of functions continuous on $[0, 2\pi]$. Now, the continuous functions are dense in $L^2[0, 2\pi]$, since the step functions are dense in L^2 and any step function is arbitrarily closely approximated by a continuous function (in the L_2-norm). (See [7.1], p. 37.]

3.34 Let $A = (a_{ij})$ be an $n \times n$ matrix.

a) Prove that the field of values $\{\langle Ax, x \rangle : \|x\| = 1\}$ is a closed, bounded convex set.

b) Show that a_{ii} is in the field of values. Hence,
$$\lambda_{\max} \geq \max\{a_{ii}\} \quad \text{and} \quad \lambda_{\min} \leq \min\{a_{ii}\}.$$

[*Hint*: For A and x real, $\langle Ax, x \rangle$ is real. For $x_1 \neq -x_2$, let $x_t = tx_1 + (1 - t)x_2$. $\langle Ax_t, x_t \rangle / \|x_t\|^2$ is continuous on $[0, 1]$, hence takes all values between $\langle Ax_1, x_1 \rangle$ and $\langle Ax_2, x_2 \rangle$. See [3.11] for complex case.]

3.35 Let $p(\lambda) = \lambda^n + a_1 \lambda^{n-1} + \cdots + a_n$ be an arbitrary polynomial with complex coefficients. The $n \times n$ matrix

$$B = \begin{vmatrix} 0 & 1 & 0 & \cdots & 0 \\ 0 & 0 & 1 & & \vdots \\ & & & \ddots & \\ & & & 1 & 0 \\ 0 & \cdots & & 0 & 1 \\ -a_n & -a_{n-1} & & & -a_1 \end{vmatrix}$$

is called the *companion matrix* of $p(\lambda)$. Prove that $p(\lambda)$ is the characteristic polynomial of B. [*Hint:* Evaluate $\det(\lambda I - B)$ by multiplying column n by λ and adding it to column $n - 1$. Then multiply column $n - 1$ of the new matrix by λ and add it to column $n - 2$, etc. The result is of the form

$$\begin{vmatrix} 0 & -1 & 0 & \cdots & & \cdots & 0 \\ 0 & 0 & -1 & 0 & & \cdots & 0 \\ \vdots & & & \ddots & & & \\ 0 & & & & 0 & & -1 \\ p(\lambda) & & & & & \cdots & \end{vmatrix}$$

which has as determinant $\pm p(\lambda)$.]

3.36 a) Extend the theorem in Exercise 1.3 to the case where f is a vector-valued function with $f(x) \in V^n$.

b) In the Banach space $C[a, b]$ with $\|f\|_\infty$, show that any bounded equicontinuous set E of functions is a compact subset of E. (a, b are finite.)

3.37 Consider the space s of infinite sequences given in Example 1 of Section 2.2. Show that a necessary condition that $X_1 = (\xi_i^{(1)}), \ldots, X_n = (\xi_i^{(n)})$ be a linearly dependent set in s is that each of the determinants

$$w_i = \begin{vmatrix} \xi_i^{(1)} & \cdots & \xi_i^{(n)} \\ \xi_{i-1}^{(1)} & \cdots & \xi_{i-1}^{(n)} \\ \xi_{i-n+1}^{(1)} & \cdots & \xi_{i-n+1}^{(n)} \end{vmatrix}$$

should be zero. [*Hint:* $\sum_1^n a_k X_k = 0$ implies that $\sum_{k=1}^n a_k \xi_i^{(k)} = 0$ for all i. Thus, the a_k are a nontrivial solution of a system of linear equations.]

3.38 Let $\mathscr{A}$ be a Banach algebra. For $A \in \mathscr{A}$, define

$$e^A = I + \sum_{n=1}^\infty A^i/i!.$$

a) Show that the exponential series converges absolutely for all A.

b) Prove that $e^{A+B} = e^A e^B$ for any $A, B \in \mathscr{A}$ such that $AB = BA$. Give an example to show that this relation can fail if A and B do not commute.

c) If $\mathscr{A}$ is a Banach algebra $A^{(n)}$ of $n \times n$ matrices, show that $\det e^A = e^{\operatorname{tr} A}$, where $\operatorname{tr} A = \sum_1^n a_{ii}$ is the trace of $A = (a_{ij})$. Hence, verify that e^A is nonsingular for all A.

d) Show that if A is an $n \times n$ matrix and B is a nonsingular $n \times n$ matrix, then $Be^A B^{-1} = e^{BAB^{-1}}$. [*Hint:* $(BAB^{-1})^k = BA^k B^{-1}$.]

e) Let

$$J = \begin{pmatrix} \lambda & 1 & & 0 \\ & \ddots & \ddots & \\ \vdots & & \ddots & 1 \\ 0 & & & \lambda \end{pmatrix}$$

be an $n \times n$ Jordan matrix. (See Exercises 3.22a, 3.23.) Let U be the upper triangular matrix defined by $J = \lambda I + U$. Show that $e^{tJ} = e^{t\lambda} e^{tU}$ for any scalar t and

$$e^{tU} = \begin{vmatrix} 1 & t & t^2/2! & \cdots & t^{n-1}/(n-1)! \\ 0 & 1 & t & \cdots & t^{n-2}/(n-2)! \\ 0 & 0 & 1 & & \vdots \\ \vdots & & & \ddots & \vdots \\ 0 & \cdots & & & 1 \end{vmatrix}$$

[*Hint*: $U^n = 0$.]

f) In part (e), let $\lambda = \alpha + i\beta$ and $\alpha < 0$. Show that there exist positive constants K and σ such that for all real $t < \infty$,

$$\|e^{tJ}\| \leq K e^{-\sigma t}.$$

[*Hint*: Take $|\sigma| < |\alpha|$.]

g) Let $A = BJB^{-1}$ where as in Exercise 3.22(a), J is the Jordan canonical form of A. Show that $e^A = Be^J B^{-1}$ where

$$e^J = \begin{vmatrix} e^{J_1} & & \\ & \ddots & \\ & & e^{J_p} \end{vmatrix}$$

3.39 Use Theorem 3.6.7 to give a simple proof of the first part of Theorem 3.6.2, that

$$r_\sigma(T) \leq \lim_{n \to \infty} \|T^n\|^{1/n}.$$

[*Hint*: $|\lambda^n| \leq \|T^n\|$ for any $\lambda \in \sigma(T)$ by the Corollary of Theorem 3.6.1.]

3.40 In Theorem 3.6.2, Corollary 2, show that the series is absolutely convergent. [*Hint*: By the proof of the theorem, for large m

$$\|T^m\| = (r_\sigma(T) + \varepsilon_m)^m < d^m,$$

where $d < 1$.]

3.41 Let $f(\lambda) = \det(\lambda I - A)$ be the characteristic polynomial of A. By the Cayley–Hamilton theorem (Exercise 3.18), $f(A) = 0$. Hence, there exists a monic polynomial

$$m(\lambda) = \lambda^q + a_1 \lambda^{q-1} + \cdots + a_q$$

of minimum degree such that $m(A) = 0$. The polynomial $m(\lambda)$ is called the *minimum polynomial* of A. Prove:

a) For any polynomial $p(\lambda)$ (with scalar coefficients) such that $p(A) = 0$, $m(\lambda)$ divides $p(\lambda)$. [*Hint*: The degree of $p(\lambda)$ is $\geq q$. Thus, $p(\lambda) = g(\lambda)m(\lambda) + r(\lambda)$, where either $r(\lambda) = 0$ or is of degree $< q$.]

b) If A is nonsingular, then the constant term a_q in $m(\lambda)$ is nonzero. [*Hint*: Otherwise, $m(A) = A(A^{q-1} + a_1 A^{q-2} + \cdots + a_{q-1}) = 0$.]

c) If A is nonsingular, then A^{-1} is expressible as a polynomial in A. [*Hint:*
$$A^{-1} = -(A^{q-1} + a_1 A^{q-2} + \cdots + a_{q-1} I)/a_q.\ ]$$

d) Let $p(\xi_1, \ldots, \xi_k)$ and $q(\xi_1, \ldots, \xi_k)$ be two polynomials (with complex coefficients) in the variables $\xi_1, \ldots, \xi_n$. Let $A_1, \ldots, A_k$ be k $n \times n$ matrices which commute in pairs (i.e., $A_i A_j = A_j A_i$, all i, j). Let $r(\xi_1, \ldots, \xi_n) = pq$. If $q(A_1, \ldots, A_k)$ is nonsingular, then $r(A_1, \ldots, A_k)$ is defined and can be expressed as a polynomial in $A_1, \ldots, A_k$. [*Hint:* $(q(A_1, \ldots, A_k))^{-1}$ is a polynomial in $q(A_1, \ldots, A_k)$.]

e) Let $p, q, r, A_1, \ldots, A_k$ be as part (d). The eigenvalues $\lambda_{11}, \ldots, \lambda_{1n}$ of A_1, $\lambda_{21}, \ldots, \lambda_{2n}$ of $A_2, \ldots$, can be ordered so that the eigenvalues of $r(A_1, \ldots, A_k)$ are
$$r(\lambda_{11}, \lambda_{21}, \ldots, \lambda_{n1}), r(\lambda_{12}, \lambda_{22}, \ldots, \lambda_{n2}), \ldots, r(\lambda_{1n}, \ldots, \lambda_{nn}).$$

[*Hint:* Let
$$m_1(\lambda) = \prod_{i=1}^{q_1} (\lambda - \lambda_{1i}), m_2(\lambda) = \prod_{i=1}^{q_2} (\lambda - \lambda_{2i}), \ldots$$

be the minimum polynomial of $A_1, A_2, \ldots$ respectively. Then if r is a polynomial

$$r(\xi_1, \ldots, \xi_n) - r(\lambda_{1i}, \lambda_{2j}, \lambda_{3k}, \ldots)$$
$$= [r(\xi_1, \lambda_{2j}, \lambda_{3k}, \ldots) - r(\lambda_{1i}, \lambda_{2j}, \lambda_{3k}, \ldots)]$$
$$+ [r(\xi_1, \xi_2, \lambda_{3k}, \ldots) - r(\xi_1, \lambda_{2j}, \lambda_{3k}, \ldots)] + \cdots$$
$$= (\xi_1 - \lambda_{1i})r_i + (\xi_2 - \lambda_{2j})r_j + \cdots.$$

Hence, taking the product over all indices $(i, j, k, \ldots)$,

$$\prod_{i,j,k,\ldots} [r(\xi_1, \ldots, \xi_n) - r(\lambda_{1i}, \lambda_{2j}, \ldots)] = g_1 m_1(\xi_1) + g_2 m_2(\xi_2) + \cdots,$$

where $1 \leq i \leq q_1, 1 \leq j \leq q_2, \ldots$, and

$$\prod_{i,j,k,\ldots} [r(A_1, \ldots, A_n) - r(\lambda_{1i}, \lambda_{2j}, \ldots)] = 0.$$

Hence, the minimum polynomial is the product of factors $\lambda - r(\lambda_{1i}, \lambda_{2j}, \ldots)$ and these include all distinct factors of the characteristic polynomial.]

3.42 Let
$$Tx = \int_a^b K(t, s) x(s)\, ds$$

be the integral operator discussed in the second example of Section 3.5. Then $T \in L_c(C[a, b], C[a, b])$. Prove that

$$\|T\| = \max_{a \leq t \leq b} \int_a^b |K(t, s)|\, ds.$$

Hint: $\int_a^b |K(t, s)|\, ds$ is continuous for t in $[a, b]$ and therefore assumes its maximum at some t^* in $[a, b]$. Let $u(s) = \operatorname{sgn} K(t^*, s)$; i.e., $u(s) = \pm 1$ depending on the sign of $K(t^*, s)$. Thus $u(s)$ is a step function which can be approximated by a continuous function $x_n = x_n(s)$ which equals $u(s)$ except on a set of measure $\varepsilon_n < 1/2Mn$. Also, $|u(s)| \leq 1$. Then

$$\|Tu - Tx_n\|_\infty \leq 1/n,$$

and

$$\int_a^b |K(t, s)u(s)|\, ds \leq \|Tu\|_\infty \leq \|Tx_n\| + 1/n \leq \|T\| + 1/n.$$

Letting $t = t^*$, we get

$$\int_a^b |K(t^*, s)|\, ds \leq \|T\| + 1/n.$$

3.43 a) Banach's *closed-graph theorem* may be stated as follows: Let X and Y be Banach spaces. If $T \in L(X; Y)$ is a closed operator, then T is continuous. Use Theorem 3.5.0 to prove the closed-graph theorem. [*Hint:* $X \times Y$ is a Banach space when viewed as the direct sum of X and Y with $\|(x, y)\| = \|x\| + \|y\|$. The graph G of T is a closed subspace of $X \times Y$, hence is a Banach space. The mapping $T_1 : (x, T(x)) \longrightarrow x$ is bijective, linear and a continuous mapping of G onto X. Hence, T_1^{-1} is continuous. The mapping $T_2 : (x, Tx) \longrightarrow Tx$ is a linear continuous operator on G onto the range of T. Hence, $T = T_2 T_1^{-1}$ is continuous.]

b) Take $X = Y = C[0, 1]$ with the uniform norm. Let $D \subset C[0, 1]$ be the subspace of continuously differentiable functions and define for $x(t) \in D$ the operator

$$T : x(t) \longrightarrow x'(t).$$

Show that $T \in L(D; X)$ is not continuous, but yet is closed. Reconcile this with the closed-graph theorem. [*Hint:* The sequence $x_n(t) = t^n$ is in D and $\|Tx_n\| = \sup |nt^{n-1}|$ is unbounded, while $\|x_n\| = 1$. If $(x_n, Tx_n) \longrightarrow (x, y)$, this implies that the sequence of derivatives $x'_n(t)$ converges uniformly to $y(t)$. Since $x_n(t) \longrightarrow x(t)$, it follows from advanced calculus that $x'(t) = y(t)$.]

3.44 Let X, Y be normed vector spaces and $L_c(X; Y)$ the space of continuous linear operators. The topology τ_u induced by the norm of an operator is called the *uniform topology* of $L_c(X; Y)$. The topology τ_s corresponding to strong convergence (Section 3.7) is called the *strong topology* or the *topology of pointwise convergence* in X. To define τ_u and τ_s it suffices to specify a base of neighborhoods in each. Show that a neighborhood base for τ_s is the family of all subsets of the form

$$N(\varepsilon; x_1, \ldots, x_m) = \{T \in L_c(X; Y) : \|Tx_i\| \leq \varepsilon, 1 \leq i \leq m\}$$

where $\varepsilon > 0$ and $\{x_1, \ldots, x_m\}$ ranges over all finite subsets of X. (See Exercise 1.9.) Likewise, a neighborhood base for τ_u is the family of all open balls

$$B(T_0; r) = \{T : \|T - T_0\| < r\}.$$

3.45 Let H be a separable infinite-dimensional Hilbert space. Let T be a bounded linear operator on H to H. Show that T can be represented by an infinite matrix (a_{ij}) such that

$$\sum_{i=1}^n \left| \sum_{j=1}^m a_{ij} \xi_j \right|^2 \leq C^2 \sum_{j=1}^m |\xi_j|^2$$

holds for some constant C and all $m, n = 1, 2, \ldots$, and any $\xi_1, \ldots, \xi_m$. [*Hint:* Choose a complete orthonormal set, $\{e_i\}$, by Theorem 3.3.10. Take $x = \sum_1^m \xi_j e_j$. As in the finite dimensional case, let $Te_j = \sum_{i=1}^\infty a_{ij} e_i$. Take $C = \|T\|$.]

Conversely, show that any infinite matrix satisfying the above inequality defines a bounded linear operator T on H with $\|T\| \leq C$. A simpler sufficient condition that (a_{ij}) define a bounded linear operator is that

$$\sum_{i=1}^{\infty} \sum_{j=1}^{\infty} |a_{ij}|^2 < \infty.$$

Verify this.

3.46 Let $T \in L(X, Y)$, where X and Y are pre-Hilbert spaces. Show that T^* is a closed operator. [*Hint:* Let (y_n) be a sequence in the domain $D(T^*) \subset Y$ such that $y_n \longrightarrow y$ and $T^* y_n \longrightarrow u$. Since $\langle Tx, y \rangle = \lim \langle Tx, y_n \rangle = \lim \langle x, T^* y_n \rangle = \langle x, u \rangle, T^* y = u$.]

CHAPTER 4

NUMERICAL SOLUTION OF LINEAR ALGEBRAIC SYSTEMS

4.1 INTRODUCTION

In this chapter, we begin our study of numerical analysis with a subject that appears to be perfectly straightforward, the numerical solution of systems of linear algebraic equations—a subject sometimes referred to as "computational linear algebra." The mathematical theory of linear algebra has been thoroughly worked out and is well understood. It is reviewed briefly in Chapter 2. That part of the theory which deals with the solution of the linear equation $Ax = b$, A a matrix and b a vector, is complete and is summarized in Theorem 2.1.6. It is, of course, based on the real (and complex) number system. Numerical analysis concerns itself with the constructive aspects of this theory. Thus, we are interested in algorithms for obtaining a solution when one exists. At first glance, this would not appear to pose any serious problems. The Gaussian elimination procedure, familiar to secondary-school students, provides us with an algorithm. However, there are two questions which arise. First, is Gaussian elimination the most efficient algorithm in all cases? Second, what is the magnitude of the error caused by rounding? The answers to these questions are found only after careful analysis. The first question can be analyzed within the framework of the real numbers if by "efficiency" we mean simply the number of operations required to obtain the solution. However, it is also important to consider approximate solutions. There are iterative methods which generate approximate solutions which are useful for certain kinds of matrices A. For such methods "efficiency" may be interpreted as the number of operations required to achieve an approximate solution of a prescribed accuracy. This leads to an analysis of the rate of convergence.

The second question, rounding errors, introduces further complications and is related to the first, since Gaussian elimination can produce only approximate solutions when rounding errors occur. In practice, rounding errors always occur because computers are finite machines and real numbers (such as $\sqrt{2}$) cannot be represented by their infinite decimal (or binary) expansions. A real number must be approximated by a finite sequence of digits. The solution of $Ax = b$ must take place within a subset of the real number system. This subset consists of those numbers expressible by a finite number, s say, of digits. The integer s is usually prescribed in advance and held fixed throughout the computation. The real numbers a_{ij} of the matrix A must be approximated by s-digit numbers $\bar{a}_{ij}$ and similarly for the components of the vector b. (Since this chapter is concerned almost exclusively with real matrices and vectors, no confusion should result from the use of the bar in $\bar{a}_{ij}$ to denote an s-digit real number which approximates the real number a_{ij}.) In place of the original system, we have the system $\bar{A}x = \bar{b}$. Furthermore, the exact arithmetic operations $(+, -, \cdot, /)$ on real numbers are replaced by operations with rounding to allow the results to be represented as s-digit numbers. (The product of two s-digit numbers is a $2s$-digit number which must be rounded off to s digits.) Therefore, we obtain an approximate solution $\bar{x}$ having s-digit numbers as components. It is assumed that the reader has had some experience with practical computation and has an intuitive understanding of rounding errors. We shall not take the time here for a lengthy discussion of the various forms possible for the finite representation of real numbers; e.g., *fixed point, floating point, normalized floating point*, etc., and the methods of rounding. We shall proceed as if the reader is acquainted with the material in [4.22].

Unfortunately, the subset of s-digit numbers does not have desirable algebraic properties with respect to the *approximate arithmetic operations* (i.e., operations with rounding). For example, the associative law for approximate multiplication does not hold in general, nor do most of the other algebraic laws of the real numbers. Indeed, the ordering of the operations in a computation with s-digit numbers can affect the final result. Since the s-digit numbers have such poor algebraic properties, the analysis of rounding error is best done by embedding them in the real numbers, treating all operations as exact operations on real numbers and adding on certain error terms to represent the rounding errors. Then we shall be able to use much of the theory of Chapter 2. However, to measure the rounding error, we shall need a topology, and this will take us into the theory developed in Chapter 3. The analysis of iterative procedures which generate a sequence of vectors converging to a solution also necessitates the introduction of norms, even when rounding errors are excluded.

There are two general classes of methods for solving the equation $Ax = b$, where A is an $n \times n$ real matrix and x and b are n-dimensional real column vectors. In one class are the *direct* methods and in the other the *iterative* methods. The term *direct* arises from considerations based on real numbers and exact arithmetic operations on real numbers. As we shall see, the distinction between the two types of methods diminishes when s-digit numerals and rounding errors are involved. A *direct* method is one which yields a solution, x, in a finite number of steps (real operations), as in Gaussian elimination. An *iterative* method, as the name implies, usually requires an infinite number of repetitions of some finite sequence of operations. When rounding accompanies the arithmetic operations, as it does in actual computers, the direct methods generally will not yield the exact solution, x, and iterations of one kind or another are required to improve the accuracy of the computed value.

In describing a direct method, we shall present the algorithm for the *exact solution*, that is, the real vector x which would be obtained by exact operations (without rounding errors) performed on real numbers of *infinite precision*. Then we shall discuss the approximate solution, $\bar{x}$, obtained when rounding errors are generated by the corresponding computer operations performed on numerals of *finite precision*, that is, on numerals which represent real numbers with a prescribed finite number of digits.

Although our discussion assumes that A is $n \times n$, in principle this entails no loss in generality. We recall that in Theorem 2.1.6 the necessary and sufficient condition for a solution to exist applies to $m \times n$ matrices. Let $r = \text{rank } A$. We select r linearly independent columns and rows and apply the algorithm to the $r \times r$ submatrix consisting of those rows and columns, solving for r unknown in terms of the others.

In this way, the solution of the $m \times n$ case reduces to the $n \times n$ case, at least in theory. In practice, rounding errors make this reduction troublesome and impractical for large n. In Exercise 4.1., it is shown how the rank of A can be computed. This procedure depends on computing exact zero's. In the presence of rounding, zero's are replaced by "small" numbers and the rank may be difficult to determine. Therefore, it may be necessary to resort to other techniques such as least-squares approximation. This is discussed in Sections 4.6 and 7.4. In this chapter, we are concerned mainly with nonsingular matrices. Nevertheless, to keep the presentation as general as possible, we do not consider that A is nonsingular, unless explicitly stated.

4.2 GAUSSIAN ELIMINATION WITH REAL NUMBERS

There are several direct methods which are feasible for computers but since they are variants of the familiar Gaussian elimination method, we shall confine our attention to this one basic procedure. (See Exercise 4.3.)

Let us consider Eqs. (2.1.1) in the real case. (The complex case is treated by splitting each equation into its real and imaginary parts as in Exercise 4.2.) The real numbers a_{ij} and b_i are represented by real numerals of infinite precision; e.g., by infinite decimal or binary expansions. We seek real values of the x_i, $i = 1, \ldots, n$, which satisfy the equations. The Gaussian elimination procedure involves the successive elimination of variables as follows. Assume for the moment that $a_{11} \neq 0$. We form the *multipliers* $m_{i1} = a_{i1}/a_{11}$, $i = 2, \ldots, n$. To eliminate x_1 from equation i ($i = 2, \ldots, n$), we multiply equation 1 by m_{i1} and subtract the result from equation i. We then eliminate x_2 from equations $3, \ldots, n$ using a new set of multipliers and so on for $x_3, \ldots, x_{n-1}$.

More precisely, suppose x_1 has been eliminated from equation $2, \ldots, n$, and x_1, x_2 from equations $3, \ldots, n$ and so on up to $x_1, \ldots, x_{k-1}$ from equations $k, \ldots, n$. (For $k = 1$, no variables have been eliminated.) In the elimination process for x_{k-1} new coefficients, $a_{ij}^{(k)}$, are computed for equations $k, \ldots, n$ so that these equations appear as

$$\sum_{j=k}^{n} a_{ij}^{(k)} x_j = b_i^{(k)}, \qquad i = k, \ldots, n. \tag{4.2.1}$$

To eliminate x_k from equations $k + 1, \ldots, n$, we form the multipliers, $m_{ik} = a_{ik}^{(k)}/a_{kk}^{(k)}$, $i = k + 1, \ldots, n$, assuming that $a_{kk}^{(k)} \neq 0$. We multiply equation k by m_{ik} and subtract the result from equation i, $i = k + 1, \ldots, n$. This yields a new set of equations,

$$\sum_{j=k+1}^{n} a_{ij}^{(k+1)} x_j = b_i^{(k+1)}, \qquad i = k + 1, \ldots, n, \tag{4.2.2}$$

where for $k = 1, \ldots, n - 1$,

$$a_{ij}^{(k+1)} = a_{ij}^{(k)} - a_{kj}^{(k)}(a_{ik}^{(k)}/a_{kk}^{(k)}), \qquad i, j = k + 1, \ldots, n. \tag{4.2.3}$$

$$b_i^{(k+1)} = b_i^{(k)} - b_k^{(k)}(a_{ik}^{(k)}/a_{kk}^{(k)}), \qquad (i = k+1, \ldots, n). \tag{4.2.4}$$

If all $a_{kk}^{(k)} \neq 0$, $k = 1, \ldots, n$, this process yields an upper triangular matrix of coefficients in the system of n equations,

$$\sum_{j=k}^{n} a_{kj}^{(k)} x_j = b_k^{(k)}, \qquad k = 1, \ldots, n. \tag{4.2.5}$$

Here, $a_{1j}^{(1)} = a_{1j}$ are the original elements of the first row of A. This triangular system is then solved by the process known as *back substitution*, defined by the equations

$$\left.\begin{array}{l} x_n = b_n^{(n)}/a_{nn}^{(n)}, \\[6pt] x_k = \dfrac{1}{a_{kk}^{(k)}}\left(b_k^{(k)} - \displaystyle\sum_{j=k+1}^{n} a_{kj}^{(k)} x_j\right), \qquad k = n-1, \ldots, 1. \end{array}\right\} \tag{4.2.6}$$

Equations (4.2.3), (4.2.4) and (4.2.6) define the Gaussian elimination procedure.

We now come back to the requirement that $a_{kk}^{(k)} \neq 0$ at each elimination step. To insure that $a_{kk}^{(k)} \neq 0$ whenever possible, a process we shall call *pivoting for maximum size* can be performed prior to the elimination of x_k. This requires a search for an element of maximum absolute value, that is, we seek max $\{|a_{ij}^{(k)}|, i, j = k, \ldots, n\}$. Suppose the element in row i_k and column j_k has the maximum value. Pivoting for maximum size consists of the elementary matrix operation of interchanging rows k and i_k followed by the elementary matrix operation of interchanging columns k and j_k. This brings the element $a_{i_k j_k}$ into the (k, k) or *pivot* position; i.e., the largest element of the submatrix $(a_{ij}^{(k)})$, $i \geq k, j \geq k$, is moved into the pivot position. If A is nonsingular, pivoting for maximum size always yields a nonzero pivot element. (See Theorem 2.1.8.) Hence, the elimination process defined by Eqs. (4.2.3) and (4.2.4) and back substitution defined by Eq. (4.2.6) can always be carried through to completion in the nonsingular case. To show this in detail, we recall (Definition 2.1.18 ff.) that each of the operations in the elimination and pivoting processes can be accomplished by multiplying A by an elementary row or column matrix. At the kth stage in the process, that is, prior to eliminating x_k, we have $E_R A E_C = A^{(k)}$, where E_R is a product of elementary row matrices, E_C is a product of elementary column matrices and $A^{(k)}$ has the form

$$A^{(k)} = \begin{bmatrix} a_{11}^{(1)} & a_{12}^{(1)} & \cdots & & & a_{1n}^{(1)} \\ 0 & a_{22}^{(2)} & \cdots & & & a_{2n}^{(2)} \\ 0 & 0 & \ddots & & & \vdots \\ \vdots & \vdots & & a_{kk}^{(k)} & \cdots & a_{kn}^{(k)} \\ & & & \vdots & & \vdots \\ 0 & 0 & & a_{nk}^{(k)} & \cdots & a_{nn}^{(k)} \end{bmatrix}. \tag{4.2.7}$$

Also $\det E_R = \det E_C = \pm 1$ so that $\det A = \pm \det A^{(k)}$. If $\det A \neq 0$, then clearly $\max \{|a_{ij}^{(k)}|; i, j = k, \ldots, n\} \neq 0$. Conversely, if $\max_{i,j \geq k} \{|a_{ij}^{(k)}|\} \neq 0$ at each stage, $k = 1, \ldots, n$, then A must be nonsingular since at the last step we have $E_R A E_C = U$, where U is the upper triangular matrix of system (4.25). Clearly, $\det U = \Pi_{k=1}^n a_{kk}^{(k)} \neq 0$ in this case and $\det A = \pm \det U$.

If A is singular, the elimination process with pivoting for maximum size must produce a matrix $A^{(k)}$ of the form in (4.2.7) with $a_{ij}^{(k)} = 0$ for $i, j = k, \ldots, n, (k \leq n)$. Matrix A has rank $k - 1$ and nullity $n - k + 1$ in this case. If the transformed vector $E_R b$ has zero coordinates $b_k^{(k)}, \ldots, b_n^{(k)}$, then the system has $n - k + 2$ linearly independent solutions. These may be found by letting $(x_k, \ldots, x_n)$ have the values $(0, 0, \ldots, 0), (1, 0, \ldots, 0), (0, 1, 0, \ldots, 0), \ldots, (0, \ldots, 0, 1)$ successively and solving (4.2.5) by back substitution starting with equation $k - 1$. Obviously, if not all $b_k^{(k)}, \ldots, b_n^{(k)}$ are zero, the singular system has no solutions.

It is clearly not necessary to have the maximum element in the pivot position at each stage. It is sufficient that the pivot element be nonzero. Furthermore, interchanges of rows and columns are operations which must be carried out by the computer and increase the computation time. Obviously, one can modify the maximum-size pivoting procedure so that it is used only if the existing pivot element is zero. However, there is some advantage to having the multipliers, $a_{ik}^{(k)}/a_{kk}^{(k)}$, be less than 1 in absolute value. This can be achieved by a *partial pivoting* process in which, prior to each elimination, we find $\max \{|a_{ik}^{(k)}|, i = k, \ldots, n\}$ and interchange rows to bring this largest element into the pivot position. This reduces the amount of searching to find the largest element, since only one subcolumn has to be searched. It also eliminates column interchanges. If $\max_{k \leq i \leq n} \{|a_{ik}^{(k)}|\} = 0$, then a column interchange should be performed before partial pivoting is continued. In this procedure, the final upper triangular matrix either has no zero's on the diagonal or all zero's occur in a block at the end as in pivoting for maximum size. Solutions, if any, are obtained as explained for that case.

The choice of a method of pivoting is primarily a question of avoiding division by zero when we are dealing with exact computations. However, when finite precision and rounding considerations enter, the choice of pivoting method should be such as to reduce rounding errors without increasing the number of operations excessively. A combination of partial pivoting and scaling appears to be an effective compromise between accuracy and speed of computation for most practical problems. We shall pursue this question further in our analysis of rounding errors.

To complete our discussion of Gaussian elimination, we observe that if no row and column interchanges are required, then the elimination process can be expressed in matrix form by multiplying $Ax = b$ on the left by a lower triangular matrix, L_1, to obtain the matrix form of (4.2.5),

$$L_1(Ax) = L_1 b = c, \qquad Ux = c, \qquad U = L_1 A, \qquad (4.2.8)$$

where U is upper triangular and L_1 is the product of lower triangular *multiplier* matrices of the form,

$$E_k = \begin{bmatrix} 1 & 0 & 0 \cdots & & & & 0 & 0 \\ 0 & 1 & 0 \cdots & & & & & \\ 0 & 0 & 1 & & & & \vdots & \vdots \\ \vdots & \vdots & & \ddots & & & & \\ & & & & 1 & & & \\ & & & & -m_{k+1,k} & 1 & & \\ & & & & \vdots & 0 & \ddots & \\ & & & & & & 1 & 0 \\ & & & & & & \vdots & \\ 0 & 0 \cdots & & -m_{nk} & & 0 \cdots & 0 & 1 \end{bmatrix}. \quad (4.2.9)$$

Each E_k has 1's on the diagonal, $-m_{ik} = -a_{ik}^{(k)}/a_{kk}^{(k)}$ in column k below the diagonal and zero's everywhere else. The product L_1 is

$$L_1 = E_{n-1}E_{n-2} \cdots E_2 E_1. \quad (4.2.10)$$

We observe for future reference that the inverse of E_k is simply

$$E_k^{-1} = \begin{bmatrix} 1 & 0 & 0 \cdots & & & & 0 & 0 \\ 0 & 1 & 0 \cdots & & & & & \\ 0 & 0 & 1 & & & & \vdots & \vdots \\ \vdots & \vdots & & \ddots & & & & \\ & & & & 1 & & & \\ & & & & m_{k+1,k} & 1 & & \\ & & & & \vdots & 0 & \ddots & \\ & & & & & & 1 & 0 \\ & & & & & & \vdots & \\ 0 & 0 \cdots & & m_{nk} & & 0 \cdots & 0 & 1 \end{bmatrix}.$$

Since $A = L_1^{-1}U$ by (4.2.8), we have

$$A = LU, \quad (4.2.11)$$

where $L = L_1^{-1} = E_1^{-1}E_2^{-1} \cdots E_{n-1}^{-1}$. Carrying out these matrix multiplications (or observing that multiplication by each E_i^{-1} effects the corresponding elementary row operations on $E_{i+1}^{-1} \cdots E_{n-1}^{-1}$, $i = n - 2, \ldots, 1,$), one easily finds that

$$L = \begin{bmatrix} 1 & 0 & & & 0 \\ m_{21} & 1 & & & \vdots \\ m_{31} & m_{32} & 1 & & \\ \vdots & \vdots & & \ddots & \\ m_{n1} & m_{n2} & \cdots & m_{n,n-1} & \end{bmatrix}. \quad (4.2.12)$$

The same result can be obtained directly by summing Eqs. (4.2.3) for $k = 1, \ldots, i - 1$. Remembering that $a_{ij}^{(1)} = a_{ij}$, we obtain

$$a_{ij} = a_{ij}^{(i)} + \sum_{k=1}^{i-1} \left(\frac{a_{ik}^{(k)}}{a_{kk}^{(k)}}\right) a_{kj}^{(k)} = \sum_{k=1}^{i} \left(\frac{a_{ik}^{(k)}}{a_{kk}^{(k)}}\right) a_{kj}^{(k)}. \quad (4.2.13)$$

Since $m_{ik} = a_{ik}^{(k)}/a_{kk}^{(k)}$, $U = (a_{kj}^{(k)})$, Eq. (4.2.13) expresses the fact that $A = LU$.

Equations (4.2.10) and (4.2.11) were derived on the assumption that no pivoting takes place. If partial pivoting occurs, then there will be nontriangular row-permutation matrices in the factor E_R, where $U = E_R A$. Let P be the product of all the permutation matrices in the order in which they occur in E_R in the pivoting process. Then PA is a matrix obtained from A by performing all the row interchanges which are necessary for partial pivoting. The equation $(PA)x = Pb$ has the same solution as $Ax = b$ and the elimination process can be carried out on it with no pivoting. The multipliers for PA will be the same as those which occur in the elimination with pivoting. Although P is not known a priori and therefore cannot be used in the computation, we shall make use of PA in doing our error analysis.

4.2a OPERATIONS COUNT IN GAUSS ELIMINATION

Let us now determine the number of arithmetic operations required to produce U and c. The elimination of x_1 requires $(n - 1)^2$ multiplications and the same number of additions; i.e., for each multiplier a_{i1}/a_{11}, $i = 2, \ldots, n$, there are $n - 1$ multiplications and additions. (We do not compute known zeros.)

In general, to eliminate x_j we perform $(n - j)^2$ multiplications and additions. The total is given by

$$\sum_{j=1}^{n-1} (n - j)^2 = \sum_{i=1}^{n-1} i^2 = (2n - 1)(n)(n - 1)/6 = (n^3/3) - (n^2/2) + (n/6).$$

Each multiplier requires one division. Hence, there are $n(n - 1)/2$ divisions in the elimination process.

The back-substitution process requires $n - i$ multiplications and subtractions and one division for each x_i, $i = 1, \ldots, n$. This is a total of $(n - 1)n/2$ multiplications and subtractions and n divisions. The application of multipliers to the right-hand side takes $n(n - 1)/2$ adds and multiply's.

Adding these to the totals for the elimination process, we get

Solution of $n \times n$ system (no scaling)
Total multiplications $= \dfrac{n^3}{3} + \dfrac{n^2}{2} - \dfrac{5}{6}n = \mu(n)$.
Total adds and subtracts $= \mu(n)$.
Total divisions $= n^2/2 + n/2$.
We write $\mu(n) = n^3/3 + O(n^2)$.

This shows that the number of arithmetic operations goes up roughly as the cube of the order of the system. For further discussion, see Exercise 4.22 and 4.25.

4.2b COMPUTATION OF A^{-1} BY GAUSSIAN ELIMINATION

To compute the inverse of an $n \times n$ matrix A, we solve the matrix equation

$$AX = I, \qquad (4.2.14)$$

where I is the $n \times n$ identity matrix. Denoting the columns of the matrix X by X_i, we have $A^{-1} = (X_1, \ldots, X_n)$. The column vectors, X_i, are determined by solving the n equations, $AX_i = \delta_i, i = 1, \ldots, n$, where δ_i is the ith column of I. This is done by Gaussian elimination. First we compute $U = E_R A$ by the elimination procedure and then solve $E_R AX = E_R$, that is

$$UX = E_R, \qquad (4.2.15)$$

where E_R is a product of elementary row matrices as before. We solve (4.2.15) by performing n back substitutions with the matrix U and the n columns of E_R as the successive right-hand sides. From the previous section, we see that the n back substitutions require $n^2(n-1)$ multiplications and additions and n^2 divisions. Adding these numbers to the number of operations involved in the elimination process of forming U, we have

Inversion of $n \times n$ matrix (no scaling)
Total number of multiplications $= \tfrac{4}{3}n^3 - \tfrac{3}{2}n^2 + n/6$.
Total number of adds and subtracts $= \tfrac{4}{3}n^3 - \tfrac{3}{2}n^2 + n/6$.
Total number of divisions $= \tfrac{3}{2}n^2 - \tfrac{1}{2}n$.

(Note: This can be reduced to n^3 operations by taking account of the zero's in the special column vectors of I. See Exercise 4.22.)

4.2c GAUSSIAN ELIMINATION WITH FINITE-PRECISION NUMBERS

In an actual computer, the real matrix $A = (a_{ij})$ is replaced by a matrix $\bar{A} = (\bar{a}_{ij})$, where $\bar{a}_{ij}$ is an approximation to a_{ij} which can be finitely represented. Similarly, the vector b is approximated by $\bar{b}$. We shall assume that the floating-point mode is used. A complete discussion of this mode of representing real numbers is given in [4.22].

We summarize the main features here. A real number y is approximated by a floating-point number of $\bar{y}$ of finite precision if $\bar{y} = \pm 0.n_1 n_2 \cdots n_s \times 10^e$, where the n_i are decimal digits. The integer s is called the *precision* of $\bar{y}$. Normally, the *exponent* e is adjusted so that $n_1 \neq 0$. We shall assume this to be the case. In general, y requires an infinite number of digits d_i to represent it exactly. Thus, $y = \pm d_1 d_2 \cdots \times 10^e$ and $\bar{y}$ is obtained by rounding off the digits $d_{s+1}, d_{s+2} \cdots$. The number $.n_1 n_2 \cdots n_s$ is called the *mantissa* or *fractional part* of $\bar{y}$. There are several ways of rounding. We shall assume the mantissa is obtained by *half-adjusting*, that is, by taking $n_1 n_2 \cdots n_s = [d_1 d_2 \cdots d_s . d_{s+1} + 0.5]$, where $[z]$ denotes the integral part of z. In a computer, the precision s is limited by the "word size," the number of digits in a memory cell. We call $\bar{y}$ a *single-precision* number. By using two memory cells, we can represent y as a number $\bar{\bar{y}}$ with a $2s$-digit mantissa. In this case $\bar{\bar{y}}$ is called a *double-precision* number. Then, $\bar{\bar{y}} = \pm .n_1 n_2 \cdots n_s n_{s+1} \cdots n_{2s} \times 10^e$ and $n_{s+1}, \ldots, n_{2s}$ are called the *low-order* digits of $\bar{\bar{y}}$.

From the definition of $\bar{y}$ it follows that

$$|y - \bar{y}| \leq (1/2)10^{e-s} = (1/2)10^{e-s} \left|\frac{y}{y}\right| \leq 5|y| \cdot 10^{-s}, \quad \text{since} \quad |y| \geq 10^{e-1}$$

when $d_1 \neq 0$. Most computers have a $2s$-digit register to hold the result of an arithmetic operation performed on two single-precision operands. When the result is stored in a single-precision memory cell, it is rounded (half-adjusted) to a single-precision number. Thus, for any two single-precision floating point numbers a, b having s-digit mantissas, we obtain

$$\left.\begin{aligned}\overline{a \pm b} &= (a \pm b)(1 + \rho 10^{-s}), \\ \overline{ab} &= ab(1 + \rho 10^{-s}), \\ \overline{(a/b)} &= \left(\frac{a}{b}\right)(1 + \rho 10^{-s}),\end{aligned}\right\} \quad 0 \leq |\rho| \leq 5$$

where $\overline{a + b}$ is the single-precision result corresponding to the exact sum $a + b$ and similarly for the other quantities.

In the Gaussian elimination algorithm, we must estimate the rounding error in a sum of products

$$S_n = \sum_{j=1}^{n} a_j b_j,$$

where each product is rounded to a single-precision number. We have

$$\bar{S}_k = [\bar{S}_{k-1} + a_k b_k (1 + \mu_k 10^{-s})](1 + \alpha_k 10^{-s}), \quad k \geq 1, \tag{1}$$

where $|\mu_k| \leq 5$, $|\alpha_k| \leq 5$, and $\alpha_1 = 0$. Using (1) recursively, starting with $k = n$, we obtain

$$\bar{S}_n = [\bar{S}_{n-1} + a_n b_n (1 + \mu_n 10^{-s})](1 + \alpha_n 10^{-s})$$

4.2c Gaussian Elimination with Finite-precision Numbers

$$= ([\bar{S}_{n-2} + a_{n-1}b_{n-1}(1 + \mu_{n-1}10^{-s})](1 + \alpha_{n-1}10^{-s}) + a_n b_n(1 + \mu_n 10^{-s}))$$
$$\times (1 + \alpha_n 10^{-s})$$
$$= [\bar{S}_{n-2} + a_{n-1}b_{n-1}(1 + \mu_{n-1}10^{-s})](1 + \alpha_{n-1}10^{-s})(1 + \alpha_n 10^{-s})$$
$$+ a_n b_n(1 + \mu_n 10^{-s})(1 + \alpha_n 10^{-s}),$$
$$\vdots$$
$$\bar{S}_n = \sum_{j=1}^{n} a_j b_j (1 + \mu_j 10^{-s}) \prod_{i=j}^{n} (1 + \alpha_i 10^{-s}) = \sum_{j=1}^{n} a_j b_j (1 + \delta_j),$$

where

$$1 + \delta_j = (1 + \mu_j 10^{-s}) \prod_{i=j}^{n} (1 + \alpha_i 10^{-s}).$$

Since $\alpha_1 = 0$, we find that

$$(1 - 5 \cdot 10^{-s})^{n-j+2} \leq 1 + \delta_j \leq (1 + 5 \cdot 10^{-s})^{n-j+2} \quad \text{for} \quad j \geq 2,$$
$$(1 - 5 \cdot 10^{-s})^n \leq 1 + \delta_1 \leq (1 + 5 \cdot 10^{-s})^n.$$

Hence, letting $\varepsilon = 5 \cdot 10^{-s}$, and noting that

$$-n\varepsilon - \binom{n}{2}\varepsilon^2 \cdots = 1 - (1 + \varepsilon)^n \leq (1 - \varepsilon)^n - 1 = -n\varepsilon + \binom{n}{2}\varepsilon^2 - \cdots,$$

we obtain

$$|\delta_1| \leq (1 + \varepsilon)^n - 1$$
$$|\delta_j| \leq (1 + \varepsilon)^{n-j+2} - 1, \quad j \geq 2.$$

Now, assume that $n \cdot 10^{1-s} \leq 1$. Then for $p \leq n$, $p\varepsilon \leq \frac{1}{2}$ and

$$(1 + \varepsilon)^p - 1 = p\varepsilon \left(1 + \frac{p-1}{2}\varepsilon + \frac{(p-1)(p-2)}{2 \cdot 3}\varepsilon^2 + \cdots\right)$$
$$\leq p\varepsilon(1 + \tfrac{1}{2} + (\tfrac{1}{2})^2 + \cdots)$$
$$\leq 2p\varepsilon = p \cdot 10^{1-s}.$$

Therefore,

$$|\delta_j| \leq (n - j + 2)10^{1-s}, \quad j \geq 2$$
$$|\delta_1| \leq n \cdot 10^{1-s}.$$

Setting $\delta a_j = a_j \delta_j$, we finally have

$$\bar{S}_n = \sum_{1}^{n} a_j b_j (1 + \delta_j) = \sum_{1}^{n} (a_j + \delta a_j) b_j. \tag{4.2.16}$$

The Gaussian elimination procedure with finite-precision floating-point numbers is defined by (4.2.3), (4.2.4) and (4.2.6), as with real numbers, except that the operation symbols ($+$, $-$, $\cdot$, $/$) now denote computer operations with rounding. Also, certain other operations, such as *scaling* of the matrix elements, are performed to normalize the rows and possibly the columns of the matrix A. The rigorous mathematical analysis of the effect of scaling remains a subject for research. (See [4.13] and [4.14], for example.) Various schemes are used in practice. One method, called *equilibration* [4.4], consists of multiplying each row and column of A by an appropriate constant, the *scale factor* for that row or column, so as to *normalize* the row and column vectors; i.e., make their l_∞-norms equal to unity. In the floating-point mode, this can be done approximately by changing the exponents, that is, all scale factors are of the form 10^e, where 10^{-e} is the largest exponent in the row (or column). This choice of scale factor introduces no rounding errors. A simpler method of scaling is to multiply each row of the augmented matrix by a scale factor which makes the l_∞-norm of the row vector of A equal to unity. Again, this can be done approximately by changing exponents. We shall assume that this method of scaling is used.

Unless otherwise stated, the precision s is to be regarded as fixed throughout the computation. Thus, the elements $\bar{a}_{ij}$ of $\bar{A}$ and the components $\bar{b}_i$ of $\bar{b}$ are single-precision floating-point numbers having s-digit mantissas. Likewise, the result of each arithmetic operation is a (rounded) s-digit floating-point number. (It is sometimes necessary to use double-precision in critical parts of the computation as, for example, in the formation of vector inner products. This decreases the effect of rounding errors appreciably, as the error analysis of the next section will show.) If we let d_i be the scale factor for row i ($i = 1, \ldots, n$) and D the diagonal matrix having d_i as its ith diagonal element, then the system $\bar{A}x = \bar{b}$ is equivalent to the scaled system $D\bar{A}x = D\bar{b}$. Letting $\bar{A}^{(1)} = D\bar{A}$ and $\bar{b}^{(1)} = D\bar{b}$, we consider Gaussian elimination with partial pivoting applied to the system $\bar{A}^{(1)}x = \bar{b}^{(1)}$. As a practical matter, it is recommended that the elimination procedure be applied separately to $\bar{A}^{(1)}$ and $\bar{b}^{(1)}$. An algorithm so constructed can be used conveniently to solve many systems with matrix $\bar{A}^{(1)}$ and different right-hand vectors. In particular, the algorithm can be used to obtain the inverse of $\bar{A}^{(1)}$ by solving the matrix equation $\bar{A}^{(1)}X = I$, as explained in the previous section.

The elimination procedure for $\bar{A}^{(1)}$ is defined by Eq (4.2.3), where all operations are to be interpreted now as computer operations in the floating-point mode with s-digit precision. Therefore, rounding errors occur and the actual computed quantities, $\bar{a}_{ij}^{(k)}$, differ from the real numbers $a_{ij}^{(k)}$ in (4.2.3). In terms of exact operations on real numbers, the computed values are given by the modified equations,

$$\left.\begin{aligned} u_{ij} &= \bar{a}_{p_i,j}^{(i)}, & 1 \leq i \leq j \leq n & \\ l_{ij} &= \begin{cases} \bar{a}_{ij}^{(j)}/u_{jj} + \mu_{ij}, & \text{if } u_{jj} \neq 0 \\ 0, & \text{if } u_{jj} = 0, \ n \geq i > j \geq 1 \end{cases} & \\ \bar{a}_{ij}^{(k+1)} &= \bar{a}_{ij}^{(k)} - l_{ik}u_{kj} + \varepsilon_{ij}^{(k)}, & i,j \geq k+1, \ 1 \leq k \leq n-1, \end{aligned}\right\} \quad (4.2.17)$$

where μ_{ij} and $\varepsilon_{ij}^{(k)}$ are the errors caused by rounding and p_i is an integer indicating the row interchanged with row i in the partial pivoting process. The l_{ij} are the computed multipliers. (We could have denoted these values by $\bar{m}_{ij}$, but we prefer to abbreviate our notation. For the same reason, we denote the final results of the elimination by u_{ij}.)

In a computer, the multipliers would be recorded as the elements of a lower triangular matrix $\bar{L} = (l_{ij})$, corresponding to the exact matrix L in (4.2.12). The $\bar{a}_{ij}^{(k+1)}$ are stored as columns adjacent to the multipliers at each step k, the old values $\bar{a}_{ij}^{(k)}$ ($i, j \geq k + 1$) being erased as the new values $\bar{a}_{ij}^{(k+1)}$ are written over them. Any row interchange required by partial pivoting is performed both on $\bar{L}$ and the $a_{ij}^{(k)}$. The interchange is recorded as an integer p_i, indicating that row p_i was interchanged with i. Since the diagonal elements of $\bar{L}$ are 1's, these need not be recorded. In their place, the scale factors d_i are recorded. At the end of the elimination computation for $\bar{A}^{(1)}$, there are $n(n + 2)$ memory cells containing the column vectors of the matrix,

$$B = \begin{bmatrix} d_1 & u_{11} & u_{12} & \cdots & u_{1n} & p_1 \\ l_{21} & d_2 & u_{22} & \cdots & u_{2n} & p_2 \\ l_{31} & l_{32} & d_3 & \cdots & u_{3n} & p_3 \\ \vdots & \vdots & \vdots & \ddots & \vdots & \vdots \\ l_{n1} & l_{n2} & l_{n3} & d_n & u_{nn} & p_n \end{bmatrix}. \quad (4.2.18)$$

The original matrix $\bar{A}$ should be retained in memory for future use.

Next, the computations defined by Eqs. (4.2.4) are performed on the vector b, using the matrix B. The components $\bar{b}_i$ are first multiplied by the scale factors d_i and the results $\bar{b}_i^{(1)}$ are permuted as indicated by the p_i. However, rather than follow (4.2.4), we note that the desired right-hand vector c is given by (4.2.8) as the solution of $Lc = b$. Since the entire matrix $\bar{L}$ is available, we can compute the components of c directly. Thus, calling the computed vector $\bar{c}$, we have

$$\left. \begin{array}{l} \bar{c}_1 = \bar{b}_{p_1}^{(1)}, \\ \bar{c}_i = \bar{b}_{p_i}^{(1)} - \sum_{k=1}^{i-1} l_{ik}\bar{c}_k + \beta_i, \quad (2 \leq i \leq n), \end{array} \right\} \quad (4.2.19)$$

where the β_i are the total rounding errors resulting from the finite-precision operations.

Finally, the back-substitution computation defined by (4.2.6) is carried out. The computed vector $\bar{x} = (\bar{x}_1, \ldots, \bar{x}_n)$ is given by

$$\bar{x}_k = (1/u_{kk})(\bar{c}_k - \sum_{j=k+1}^{n} u_{kj}\bar{x}_j) + \xi_k, \quad k = n, n-1, \ldots, 1, \quad (4.2.20)$$

where ξ_k is the total rounding error resulting from the computer operations corresponding to the exact operations in (4.2.20). Here, we have assumed that $u_{kk} \neq 0$, $k = 1, \ldots, n$. If A is nonsingular, so is U. Our assumption is valid if $\bar{U} = (u_{ij})$ is sufficiently close to U.

4.3 ESTIMATES OF ERROR IN GAUSSIAN ELIMINATION

The computed solution, $\bar{x}$, given by the Gaussian elimination single-precision algorithm described in the previous section, differs from the theoretical exact solution,

x, of $Ax = b$ by an *error* Δx. Thus, $\bar{x} = x + \Delta x$. The error Δx may be considered to arise from two sources: (1) *initial errors* caused by rounding A to $\bar{A}$ and b to $\bar{b}$; (2) rounding errors in the elimination procedure. Now let $\bar{A} = A + \Delta A, \bar{b} = b + \Delta b$ and let $\Delta_1 x$ be the error caused by ΔA and Δb when no rounding errors occur in the elimination procedure, that is, $\Delta_1 x$ is the part of the total error ascribable to initial errors. (We may consider ΔA and Δb to include initial errors in the given values arising from sources other than rounding. For example, if the solution of $Ax = b$ is part of a larger problem, the coefficients a_{ij} and b_i may be determined by some method which generates errors.) A bound on $\|\Delta_1 x\|$ is obtained easily as in Theorem 3.5.3. By the definition of $\Delta_1 x$, we have (using exact operations),

$$(A + \Delta A)(x + \Delta_1 x) = b + \Delta b.$$

As in the proof of Theorem 3.5.3, we obtain

$$\Delta_1 x = -(A + \Delta A)^{-1}(\Delta A)x + (A + \Delta A)^{-1}\Delta b,$$

$$\|\Delta_1 x\| \leq \frac{\|A^{-1}\|}{1 - \|A^{-1}\|\|\Delta A\|}(\|\Delta A\|\|x\| + \|\Delta b\|),$$

$$\frac{\|\Delta_1 x\|}{\|x\|} \leq \frac{\gamma}{1 - \gamma\frac{\|\Delta A\|}{\|A\|}}\left(\frac{\|\Delta A\|}{\|A\|} + \frac{\|\Delta b\|}{\|b\|}\right), \tag{4.3.1}$$

since $\|b\| \leq \|A\|\|x\|$. As before, $\gamma = \|A\|\|A^{-1}\|$ is the condition number of A and we are assuming that $\|\Delta A\| \leq 1/\|A^{-1}\|$. As a practical estimation formula, (4.3.1) is of limited usefulness, since it requires an estimate of $\|A^{-1}\|$, A^{-1} being generally unknown. (An approximation to A^{-1} can be computed by Gaussian elimination, but we would then have the problem of estimating the error again.) Nevertheless, (4.3.1) sheds light on the effect of initial errors. If the condition number is much larger than unity, then we might expect the *relative error*, $\|\Delta_1 x\|/\|x\|$, in the solution to be much larger than the initial relative errors in A and b. In this case, A and equation $Ax = b$ are called *ill-conditioned*. If it is known that γ is large (say $\gamma \geq 1000$), then the matrix A and the vector b should be determined very accurately.

We have called $\|\Delta_1 x\|/\|x\|$ the *relative error*. This quantity roughly measures the number of correct digits in the mantissa of the computed solution. Note, however, that $\|\Delta_1 x\|/\|x\|$ being small, say $< 10^{-7}$, does not imply that all components of x have been determined with 7 correct decimal digits. For example, if we use the l_∞-norm, then we can have

$$\max_i |\Delta x_i|/\max_i |x_i| < 10^{-7},$$

while $|\Delta x_1| = 10^{-8}, |x_1| = 10^{-5}$, $\max_i |x_i| = 1$ and $\max_i |\Delta x_i| < 10^{-7}$. Thus, $|\Delta x_1|/|x_1| = 10^{-3}$, so that x_1 has been determined with only 3 correct digits. A more realistic measure of the relative error is given by the quantity $\max_i \{|\Delta x_i|/|x_i|\}$. However, this quantity cannot be expressed in terms of norms and is therefore difficult to estimate.

4.3 Estimates of Error in Gaussian Elimination

Let us now turn to the second source of error, the rounding errors in the elimination procedure. To simplify the analysis of the error due to this source, we shall assume that there are no initial errors and $A = \bar{A}^{(1)}$. Furthermore, we shall assume that all permutations required for partial pivoting are somehow known and performed prior to any computation. Thus, the rows of (A, b) are assumed to be so positioned that the maximum column pivots are already in position at each step of the elimination procedure. This assumption simply allows us to write A instead of PA, where P is some permutation matrix.

Now, let x be the exact solution and $\bar{x} = x + \delta x$ be the solution obtained by the single-precision floating-point Gaussian elimination procedure. Then $Ax = b$ and $A(x + \delta x) = b + r$, where r is the *exact residual vector*. Since A, b and $\bar{x}$ are all single-precision quantities, $r = A\bar{x} - b$ is computable provided that all operations are carried out with a sufficiently high (but finite) precision. Suppose also that an approximate inverse, B, of A is computed, say by solving $AX = I$ by the single-precision elimination procedure. Let $R = AB - I$. From Corollary 3 of Theorem 3.5.2 it follows that if $\|R\| < 1$, then

$$\|A^{-1}\| \le \frac{\|B\|}{1 - \|R\|}.$$

In fact, we compute an approximation $\bar{R} = R + \delta R$, but $\|\delta R\|$ can be bounded easily by analyzing the rounding error in the matrix multiplication AB. (The main computation is of the inner product $\sum_{k=1}^{n} a_{ik}b_{kj}$ and we have already obtained estimates of the rounding error in an inner product. See formula (4.2.16).) Thus $\|R\| \le \|\bar{R}\| + \|\delta R\|$, which yields as a bound for $\|A^{-1}\|$,

$$\|A^{-1}\| \le \frac{\|B\|}{1 - \|\bar{R}\| - \|\delta R\|}.$$

Since $\bar{x} = A^{-1}(r + b)$ and $x = A^{-1}b$, we have

$$\|\delta x\| = \|\bar{x} - x\| \le \|A^{-1}\| \|r\| \le \frac{\|B\| \|r\|}{1 - \|\bar{R}\| - \|\delta R\|}.$$

Since the computation of r usually requires double precision at least, it is worthwhile to point out that the single-precision residual $\bar{r} = A\bar{x} - b + \delta r$ can be used. The norm of the rounding error $\|\delta r\|$ can be estimated, since the single-precision product $A\bar{x}$ involves the inner products $\sum_{k=1}^{n} a_{ik}\bar{x}_k$. Hence, $\|r\| \le \|\bar{r}\| + \|\delta r\|$, which yields the error bound,

$$\|\delta x\| \le \frac{\|B\|(\|\bar{r}\| + \|\delta r\|)}{1 - \|\bar{R}\| - \|\delta R\|}, \tagright{(4.3.2)}$$

in which all quantities are computable from results of the elimination computation. Therefore, (4.3.2) is called *a posteriori* error bound. It can be used to gain some idea of the accuracy of the computed solution. However, it does not throw any light on such questions as how the precision s affects the accuracy. To gain some insight into this matter, we shall now develop *a priori* error bounds.

If we "perturb" the nonsingular matrix A, that is, add a matrix δA of small norm to A, we obtain a matrix $A + \delta A$ which is also nonsingular provided $\|\delta A\| < 1/\|A^{-1}\|$. (Theorem 3.5.2.) Now, $x = A^{-1}b$ depends continuously on A, since inversion is a continuous operation by Theorem 3.5.2, Corollary 1, and matrix multiplication is continuous. Therefore, the solution $(A + \delta A)^{-1}b$ of the perturbed system should be close to x. This suggests that, conversely, the computed solution $\bar{x} = x + \delta x$ might be the exact solution of some perturbed system. If this is the case, then

$$(A + \delta A)(x + \delta x) = b, \qquad (4.3.3)$$

and inequality (3.5.1) can be used provided that a bound for $\|\delta A\|$ can be estimated. We shall show that such a perturbation exists, although it is not unique. At the same time, we shall obtain an a priori bound for $\|\delta A\|$ in terms of the precision s and the order n of the matrix A.

As with the a posteriori error bound, we assume that $A = \bar{A}^{(1)}$ and all required permutations have been performed so that no pivoting is needed in the elimination procedure. In that case, $A = LU$ by (4.2.11). The lower triangular matrix $\bar{L} = (l_{ij})$, where l_{ij} is given in (4.2.17) for $i > j$ and $l_{ii} = 1$ ($l_{ij} = 0$ for $i < j$), is the computed approximation to the exact matrix L. Likewise, the upper triangular matrix $\bar{U} = (u_{ij})$, where u_{ij} is given by (4.2.17), is the computed approximation to U. This suggests that

$$\bar{L}\bar{U} = A + E, \qquad (4.3.4)$$

where $E = (e_{ij})$ may be thought of as an error matrix resulting from the rounding errors in the approximations $\bar{L}$ and $\bar{U}$. Actually, E is defined by (4.3.4) regardless of how we may interpret it.

Referring to (4.2.8), we see that $Lc = b$, since $L = L_1^{-1}$. In the single-precision computation, L is replaced by $\bar{L}$ and c is replaced by a computed approximation, $c + \delta c$, given by (4.2.19) (with all $p_i = i$ under our assumption). In view of the rounding errors in (4.2.19), we do not have $\bar{L}(c + \delta c) = b$, but rather expect that $c + \delta c$ is the solution of a perturbed system. Hence, we define the matrix $\delta \bar{L}$ to be such that

$$(\bar{L} + \delta \bar{L})(c + \delta c) = b. \qquad (4.3.5)$$

To determine the perturbation $\delta \bar{L} = (\delta l_{ij})$ we note that

$$\delta \bar{L}(c + \delta c) = b - \bar{L}(c + \delta c) = v, \qquad (4.3.6)$$

where v is a known vector, since b, $\bar{L}$ and the computed vector $c + \delta c$ are known. The ith equation of (4.3.6) is

$$\sum_{j=1}^{n} \delta l_{ij}(c_j + \delta c_j) = v_i. \qquad (4.3.7)$$

4.3 Estimates of Error in Gaussian Elimination   117

If $c + \delta c \neq 0$, then some component $c_k + \delta c_k \neq 0$. Hence, we can take $\delta l_{ik} = v_i/(c_k + \delta c_k)$ and $\delta l_{ij} = 0$ for $j \neq k$. If $c + \delta c = 0$, we must have $v = b$. If $b = 0$ also, then $\delta \bar{L}$ can be chosen arbitrarily and (4.3.5) will be satisfied. If $b \neq 0$ while $c + \delta c = 0$, then there is no $\delta \bar{L}$ which satisfies (4.3.5). However, if $b \neq 0$ and $\bar{L}$ is nonsingular, in general $c + \delta c$ will also be nonzero since it is close to the exact solution $\bar{L}^{-1}b$. In any event, we shall assume that $c + \delta c = 0$ and $b \neq 0$ do not occur simultaneously. Therefore, a perturbation $\delta \bar{L}$ satisfying (4.3.5) exists. It is clear from (4.3.7) that $\delta \bar{L}$ is not unique.

Referring to (4.2.8) again, we observe that the computed solution $x + \delta x$ is the result of trying to solve $\bar{U}y = c + \delta c$ exactly. Since rounding errors occur, as shown in (4.2.20), we again seek a perturbation, $\delta \bar{U}$, such that

$$(\bar{U} + \delta \bar{U})(x + \delta x) = c + \delta c. \tag{4.3.8}$$

Following the same line of reasoning as with $\delta \bar{L}$, one can prove that such a perturbation $\delta \bar{U}$ exists but is not unique.

Now, multiplying (4.3.8) by the matrix $\bar{L} + \delta \bar{L}$ and using (4.3.5), we obtain

$$(\bar{L} + \delta \bar{L})(\bar{U} + \delta \bar{U})(x + \delta x) = b. \tag{4.3.9}$$

Since $\bar{L}\bar{U}$ is approximately A (if $\|E\|$ in (4.3.4) is small), this suggests that the desired perturbation δA be defined by the equation

$$A + \delta A = (\bar{L} + \delta \bar{L})(\bar{U} + \delta \bar{U}), \tag{4.3.10}$$

in order to get the perturbed system (4.3.3). From (4.3.10) and (4.3.4) it follows that

$$\delta A = E + \bar{L}\,\delta U + \delta \bar{L}\, U + \delta \bar{L}\, \delta \bar{U}. \tag{4.3.11}$$

Conversely, a δA given by (4.3.11), with E determined by (4.3.4), $\delta \bar{L}$ by (4.3.5) and $\delta \bar{U}$ by (4.3.8) must satisfy (4.3.3). Since

$$\|\delta A\| \leq \|E\| + \|\bar{L}\|\,\|\delta \bar{U}\| + \|\delta \bar{L}\|\,\|\bar{U}\| + \|\delta \bar{L}\|\,\|\delta \bar{U}\|,$$

it suffices to determine bounds for the norms of E, $\bar{L}$, $\bar{U}$, $\delta \bar{L}$ and $\delta \bar{U}$. These bounds will now be calculated by a detailed analysis of the rounding errors in Eqs. (4.2.17), (4.2.19) and (4.2.20).

Lemma 4.3.1 Let A be a nonsingular $n \times n$ matrix and let $\bar{a}_{ij}^{(k)}$ be given by (4.2.17) in the Gaussian elimination procedure with partial pivoting and floating-point precision s. Let

$$a = \max_{1 \leq i \leq j \leq n}\{|a_{ij}|\}.$$

There exists a quantity g_n such that

$$\left.\begin{array}{l}|\bar{a}_{ij}^{(k)}| \leq g_n a, \\ g_n \leq 2^{n-1} + O((n-1)10^{1-s}).\end{array}\right\} \tag{4.3.12}$$

Proof. For the exact multipliers, $|m_{ij}| \leq 1$ when partial pivots are used. This is also true for the computed multipliers l_{ij} in (4.2.17), since $|\bar{a}_{ij}^{(j)}| \leq |u_{jj}|$ implies that the absolute value of the mantissa of $\bar{a}_{ij}^{(j)}$ is less than or equal to the absolute value of the mantissa of u_{jj}. The division operation with rounding must produce $|l_{ij}| \leq 1$. (Since A is nonsingular, $u_{jj} \neq 0$.) The rounding error in (4.2.17) consists of two rounding errors. Thus, we have

$$\bar{a}_{ij}^{(k+1)} = (\bar{a}_{ij}^{(k)} - l_{ik}u_{kj}(1 + \mu_{ij}^{(k)} \cdot 10^{-s}))(1 + \alpha_{ij}^{(k)} \cdot 10^{-s}), \qquad (4.3.13)$$

where $|\mu_{ij}^{(k)}| \leq 5$ and $|\alpha_{ij}^{(k)}| \leq 5$. Since $|l_{ik}| \leq 1$ and $|\bar{a}_{ij}^{(1)}| = |a_{ij}| \leq a$, we find that

$$|\bar{a}_{ij}^{(2)}| \leq a(1 + (1 + 5 \cdot 10^{-s}))(1 + 5 \cdot 10^{-s}), \qquad (i, j \geq 2).$$

Continuing with this bound for $|\bar{a}_{ij}^{(2)}|$, we get

$$|\bar{a}_{ij}^{(3)}| \leq a(1 + (1 + 5 \cdot 10^{-s}))^2(1 + 5 \cdot 10^{-s})^2,$$

and so on until $k = n - 1$. Hence,

$$g_n \leq (1 + (1 + 5 \cdot 10^{-s}))^{n-1}(1 + 5 \cdot 10^{-s})^{n-1} = 2^{n-1} + O((n-1)10^{1-s}). \blacksquare$$

Using g_n we can obtain an upper bound for the norm of the matrix E defined in (4.3.4).

Lemma 4.3.2 From the same hypotheses as in Lemma 4.3.1 it follows that the elements e_{ij} of the matrix E in (4.3.4) must satisfy the inequalities

$$|e_{ij}| \leq 2(i-1)g_n a \cdot 10^{-s}, \qquad \text{for } i \leq j,$$
$$|e_{ij}| \leq 2j g_n a \cdot 10^{1-s}, \qquad \text{for } i > j.$$

Thus,
$$|e_{ij}| \leq 2n g_n a \cdot 10^{1-s}.$$

Proof. From (4.3.13) and (4.2.17) it follows that

$$\bar{a}_{ij}^{(k+1)} = \bar{a}_{ij}^{(k)} - l_{ik}u_{kj} + \varepsilon_{ij}^{(k+1)}, \qquad (i, j \geq k+1), \qquad (4.3.14)$$

and

$$\varepsilon_{ij}^{(k+1)} = \alpha_{ij}^{(k)} \cdot 10^{-s}\bar{a}_{ij}^{(k)} - l_{ik}u_{kj}(\mu_{ij}^{(k)} \cdot 10^{-s} + (1 + \mu_{ij(k)}10^{-s})\alpha_{ij}^{(k)} \cdot 10^{-s}).$$

Applying (4.3.12) and recalling that $|l_{ik}| \leq 1$, we obtain

$$|\varepsilon_{ij}^{(k+1)}| \leq 5 \cdot 10^{-s}g_n a(3 + 5 \cdot 10^{-s}) \leq 2g_n a \cdot 10^{1-s}.$$

Now, we note that the (i, j)-entry of the matrix $\bar{L}\bar{U}$ is given by

$$\sum_{k=1}^{\min\{i,j\}} l_{ik}u_{kj}.$$

Summing (4.3.14) over k from 1 to $j - 1$ for $i > j$, we get

$$\sum_{k=1}^{j-1} \bar{a}_{ij}^{(k+1)} = \sum_{k=1}^{j-1} \bar{a}_{ij}^{(k)} - \sum_{k=1}^{j-1} l_{ik}u_{kj} + \sum_{k=1}^{j-1} \varepsilon_{ij}^{(k+1)},$$

whence
$$\bar{a}_{ij}^{(j)} = \bar{a}_{ij}^{(1)} - \sum_{k=1}^{j-1} l_{ik}u_{kj} + \sum_{k=1}^{j-1} \varepsilon_{ij}^{(k+1)}. \qquad (4.3.15)$$

From (4.2.17) we obtain for $i > j$,

$$l_{ij} = \frac{\bar{a}_{ij}^{(j)}}{u_{jj}}(1 + \sigma_{ij} \cdot 10^{-s}),$$

which may be rewritten as

$$\bar{a}_{ij}^{(j)} - l_{ij}u_{jj} + \varepsilon_{ij}^{(j+1)} = 0,$$

where
$$|\varepsilon_{ij}^{(j+1)}| = |\sigma_{ij}\bar{a}_{ij}^{(j)} \cdot 10^{-s}| \leq 2g_n a \cdot 10^{1-s},$$
since $|\sigma_{ij}| \leq 5$. Combining the last equation with (4.3.15), we find that
$$\bar{a}_{ij}^{(1)} - \sum_{k=1}^{j} l_{ik} u_{kj} + \sum_{k=1}^{j} \varepsilon_{ij}^{(k+1)} = 0.$$
For $i \leq j$, we sum (4.3.14) over k from 1 to $i-1$. Observing that $l_{ii} = 1$ and $u_{ij} = \bar{a}_{ij}^{(i)}$ (see 4.2.17), we obtain
$$\bar{a}_{ij}^{(1)} - \sum_{k=1}^{i} l_{ik} u_{kj} + \sum_{k=1}^{i-1} \varepsilon_{ij}^{(k+1)} = 0.$$
Since we are assuming $A = (\bar{a}_{ij}^{(1)})$, the last two equations are the same as (4.3.4) with
$$e_{ij} = \begin{cases} \sum_{k=1}^{j} \varepsilon_{ij}^{(k+1)}, & i > j, \\ \sum_{k=1}^{i-1} \varepsilon_{ij}^{(k+1)}, & i \leq j. \end{cases}$$
Using the bound for $|\varepsilon_{ij}^{(k+1)}|$ obtained above, we obtain the result immediately.

We now seek a bound for the norm of the matrix $\delta \bar{L}$ defined by (4.3.5).

Lemma 4.3.3 There exists a matrix $\delta \bar{L} = (\delta l_{ij})$ satisfying (4.3.5) and such that
$$|\delta l_{ij}| \leq n |l_{ij}| \cdot 10^{1-s}$$
provided that $n \cdot 10^{1-s} \leq 1$.

Proof. Let $\bar{c} = c + \delta c$ be the vector resulting from the elimination procedure applied to b. Then $\bar{c}_i$ is given by (4.2.19) with $b_i = \bar{b}_i^{(1)}$ and $p_i = i$ (since we are assuming that all permutations and scaling have been done in advance). The total rounding error β_i may be analyzed into individual rounding errors associated with each operation in computing the $\bar{c}_i$. Thus, (4.2.19) becomes

$$\left. \begin{aligned} \bar{c}_1 &= (b_1/l_{11})(1 + d_1 \cdot 10^{-s}), \\ \bar{c}_i &= \frac{(b_i - \bar{\Sigma}_i)}{l_{ii}}(1 + \alpha_i \cdot 10^{-s})(1 + d_i \cdot 10^{-s}), \quad i \geq 2, \end{aligned} \right\} \quad (4.3.16)$$

where $\bar{\Sigma}_i$ is the computed value of the sum $\sum_{j=1}^{i-1} l_{ij} \bar{c}_j$ using single-precision floating-point arithmetic, α_i is the rounding error in the subtraction and d_i is the rounding error in the division. In fact, since all $l_{ii} = 1$, there is no division and $d_i = 0$. However, we shall need the generality of (4.3.16), which applies to the solution of any triangular system.

Using (4.2.16), we may write
$$\bar{\Sigma}_i = \sum_{j=1}^{i-1} (l_{ij} + \delta l_{ij}) \bar{c}_j,$$
where
$$|\delta l_{ij}| \leq (i - j + 1) |l_{ij}| \cdot 10^{1-s}, \quad 2 \leq j \leq i - 1$$
$$|\delta l_{ij}| \leq n |l_{ij}| \cdot 10^{1-s}, \quad j = 1, i \geq 2.$$

Equation (4.3.16) then becomes

$$l_{11}(1 + d_1 \cdot 10^{-s})^{-1} \bar{c}_1 = b_1,$$

$$\sum_{j=1}^{i-1} (l_{ij} + \delta l_{ij})\bar{c}_j + l_{ii}(1 + \alpha_i \cdot 10^{-s})^{-1}(1 + d_i \cdot 10^{-s})^{-1} \bar{c}_i = b_i,$$

which is of the form of system (4.3.5) with $|\delta l_{ii}| \leq 2|l_{ii}| \cdot 10^{1-s}$, $|\delta l_{11}| \leq |l_{11}| \cdot 10^{1-s}$. This completes the proof. ∎

The preceding lemma applies to the solution of any triangular system effected by the same sequence of floating-point operations as the computation of $\bar{c}$. In particular, it applies to the computed solution $\bar{x} = x + \delta x$ obtained by the back-substitution procedure in (4.2.20). Therefore, we may state the following corollary.

Corollary Let $x + \delta x$ be the s-digit floating-point solution determined by (4.2.20). There exists a matrix $\delta \bar{U} = (\delta u_{ij})$ such that (4.3.8) is satisfied and

$$|\delta u_{ij}| \leq n|u_{ij}| \cdot 10^{1-s}.$$

Combining the preceding results, we obtain an upper bound for the perturbation δA in (4.3.11).

Theorem 4.3.1 Let $x + \delta x$ be the approximate solution of $Ax = b$ (A nonsingular) computed by the floating-point Gaussian elimination procedure with partial pivoting and a precision of s digits, where $n \cdot 10^{1-s} < 1$, n being the order of A. There exists a matrix $\delta A = (\delta a_{ij})$ such that $(A + \delta A)(x + \delta x) = b$ and

$$|\delta a_{ij}| \leq (3n^2 + 2n)g_n a \cdot 10^{1-s}, \tag{4.3.17}$$

where g_n and a are given in Lemma 4.3.1.

Proof. From (4.3.11) we have

$$\delta a_{ij} = e_{ij} + \sum_{k=1}^{\min\{i,j\}} (l_{ik} \delta u_{kj} + \delta l_{ik} u_{kj} + \delta l_{ik} \delta u_{kj}).$$

Since $|l_{ik}| \leq 1$ and $|u_{kj}| \leq g_n a$, it follows from the preceding lemmas and corollary that

$$|\delta a_{ij}| \leq (2n + 2n^2 + n^3 \cdot 10^{1-s})g_n a \cdot 10^{1-s}.$$

Since we are assuming s is sufficiently large so that $n \cdot 10^{1-s} \leq 1$, the result is immediate. ∎

Computational Aspects

Let us examine some of the consequences of the last theorem. Since the rows of A are assumed to have been scaled, we may take $a = 1$. The precision s is to be taken sufficiently large so that $n \cdot 10^{1-s} \leq 1$. Therefore, we may approximate and take $g_n = 2^{n-1}$. The theorem then yields $|\delta a_{ij}| \leq K_n$, where

$$K_n = (3n^2 + 2n)2^{n-1} \cdot 10^{1-s}.$$

4.3 Estimates of Error in Gaussian Elimination

Let us use the l_∞ vector norm and the induced *maximum row sum* matrix norm (Section 3.5). We have $1 \leq \|A\|_\infty \leq n$ and $\|\delta A\|_\infty \leq nK_n$. Hence, $\|\delta A\|_\infty / \|A\|_\infty \leq nK_n$. In order to use inequality (3.5.1) as an error bound (and to insure the existence of $(A + \delta A)^{-1}$) we must have $\gamma n K_n < 1$. Letting $\gamma = 10^q$, this requires that $(3n^3 + 2n^2)2^{n-1} < 10^{s-q-1}$. As an example, suppose $n = 10$. Then we must have $3200 \cdot 2^q < 10^{s-q-1}$ or $2^{14} < 10^{s-q-3}$. Thus, we must have $s > q + 3 + 14/\log_2 10 > q + 7$. For a *well-conditioned* matrix with $q = 0$, this implies $s > 7$. Taking $s = 8$, we obtain $\gamma n K_n = 0.16384$. Therefore, $\|\delta x\|_\infty / \|x\|_\infty \leq 0.164/0.836 \leq .2$. This would seem to indicate that we can guarantee at most one correct digit in the worst component of the solution and suggests that a precision of $s = 15$ digits is needed to obtain 7 correct digits in all components of the solution. For ill-conditioned matrices (e.g., $\gamma = 10^3$), about 18-digit precision seems to be indicated. For $n = 10^2$, a similar computation yields $s > q + 37$. For $q = 0$ and $s = 38$, we have $\gamma n K_n = 0.1503$ and an error bound of approximately 0.18. Thus, if we were to rely on (3.5.1) for an a priori estimate, a precision of roughly 45 digits would be indicated to ensure an accuracy of 7 correct digits in the solution. In the early days of computing [4.1], it was feared that the solution of very high-order linear systems ($n \geq 100$, say) would not be feasible with the precision available on computers. (In current computers, generally, s is somewhere in the range of 8 to 16 decimal digits.) However, experience has demonstrated that for well-conditioned matrices satisfactory accuracy can be obtained with a much smaller precision than in the above estimates. Therefore, the error bounds given by (3.5.1), (4.3.12) and (4.3.17) must be considered to be extremely conservative. This is to be expected for several reasons. First, the bound is general, that is, it applies to all nonsingular matrices. In particular, the upper bound for the quantity g_n is much too large for many matrices. For example, for a symmetric positive definite matrix (Section 3.6) it can be shown that $g_n = 1$ for exact operations. (See [4.1] and Exercise 4.4.) For g_n of order 1 and $n = 10^2$ the a priori estimate shows that $s = q + 15$ will produce a relative error bound of 10^{-7}. Smaller values of g_n can be achieved for arbitrary matrices by using maximal pivoting. For maximal pivoting it can be shown that with no rounding errors, the growth of the pivots is restricted according to the inequality,

$$|a_{kk}^{(k)}| < k^{1/2}(2 \cdot 3^{1/2} \cdot 4^{1/3} \cdots k^{1/k-1})^{1/2} a.$$

(See [4.4], Section 16.) Since $|a_{ij}^{(k)}| \leq |a_{kk}^{(k)}|$ for $i, j \geq k$, if we assume that rounding errors do not appreciably affect the pivots, then we may take

$$g_n < n^{1/2}(2 \cdot 3^{1/2} \cdots n^{1/(n-1)})^{1/2},$$

which reduces appreciably the estimates of precision required.

A second reason why the bound is too large in general is that the above error analysis is based on maximum rounding errors, that is, on the worst case in each operation. In practice, the rounding errors are distributed in some random fashion between -5 and 5. (A common assumption is that the distribution is uniform.) Hence, there is some cancellation of errors and the probable error is smaller than the maximum error bound. An error analysis based on this probabilistic approach to rounding error is given in [4.2].

If the precision required is, in fact, excessive, the error bound and the error itself can be reduced by using *double-precision accumulation* to compute inner products. Thus, to compute $\sum_{j=1}^{n} a_j b_j$, where the a_j, b_j have s-digit mantissas, we compute the products $a_j b_j$ as double-precision numbers with $2s$-digit mantissas and add two such numbers by rounding the sum, if necessary, to a $2s$-digit number. (Most computers have a $2s$-digit *accumulator* register for products and summands.) The final sum of the n double-precision numbers is then rounded to an s-digit mantissa. Referring to our earlier analysis of the rounding error in a sum of products, we see that

$$\bar{\bar{S}}_n = (\bar{\bar{S}}_{n-1} + a_n b_n)(1 + \alpha_n \cdot 10^{-2s})$$

where $\bar{\bar{S}}_n$ is the double-precision sum from 1 to n and $\bar{\bar{S}}_{n-1}$ is the double-precision sum from 1 to $n - 1$. As before, $|\alpha_n| \leq 5$ and we find that

$$\bar{\bar{S}}_n = \sum_{j=1}^{n} a_j b_j (1 + \bar{\bar{\delta}}_j) = \sum_{j=1}^{n} (a_j + \bar{\bar{\delta}} a_j) b_j,$$

where

$$|\bar{\bar{\delta}} a_j| \leq (n - j + 1) 10^{1-2s} |a_j|.$$

The final rounding yields a single-precision result $\bar{S}_n = \bar{\bar{S}}_n (1 + \alpha \cdot 10^{-s})$, $|\alpha| \leq 5$. Therefore, $\bar{S}_n = \sum_{j=1}^{n} (a_j + \delta a_j) b_j$, where $|\delta a_j| \leq \delta_j |a_j|$ and we may take $\delta_j = 10^{-s}$, since $n \cdot 10^{1-2s}$ is usually negligible compared to 10^{-s}. Applying these bounds to the proof of the last theorem, we eliminate a factor of n in δu_{kj} and δl_{ik}, obtaining

$$|\delta a_{ij}| \leq \left(2n + \frac{2n}{10} + n \cdot 10^{-s-2}\right) g_n a \cdot 10^{1-s} \leq 3n g_n a \cdot 10^{1-s},$$

which is an improvement over the single-precision bound by a factor of n. Thus, for $n = 10$, the a priori error bound would predict one more correct digit in the computed solution using double-precision accumulation. For $n = 10^2$, two additional correct digits may be achieved. In practice, results may be considerably better. Double-precision accumulation may be used in the elimination procedure for A as well if the Crout variant of Gaussian elimination is used. (See Exercise 4.3.)

Observe that a large condition number $\|A^{-1}\| \|A\|$ does not of itself signify that the solution of the system $Ax = b$ is ill-determined. For example, if A is a diagonal matrix, the system is well-determined in the sense that the relative error in x is of the same order of magnitude as the relative errors in A. (See Exercise 4.16) Now, suppose that A has been scaled properly so that $|d_{ii}| = 1$ for all i. Then $\|A^{-1}\|_\infty \cdot \|A\|_\infty = 1$. However, if we multiply the first equation of the system by an arbitrary small factor ε, we obtain a matrix B such that $\|B\|_\infty = 1$ but $\|B^{-1}\|_\infty = 1/\varepsilon$. Thus, B can be made to have an arbitrarily large condition number while the resulting system, being diagonal, has a well-determined solution. It seems clear that some sort of scaling which normalizes the rows and/or columns of A must be performed before computing the condition number to be used in formula (4.3.1) or other error estimates. See [4.15] and Exercises 4.19 and 4.24 for further discussion.

Another method for improving the accuracy of the computed result involves an iterative procedure of successive corrections based on Gaussian elimination. We shall discuss this in the next section.

4.4 ITERATIVE METHODS

Iteration on the Residuals

We have seen that the Gaussian elimination procedure, when carried out in finite precision, generally does not yield the exact solution x of $Ax = b$. The computed approximation $\bar{x} = x + \delta x$ is such that the residual $r = b - A\bar{x}$ is nonzero. Since $r = b - A(x + \delta x) = -A\,\delta x$, we have $x = \bar{x} + A^{-1}r$. This suggests that the approximation $\bar{x}$ can be improved by adding on a "correction," Δx, which is obtained by solving the system,

$$A\,\Delta x = r. \tag{4.4.1}$$

Indeed, if we could solve (4.4.1) exactly and perform the sum $\bar{x} + \Delta x$ exactly, the sum would be the exact solution. However, rounding errors are unavoidable in most cases. Any method of numerical solution, such as Gaussian elimination, will produce an approximate correction only. Nevertheless, Gaussian elimination would be a wise choice, since we have already done the computational work to reduce A to triangular form. Having organized the algorithm in two parts, we can apply the previously computed matrices $\bar{L}$ and $\bar{U}$ to the new right-hand vector r in (4.4.1). If there were no further rounding errors, an approximate correction $\overline{\Delta x}$ would be obtained by solving the triangular systems $\bar{L}c = r$ and $\bar{U}\,\overline{\Delta x} = c$. In fact, there are rounding errors committed in the computer solution of these triangular systems. Furthermore, there is an initial error in computing the residual and we actually start with a single-precision approximation $\bar{r}$. As a consequence of these errors, the computation yields an approximation, $\overline{\Delta x^{(1)}}$, to the exact correction. Nevertheless, we would still expect $\bar{x}^{(1)} = x + \overline{\Delta x^{(1)}}$ to be closer to the exact solution. It seems plausible, therefore, to repeat the procedure with $\bar{x}^{(1)}$ replacing $\bar{x}$ and continue to iterate the process until some desired accuracy is achieved. This gives rise to an iterative procedure which generates a sequence of vectors $(\bar{x}^{(m)})$ starting with $\bar{x}^{(0)} = \bar{x}$. We shall call this procedure *iteration on the residuals*. To describe it we shall first assume that there are no rounding errors in the residual r and none in the computation of the correction. In this case, we obtain a sequence of real vectors $x^{(m)}$ defined in terms of exact operations as follows:

$$\left.\begin{aligned}
x^{(0)} &= \bar{x}, \\
r^{(m)} &= b - Ax^{(m)}, \quad m = 0, 1, 2, \ldots, \\
\bar{L}\bar{U}\,\Delta x^{(m)} &= r^{(m)}, \\
x^{(m+1)} &= x^{(m)} + \Delta x^{(m)}.
\end{aligned}\right\} \tag{4.4.2}$$

If $\bar{L}\bar{U}$ were equal to A, then one such iteration would yield the exact solution; i.e., $x^{(1)} = x$. However, according to (4.3.4), $\bar{L}\bar{U} = A + E$, where E is generally not the zero matrix. Therefore, we must expect to iterate indefinitely, hoping to obtain the exact solution as the limit of the sequence $(x^{(m)})$. Although this could not be done in practice, it is worthwhile to investigate the question of whether $x^{(m)}$ converges to x, for if $(x^{(m)})$ does not converge to x, it seems unlikely that $(\bar{x}^{(m)})$ will converge to $\hat{x}$, where $\hat{x}$ is the result of rounding x to s digits. On the other hand, if $\lim x^{(m)} = x$ and rounding errors are kept sufficiently small, it is likely that $\lim \bar{x}^{(m)} = \hat{x}$, which is all we can hope to achieve with s-digit arithmetic.

Referring to (4.4.2), we see that $b - Ax^{(m)} = \bar{L}\bar{U} \Delta x^{(m)} = \bar{L}\bar{U}(x^{(m+1)} - x^{(m)})$, whence

$$\bar{L}\bar{U}x^{(m+1)} = (\bar{L}\bar{U} - A)x^{(m)} + b. \tag{4.4.3}$$

Equation (4.4.3) has a form common to other iterative procedures to be discussed below, namely, the form

$$Mx^{(m+1)} = Nx^{(m)} + b, \tag{4.4.4}$$

where

$$A = M - N. \tag{4.4.5}$$

(This is a *splitting* of A.)

Given $x^{(m)}$, a unique solution $x^{(m+1)}$ of (4.4.4) exists if and only if M^{-1} exists. In that case, we have

$$x^{(m+1)} = M^{-1}Nx^{(m)} + M^{-1}b = Qx^{(m)} + d, \tag{4.4.6}$$

where we have set $Q = M^{-1}N$ and $d = M^{-1}b$. By virtue of (4.4.5), the exact solution of $Ax = b$ satisfies

$$Mx = Nx + b,$$

so that

$$x = M^{-1}Nx + M^{-1}b = Qx + d. \tag{4.4.7}$$

Now let $e^{(m)} = x^{(m)} - x$ be the error in the mth approximation. Subtracting (4.4.7) from (4.4.6), we find that

$$e^{(m+1)} = Qe^{(m)}, \tag{4.4.8}$$

from which it follows that

$$e^{(m)} = Q^m e^{(0)}. \tag{4.4.9}$$

The sequence $(x^{(m)})$ defined by (4.4.6) converges to x for any starting value $x^{(0)}$ if and only if $e^{(m)}$ converges to 0 for any initial $e^{(0)}$. We shall now prove that $e^{(m)} \to 0$ for arbitrary $e^{(0)}$ if and only if $Q^m \to 0$ (the zero operator). We recall (Section 3.6) that Q is said to be quasinilpotent if $Q^m \to 0$ and Theorem 3.6.2, Corollary 1, gives a necessary and sufficient condition for Q to be quasinilpotent. (A quasinilpotent operator is frequently called *convergent*.)

Theorem 4.4.1 Let (x^m) be a sequence of n-dimensional vectors satisfying (4.4.4), where M is a real nonsingular $n \times n$ matrix and N is a real $n \times n$ matrix such that

$A = M - N$ is nonsingular. Let $x = A^{-1}b$. Choose any vector norm topology and the induced matrix norm topology. Then

$$\lim_{m \to \infty} \|x^{(m)} - x\| = 0$$

for all initial values $x^{(0)}$ if and only if $Q = M^{-1}N$ is quasinilpotent.

Proof. First, suppose Q is quasinilpotent. Then $\|Q^m\| \to 0$. By (4.4.9),

$$\|e^{(m)}\| \le \|Q^m\| \, \|e^{(0)}\| \to 0.$$

To prove the converse, suppose Q is not quasinilpotent. Then by Theorem 3.6.2, Corollary 1, the spectral radius $r_\sigma(Q) \ge 1$, so that there exists an eigenvalue, λ, of Q with $|\lambda| \ge 1$. Let u be an eigenvector belonging to λ and normalized so that $\|u\|_\infty = 1$. If λ is real, then u is real and $\|Q^m u\|_\infty = |\lambda|^m \|u\|_\infty = |\lambda|^m \ge 1$. Therefore, if we take $x^{(0)}$ such that $e^{(0)} = x - x^{(0)} = u$, we cannot have $\|e^{(m)}\|_\infty = \|Q^m u\|_\infty$ converging to zero. (Since all norms in V^n are equivalent by Theorem 3.2.3, this holds for any norm.) If Im $(\lambda) \ne 0$, then the complex conjugate $\bar{\lambda}$ is also an eigenvalue of Q and $Q\bar{u} = \bar{\lambda}\bar{u}$. (See the Example following Definition 3.6.1.) Hence, $Q^m(u + \bar{u}) = \lambda^m u + \bar{\lambda}^m \bar{u} = 2 \, \text{Re}\, (\lambda^m u)$. Taking $e^{(0)} = u + \bar{u}$, we have

$$\|e^{(m)}\|_\infty = \|Q^m e^{(0)}\|_\infty = \max_{1 \le j \le n} \{|2 \, \text{Re}\, \lambda^m u_j|\},$$

where the u_j are the components of u. Writing $\lambda = |\lambda| \exp(i\theta)$, $0 < \theta < 2\pi$, $\theta \ne \pi$, and $u_j = \rho_j \exp(i\alpha_j)$, we have

$$\|e^{(m)}\|_\infty = \max_{1 \le j \le n} \{2 |\lambda|^m \rho_j |\cos(m\theta + \alpha_j)|\}.$$

We shall show that $e^{(m)} \ne 0$ for all m. Suppose $e^{(m)} = 0$. This implies $\rho_j = 0$ or $\cos(m\theta + \alpha_j) = 0$ for all $j = 1, \ldots, n$. Since $\|u\|_\infty = \max_{1 \le j \le n} \rho_j = 1$, some ρ_j (say ρ_k) is not zero. Then we must have $\cos(m\theta + \alpha_k) = 0$ and $m\theta + \alpha_k = (2q + 1)\pi/2$ for some integer q. However, $e^{(m)} = 0$ implies that $e^{(m+1)} = Qe^{(m)} = 0$. Hence, $(m + 1)\theta + \alpha_k = (2q' + 1)\pi/2$. By subtraction, $\theta = (q' - q)\pi$. But this implies that Im $(\lambda) = 0$, which is a contradiction. Hence, $e^{(m)} \ne 0$ for all m. Furthermore, $e^{(m)}$ cannot converge to zero, since $\lim_{m \to \infty} (m\theta + \alpha_k) \ne \pm\pi/2$ or $3\pi/2$. Thus, we have shown that if Q is not quasinilpotent, then there is some initial value $x^{(0)}$ for which the sequence $x^{(m)}$ does not converge. This completes the proof. ∎

Returning to (4.4.3), we see that exact iteration on the residuals converges for arbitrary $\bar{x}^{(0)}$ if and only if $Q = (\bar{L}\bar{U})^{-1}E$ is quasinilpotent. A sufficient condition for this is that $\|Q\| < 1$. Now,

$$\|Q\| \le \|(A + E)^{-1}E\| \le \|(A + E)^{-1}\| \, \|E\| \le \|A^{-1}\| \, \|E\|/(1 - \|A^{-1}\| \, \|E\|).$$

Therefore, if $\|A^{-1}\| \cdot \|E\| < \frac{1}{2}$, then $\|Q\| < 1$. Using infinity norms and Lemma 4.3.2, we see that $\|E\| < 2n^2 \cdot 10^{1-s} \cdot g_n$. A reasonable practical value for g_n is n (see [4.4]). Thus, if $\|A^{-1}\| < 10^{s-1}/4n^3$, Q is quasinilpotent. It seems likely that this condition will be satisfied by most matrices A encountered in practical problems.

Furthermore, if $g_n = 1$, the condition is $\|A^{-1}\| < 10^{s-1}/4n^2$, which is even less stringent. Therefore, we can expect that $x^{(m)}$ will converge to the exact solution x.

Since rounding errors occur in each iteration of the finite-precision computation, we cannot expect that the computed iterants $\bar{x}^{(m)}$ will converge to x even when Q is quasinilpotent. However, by using double precision in the computation of the residuals, we can often approach $\hat{x}$, the exact solution rounded to s digits. Double precision is used in the computation of $A\bar{x}^{(m)}$ to accumulate the sums $\sum_{k=1}^{m} a_{ik}\bar{x}_k^{(m)}$, $1 \leq i \leq n$. The remaining operations are done in single precision. In place of (4.4.2) we have for $m = 0, 1, 2, \ldots$,

$$\left.\begin{aligned} \bar{x}^{(0)} &= \bar{x}, \\ \bar{r}^{(m)} &= b - A\bar{x}^{(m)} + \rho^{(m)}, \\ L_m U_m \overline{\Delta x}^{(m)} &= \bar{r}^{(m)}, \\ \bar{x}^{(m+1)} &= \bar{x}^{(m)} + \overline{\Delta x}^{(m)} + \alpha^{(m)}, \end{aligned}\right\} \quad (4.4.10)$$

where $\rho^{(m)}$ is the total rounding error due to the rounding errors in the double-precision accumulation and the single-precision subtraction, and $\alpha^{(m)}$ is the rounding error in the single-precision addition. The matrices L_m, U_m, ($m = 0, 1, 2, \ldots$,) are the perturbed matrices for which $\overline{\Delta x}^{(m)}$ is an exact solution, that is $\overline{\Delta x}^{(m)}$ is computed using $\bar{L}\bar{U}$ as in (4.4.2) but with single-precision arithmetic. Therefore, it does not satisfy the third equation of (4.4.2) but rather the perturbed equation in (4.4.10), where $L_m = \bar{L} + \delta L_m$ and $U_m = \bar{U} + \delta U_m$. The perturbations δL_m, δU_m exist by Lemma 4.3.3. We assume that $\|\delta L_m\| < \|\bar{L}^{-1}\|^{-1}$ and $\|\delta U_m\| < \|\bar{U}^{-1}\|^{-1}$.

Now, let us set $A_m = L_m U_m$ and $Q_m = I - A_m^{-1} A$. We shall assume that $\|Q_m\| < d < 1$ for all m. Then in place of (4.4.3) we have

$$A_m \bar{x}^{(m+1)} = (A_m - A)\bar{x}^{(m)} + A_m \alpha^{(m)} + b + \rho^{(m)}. \quad (4.4.11)$$

The exact solution $x = A^{-1}b$ satisfies the equation,

$$A_m x = (A_m - A)x + b.$$

Subtracting and putting $\bar{e}^{(m)} = \bar{x}^{(m)} - x$, we get

$$\bar{e}^{(m+1)} = (I - A_m^{-1} A)\bar{e}^{(m)} + \alpha^{(m)} + A_m^{-1} \rho^{(m)}. \quad (4.4.12)$$

Referring to (4.4.10), we note that $\rho^{(m)}$ is the result of the errors in forming $A\bar{x}^{(m)}$ using double-precision accumulation, then rounding and subtracting in single precision. Applying (4.2.16), we obtain

$$\bar{r}_i^{(m)} = \left(b_i - \left(\sum_{k=1}^{n}(a_{ik} + \delta_{ik}^{(m)} a_{ik})\bar{x}_k^{(m)}\right)(1 + \sigma_i^{(m)})\right)(1 + \beta_i^{(m)}), \quad (4.4.13)$$

where

$$|\delta_{ik}^{(m)}| \leq n \cdot 10^{1-2s}, \quad |\sigma_i^{(m)}| \leq 5 \cdot 10^{-s}, \quad |\beta_i^{(m)}| \leq 5 \cdot 10^{-s}.$$

For convenience, set $h = 10^{1-s}$. Henceforth, we shall assume that $nh < 1$. Letting $\Delta A^{(m)} = (\delta_{ij}^{(m)} a_{ij})$, and using either the l_∞- or l_1-norm and the induced operator norm,

$\|\Delta A^{(m)}\| \leq h \|A\|$ and $\|A + \Delta A^{(m)}\| \leq (1 + h) \|A\|$. Equation (4.4.13) may be rewritten in vector form as

$$\bar{r}^{(m)} = b - (A + \Delta A^{(m)})\bar{x}^{(m)} + \gamma^{(m)}. \quad (4.4.14)$$

Hence,

$$\rho^{(m)} = -\Delta A^{(m)}\bar{x}^{(m)} + \gamma^{(m)},$$

where

$$\|\gamma^{(m)}\| \leq h \|A\| \|\bar{e}^{(m)}\| + h \|\Delta A^{(m)}\| \|\bar{x}^{(m)}\| + h(1 + h)\|A + \Delta A^{(m)}\| \|\bar{x}^{(m)}\|$$

$$\leq h \|A\| \|\bar{e}^{(m)}\| + h \|A\| \|\bar{x}^{(m)}\|(1 + 3h + h^2)$$

$$\|\gamma^{(m)}\| \leq h(2 + 3h + h^2)\|A\| \|\bar{e}^{(m)}\| + h(1 + 3h + h^2) \|A\| \|x\| \quad (4.4.15)$$

since $\|\bar{x}^{(m)}\| \leq \|\bar{x}^{(m)} - x\| + \|x\| = \|\bar{e}^{(m)}\| + \|x\|$. Therefore,

$$\|\rho^{(m)}\| \leq h \|A\|(3 + 3h + h^2) \|\bar{e}^{(m)}\| + h \|A\| (2 + 3h + h^2) \|x\|. \quad (4.4.16)$$

Now, from (4.4.10), (4.4.14), $\overline{\Delta x}^{(m)} = A_m^{-1} A \bar{e}^{(m)} - A_m^{-1} \Delta A \bar{x}^{(m)} + A_m^{-1} \gamma^{(m)}$.

$$\|\overline{\Delta x}^{(m)}\| \leq \|A_m^{-1}\| \|A\| \|\bar{e}^{(m)}\| + h \|A_m^{-1}\| \|A\| \|\bar{x}^{(m)}\| + \|A_m^{-1}\| \|\gamma^{(m)}\|$$

$$\leq \|A_m^{-1}\| \|A\| \|\bar{e}^{(m)}\| (1 + h) + h \|A_m^{-1}\| \|A\| \|x\| + \|A_m^{-1}\| \|\gamma^{(m)}\|. \quad (4.4.17)$$

Since $\|\alpha^{(m)}\| \leq h(\|\bar{x}^{(m)}\| + \|\overline{\Delta x}^{(m)}\|)$ and noting that $\|A_m^{-1}\| \leq \|I - Q_m\| \|A^{-1}\| \leq (1 + d) \|A^{-1}\|$, we collect terms in (4.4.15), (4.4.17) to obtain

$$\|\alpha^{(m)}\| \leq h(1 + d) \|A^{-1}\| \|A\| (1 + O(h)) \|\bar{e}^{(m)}\| + h \|\bar{e}^{(m)}\|$$
$$+ \|x\| h^2 (1 + d) \|A^{-1}\| \|A\| (2 + O(h)) + h \|x\|, \quad (4.4.18)$$

where $O(h)$ denotes a term $< 10h$. Finally, combining (4.4.12) (4.4.16), (4.4.18), we obtain

$$\|\bar{e}^{(m+1)}\| \leq \|Q_m\| \|\bar{e}^{(m)}\| + h \|A\| \|A^{-1}\| (5 + d + O(h)) \|\bar{e}^{(m)}\|$$
$$+ h \|A\| \|A^{-1}\| (3 + O(h)) \|x\|,$$

which leads to the inequality

$$\|\bar{e}^{(m+1)}\| \leq (\|Q_m\| + 7h \|A\| \|A^{-1}\|) \|\bar{e}^{(m)}\| + 3h \|A^{-1}\| \|A\| \|x\|.$$

Let $\gamma_A = \|A\| \|A^{-1}\|$ be the condition number of A. Then $\|Q_m\| + 7h \|A\| \|A^{-1}\| \leq d + 7h\gamma_A$ and the preceding inequality yields

$$\|\bar{e}^{(m)}\| \leq (d + 7h\gamma_A)^m \|\bar{e}^{(0)}\| + 3h\gamma_A \|x\| \sum_{j=0}^{m-1} (d + 7h\gamma_A)^j. \quad (4.4.19)$$

If $7h\gamma_A < 1 - d$, then $(d + 7h\gamma_A)^m \to 0$ and we have

$$\limsup_{m \to \infty} \frac{\|\bar{e}^{(m)}\|}{\|x\|} \leq \frac{3h\gamma_A}{1 - d - 7h\gamma_A}.$$

Thus, if d is not close to 1 (say $d \leq \frac{1}{2}$), and γ_A is of order 1, this is an error in the sth significant digit, that is, $\bar{x}^{(m)}$ approaches the rounded value $\hat{x}$ with an error in the last digit.

To show that $\bar{x}^{(m)}$ need not converge, let
$$f^{(m)} = \alpha^{(m)} + A_m^{-1}\rho^{(m)}.$$

Then from (4.4.12), we get the difference equation
$$\bar{e}^{(m+1)} = Q_m\bar{e}^{(m)} + f^{(m)},$$

which has the solution (as one can verify directly),
$$\bar{e}^{(m)} = \left(\prod_{j=1}^{m-1} Q_j\right)\bar{e}^{(0)} + \sum_{j=1}^{m}\left(\prod_{k=1}^{m-j} Q_k\right)f^{(j-1)}, \qquad (4.4.20)$$

where
$$\prod_{k=1}^{m-j} Q_k = Q_{m-j}\cdots Q_2 Q_1 \quad \text{and} \quad \prod_{k=1}^{0} Q_k = I.$$

Since $\|\bar{e}^{(m)}\|$ is bounded, as we have seen, so are the $\|\rho^{(m)}\|$ and $\|\alpha^{(m)}\|$, by (4.4.16) and (4.4.18). In fact, they are of order $h\gamma_A$. Under our assumption that $\|Q_m\| < d$, the first term on the right of (4.4.20) is bounded by $d^m\|\bar{e}^{(0)}\|$, which approaches zero. The second term is a sum which is absolutely bounded by the sum
$$\sum_{j=1}^{m} \|f^{(j-1)}\| d^{m-j} \leq K/(1-d),$$

where K is such that $\|f^{(m)}\| \leq \|\alpha^{(m)}\| + (1+d)\|A^{-1}\|\|\rho^{(m)}\| < K$. Hence, the error norms $\|\bar{e}^{(m)}\|$ become $\leq K/(1-d)$ in the limit. However, $\bar{e}^{(m)}$ itself will usually exhibit an oscillatory behavior owing to the random character of $f^{(m)}$.

We summarize these results as a theorem.

Theorem 4.4.2 Let $(\bar{x}^{(m)})$ be the sequence of vectors obtained by iteration on the residuals as defined by (4.4.10) and using double-precision accumulation to form $A\bar{x}^{(m)}$. Let $n \cdot 10^{1-s} < 1$. Let $\|I - (L_m U_m)^{-1}A\|_\infty < d < 1$ for all m. If
$$7 \cdot 10^{1-s} \|A\|_\infty \|A^{-1}\|_\infty < 1 - d,$$
then
$$\limsup_{m\to\infty} \frac{\|\bar{x}^{(m)} - x\|_\infty}{\|x\|_\infty} \leq \frac{3\cdot 10^{1-s}\|A\|_\infty\|A^{-1}\|_\infty}{1 - d - 7\cdot 10^{1-s}\|A\|_\infty\|A^{-1}\|_\infty}.$$

As in the case of iteration with no rounding error, we may obtain a condition on $\|A^{-1}\|$ to insure that $\|Q_m\| < d < 1$. Again, we write $Q_m = I - (L_m U_m)^{-1}A = A_m^{-1}(A_m - A)$, where $A_m = L_m U_m = (\bar{L} + \delta L_m)(\bar{U} + \delta U_m)$ and δL_m is given by Lemma 4.3.3 and δU_m by its corollary. Thus, $A_m = A + E_m$, where
$$E_m = E + \delta L_m \bar{U} + \bar{L}\,\delta U_m + \delta L_m\,\delta U_m,$$
and
$$\|E_m\|_\infty \leq (3n^3 + 2n^2)g_n \cdot 10^{1-s}$$

by Theorem 4.3.1. Therefore, using infinity operator norms,

$$\|Q_m\| \leq \|A^{-1}\| \|E_m\|/(1 - \|A^{-1}\| \|E_m\|) < 1 \quad \text{if} \quad \|A^{-1}\| \|E_m\| < \tfrac{1}{2}.$$

If $g_n = n$, then

$$\|A^{-1}\| < \frac{10^{s-2}}{n^4}$$

is a sufficient condition for $\|Q_m\| < 1$.

Computational Aspects

From (4.4.19) we can conclude that the rate of convergence depends on the factor $d + 7 \|A\|_\infty \|A^{-1}\|_\infty 10^{1-s}$. In practice, A^{-1} is unknown. However, (4.4.19) suggests that the error will be reduced at least by this factor after each iteration. Therefore, $\bar{x}^{(m+1)}$ should soon agree with $\bar{x}^{(m)}$ up to a certain number $q < s$ digits. This can be observed and the iteration procedure may usually be terminated when the criterion

$$\max_{1 \leq i \leq n} \frac{|\bar{x}_i^{(m+1)} - \bar{x}_i^{(m)}|}{|\bar{x}_i^{(m)}|} < 10^{-q},$$

is satisfied, where q is the desired number of correct digits. As a practical matter, 6 to 10 iterations will usually produce $s - 1$ correct digits. For example, if $\|A^{-1}\|_\infty < 10^{s-2}/4n^4$, then $d \leq \tfrac{1}{7}$ and $d + 7 \|A\|_\infty \|A^{-1}\|_\infty \cdot 10^{1-s} < \tfrac{1}{7} + \tfrac{1}{5}n^{-4}$. (See Exercise 4.23.)

Jacobi Iteration

A specific instance of (4.4.5), is obtained by taking $M = D$, where D is the diagonal matrix

$$D = \begin{bmatrix} a_{11} & 0 & \cdots & 0 \\ 0 & a_{22} & & \vdots \\ \vdots & & \ddots & 0 \\ 0 & \cdots & 0 & a_{nn} \end{bmatrix}.$$

This leads to the system of equations

$$\left. \begin{aligned} x_i^{(m+1)} &= \frac{1}{a_{ii}}\left(b_i - \sum_{\substack{j=1 \\ j \neq i}}^{n} a_{ij} x_j^{(m)}\right), \quad i = 1, \ldots, n, \\ \text{or} \\ x^{(m+1)} &= D^{-1}(D - A)x^{(m)} + D^{-1}b, \quad m \geq 0. \end{aligned} \right\} \quad (4.4.21)$$

The procedure defined by (4.4.21) is known as the *Jacobi iteration*, and it is well defined provided that $a_{ii} \neq 0$ for all $1 \leq i \leq n$. If A is nonsingular, we can permute the rows of (A, b) by a permutation matrix P such that PA has no zero's on the diagonal. (See Theorem 2.1.9, Corollary.) Then we solve the equivalent system $PAx = Pb$. We could, of course, permute columns as well, as in a complete pivoting procedure. We shall assume henceforth that all $a_{ii} \neq 0$. Convergence is assured if $\|Q_J\| = \|I - D^{-1}A\| < 1$ for any operator norm, for example, if

$$\|Q_J\|_\infty = \|I - D^{-1}A\|_\infty = \max_i \left\{ \sum_{\substack{j=1 \\ j \neq i}}^n \left| \frac{a_{ij}}{a_{ii}} \right| \right\} = c < 1. \tag{4.4.22}$$

This norm is computed easily. If the quantity c is less than 1, then A is called *strictly diagonally dominant*.

The error in the mth Jacobi iterant is given by (4.4.9), where now $Q = I - D^{-1}A$. If $\|Q\| < 1$ for some operator norm, then the error is reduced by the factor $\|Q\|$, at least, in each iteration. More generally, however, we have convergence if $\|Q^m\| \to 0$; i.e., if Q is quasinilpotent. In this case, we have for m sufficiently large,

$$\frac{\|e^{(m)}\|}{\|e^{(0)}\|} \leq \|Q^m\| < 1,$$

so that the error (and the relative error as well) is reduced by the factor $\|Q^m\|$ in m iterations. Therefore, on the average, in each iteration the error is reduced by the factor $\|Q^m\|^{1/m}$. If we wish to reduce the error by a specified factor of 10^{-t}, then we will need N iterations, where N is the smallest integer such that

$$(\|Q^m\|^{1/m})^N \leq 10^{-t}.$$

Taking logarithms,

$$N \geq \frac{t}{-\log \|Q^m\|^{1/m}}.$$

Thus, N varies inversely as

$$R(Q^m) = -\log \|Q^m\|^{1/m}$$

and, therefore, this quantity is sometimes called the *average rate of convergence* for m iterations. Since

$$\lim_{m \to \infty} \|Q^m\|^{1/m} = r_\sigma(Q)$$

by Theorem 3.6.2, this suggests that

$$R(Q) = -\log r_\sigma(Q) = \log \frac{1}{r_\sigma(Q)}$$

be called the *asymptotic rate of convergence*. (See [4.8].) Since $r_\sigma(Q) \leq \|Q\|$ for any operator norm, we have the lower bound

$$R(Q) \geq \log \frac{1}{\|Q\|}.$$

To obtain a rough comparison of the efficiency of the Jacobi method with the Gaussian elimination method we note that there are n^2 multiplications and additions in one Jacobi iteration (see (4.4.21)). To reduce the initial error by a factor of 10^{-t} requires about $t/R(Q_J)$ iterations. If we take $t = s - 1$, where s is the precision, and if we assume that Gaussian elimination without iteration on the residuals yields a relative error of less than 10^{1-s} in the computed solution, then to be of comparable efficiency the Jacobi method should satisfy

$$\frac{(s-1)}{R(Q_J)} n^2 \leq \frac{n^3}{3}$$

or

$$R(Q_J) \geq \frac{3(s-1)}{n}.$$

In terms of norms, we should have

$$\log \frac{1}{\|Q_J\|} \geq \frac{3(s-1)}{n}$$

or

$$\|Q_J\| \leq 10^{3(1-s)/n}.$$

This is a rather stringent requirement on Q_J for small n ($n < 10$) and would only be satisfied for $\|Q_J\|_\infty$ if the diagonal elements a_{ii} are large in modulus compared to the off-diagonal elements. It could be ameliorated somewhat if the initial guess, $x^{(0)}$, is close to the exact solution. Observe also that the situation improves as n gets larger. Furthermore, for certain special matrices, called *band* matrices, which have zero entries everywhere except in a narrow band along the main diagonal, the number of operations is reduced by taking advantage of the zero's. Thus, in (4.4.21), if $a_{ij} = 0$, the multiplication by a_{ij} can be omitted. For example, in the numerical solution of partial differential equations (Chapter 11), matrices A arise which are of large order ($n = 10^2$ or 10^3) and which have zero's everywhere except on the main diagonal, the subdiagonal ($a_{i,\,i-1}$) and the superdiagonal ($a_{i,\,i+1}$). For such a matrix there would be $4n$ operations per iteration and we would need

$$\|Q_J\| \leq 10^{12(1-s)/n^2}.$$

For example, if $s = 11$ and $n \geq 20$, then $\|Q\| \leq \frac{1}{2}$ would suffice. Conversely, suppose $\|Q_J\| \leq \frac{1}{2}$. Then to reduce the initial error by a factor of 10^{1-s}, it suffices to use a number N of iterations

$$N \geq \frac{s-1}{\log(1/\|Q_J\|)} \geq \frac{s-1}{\log 2} = 3.33(s-1).$$

For $s = 11$, $N \geq 34$ iterations suffice. The total number of operations is $34n^2$, which is to be compared with $n^3/3$ for Gaussian elimination. If $n/3 > 34$, the Jacobi method may be more efficient in these cases ($\|Q_J\| \leq \frac{1}{2}$ or $r_\sigma(Q_J) \leq \frac{1}{2}$).

The effect of rounding errors in many iterative methods is not serious. If a method is convergent, it is usually *stable* with respect to rounding errors. The Jacobi method illustrates this point very well. In a single-precision computation we have, in place of (4.4.21),

$$\bar{x}_i^{(m+1)} = \frac{1}{a_{ii}}\left(b_i - \sum_{\substack{j=1 \\ j \neq i}}^{n} a_{ij}\bar{x}_j^{(m)}\right) + \rho_i^{(m)},$$

or

$$\bar{x}^{(m+1)} = Q_J \bar{x}^{(m)} + D^{-1}b + \rho^{(m)},$$

(4.4.23)

where ρ_i is the total rounding error in computing $\bar{x}_i^{(m+1)}$. (We are assuming for the moment that there are no rounding errors in A or b and that any errors in forming $Q_J = I - D^{-1}A$ are included in $\rho^{(m)}$. The latter assumption is quite reasonable if the computation of $\bar{x}_i^{(m+1)}$ is performed as indicated in (4.4.23)) If s-digit single-precision floating-point arithmetic is used throughout, then

$$\|\rho^{(m)}\|_\infty \leq n \cdot 10^{1-s} \|A\|_\infty \|\bar{x}^{(m)}\|_\infty + 2 \cdot 10^{1-s} \|Q_J\|_\infty \|\bar{x}^{(m)}\|_\infty.$$

Let $r^{(m)} = x^{(m)} - \bar{x}^{(m)}$ be the rounding error in the mth iterant $\bar{x}^{(m)}$. Then

$$\|\bar{x}^{(m)}\| \leq \|x^{(m)} - \bar{x}^{(m)}\| + \|x^{(m)}\|$$

and

$$\|\rho^{(m)}\|_\infty \leq (n \cdot 10^{1-s} \|A\|_\infty + 2 \cdot 10^{1-s}\|Q_J\|_\infty)\|r^{(m)}\| + K\|x^{(m)}\|$$

where K is a constant of order 10^{1-s}, since $\|x^{(m)}\|$ is bounded. Subtracting (4.4.23) from (4.4.21), we get

$$\|r^{(m+1)}\|_\infty \leq ((1 + 2 \cdot 10^{1-s})\|Q_J\|_\infty + n \cdot 10^{1-s} \|A\|_\infty)\|r^{(m)}\|_\infty + K\|x^{(m)}\|.$$

For any reasonable value of s, the quantity in parenthesis is less than some constant $d < 1$. Hence,

$$\limsup_{m \to \infty} \frac{\|r^{(m)}\|}{\|x\|} \leq \frac{K}{1-d}.$$

Since $K = O(10^{1-s})$, we can expect $\bar{x}^{(m)}$ to converge to $\hat{x}$, the exact solution rounded to s digits, provided d is not too close to 1. Furthermore, as the precision s is increased, $\|\rho^{(m)}\|$ decreases and $\lim_{m \to \infty} \bar{x}^{(m)}$ approaches x and the sequence $(\bar{x}^{(m)})$ approaches the sequence $(x^{(m)})$. We may make this precise by introducing the concept of a *stable solution* of any iterative scheme given by a difference equation,

$$x^{(m+1)} = f(x^{(m)}),$$

(4.4.24)

where f is any mapping of V^n into itself. By a *solution* of (4.4.24) we mean a sequence of vectors $(x^{(m)})$ which has a specified initial vector $x^{(0)} = u$, called the *starting value* (or *initial value*) and which satisfies (4.4.24) for all $m = 0, 1, 2, \ldots$. If (4.4.24) represents an iterative method, the sequence $(x^{(m)})$ should converge to a vector x, which is the solution of the problem to be solved by iteration. Hence, we may restrict our attention to bounded sequences. We can introduce a norm topology into the space of bounded sequences by taking (Section 3.1, Example 2)

$$\|(x^{(m)})\| = \sup_{0 \le m < \infty} \{\|x^{(m)}\|\}.$$

We may then measure the effect on the sequence of a change in the initial value and the effect of a *perturbation* in the Eq. (4.4.24). By a *perturbation* we shall mean a bounded sequence $(\rho^{(m)})$ which is introduced as an additive term in (4.4.24). Thus, we consider the *perturbed equation* and initial value,

$$\left. \begin{array}{l} x^{(m+1)} = f(x^{(m)}) + \rho^{(m)}, \quad m \ge 0, \\ x^{(0)} = u. \end{array} \right\} \quad (4.4.25)$$

Let $(x^{(m)})$ be a sequence satisfying (4.4.25). Then $(x^{(m)})$ depends on $(\rho^{(m)})$ and u and we may write $(x^{(m)}) = \chi((\rho^{(m)}), u)$, where χ is the mapping defined by (4.4.25). We regard $((\rho^{(m)}), u)$ as an element in the space of bounded sequences and define a norm topology in this space by

$$\|((\rho^{(m)}), u)\| = \sup_m \{\|\rho^{(m)}\|, \|u\|\}.$$

We may now define *stability* for the method (4.4.24).

Definition 4.4.1 A *solution* $x^{(m)} = \chi((\rho^{(m)}), u)$ of (4.4.25) is said to be *stable* if the mapping χ is continuous at the point $((\rho^{(m)}), u)$. Otherwise, the solution is said to be *unstable*. If χ is continuous at all points $(0, u)$, where 0 is the perturbation $\rho^{(m)} = 0$ for all m, then we say that the *iterative method* (4.4.24) is *stable*, and otherwise that it is *unstable*.

According to this definition, a stable sequence $(x^{(m)})$ is one which changes very little when $((\rho^{(m)}), u)$ is changed by a small amount. Hence, for $\|((\rho^{(m)}), u)\|$ sufficiently small, the change in $\|(x^{(m)})\|$ can be made less than an arbitrary $\varepsilon > 0$. If $\rho^{(m)}$ is interpreted as a rounding error, as in the Jacobi iterative method, stability implies that two different computers employing different rounding techniques will produce sequences which are close provided that the rounding errors are of small magnitude. (This will be the case if the precisions are high enough.) The perturbation $\rho^{(m)}$ may also be interpreted as including errors in b and A (the given data in this case). If the method is stable, then a computed sequence $(\bar{x}^{(m)})$ can be made to be arbitrarily close to the exact solution $(x^{(m)})$ by making $\|(\rho^{(m)})\|$ sufficiently small; i.e., by using a sufficiently high precision.

We can apply these notions to the general linear iterative method given in (4.4.4), (4.4.5). The perturbed equation appears as

$$x^{(m+1)} = Qx^{(m)} + M^{-1}b + \rho^{(m)}, \quad m \ge 0. \quad (4.4.26)$$

This method is convergent if and only if Q is quasinilpotent (Theorem 4.4.1). We shall now show that this condition is also necessary and sufficient for stability of the method.

Theorem 4.4.3 If Q is quasinilpotent, then all solutions of (4.4.26) are stable.

Proof. Let $(x^{(m)})$ be a solution corresponding to the initial value u and perturbation $(\rho^{(m)})$. Let $(y^{(m)})$ be a solution corresponding to the initial value v and perturbation $(\eta^{(m)})$, that is,

$$y^{(m+1)} = Qy^{(m)} + M^{-1}b + \eta^{(m)}, \quad m \geq 0.$$

Subtracting this equation from (4.4.26), we obtain

$$x^{(m+1)} - y^{(m+1)} = Q(x^{(m)} - y^{(m)}) + \rho^{(m)} - \eta^{(m)},$$

whence

$$x^{(m+1)} - y^{(m+1)} = Q^{m+1}(u - v) + \sum_{j=0}^{m} Q^j(\rho^{(j)} - \eta^{(j)}), \quad (4.4.27)$$

$$\|x^{(m+1)} - y^{(m+1)}\| \leq \|Q^{m+1}\| \|u - v\| + \left(\sum_{j=0}^{m} \|Q^j\|\right) \sup_{0 \leq j < \infty} \|\rho^{(j)} - \eta^{(j)}\|.$$

By Exercise 3.40, $\sum_0^\infty \|Q^j\| < K_2 < \infty$. Since $\|Q^m\| \to 0$, we also have $\|Q^m\| < K_1 < \infty$. Hence,

$$\|x^{(m+1)} - y^{(m+1)}\| \leq K_1 \|u - v\| + K_2 \|(\rho^{(m)}) - (\eta^{(m)})\|_\infty,$$

which shows that the mapping $\chi : ((\rho^{(m)}), u) \longmapsto (x^{(m)})$ defined by (4.4.26) is (Lipschitz) continuous. Therefore, all solutions are stable. ∎

We have the following strong converse type of statement.

Theorem 4.4.4 If Q is not quasinilpotent, then every solution of (4.4.26) is unstable.

Proof. $r_\sigma(Q) \geq 1$ implies that Q has an eigenvalue λ such that $|\lambda| \geq 1$.

CASE 1. If $|\lambda| > 1$ or $\lambda = 1$, let $\rho^{(j)} - \eta^{(j)} = w$, where w is an eigenvector belonging to λ. Then the sum in (4.4.27) becomes $\sum_0^m Q^j w = (\sum_0^m \lambda^j) w$. The geometric series $\sum_0^m \lambda^j$ diverges to ∞ in this case. Hence, $\|y^{(m+1)} - z^{(m+1)}\|$ is unbounded no matter how small $\|w\| > 0$ is chosen.

CASE 2. If all $|\lambda| \leq 1$ and $\lambda \neq 1$, then some $\lambda = e^{i\theta}$, $\theta \neq 0$. In this case take $\rho^{(j)} - \eta^{(j)} = w^{(j)} = w + \bar{w}$, where $\bar{w}$ is an eigenvector belonging to $\bar{\lambda}$. Hence,

$$\sum_0^m Q^j w^{(j)} = \left(\sum_0^m 2 \cos j\theta\right)(w + \bar{w}),$$

does not converge (see Exercise 7.22). ∎

We have said that the effect of rounding errors in iterative methods is "usually not serious." This is somewhat misleading from a practical standpoint. We have proven that the iterative methods (4.4.4) are stable if and only if they converge. Stability implies that rounding errors do not grow in the course of iteration. This would seem to imply that ill-conditioning can be overcome by using an iterative technique. However, this is not the case. We have shown above that in the Jacobi method, the rounding error $r^{(m)} = x^{(m)} - \bar{x}^{(m)}$ satisfies an inequality

$$\lim_{m \to \infty} \|r^{(m)}\| \leq K/(1-d).$$

However, $d \geq (1 + 2 \cdot 10^{1-s}) \|Q_J\|_\infty + n \cdot 10^{1-s} \|A\|_\infty$. If $\|Q_J\|_\infty$ is close to 1, then $1/(1-d)$ is large and although $K = C \cdot 10^{1-s}$ where $0 < C < 10$, the bound for $\|r^{(m)}\|$ can be several orders of magnitude greater than 10^{1-s}. Indeed, since $A^{-1} = (I - Q)^{-1} M^{-1}$, we have

$$\|A^{-1}\| \leq \frac{\|M^{-1}\|}{1 - \|Q\|}.$$

Let A be scaled so that $\max |a_{ij}| = 1$ for all i. If $\|Q_J\|_\infty < 1$, then A is strictly diagonally dominant and $\|D\|_\infty = 1$. In fact, $|a_{ii}| = 1$ for all i. Hence, $\|M^{-1}\|_\infty = \|D^{-1}\|_\infty = 1$. If $\|A^{-1}\|_\infty$ is large, this implies that $1/(1 - \|Q_J\|_\infty)$ is large; i.e., $\|Q_J\|_\infty$ is close to 1. Thus, we might have slow convergence and the limit may be subject to relatively large rounding errors (compared to 10^{1-s}).

Gauss-Seidel Iteration

Another well-known specific instance of Eqs. (4.4.4), (4.4.5) is the *Gauss-Seidel method* in which M is taken to be the lower triangular part of A. Thus,

$$M_G = \begin{bmatrix} a_{11} & 0 & 0 \\ a_{21} & a_{22} & 0 \\ \vdots & & \ddots \\ a_{n1} & \cdots & a_{nn} \end{bmatrix}$$

and $N_G = M_G - A$. In component form, (4.4.4) becomes

$$\sum_{j=1}^{i} a_{ij} x_j^{(m+1)} = -\sum_{j=i+1}^{n} a_{ij} x_j^{(m)} + b_i, \quad i = 1, \ldots, n.$$

Starting with an arbitrary initial guess $x^{(0)}$, we solve for $x_1^{(1)}$ in the first equation, using the "old" values $x_2^{(0)}, \ldots, x_n^{(0)}$. Then in Eq. (2), we solve for $x_2^{(1)}$ using the "new" value $x_1^{(1)}$ and the old values $x_3^{(0)}, \ldots, x_n^{(0)}$ and so on for Eqs. (3), ..., (n). Thus, in each equation, the "most recent" values of the components are used. This completes one iteration. $x^{(1)}$ then becomes the "old" vector and the process is repeated to find the next iterant $x^{(2)}$ and so on, generating a sequence $(x^{(m)})$.

The iteration can be carried out provided that $a_{ii} \neq 0$, $1 \leq i \leq n$, since the solution is given by

$$x_i^{(m+1)} = (1/a_{ii}) \left(-\sum_{j=1}^{i-1} a_{ij} x_j^{(m+1)} - \sum_{j=i+1}^{n} a_{ij} x_j^{(m)} + b_i \right), \quad (4.4.28)$$

where the first sum is taken to be zero for $i = 1$.

As in the general case (4.4.4), the Gauss-Seidel iterants converge to $x = A^{-1}b$ if and only if $Q_G = M_G^{-1}N_G$ is quasinilpotent. To check this condition computationally we would have to obtain bounds on the eigenvalues of Q_G. In general, this is a difficult computational problem. A simply calculable sufficient condition for convergence is the following.

Theorem 4.4.5 Let

$$c_i = (1/|a_{ii}|) \sum_{\substack{j=1 \\ j \neq i}}^{n} |a_{ij}|$$

and $c = \max_{1 \leq i \leq n}\{c_i\}$. If $c < 1$, the Gauss-Seidel sequence (4.4.28) converges to $A^{-1}b$ for arbitrary starting values. Also, $\|Q_G\|_\infty \leq c$ if $c \leq 1$.

Proof. We note that although $c < 1$ is the same sufficient condition as in (4.4.22), now it does not imply that $\|Q_G\|_\infty = c$. However, we shall show that $c \leq 1$ implies $\|Q_G\|_\infty \leq c$, which will establish convergence when $c < 1$.

By Definition 3.5.1,

$$\|Q_G\|_\infty = \sup_{\|x\|_\infty = 1} \|Q_G x\|_\infty.$$

Let x be an arbitrary vector such that $\|x\|_\infty = 1$. Let $y = Q_G x = M_G^{-1} N_G x$. We shall prove that $\|y\|_\infty \leq c$, which implies

$$\sup_{\|x\|_\infty = 1} \|Q_G x\|_\infty \leq c.$$

We have $M_G y = N_G x$. Hence,

$$|y_1| \leq (1/|a_{11}|) \sum_{j=2}^{m} |a_{1j}| |x_j| \leq c_1 \leq c.$$

Now, assume that $|y_k| \leq c \leq 1$ for $1 \leq k \leq i - 1$. Then

$$|y_i| \leq (1/|a_{ii}|) \left(\sum_{j=1}^{i-1} |a_{ij}| |y_j| + \sum_{j=i+1}^{n} |a_{ij}| |x_j| \right)$$

$$\leq (1/|a_{ii}|) \left(c \sum_{j=1}^{i-1} |a_{ij}| + \sum_{j=i+1}^{n} |a_{ij}| \right) \leq c_i \leq c.$$

Therefore, by induction on i, $|y_i| \leq c$ for $1 \leq i \leq n$, that is $\|y\|_\infty \leq c$, as we wished to show. ∎

Diagonal Dominance and Irreducible Matrices

If the condition $c < 1$ of the preceding theorem is satisfied, the matrix A is said to be *strictly diagonally dominant*. If $c \leq 1$, then

$$\sum_{\substack{j=1 \\ j \neq i}}^{n} |a_{ij}| \leq |a_{ii}|, \qquad i = 1, \ldots, n,$$

and the matrix is said to be *diagonally dominant*. As we shall see, for certain matrices A, diagonal dominance suffices for convergence of both the Jacobi and Gauss-Seidel methods.

Definition 4.4.2 A matrix A of order $n \geq 2$ is said to be *reducible* if there exists a permutation matrix P or order n such that

$$PAP^T = \begin{bmatrix} A_1 & B \\ O & C \end{bmatrix},$$

where A_1 is an rth order matrix and O is the zero $(n - r) \times r$ matrix. If no such P exists, A is called *irreducible*.

(For an example, see Exercise 4.7. Also see Exercise 4.8.)

Remark If a matrix is irreducible, then given any indices i, j there must exist a sequence of nonzero entries of the form $(a_{ir}, a_{rs}, a_{st}, \ldots, a_{uj})$, i.e., the row index of an entry agrees with the column index of the previous member of the sequence. (Exercise 4.8(b).) This condition is also sufficient for irreducibility.

Theorem 4.4.6 Let A be an irreducible nth order matrix. If there exists a row q such that $\sum_{j=1}^{n} |a_{qj}| < \|A\|_\infty$, then $r_\sigma(A) < \|A\|_\infty$. Similarly, if $\sum_{i=1}^{n} |a_{iq}| < \|A\|_1$ for some q, then $r_\sigma(A) < \|A\|_1$.

Proof. Let λ be an eigenvalue and x a corresponding eigenvector with $\|x\|_\infty = 1$. Let

$$C_i = \{z : |z - a_{ii}| \leq \sum_{\substack{j=1 \\ j \neq i}}^{n} |a_{ij}| = r_i\}, \qquad 1 \leq i \leq n.$$

Then $\lambda \in \bigcup_{i=1}^{n} C_i$. In proof, note that $|x_k| = 1$ for some k. Since $\lambda x = Ax$, we have

$$|\lambda - a_{kk}| = \sum_{\substack{j=1 \\ j \neq k}}^{n} a_{kj} x_j.$$

Hence,

$$|\lambda - a_{kk}| \leq \sum_{\substack{j=1 \\ j \neq k}}^{n} |a_{kj}| = r_k.$$

(This is (Hadamard's) Theorem 6.2.1.) Now, suppose λ is on the boundary of $\bigcup C_i$. We shall show that this implies $|\lambda - a_{ii}| = r_i$ for all i. First, it is clear that λ must be on the boundary of C_k. For any i, there exists a connecting sequence of nonzero elements $(a_{ks}, a_{st}, \ldots, a_{mi})$, by the Remark above. From

$$r_k = |\lambda - a_{kk}| \leq \sum_{j \neq k} |a_{kj}| |x_j| \leq r_k,$$

it follows that $|x_s| = 1$, since $a_{ks} \neq 0$. Hence, as before, $|\lambda - a_{ss}| \leq r_s$ and since λ is on the boundary of $\bigcup C_i$, $|\lambda - a_{ss}| = r_s$. Continuing through the sequence with $a_{tt}, \ldots, a_{ii}$, we obtain $|\lambda - a_{ii}| = r_i$.

Now, $z \in C_i$ implies $|z| \le |a_{ii}| + r_i \le \|A\|_\infty$. Thus, each C_i is contained in $C = \{z : |z| \le \|A\|_\infty\}$. By hypothesis, $|a_{qq}| + r_q < \|A\|_\infty$, so that C_q is contained in the interior of C. If there were an eigenvalue λ with $|\lambda| = \|A\|_\infty$, it would have to be on the boundary of $\bigcup_1^n C_i$. By the preceding analysis, this implies $|\lambda - a_{ii}| = r_i$ for all i. Hence, $|\lambda| \le |a_{qq}| + r_q < \|A\|_\infty$, which is a contradiction. Therefore $|\lambda| < \|A\|_\infty$. ∎

Corollary 1 If A is irreducible and diagonally dominant and if for some q,

$$|a_{qq}| > \sum_{\substack{i=1 \\ i \ne q}}^n |a_{iq}|,$$

then A is nonsingular and $a_{ii} \ne 0$ for all i. If all $a_{ii} > 0$, then any eigenvalue λ has a positive real part.

Proof. As a consequence of diagonal dominance, for each of the disks C_i in the proof of the theorem, we have $|z - a_{ii}| \le r_i \le |a_{ii}|$. Hence, if 0 were an eigenvalue, it would be on the boundary of $\bigcup C_i$ and this would imply $|0 - a_{qq}| = r_q < |a_{qq}|$, which is a contradiction. Since row i has at least one nonzero off-diagonal element, $|a_{ii}| > 0$. ∎

Corollary 2 If A is an arbitrary strictly diagonally dominant matrix, then A is nonsingular.

Proof. As in Corollary 1, $|z - a_{ii}| < |a_{ii}|$, so that $0 \notin \bigcup C_i$. Hence, 0 is not an eigenvalue. ∎

Theorem 4.4.7 Let A be irreducible and diagonally dominant and let

$$|a_{KK}| > \sum_{\substack{K \ne j \\ j=1}}^n |a_{Kj}|$$

for some K. Then the Jacobi method converges.

Proof. The matrix $Q_J = (q_{ij})$ has elements $q_{ij} = -a_{ij}/a_{ii}$, $i \ne j$, and $q_{ii} = 0$ for all i. Hence, by the diagonal dominance, $\|Q_J\|_\infty \le 1$. If $\|Q_J\|_\infty < 1$, Q_J is quasinilpotent. If $\|Q_J\|_\infty = 1$, then $\sum_{i=1}^n |q_{Kj}| < \|Q_J\|_\infty$. Since Q_J is also irreducible, $r_\sigma(Q_J) < 1$ by Theorem 4.4.6. (Q_J is clearly irreducible, since for any i, j there is a connecting sequence of nonzero entries $(q_{is}, q_{st}, \ldots, q_{mj})$ obtained from $(a_{is}, a_{st}, \ldots, a_{mj})$.) ∎

The same result holds for the Gauss-Seidel method.

Theorem 4.4.8 Let A be irreducible and diagonally dominant and let

$$|a_{qq}| > \sum_{\substack{j=1 \\ j \ne q}}^n |a_{qj}|$$

for some q. Then the Gauss-Seidel method converges.

Proof. Let $L = (l_{ij})$, where $l_{ij} = -a_{ij}$ for $i > j$ and $l_{ij} = 0$ for $i \leq j$. Let $U = (u_{ij})$ where $u_{ij} = -a_{ij}$ for $i < j$ and $u_{ij} = 0$ for $i \geq j$. (Thus, $-L$ is the *lower triangular part* of A and $-U$ is the *upper triangular part*, not including the diagonal.) As above, let D be the diagonal matrix whose entries are a_{ii}. Then from (4.4.28) we see that for the Gauss-Seidel procedure,

$$Q_G = M_G^{-1} N_G = (D - L)^{-1} U = (I - D^{-1}L)^{-1} D^{-1} U.$$

Now, by Theorem 4.4.5, since A is diagonally dominant, it follows that $c \leq 1$ and $\|Q_G\|_\infty \leq c \leq 1$. Hence, if λ is an eigenvalue of Q_G, then $|\lambda| \leq 1$. Suppose $|\lambda| = 1$. We shall show this leads to a contradiction.

Let x be an eigenvector belonging to λ. Then

$$(I - D^{-1}L)^{-1} D^{-1} U x = \lambda x$$
$$D^{-1} U x = -\lambda D^{-1} L x + \lambda x$$
$$\left(D^{-1} L + \frac{1}{\lambda} D^{-1} U \right) x = x.$$

Thus, x is an eigenvector of the matrix $B = D^{-1}L + (1/\lambda)D^{-1}U$ belonging to the eigenvalue 1. However, the matrix B has elements $b_{ij} = -a_{ij}/a_{ii}$ for $i > j$, $b_{ij} = -a_{ij}/\lambda a_{ii}$ for $i > j$ and $b_{ii} = 0$ on the diagonal. Since A is irreducible, so is B. (See proof of Theorem 4.4.7.) Furthermore, by the diagonal dominance, if $|\lambda| = 1$, then

$$\sum_{j=1}^{n} |b_{ij}| = \sum_{j=1}^{i-1} \frac{|a_{ij}|}{|a_{ii}|} + \sum_{j=i+1}^{n} \frac{|a_{ij}|}{|\lambda| |a_{ii}|} = \sum_{\substack{j=1 \\ j \neq i}}^{n} \frac{|a_{ij}|}{|a_{ii}|} \leq 1.$$

Thus, $\|B\|_\infty \leq 1$. If

$$c_i = \sum_{\substack{j=1 \\ j \neq i}}^{n} |a_{ij}|/|a_{ii}| < 1 \qquad \text{for all } i,$$

then the Gauss-Seidel method converges by Theorem 4.4.5. Hence, we may suppose that $c_i = 1$ for some i, which implies that $\|B\|_\infty = 1$. But $\sum_{j=1}^{n} |b_{qj}| = c_q < 1$ by hypothesis. By Theorem 4.4.6, $r_\sigma(B) < \|B\|_\infty = 1$. However, we have seen that 1 is an eigenvalue of B. This is a contradiction. Therefore, $|\lambda| < 1$ for every eigenvalue λ of Q_G; i.e., $r_\sigma(Q_G) < 1$ and Q_G is quasinilpotent. Thus, the Gauss-Seidel method converges. ∎

Theorem 4.4.9 Let A be irreducible. Let its Jacobi matrix Q_J be nonnegative. Then exactly one of the following cases holds:

$$0 < r_\sigma(Q_G) < r_\sigma(Q_J) < 1; \tag{1}$$

$$1 = r_\sigma(Q_J) = r_\sigma(Q_G); \tag{2}$$

$$1 < r_\sigma(Q_J) < r_\sigma(Q_G); \tag{3}$$

$$r_\sigma(Q_J) = r_\sigma(Q_G) = 0. \tag{4}$$

Thus, for such a matrix A, the Jacobi and Gauss-Seidel methods either both converge or both diverge. If both converge, the Gauss-Seidel has a higher asymptotic rate of convergence. To prove the theorem we shall establish several results about irreducible nonnegative matrices. For convenience, we write $A \geq 0$ to indicate that $a_{ij} \geq 0$ for all elements of A. Similarly, for a vector, $x \geq 0$ indicates that all coordinates are nonnegative. $A \geq B$ indicates that $A - B \geq 0$. (See Exercise 2.16.)

Lemma 4.4.1 Let A be a nonnegative irreducible $n \times n$ matrix. Then the matrix $B = I + A + A^2 + \cdots + A^{n-1}$ has only positive entries.

Proof. Consider the canonical basis $\{e_1, \ldots, e_n\}$ of (2.1.9). Since A is irreducible, it cannot leave invariant any subspace spanned by a subset of the $\{e_i\}$. (Exercise 4.8a.) Consider $e_1 = Ie_1$. Then $Ae_1 = \sum_{i=1}^n a_{i1} e_i$ and $a_{i1} \neq 0$ for at least one $i \geq 2$. (Otherwise, A leaves invariant the subspace spanned by $\{e_1\}$.) Let $\{e_{i_1}, \ldots, e_{i_q}\}$ be the vectors corresponding to the a_{i1} which are nonzero. Let V_1 be the subspace spanned by these vectors and e_1. Now, $A^2 e_1 = \sum_{j=1}^q a_{i_j 1} A e_{i_j}$ and the vector $Ae_{i_1}, \ldots, Ae_{i_q}$ do not all lie in the subspace V_1. (Otherwise $AV_1 \subset V_1$.) Therefore, $A^2 e_1 = \sum_{i=1}^n \alpha_{2i} e_i$, where $\alpha_{2i} \neq 0$ for some index i other than $\{1, i_1, \ldots, i_q\}$. Continuing in this way with $A^3 e_1, \ldots, A^{n-1} e_1$, we conclude that every vector e_i, $1 \leq i \leq n$, occurs at least once with a nonzero coordinate in one of the vectors $e_1, Ae_1, \ldots, A^{n-1} e_1$. Since A is nonnegative, so is A^i for all i. Therefore,

$$Be_1 = \left(\sum_{i=0}^{n-1} A^i\right) e_1 = \sum_{i=0}^{n-1} A^i e_1 = \sum_{i=1}^n b_{i1} e_i,$$

and $b_{i1} > 0$ for all i. Applying the same argument to $e_2, \ldots, e_n$, we conclude that $b_{ij} > 0$ for every entry of B. ∎

Corollary If A is a nonnegative and irreducible matrix of order n, then $(I + A)^{n-1}$ has only positive entries.

Proof. Consider the expansion

$$(I + A)^{n-1} = \sum_{i=0}^{n-1} \binom{n}{i} A^i,$$

where $\binom{n}{i}$ are the binomial coefficients. Since the entries of $\sum_0^{n-1} A^i$ are all positive by Lemma 4.4.1, so are the entries of $(I + A)^{n-1}$. ∎

Lemma 4.4.2 Let A be an irreducible nonnegative matrix of order n. For any nonnegative vector $x \in V^n$, let

$$\rho_x = \sup\{\lambda \geq 0 : Ax \geq \lambda x\}$$

and

$$\rho = \sup\{\rho_x : x \text{ nonnegative and } x \neq 0\}.$$

Then $\rho > 0$. If u is a nonnegative nonzero vector such that $Au \geq \rho u$, then $Au = \rho u$ and $u > 0$; i.e., u is an eigenvector belonging to the eigenvalue ρ. There exists at least one such u.

Proof. Note that $\rho_x = \rho_{\alpha x}$ for any positive scalar α. Also,

$$\rho_x = \min_{x_i > 0} \left\{\frac{\sum_{j=1}^n a_{ij} x_j}{x_i}\right\}. \qquad (4.4.29)$$

For $x = (1, 1, \ldots, 1)$,

$$\rho_x = \min_i \left\{ \sum_{j=1}^n a_{ij} \right\} > 0$$

by the irreducibility. Hence $\rho > 0$. Since $\|Ax\|_\infty \le \|A\|_\infty$ for $\|x\|_\infty = 1$, we see that $\rho_x \le \|A\|_\infty$ and therefore $\rho \le \|A\|_\infty$. Let $S = \{x : \|x\|_\infty = 1 \text{ and } x \ge 0\}$. Then $\rho = \sup\{\rho_x : x \in S\}$. Now $(I + A)^{n-1}A = A(I + A)^{n-1}$ and $(I + A)^{n-1} > 0$ by Lemma 4.4.1. For any $x \in S$ $Ax \ge \rho_x x$ and multiplying by $(I + A)^{n-1}$, we have

$$A(I + A)^{n-1}x \ge \rho_x(I + A)^{n-1}x.$$

Let $E = \{(I + A)^{n-1}x, x \in S\}$. For $y \in E$ we have shown $Ay \ge \rho_x y$. By definition of ρ_y, this implies $\rho_y \ge \rho_x$. Hence, $\rho = \sup\{\rho_y : y \in E\}$. Furthermore, since $y = (I + A)^{n-1}x$, where $(I + A)^{n-1} > 0$ and $x \ne 0$, we have $y > 0$.

Now, S is a closed bounded set in V^n, therefore compact. E is a continuous image of S under the mapping $(I + A)^{n-1}$. Thus, E is compact (by Theorem 1.4.1). By (4.4.29) ρ_y is continuous in y. Hence, ρ_y assumes its supremum value, ρ, on E; i.e., there exists $u \in E$ such that $\rho = \rho_u$. Therefore, $Au \ge \rho_u u = \rho u$.

Now, suppose $Au \ge \rho u$ for any $u \ge 0$, $u \ne 0$. If $Au - \rho u \ne 0$, then some component is positive. Hence, $(I + A)^{n-1}(Au - \rho u) > 0$, which implies $Az > \rho z$ for $z = (I + A)^{n-1}u$. But then $\rho_z > \rho$, which contradicts the definition of ρ. Hence, $Au = \rho u$ and $z = (1 + \rho)^{n-1}u$, which implies that $u > 0$ (since $z > 0$). ∎

Lemma 4.4.3 Let A and ρ be as in Lemma 4.4.2. Let $B = (b_{ij})$ be any $n \times n$ matrix such that $|b_{ij}| \le a_{ij}$ for all i and j. If λ is an eigenvalue of B, then $|\lambda| \le \rho$. Furthermore, $|\lambda| = \rho$ only if $|b_{ij}| = a_{ij}$ for all i and j.

Proof. Let $\lambda x = Bx$, where $x \ne 0$ is an eigenvector belonging to λ. Writing $|B|$ for the matrix $(|b_{ij}|)$ and $|x|$ for the vector $(|x_1|, \ldots, |x_n|)$, we have

$$|\lambda| |x_i| \le \sum_{j=1}^n |b_{ij}| |x_j| \le \sum_{j=1}^n a_{ij} |x_j|,$$

so that

$$A|x| \ge |B| |x| > |\lambda| |x|.$$

This implies $|\lambda| \le \rho_{|x|} \le \rho$. If $|\lambda| = \rho$, then by Lemma 4.4.2, $|x|$ is an eigenvector, so that $A|x| = |B||x| = |\lambda||x|$. Since $|x| > 0$ by Lemma 4.4.2, and $a_{ij} \ge |b_{ij}|$ by hypothesis, we cannot have $a_{ij} > |b_{ij}|$ for any i, j. Therefore, $a_{ij} = |b_{ij}|$ for all i, j. ∎

These lemmas lead to the following result, due to Perron and Frobenius.

Theorem 4.4.10 If A is a nonnegative irreducible matrix of order n, then

i) The quantity ρ of Lemma 4.4.2 is equal to the spectral radius $r_\sigma(A)$ so that $r_\sigma(A)$ is an eigenvalue which has an eigenvector $u > 0$;

ii) $r_\sigma(A)$ increases if any entry of A increases.

Proof. In Lemma 4.4.3, take $B = A$ to obtain (i). To prove (ii), if some a_{ij} is increased, we obtain a nonnegative matrix $A' \ge A$. By Lemma 4.4.3, and (i), $r_\sigma(A) < r_\sigma(A')$; (i.e., for any eigenvalue λ of A, we must have $|\lambda| \le \rho' = r_\sigma(A')$ and $|\lambda| \ne \rho'$ since $A \ne A'$.) ∎

Lemma 4.4.4 Let A be a nonnegative and irreducible $n \times n$ matrix. For any positive vector $x > 0$, either

$$\min_{1 \leq i \leq n} \left\{ \frac{\sum_{j=1}^{n} a_{ij} x_j}{x_i} \right\} < r_\sigma(A) < \max_{1 \leq i \leq n} \left\{ \frac{\sum_{j=1}^{n} a_{ij} x_j}{x_i} \right\} \tag{4.4.30}$$

or

$$\frac{\sum_{j=1}^{n} a_{ij} x_j}{x_i} = r_\sigma(A), \quad 1 \leq i \leq n. \tag{4.4.31}$$

Also, either

$$\sum_{j=1}^{n} a_{ij} = r_\sigma(A), \quad 1 \leq i \leq n$$

or

$$\min_{1 \leq i \leq n} \left\{ \sum_{j=1}^{n} a_{ij} \right\} < r_\sigma(A) < \max_{1 \leq i \leq n} \left\{ \sum_{j=1}^{n} a_{ij} \right\} = \|A\|_\infty. \tag{4.4.32}$$

Proof. If all row sums $\sum_{j=1}^{n} a_{ij} = \|A\|_\infty$, then taking $u = (1, 1, \ldots, 1)$, we have $Au = \|A\|_\infty u$. Hence, $\|A\|_\infty \leq r_\sigma(A)$. But $r_\sigma(A) \leq \|A\|_\infty$. Thus, $r_\sigma(A) = \sum_{j=1}^{n} a_{ij}$.

If

$$\min_{1 \leq i \leq n} \sum_{j=1}^{n} a_{ij} < \|A\|_\infty,$$

there exists a nonnegative matrix $B = (b_{ij})$ such that $b_{ij} \leq a_{ij}$ and

$$\sum_{j=1}^{n} b_{ij} = \min_{1 \leq i \leq n} \sum_{j=1}^{n} a_{ij}$$

for all i. The matrix B is obtained from A by decreasing suitable positive a_{ij}. By Theorem 4.4.10(ii), $r_\sigma(B) < r_\sigma(A)$. Since $r_\sigma(B) = \sum_{j=1}^{n} b_{ij}$ by the preceding paragraph, this establishes the first inequality in (4.4.32). Similarly, by increasing certain elements of A we obtain a matrix B' with $\sum_{j=1}^{n} b'_{ij} = \|A\|_\infty$ for all i. Reasoning as before, we obtain $r_\sigma(A) < r_\sigma(B')$, which is the second inequality in (4.4.32).

To establish that (4.4.30), (4.4.31) for $x = (x_1, \ldots, x_n)$ where $x_i > 0$, we let D be the diagonal matrix with $d_{ii} = x_i$. Then $B = D^{-1}AD$ is clearly nonnegative and irreducible. Since $b_{ij} = a_{ij} x_j / x_i$, the result follows directly from the previous paragraphs. ∎

Observe that the nonnegative irreducible matrices are dense in the set of all nonnegative matrices of order n. For suppose $A = (a_{ij})$ is nonnegative. For every $a_{ij} = 0$ we define $b_{ij} = \varepsilon > 0$ and otherwise let $b_{ij} = a_{ij}$. Then B is irreducible and $\|B - A\|_\infty \leq n\varepsilon$. Since ε is arbitrary, there is an irreducible matrix in every neighborhood of A.

Theorem 4.4.11 If A is a nonnegative matrix of order n, then A has a nonnegative real eigenvalue equal to its spectral radius $r_\sigma(A)$. There is a nonnegative eigenvector, $x \geq 0$ belonging to the eigenvalue $r_\sigma(A)$.

Proof. Let (A_m) be a sequence of irreducible nonnegative matrices converging to A (in the induced operator norm). By Theorem 4.4.10, there exist positive vectors u_m such that $A_m u_m = r_\sigma(A_m) u_m$, with $\|u_m\| = 1$. Since the unit sphere is compact in V^n (Theorem 3.2.3, Corollary 4), there is a convergent subsequence (u_{m_i}) converging to a unit vector u. Then

$$\|A_{m_i} u_{m_i} - Au\| \leq \|A_{m_i} u_{m_i} - Au_{m_i}\| + \|Au_{m_i} - Au\|$$
$$\leq \|A_{m_i} - A\| \, \|u_{m_i}\| + \|A\| \, \|u_{m_i} - u\|.$$

Since $\|A_{m_i} - A\| \to 0$ and $\|u_{m_i} - u\| \to 0$, it follows easily that $A_{m_i}u_{m_i} \to Au$. Hence, $r_\sigma(A_{m_i})u_{m_i} \to Au$ also. We see that $(r_\sigma(A_{m_i}))$ converges, say to λ, and $u_{m_i} \to u$. Therefore, $Au = \lambda u$. (From formula (3.6.3), it is clear that $r_\sigma(A)$ is a continuous function of A in the operator norm topology. One can also see this from the fact that the eigenvalues of a matrix A are zeros of the characteristic polynomial. See the last part of Section 6.2.) Therefore $\lambda = r_\sigma(A) = \lim r_\sigma(A_{m_i})$. Finally, since $r_\sigma(A_{m_i}) > 0$ by Theorem 4.4.10, $r_\sigma(A) \geq 0$. Similarly, $u \geq 0$, since $u_{m_i} > 0$. This completes the proof. ∎

We can now give a proof of Theorem 4.4.9.

Proof (Theorem 4.4.9). We shall set the Jacobi matrix $Q_J = L' + U'$, where $L' = (l'_{ij})$, $U' = (u'_{ij})$, $l'_{ij} = -a_{ij}/a_{ii}$ for $i > j$ and is zero otherwise, and $u'_{ij} = -a_{ij}/a_{ii}$ for $i < j$ and is zero otherwise. Since A is an irreducible matrix, so is $L' + U'$.

Now, the Gauss-Seidel matrix $Q_G = (I - L')^{-1}U'$. We see that $(I - L')^{-1} = \sum_{j=0}^{n-1} L'^j$. Since L' and U' are nonnegative, this shows that Q_G is nonnegative. By Theorem 4.4.11, there exists a unit vector $u \geq 0$ such that $(I - L')^{-1}U'u = \lambda u$, where $\lambda = r_\sigma(Q_G) \geq 0$. It follows that

$$(\lambda L' + U')u = \lambda u.$$

In fact, the matrices $B_\alpha = \alpha L' + U'$ are nonnegative and irreducible for all $\alpha > 0$. It follows that $\lambda > 0$. (See Exercise 4.9.) Since B_λ is irreducible and nonnegative, $(I + B_\lambda)^{n-1}$ is positive. Hence, $(I + B_\lambda)^{n-1}B_\lambda u = \lambda(1 + \lambda)^{n-1}u$ is positive, which implies that $u > 0$.

Let $C_\alpha = L' + (1/\alpha)U'$, $\alpha > 0$. We see that $C_\lambda u = (L' + (1/\lambda)U')u = u$. Hence, by Lemma 4.4.4, $r_\sigma(C_\lambda) = 1$. Similarly, $r_\sigma(B_\lambda) = \lambda$. For $\alpha = 1$,

$$r_\sigma(B_1) = r_\sigma(C_1) = r_\sigma(Q_J).$$

From Theorem 4.4.10 (ii) it follows that $r_\sigma(B_\alpha)$ increases and $r_\sigma(C_\alpha)$ decreases as α increases.

CASE 1. If $0 < r_\sigma(Q_J) < 1$, then $r_\sigma(C_1) < 1$. Since $r_\sigma(C_\lambda) = 1$, in view of the decreasing behavior of $r_\sigma(C_\alpha)$, we must have $0 < \lambda < 1$. But then

$$r_\sigma(Q_G) = r_\sigma(B_\lambda) < r_\sigma(B_1) = r_\sigma(Q_J),$$

as we wished to prove.

CASE 2. If $r_\sigma(Q_J) = 1$, then $r_\sigma(C_1) = 1$ and since $r_\sigma(C_\alpha)$ is strictly decreasing and $r_\sigma(C_\lambda) = 1$, we must have $\lambda = 1$; i.e., $r_\sigma(Q_G) = 1$. Conversely, $\lambda = 1$ implies $r_\sigma(Q_J) = r_\sigma(B_1) = 1$.

CASE 3. If $r_\sigma(Q_J) > 1$, then $r_\sigma(C_1) > 1$, which with $r_\sigma(C_\lambda) = 1$ implies $\lambda > 1$. But then we also have $r_\sigma(Q_G) = r_\sigma(B_\lambda) > r_\sigma(B_1) = r_\sigma(Q_J)$. ∎

The self-adjoint case

If A is a (complex) Hermitian $n \times n$ matrix, we can write $A = D - L - L^*$, where D is a real diagonal matrix and $L = (l_{ij})$ with $l_{ij} = -a_{ij}$ for $i > j$ and $l_{ij} = 0$ for $i \leq j$. In this case, for the Gauss-Seidel method we have $M = D - L$ and $N = L^*$. We shall show that convergence depends on whether or not A is positive definite (Section 3.6).

Suppose λ is an eigenvalue of $Q = M^{-1}N = I - M^{-1}A$. Then for any eigenvector u belonging to λ,

$$(I - M^{-1}A)u = \lambda u,$$
$$Au = (1 - \lambda)Mu. \qquad (4.4.33)$$

Assuming that A is nonsingular, we must have $\lambda \neq 1$. Now, observe that

$$\langle Au, u \rangle = (1 - \lambda)\langle Mu, u \rangle. \qquad (4.4.34)$$

Taking complex conjugates and remembering that $\langle Au, u \rangle$ is real we also have

$$\langle Au, u \rangle = (1 - \bar{\lambda})\overline{\langle Mu, u \rangle} = (1 - \bar{\lambda})\langle u, Mu \rangle = (1 - \bar{\lambda})\langle M^*u, u \rangle. \qquad (4.4.35)$$

Finally, noting that $M + M^* = D - L + D - L^* = A + D$, we obtain by addition of (4.4.34) and (4.4.35),

$$\left(\frac{1}{1-\lambda} + \frac{1}{1-\bar{\lambda}}\right)\langle Au, u \rangle = \langle(M + M^*)u, u \rangle = \langle Au, u \rangle + \langle Du, u \rangle,$$

or

$$\left(\frac{1}{1-\lambda} + \frac{1}{1-\bar{\lambda}} - 1\right)\langle Au, u \rangle = \langle Du, u \rangle. \qquad (4.4.36)$$

But

$$\alpha = \frac{1}{1-\lambda} + \frac{1}{1-\bar{\lambda}} - 1 = \frac{1 - |\lambda|^2}{1 - 2\,\text{Re}\,\lambda + |\lambda|^2}$$

and

$$1 + |\lambda|^2 - 2\,\text{Re}\,\lambda \geq 1 + |\lambda|^2 - 2|\lambda| = (1 - |\lambda|)^2 \geq 0.$$

Hence, $|\lambda| < 1$ if and only if $\alpha > 0$.

In many of the practical cases of interest, the diagonal elements of A are all positive. Therefore, $\langle Du, u \rangle > 0$. In fact, D is positive definite. If A is also positive definite, then by (4.4.36), $\alpha > 0$ and $|\lambda| < 1$. Therefore, this is a sufficient condition for Q to be quasinilpotent. It is also necessary. To prove this, suppose A is not positive definite. Then there exists a vector v_0 such that $\langle Av_0, v_0 \rangle \leq 0$. Let $v_m = Q^m v_0$. Then $v_{m+1} = Qv_m$, so that $Mv_{m+1} = Nv_m$ or

$$(D - L)v_{m+1} = L^*v_m. \qquad (4.4.37)$$

Subtracting $(D - L)v_m$ from both sides, we get

$$(D - L)\delta_m = -(D - L - L^*)v_m = -Av_m,$$

where $\delta_m = v_{m+1} - v_m$. Subtracting L^*v_{m+1} from both sides, we find that

$$Av_{m+1} = -L^*\delta_m.$$

From the last two equations it follows that
$$\langle (D - L) \delta_m, v_m \rangle = -\langle Av_m, v_m \rangle,$$
and
$$\langle Av_{m+1}, v_{m+1} \rangle = -\langle L^* \delta_m, v_{m+1} \rangle.$$
Hence,
$$\langle Av_{m+1}, v_{m+1} \rangle - \langle Av_m, v_m \rangle = \langle (D - L) \delta_m, v_m \rangle - \langle L^* \delta_m, v_{m+1} \rangle.$$
$$= \langle D \delta_m, v_m \rangle - \langle L \delta_m, v_m \rangle - \langle L^* \delta_m, v_{m+1} \rangle$$
$$= -\langle D \delta_m, \delta_m \rangle + \langle D \delta_m, v_{m+1} \rangle - \langle L^* \delta_m, v_{m+1} \rangle - \langle \delta_m, L^* v_m \rangle.$$
Using (4.4.37) again, we obtain
$$\langle \delta_m, L^* v_m \rangle = \langle \delta_m, (D - L) v_{m+1} \rangle = \langle (D - L^*) \delta_m, v_{m+1} \rangle.$$
Hence,
$$\langle Av_{m+1}, v_{m+1} \rangle = \langle Av_m, v_m \rangle - \langle D \delta_m, \delta_m \rangle < \langle Av_m, v_m \rangle,$$
since D is positive definite. Since $\langle Av_0, v_0 \rangle \leq 0$, $\langle Av_m, v_m \rangle < 0$ for $m = 1, 2, \ldots$. Hence, we cannot have $\lim v_m = 0$. Therefore, $\lim Q^m \neq 0$; i.e., Q is not quasinilpotent. We have proven the following theorem.

Theorem 4.4.12 Let $A = D - L - L^*$, where D is a real diagonal matrix and L is a lower triangular matrix with zero's on the diagonal. Suppose D is positive definite. Then the Gauss-Seidel method converges if and only if A is positive definite.

As a generalization we have

Theorem 4.4.13 Let $A = A^*$. Let $A = M - N$, where M^{-1} exists and $M + M^* - A$ is positive definite. Then $Q = M^{-1}N$ is quasinilpotent if and only if A is positive definite.

Proof. The proof of the previous theorem carries over with $D - L$ replaced by M and L^* by N. ∎

Corollary Let $A = D - L - L^*$, where D is a real nonsingular diagonal matrix and L is lower triangular with zero's on the diagonal. Suppose $D + L + L^*$ is positive definite. Then the Jacobi method converges if and only if A is positive definite.

4.5 RELAXATION METHODS, ACCELERATION OF CONVERGENCE

The Gauss-Seidel procedure (4.4.28) may be restated as follows. Let
$$r_i^{(m)} = b_i - \sum_{j=1}^{i-1} a_{ij} x_j^{(m+1)} - \sum_{j=i}^{n} a_{ij} x_j^{(m)},$$
be the residual of the ith equation prior to solving (4.4.28) for $x_i^{(m+1)}$. In general, $r_i^{(m)} \neq 0$. From (4.4.28) we see that
$$x_i^{(m+1)} = x_i^{(m)} + \frac{1}{a_{ii}} r_i^{(m)}. \qquad (4.5.1)$$

Let $u^{(i)}$ be the vector having components $u_j = x_j^{(m+1)}$, $1 \le j \le i-1$ and $u_j = x_j^{(m)}$ for $i \le j \le n$. Let $u^{(i+1)} = u^{(i)} + a_{ii}^{-1} r_i^{(m)} e_i$, where e_i is the ith canonical basis vector (2.1.9). According to (4.5.1), $u^{(i+1)}$ is the vector obtained after the ith step in one "cycle" of a Gauss-Seidel iteration, a cycle consisting of n applications of (4.5.1), $1 \le i \le n$. (Strictly speaking we should write $u^{(i)} = u^{(m,i)}$, but we suppress the iteration index m for convenience.)

Now, consider the quadratic functional,

$$f(x) = \langle Ax, x \rangle - 2\langle b, x \rangle, \tag{4.5.2}$$

and assume that $A = A^*$ and A is positive definite. We have

$$f(x) = \langle A(x - A^{-1}b), (x - A^{-1}b) \rangle - \langle A^{-1}b, b \rangle,$$

which shows that f is minimized by taking $x = A^{-1}b$. Hence, solving $Ax = b$ is equivalent to minimizing $f(x)$ in the positive self-adjoint case. A simple calculation yields

$$\begin{aligned} f(u^{(i)} + se_i) &= f(u^{(i)}) + 2s\langle Au^{(i)} - b, e_i \rangle + s^2 \langle Ae_i, e_i \rangle \\ &= f(u^{(i)}) - 2sr_i^{(m)} + a_{ii}s^2. \end{aligned} \tag{4.5.3}$$

Thus, $f(u^{(i)} + se_i)$ is a quadratic function of s with $a_{ii} > 0$. To minimize it, we differentiate with respect to s and set the derivative equal to zero. This yields $s = r_i^{(m)}/a_{ii}$, so that $u^{(i+1)}$ is the vector which minimizes f along the line $u^{(i)} + se_i$. Since we are trying to obtain the absolute minimum of f, the Gauss-Seidel iteration can be viewed as a minimization procedure which minimizes f successively with respect to individual components. In fact, it is sometimes called the *method of successive displacements*. However, from (4.5.3) we see that

$$f(u^{(i)}) - f(u^{(i)} + \omega r_i^{(m)} a_{ii}^{-1} e_i) = \omega(2 - \omega)(r_i^{(m)})^2/a_{ii},$$

showing a decrease for any $s = \omega r_i^{(m)}/a_{ii}$ provided that $0 < \omega < 2$. Taking $\omega = 1$ yields the Gauss-Seidel method. It is conceivable that other choices of ω might improve the rate of convergence. The factor ω is called the *relaxation factor*. Choosing $0 < \omega < 1$ gives rise to *underrelaxation methods* and taking $1 < \omega < 2$ gives rise to *successive overrelaxation methods*.

In the general case, for any matrix A, these methods are defined by the equations

$$x_i^{(m+1)} = x_i^{(m)} + \frac{\omega}{a_{ii}} \left(b_i - \sum_{j=1}^{i-1} a_{ij} x_j^{(m+1)} - \sum_{j=i}^{n} a_{ij} x_j^{(m)} \right), \quad 1 \le i \le n.$$

Using our previous notation, we write $A = D - L - U$, and in matrix form the above becomes

$$(D - \omega L)x^{(m+1)} = ((1 - \omega)D + \omega U)x^{(m)} + \omega b. \tag{4.5.4}$$

Letting $M_\omega = (D - \omega L)$ and $N_\omega = (1 - \omega)D + \omega U$, we see that $(1/\omega)M_\omega - (1/\omega)N_\omega = A$, which is a "splitting" of A of the type given in the general iterative procedure (4.4.4). The corresponding Q-matrix is

$$Q_\omega = (D - \omega L)^{-1}((1 - \omega)D + \omega U). \tag{4.5.5}$$

4.5 Relaxation Methods, Acceleration of Convergence   147

For $\omega = 1$ we obtain the Gauss-Seidel case and the possibility arises of choosing ω to reduce the spectral radius below $r_\sigma(Q_1) = r_\sigma(Q_G)$. In fact, we would like to choose ω so as to minimize the spectral radius $r_\sigma(Q_\omega)$. We shall consider this problem for matrices having a special form first studied by Frankel [4.12] and Young [4.11]. (See also [4.8], [4.9], [4.10].)

A matrix A of order n is said to be *block tridiagonal* if it is of the form

$$A = \begin{bmatrix} D_1 & U_1 & 0 & \cdots & & & 0 \\ L_2 & D_2 & U_2 & 0 & \cdots & & 0 \\ 0 & L_3 & D_3 & U_3 & 0 & & \vdots \\ \vdots & & \ddots & \ddots & \ddots & \ddots & \\ 0 & 0 & & & & & 0 \\ \vdots & & \ddots & & L_{p-1} & D_{p-1} & U_{p-1} \\ 0 & \cdots & & & 0 & L_p & D_p \end{bmatrix} \quad (4.5.6)$$

where the D_i are square matrices. Here, $p > 2$. If each D_i is a diagonal matrix, A is said to be *diagonally block tridiagonal*. A matrix A is said to have (Young's) *property (A)* if there is a permutation matrix P such that PAP^T is diagonally block tridiagonal; i.e., some permutation of the canonical basis vectors (2.1.9) is a basis with respect to which the operator A has a matrix representation in diagonally block tridiagonal form. We shall suppose that the order of solving for the x_i in one cycle of successive overrelaxation agrees with the ordering of the permuted variables; i.e., we solve for the vector Px in the system $(PAP^T)Px = Pb$. (Recall that $P^T = P^{-1}$.) Such an ordering is said to be *consistent* with the block tridiagonal form PAP^T. (There may be several such forms obtainable by applying permutation matrices to A.) In the remainder of this section, we shall assume that A is as shown in (4.5.6) with all D_i diagonal matrices. It will be seen that there is no loss in generality in proceeding this way, since all determinants to be considered are invariant under the similarity transformation PAP^T.

The eigenvalues of Q_ω are roots of the characteristic equation $\det(\lambda I - Q_\omega) = 0$. From (4.5.5) we see that this equation is equivalent to the equation,

$$q_\omega(\lambda) = \det\left(\frac{(\lambda - 1 + \omega)}{\omega} D - \lambda L - U\right) = 0. \quad (4.5.7)$$

Now consider the eigenvalues of the Jacobi matrix $Q_J = D^{-1}(L + U)$. These are roots of the equation $p(\lambda) = \det(\lambda I - D^{-1}(L + U)) = \det(\lambda D - L - U) = 0$. From (4.5.6) we see that $B = \lambda D - L - U$ has the form

$$B = \begin{bmatrix} \lambda D_1 & U_1 & 0 & \cdots & & & 0 \\ L_2 & \lambda D_2 & U_2 & 0 & \cdots & & \vdots \\ 0 & L_3 & \lambda D_3 & U_3 & & & \\ \vdots & & \ddots & \ddots & \ddots & & 0 \\ & & & & L_{p-1} & \lambda D_{p-1} & U_{p-1} \\ 0 & & \cdots & & 0 & L_p & \lambda D_p \end{bmatrix}$$

Let T be the diagonal matrix

$$\begin{bmatrix} -I_1 & & & & \\ & I_2 & & & \\ & & -I_3 & & \\ & & & \ddots & \\ & & & & (-1)^p I_p \end{bmatrix},$$

where I_j is the identity matrix of order equal to the order of D_j, $1 \le j \le p$. Since $T = T^{-1}$, $p(\lambda) = \det B = \det(TBT)$. But

$$TBT = \begin{bmatrix} \lambda D_1 & -U_1 & & \cdots & 0 \\ -L_2 & \lambda D_2 & -U_2 & & \vdots \\ \vdots & \ddots & \ddots & \ddots & \\ & & & & -U_{p-1} \\ 0 & \cdots & & -L_p & \lambda D_p \end{bmatrix}.$$

Replacing λ by $-\lambda$ in TBT, we see that -1 can be factored out of each row. Hence, $p(\lambda) = \det(TBT) = (-1)^n p(-\lambda)$, which shows that the nonzero eigenvalues of Q_J occur in pairs $\pm\mu_i$. Now, consider (4.5.7) again. Letting

$$\eta = (\lambda - 1 + \omega)/\omega \tag{4.5.8}$$

and $C = \eta D - \lambda L - U$, it follows from (4.5.6) that

$$C = \begin{bmatrix} \eta D_1 & U_1 & & & \\ \lambda L_2 & \eta D_2 & U_2 & & \\ & \ddots & \ddots & \ddots & \\ & & \lambda L_{p-1} & \eta D_{p-1} & U_{p-1} \\ & & & \lambda L_p & \eta D_p \end{bmatrix}.$$

Let M_R be the diagonal matrix,

$$M_R = \begin{bmatrix} \lambda^{-1/2} I_1 & & & & \\ & I_2 & & & \\ & & \lambda^{1/2} I_3 & & \\ & & & \ddots & \\ & & & & \lambda^{(p-2)/2} I_p \end{bmatrix}$$

and let

$$M_L = \begin{bmatrix} I_1 & & & & \\ & \lambda^{-1/2} I_2 & & & \\ & & \lambda^{-1} I_3 & & \\ & & & \ddots & \\ & & & & \lambda^{-(p-1)/2} I_p \end{bmatrix}.$$

4.5 Relaxation Methods, Acceleration of Convergence 149

Multiplying C on the right by M_R and on the left by M_L, we obtain the matrix

$$M_L C M_R = \begin{bmatrix} \lambda^{-1/2}\eta D_1 & U_1 & & & \\ L_2 & \lambda^{-1/2}\eta D_2 & U_2 & & \\ & \ddots & \ddots & \ddots & \\ & & & \lambda^{-1/2}\eta D_{p-1} & U_{p-1} \\ & & & L_{p-1} & \lambda^{-1/2}\eta D_p \end{bmatrix}$$

Comparing this with B, we see that $\det(M_L C M_R) = p(\lambda^{-1/2}\eta)$. Since $\det(M_R M_L) = \lambda^{-n/2}$, we find that

$$q_\omega(\lambda) = \det C = \lambda^{n/2} \det(M_L C M_R) = \lambda^{n/2} p(\lambda^{-1/2}\eta). \tag{4.5.9}$$

This establishes the following theorem.

Theorem 4.5.1 Let $A = D - L - U$ be a matrix in the diagonally block tridiagonal form of (4.5.6). Let λ satisfy the relation

$$\frac{\lambda - 1 + \omega}{\omega} = \pm \lambda^{1/2} \mu \tag{4.5.10}$$

where $\omega \neq 0$. If μ is an eigenvalue of the Jacobi matrix $Q_J = D^{-1}(L + U)$, then λ is an eigenvalue of the successive overrelaxation matrix Q_ω of (4.5.5), and conversely.

Proof. μ is an eigenvalue of Q_J if and only if $p(\mu) = 0$. λ is an eigenvalue of Q_ω if and only if $q_\omega(\lambda) = 0$. The result then follows from (4.5.8), (4.5.9). ∎

The following result [4.8] shows that we may restrict our attention to $0 < \omega < 2$.

Lemma 4.5.1 Let Q_ω be given by (4.5.5). For all real ω,

$$r_\sigma(Q_\omega) \geq |\omega - 1|.$$

Hence, if $r_\sigma(Q_\omega) < 1$, then $0 < \omega < 2$.

Proof. The product of the eigenvalues λ_i of Q_ω is given by dividing the constant term of $q_\omega(\lambda)$ in (4.5.7) by the leading coefficient. The constant term is

$$q_\omega(0) = \det\left(\left(\frac{\omega-1}{\omega}\right)D - U\right) = \left(\frac{\omega-1}{\omega}\right)^n \prod_{i=1}^n a_{ii},$$

since U is upper triangular with zero's on the diagonal. The leading coefficient is $(\prod_{i=1}^n a_{ii})/\omega^n$. Hence, $\prod_{i=1}^n (-\lambda_i)^n = (\omega - 1)^n$ and it follows that $|\omega - 1| \leq \max_i |\lambda_i| = r_\sigma(Q_\omega)$. ∎

We shall now determine the value of ω which minimizes $r_\sigma(Q_\omega)$ when all eigenvalues of Q_J are real and $r_\sigma(Q_J) < 1$. This is an important special case. For example, if A is self-adjoint, we have seen that $Q_J = D^{-1}(L + L^*)$ is also self-adjoint, and therefore has real eigenvalues. The condition $r_\sigma(Q_J) < 1$ assures us that $r_\sigma(Q_1) = r_\sigma(Q_G) < 1$ also (Theorem 4.4.9). (We would certainly wish to require that convergence obtains before attempting to accelerate it.)

Theorem 4.5.2 Let A be as in Theorem 4.5.1. Suppose that all eigenvalues of the Jacobi matrix Q_J are real and $r_\sigma(Q_J) < 1$. Then $r_\sigma(Q_\omega)$ is minimized by taking the overrelaxation factor $\omega = \omega_0$, where

$$\omega_0 = \frac{2}{1 + \sqrt{1 - (r_\sigma(Q_J))^2}}. \tag{4.5.11}$$

The resulting successive overrelaxation procedure is convergent and $r_\sigma(Q_{\omega_0}) = \omega_0 - 1$.

Proof. Consider (4.5.10) as a quadratic in $\lambda^{1/2}$. (We take the positive root. If the negative root is taken the same value of λ is obtained.) Rewriting (4.5.10), we have

$$\lambda - \omega\mu\lambda^{1/2} + (\omega - 1) = 0. \tag{4.5.12}$$

Solving for $\lambda^{1/2}$, we obtain

$$\lambda^{1/2} = \frac{\omega\mu \pm \sqrt{\omega^2\mu^2 - 4(\omega - 1)}}{2}. \tag{4.5.13}$$

If $\mu = 0$, then $\lambda = -(\omega - 1)$. If $\mu_i \neq 0$ is an eigenvalue of Q_J, then so is $-\mu_i$, since the nonzero eigenvalues of Q_J were shown to occur in pairs. Therefore, for the purposes of this analysis, we may take $0 < \mu < 1$ as the condition satisfied by a nonzero eigenvalue μ of Q_J. ($-\mu$ would yield the same values of λ determined from (4.5.13).) For $0 < \mu < 1$ it is clear that the eigenvalue of larger modulus corresponds to the plus sign in (4.5.13). Thus, we consider

$$\lambda^{1/2} = \frac{\omega\mu + \sqrt{\omega^2\mu^2 - 4(\omega - 1)}}{2}. \tag{4.5.14}$$

Let $\lambda(\omega)$ be the eigenvalue for a given μ given by this function of ω. We see from (4.5.14) that for $0 < \omega < 1$, $|\lambda(\omega)| > |\lambda(1)| = \mu^2$; i.e., underrelaxation increases the spectral radius. Differentiating (4.5.14) yields

$$\left.\frac{d\lambda^{1/2}}{d\omega}\right|_{\omega=1} = \frac{1+\mu}{2} - \frac{1}{\mu} < 0$$

for $0 < \mu < 1$.) Hence, we must consider overrelaxation. For $1 \leq \omega \leq 2$, the quantity $\omega^2\mu^2 - 4(\omega - 1)$ decreases from its value μ^2 at $\omega = 1$ as ω increases. (To see this, take the derivative with respect to ω and note that it is negative.) This causes the modulus $|\lambda(\omega)|$ to decrease, as can be seen from (4.5.14), until ω reaches a value such that

$$\mu^2\omega^2 - 4\omega + 4 = 0, \tag{4.5.15}$$

after which $|\lambda(\omega)|$ increases. Therefore, for a given eigenvalue $0 < \mu_i < 1$ of Q_J, the corresponding eigenvalue $\lambda_i(\omega)$ of Q_ω has minimum modulus when $\omega = \omega_i$, where

$$\omega_i = \frac{2(1 - \sqrt{1 - \mu_i^2})}{\mu_i^2} = \frac{2}{1 + \sqrt{1 - \mu_i^2}}. \tag{4.5.16}$$

For this value of ω, (4.5.12) has a double root, $\lambda_i^{1/2} = \omega_i\mu/2$. From (4.5.15) we have $(\omega_i\mu)^2 = 4(\omega_i - 1)$, so that $\lambda_i = \omega_i - 1$. Now, as ω increases beyond ω_i, the roots $\lambda_i^{1/2}$ are complex conjugates. Since the product of the roots of (4.5.12) is $\omega - 1$, we see that $|\lambda_i| = \omega - 1$ for $\omega > \omega_i$. The largest value of ω_i is obtained by setting $\mu_i = \mu_0 = r_\sigma(Q_J)$ in (4.5.16). This yields the value ω_0 in (4.5.11). For $\omega = \omega_0$ the root λ_0 corresponding to μ_0 has its smallest modulus $|\lambda_0| = \omega_0 - 1$. All other $|\lambda_i| = \omega_0 - 1$. For $\omega_i < \omega < \omega_0$, $|\lambda_0| > \omega_0 - 1$ and $|\lambda_i| = \omega - 1$. Hence, the choice $\omega = \omega_0$ minimizes the maximum modulus of all λ_i; i.e., it minimizes $r_\sigma(Q_\omega)$, which then has the minimum value $\omega_0 - 1$. ∎

Successive overrelaxation may be viewed as a method of accelerating the rate of convergence of the Gauss–Seidel procedure. In Theorem 4.5.2, it is shown that the spectral radius $r_\sigma(Q_\omega)$ can be made less than $r_\sigma(Q_G) = r_\sigma(Q_1)$ by choosing $\omega = \omega_0 > 1$. As defined by (4.5.4),

4.5 Relaxation Methods, Acceleration of Convergence

overrelaxation is derived from a splitting, $A = (1/\omega)M_\omega - (1/\omega)N_\omega$, of A into two one-parameter families of matrices. If we let $\alpha = 1/\omega$, we see that the parameter α enters linearly. This suggests other possible splittings involving parameters with the object of choosing the parameter to accelerate convergence by minimizing the spectral radius.

For example, we can define $M_\alpha = (1 + \alpha)M$ and $N_\alpha = M_\alpha - A = N + \alpha M$, where $A = M - N$ is a given splitting giving rise to an iterative method of the type (4.4.4); e.g., $M = D$ in the Jacobi method and $M = D - L$ in the Gauss–Seidel method. The method

$$M_\alpha x^{(m+1)} = N_\alpha x^{(m)} + b \tag{4.5.17}$$

will converge at an asymptotic rate depending on $r_\sigma(Q_\alpha)$, where $Q_\alpha = M_\alpha^{-1} N_\alpha$. Thus, letting $M = M_0$, $N = N_0$ and $Q = Q_0$, we have

$$Q_\alpha = \frac{\alpha}{1+\alpha} I + \frac{1}{1+\alpha} Q_0. \tag{4.5.18}$$

If we set $s = 1/(1 + \alpha)$, (4.5.18) becomes

$$Q_s = (1 - s)I + sQ_0.$$

By Theorem 3.6.7, μ_i is an eigenvalue of Q_0 if and only if

$$\lambda_i = \lambda_i(s) = (1 - s) + s\mu_i = (\mu_i - 1)s + 1 \tag{4.5.19}$$

is an eigenvalue of Q_s. If we assume that all eigenvalues μ of Q are real and $r_\sigma(Q) < 1$, then $\mu_i - 1 < 0$ and (4.5.19) defines λ_i as a linear function of s, having negative slope and intercept 1. If μ_1 and μ_n are the largest and smallest eigenvalues respectively, then for any given $s > 0$,

$$(\mu_n - 1)s + 1 \le \lambda_i(s) \le (\mu_1 - 1)s + 1.$$

(For $s < 0$, $|\lambda_i(s)| > 1$.) Hence, to minimize $\max_{1 \le i \le n} \{|\lambda_i(s)|\}$, we must choose s so that

$$|(\mu_n - 1)s + 1| = |(\mu_1 - 1)s + 1|,$$

or

$$(1 - \mu_n)s - 1 = (\mu_1 - 1)s + 1,$$

which yields

$$s_0 = \frac{2}{2 - \mu_1 - \mu_n}.$$

The corresponding spectral radius is

$$r_\sigma(Q_{s_0}) = \frac{\mu_1 - \mu_n}{2 - \mu_1 - \mu_n}.$$

If all eigenvalues of Q are positive, it is easy to see that $r_\sigma(Q_{s_0}) < r_\sigma(Q) = \mu_1$. We can rewrite (4.5.17) as

$$x^{(m+1)} = Q_\alpha x^{(m)} + M_\alpha^{-1} b,$$

$$= \frac{\alpha}{1+\alpha} x^{(m)} + \frac{1}{1+\alpha} (Q_0 x^{(m)} + M_0^{-1} b).$$

Letting

$$y^{(m+1)} = Q_0 x^{(m)} + M_0^{-1} b, \tag{4.5.20}$$

we have

$$x^{(m+1)} = (1 - s)x^{(m)} + sy^{(m+1)}. \tag{4.5.21}$$

Therefore, (4.5.17) can be interpreted as a modification of the iterative method given by the splitting $A = M - N$ in which the vector $y^{(m+1)}$ given by the basic iteration is replaced by a vector on the line segment joining the previous iteration $x^{(m)}$ to $y^{(m+1)}$. Letting $e^{(m+1)} = x^{(m+1)} - x$ and $\hat{e}^{(m+1)} = y^{(m+1)} - x$, we obtain by subtracting x from both sides of (4.5.21),

$$e^{(m+1)} = (1-s)e^{(m)} + s\hat{e}^{(m+1)}.$$

But, since $x = Q_0 x + M_0^{-1} b$, (4.5.20) yields

$$\hat{e}^{(m+1)} = Q_0 e^{(m)}.$$

Hence,

$$e^{(m+1)} = ((1-s)I + sQ_0)e^{(m)}. \tag{4.5.22}$$

This suggests a generalization in which we perform k iterations of the type

$$y^{(j+1)} = Qy^{(j)} + M^{-1}b, \qquad 0 \le j \le k, \tag{4.5.23}$$

starting with some initial value $y^{(0)} = x^{(0)}$ and then compute

$$x^{(1)} = \sum_{j=0}^{k+1} s_j y^{(j)}.$$

This procedure is then repeated, using $x^{(1)}$ as a new starting value $y^{(0)}$. Thus, we have

$$x^{(m+1)} = \sum_{j=0}^{k+1} s_j y^{(j)}, \qquad y^{(0)} = x^{(m)}. \tag{4.5.24}$$

If $y^{(0)} = x$, we would obtain $y^{(j)} = x$ for all j and would wish $x^{(m)}$ to equal x. Therefore, we require that $\sum_{j=0}^{k+1} s_j = 1$. (In the linear case, we had $s_0 = 1 - s$, $s_1 = s$.) As before, we obtain for $\hat{e}^{(j)} = y^{(j)} - x$,

$$\hat{e}^{(j+1)} = Q\hat{e}^{(j)} = Q^{j+1}\hat{e}^{(0)}.$$

From (4.5.24) we get for $e^{(m+1)} = x^{(m+1)} - x$,

$$e^{(m+1)} = \sum_{j=0}^{k+1} s_j \hat{e}^{(j)} = \left(\sum_{j=0}^{k+1} s_j Q^j\right)\hat{e}^{(0)}, \tag{4.5.25}$$

which generalizes (4.5.22) (with $k = 0$). Now, let

$$p(t) = \sum_{j=0}^{k+1} s_j t^j.$$

Then (4.5.25) can be rewritten as

$$e^{(m+1)} = p(Q)\hat{e}^{(0)} = p(Q)e^{(m)}.$$

The method will converge if and only if $p(Q) = \sum_{j=0}^{k+1} s_j Q^j$ is quasinilpotent and the asymptotic rate of convergence is maximized by minimizing $r_\sigma(p(Q))$. By Theorem 3.6.7, the spectrum of $p(Q)$ is $\{p(\mu_i) : \mu_i \in \sigma(Q)\}$. We shall assume that $r_\sigma(Q) < 1$, since we are trying to accelerate the convergence of (4.5.23). We would like to minimize $\max_i \{|p(\mu_i)|\}$ by choosing k and the coefficients s_j appropriately. If we take p to be the characteristic polynomial of Q, then $p(\mu_i) = 0$. However, this requires that we determine all the eigenvalues of Q, since then $p(t) = \prod_{i=1}^n (t - \mu_i)$. If n is large, this is a large and difficult computational problem, as we shall see in Chapter 6. Therefore, rather than try to minimize $\max_i \{|p(\mu_i)|\}$, we seek to determine p such that

$$\max_{-1 < a \le t \le b < 1} \{|p(t)|\} = \|p\|_\infty$$

is a minimum for all polynomials of a specified maximum degree k, where a and b are lower and upper bounds respectively for the eigenvalues μ_i. We have already seen how such bounds may be obtained; e.g., in the general case we might try $\|Q\|_1$ or $\|Q\|_\infty$, or use Lemma 4.4.4 in special cases.

The problem of finding a polynomial p of degree $\leq k$ which has the smallest ∞-norm on a given closed interval $[a, b]$ is treated in Chapter 7. The solution for the interval $[-1, 1]$ is given by the Chebyshev polynomials (Definition 7.6.1). For further details and other accelaration methods see [4.8].

4.6 SOLUTION OF SINGULAR LINEAR SYSTEMS

In Section 4.1, we referred briefly to the case when A is an $m \times n$ rectangular matrix. The techniques of the previous sections are not easily adapted to this case or to the case of a square singular matrix. To solve $Ax = b$ in these cases it is customary to seek a *least squares* solution. This is discussed fully in Section 7.4(a) as a *closest approximation* problem. For the matrix case at hand, we remark briefly that a least squares solution is one which minimizes $\|Ax - b\|_2$. A necessary and sufficient condition for x^* to be a least squares solution is that it satisfy the normal equations, $A^T A x = A^T b$. (See Section 7.4.*a*.) If $A^T A$ is nonsingular and well-conditioned, the normal equations can be solved by one of the methods of the preceding sections. If $A^T A$ is singular, a solution can be obtained by using the pseudoinverse, $A^\dagger$, of A. The pseudoinverse is defined and its properties developed in Exercise 2.13. An alternative definition is given in Exercise 7.31, where it is shown that $A^\dagger b$ is the least squares solution having minimum l_2-norm. (See also Exercise 4.21) In this section, we discuss one method of computing $A^\dagger$ given in [2.9], [4.17]. For another method see [2.8].

Let A be an $m \times n$ matrix of rank r. Referring to Exercise 2.13(j), we see that $A = BC$, where B is $m \times r$ and C is $r \times n$, both having rank r. The pseudoinverse of B is given by

$$B^\dagger = (B^T B)^{-1} B^T \tag{4.6.1}$$

and

$$A^\dagger = C^T (CC^T)^{-1} B^\dagger. \tag{4.6.2}$$

Also, since $B^\dagger B = I$, we see that

$$C = B^\dagger A. \tag{4.6.3}$$

The matrix B is any matrix consisting of r linearly independent columns of A. To compute B and $(B^T B)^{-1}$ the following recursive procedure can be used.

Denote the columns of A by A_i, $1 \leq i \leq n$. Choose A_1. Now, suppose k linearly independent columns have been selected and we have the matrix B_k consisting of these columns. Also suppose the inverse matrix $(B_k^T B_k)^{-1}$ has been computed. Select a column, say A_j, from the remaining columns of A. Using (4.6.1) applied to B_k, we compute

$$v_k = A_j - B_k B_k^\dagger A_j. \tag{4.6.4}$$

By Exercise 2.13k, we know that $B_k B_k^\dagger$ is the projection operator which maps a vector onto its projection in $R(B_k)$, the range of B_k. Hence, $v_k = 0$ if and only if $A_j \in R(B_k)$. If $\|v_k\|^2 > \varepsilon$, where ε is a parameter depending on the precision, then A_j is accepted as the next linearly independent column to be adjoined to B_k, thereby forming B_{k+1}. Using the bordering method (Exercise 4.20), we can compute

$$(B_{k+1}^T B_{k+1})^{-1} = \begin{bmatrix} B & w \\ z^T & c \end{bmatrix},$$

where $B = (B_k^T B_k)^{-1} + c u_k u_k^T$, $w = -c u_k$, $z^T = -c u_k^T$, $1/c = \|v_k\|^2$ and $u_k = B_k^\dagger A_j$. The procedure is continued with B_{k+1}. If $\|v_k\|^2 < \varepsilon$, the column A_j is rejected and one of the remaining columns is selected for testing. In this way, r columns are selected to form matrix B. C is computed by using (4.6.3) and $A^\dagger$ by (4.6.2). Finally, the solution is computed as $A^\dagger b$.

The practical difficulties in the preceding method center on the choice of a value for the parameter ε. For well-conditioned matrices A, a value of 10^{-4} is suggested [4.18]. For ill-conditioned matrices a value of .05 has been found to be effective. However, the normal equations are frequently ill-conditioned (see Exercise 7.19). Therefore, alternative techniques may be required. See, for example, [4.19], [4.20] and [4.21]. In most cases, special attention and care must be exercised in carrying out the computations.

REFERENCES

4.1. J. von Neumann and H. H. Goldstine, "Numerical inverting of matrices of high order," *Bull. Amer. Math. Soc.* **53**, 1021–1099 (Nov. 1947)

4.2. J. von Neumann and H. H. Goldstine, "Numerical inverting of matrices of high order, Part II," *Proc. Amer. Math. Soc.* **2**, 188–202 (1951)

4.3. J. H. Wilkinson, "Error analysis of direct methods of matrix inversion," *J.A.C.M.* **8**, 281–330 (1961)

4.4. J. H. Wilkinson, *Rounding Errors in Algebraic Processes*, Prentice-Hall, Englewood Cliffs, N.J. (1963)

4.5. J. H. Wilkinson, *The Algebraic Eigenvalue Problem*, Oxford University Press, London (1965)

4.6. J. Todd, "The condition of the finite segments of the Hilbert matrix," in *Contributions to the Solution of Systems of Linear Equations and the Determination of Eigenvalues*, National Bureau of Standards, Applied Mathematics Series, Vol. 39, 109–116 (1954)

4.7. A. S. Householder, *The Theory of Matrices in Numerical Analysis*, Blaisdell, New York (1964)

4.8. R. S. Varga, *Matrix Iterative Analysis*, Prentice-Hall, Englewood Cliffs, N.J. (1962)

4.9. B. Friedman, *The Iterative Solution of Elliptic Difference Equations*, Report NYO-7698, Institute of Mathematical Sciences, New York University (1957)

4.10. A. S. Householder, "The approximate solution of matrix problems," *J.A.C.M.* **5**, 205–243 (1958)

4.11. D. M. Young, "Iterative methods for solving partial difference equations of elliptic type," *Trans. Amer. Math. Soc.* **76**, 92–111 (1954)

4.12. S. P. Frankel, "Convergence rates of iterative treatments of partial differential equations," *Mathematics of Computation* (formerly *Math Tables*) **4**, 65–75 (1950)

4.13. F. L. BAUER, "Optimally scaled matrices," *Numerische Mathematik* **5**, 73–87 (1963)
4.14. E. E. OSBORNE, "On pre-conditioning of matrices," *J.A.C.M.* **7**, 338–345 (1960)
4.15. G. E. FORSYTHE, "Today's computational methods of linear algebra," *SIAM Review* **9**, 489–515 (July 1967)
4.16. G. E. FORSYTHE and C. B. MOLER, *Computer Solution of Linear Algebraic Systems*, Prentice-Hall, Englewood Cliffs, N.J. (1967)
4.17. J. B. ROSEN, "Minimum and basic solutions to singular linear systems," *J. SIAM* **12**, No. 1 (March 1964)
4.18. V. PEREYRA and J. B. ROSEN, *Computation of the Pseudoinverse of a Matrix of Unknown Rank*, Technical Report CS 13, Computer Science Division, Stanford University (September 1964)
4.19. G. GOLUB, *Matrix Decompositions and Statistical Calculations*, Technical Report CS 124, Computer Science Department, Stanford University (March 1969)
4.20. A. BJORCK, "Iterative refinement of linear least squares solutions II," *BIT* **8**, 8–30 (1968)
4.21. G. H. GOLUB and W. KAHAN, Calculating the singular values and pseudoinverse of a matrix," *J. SIAM Numer. Anal. Ser. B* **2**, 205–224 (1965)
4.22. D. KNUTH, *Semi-Numerical Algorithms*, Addison-Wesley, Reading, Mass. (1969)
4.23. S. WINOGRAD, "On the number of multiplications required to compute certain functions," *Proc. Nat. Acad. Science* **58**, 1840–1842 (1967)
4.24. S. WINOGRAD, "On the number of multiplications necessary to compute certain functions," *Comm. Pure Appl. Math.* **23**, 165–179 (1970)
4.25. V. STRASSEN, "Gaussian elimination is not optimal," *Numer. Math.* **13**, 4, 354–356 (August 1969)

EXERCISES

4.1 Let A be an $m \times n$ matrix. Show that the elimination procedure with maximal pivoting (using column and row interchanges) reduces the matrix to *upper triangular* form U (all elements $u_{ij} = 0$ for $i > j$). Prove that the rank of A is the number of nonzero "diagonal" elements u_{ii}, $i = 1, \ldots, \min(m, n)$.

4.2 Given that A is a complex $n \times n$ matrix and b is a complex n-dimensional vector, show that $Ax = b$ is equivalent to $(C + iD)(y + iz) = c + id$, where C, D, y, z, c, d are real. Hence, $Ax = b$ is equivalent to the real system of order $2n$,

$$\begin{pmatrix} C & -D \\ D & C \end{pmatrix} \begin{pmatrix} y \\ z \end{pmatrix} = \begin{pmatrix} c \\ d \end{pmatrix}. \tag{1}$$

Show that (1) is also equivalent to the two nth order systems, having the same matrix,

$$(D^{-1}C + C^{-1}D)y = D^{-1}c + C^{-1}d,$$
$$(D^{-1}C + C^{-1}D)z = D^{-1}d - C^{-1}c.$$

[*Hint*: Use "block" elimination on system (1).]

4.4 Let $A = (a_{ij})$ be a symmetric positive definite matrix of order n. (See Section 3.6.) Prove the following:

a) $a_{ii} > 0$, $1 \leq i \leq n$;

b) $\max_{i,j} \{|a_{ij}|\} = \max_{i} \{|a_{ii}|\}$.

[*Hint*: Let $|a_{pq}| = \max\{|a_{ij}|\}$. Let $x = (x_1, \ldots, x_n)$ be such that $x_i = 0$ for $i \neq p, q$. Then $\langle Ax, x \rangle = a_{pp}x_p^2 + 2a_{pq}x_px_q + a_{qq}x_q^2$. If $\max\{|a_{ii}|\} < \max\{|a_{ij}|\}$, then $|a_{pp}a_{qq}| < |a_{pq}|^2$. This implies $\langle Ax, x \rangle = 0$ for some $x \neq 0$.]

156 Numerical Solution of Linear Algebraic Systems

c) The submatrices $(a_{ij}^{(k)})$, $k \leq i, j \leq n$, in Gaussian elimination are all symmetric positive definite. [*Hint:* Induction on k. Use Eq. (4.2.3) to prove symmetry. Then note that

$$\sum_{i,j=k+1}^{n} a_{ij}^{(k+1)} x_i x_j = \sum_{i,j=k}^{n} a_{ij}^{(k)} x_i x_j - a_{kk}^{(k)} \left(x_k + \sum_{i=k+1}^{n} \frac{a_{ik}^{(k)}}{a_{kk}^{(k)}} x_i \right)^2$$

Take

$$x_k = -\sum_{i=k+1}^{n} \left(\frac{a_{ik}^{(k)}}{a_{kk}^{(k)}} \right) x_i.$$

If $(a_{ij}^{(k+1)})$ is not positive definite, then neither is $(a_{ij}^{(k)})$.]

d) $a_{ii}^{(k)} \leq a_{ii}^{(k-1)}$ for $k \leq i \leq n$. Hence,

$$\max_{i,j} \{|a_{ij}^{(k)}|\} \leq \max_{i,j} \{|a_{ij}|\}.$$

4.3 (Crout's variant of Gaussian elimination.)
Suppose that $A = (a_{ij})$ is an $n \times n$ nonsingular matrix such that no column or row interchanges are required in the Gaussian elimination procedure. Since only the elements $a_{ij}^{(k)}$ for which $j \geq i$ and $i \leq k$ are required for the back-substitution procedure, we need not compute the others explicitly. Thus, we seek the factorization $LU = A$, with $l_{ii} = 1$ and $l_{ij} = m_{ij}$. Show that u_{ij} are given by the equations

$$u_{ij} = a_{ij} - \sum_{k=1}^{i-1} m_{ik} u_{kj}, \qquad i \leq j$$

$$m_{ik} = \frac{1}{u_{kk}} \left(a_{ik} - \sum_{q=1}^{k-1} m_{iq} u_{qk} \right), \qquad i > k.$$

4.4 Define $u_{1j} = a_{1j}$, $1 \leq j \leq n$, and $m_{i1} = a_{i1}/a_{11}$, $2 \leq i \leq n$. Then show that the u_{ij} and m_{ik} are determined recursively by computing the ith row of U and then the ith column of L, $i = 2, \ldots, n$.

Show that $L^{-1}b$ is obtained by considering b to be the $(n+1)$-st column of A and defining $u_{1,n+1} = b_1$. Then $L^{-1}b$ is computed as $u_{i,n+1}$, $i = 1, 2, \ldots, n$. (Note that double-precision accumulation can be used to form the sums.)

4.5 a) Consider the matrix

$$A = \begin{pmatrix} \lambda & 0 \\ 2 & \lambda \end{pmatrix}.$$

Verify that $r_\sigma(A) = |\lambda|$ and therefore that $A^m \to 0$ if and only if $|\lambda| < 1$. Show that $\|A\|_\infty$, $\|A\|_1$ and $\|A\|_s$ are all greater than 1. Verify that

$$A^m = \begin{pmatrix} \lambda^m & 0 \\ 2m\lambda^{m-1} & \lambda^m \end{pmatrix}.$$

b) Prove that if $r_\sigma(T) \geq 1$ for any operator T in $L_c(X; X)$, where X is a finite-dimensional normed vector space, then the series $\sum_{j=0}^{\infty} T^j$ cannot converge (in the operator norm topology) to an element of $L_c(X; X)$. [*Hint:* Let λ be an eigenvalue such that $|\lambda| \geq 1$ and u an eigenvector belonging to λ. Then

$$v_m = \left(\sum_{j=0}^{m} \cdot T^j \right) u = \left(\sum_{0}^{m} \lambda^j \right) u = \alpha_m u$$

where

$$\alpha_m = (1 - \lambda^{m+1})/(1 - \lambda) \quad \text{if } \lambda \neq 1$$

and

$$\alpha_m = (m + 1) \quad \text{if } \lambda = 1.$$

If $|\lambda| > 1$, $\lambda^{m+1} \to \infty$. If $\lambda = e^{i\theta}$, then $\lambda^m = e^{im\theta}$ cycles around the unit circle. Hence, v_m cannot converge.]

c) Prove that if $r_\sigma(A) = 1$ for any nondiagonalizable $n \times n$ matrix A, then there exists a vector x such that $\lim_{m \to \infty} \|A^m x\| = \infty$ or 1. [*Hint:* Use Exercise 3.23, where $|\lambda| = 1$ and choose $x = (0, 1, 0, \ldots, 0)$.]

d) Prove that if A is a diagonalizable $n \times n$ matrix with $r_\sigma(A) \leq 1$, then

$$\limsup_{m \to \infty} \|A^m x\| < \infty$$

for all x in V^n.

4.6 Show that for any 2×2 matrix, both the Gauss-Seidel and Jacobi methods have the same convergence behavior, i.e., either both converge or both diverge.

4.7 Matrices of the following form arise in the numerical solution of differential equations.

$$A = \begin{bmatrix} 2 & -1 & 0 & 0 & 0 \\ -1 & 2 & -1 & 0 & 0 \\ 0 & -1 & 2 & -1 & 0 \\ 0 & 0 & -1 & 2 & -1 \\ 0 & 0 & 0 & -1 & 2 \end{bmatrix}.$$

a) Compute the matrices Q for the Jacobi and Gauss-Seidel methods.

b) Prove that the Jacobi Q matrix is nonnegative, irreducible and quasinilpotent.

c) Prove that the Gauss-Seidel Q is nonnegative, reducible and quasinilpotent.

d) Prove that all eigenvalues of A are positive. [*Hint:* Theorem 4.4.6, Corollary 1 and Exercise 4.8(c). Or use Lemma 3.6.1.]

4.8 a) Prove: A matrix A of order $n \geq 2$ is reducible if and only if the operator A leaves invariant some subspace spanned by a subset of the canonical basis $\{e_j\}$ (see 2.1.9). [*Hint:* The matrix of the operator A relative to the basis $\{e_j\}$ is A itself. Suppose A leaves invariant the subspace V^r spanned by $\{e_{k_1}, \ldots, e_{k_r}\}$; i.e., $AV^r \subset V^r$. Let $e_{k_{r+1}}, \ldots, e_{k_n}$ be the remaining canonical vectors. Let $v_j = e_{k_j}$. Then $v_j = P^T e_j$, where P^T is the permutation matrix which permutes columns $(1, \ldots, n)$ into columns $\{k_1, \ldots, k_n\}$. Apply Theorem 2.1.7 to show that $B = PAP^T$ is the matrix of the operator A relative to the $\{v_j\}$ basis. Since $Bv_j = \sum_{i=1}^n b_{ij} v_i$, $b_{ij} = 0$ for $1 \leq j \leq r$ and $i > r$.]

b) Let A be irreducible. Show that for any pair $1 \leq i, j \leq n$, there exists a "connecting" sequence of nonzero entries of the form $(a_{ir}, a_{rs}, \ldots, a_{tj})$; i.e., the row index of any member of the sequence equals the column index of the preceding member.
[*Hint:* Prove the contrapositive. Choose j such that for some i there is no connecting sequence from i to j. Then column j has at least one zero entry, a_{ij}. Consider all nonzero entries in column j. Let these be in rows $j_1, \ldots, j_q$. Consider columns $j_1, \ldots, j_q$ and all rows which have nonzero entries. (In each column j_k the element $a_{ij_k} = 0$.) Continue in this manner until no new column indices appear. Not all column indices appear, since i cannot be connected to j. We have generated all connecting sequences ending

with column index j.) Let these columns be $A_{i_1}, \ldots, A_{i_r}$. Then each column must have zero's in all rows except (possibly) $i_1, \ldots, i_r$. Hence, the canonical vectors $\{e_{i_1}, \ldots, e_{i_r}\}$ span an invariant subspace. Prove that the converse is also true.]

c) Verify that the matrix of Exercise 4.7 is irreducible. [*Hint:* For $i < j$, use $(a_{i,i+1}, a_{i+1,i+2}, \ldots, a_{j-1,j})$ and for $j < i$, take $(a_{i,i-1}, \ldots, a_{j+1,j})$.]

4.9 Let the Jacobi matrix $Q_J = L' + U'$ be nonnegative and irreducible. Prove that the Gauss-Seidel matrix $Q_G = (I - L')^{-1}U'$ has a positive spectral radius. (Here, L' has zero entries on and above the diagonal while U' has zero entries on and below the diagonal.) [*Hint:* See Theorem 4.4.10.]

4.10 Solve $Ax = b$, where $b^* = (1, 10, 7, -3, 2)$ and

$$A = \begin{pmatrix} 1 & -1 & 2 & -1 & 1 \\ 1 & -4 & 4 & -1 & 1 \\ 3 & -4 & 6 & -3 & 2 \\ 2 & 2 & -4 & 1 & 1 \\ 2 & -1 & 1 & 1 & 1 \end{pmatrix}$$

Use finite-precision arithmetic of different precisions. Then solve exactly using rational numbers.

4.11 Compute A^{-1}, where

$$A = \begin{pmatrix} .2641 & .1735 & .8642 \\ .9411 & -.0175 & .1463 \\ -.8641 & -.4243 & .0711 \end{pmatrix}$$

Also compute the condition number of A and show that A is not ill-conditioned. Estimate $\|A^{-1}\|$ using Theorem 3.5.2, Corollary 3.

4.12 Given

$$A = \begin{pmatrix} .05 & .07 & .06 & .05 \\ .07 & .10 & .08 & .07 \\ .06 & .08 & .10 & .09 \\ .05 & .07 & .09 & .10 \end{pmatrix},$$

$$b_1^* = (.23, .32, .33, .31),$$
$$b_2^* = (-.01, -.02, -.01, -.03)$$
$$b_3^* = (.03, .04, .03, .01),$$

solve $Ax = b_i$, $i = 1, 2, 3$. Compute A^{-1} and the condition number. Give *a priori* and *a posteriori* error estimates for the computed solutions. Verify that A is ill-conditioned.

4.13 (Wilson's matrix) Let

$$A = \begin{pmatrix} 10 & 7 & 8 & 7 \\ 7 & 5 & 6 & 5 \\ 8 & 6 & 10 & 9 \\ 7 & 5 & 9 & 10 \end{pmatrix} \quad \text{and} \quad b = \begin{pmatrix} 32 \\ 23 \\ 33 \\ 31 \end{pmatrix}.$$

a) Compute the condition number. Solve $Ax = b$. Perturb the elements of A by $+.01$ and solve again. Verify that A is ill-conditioned. Apply Gaussian elimination using single and double precision. [*Hint:* The exact solution is (1, 1, 1, 1).]

b) Compute A^{-1}. *Hint:* The exact inverse is

$$A^{-1} = \begin{bmatrix} 25 & -41 & 10 & -6 \\ -41 & 68 & -17 & 10 \\ 10 & -17 & 5 & -3 \\ -6 & 10 & -3 & 2 \end{bmatrix}$$

c) Solve $Ax = b$ by the Gauss-Seidel method, using both single and double precision. Verify that A is positive definite (Lemma 3.6.1) and use Theorem 4.4.12 to prove convergence. Explain the very slow convergence.

4.14 Show by an operation count that it is more costly to solve $Ax = b$ by computing A^{-1}, using Gaussian elimination, and then forming the product $A^{-1}b$ (as opposed to the algorithm in Section 4.2 in which the elimination is done first on A, then on b, and x is found by back substitution). Show that this is true even if we have to solve $Ax = b$ for many different b vectors. (See Exercise 4.25 also.)

4.15 Give an alternative proof of Theorem 4.4.6, Corollary 2, using Theorem 3.5.1, Corollary 2. [*Hint:* Let $D = (a_{ij} \delta_{ij})$ be the diagonal part of A. If A is strictly diagonally dominant,

$$\|D^{-1}(A - D)\|_\infty < 1.]$$

4.16 Show that a diagonal matrix D always gives rise to a "well-determined" system in the sense that small perturbations of the diagonal elements produce correspondingly small relative errors in the solution of $Dx = b$. [*Hint:* Let d_{ii} be the ith diagonal element of D and Δd_{ii} a perturbation. Then

$$(d_{ii} + \Delta d_{ii})(x_i + \Delta x_i) = b_i \quad \text{and} \quad d_{ii}x_i = b_i.$$

$$\frac{\Delta x_i}{x_i} = -\left(\frac{\Delta d_{ii}}{d_{ii}}\right)\left(\frac{1}{1 + \Delta d_{ii}/d_{ii}}\right). \ ]$$

4.17 a) Let A be a tridiagonal matrix; i.e., $a_{ij} = 0$ except for $i = j, j = i - 1$ and $j = i + 1$. Show that the solution of $Ax = b$ by Gaussian elimination without pivoting can be done using only $3(n - 1)$ operations for the elimination procedure and $3(n - 1) + 1$ operations in the back-substitution procedure. [*Hint:* Taking advantage of the zero's, note that only the subdiagonal element must be reduced to zero. This determines the multiplier $m_i = a_{i+1,j}/a_{ii}^{(i)}$. It is only necessary to compute

$$a_{i+1,i+1}^{(i+1)} = a_{i+1,i+1}^{(i)} - m_i a_{i,i+1}^{(i)}$$

in the ith stage of the elimination. Thus, only the diagonal elements are changed.]

b) With reference to part (a), show that partial pivoting results in a triangular matrix with 3 nonzero diagonals instead of 2. Hence, back substitution requires $5(n - 2) + 4$ operations.

4.18 Let A be a strictly diagonally dominant matrix. Prove that

$$\|A^{-1}\|_\infty \le \frac{1}{\min_i \left(|a_{ii}| - \sum_{\substack{j=1 \\ j \ne i}}^{n} |a_{ij}|\right)}.$$

[*Hint:* $A^{-1}y = x$ implies $y = Ax$. Let $\|x\|_\infty = |x_k|$. Then

$$\|y\|_\infty = \|Ax\|_\infty = \max_i \sum_{}^{n} |a_{ij}x_j| \geq \sum_{j=1}^{n} |a_{kj}x_j|$$

$$\geq |a_{kk}|\,\|x\|_\infty - \sum_{j \neq k} |a_{kj}|\,|x_j|$$

$$\geq \left(|a_{kk}| - \sum_{j \neq k} |a_{kj}|\right)\|x\|_\infty$$

$$\geq \min_i \left(|a_{ii}| - \sum_{j \neq i} |a_{ij}|\right)\|x\|_\infty. \;]$$

4.19 A well-known example of an ill-conditioned matrix is the *n*th order *Hilbert matrix* $H_n = (h_{ij})$, where $h_{ij} = 1/(i + j - 1)$ for $1 \leq i, j \leq n$.

a) Verify that the condition number γ_n has the following approximate values:

$$\gamma_2 = 19.3, \qquad \gamma_3 = 524.0, \qquad \gamma_6 = 1.5 \times 10^7.$$

b) Verify the effect of ill-conditioning by solving $H_n x = b$ for $n = 2, 3, 6$ by Gaussian elimination using single and double-precision values for the elements h_{ij}.

c) Show that $h_{ij} = \int_0^1 x^{i+j-2}\,dx$. Compare this with the matrix (b_{ij}) of Exercise 7.19, which arises in discrete least squares polynomial approximation.

4.20 Derive the *bordering method* of inverting a matrix A_{n+1} of order $n + 1$ by expressing A_{n+1}^{-1} in terms of A_n^{-1}, where A_n is the submatrix consisting of the first *n* rows and columns of A_{n+1}. (Assume that A_n^{-1} exists.) [*Hint:* Write A_{n+1} as a *bordered matrix*,

$$A_{n+1} = \begin{bmatrix} A_n & u \\ v^T & a \end{bmatrix}, \qquad a = a_{n+1,n+1},$$

$$v^T = (a_{n+1,1}, \ldots, a_{n+1,n}), \qquad u^T = (a_{1,n+1}, \ldots, a_{n,n+1}).$$

Write A_{n+1}^{-1} as a bordered matrix,

$$A_{n+1}^{-1} = \begin{bmatrix} B & w \\ z^T & c \end{bmatrix}.$$

$$A_{n+1}A_{n+1}^{-1} = \begin{bmatrix} A_n B + u z^T & A_n w + cu \\ v^T B + a z^T & v^T w + ac \end{bmatrix} = \begin{bmatrix} I_n & 0 \\ 0 & 1 \end{bmatrix}.$$

$$A_n B + u z^T = I, \qquad v^T B + a z^T = 0,$$
$$A_n w + cu = 0, \qquad v^T w + ac = 1.$$

It follows that

$$w = -c A_n^{-1} u, \qquad \frac{1}{c} = a - v^T A_n^{-1} u,$$

$$z^T = -c v^T A_n^{-1}, \qquad B = A_n^{-1} + c A_n^{-1} u v^T A_n^{-1}. \;]$$

4.21 Let A be an $m \times n$ matrix of rank r. In Exercise 2.13(j), it is shown that $A = BC$ and the pseudoinverse is given by formula (5). (See also Eqs. 4.6.1 and 4.6.2.) Let $x^* = A^\dagger b$. Verify that x^* is a solution of the normal equations $A^T A z = A^T b$.
[*Hint:* $(BC)A^\dagger = BB^\dagger$ and $C^T B^T (BB^\dagger) = C^T B^T$.]

4.22 a) In [4.24], the following procedure for computing the inner product of two vectors $x = (x_1, \ldots, x_n)$ and $y = (y_1, \ldots, y_n)$ is given: Compute the numbers

$$\xi = \sum_{j=1}^{[n/2]} x_{2j-1} x_{2j} \qquad \eta = \sum_{j=1}^{[n/2]} y_{2j-1} y_{2j},$$

where $[n/2]$ is the integral part of $n/2$. Then, if n is even,

$$\langle x, y \rangle = \sum_{j=1}^{[n/2]} (x_{2j-1} + y_{2j})(x_{2j} + y_{2j-1}) - \xi - \eta$$

and, if n is odd, we should add $x_n y_n$ to the right-hand side. Now, suppose Q vectors of dimension n are given and it is required to compute R inner products. Show that the total multiplications are $Q[n/2] + R[(n+1)/2] = Qn + (R-Q)[(n+1)/2]$ compared to only $R(n-1)$ in the straightforward procedure.

b) Let A be an $m \times n$ and B an $n \times p$ matrix. Computation of AB requires $R = mp$ inner products, given $Q = m + p$ vectors. Verify that: Winograd's algorithm requires $(m+p)n + (mp - m - p)[(n+1)/2]$ multiplications compared to mpn in the usual method; the number of additions is $(m+p)([n/2] - 1) + mp(n + [n/2] + 1)$ compared to $mp(n-1)$; for m, n, and p large the algorithm requires on the order of $mnp/2$ multiplications and $(3/2) mnp$ additions compared to mnp in the usual method.

c) Show that A^{-1} can be computed with the order of $n^3/2$ multiplications and $(3/2)n^3$ additions using Winograd's procedure, compared to n^3 by the method in Section 4.2. Similarly, show that $Ax = b$ can be solved using on the order of $n^3/6$ multiplications and $n^3/2$ additions.

4.23 The criterion for stopping iteration on the residuals given under Computational Aspects after Theorem 4.4.2 involves comparing successive iterates. It is suggested that if $\bar{x}^{(m+1)}$ agrees with $\bar{x}^{(m)}$ up to q digits then $\bar{x}^{(m)}$ agrees with the exact solution x up to q digits. Show that this is often justified in practice. [*Hint:* If the stopping criterion is satisfied, then

$$\bar{x}^{(m+1)} = \bar{x}^{(m)} + \delta_m, \qquad \text{where} \quad \delta_m = D\bar{x}^{(m)} \tag{1}$$

and D is a diagonal matrix with entries $d_{ii} \leq 10^{-q}$. From (1), $\bar{e}^{(m+1)} = \bar{e}^{(m)} + \delta_m$. From Section (4.4), $\bar{e}^{(m+)} = Q_m \bar{e}^{(m)} + f^{(m)}$. Hence, $(I - Q_m)\bar{e}^{(m)} = f^{(m)} - \delta_m = f^{(m)} - D\bar{x}^{(m)}$, $(I - Q_m + D)\bar{x}^{(m)} = f^{(m)} + (I - Q_m)x$ and $(I - Q_m + D)(\bar{x}^{(m)} - x) = f^{(m)} - Dx$ so that

$$\|\bar{x}^{(m)} - x\| \leq \frac{\|D\| \|x\| + \|f^{(m)}\|}{1 - \|Q_m\| - \|D\|}$$

If l_∞-norms are used, $\|D\|_\infty \leq 10^{-q}$. We assume $\|Q_m\| \leq d < 1$. Then

$$\frac{\|\bar{x}^{(m)} - x\|}{\|x\|} \leq \frac{10^{-q}}{1 - d - 10^{-q}} + \frac{\|f^{(m)}\|}{\|x\|(1 - d - 10^{-q})}.$$

If d is not too close to 1, the relative error in $\bar{x}^{(m)}$ is approximately 10^{-q}, since, by Section 4.4, the second term is of order $\gamma_A 10^{1-s}$.]

4.24 Let A be a nonsingular matrix, b a fixed vector, and $x = A^{-1}b$. Consider x to be a function of A and write $x + \Delta x = (A + \Delta A)^{-1}b$ when $\|\Delta A\| < 1/\|A^{-1}\|$.
a) Verify that $\Delta x = -A^{-1} \Delta A A^{-1} b + \varepsilon$, where $\|\varepsilon\| = O(\|\Delta A\|^2)$. (In Chapter 5, we shall see that $dx = -A^1 \Delta A A^{-1} b$ is the differential of x.)

b) Assume that $\|x\| = \|A^{-1}b\| = 1$. Let x_0 be a unit vector such that $\|Ax_0\| = 1/\|A^{-1}\| = \inf\{\|Ay\| : \|y\| = 1\}$. There exists an orthogonal matrix U such that $Ux = Ax_0/\|Ax_0\|$. Take $\Delta A = (\alpha/\|A^{-1}\|)U$. For $\|\Delta A\|$ sufficiently small, show that this perturbation of A yields a solution $x + \Delta x$ having a relation error $\|\Delta x\|/\|x\|$ approximately equal to the condition number of A times the relative error $\|\Delta A\|/\|A\|$. [*Hint:* For $\|\Delta A\|$ small, $\Delta x = dx + O(\|\Delta A\|^2)$.]

c) Apply part (b) to the diagonal matrix

$$A = \begin{bmatrix} a & 0 \\ 0 & 1 \end{bmatrix},$$

where $|a|$ is small compared to 1; i.e. perturb A by the matrix

$$\Delta A = a^2 \begin{bmatrix} 0 & 1 \\ 1 & 0 \end{bmatrix}$$

and show that the relative error $|\Delta x_2|/|x_2| = O(\gamma_A \|\Delta A\|/\|A\|)$. Thus, a large condition number (small $|a|$) does mean that the relative error in the solution can be large even when the relative error in A is small. (Note, however, that ΔA is not diagonal as in Exercise 4.16.)

4.25 **a)** In [4.25] the following algorithm for multiplying two matrices A and B of order n is given:

First, suppose that $n = m2^k$ and define algorithms α_{mk} by induction on k. Take α_{m0} to be the usual algorithm for matrix multiplication (using m^3 multiplications and $m^2(m-1)$ additions). Assume that $\alpha_{m,k-1}$ is known and define α_{mk} in terms of $\alpha_{m,k-1}$ by subdividing A and B into blocks of order $n/2$. Thus

$$A = \begin{bmatrix} A_{11} & A_{12} \\ A_{21} & A_{22} \end{bmatrix}, \quad B = \begin{bmatrix} B_{11} & B_{12} \\ B_{21} & B_{22} \end{bmatrix}, \quad AB = \begin{bmatrix} C_{11} & C_{12} \\ C_{21} & C_{22} \end{bmatrix}.$$

Compute

$$\begin{aligned}
D_1 &= (A_{11} + A_{22})*(B_{11} + B_{22}), \\
D_2 &= (A_{21} + A_{22})*B_{11}, \\
D_3 &= A_{11}*(B_{12} - B_{22}), \\
D_4 &= A_{22}*(B_{21} - B_{11}), \\
D_5 &= (A_{11} + A_{12})*B_{22}, \\
D_6 &= (A_{21} - A_{11})*(B_{11} + B_{12}), \\
D_7 &= (A_{12} - A_{22})*(B_{21} + B_{22}), \\
C_{11} &= D_1 + D_4 - D_5 + D_7, \\
C_{21} &= D_2 + D_4, \\
C_{12} &= D_3 + D_5, \\
C_{22} &= D_1 + D_3 - D_2 + D_6,
\end{aligned}$$

where * denotes multiplication, using algorithm $\alpha_{m,k-1}$, and + is the usual addition of matrices.

Prove by induction on k that α_{mk} requires $m^3 7^k$ multiplications and $(5 + m)m^2 7^k - 6(m2^k)^2$ additions.

b) In [4.25] it is proved that for arbitrary order n, the product AB can be computed in less than $4.7 n^{\log_2 7}$ arithmetic operations ($\log_2 7 < 2.81$). This is done by imbedding

A and B into matrices A' and B' of order $m2^k$, where $k = [\log n - 4]$ and $m = [n2^{-k}] + 1$. ($[\cdot]$ denotes the integral part.) Verify this by showing that $n \leq m2^k$, and that α_{mk} requires a number of operations given by

$$(5 + 2m)m^2 7^k - 6(m2^k)^2 < (5 + 2(n2^{-k} + 1))(n2^{-k} + 1)^2 7^k$$
$$< 2n^3(7/8)^k + 12.03n^2(7/4)^k \text{ (since } 16 \cdot 2^k \leq n)$$
$$= (2(8/7)^{\log n - k} + 12.03(4/7)^{\log n - k})n^{\log 7}$$
$$\leq \max_{4 \leq t \leq 5} (2(8/7)^t + 12.03(4/7)^t)n^{\log 7}.$$

[*Hint:* Use a convexity argument.]

c) Apply α_{mk} to compute A^{-1} for $n = m2^k$ by algorithm β_{mk} as follows. β_{m0} is the ordinary Gauss elimination method. Let

$$A^{-1} = \begin{bmatrix} G_{11} & G_{12} \\ G_{21} & G_{22} \end{bmatrix}$$

where the G's are of order $n/2$.

$E_1 = A_{11}^{-1}$, $\quad E_2 = A_{21}*E_1$, $\quad E_3 = E_1*A_{12}$,
$E_4 = A_{21}*E_3$, $\quad E_5 = E_4 - A_{22}$, $\quad E_6 = E_5^{-1}$,
$G_{12} = E_3*E_6$, $\quad G_{21} = E_6*E_2$, $\quad G_{11} = E_1 - E_3*C_{21}$, $\quad G_{22} = -E_6$.

Here $*$ denotes $\alpha_{m,k-1}$ and inversion denotes $\beta_{m,k-1}$. Prove by induction on k that β_{mk} requires $m2^k$ divisions, at most $(6/5)m^3 7^k - m2^k$ multiplications, and at most $(6/5)(5 + m)m^2 7^k - 7(m2^k)^2$ additions. Hence A^{-1} can be computed with less than $5.64n^{\log 7}$ arithmetic operations.

d) Analyze the nonarithmetic operations required in programming the preceding algorithms for a modern computer.

CHAPTER 5

SOLUTION OF NONLINEAR EQUATIONS

5.1 FIXED POINTS AND CONTRACTIVE MAPPINGS

Let $f: X \longrightarrow X$ be a mapping of a topological vector space X into itself. We consider the problem of determining all $x^* \in X$ such that $f(x^*) = 0$.

Writing $g(x) = x + f(x)$, we see that $g(x^*) = x^*$ if and only if $f(x^*) = 0$. Thus, a zero of $f(x)$ is a *fixed point* of $g(x)$. The simplest method of determining fixed points is the method of *successive approximation* defined by the iterative scheme

$$x_{m+1} = g(x_m), \quad m = 0, 1, 2, \ldots, \tag{5.1.1}$$

where x_0 is chosen arbitrarily. The simplest convergence result is obtained when g is a contractive mapping (Definition 1.3.4).

Theorem 5.1.1 Let $g: X \longrightarrow X$ be a contractive mapping of a complete metric space X into itself. Then g has a unique fixed point $x^* \in X$ and $x^* = \lim_{m \to \infty} x_m$, where x_m is given by (5.1.1) for any choice of x_0. For such a sequence, the rate of convergence is governed by the inequality

$$d(x_m, x^*) \leq \left(\frac{L^m}{1-L}\right) d(x_1, x_0), \tag{5.1.2}$$

where $L < 1$ is a Lipschitz constant of g. We also have

$$d(x_m, x^*) \leq L^m d(x_0, x^*). \tag{5.1.3}$$

Proof. We have $x_{m+1} = g(x_m)$ and $x_m = g(x_{m-1})$. Hence,

$$d(x_{m+1}, x_m) = d(g(x_m), g(x_{m-1})) \leq L\, d(x_m, x_{m-1}).$$

By induction, we obtain $d(x_{m+1}, x_m) \leq L^m d(x_1, x_0)$. Therefore, by the generalized triangle inequality,

$$d(x_{m+p}, x_m) \leq d(x_{m+p}, x_{m+p-1}) + \cdots + d(x_{m+1}, x_m)$$

$$\leq \sum_{j=m}^{m+p-1} L^j d(x_1, x_0) \leq \left(\frac{L^m}{1-L}\right) d(x_1, x_0). \tag{5.1.4}$$

Since $L^m \to 0$ as $m \to \infty$, (x_m) is a Cauchy sequence. Since X is complete, $\lim_{m \to \infty} x_m = x^*$ exists in X. Allowing $p \to \infty$ in the above inequality, we obtain (5.1.2). Now, $d(x_{m+1}, g(x^*)) = d(g(x_m), g(x^*)) \leq L\, d(x_m, x^*) \to 0$ as $m \to \infty$. Hence, $x^* = \lim_{m \to \infty} x_{m+1} = g(x^*)$, showing that x^* is a fixed point of g. This also shows that

$$d(x_{m+1}, x^*) \leq L\, d(x_m, x^*) \tag{5.1.5}$$

from which (5.1.3) follows easily.

To show uniqueness, suppose $g(y) = y$. Then
$$d(x^*, y) = d(g(x^*), g(y)) \leq L\, d(x^*, y),$$
whence $(1 - L)\, d(x^*, y) \leq 0$. Since $L < 1$, $d(x^*, y) = 0$ and $x^* = y$. ∎

Corollary Let $g(x) = Tx + c$, where $T \in L_c(X; X)$ and X is a Banach space. If $\|T\| < 1$, then the sequence of successive approximations (5.1.1) converges to a fixed point of g for any $x_0 \in X$.

Proof. Since
$$\|g(x) - g(y)\| = \|Tx - Ty\| \leq \|T\|\, \|x - y\|,$$
g is contractive. ∎

Consider the iterative matrix methods of Section 4.4, given by $x_m = M^{-1}Nx_{m-1} + M^{-1}b$. If we take $T = M^{-1}N = Q$ and $c = M^{-1}b$, we see that these methods are special cases of the successive approximation method. The contractive condition appears as $\|T\| < 1$, which we proved to be a sufficient condition for convergence by using the fact that this implies $r_\sigma(T) < 1$. The above corollary proof is independent of spectral considerations and uses only the contractive property of T (and its linearity). The linear case also shows that the contractive condition is not a necessary condition for convergence. Indeed, consider the matrix

$$T = \begin{pmatrix} \lambda & 1 \\ 0 & \lambda \end{pmatrix}$$

with $0 < \lambda < 1$. For $x = (\xi_1, \xi_2)$, $\|Tx\|_2^2 - \|x\|_2^2 = (\lambda^2 - 1)\xi_1^2 + 2\lambda\xi_1\xi_2 + \lambda^2\xi_2^2 > 0$ for $\xi_2 > 0$ sufficiently large. Hence, $\|T\|_2 > 1$ and $r_\sigma(T) < 1$. Therefore, by Section 4.4, the successive approximations $x_m = Tx_{m-1} + c$ converge and $Tx + c$ has a fixed point.

If g is not contractive, it need not have a fixed point. For example, let $g(x) = \pi/2 + x - \arctan x$, $x \in R$. Then

$$|g(x) - g(x')| = |g'(\xi)|\, |x - x'|, \quad \text{where } x < \xi < x'.$$

Since $|g'(\xi)| = |1 - 1/(1 + \xi^2)| \to 1$ as $\xi \to \infty$, the smallest Lipschitz constant for $g(x)$ is 1. Thus, although $|g(x) - g(x')| < |x - x'|$, g is not contractive. $g(x^*) = x^*$ implies $\arctan x^* = \pi/2$ and $x^* = \infty$. Hence, g has no fixed point in R.

Theorem 5.1.1 is a *global* result. A *local* result which also sheds light on the case $L \leq 1$ is given in the next theorem [5.15].

Theorem 5.1.2 Let X be a metric space. Let g map a subset $D \subset X$ into a compact subset $E \subset D$. If g is such that $d(g(x), g(y)) < d(x, y)$ for any two distinct points $x, y \in D$, there exists a unique fixed point x^* in D and it is the limit of the sequence of successive approximations with any initial guess $x_0 \in E$.

Proof. Since $g(E) \subset E \subset D$, $x_m = g(x_{m-1}) \in E$ for $m \geq 1$ and any $x_0 \in D$. Since E is sequentially compact (Remark 2 of Section 1.4), the sequence (x_m) has a convergent subsequence (x_{m_i}) with limit point $x^* \in E$. We shall prove that $g(x^*) = x^*$.

First, observe that for all m,
$$d(x_{m+1}, x_m) = d(g(x_m), g(x_{m-1})) < d(x_m, x_{m-1}). \tag{1}$$
Now, suppose that $g(x^*) \neq x^*$. Then
$$d(g(g(x^*)), g(x^*)) = c_0\, d(g(x^*), x^*), \quad \text{where } c_0 < 1.$$
By continuity at x^*, for any c with $c_0 < c < 1$, there exists a neighbourhood U of x^* such that for $x \in U \cap E$, we have
$$d(g(g(x)), g(x)) \leq c\, d(g(x), x).$$
Hence, for $i > N$ sufficiently large,
$$d(g(g(x_{m_i})), g(x_{m_i})) \leq c\, d(g(x_{m_i}), x_{m_i}).$$
Remembering that $x_{m_i+1} = g(x_{m_i})$, we deduce that
$$d(x_{m_i+2}, x_{m_i+1}) = d(g(x_{m_i+1}), g(x_{m_i})) \leq c\, d(x_{m_i+1}, x_{m_i})$$
If $m_{i+1} > m_i + 1$, we can apply inequality (1) repeatedly to obtain
$$d(x_{m_{i+1}+1}, x_{m_{i+1}}) < \cdots < d(x_{m_i+2}, x_{m_i+1}) \leq c\, d(x_{m_i+1}, x_{m_i}). \tag{2}$$
Since (2) holds for all $i > N$, we have for $j > i$,
$$d(x_{m_j+1}, x_{m_j}) < c^{j-i}\, d(x_{m_i+1}, x_{m_i}).$$
It follows that $d(g(x_{m_j}), x_{m_j}) \to 0$ as $j \to \infty$. But this implies $d(g(x^*), x^*) = 0$, which contradicts the assumption that $g(x^*) \neq x^*$. Hence, $g(x^*) = x^*$.

Consequently, $d(x_m, x^*) = d(g(x_{m-1}), g(x^*)) < d(x_{m-1}, x^*)$. Thus, $d(x_m, x^*)$ is a monotone decreasing sequence and since $d(x_{m_i}, x^*) \to 0$, it follows that $d(x_m, x^*) \to 0$ also, as we wished to prove. Uniqueness is proved as in Theorem 5.1.1. ∎

A mapping g such that $d(g(x), g(y)) \leq d(x, y)$ is called *nonexpansive*. If strict inequality holds, as in the preceding theorem, g is called *strictly nonexpansive*. We shall prove that under certain conditions a nonexpansive mapping g also has a fixed point. Here, we must require that the domain of g be a vector space, since the condition of convexity is required. Perhaps, the best known theorem on fixed points was proved by Brouwer (in 1912).

Brouwer Fixed-Point Theorem Let g be a mapping of a convex compact subset $E \subset R^n$ into itself; i.e. $g(E) \subset E$. If g is continuous on E, then g has a fixed point in E.

We shall not give a proof of this theorem. (See [5.14].) Known proofs are based on nonconstructive concepts, probably because of the generality of g. (It need only be continuous.) Certainly, successive approximation will not converge for all such g. However if we assume that g is nonexpansive, it is possible to give a proof based on the contractive mapping theorem, 5.1.1.

Theorem Let V be a normed vector space. Let g be a mapping of a convex compact set $E \subset V$ into itself. If g is nonexpansive on E, then g has a fixed point in E.

Proof. For $0 < s < 1$ and some fixed $u \in E$ define the mapping g_s by $g_s(x) = sg(x) + (1 - s)u$. Since E is convex and $g(x) \in E$ for all $x \in E$, we see that $g_s(E) \subset E$. Also, by nonexpansiveness.

$$\|g_s(x) - g_s(y)\| = s \|g(x) - g(y)\| \le s \|x - y\|.$$

This holds for all $x, y \in E$. Hence, g_s is contractive on E. Since E is compact, it is a complete metric space and by Theorem 5.1.1, g_s has a unique fixed point, $x_s \in E$. Further, $x_s - sg(x_s) = g_s(x_s) - sg(x_s) = (1 - s)u$. Hence, dividing by s, we have

$$\lim_{s \to 1} \left(\frac{1}{s} x_s - g(x_s)\right) = \lim_{s \to 1} \left(\frac{1 - s}{s}\right) u = 0.$$

Now, choose any sequence $s_m \to 1$ and let x_m be the fixed point of g_{s_m}. By the compactness of E (here is the nonconstructive step), there exists a convergent subsequence (x_{m_i}) with limit point x^*. Since $s_{m_i} \to 1$,

$$\lim_{i \to \infty} \frac{1}{s_{m_i}} x_{m_i} = x^* = \lim_{i \to \infty} g(x_{m_i}) = g(x^*),$$

by continuity of g. Hence, x^* is a fixed point of g. ∎

Another type of *local* fixed-point theorem is the following.

Theorem 5.1.3 Let G be a mapping of the closed ball $\bar{B} = \bar{B}(x_0, r)$ of a complete metric space X into X. Suppose
i) g is contractive on $\bar{B}$ with Lipschitz constant $L < 1$, and
ii) $d(x_0, g(x_0)) \le (1 - L)r$.
Then the sequence (x_m) of successive approximations given by (5.1.1) with x_0 as starting value is defined and converges to a point $x^* \in \bar{B}$. x^* is the unique fixed point of g in $\bar{B}$ and

$$d(x_m, x^*) \le L^m r. \tag{5.1.6}$$

Proof. By induction, we prove that $x_m \in \bar{B}$, $m = 1, 2, \ldots$. First, $d(x_1, x_0) = d(g(x_0), x_0) \le (1 - L)r$. Now, assume that $x_1, \ldots, x_q \in \bar{B}$. Then

$$d(x_{j+1}, x_j) = d(g(x_j), g(x_{j-1})) \le L\, d(x_j, x_{j-1}) \qquad \text{for } j = 1, \ldots, q.$$

Hence, $d(x_{j+1}, x_j) \le L^j d(x_1, x_0)$.
By the triangle inequality,

$$d(x_{q+1}, x_0) \le d(x_{q+1}, x_q) + \cdots + d(x_1, x_0)$$

$$\le \sum_{j=0}^{q} L^j d(x_1, x_0) \le \frac{d(x_1, x_0)}{1 - L} \le r. \tag{5.1.7}$$

Hence, $x_{q+1} \in \bar{B}$ and the induction is complete. Therefore, $g(x_m)$ is defined for all m. By (5.1.4), (5.1.2) and condition (ii) of the theorem, we deduce (5.1.6). The remaining conclusions follow directly as in the proof of Theorem 5.1.1. ∎

168   Solution of Nonlinear Equations

Example. In the one-dimensional case where $g(x)$ is a real function defined on the closed interval $[x_0 - r, x_0 + r]$, the geometric interpretation of the conditions in Theorem 5.1.3 is apparent in Fig. 5.1.

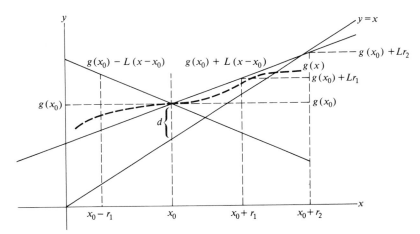

Figure 5.1

Since $d(g(x), g(x_0)) = |g(x) - g(x_0)| \le L|x - x_0|$, we see that the graph of $g(x)$ must lie in the shaded triangles. For $g(x_0) + Lr_1 > x_0 + r_1$ (i.e., $g(x_0) - x_0 >$ $(1 - L)r_1$) as shown, there need not be a fixed point in $[x_0 - r_1, x_0 + r_1]$. For $g(x_0) + Lr_2 < x_0 + r_2$ (i.e., $g(x_0) - x_0 < (1 - L)r_2$), the graph of $g(x)$ must cross the line $y = x$ at least once in $[x_0 - r_2, x_0 + r_2]$.

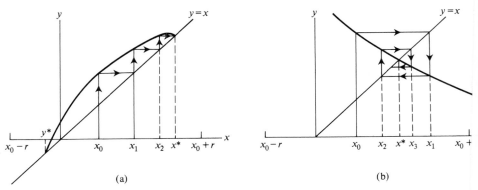

Figure 5.2

In case g is differentiable in the open interval, $B(x_0, r) = (x_0 - r, x_0 + r)$, and continuous in the closed interval, the mean-value theorem yields for $x_0 - r$ $x_1 < x_2 \le x_0 + r$,

$$g(x_1) - g(x_2) = g'(\xi)(x_1 - x_2),$$

where $x_1 < \xi < x_2$. If $|g'(\xi)| \leq L < 1$ for all $\xi \in B(x_0, r)$, then g is contractive in $\bar{B}(x_0, r)$ with Lipschitz constant L. If condition (ii) of Theorem 5.1.3 holds also, then g has a fixed point in $\bar{B}$. If $g'(\xi) > 0$ in B, the convergence of x_n to x^* is monotone as shown in Fig. 5.2(a). If $g'(\xi) < 0$, the sequence x_n oscillates in magnitude as it converges to x^* (Fig. 5.2b). Figure 5.2(a) also illustrates nonconvergence at the fixed point y^*.

In the previous theorems, the fixed point was proved to exist and shown to be the limit of the sequence of successive approximations. Suppose that the existence of a fixed point has already been established by some other method. What can be said about the successive approximations if the mapping is not contractive?

Theorem 5.1.4 Let X be a metric space and $x^* \in X$ be such that $g(x^*) = x^*$. If there exists a ball $B = B(x^*, r)$ such that for all $x \in B$,

$$d(g(x), g(x^*)) \leq L\, d(x, x^*) \tag{5.1.8}$$

with $L < 1$, then for any $x_0 \in B$, all the successive approximations x_m are in B, $\lim x_m = x^*$ and x^* is the unique fixed point of g in B.

Proof. We use induction again to show $x_m \in B$. $x_0 \in B$ by hypothesis. If $x_1, \ldots, x_q \in B$, then for $j = 1, \ldots, q$,

$$d(x_{j+1}, x^*) = d(g(x_j), g(x^*)) \leq L\, d(x_j, x^*) \leq L^{j+1} d(x_0, x^*)$$

$$\leq L^{j+1} r < r.$$

Hence, $d(x_{q+1}, x^*) \leq L^{q+1} r$. Letting $q \to \infty$, we obtain $x_q \to x^*$. Uniqueness is proved as in Theorem 5.1.3. ∎

Note that inequality (5.1.8) does not imply that g is contractive in B, since one of the two points must be x^*. In fact, g need not be continuous in B. In the one-dimensional case, (5.1.8) is satisfied if $|g'(x^*)| < 1$, since

$$|g(x) - g(x^*)| \leq |g'(x^*)|\,|x - x^*| + |\varepsilon|,$$

where $|\varepsilon|/|x - x^*| \to 0$ as $x \to x^*$. This proves

Corollary 1 Let $g(x)$ be a real-valued function of the real variable x and suppose $g(x^*) = x^*$. If $|g'(x^*)| < 1$ and g is defined in a neighbourhood of x^*, then there exists a ball $B(x^*, r)$ in which (5.1.8) holds and the successive approximations converge.

Corollary 2 Let $g(x)$ be a real function of the real variable x such that $g(x^*) = x^*$. If $g'(x^*) = 0$, there exists a ball $B = B(x^*, r)$ such that for $x_0 \in B$, the successive approximations $x_m = g(x_{m-1})$, $m = 1, 2, \ldots$, converge to x^*. Further, if $g''(x)$ exists and is bounded and $g'(x)$ is continuous in $\bar{B}$, then

$$|x_{m-1} - x^*| \leq C\,|x_m - x^*|^2.$$

Proof. By the definition of the derivative, we have

$$g(x) - g(x^*) = g'(x^*)(x - x^*) + \varepsilon(x),$$

where $\lim_{x \to x^*} \varepsilon(x)/(x - x^*) = 0$. Hence, for $|x - x^*| < r$ and r sufficiently small, $|g(x) - g(x^*)| = |\varepsilon(x)| \le L|x - x^*|$, with $L < 1$. Thus, g is contractive in B and the theorem applies to yield convergence. Now, if $g''(x)$ exists in some ball $B(x^*, r)$, Taylor's formula yields

$$g(x_m) - g(x^*) = g'(x^*)(x_m - x^*) + g''(\xi)(x_m - x^*)^2/2.$$

Since $g'(x^*) = 0$, and $|g''(\xi)| \le 2C$ we have

$$|x_{m+1} - x^*| = |g''(\xi)| |(x_m - x^*)|^2/2 \le C|x_m - x^*|^2,$$

as was to be proved. ∎

Computational Aspects (Rounding Errors, Rates of Convergence)

If an iterative method is such that $x_m \to x^*$ and

$$d(x_{m+1}, x^*) \le L[d(x_m, x^*)]^2$$

or

$$\|x_{m+1} - x^*\| \le L\|x_m - x^*\|^2,$$

as in Corollary 2, then we say that it is a *second-order* method and *converges quadratically*. The "error" at the $(m + 1)$-st iteration behaves like the square of the error at the mth iteration. Thus, if $\|x_m - x^*\| = .01$, then $\|x_{m+1} - x^*\| \le L \cdot 10^{-4}$. Quadratic convergence is quite rapid from a computational standpoint, as we shall see when we consider Newton's method in the next section.

In the general case, the sequence of successive approximation does not converge quadratically. Instead, according to (5.1.5), the error at the $(m + 1)$st iteration is less than a constant multiple, L, of the error at the mth iteration. For example, if $L = \frac{1}{2}$, then we can expect the error to be reduced by $\frac{1}{2}$, at least, every iteration. An iterative method in which the error behaves according to (5.1.5) with $L < 1$ is said to be a *first-order* method and the convergence is said to be *linear*. Since $\|x_m - x^*\| \le L^m \|x_0 - x^*\|$, the number of iterations to insure a reduction in the initial error by a factor of 10^{-q} is obtained by setting $L^m \le 10^{-q}$. Taking logarithms, we find that

$$m \ge q/\log_{10}(1/L).$$

Thus, the number of iterations required is roughly proportional to q. By comparison, quadratic convergence yields (assuming $L\|x_0 - x^*\| < 1$)

$$\|x_m - x^*\| \le L^{2^m - 1} \|x_0 - x^*\|^{2^m}.$$

To obtain a reduction by a factor of 10^{-q}, we set

$$L^{2^m - 1} \|x_0 - x^*\|^{2^m} \le 10^{-q} \|x_0 - x^*\|$$

and taking logarithms as before, we find

$$m \ge \log_2 q - \log_2 \log_{10}(1/L\|x_0 - x^*\|).$$

Thus, m is roughly proportional to $\log_2 q$. For example, if $q = 8$, then $m \ge 3$ iterations will generally suffice. (We are assuming that $\|x_0 - x^*\|$ has order of magnitude 1 and $L < 1$.) For $L = \frac{1}{2}$, a first-order method would require roughly 27 iterations. See Exercise 5.25 for a practical error criterion for stopping the iteration.

In actual computations, rounding errors are committed in evaluating $g(x_m)$. Suppose X is a Banach space. Then the computed successive approximations $\bar{x}_m$ are defined by

$$\bar{x}_0 = a + \Delta a,$$
$$\bar{x}_{m+1} = g(\bar{x}_m) + \delta_m, \quad m \geq 0. \qquad (5.1.9)$$

where δ_m is the rounding error and $a = x_0$. We cannot expect $\bar{x}_m$ to converge to the fixed point x^*. However, if g is contractive and if $\|\delta_m\| < \delta$, then the $\bar{x}_m$ will all lie in some ball with center x^* and radius determined by δ.

Theorem 5.1.5 Let $g : X \to X$ be a contractive mapping of a Banach space X into itself with fixed point x^*. Let $\bar{x}_m$ be the sequence defined by (5.1.9), where $x_0 \in X$ is arbitrary. If $\|\Delta a\| < \delta$ and $\|\delta_m\| < \delta$ for all m, then

$$\|\bar{x}_m - x^*\| \leq \delta/(1-L) + \frac{L^m}{1-L}((L+1)\delta + \|\bar{x}_1 - \bar{x}_0\|), \qquad (5.1.10)$$

where $0 < L < 1$ is a Lipschitz constant of g.

Proof. We have (using $\|\delta_m\| < \delta$),

$$\|\bar{x}_{m+1} - x_{m+1}\| \leq \|g(\bar{x}_m) - g(x_m)\| + \|\delta_m\| \leq L\|\bar{x}_m - x_m\| + \|\delta_m\|$$
$$\leq L^{m+1}\|\Delta a\| + L^m\|\delta_0\| + \cdots + \|\delta_m\| \leq \delta/(1-L). \qquad (5.1.11)$$

Also, $\|x_1 - x_0\| \leq \|x_1 - \bar{x}_1\| + \|\bar{x}_1 - \bar{x}_0\| + \|\Delta a\| \leq (L+1)\delta + \|\bar{x}_1 - \bar{x}_0\|$. By Theorem 5.1.1, $\|x_m - x^*\| \leq L^m\|x_1 - x_0\|/(1-L)$. Since $\|\bar{x}_m - x^*\| \leq \|\bar{x}_m - x_m\| + \|x_m - x^*\|$, the inequality (5.1.10) follows. ∎

Since $L^m \to 0$ as $m \to \infty$, Theorem 5.1.5 asserts that we can approximate x^* with an error of the same order as δ. Thus, if $\delta = 10^{-s}$ say, where s is the precision of the arithmetic used in the computation of $g(\bar{x})$, then for m sufficiently large, the error will be $\leq 10^{-s}/(1-L) + \varepsilon_m$, where $\varepsilon_m > 0$ and $\varepsilon_m \to 0$. This result is somewhat deceptive, since it seems to suggest that errors due to rounding are small provided g is contracting. However, we must not overlook the presence of the factor $1/(1-L)$. If g is contractive with constant $L = 1 - \varepsilon$, then $1/(1-L) = 1/\varepsilon$ can be large if ε is small. The error in this case is $10^{-s}/\varepsilon$ and if $\varepsilon = 10^{-s}$ say, then the error in any iterate is of magnitude 1. For example, if $g(x) = (1-\varepsilon)x$, the fixed point is 0. If $\varepsilon = 10^{-s-1}$ where s is the precision, then the rounded value of $1 - \varepsilon$ is 1 (assuming half-adjust rounding). Thus, we actually compute $\bar{x}_{m+1} = \overline{g(\bar{x}_m)} = \bar{x}_m = g(\bar{x}_m) + \varepsilon\bar{x}_m$. Therefore, the quantity δ_m in (5.1.9) is here given by $\delta_m = \varepsilon\bar{x}_m$. If $\bar{x}_0 = 1$, then $\bar{x}_m = 1$ for all m and (5.1.10) yields $|\bar{x}_m - x^*| = |\bar{x}_m| \leq 10^{-s-1}/\varepsilon = 1$.

Rounding errors cause us to modify Theorem 5.1.3 as follows.

Theorem 5.1.6 Let g be a mapping of the closed ball $\bar{B} = \bar{B}(x_0, r)$ of a Banach space X into X. Suppose (i) g is contractive on $\bar{B}$ with Lipschitz constant $0 \leq L < 1$, and

(ii) $\|x_0 - g(x_0)\| \leq (1 - L)r - \delta$. Let $\bar{x}_m$ be determined by (5.1.9) with $\|\delta_m\| < \delta$ and $\|\Delta a\| < \delta$. If $\delta < (1 - L)r$, then $\bar{x}_m \in \bar{B}$ and
$$\|\bar{x}_m - x^*\| \leq \delta/(1 - L) + L^m(r - \delta/(1 - L)).$$

Proof. We have $\|\bar{x}_0 - x_0\| \leq \|\Delta a\| < \delta < r$. Thus, $\bar{x}_0 \in B$. Observe that condition (ii) above implies condition (ii) of Theorem 5.1.3. Hence, $x_m \in \bar{B}$ and $\|x_m - x_0\| \leq r - \delta/(1 - L)$ by (5.1.7). Now suppose $\bar{x}_m \in \bar{B}$ for $m \leq q$. Then
$$\|\bar{x}_{q+1} - x_0\| \leq \|\bar{x}_{q+1} - x_{q+1}\| + \|x_{q+1} - x_0\|$$
$$\leq \delta/(1 - L) + r - \delta/(1 - L) = r,$$
by (5.1.11) of Theorem 5.1.5. Hence, $\bar{x}_m \in \bar{B}$ for all m. Using (5.1.6) and (5.1.11) we obtain the result,
$$\|\bar{x}_m - x^*\| \leq \|\bar{x}_m - x_m\| + \|x_m - x^*\| \leq \delta/(1 - L) + L^m\left(r - \frac{\delta}{1 - L}\right). \blacksquare$$

Stability of Successive Approximations

Equation (5.1.1) is an example of a *difference equation*. Let us define the operator $\mathscr{L}$ by
$$\mathscr{L}(x_m) = (y_m),$$
$$y_m = x_{m+1} - g(x_m), \qquad m = 0, 1, \ldots.$$

$\mathscr{L}$ is called a difference operator. It maps a sequence (x_m) into another sequence (y_m). For $m = 0, 1, 2, \ldots$, the terms of the sequences, x_m and y_m, are elements of a normed vector space X. The mapping g is an arbitrary mapping of X into itself. The sequence of successive approximations (x_m) defined by (5.1.1) is a solution of the *homogeneous difference equation*,
$$\mathscr{L}(x_m) = (0), \qquad (5.1.12)$$
where (0) is the sequence consisting of all zero's. The sequence (x_m) may be regarded as a vector in the space s_x of sequences of elements of X. (Compare this with s, the space of complex sequences, Example 1 of Chapters 2 and 3. Addition and scalar multiplication of sequences in s_x are defined as in s.)

Equation (5.1.12) has many solutions. However, if we specify the initial value $x_0 = a$, we obtain an *initial-value problem*,
$$x_0 = a, \qquad \mathscr{L}(x_m) = 0, \qquad (5.1.13)$$
We shall now generalize the discussion in Section 4.4. (See Definitions 4.4.1 ff.)

The problem defined by (5.1.13) has a unique solution, namely, the sequence of successive approximations with starting value a. Suppose the starting value is "perturbed" by a small amount Δa, where $\|\Delta a\| = \delta$. Does the sequence (x_m) change much? To answer this question, of course, we must first introduce a topology into the space of sequences. This can be done as in Chapter 3, Example 1, or we may restrict our attention to the subspace of bounded sequences, as in Example 2 of Chapter 3. We shall denote this subspace by m_x. For any sequence $(x_m) \in m_x$ we denote the norm by $\|(x_m)\|_\infty$. Thus, $\|(x_m)\|_\infty = \sup_m\{\|x_m\|\}$. If we assume that g satisfies a Lipschitz condition with constant L, then the operator $\mathscr{L}$ maps the space m_x into itself. To verify this, suppose $(x_m) \in m_x$. Then for $(y_m) = \mathscr{L}(x_m)$ we have

$$\|y_m\| \leq \|x_{m+1}\| + \|g(x_m)\| \leq (1 + L)\|(x_m)\|_\infty + \|g(0)\|,$$

since

$$\|g(x_m)\| - \|g(0)\| \leq \|g(x_m) - g(0)\| \leq L\|x_m\| \leq L\|(x_m)\|_\infty.$$

If we assume further that g is a contracting map, then any solution of the initial-value problem (5.1.13) must be a Cauchy sequence (Theorem 5.1.1) and, therefore, a bounded sequence.

Now, let (u'_m) be a solution of the perturbed initial-value problem

$$x_0 = a + \Delta a, \qquad \mathscr{L}(x_m) = 0,$$

and let (u_m) be a solution of (5.1.13). Assume that both sequences are bounded. Then

$$u'_{m+1} - g(u'_m) = u_{m+1} - g(u_m) = 0,$$
$$\|u'_{m+1} - u_{m+1}\| = \|g(u'_m) - g(u_m)\| \leq L\|u'_m - u_m\|,$$

and it follows that for $m = 0, 1, 2, \ldots$,

$$\|u'_m - u_m\| \leq L^m \|u'_0 - u_0\| = L^m \|\Delta a\| \leq L^m \delta.$$

Hence, if $L \leq 1$, then $\|(u'_m - u_m)\|_\infty \leq \delta$. This answers the question raised earlier. We see that if δ is sufficiently small, then the two sequences of successive approximations will be close, provided they are bounded and g has a Lipschitz constant $L \leq 1$.

Now, let us consider Eq. (5.1.9) from this new standpoint. We may regard the sequence (δ_m) as a perturbation of the difference equation (5.1.12) (or of the operator $\mathscr{L}$). Thus, we have a new initial-value problem,

$$x_0 = a + \Delta a = a', \qquad \mathscr{L}(x_m) = (\delta_m). \qquad (5.1.14)$$

For any pair $(a', (\delta_m)) \in X \times s_x$, the initial-value problem (5.1.14) has a unique solution $(\bar{x}_m)$ given by

$$\bar{x}_0 = a',$$
$$\bar{x}_{m+1} = g(\bar{x}_m) + \delta_m, \qquad m \geq 0. \qquad (5.1.15)$$

This defines a mapping $G : X \times s_x \longrightarrow s_x$, where $(a', (\delta_m)) \longrightarrow (\bar{x}_m)$ and $(\bar{x}_m)$ is given by (5.1.15). In practice, (δ_m) and Δa are perturbations caused by rounding errors. We would like to know whether a "small" perturbation $(\Delta a, (\delta_m))$ produces a small change in the solution $(\bar{x}_m)$. To make this precise, we topologize the domain and range of G and consider the continuity of G.

Suppose g is contractive. If $(\delta_m) \in m_x$, then $(\bar{x}_m) \in m_x$. This follows from inequality (5.1.11). In this case, we may consider the restriction of G to $X \times m_x$. The results in Theorem 5.1.5 may be restated as follows:

Theorem 5.1.7 The mapping $G : X \times m_x \longrightarrow m_x$ defined by (5.1.15) with g contractive is continuous at the point $(a, (\delta_m))$.

Proof. $G(a, (\delta_m)) = (x_m)$, where (x_m) is a solution of the initial-value problem (5.1.14) with $\Delta a = 0$. For any $a' = a + \Delta a$ and (δ'_m) with $\|\Delta a\| < \delta$ and $\|\delta'_m - \delta_m\| < \delta$ for all m, an inequality like (5.1.11) yields

$$\|\Delta G\| = \|G(a'\,(\delta'_m)) - G(a, (\delta_m))\| = \|(\bar{x}_m - x_m)\|_\infty \le \delta/(1 - L).$$

Thus, $\|\Delta G\| < \varepsilon$ if δ is chosen sufficiently small; i.e., G is continuous at $(a, (\delta_m))$. ∎

As in Definition 4.4.1, if a solution of a difference equation behaves continuously with respect to perturbations, we shall say that the solution is *stable*. If G is continuous at $(a, (0))$ for every a, then we say that the difference operator $\mathscr{L}$ is *stable* and that the difference method of generating the sequence (x_n) is a *stable method*. Stability in any of these forms is obviously a desirable property in numerical computations. A method which is *unstable* would not be practical, since two different computers using slightly different rounding procedures could produce widely differing results. In fact, this indicates that we would like to have continuity in a neighbourhood of the zero perturbation, rather than just at zero. Indeed, we have this for the method of successive approximations, provided that g is contractive. Hence, two computers using rounding techniques which produce rounding errors of the same order of magnitude should produce results which are close to each other.

We have limited our attention to what are called *first-order* difference operators and equations. We could also consider a difference equation of order n,

$$x_{m+1} = g(x_m, x_{m-1}, \ldots, x_{m-n+1}).$$

As we shall see in Chapter 10, these nth-order difference equations can be transformed into first-order equations.

5.2 NEWTON'S METHOD—THE ONE-DIMENSIONAL CASE

We have seen that the method of successive approximations converges at a linear rate. For many applications this would involve an excessive number of iterations to achieve desired accuracy. Even with high-speed computers at our disposal it may be necessary—and certainly is desirable—to have faster methods of solving the equation $f(x) = 0$. As we know, unless f is a very simple function, we shall have to use an iterative method. (For the general polynomial in x of degree ≥ 5, Galois proved that the roots cannot be computed by a finite number of algebraic operations, a fortiori not by a finite number of arithmetic operations $(+, -, \cdot, /)$.) Even for linear problems, iterative methods are sometimes called for, as we saw in Chapter 4. When we say that we would like to have a "faster" method, we mean an iterative method for which the rate of convergence is greater than linear. The best known and most widely used method which achieves a quadratic rate of convergence in many cases is the one called *Newton's method* or sometimes the *Newton–Raphson* method. As is to be expected, quadratic convergence is attained at the expense of more computation per iteration. However, the marked decrease in the number of iterations required more than compensates for the additional computation in each iteration. Thus, in general, Newton's

5.2 Newton's Method—The One-Dimensional Case

method is more efficient than the successive approximation method in that fewer computations are required to obtain a root of $f(x) = 0$ with a prescribed accuracy. As with the method of successive approximations, Newton's method can be applied to vector functions f of a vector variable x. However, it is instructive to consider first the one-dimensional case of a real-valued function f of a real variable x. We do this in this section. The treatment of the vector case is deferred to Section 5.4 after we have developed the requisite generalization of differential calculus in Section 5.3.

Suppose that we wish to find a zero of a real function $f(x)$ having a graph as shown in Fig. 5.3.

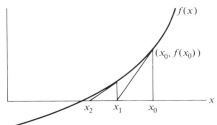

Figure 5.3

We choose a starting value x_0 and compute $f(x_0)$. If f were linear, its graph would be a straight line through the point $(x_0, f(x_0))$ and a zero could be found by setting $f(x_0) + A(x - x_0) = 0$, A being the slope. Solving for x, we get $x = x_0 - A^{-1}f(x_0)$. If f is not linear, we may approximate it by a linear function provided that its derivative $f'(x_0)$ exists. From the definition of the derivative, we have

$$f(x) = f(x_0) + f'(x_0)(x - x_0) + \varepsilon(x), \tag{5.2.0}$$

where $\varepsilon(x)/|x - x_0| \to 0$ as $|x - x_0| \to 0$. The quantity $f(x_0) + f'(x_0)(x - x_0)$ is called the *linear part* of $f(x)$ at x_0. Setting the linear part of $f(x)$ equal to zero, we obtain

$$x_1 = x_0 - \frac{f(x_0)}{f'(x_0)}$$

as an approximation to a zero of $f(x)$. Geometrically, as Fig. 5.3 shows, x_1 is the x-intercept of the tangent line at x_0. This process of *linearization* can be iterated to yield the sequence (x_m) given by

$$x_{m+1} = x_m - (f'(x_m))^{-1}f(x_m). \tag{5.2.1}$$

The procedure defined by (5.2.1) is known as *Newton's method* (also Newton–Raphson). Note that an initial value, x_0, is required. If the Newton iterates x_m converge to a point x^* and if f' exists and is continuous at x^* and $f'(x^*) \neq 0$, then passing to the limit in (5.2.1), we obtain

$$x^* = x^* - (f'(x^*))^{-1}f(x^*).$$

This implies that $f(x^*) = 0$. In fact, we see that x^* is a fixed point of the function

$$g(x) = x - (f'(x))^{-1}f(x) \tag{5.2.2}$$

and the Newton iterates are given by $x_{m+1} = g(x_m)$. Therefore, Newton's method can be viewed as a generalization of the idea of finding a zero of $f(x)$ by finding a fixed point of $x - f(x)$. To develop this idea further, let us consider

$$g(x) = x - \varphi(x)f(x). \tag{5.2.3}$$

We take $\varphi(x) = (f'(x))^{-1}$ to obtain Newton's method and $\varphi(x) = 1$ to obtain the successive approximation scheme. Since $(f'(x))^{-1}$ must exist in Newton's method, we usually require that $f'(x) \neq 0$ for all $x \neq x^*$ in some ball $B(x^*, r)$.

Formula (5.2.1) involves an evaluation of the derivative $f'(x_m)$ in each iteration. An alternative procedure is to approximate $f'(x_m)$ by the difference quotient,

$$\frac{f(x_m) - f(x_{m-1})}{x_m - x_{m-1}}.$$

This yields the iterative procedure,

$$x_{m+1} = x_m - f(x_m)\left(\frac{x_m - x_{m-1}}{f(x_m) - f(x_{m-1})}\right). \tag{5.2.4}$$

known as the *method or rule of false position* (or *regula falsi*). The procedure (5.2.4) cannot be viewed as successive approximation applied to a real function $g(x)$, since it requires two preceding values. Also, note that (5.2.4) requires two starting values, x_0 and x_1.

A variation of Newton's method is obtained by assuming that $f'(x_m)$ does not change appreciably as x_m approaches x^*. This suggests the formula (sometimes called the *chord method*),

$$x_{m+1} = x_m - (f'(x_0))^{-1}f(x_m). \tag{5.2.5}$$

Here, if the limit x^* exists, it must be a fixed point of

$$g(x) = x - (f'(x_0))^{-1}f(x),$$

that is, the function $\varphi(x)$ in (5.2.3) is the constant function $(f'(x_0))^{-1}$.

A simple convergence theorem for Newton's method is the following.

Theorem 5.2.1 Let $f(x)$ be a real function of the real variable x such that $f(x^*) = 0$. If $f'(x)$ and $f''(x)$ exist and $f'(x) \neq 0$ in a neighborhood of x^*, then there is a ball $B = B(x^*, r)$ such that for any initial value $x_0 \in B$, the Newton iterates converge to x^*. Further, if $f''(x)$ is continuous in the closed ball $\bar{B}$ and $f'''(x)$ exists and is bounded for $x \in B$, then the convergence is quadratic.

Proof. Let $g(x) = x - (f'(x))^{-1}f(x)$. Differentiating, we obtain

$$g'(x^*) = 1 - (f'(x^*))^{-1}f'(x^*) + (f'(x^*))^{-2}f''(x^*)f(x^*) = 0,$$

since $f(x^*) = 0$. The conclusions then follow directly from Corollary 2 of Theorem 5.1.4. ∎

5.2 Newton's Method—The One-Dimensional Case

One of the most familiar applications of Newton's method is to the computation of square roots of real numbers. Let $f(x) = x^2 - a$. The solutions of $f(x) = 0$ are $\pm\sqrt{a}$. Since $f'(x) = 2x$, Eq. (5.2.1) becomes in this case

$$x_{m+1} = x_m - \frac{(x_m^2 - a)}{2x_m} = \frac{1}{2}\left(x_m + \frac{a}{x_m}\right). \tag{5.2.6}$$

Since $f'(x) \neq 0$ for $x \neq 0$ and $f''(x) = 2$, Theorem 5.2.1 applies and we conclude that (5.2.6) yields a sequence (x_m) converging to $\sqrt{a}$ for any x_0 in a neighborhood of $\sqrt{a}$. In fact, it is clear from Fig. 5.4 that any positive initial value x_0 will yield such a convergent sequence.

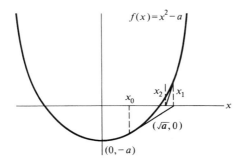

Figure 5.4

In Theorem 5.2.1, we required that $f'(x) \neq 0$ for all x in some neighborhood of x^*. It sometimes happens that $f'(x^*) = 0$ and $f'(x) \neq 0$ for $x \neq x^*$. In this case, Newton's method can be modified by choosing a different function φ in (5.2.3) to achieve quadratic convergence. Let $g(x) = x - (f'(x))^{-1}f(x)$ for $x \neq x^*$ and $g(x^*) = x^*$. Suppose that $f(x^*) = f'(x^*) = \cdots = f^{(m-1)}(x^*) = 0$ and $f^{(m)}(x^*) \neq 0$.
By Taylor's formula, with $x = x^* + h$, we have

$$f(x^* + h) = \frac{f^{(m)}(x^*)h^m}{m!}(1 + O(h)),$$

and

$$f'(x^* + h) = \frac{f^m(x^*)h^{m-1}}{(m-1)!}(1 + O(h)).$$

Hence,

$$\frac{f(x^* + h)}{f'(x^* + h)} = \frac{h}{m}(1 + O(h)) = \frac{h}{m} + O(h^2),$$

and for $f'(x^* + h) \neq 0$, $g(x^* + h) = x^* + h - h/m + O(h^2)$.

Therefore, since $g(x^*) = x^*$,

$$g'(x^*) = \lim_{h \to 0} \frac{g(x^* + h) - g(x^*)}{h}$$

$$= \lim_{h \to 0} \frac{h - h/m + O(h^2)}{h} = 1 - \frac{1}{m} < 1.$$

By Corollary 1 of Theorem 5.1.4, Newton's method converges. However, if $m > 1$, $g'(x^*) \neq 0$ and we cannot be assured of quadratic convergence. To obtain quadratic convergence, we take $\varphi(x) = -m/f'(x)$; i.e. $g(x) = x - m(f'(x))^{-1}f(x)$. From the preceding analysis, it is easy to see that for this choice of φ, $g'(x^*) = 0$.

The underlying idea in Newton's method is the linearization of $f(x)$, that is, the replacement of $f(x)$ by its linear part. The latter notion requires the concept of the derivative. Therefore, to extend Newton's method to the case where f is a mapping of a normed vector space X into itself, we must generalize the derivative concept. The next section is devoted to a development of differential calculus in abstract vector spaces which includes a generalization of the derivative, Taylor expansions etc. Since we have other applications in mind, we shall carry this development out with greater scope than is required for Newton's method. (See [5.21].)

5.3 DIFFERENTIAL CALCULUS OF VECTOR FUNCTIONS

The derivative concept arises from a limiting process applied to the rate of change (difference quotient) of a function. However, much of its importance as a mathematical entity stems from its role as a linear operator. In the one-dimensional case, linearity is a consequence of the limit-process definition. In the general case, this is no longer true. For numerical purposes, we are most often interested in the derivative when it is a linear operator. Nevertheless, we shall first give a definition based on the difference quotient and then make further assumptions to achieve linearity.

Throughout this section, X and Y denote real normed vector spaces and f is a partial mapping from X into Y; i.e., $f : X \longrightarrow Y$.

Definition 5.3.1(a) Let $u, h \in X$ be fixed. Let s be a scalar. If f is defined on the set $\{u + sh : 0 \leq s < \delta\}$ and the limit

$$\lim_{s \to 0^+} \frac{f(u + sh) - f(u)}{s} = df(u; h^+)$$

exists, it is called the *one-sided directional differential of f at u with increment h.*

Definition 5.3.1(b) Let f be defined on the set $\{u + sh : 0 \leq |s| < \delta\}$. If

$$\lim_{|s| \to 0} \frac{f(u + sh) - f(u)}{s} = \left. \frac{df(u + sh)}{ds} \right|_{s=0}$$

exists, it is called the *directional differential* of f at u with increment h.

5.3 Differential Calculus of Vector Functions

We observe that in the case where f is a real functional (i.e., $Y = R$) $f(u + sh)$ is an ordinary real function of the real variable s and $df(u + sh)/ds|_{s=0}$ is the ordinary derivative at $s = 0$. Similarly, $df(u; h^+)$ is a one-sided derivative.

The quantity $df(u; h^+)$ is an element of Y, since it is the limit of elements of Y. The limit is, of course, taken in the norm topology of Y.

Definition 5.3.2 Let $u \in X$ be fixed. Suppose that for every $h \in X$ there exists $\delta = \delta(h) > 0$ such that f is defined on the "segment" $\{u + sh : 0 \le |s| < \delta\}$. If

$$\left.\frac{df(u + sh)}{ds}\right|_{s=0}$$

exists for all h, we say that f has a *weak* (or *Gâteaux*) *differential* at u. We write

$$df(u; h) = \frac{d}{ds} f(u + sh)\Big|_{s=0} = \lim_{s \to 0} \frac{f(u + sh) - f(u)}{s} \qquad (5.3.1)$$

in this case and call $df(u; h)$ the *weak* (or *Gâteaux*) *differential* of f at u with increment h.

Remark 1 Observe that for all real t

$$df(u; th) = t\, df(u; h), \qquad (5.3.2)$$

since

$$\lim_{|s| \to 0} \frac{f(u + sth) - f(u)}{s} = t \lim_{|st| \to 0} \frac{f(u + sth) - f(u)}{st}.$$

However, the weak differential is not always additive in h and, therefore, is not always linear in h. As a simple example, consider the real function, φ, defined on $X = V^2$ by

$$\varphi(\xi, \eta) = \begin{cases} (\xi^2 + \eta^2)^{1/2} & \text{for } \eta > 0 \text{ and on the positive } \xi\text{-axis} \\ -(\xi^2 + \eta^2)^{1/2} & \text{for } \eta < 0 \text{ and on the negative } \xi\text{-axis} \end{cases}$$

Let $u = (0, 0)$ be the origin and $h = (h_1, h_2)$ be an arbitrary two-dimensional vector. Then

$$d\varphi(0; h) = \lim_{s \to 0} (\varphi(u + sh) - \varphi(u))/s = \lim_{s \to 0} \varphi(sh_1, sh_2)/s = \pm(h_1^2 + h_2^2)^{1/2},$$

which is clearly not additive in h.

From advanced calculus we know that for any real function $f(\xi_1, \ldots, \xi_n)$ of n real variables, if f has partial derivatives $\partial f/\partial \xi_i$, $1 \le i \le n$, which exist and are continuous in a neighborhood $u = (\xi_1, \ldots, \xi_n)$, then

$$f(u + h) - f(u) = \langle \operatorname{grad} f(u), h \rangle + \varepsilon(u, h), \qquad (5.3.3)$$

where $\varepsilon/\|h\|_2 \to 0$ as $\|h\|_2 \to 0$. Here, $\operatorname{grad} f(u) = (\partial f/\partial \xi_1, \ldots, \partial f/\partial \xi_n)$. (See Exercise 5.2.) This shows that $df(u; h) = \langle \operatorname{grad} f(u), h \rangle$, which is linear in h. Furthermore, since the inner product is continuous in h, we also obtain continuity of $df(u; h)$.

When f is real-valued, so is $df(u; h)$. Hence, for fixed u, $df(u; h)$ is a real functional (mapping h into R). Since any linear functional on a finite-dimensional space is bounded, it is to be expected that $df(u; h)$ is continuous in h whenever it is linear in h. In the infinite-dimensional case, we cannot expect continuity to be a concomitant of linearity. However, as a generalization of the advanced calculus result, we might expect some kind of continuity condition on $df(x; h)$ with respect to x to be sufficient for linearity in h.

Lemma 5.3.1 Let $f: X \longrightarrow R$. If the weak differential $df(x; h)$ exists for all x in a ball B centered at $u_0 \in X$ and is continuous at u_0 (as a function of x for arbitrary fixed h), then $df(u_0; h)$ is linear in h.

Proof. In virtue of (5.3.2), it remains to establish the additivity of $df(u_0; h)$; i.e., to show that

$$df(u_0; h_1 + h_2) = df(u_0; h_1) + df(u_0; h_2) \tag{5.3.4}$$

for arbitrary $h_1, h_2 \in X$. Let $g(t_1, t_2) = f(u_0 + t_1 h_1 + t_2 h_2)$. g is a real function of the scalars t_1 and t_2. From (5.3.1) with $u = u_0 + t_1 h_1 + t_2 h_2$ and $h = h_1$ we obtain

$$\frac{\partial g}{\partial t_1}(t_1, t_2) = df(u_0 + t_1 h_1 + t_2 h_2; h_1)$$

for $|t_1| + |t_2|$ sufficiently small so that $u_0 + t_1 h_1 + t_2 h_2 \in B$. Similarly, with $h = h_2$, we find

$$\frac{\partial g}{\partial t_2} = df(u_0 + t_1 h_1 + t_2 h_2; h_2)$$

Since $df(x; h)$ is continuous at u_0, it follows that $\partial g/\partial t_1$ and $\partial g/\partial t_2$ exist and are continuous functions of (t_1, t_2) at the origin $(0, 0)$. Therefore (see Exercise 5.2c),

$$g(t_1, t_2) = g(0, 0) + t_1 \left.\frac{\partial g}{\partial t_1}\right|_{(0,0)} + t_2 \left.\frac{\partial g}{\partial t_2}\right|_{(0,0)} + \varepsilon,$$

where $\varepsilon/(t_1^2 + t_2^2)^{1/2} \to 0$ as $(t_1, t_2) \to (0, 0)$ along a ray through the origin. This can be rewritten as

$$f(u_0 + t_1 h_2 + t_2 h_2) - f(u_0) = t_1 df(u_0; h_1) + t_2 df(u_0; h_2) + \varepsilon.$$

Taking $t_1 = t_2 = t$, we obtain

$$\lim_{t \to 0} (f(u_0 + t(h_1 + h_2)) - f(u_0))/t = df(u_0; h_1) + df(u_0; h_2).$$

Since the left member of this equation is $df(u_0; h_1 + h_2)$, this establishes (5.3.4). ∎

To obtain continuity in h as well as linearity we must simply specify that $df(u; h)$ be continuous at $h = 0$, since any linear operator which is continuous at 0 is bounded and therefore continuous everywhere. A general result which insures that $df(u; h)$ is both linear and continuous in h is the following.

Theorem 5.3.1 Let $f: X \longrightarrow Y$, where X and Y are normed vector spaces. If (1) $df(x; h)$ exists in a neighborhood of $u \in X$ and is continuous in x at the point u for arbitrary fixed h, and (2) $df(u; h)$ is continuous in h at $h = 0$, then $df(u; h)$ is linear (and therefore continuous) in h.

Proof. See [5.1]. ∎

Definition 5.3.3 Let $df(u; h)$ be linear and continuous in h for a fixed $u \in X$. Then $df(u; h)$ defines a linear operator $f'_w(u) \in L_c(X; Y)$ called the *weak* (or *Gâteaux*) *derivative of f at u*; i.e., $df(u; h) = f'_w(u) \cdot h$. Thus, for any $x \in X$ such that $f'_w(x)$ exists,

$$\lim_{s \to 0} \frac{f(x + sh) - f(x)}{s} = f'_w(x) \cdot h. \tag{5.3.5}$$

The mapping $f'_w : X \longrightarrow L_c(X; Y)$ which maps x onto $f'_w(x)$ is called the *G-derivative* (or *weak derivative*) of f.

To illustrate the notion of weak derivative, we let $X = V^n$ and $Y = V^m$. Then $x = (\xi_1, \ldots, \xi_n)$ and $f(x) = (f_1(\xi_1, \ldots, \xi_n), \ldots, f_m(\xi_1, \ldots, \xi_n))$. If $f'_w(x)$ exists, then $f'_w(x) \in L(V^n; V^m)$. Since every linear operator from V^n to V^m is represented by an $m \times n$ matrix, there exists an $m \times n$ matrix $A = (a_{ij})$, $a_{ij} = a_{ij}(x)$, such that $f'_w(x) = A$. By (5.3.5), for any $h = (h_1, \ldots, h_n)$ we must have

$$\lim_{s \to 0} \frac{f(x + sh) - f(x)}{s} = Ah.$$

This implies that for $1 \leq i \leq m$,

$$\lim_{s \to 0} \frac{f_i(\xi_1 + sh_1, \ldots, \xi_n + sh_n) - f_i(\xi_1, \ldots, \xi_n)}{s} = \sum_{k=1}^{n} a_{ik} h_k.$$

Taking $h_k = 0$ for $k \neq j$ and $h_j = 1$ and passing to the limit, we obtain for $1 \leq j \leq n$,

$$\frac{\partial f_i}{\partial \xi_j} = a_{ij}.$$

Therefore, $f'_w(x) = J(x) = (\partial f_i / \partial \xi_j)$; i.e., $f'_w(x)$ is the Jacobian of f evaluated at x.

If $X = V^n$ and $Y = R$, the preceding result reduces to

$$f'_w(x) = \left(\frac{\partial f}{\partial \xi_1}, \ldots, \frac{\partial f}{\partial \xi_n} \right) = \operatorname{grad} f(x).$$

However, as Exercise 5.2b demonstrates, the existence of the partials is not sufficient for f to have a weak derivative or a uniform linear approximation as in (5.3.3). As we know, a sufficient condition is that $\operatorname{grad} f(y)$ exist and be continuous for all y in a neighborhood of x. (See Exercise 5.2.)

As a consequence of (5.3.5), whenever $f'_w(x)$ exists we have $f(x + sh) - f(x) = f'_w(x) \cdot sh + \varepsilon(x, h, s)$, where $\varepsilon/s \to 0$ as $s \to 0$, h and x being held fixed. However, the convergence may not be uniform with respect to h. (See Exercise 5.2d.) In that case, f cannot be approximated by a linear function with uniform accuracy in a neighborhood of x and we cannot replace f by a linear part as required in Newton's method. In fact, Definition 5.3.2 does not require f to be defined in a neighborhood of x.

Definition 5.3.4 Let f be defined in a neighborhood of x. If the convergence in (5.3.5) is uniform with respect to all h with $\|h\| = 1$, then the weak derivative is called the *strong derivative* (or *Fréchet derivative*) of f at x. We write $f'(x) = f'_w(x)$ in this case.

Setting $\Delta x = sh$ in (5.3.5), we have

$$\lim_{s \to 0} \frac{f(x + \Delta x) - f(x) - f'_w(x) \Delta x}{s} = 0.$$

If the convergence is uniform for all $\|h\| = 1$, this can be written as

$$\lim_{\|\Delta x\| \to 0} \frac{\|f(x + \Delta x) - f(x) - f'(x) \Delta x\|}{\|\Delta x\|} = 0, \qquad (5.3.6)$$

since $\|\Delta x\| = \|sh\| = |s|$ when $\|h\| = 1$.

From (5.3.6) it follows that

$$f(x + \Delta x) = f(x) + f'(x) \Delta x + \varepsilon, \qquad (5.3.7)$$

where $\varepsilon = \varepsilon(\Delta x)$ is a function of Δx such that

$$\lim_{\|\Delta x\| \to 0} \frac{\|\varepsilon\|}{\|\Delta x\|} = 0. \qquad (5.3.8)$$

Equation (5.3.7) is a generalization of (5.2.0) and shows that the existence of the Fréchet derivative $f'(x)$ permits us to approximate $f(x + \Delta x)$ by its linear part $f(x) + f'(x) \Delta x$ with uniform accuracy ε in a neighborhood of x; i.e., for all $\|\Delta x\| < \delta$, δ sufficiently small.

We call attention to the fact that $f'_w(u)$ can exist without f being continuous at u. (See Exercise 5.2f.) However, if $f'(u)$ exists, then f must be continuous at u. This follows from (5.3.7), since

$$\|f(x + \Delta x) - f(x)\| \leq \|f'(x)\| \|\Delta x\| + \|\varepsilon\|,$$

and for $\|\Delta x\| \leq r$ sufficiently small, the right side of the inequality can be made arbitrarily small. Note that $f(u + sh)$ is continuous in s at $s = 0$ when $f'_w(u)$ exists.

By definition, if the strong derivative exists, then so does the weak derivative. Exercise 5.2(d) shows that the converse is not true. However, by considering the behavior of the G-derivative f'_w, we can state the following useful result.

Theorem 5.3.2 If the G-derivative f'_w is defined for all x in a neighborhood of $u \in X$ and is continuous at u, then the strong derivative $f'(u)$ exists.

Proof. (See Exercise 5.8.) ∎

Thus, if $f'_w(x)$ exists in a neighborhood of u and is continuous (in the norm topology of $L_c(X; Y)$), then $f'(u)$ exists. Since $f'_w(x)$ is easier to calculate in most cases, we need only establish its continuity to show that we have also calculated the strong derivative. For example, if $f : V^n \longrightarrow R$, then $f'_w(x) = \operatorname{grad} f(x)$. If we can show that the partials $\partial f/\partial \xi_i$, $1 \le i \le n$ are continuous at u, then we know that $f'(u) = \operatorname{grad} f(u)$. Theorem 5.3.2 generalizes and sharpens the classical result of calculus referred to earlier, and given in Exercise 5.2, which permits us to write (5.3.3) when $\operatorname{grad} f(x)$ is continuous in a neighborhood of u. Since Newton's method is based on the replacement of f by its linear part, the existence of the strong derivative seems to be what is required to obtain a useful generalization of the one-dimensional case. The existence of the weak derivative alone does not seem to provide a good enough linear approximation to insure convergence. Indeed, as we have pointed out above, it allows f to be discontinuous. It should be observed that conceptually the formulation of the general Newton's method could be done in terms of the weak derivative. However, we shall confine our treatment to the strong derivative. For example, suppose $f: V^n \longrightarrow V^n$ and that $f'(x)$ exists. As we have seen above, $f'(x) = J(x)$, where J is the Jacobian of f. In this instance J is an $n \times n$ matrix and we may consider its inverse $J^{-1}(x)$, when it exists. This suggests that the generalization of Newton's method, as given in (5.2.1) for the one-dimensional case, should be, for the n-dimensional case,

$$x_{m+1} = x_m - J^{-1}(x_m) \cdot f(x_m). \tag{5.3.9}$$

Here, $\cdot$ refers to the matrix $J^{-1}(x_m)$ operating on the vector $f(x_m)$. In the infinite-dimensional case, we revert to (5.2.1) as the defining equation of Newton's method except that $(f'(x_m))^{-1}$ is to be interpreted as the inverse of the linear operator $f'(x_m)$.

In order to obtain convergence, we shall have to impose further conditions on $f'(x)$. To state these conditions we must first develop certain aspects of the differential calculus of vector functions.

Lemma 5.3.2 Let $\varphi, f,$ and g be mappings defined on a subset of X to Y and such that $\varphi(x) = \alpha f(x) + \beta g(x)$. If $f'_w(u)$ and $g'_w(u)$ exist, then $\varphi'_w(u)$ exists and $\varphi'_w(u) = \alpha f'_w(u) + \beta g'_w(u)$. Similarly if $f'(u)$ and $g'(u)$ exist, then $\varphi'(u)$ exists and $\varphi'(u) = \alpha f'(u) + \beta g'(u)$.

Proof. Immediate from the definition of derivative. ∎

Lemma 5.3.3 If $f: X \longrightarrow Y$ is a bounded linear mapping (i.e., $f \in L_c(X; Y)$), then for all $x \in X$ $f'(x)$ exists and $f'(x) = f$.

Proof. $f(x + sh) = f(x) + sf(h)$ by the linearity of f. Hence,

$$\frac{f(x + sh) - f(x)}{s} = f(h),$$

showing that the weak derivative exists. Furthermore, the convergence is clearly uniform in h; i.e.,

$$\lim_{s \to 0} \frac{f(x + sh) - f(x)}{s} = \lim_{s \to 0} f(h) = f(h).$$

Thus, $f'(x)$ exists and $f'(x)h = f(h)$; i.e., $f'(x)$ and f are the same linear operator. ∎

As in ordinary calculus, we have a *composite function or chain rule for differentiation*.

Lemma 5.3.4 Let $f: X \longrightarrow Y$ and $g: Y \longrightarrow Z$. Let $\varphi = g \cdot f : X \longrightarrow Z$ be the composition of f and g (defined for those $x \in X$ such that $f(x)$ is in the domain of g). If $f'_w(u)$ exists and $g'(v)$ exists, where $v = f(u)$, then $\varphi'_w(u)$ exists and $\varphi'_w(u) = g'(f(u)) \cdot f'_w(u)$. Furthermore, if $f'(u)$ exists, so does $\varphi'(u)$.

Proof. Choose $h \in X$ and a scalar s sufficiently small so that $f(u + sh)$ is defined. Let $\Delta v = f(u + sh) - f(u)$; i.e., $f(u + sh) = v + \Delta v$. Then

$$\frac{\varphi(u + sh) - \varphi(u)}{s} = \frac{g(f(u + sh)) - g(f(u))}{s}$$

$$= \frac{g(v + \Delta v) - g(v)}{s} = \frac{g'(v) \Delta v + \varepsilon}{s},$$

where $\|\varepsilon\|/\|\Delta v\| \to 0$ as $\|\Delta v\| \to 0$. Note that $\Delta v \to 0$ as $s \to 0$ by the continuity of $f(u + sh)$ with respect to s. Now (when $\Delta v \neq 0$),

$$\frac{g'(v) \cdot \Delta v + \varepsilon}{s} = g'(v)\left(\frac{f(u + sh) - f(u)}{s}\right) + \frac{\varepsilon}{\|\Delta v\|} \frac{\|\Delta v\|}{s}.$$

Since $\lim_{s \to 0} \Delta v / s = f'(u)h$ and $\varepsilon / \|\Delta v\| \to 0$ as $s \to 0$, we obtain, using the continuity of the linear operator $g'(v)$,

$$\varphi'_w(u)h = \lim_{s \to 0} \frac{\varphi(u + sh) - \varphi(u)}{s} = g'(v) \cdot f'_w(u)h.$$

Since this holds for all $h \in X$, $\varphi'_w(u) = g'(v) \cdot f'_w(u)$, where $\cdot$ denotes composition of the linear operators $f'_w(u) \in L_c(X; Y)$ and $g'(v) \in L_c(Y; Z)$. Clearly, if $f'(u)$ exists, then the convergence is uniform with respect to h and $\varphi'(u)$ exists. This completes the proof. ∎

In ordinary calculus, the mean-value theorem asserts that $f(x + h) - f(x) = f'(x + \theta h)h$, $0 < \theta < 1$, whenever the derivative exists in the interval $(x, x + h)$ and f is continuous on $[x, x + h]$. Hence,

$$|f(x + h) - f(x)| \leq \sup_{0 < \theta < 1} |f'(x + \theta h)| |h|.$$

When f is a vector function, the generalization of this result takes the form of the inequality rather than the equality.

(Mean-Value) Theorem 5.3.3 Let $f: X \longrightarrow Y$ be defined in a neighborhood of $x_0 \in X$. Let $h \in X$ be such that $f'_w(x)$ exist for all $x = x_0 + sh$, $0 \leq s \leq 1$. Then

$$\|f(x_0 + h) - f(x_0)\| \leq \sup_{0 < \theta < 1} \|f'_w(x_0 + \theta h)\| \|h\|, \tag{5.3.10}$$

and

$$\|f(x_0 + h) - f(x_0) - f'_w(x_0)h\| \leq \sup_{0 < \theta < 1} \|f'_w(x_0 + \theta h) - f'_w(x_0)\| \|h\|. \tag{5.3.11}$$

5.3 Differential Calculus of Vector Functions

Proof. Let Y' be the conjugate space of Y (Section 3.5) and let $g \in Y'$ be an arbitrary bounded linear functional. Define the real function $\varphi(s) = g(f(x_0 + sh)), 0 \leq s \leq 1$. Then

$$\varphi'(s) = \lim_{\Delta s \to 0} \frac{\varphi(s + \Delta s) - \varphi(s)}{\Delta s} = \lim_{\Delta s \to 0} g\left(\frac{f(x_0 + sh + \Delta sh) - f(x_0 + sh)}{\Delta s}\right)$$

$$= g\left(\lim_{\Delta s \to 0} \frac{f(x_0 + sh + \Delta sh) - f(x_0 + sh)}{\Delta s}\right)$$

$$= g(f'_w(x_0 + sh)h).$$

(We have used both the linearity and continuity of g.) This shows that $\varphi'(s)$ exists for all $0 \leq s \leq 1$. Applying the mean-value theorem for real functions, we have

$$\varphi(1) - \varphi(0) = \varphi'(\theta), \qquad 0 < \theta < 1,$$

or

$$g(f(x_0 + h) - f(x_0)) = g(f'_w(x_0 + \theta h)h).$$

Note that θ depends upon g.

Hence,

$$|g(f(x_0 + h) - f(x_0))| \leq \|g\| \sup_{0 < \theta < 1} \|f'_w(x_0 + \theta h)\| \|h\|. \tag{5.3.12}$$

By the Hahn–Banach theorem (Appendix II) there exists $g_1 \in Y'$, $g_1 \neq 0$, such that $\|g_1\| = 1$ and

$$g_1(f(x_0 + h) - f(x_0)) = \|f(x_0 + h) - f(x_0)\|.$$

Using this particular functional in (5.3.12), we obtain (5.3.10).

Now, consider the function $\psi(x) = f(x) - f'_w(x_0)x$. Observe that $\psi(x + sh) = f(x + sh) - f'_w(x_0)(x + sh)$ and by Lemmas 5.3.2, 5.3.3,

$$\psi'_w(x) = f'_w(x) - f'_w(x_0).$$

Therefore,

$$\psi(x_0 + h) = f(x_0 + h) - f'_w(x_0)(x_0 + h),$$
$$\psi(x_0) = f(x_0) - f'_w(x_0)x_0,$$

and

$$\psi'_w(x_0 + \theta h) = f'_w(x_0 + \theta h) - f'_w(x_0).$$

Applying (5.3.10) to the function ψ, we obtain

$$\|f(x_0 + h) - f(x_0) - f'_w(x_0)h\| \leq \sup_{0 < \theta < 1} \|\psi'(x_0 + \theta h)\| \|h\|$$

and (5.3.11) follows. ∎

186 Solution of Nonlinear Equations

It is worth pointing out that in the special case where f is a real functional (i.e. $f\colon X \longrightarrow R$) the classical mean-value theorem remains valid.

Lemma 5.3.5 Let $f\colon X \longrightarrow R$ have a weak differential at all points of the line segment $\{x + th\colon 0 < t < 1\}$ and be continuous at the points x and $x + h$. Then there exists $0 < \theta < 1$ such that
$$f(x + h) - f(x) = df(x + \theta h; h).$$

Proof. Let $g(t) = f(x + th)$. Then
$$g'(t) = \lim_{s \to 0} \frac{f(x + th + sh) - f(x + th)}{s} = df(x + th; h).$$

Hence, by the classical mean-value theorem,
$$f(x + h) - f(x) = g(1) - g(0) = g'(\theta) = df(x + \theta h; h). \blacksquare$$

Another instance in which the classical mean-value theorem holds is the following:

Lemma 5.3.6 Let $f\colon X \longrightarrow Y$ have a weak differential on the line segment $\{x + th\colon 0 < t < 1\}$ and be continuous at x and $x + h$. Let $g^* \in Y'$ be a bounded linear functional on Y. Then there exists $0 < \theta < 1$ such that
$$\langle f(x + h) - f(x), g^* \rangle = \langle df(x + \theta h; h), g^* \rangle.$$

(Note that θ depends on g^*.)

Proof. Let $\varphi(t) = \langle f(x + th), g^* \rangle$. Thus, φ is a real function continuous for $0 \le t \le 1$. Furthermore,
$$\varphi'(t) = \lim_{s \to 0} \frac{\langle f(x + th + sh), g^* \rangle - \langle f(x + th), g^* \rangle}{s}$$
$$= \lim_{s \to 0} \frac{\langle f(x + th + sh) - f(x + th), g^* \rangle}{s}$$
$$= \langle df(x + th; h), g^* \rangle,$$

by the continuity of g^*. Applying the ordinary mean-value theorem as in the previous lemma, we obtain the result. $\blacksquare$

In Lemma 5.3.2, we generalized to vector functions the theorem of elementary calculus which states that the derivative of the sum of two functions is the sum of their derivatives. Now, consider the theorem on the differential of the product of two real functions, which we may state in the form $d(gf) = (dg)f + g(df)$. How should this be extended to vector functions? Since the operation of real multiplication by g is a linear one (i.e., $g(\alpha f_1 + \beta f_2) = \alpha g f_1 + \beta g f_2$), it seems reasonable to consider the situation in which the product is replaced by an arbitrary linear operation. Thus, in place of $g(x)$ we consider $T(x)$, where $T(x)$ is a bounded linear operator on Y to Z (Y and Z normed vector spaces); i.e., $T\colon X \longrightarrow L_c(Y; Z)$ is defined on a subset D of the normed vector space X and for each $x \in D$, $T(x)$ is an element of the normed vector space of bounded linear operators $L_c(Y; Z)$. The norm, as usual, is the operator norm. For any $f: X \longrightarrow Y$ having domain $D_1 \subset X$, we can consider the application of the linear operator $T(x)$ to the vector $f(x) \in Y$. The result is a vector $z(x) = T(x)f(x)$ defined for all $x \in D \cap D_1$. ($z(x) \in Z$.) Hence, $z: X \longrightarrow Z$ is a vector function and we may seek to determine its differential $dz(x; h)$.

Lemma 5.3.7 Let $f: X \longrightarrow Y$ and $T: X \longrightarrow L_c(Y; Z)$ be defined on a subset $D \subset X$. Suppose that $f'_w(x)$ and $T'_w(x)$ exist. Then $z(x) = T(x)f(x)$ has a weak derivative and for all $h \in X$

$$dz(x; h) = dT(x; h)f(x) + T(x)df(x; h) \tag{5.3.13}$$

or

$$z'_w(x)h = (T'_w(x)h)f(x) + T(x)(f'_w(x)h). \tag{5.3.14}$$

Proof. Let $\Delta z = z(x + sh) - z(x)$. Then $\Delta z = T(x + sh) \cdot f(x + sh) - T(x) \cdot f(x)$ and

$$\frac{\Delta z}{s} = \left(\frac{T(x + sh) - T(x)}{s}\right) f(x + sh) + T(x) \left(\frac{f(x + sh) - f(x)}{s}\right). \tag{5.3.15}$$

Now, $\lim_{s \to 0} (T(x + sh) - T(x))/s = T'_w(x) \cdot h$ exists and is a linear operator on Y to Z for every $h \in X$. The limit is taken in the operator norm topology. Also, $\lim_{s \to 0} f(x + sh) = f(x)$. By the continuity of $T(x)$, we have

$$\lim_{s \to 0} T(x) \cdot \left(\frac{f(x + sh) - f(x)}{s}\right) = T(x) \cdot \lim_{s \to 0} \frac{f(x + sh) - f(x)}{s}.$$

Therefore, passing to the limit as $s \to 0$ in (5.3.15), we obtain (using Exercise 5.6) the relations (5.3.13) and (5.3.14). ∎

Corollary If the strong derivatives $T'(x)$ and $f'(x)$ exist, then

$$z'(x)\Delta x = (T'(x)\Delta x)f(x) + T(x)(f'(x)\Delta x). \tag{5.3.16}$$

Proof.

$$\Delta z = (T(x + \Delta x) - T(x)) \cdot f(x + \Delta x) + T(x) \cdot (f(x + \Delta x) - f(x))$$
$$= (T'(x)\Delta x + \varepsilon_1) \cdot f(x + \Delta x) + T(x)(f'(x)\Delta x + \varepsilon_2).$$

Since $f(x + \Delta x) = f(x) + \varepsilon_3$,

$$\|z(x + \Delta x) - z(x) - (T'(x)\Delta x \cdot f(x) + T(x) \cdot f'(x)\Delta x)\| \leq$$
$$(\|T'(x)\| \|\Delta x\| + \|\varepsilon_1\|) \|\varepsilon_3\| + \|\varepsilon_1\| \|f(x)\| + \|T(x)\| \|\varepsilon_2\|.$$

Dividing by $\|\Delta x\|$ and noting that $\|\varepsilon_1\|/\|\Delta x\| \to 0$ and $\|\varepsilon_2\|/\|\Delta x\| \to 0$ as $\|\Delta x\| \to 0$, by definition of the strong derivative, and $\|\varepsilon_3\| \to 0$ by continuity of f at x, we see that (5.3.16) holds for every Δx. ∎

Remark. In both (5.3.14) and (5.3.16) we have written the differential rather than the derivative. The latter could be denoted by the notation

$$z'(x) = (T'(x) -)\cdot f(x) + T(x)\cdot(f'(x) -)$$

where the $|-|$ denotes the position of the operand. This is awkward and serves to emphasize the hierarchical structure of the mappings f, T, f' and T'. (One could resort to left application of operators and write $f(x)*T'(x) \Delta x = (T'(x) \Delta x)\cdot f(x)$, where $*$ denotes that the operator is written to its right.) For one case of particular interest in Chapter 10, these difficulties do not arise. This is the case $X = R$; i.e., f is a vector function of a real variable x and $T = T(x)$ is a one-parameter family of operators defined for real x in some subset of R. Since Δx is a scalar and $df(x; \Delta x) = f'(x) \Delta x$ is a vector in Y, it follows that $f'(x) = (1/\Delta x) df(x; \Delta x)$ is also a vector in Y and the linear operation is simply scalar multiplication, that is, $f'(x) \Delta x = \Delta x f'(x)$. Similarly, $T'(x) \Delta x = \Delta x T'(x)$ and $T'(x)$ is an element of $L_c(Y; Z)$. In this case, (5.3.16) becomes

$$\Delta x z'(x) = (\Delta x T'(x))f(x) + T(x)\cdot(\Delta x f'(x))$$
$$= \Delta x(T'(x)f(x) + T(x)f'(x)),$$

by the linearity of $T(x)$. Therefore, $z'(x) = T'(x)f(x) + T(x)f'(x)$. To emphasize that this is only valid when $X = R$ we shall use t instead of x and write

$$(T(t)\cdot f(t))' = T'(t)\cdot f(t) + T(t)\cdot f'(t), \qquad (5.3.17)$$

which looks exactly like the elementary calculus formula except that "$\cdot$" denotes the application of a linear operator. (We usually omit "$\cdot$", as we have above.)

For a vector function $f: R \longrightarrow Y$, where $f(t)$ is defined for t in the closed bounded interval $[a, b]$, we can define a Riemann integral in the same way as for real functions. We choose a subdivision d of $[a, b]$ into subintervals having end points $a = t_0 < t_1 < \cdots < t_m = b$. Then we form the sum

$$S_d = \sum_{i=0}^{m-1} (t_{i+1} - t_i)f(t_i).$$

Let $\|d\| = \max_{0 \le i \le m-1}|t_{i+1} - t_i|$. If for any sequence of subdivisions (d_i) such that $\|d_i\| \to 0$ as $i \to \infty$, $\lim_{i\to\infty} S_{d_i} = S$ exists and S is independent of the sequence (d_i), then it is called the *(Riemann) integral* of f on $[a, b]$ and is denoted by $\int_a^b f(t)\,dt$. Since $S_{d_i} \in Y$ for every subdivision d_i, $\lim_i S_{d_i}$ is also a vector in Y; i.e., $\int_a^b f(t)\,dt \in Y$, the limit being taken in the norm topology of Y.

Theorem 5.3.4 Let $f: [a, b] \longrightarrow Y$ be a continuous function on the closed bounded interval to the Banach space Y. Then $\int_a^b f(t)\,dt$ exists.

Proof. See Exercise 5.7a. ∎

It is easy to verify that

$$\int_a^b [f(t) + g(t)]\,dt = \int_a^b f(t)\,dt + \int_a^b g(t)\,dt$$

and for any real scalar α,

$$\int_a^b \alpha f(t)\, dt = \alpha \int_a^b f(t)\, dt.$$

Likewise, from the definition one easily establishes that

$$\left\| \int_a^b f(t)\, dt \right\| \leq \int_a^b \| f(t) \|\, dt.$$

Higher-Order Derivatives and Differentials

For an ordinary real function $f: R \longrightarrow R$, to obtain an approximation to $f(x)$ which is locally more accurate than the linear part of f at x we may use the next term in the Taylor series expansion. Thus, we have

$$f(x+h) - f(x) = f'(x)h + \frac{f''(x)}{2!} h^2 + \varepsilon,$$

where $\varepsilon/h^2 \to 0$ as $h \to 0$. (As in Section 1.8, this is sometimes indicated by writing $\varepsilon = o(h^2)$. In general, $\varepsilon = o(\xi)$ means that $\lim_{\xi \to 0} \varepsilon/\xi = 0$.) In the case of a real function f of two real variables ξ and η, the Taylor expansion of order 2 is

$$f(\xi + h_1, \eta + h_2) = f(\xi, \eta) + \langle \nabla f(x), h \rangle$$

$$+ \frac{1}{2}\left(\frac{\partial^2 f}{\partial \xi^2} h_1^2 + \frac{2\partial^2 f}{\partial \xi \partial \eta} h_1 h_2 + \frac{\partial^2 f}{\partial \eta^2} h_2^2 \right) + \varepsilon, \qquad (5.3.18)$$

where ∇f is the gradient vector $(\partial f/\partial \xi, \partial f/\partial \eta)$, $x = (\xi, \eta)$, $h = (h_1, h_2)$, and $\varepsilon = o(\|h\|_2^2)$, under suitable conditions on the partial derivatives. As we have seen, the strong derivative $f'(x) = \nabla f(x)$ and the linear part is $f(x) + f'(x)h$, where the differential $f'(x)h = \langle \nabla f(x), h \rangle$, which is linear in h. The next term in the Taylor expansion is expressible in the form $\frac{1}{2}\langle f^{(2)}(x)h, h \rangle$, where

$$f^{(2)}(x) = \begin{pmatrix} \dfrac{\partial^2 f}{\partial \xi^2} & \dfrac{\partial^2 f}{\partial \xi \partial \eta} \\ \dfrac{\partial^2 f}{\partial \xi \partial \eta} & \dfrac{\partial^2 f}{\partial \eta^2} \end{pmatrix}. \qquad (5.3.19)$$

Thus, $f^{(2)}(x)$ is the matrix of second partial derivatives (the *Hessian*) evaluated at x and the *second-order differential* $\langle f^{(2)}(x)h, h \rangle$ is a quadratic form in h. This suggests that the generalization of the ordinary second derivative to the case of a vector function $f: X \longrightarrow Y$ should be a bilinear mapping $f^{(2)}(x): X \times X \longrightarrow Y$ and accordingly, the nth derivative of f at x should be a multilinear mapping $f^{(n)}(x): X^n \longrightarrow Y$.

In Section 2.3, we discussed the algebraic properties of multilinear mappings. For any $f \in L^n(V_1, \ldots, V_n; V_0)$ we have agreed to write $fx_1 x_2 \cdots x_n$ for $f(x_1, \ldots, x_n) \in V_0$. When the V_i, $0 \leq i \leq n$, are normed vector spaces, we may consider the continuity of f (with respect to the product topology). If f is continuous as a multilinear mapping, then we may regard f as a continuous linear operation in

$$L_c(V_1; L_c(V_2; L_c \ldots; L_c(V_{n-1}; L_c(V_n; V_0))\ldots)).$$

For example, when $n = 2$, $f \in L_c^2(V_1, V_2; V_0)$ and the continuity of f as a bilinear mapping $f: V_1 \times V_2 \longrightarrow V_0$ implies that fx_1 is a continuous linear operator on $x_2 \in V_2$ for fixed $x_1 \in V_1$, i.e., $fx_1: V_2 \longrightarrow V_0$ is a continuous linear operator in $L_c(V_2; V_0)$. Similarly, f maps x_1 onto the element fx_1 and is continuous and linear in x_1. Hence, $f \in L_c(V_1; L_c(V_2; V_0))$. Therefore, fx_1 is bounded and has norm

$$\|fx_1\| = \sup_{\|x^2\| \leq 1} \{\|fx_1 x_2\|\}.$$

Consequently,

$$\|f\| = \sup_{\|x_1\| \leq 1} \{\|fx_1\|\} = \sup_{\|x_1\| \leq 1} \{\sup_{\|x_1\| \leq 1} \|fx_1 x_2\|\} = \sup_{\substack{\|x_1\| \leq 1 \\ \|x_2\| \leq 1}} \|fx_1 x_2\|.$$

In the general case,

$$\|f\| = \sup_{\substack{\|x_i\| \leq 1 \\ 1 \leq i \leq n}} \{\|fx_1 x_2 \cdots x_n\|\},$$

and

$$\|fx_1 x_2 \cdots x_n\| \leq \|f\| \|x_1\| \cdots \|x_n\|.$$

Of particular interest for differentiability are the continuous symmetric multilinear mappings. If the continuous multilinear mapping $f: X^n \longrightarrow Y$ is symmetric (Section 2.3), then the restriction of f to n-tuples $(h, h, \ldots, h)$ is called a *homogeneous form of degree n* (on X to Y) and its values are denoted by fh^n. When $n = 2$, fh^2 is called a *quadratic map*. (Compare this with the special case $\langle Th, h \rangle$ when T is a linear operator on a Hilbert space. See Exercise 2.15 and the discussion following Theorem 3.3.10.) If f_i, $1 \leq i \leq n$ are homogeneous forms of degree i on X to Y, then the sum $p(h) = \sum_{i=1}^{n} f_i h^i$ defines a continuous mapping from X to Y called a *polynomial of degree n* in h. For example, the linear part of a function $f: X \longrightarrow Y$ is a polynomial $fx + f'(x)h$ of degree 1 in h. (In fact, if the weak derivative exists, then $f_w''(x)h$ is a polynomial of degree 1.) Equation (5.3.18) can be rewritten in polynomial form. Let $x = (\xi, \eta)$, $f_1 = f'(x) = \nabla fx$ and $f_2 = \tfrac{1}{2} f^{(2)}(x)$, as given by (5.3.19). Then (5.3.18) becomes

$$f(x + h) = f(x) + f_1 h + f_2 h^2 + \varepsilon,$$

or

$$f(x + h) - f(x) = f_1 h + f_2 h^2 + \varepsilon.$$

This is a generalization of the Taylor expansion of order 2 for ordinary real functions. To apply it to a vector function $f: X \longrightarrow Y$ we must define the second derivative of such a function.

Definition 5.3.5 Let $f: X \longrightarrow Y$ and let the strong derivative of f exist in a neighborhood N of $x \in X$; i.e., $f': X \longrightarrow L_c(X; Y)$ is defined for all $x \in N$. If the strong derivative of f' at x exists, it is called the *strong* (or *Frechet*) *second derivative of f at x* and is denoted by $f''(x)$ or $f^{(2)}(x)$. In general, we define $f^{(n)}(x), n \geq 1$, by induction to be the *strong derivative* of $f^{(n-1)}$ at x, $(f^{(0)} = f)$.

Since $f': X \longrightarrow L_c(X; Y)$, for each x, $f''(x)$ is a continuous linear operator mapping each $h_1 \in X$ onto an element of $L_c(X; Y)$; i.e., $f''(x) h_1 \in L_c(X; Y)$. Hence, $f''(x) h_1 h_2 \in Y$ for all $h_2 \in X$ and we see that $f''(x) \in L_c(X; L_c(X; Y)) = L_c^2(X, X; Y)$, that is, $f''(x)$ is a continuous bilinear mapping on X to Y. From Definitions 5.3.3 and 5.3.4 it follows that

$$\lim_{s \to \infty} \frac{f'(x + sh_1) - f'(x)}{s} = f''(x) h_1, \qquad (5.3.20)$$

for all $h_1 \in X$ and the convergence is uniform with respect to h_1 for $\|h_1\| = 1$. The topology is that of the operator norm of $L_c(X; Y)$. Letting $T_s = T_s(h_1) = (f'(x + sh_1) - f'(x))/s$, we observe that $T_s \in L_c(X; Y)$ and (5.3.20) can be rewritten as $\|T_s - f''(x)h_1\| \to 0$ as $s \to 0$, the convergence being uniform with respect to all h_1 with $\|h_1\| = 1$. Therefore, for any $h_2 \in X$, $T_s h_2 \to f''(x) h_1 h_2$ as $s \to 0$, that is

$$\lim_{s \to 0} \frac{f'(x + sh_1) h_2 - f'(x) h_2}{s} = f''(x) h_1 h_2. \qquad (5.3.21)$$

Example Let $X = V^n$ and $Y = V^m$ be real finite-dimensional vector spaces. Let A be a bilinear mapping on V^n to V^m, i.e., $A \in L^2(X, X; Y)$. Consider the operation of A on the canonical basis $\{e_1, \ldots, e_n\}$ (Definition 2.1.9): $Ae_i e_j = A(e_i, e_j) \in V^m$. Choosing a basis in V^m, we can express $Ae_i e_j$ as a row vector, say

$$Ae_i e_j = (a_{ij}^{(1)}, \ldots, a_{ij}^{(m)}), \qquad 1 \leq i, j \leq n.$$

For any $x, y \in V^n$, $x = (\xi_1, \ldots, \xi_n)$ and $y = (\eta_1, \ldots, \eta_n)$, we have by the bilinearity of A,

$$Axy = A(x, y) = A\left(\sum_1^n \xi_i e_i, \sum_1^n \eta_j e_j\right) = \sum_{i, j=1}^n \xi_i \eta_j Ae_i e_j.$$

Since $Axy = (\zeta_1, \ldots, \zeta_m) \in V^m$, the kth component ζ_k is given by

$$\zeta_k = \sum_{i, j=1}^n a_{ij}^{(k)} \xi_i \eta_j, \qquad 1 \leq k \leq m.$$

Let $A^{(k)} = (a_{ij}^{(k)})$ be the $n \times n$ matrices $(k = 1, \ldots, m)$ determined as above. Then

$$Axy = (\langle A^{(1)} x, y \rangle, \ldots, \langle A^{(m)} x, y \rangle), \qquad (5.3.22)$$

so that we can write

$$A = (A^{(1)}, \ldots, A^{(m)}); \qquad (5.3.23)$$

i.e., any bilinear mapping A on V^n to V^m is represented by an m-dimensional vector whose components are $n \times n$ matrices as in (5.3.23) and Axy is given by (5.3.22).

Now, suppose $f: V^n \longrightarrow V^m$; i.e., $f(x) = (f_1(x_1,\ldots,x_n),\ldots,f_m(x_1,\ldots,x_n))$, $x = (x_1,\ldots,x_n)$. Suppose $f''(x)$ exists. We have seen that $f'(x) = f'_w(x) = J(x)$, the Jacobian of f at x. Since $f''(x)$ is a bilinear mapping on V^n to V^m, it must have a representation as in (5.3.23) and $f''(x)h_1h_2$ must be given by (5.3.22). Since (5.3.21) also holds, we find that the kth component of the vector $f''(x)h_1h_2$ is, with $h_1 = (\xi_1,\ldots,\xi_n)$, $h_2 = (\eta_1,\ldots,\eta_n)$, $x = (x_1,\ldots,x_n)$,

$$\langle A^{(k)}h_1, h_2 \rangle = \sum_{i,j=1}^n a_{ij}^{(k)} \xi_i \eta_j = \lim_{s \to 0} \frac{(J(x+sh_1)h_2 - J(x)h_2)_k}{s}$$

$$= \lim_{s \to 0} \frac{\sum_{i=1}^n \left(\frac{\partial f_k}{\partial x_i}\right)_{x+sh_1} \eta_i - \sum_{i=1}^n \left(\frac{\partial f_k}{\partial x_i}\right)_x \eta_i}{s}$$

$$= \sum_{i=1}^n \eta_i \lim_{s \to 0} \frac{\left[\left(\frac{\partial f_k}{\partial x_i}\right)_{x+sh_1} - \left(\frac{\partial f_k}{\partial x_i}\right)_x\right]}{s}.$$

Now, take h_1 such that $\xi_j = 1$ is its only nonzero coordinate and h_2 such that $\eta_i = 1$ is its only nonzero coordinate. The above becomes in this case,

$$a_{ij}^{(k)} = \frac{\partial^2 f_k}{\partial x_j \partial x_i}, \quad i, j = 1, \ldots, n; k = 1, \ldots, m,$$

where the second partials are evaluated at x. Thus, in this example,

$$f''(x) = \left(\left(\frac{\partial^2 f_1}{\partial x_i \partial x_j}\right), \ldots, \left(\frac{\partial^2 f_m}{\partial x_i \partial x_j}\right)\right),$$

where $(\partial^2 f_k / \partial x_i \partial x_j)$ is the $n \times n$ matrix of second partial derivatives of f_k, $1 \le i, j \le n$ ($k = 1, \ldots, m$). This generalizes (5.3.19), which holds when $m = 1, n = 2$.

Now, let us consider how to obtain the generalization of Taylor's expansion. We have defined the Riemann integral $\int_0^1 g(t)\,dt$ for a function $g: [0,1] \longrightarrow Y$. Theorem 5.3.4 shows that the integral exists if g is continuous. In Exercise 5.7b, it is proved that the generalization of the fundamental theorem of calculus,

$$\int_0^1 g'(t)\,dt = g(1) - g(0), \tag{5.3.24}$$

holds if the derivative $g'(t)$ is continuous on $[0, 1]$. We shall use this result to obtain a generalized Taylor's formula with a remainder.

Theorem 5.3.5 Let $f: X \longrightarrow Y$ have a strong derivative in a ball $B(x, r)$, $x \in X$. Let the strong second derivative $f''(x)$ exist. Then for all h with $\|h\| < r$,

$$f(x + h) = f(x) + f'(x)h + \tfrac{1}{2} f''(x)h^2 + \varepsilon, \tag{5.3.25}$$

where $\|\varepsilon\|/\|h\|^2 \to 0$ as $\|h\| \to 0$; i.e., $\|\varepsilon\| = o(\|h\|^2)$.

Proof. Let $g(t) = f(x + th)$, $\|h\| < r$. By (5.3.1), $g'(t) = f'(x + th)h$. Using (5.3.24), we get

$$g(1) - g(0) = f(x + h) - f(x) = \int_0^1 f'(x + th)h \, dt.$$

By Definition 5.3.5, $f'(x + th) = f'(x) + f''(x)th + \varepsilon_1 \|th\|$, where $\varepsilon_1 = \varepsilon_1(t) \to 0$ as $\|th\| = |t| \|h\| \to 0$. Hence,

$$f(x + h) - f(x) = \int_0^1 f'(x)h \, dt + \int_0^1 tf''(x)h^2 \, dt + \int_0^1 \varepsilon_1 |t| \|h\| \, h \, dt,$$

$$f(x + h) - f(x) = f'(x)h + \tfrac{1}{2}f''(x)h^2 + \varepsilon,$$

where $\|\varepsilon\| \le \int_0^1 \|\varepsilon_1\| |t| \|h\|^2 \, dt \le \|\varepsilon_1\| \|h\|^2$, so that $\|\varepsilon\|/\|h\|^2 \le \|\varepsilon_1\|$. As $\|h\| \to 0$ so does $|t| \|h\|$ for all $0 \le t \le 1$. Hence, $\|\varepsilon_1(t)\| \to 0$ as $\|h\| \to 0$; i.e., uniformly with respect to h, which proves the theorem. ∎

(See Exercise 12.11 for similar results with weak derivatives.)

Corollary If f has a strong nth derivative $f^{(n)}(x)$ and strong derivatives of orders 1 to $n - 1$ in a ball $B(x; r)$, then for all $\|h\| < r$,

$$f(x + h) = f(x) + f'(x)h + \frac{1}{2!}f''(x)h^2 + \frac{1}{3!}f^{(3)}(x)h^3 + \cdots + \frac{1}{n!}f^{(n)}(x)h^n + \varepsilon_n$$

where $\|\varepsilon_n\| = o(\|h\|^n)$; i.e., $\|\varepsilon_n\|/\|h\|^n \to 0$ as $\|h\| \to 0$.

Proof. Use induction on n. ∎

In ordinary calculus, if $f(x)$ is real-valued and twice differentiable on the interval $(x_0, x_0 + h)$ and $f'(x)$ is continuous on $[x_0, x_0 + h]$, then

$$f(x_0 + h) = f(x_0) + f'(x_0)h + \tfrac{1}{2}f''(x_0 + \theta h)h^2, \qquad 0 < \theta < 1. \quad (5.3.26)$$

We have seen that the generalization of the first-order case, $f(x + h) = f(x) + f'(x + \theta h)h$, which is the mean-value theorem, is given by (5.3.10). We now derive a similar generalization of the second-order Taylor's formula (5.3.26).

Theorem 5.3.6 Let $f: X \longrightarrow Y$ be such that $f'(x)$ exists and is continuous for all x in a closed ball $\bar{B}(x_0; r)$. Let h be such that $\|h\| \le r$ and $f''(x)$ exists for all $x = x_0 + sh$, $0 < s < 1$. Then there exists $0 < \theta < 1$ such that

$$\| f(x_0 + h) - f(x_0) - f'(x_0)h \| \le \tfrac{1}{2} \| f''(x_0 + \theta h)\| \|h\|^2. \quad (5.3.27)$$

Proof. As in Theorem 5.3.3, let $g \in Y'$, the conjugate space of Y. Define $\varphi(s) = g(f(x_0 + sh)), 0 \le s \le 1$. As in Theorem 5.3.3 we prove that $\varphi'(s) = g(f'(x_0 + sh)h)$. Similarly, it can be shown that $\varphi''(s) = g(f''(x_0 + sh)h^2)$. By applying Taylor's formula (5.3.26) to the real function φ, we obtain

$$g(f(x_0 + h) - f(x_0) - f'(x_0)h) = \varphi(1) - \varphi(0) - \varphi'(0) = \tfrac{1}{2}\varphi''(\theta)$$
$$= \tfrac{1}{2}g(f''(x_0 + \theta h)h^2),$$

194 Solution of Nonlinear Equations

and therefore
$$|g(f(x_0 + h) - f(x_0) - f'(x_0)h)| \leq \tfrac{1}{2} \|g\| \, \|f''(x_0 + \theta h)\| \, \|h\|^2.$$
Proceeding as in Theorem 5.3.3, we choose a $g_1 \in Y'$ such that $\|g_1\| = 1$ and
$$g_1(f(x_0 + h) - f(x_0) - f'(x_0)h) = \|f(x_0 + h) - f(x_0) - f'(x_0)h\|.$$
This yields (5.3.27). ∎

Observe that Definition 5.3.5 does not require that $f''(x)$ be symmetric as a bilinear mapping, nor was this property used in the ensuing discussion. As in ordinary calculus, to obtain symmetry it suffices to assume that f'' is continuous at x. (See Exercise 5.9e for further details.)

Theorem 5.3.7 Let $f: X \longrightarrow R$ be a real functional. Let $h \in X$ be such that $f'(x)$ exists and is continuous in an open set containing the segment $\{u + th: 0 \leq t \leq 1\}$. Further, suppose that $f''(x)$ exists for all $x = u + th$, $0 < t < 1$. Then
$$f(u + h) = f(u) + f'(u)h + \tfrac{1}{2}f''(u + \theta h)h^2, \qquad 0 < \theta < 1. \qquad (5.3.28)$$

Proof. Let $g(t) = f(x + th)$. As in Lemma 5.3.5, $g'(t) = f'(u + th)h$ for $0 \leq t \leq 1$ and $g'(t)$ is continuous on $[0, 1]$. Similarly, by (5.3.21) we have for $0 < t < 1$,
$$g''(t) = \lim_{s \to 0} \frac{f'(u + th + sh)h - f'(u + th)h}{s} = f''(u + th)h^2.$$
(5.3.28) then follows from Taylor's formula of ordinary calculus. ∎

Partial Differentiation

The generalization of the preceding concepts and results to functions of several variables proceeds in an obvious way. We illustrate with the case of a function of two variables $f(x, y)$, where $x \in X$, $y \in Y$, $f(x, y) \in V$ are elements in three arbitrary normed vector spaces. For example, definitions (5.3.2), (5.3.3) and (5.3.4) generalize as follows.

Definition 5.3.6 Let $(x, y) \in X \times Y$ be fixed. Suppose that for every $h \in X$ there exists $\delta = \delta(h) > 0$ such that f is defined on the segment $\{(x + sh, y) : 0 \leq |s| < \delta\}$. The *weak partial differential* of f with respect to x with increment h is
$$d_x f(x, y; h) = \frac{d}{ds} f(x + sh, y)\Big|_{s=0}$$
$$= \lim_{s \to 0} \frac{f(x + sh, y) - f(x, y)}{s}. \qquad (5.3.29)$$

If $d_x f(x, y; h)$ is linear and continuous in h, then
$$d_x f(x, y; h) = f_{xw}(x, y)h$$
and the linear operator $f_{xw}(x, y)$ is called the *weak partial derivative* of f with respect to x at (x, y). When the convergence is uniform with respect to h on the unit ball

$\|h\| = 1$, the weak partial derivative is called the *strong partial derivative* of f with respect to x and is denoted by $f_x(x, y)$.

The partial derivative with respect to y is defined analogously. To relate the partial differentials to the (total) differential of f, we simply form the direct sum (Definition 2.2.1) $Z = X \oplus Y$ and introduce the norm $\|(x, y)\| = \|x\| + \|y\|$. Regarding f as a mapping from Z into V, we define

$$df((x, y); (h, k)) = \lim_{s \to 0} \frac{f(x + sh, y + sk) - f(x, y)}{s} \quad (5.3.30)$$

to be the (total) weak differential of f at (x, y) with increment (h, k). The (total) weak derivative $f'_w(x, y)$ is a linear bounded operator from Z to V such that

$$df((x, y); (h, k)) = f'_w(x, y) \cdot (h, k).$$

By Exercise 2.11c, if $f'_w(x, y)$ exists, then

$$f'_w(x, y) \cdot (h, k) = f_{xw}(x, y) \cdot h + f_{yw}(x, y) \cdot k \quad (5.3.31)$$

since from the definitions it follows that $f'_w(x, y) \cdot (h, 0) = f_{xw}(x, y) \cdot h$. As Exercise 5.2b shows, the existence of the partial derivatives at (x, y) does not imply that the weak total derivative exists. However, if the strong partials exist in a neighborhood of (x, y) and are continuous at (x, y), then the strong total derivative exists. This follows in a straightforward manner from

$$f(x + h, y + k) - f(x, y) = f(x + h, y + k) - f(x, y + k) + f(x, y + k) - f(x, y),$$

$$f(x + h, y + k) - f(x, y + k) = f_x(x, y + k) \cdot h + \varepsilon_1 = (f_x(x, y) + \varepsilon_2) \cdot h + \varepsilon_1,$$

$$f(x, y + k) - f(x, y) = f_y(x, y) \cdot k + \varepsilon_3,$$

where

$$\frac{\|\varepsilon_1\|}{\|h\| + \|k\|} \to 0$$

as $\|h\| + \|k\| \to 0$ by definition of f_x. Similarly, $\|\varepsilon_3\|/(\|h\| + \|k\|) \to 0$. Finally, $\|\varepsilon_2\| \to 0$ by the continuity of f_x at (x, y). Hence,

$$\frac{\|\varepsilon_2\| \|h\|}{\|h\| + \|k\|} \to 0.$$

5.4 NEWTON'S METHOD FOR VECTOR FUNCTIONS

We return to the problem of finding a zero of a vector function. Let $f: X \to Y$, where X and Y are arbitrary normed vector spaces. Suppose that the strong derivative $f'(x)$ exists for all x in some subset $E \subset X$. Then we can approximate $f(x)$ by its linear part at any point x_0 in E. Suppose further that $f'(x_0)$ is a bijective mapping (of X onto Y), so that the inverse mapping $(f'(x_0))^{-1}$ exists. Setting the linear part equal to zero, we have $f(x_0) + f'(x_0)(x - x_0) = 0$, whence $x = x_0 - (f'(x_0))^{-1}f(x_0)$. If $f(x)$ were equal to its linear part at x_0, the point x so determined would be a zero of f. When $f(x)$ is not linear in x, we are led to Newton's procedure (5.2.1), where now, in the general case of vector functions, $(f'(x_m))^{-1}$ is to be interpreted as the inverse operator of the strong derivative $f'(x_m)$.

For example, in the finite-dimensional case when $X = Y = V^n$ the procedure is given by (5.3.9). In this case, $f'(x) = J(x)$ is the Jacobian matrix $(\partial f_i/\partial \xi_j)$, where the partial derivatives are evaluated at $x = (\xi_1, \ldots, \xi_n)$. When formulas for the partial derivatives can be obtained explicitly by differentiating the component functions $f_1, \ldots, f_m$, the Jacobian $J(x_m)$ can be computed directly by evaluating the formulas for the partials at the point $x = x_m$. Instead of computing $(J(x_m))^{-1}$, we express (5.3.9) as a linear system,

$$J(x_m)(x_{m+1} - x_m) = -f(x_m),$$

to be solved for $\Delta x_m = x_{m+1} - x_m$ (by Gaussian elimination for example). Then $x_{m+1} = x_m + \Delta x_m$. This process is iterated starting with some initial guess x_0. (See Exercise 5.10.) The question of whether the sequence (x_m) converges to a vector x^* such that $f(x^*) = 0$ will now be considered in the general setting of arbitrary normed vector spaces X and Y.

Theorem 5.2.1, which establishes convergence in the one-dimensional case, provides us with some clues as to what will be needed in the general case. In Theorem 5.2.1, we assume the existence of a zero x^* and show that the sequence of Newton iterates (x_m) converges to x^* provided that the initial guess x_0 is sufficiently close to x^* and the second derivative is continuous in a neighborhood of x^*. To obtain quadratic convergence we assumed the existence of the third derivative. This result can be generalized and improved at the same time. We shall not assume the existence of x^*, but shall prove its existence under certain conditions on f, f' and f''. We shall not require the existence of the third derivative. It will suffice to assume that the second (strong) derivative $f''(x)$ exists and is not too large in norm. We observe that if the Newton iterates x_m converge to a point x^*, then we have from (5.2.1) under suitable continuity conditions on f' and f and assuming $(f'(x^*))^{-1}$ exists,

$$x^* = \lim_{m \to \infty} x_{m+1} = \lim_{m \to \infty} x_m - \lim (f'(x_m))^{-1} f(x_m)$$
$$= x^* - (f'(x^*))^{-1} f(x^*).$$

Hence, $(f'(x^*))^{-1} f(x^*) = 0$, which implies that $f(x^*) = 0$; i.e., if the Newton iterates converge at all, they must converge to a zero of f. If $\|x_0 - x^*\| < r$ for some initial guess x_0, we would expect $\|x_m - x^*\| < r$ for all terms of the convergent sequence. We would also like to have $\|x_{m+1} - x^*\| < \|x_m - x^*\|$ for $m \geq 0$, since the successive approximation method has this property when it converges. Now, recall condition (ii) of Theorem 5.1.3. Since $x_1 - x_0 = -(f'(x_0))^{-1} f(x_0)$, this suggests that we impose the condition

$$\|(f'(x_0))^{-1} f(x_0)\| \leq (1 - k)r, \tag{5.4.1}$$

where $0 < k \le \frac{1}{2}$. This will insure that $\|x_1 - x_0\| \le r$. In fact, we wish to have this hold for all the Newton iterates. Furthermore, to carry out (5.2.1) the inverse $(f'(x_m))^{-1}$ must exist for all x_m. From Corollary 4 of Theorem 3.5.2 it follows that $(f'(x_m))^{-1}$ will exist if

$$\|f'(x_m) - f'(x_0)\| < 1/\|(f'(x_0))^{-1}\|. \tag{5.4.2}$$

Since $f''(x)$ is the derivative of $f'(x)$, we can apply the generalized mean-value theorem (Theorem 5.3.3) to $f'(x)$ to obtain (see (5.3.10)),

$$\|f'(x_m) - f'(x_0)\| \le \sup_{0 < \theta < 1} \|f''(x_0 + \theta h)\| \, \|x_m - x_0\|, \tag{5.4.3}$$

where $h = x_m - x_0$. If $\|x_m - x_0\| < r$, then for (5.4.2) to hold it suffices to impose the condition,

$$\|f''(x)\| \, \|(f'(x_0))^{-1}\| < 2k/r, \tag{5.4.4}$$

for all x in the ball $\bar{B}(x_0; r)$, where, as before, $0 < k \le \frac{1}{2}$. Clearly (5.4.4) and (5.4.3) imply (5.4.2), provided that $x_m \in \bar{B}(x_0; r)$. The latter will follow from (5.4.1) as we shall see. We shall show that (5.4.1) and (5.4.4) are sufficient conditions for convergence of the Newton iterates. (See [5.3] and [5.5] for more complete treatments.) We first give a somewhat less general result.

Theorem 5.4.1 Let $f: X \longrightarrow Y$, X a Banach space, be such that $f''(x)$ exists in some open set $U \subset X$. Suppose $x_0 \in U$ is such that $(f'(x_0))^{-1}$ exists and in some closed ball $\bar{B}(x_0, r)$ the following conditions hold for all $x \in \bar{B}(x_0, r)$ and some constant $0 < L < 1$:

i) $\|(f'(x_0))^{-1}\| \, \|f''(x)\| \le L/r$;
ii) $\|(f'(x))^{-1} f(x)\| \le (1 - L)r$;
iii) $\|(f'(x))^{-1}\| \, \|f''(x)\| \le L/r$.

Then the sequence of Newton iterates (x_m), starting with x_0, converges to a unique zero $x^* \in \bar{B}(x_0, r)$. Furthermore, the convergence is quadratic, that is, there exists a constant C such that

$$\|x^* - x_{m+1}\| \le C \|x^* - x_m\|^2.$$

Proof. For any $x = x_0 + h \in \bar{B}(x_0, r)$, the mean-value theorem and condition (i) of the theorem yield

$$\|f'(x) - f'(x_0)\| \le \sup_{0 < \theta < 1} \|f''(x_0 + \theta h)\| \, \|h\|$$

$$\le \frac{L\|h\|}{r \, \|(f'(x_0))^{-1}\|} < \frac{1}{\|(f'(x_0))^{-1}\|}.$$

Hence, $(f'(x))^{-1}$ exists.

Let $g(x) = x - (f'(x))^{-1}f(x)$. By Exercise 5.13 and Lemma 5.3.7, Corollary, we have for arbitrary $h \in X$ and $x \in \bar{B}(x_0, r)$,

$$g'(x)h = Ih + (f'(x))^{-1}(f''(x)h)(f'(x))^{-1}f(x) - (f'(x))^{-1}f'(x)h$$
$$= (f'(x))^{-1}(f''(x)h)(f'(x))^{-1}f(x).$$

Hence, by conditions (ii) and (iii),

$$\|g'(x)\| \le \|(f'(x))^{-1}\| \|f''(x)\| \|(f'(x))^{-1}f(x)\|$$
$$\le \frac{L}{r}(1-L)r = (1-L)L \le L < 1.$$

By the generalized mean-value theorem,

$$\|g(x+h) - g(x)\| \le \sup_{0 < \theta < 1} \|g'(x + \theta h)\| \|h\| < L\|h\|$$

for any x, $x + h \in \bar{B}(x_0, r)$, which shows that g is a contractive mapping on $\bar{B}(x_0, r)$.

For any $x \in \bar{B}(x_0, r)$, condition (ii) yields $\|g(x) - x\| = \|-(f'(x))^{-1}f(x)\| \le r(1 - L)$. In particular, this holds for x_0, so that g satisfies the hypotheses of Theorem 5.1.3. It follows that the sequence (x_m) of Newton iterates of f (which is also the sequence of successive approximations of g) converges to a unique fixed-point x^* of g in $\bar{B}(x_0, r)$. By the continuity of f and f' at x^*, $f(x^*) = 0$. Note that $(f'(x^*))^{-1}$ exists, since $x^* \in \bar{B}(x_0, r)$.

To establish quadratic convergence, we observe that

$$x^* - x_{m+1} = x^* - x_m + (f'(x_m))^{-1}f(x_m)$$
$$= (f'(x_m))^{-1}(f'(x_m)(x^* - x_m) + f(x_m) - f(x^*)),$$

since $f(x^*) = 0$. By Taylor's formula (5.3.27),

$$\|x^* - x_{m+1}\| \le \|(f'(x_m))^{-1}\| \frac{\|f''(x_m + \theta_m h_m)\|}{2} \|h_m\|^2,$$

where $h_m = x^* - x_m$. Now, let $(f'(x_m))^{-1} - (f'(x_m + \theta_m h_m))^{-1} = y_m$. Since x_m and $x_m + \theta_m h_m$ converge to x^*, by continuity

$$\Delta f'_m = f'(x_m) - f'(x_m + \theta_m h_m) \to 0$$

and

$$(f'(x_m + \theta_m h_m))^{-1} \to (f'(x^*))^{-1}.$$

For m sufficiently large, we have by Corollary 4 of Theorem 3.5.2,

$$\|y_m\| \le \frac{\|(f'(x_m + \theta_m h_m))^{-1}\|^2 \|\Delta f'_m\|}{1 - \|(f'(x_m + \theta_m h_m))^{-1}\| \|\Delta f'_m\|}.$$

Therefore,

$$\|(f'(x_m))^{-1}\| = \|(f'(x_m + \theta_m h_m))^{-1}\|(1 + \delta_m),$$

where $\delta_m \to 0$ as $m \to \infty$. By condition (iii),

$$\|x^* - x_{m+1}\| \leq \frac{L}{2r}(1 + \delta_m)\|x^* - x_m\|^2.$$

Hence, there exists a constant C such that for all m, $\|x^* - x_{m+1}\| \leq C\|x^* - x_m\|^2$. ∎

Now, we sharpen this result by making the hypotheses somewhat more general.

Theorem 5.4.2 Let X by a Banach space. Let $f: X \longrightarrow Y$ be such that $f''(x)$ exists for all x in some open subset $U \subset X$. Let $x_0 \in U$ be such that (i) $(f'(x_0))^{-1}$ exists; and (ii) for some closed ball $B(x_0; r)$ of radius $r > 0$ and some constant $0 < k < \frac{1}{2}$ the conditions (5.4.4) and (5.4.1) hold. Then the Newton iterates given by (5.2.1) are defined for all $m \geq 0$ and the sequence (x_m) converges to a zero x^* of f. Furthermore, $x_m \in \bar{B}(x_0; r)$ for all m, so that $x^* \in \bar{B}(x_0; r)$. The rate of convergence is governed by the inequalities

$$\|x_{m+1} - x^*\| \leq 2^m \frac{k}{r}\|x_m - x^*\|^2, \tag{5.4.5}$$

and

$$\|x_m - x^*\| < \frac{r}{2^m}(2k)^{2^m - 1}. \tag{5.4.6}$$

Proof. See [5.5]. ∎

Computational Aspects

From a practical standpoint, the results in Theorems 5.4.1 and 5.4.2 are of limited utility. In the vector case, the second derivative is a bilinear mapping. As shown in the preceding section, even in the finite-dimensional case the computation of $f''(x)$ is rather involved. Even when this computation can be carried out, the verification of (5.4.1) or condition (iii) of Theorem 5.4.1 may be difficult to effect. Perhaps even more difficult is the computation of $(f'(x))^{-1}$ and the estimate of its norm for all x in $\bar{B}(x_0, r)$, which appears to be required by condition (ii). This is avoided in (5.4.1) which only requires that we compute $\|x_1 - x_0\| = \|(f'(x_0))^{-1}f(x_0)\|$. We recall that in the finite-dimensional case this is done by solving $f'(x_0) \cdot (x_1 - x_0) = -f(x_0)$. Note that from an estimate of $\|(f'(x_0))^{-1}f(x_0)\|$ it may be possible to estimate $\|(f'(x))^{-1}f(x)\|$ using the continuity of $(f'(x))^{-1}$ and $f(x)$ at x_0.

When conditions (i), (ii), (iii) of Theorem 5.4.1 hold, the Newton iteration procedure is stable, since it is equivalent to successive approximation with a contractive mapping. See Exercise 5.25 for a discussion of rounding error and stopping criterion.

5.5 MODIFICATIONS OF NEWTON'S METHOD

The determination of $f'(x_m)$ and its inverse at each step of the Newton iterative procedure could entail a considerable amount of computation. The modification given in (5.2.5) requires only the determination of $(f'(x_0))^{-1}$. In the finite-dimensional case

$f: V^n \longrightarrow V^n$, we have seen that $f'(x) = J(x)$, the Jacobian. In this instance, (5.2.5) can be rewritten as $J(x_0) \Delta x_m = J(x_0)(x_{m+1} - x_m) = -f(x_m)$, $m = 0, 1, \ldots$. The matrix $J(x_0) = (\partial f_i/\partial \xi_j)$ is evaluated just once at x_0 and is used thereafter to solve for Δx_m with the different right-hand sides, $-f(x_m)$, where $x_{m+1} = x_m + \Delta x_m$, $m = 0, 1, 2, \ldots$. The Gaussian elimination algorithm of Chapter 4 is especially well suited to this iteration, since the elimination on $J(x_0)$ is done only once. We shall prove that (5.2.5) defines a sequence convergent to a zero x^* of f under rather simple conditions on $f'(x)$. However, we cannot expect the convergence to be quadratic in general. Thus, there are fewer computations per iteration but more iterations are required for a given accuracy as compared to (5.2.1).

Theorem 5.5.1 Let $f: X \longrightarrow Y$, X and Y real Banach spaces, have a continuous derivative $f'(x)$ for x in some open set U. Let $x_0 \in U$ be such that $(f'(x_0))^{-1}$ exists. For some L, $0 < L < 1$, let r be such that the inequality

i) $\|(f'(x_0))^{-1}\| \|f'(x) - f'(x_0)\| < L$

holds for all $x \in \bar{B}(x_0, r)$. Suppose that

ii) $\|(f'(x_0))^{-1} f(x_0)\| < r(1 - L)$.

Then there exists a unique zero x^* of f in $\bar{B}(x_0, r)$. The zero x^* is the limit of the sequence (x_m) defined by the modified Newton method (5.2.5) and

$$\|x^* - x_{m+1}\| \le L \|x^* - x_m\|$$

$$\|x^* - x_m\| \le \frac{L^m}{1 - L} \|x_1 - x_0\|.$$

Proof. First, we note that for any $0 < L < 1$, there exists $r > 0$ and a closed ball $\bar{B}(x_0, r)$ in which the inequality (i) above is satisfied. This is a direct consequence of the hypothesis that $f'(x)$ is continuous at x_0.

Now, let $g(x) = x - (f'(x_0))^{-1} f(x)$. Then $g'(x) = I - (f'(x_0))^{-1} f'(x)$. Hence, for all $x \in \bar{B}(x_0, r)$, $\|g'(x)\| \le \|(f'(x_0))^{-1}\| \|f'(x_0) - f'(x)\| < L < 1$. By the mean-value theorem it follows that g is contractive in $\bar{B}(x_0, r)$ with Lipschitz constant L. From (5.2.5) and condition (ii) of the theorem it follows that

$$\|x_1 - x_0\| \le \|(f'(x_0))^{-1} f(x)\| < r(1 - L).$$

Therefore, g satisfies the hypotheses of Theorem 5.1.3. The conclusions follow immediately from that theorem. ∎

Another modification is based on a generalization of the regula falsi given in (5.2.4) for the one-dimensional case. In certain cases, the derivative may be too difficult to compute exactly. In such cases, it may be possible to approximate $f'(x_m)$ by a bounded linear operator T_m such that for all $h \in X$ and s sufficiently small

$$T_m h = \frac{f(x_m + sh) - f(x_m)}{s}. \tag{5.5.1}$$

5.5 Modifications of Newton's Method

If s is sufficiently small and T_m satisfying (5.5.1) exists, then by Definition 5.3.4,

$$\|T_m h - f'(x_m) h\| < \varepsilon < \frac{1}{\|f'(x_m)\|^{-1}}$$

for all h with $\|h\| = 1$. Hence,

$$\|T_m - f'(x_m)\| < \varepsilon < \frac{1}{\|(f'(x_m))^{-1}\|}$$

and T_m^{-1} exists by Theorem 3.5.2, Corollary 4. Furthermore, $\|T_m^{-1} - (f'(x_m))^{-1}\|$ is small if ε is small, which is the case if $|s|$ is small. Therefore, it seems plausible that the sequence defined by

$$x_{m+1} = x_m - T_m^{-1} f(x_m) \tag{5.5.2}$$

should converge whenever the Newton iterates converge. In fact, we need not insist that T_m satisfy (5.5.1) but simply that it be close to $f'(x_m)$. Since $f'(x_m)$ is close to $f'(x_0)$ when f' is continuous and all x_m are in a sufficiently small ball $\bar{B}(x_0, r)$, it suffices to require all T_m to be close to $f'(x_0)$ as in the next theorem.

Theorem 5.5.2 Let $f: X \longrightarrow Y$, X a real Banach space, have a continuous derivative $f'(x)$ for x in some open set U. Let $x_0 \in U$ be such that $(f'(x_0))^{-1}$ exists. For $0 < L < 1$ let $r > 0$ be such that the inequality

i) $\|(f'(x_0))^{-1}\| \|f'(x) - f'(x_0)\| < L/3$

holds for all $x \in B(x_0, r)$. Suppose that

ii) $\|(f'(x_0))^{-1} f(x_0)\| < r(1 - L)$

and that there exists a sequence (T_m) of bounded linear operators on X to Y such that for all $m \geq 0$

iii) $\|(f'(x_0))^{-1}\| \|T_m - f'(x_0)\| < L/3$.

Then T_m^{-1} exists and the sequence (x_m) defined by (5.5.2) converges to a unique zero $x^* \in B(x_0, r)$ and

$$\|x^* - x_{m+1}\| \leq \frac{2L}{3 - L} \|x^* - x_m\|. \tag{5.5.3}$$

Proof. We shall show by induction that $x_m \in \bar{B}(x_0, r)$. From (5.5.2) we have

$$x_1 - x_0 = -T_0^{-1} f(x_0) = -T_0^{-1} f'(x_0) (f'(x_0))^{-1} f(x_0).$$

By condition (iii) of the theorem,

$$\|(f'(x_0))^{-1} T_0 - I\| < L/3.$$

Applying Theorem 3.5.1, Corollary 2 (with $b = (f'(x_0))^{-1} T_0$ and $b^{-1} = T_0^{-1} f'(x_0)$), we obtain

$$\|T_0^{-1} f'(x_0)\| \leq \frac{1}{1 - L/3} = \frac{3}{3 - L}.$$

Hence, using condition (ii), we find

$$\|x_1 - x_0\| \leq \frac{3}{3-L} r(1-L) < r,$$

so that $x_1 \in \bar{B}(x_0, r)$. Now, assume that $x_1, \ldots, x_q \in \bar{B}(x_0, r)$. Then from (5.5.2) it follows that for $k = 0, \ldots, q$,

$$f(x_{k+1}) = f(x_{k+1}) - f(x_k) - T_k(x_{k+1} - x_k)$$
$$= f(x_{k+1}) - f(x_k) - f'(x_0)(x_{k+1} - x_k) + (f'(x_0) - T_k)(x_{k+1} - x_k).$$

Now, by condition (i) and Exercise 5.16,

$$\|f(x_{k+1}) - f(x_k) - f'(x_0)(x_{k+1} - x_k)\| \leq \sup_{x \in B} \|f'(x) - f'(x_0)\| \|x_{k+1} - x_k\|$$
$$< \frac{L}{3} \frac{\|x_{k+1} - x_k\|}{\|(f'(x_0))^{-1}\|}.$$

By condition (iii),

$$\|(f'(x_0) - T_k)(x_{k+1} - x_k)\| \leq \frac{L}{3} \frac{\|x_{k+1} - x_k\|}{\|(f'(x_0))^{-1}\|}.$$

Hence,

$$\|f(x_{k+1})\| \leq \frac{2L}{3} \frac{\|x_{k+1} - x_k\|}{\|(f'(x_0))^{-1}\|}.$$

Using (5.5.2), we obtain for $k = 0, \ldots, q-1$,

$$\|x_{k+2} - x_{k+1}\| \leq \|T_{k+1}^{-1}\| \|f(x_{k+1})\|.$$

From condition (iii) and Theorem 3.5.2, Corollary 4, it follows that for all $m \geq 0$,

$$\|T_m^{-1}\| \leq \frac{3}{3-L} \|(f'(x_0))^{-1}\|. \tag{5.5.4}$$

Hence,

$$\|x_{k+2} - x_{k+1}\| \leq \frac{2L}{3-L} \|x_{k+1} - x_k\|. \tag{5.5.5}$$

This yields

$$\|x_{q+1} - x_0\| \leq \sum_{k=0}^{q} \left(\frac{2L}{3-L}\right)^k \|x_1 - x_0\|$$
$$< \frac{1}{1 - 2L/(3-L)} \frac{3(1-L)}{3-L} r = r,$$

which completes the induction.

It follows that (5.5.5) holds for all k, which implies that for any m, q,

$$\|x_{m+q} - x_m\| \leq \sum_{k=m}^{m+q-1} \left(\frac{2L}{3-L}\right)^k \|x_1 - x_0\|$$

$$\leq \left(\frac{2L}{3-L}\right)^m r.$$

Since $2L/(3-L) < 1$, (x_m) is a Cauchy sequence and converges to a point $x^* \in B(x_0, r)$. From $T_m(x_{m+1} - x_m) = -f(x_m)$ it follows that

$$\|f(x_m)\| \leq \|T_m\| \|x_{m+1} - x_m\| \to 0,$$

since $\|T_m\|$ is bounded by condition (iii). Since

$$\lim_{m \to \infty} f(x_m) = f(x^*),$$

by continuity, $f(x^*) = 0$.

Furthermore,

$$x^* - x_{m+1} = x^* - x_m + T_m^{-1} f(x_m)$$
$$= T_m^{-1}(f(x_m) - f(x^*) - T_m(x_m - x^*))$$
$$= T_m^{-1}(f(x_m) - f(x^*) - f'(x_0)(x_m - x^*))$$
$$+ T_m^{-1}(f'(x_0) - T_m)(x_m - x^*).$$

Again using Exercise 5.16, (5.5.4) above and condition (iii) of the theorem, we obtain (5.5.3).

If $x \in B(x_0, r)$ is such that $f(x) = 0$, then applying Exercise 5.16 once more, we have

$$\|x^* - \bar{x}\| \leq \|(f'(x_0))^{-1}\| \|f(x^*) - f(\bar{x}) - f'(x_0)(x^* - \bar{x})\|$$

$$\leq \frac{L}{3} \|x^* - \bar{x}\|,$$

which implies $\bar{x} = x^*$, establishing uniqueness. ∎

Corollary Under the hypotheses of the theorem and with conditions (i) and (ii), the Newton iterates converge to a unique zero $x^* \in B(x_0, r)$. The rate of convergence is given by (5.5.3).

Proof. Take $T_m = f'(x_m)$. Then condition (i) implies (iii) if we can show $x_m \in B(x_0, r)$ for all Newton iterates. This follows by induction just as in the proof of the theorem. ∎

The corollary shows that Newton's method converges even when we do not assume the existence of the second derivative. Of course, the rate of convergence may not be quadratic. In the case $X = Y = V^n$, the operators T_m may be chosen in such a

Solution of Nonlinear Equations

way as to produce a rate close to quadratic convergence. The choice of T_m is motivated by (5.5.1), that is, we try to use the difference quotient as an approximation to the derivative. We do this as follows.

Let U be an orthogonal $n \times n$ matrix. For example, we could take $U = I$. Let $\{e_j : 1 \leq j \leq n\}$ be an orthonormal basis in V^n. For any sequence of points $x_m \in V^n$ we define the matrices $S_m, R_m, T_m, m \geq 1$, by

$$S_m = c \, \|x_m - x_{m-1}\| \, U, \tag{5.5.6}$$

$$R_m e_j = f(x_m + S_m e_j) - f(x_m), \quad 1 \leq j \leq n, \tag{5.5.7}$$

$$T_m = R_m S_m^{-1}, \tag{5.5.8}$$

where c is a constant to be specified. Equation (5.5.7) specifies the column vectors of R_m. For $h = e_j$ and $U = I$, we have

$$T_m h = T_m e_j = \frac{f(x_m + c \, \|x_m - x_{m-1}\| \, e_j) - f(x_m)}{c \, \|x_m - x_{m-1}\|}.$$

Thus, $T_m e_j$ is an approximation to the differential $df(x_m; e_j) = f'(x_m) e_j$, $1 \leq j \leq m$. Now, for arbitrary $h = \sum_{j=1}^{n} h_j e_j$ we have

$$\|T_m h - f'(x_m) h\| \leq \sum_{j=1}^{n} |h_j| \, \|T_m e_j - f'(x_m) e_j\| \leq \|h\|_1 \max_j \|T_m e_j - f'(x_m) e_j\|.$$

which means that

$$\|T_m - f'(x_m)\| \leq \max_j \|T_m e_j - f'(x_m) e_j\|.$$

Therefore, if c is chosen so that $T_m e_j$ is a good approximation to $f'(x_m) e_j$ for $1 \leq j \leq m$, then T_m will be a good approximation to $f'(x_m)$ in the operator norm topology and the sequence (x_m) determined by

$$x_{m+1} = x_m - T_m^{-1} f(x_m), \quad m = 1, 2, \ldots, \tag{5.5.9}$$

should be close to the Newton iterates and converge at nearly the same rate.

Theorem 5.5.3 Let $f : V^n \longrightarrow V^n$ be such that $f''(x)$ exists in some open subset E and the Jacobian $J(x_1)$ is nonzero for some $x_1 \in E$. Suppose there exists a constant $0 < L \leq \frac{1}{8}$ and a closed ball $\bar{B}(x_1, r) \subset E$ such that

i) $\|J^{-1}(x_1) f(x_1)\| \leq \rho_L (1 - L) r$,

ii) $\|J^{-1}(x_1)\| \, \|f''(x)\| \leq L / r \rho_L$,

for all $x \in B(x_1, r)$, where

$$\rho_L = 1 - \frac{3L}{2(1 - L)^2}.$$

For any $x_0 \in E$ with

$$0 < \|x_1 - x_0\| \leq \frac{\rho_L r}{1 - \rho_L},$$

5.5 Modifications of Newton's Method

the sequence (x_m) determined by (5.5.9) and x_0, where T_m is defined in (5.5.6), (5.5.7), (5.5.8) by choosing

$$c = 2 \min\left\{1, \frac{\sqrt{n}}{2}\right\} \frac{(1 - \rho_L)}{\sqrt{n}},$$

lies in $\bar{B}(x_1, r)$ and converges to a unique zero x^* of f in this ball. Further, there exist constants C_1 and C_2 such that

$$\|x^* - x_{m+1}\| \leq C_1 \|x^* - x_m\|^2 + C_2 \|x^* - x_m\| \|x^* - x_{m-1}\|. \quad (5.5.10)$$

Proof. We shall prove by induction that the following relations hold for every m:

$$\bar{B}(x_m, r_m) \subset E, \quad (5.5.11)$$

where $r_1 = r$ and $r_{m+1} = (1 - \rho_L)r_m$, $m \geq 1$;

$$\|x_m - x_{m-1}\| \leq \frac{\rho_L r_m}{1 - \rho_L}; \quad (5.5.12)$$

$J(x_m)$ is a nonsingular matrix; $\quad (5.5.13)$

$$\|J^{-1}(x_m)f(x_m)\| \leq r_m \rho_L (1 - L); \quad (5.5.14)$$

$$\|J^{-1}(x_m)\| \|f''(x)\| \leq \frac{L}{r_m \rho_L} \quad (5.5.15)$$

for all $x \in \bar{B}(x_m, r_m)$. (All norms are l_2-vector norms and their induced operator norms.)

The relations clearly hold for $m = 1$ as a direct consequence of the hypotheses of the theorem. Suppose they hold for $m = k$. From (5.5.6),

$$\|S_k e_j\| \leq c \|x_k - x_{k-1}\| \quad \text{for } 1 \leq j \leq n.$$

Using Taylor's formula (5.3.27), we obtain

$$\|f(x_k + S_k e_j) - f(x_k) - f'(x_k)S_k e_j\| \leq \tfrac{1}{2} \|f''(y)\| \|S_k e_j\|^2,$$

where $y = x_k + \theta S_k e_j$. Applying (5.5.7), (5.5.12) and (5.5.15) with $m = k$, we get

$$\|(R_k - J(x_k)S_k)e_j\| \leq \frac{c^2 \|x_k - x_{k-1}\|^2 L}{2 r_k \rho_L \|J^{-1}(x_k)\|}.$$

Since

$$\|R_k - J(x_k)S_k\| = \|(R_k S_k^{-1} - J(x_k))S_k\| = \|R_k S_k^{-1} - J(x_k)\| \, c \, \|x_k - x_{k-1}\|,$$

the preceding inequality and (5.5.12) imply that

$$\|R_k S_k^{-1} - J(x_k)\| \leq \frac{n^{1/2} c \|x_k - x_{k-1}\| L}{2 \rho_L r_k \|J^{-1}(x_k)\|} \leq \frac{L}{\|J^{-1}(x_k)\|}. \quad (5.5.16)$$

Since $L < 1$, it follows that $T_k = R_k S_k^{-1}$ has an inverse. Furthermore, by the preceding inequality $\|J^{-1}(x_k)T_k - I\| \leq L$. It follows from Theorem 3.5.1, Corollary 2, that $\|T_k^{-1}J(x_k)\| \leq 1/(1 - L)$. Also, from Theorem 3.5.2, Corollary 4, $\|T_k^{-1}\| \leq \|J^{-1}(x_k)\|/(1 - L)$. Now, from (5.5.9) we have

$$\|x_{k+1} - x_k\| \leq \|T_k^{-1}J(x_k)\| \|J^{-1}(x_k)f(x_k)\|.$$

206   Solution of Nonlinear Equations

Applying (5.5.14) with $m = k$, we obtain
$$\|x_{k+1} - x_k\| \leq r_k \rho_L = \frac{r_k(1 - \rho_L)\rho_L}{1 - \rho_L} = \frac{r_{k+1}\rho_L}{1 - \rho_L}.$$
This establishes the induction step for (5.5.12). It follows immediately that for $x \in \bar{B}(x_{k+1}, r_{k+1})$,
$$\|x - x_k\| \leq \|x - x_{k+1}\| + \|x_{k+1} - x_k\| \leq r_{k+1}\left(1 + \frac{\rho_L}{1 - \rho_L}\right) = r_k.$$
Thus, $\bar{B}(x_{k+1}, r_{k+1}) \subset \bar{B}(x_k, r_k) \subset E$. Also, from the mean-value theorem applied to $f'(x)$ and (5.5.15) with $m = k$ it follows that
$$\|J(x_{k+1}) - J(x_k)\| \leq \frac{L\|x_{k+1} - x_k\|}{r_k \rho_L \|J^{-1}(x_k)\|} \leq \frac{L}{\|J^{-1}(x_k)\|}, \qquad (5.5.17)$$
which shows that $J^{-1}(x_{k+1})$ exists. This completes the induction on (5.5.13). To establish (5.5.14) we note that $f(x_k) = -T_k(x_{k+1} - x_k)$ and
$$f(x_{k+1}) = f(x_{k+1}) - f(x_k) - f'(x_k)(x_{k+1} - x_k) + (f'(x_k) - T_k)(x_{k+1} - x_k).$$
Hence, applying Taylor's formula and (5.5.16), we get
$$\|J^{-1}(x_{k+1})f(x_{k+1})\| \leq \|J^{-1}(x_{k+1})\| \left(\frac{\|f''(x)\|}{2}\|x_{k+1} - x_k\|^2 + \frac{L\|x_{k+1} - x_k\|}{\|J^{-1}(x_k)\|}\right).$$
But (5.5.17) and Theorem 3.5.2, Corollary 4 imply that
$$\|J^{-1}(x_{k+1})\| \leq \|J^{-1}(x_k)\|/(1 - L).$$
Hence
$$\|J^{-1}(x_{k+1})f(x_{k+1})\| \leq \frac{\|J^{-1}(x_k)\| \|f''(x)\| \|x_{k+1} - x_k\|^2}{2(1 - L)} + \frac{L\|x_{k+1} - x_k\|}{1 - L}.$$
By the induction hypothesis on (5.5.15),
$$\|J^{-1}(x_{k+1})f(x_{k+1})\| \leq \frac{L}{(1 - L)}\|x_{k+1} - x_k\|\left(\frac{\|x_{k+1} - x_k\|}{2\rho_L r_k} + 1\right).$$
But (5.5.12) holds for $m = k + 1$ so that the term on the right is
$$\leq \frac{L}{(1 - L)} \frac{\rho_L r_{k+1}}{(1 - \rho_L)}\left(\frac{1}{2} + 1\right) = \frac{3}{2} \frac{L}{(1 - L)^2} \frac{\rho_L r_{k+1}(1 - L)}{(1 - \rho_L)}$$
$$= r_{k+1}(1 - L)\rho_L,$$
by the definition of ρ_L. This completes the induction on (5.5.14).
Similarly,
$$\|J^{-1}(x_{k+1})\| \|f''(x)\| \leq \frac{\|J^{-1}(x_k)\| \|f''(x)\|}{1 - L} \leq \frac{L}{(1 - L)r_k \rho_L}$$
$$= \frac{L(1 - \rho_L)}{(1 - L)r_{k+1}\rho_L} = \frac{3L}{2(1 - L)^3} \frac{L}{r_{k+1}\rho_L},$$
and (5.5.15) follows, since $3L/2(1 - L)^3 < 1$ for $0 < L \leq \frac{1}{8}$. This completes the induction. It follows that T_m^{-1} exists for all m and the sequence (x_m) is defined.

Since (5.5.12) holds for all m, we have $\|x_m - x_{m-1}\| \le r_m = (1 - \rho_L)^m r_1$, which implies that (x_m) is a Cauchy sequence. Therefore, there exists a limit $x^* \in \bar{B}(x_1, r)$. But by (5.5.9), $T_m(x_{m+1} - x_m) = -f(x_m)$ and $(\|T_m\|)$ is a bounded sequence as a consequence of (5.5.15), (5.5.16). Hence, $x_m \to x^*$ implies that $f(x^*) = 0$.

It remains to establish (5.5.10). Since $f(x^*) = 0$, we have by (5.5.9)

$$x^* - x_{m+1} = x^* - x_m + T_m^{-1} f(x_m)$$
$$= -T_m^{-1}(f(x^*) - f(x_m) - f'(x_m)(x^* - x_m))$$
$$+ (I - T_m^{-1} f'(x_m))(x^* - x_m).$$

Applying Taylor's formula, we find

$$\|x^* - x_{m+1}\| \le \|T_m^{-1}\| \frac{K}{2} \|x^* - x_m\|^2 + \|I - T_m^{-1} J(x_m)\| \|x^* - x_m\|,$$

where $K = L/r\rho_L \|J^{-1}(x_1)\|$ by condition (ii). Again, by condition (ii) and the mean-value theorem, for any $x \in \bar{B}(x_1, r)$,

$$\|J(x) - J(x_1)\| \le \frac{L \|x - x_1\|}{r \rho_L \|J^{-1}(x_1)\|} \le \frac{L}{\rho_L \|J^{-1}(x_1)\|}.$$

By Theorem 3.5.2, Corollary 4,

$$\|J^{-1}(x)\| \le \frac{\rho_L}{\rho_L - L} \|J^{-1}(x_1)\|.$$

Hence,

$$\|T_m^{-1}\| \le \frac{\|J^{-1}(x_m)\|}{1 - L} \le \frac{\rho_L \|J^{-1}(x_1)\|}{(1 - L)(\rho_L - L)}.$$

Also, by (5.5.16)

$$\|I - T_m^{-1} J(x_m)\| \le \|T_m^{-1}\| \|T_m - J(x_m)\|$$
$$\le \frac{L}{1 - L} \frac{\|x_m - x_{m-1}\|(1 - \rho_L)}{2 \rho_L r_m}.$$

Therefore,

$$\|x^* - x_{m+1}\| \le \left(\frac{1}{2r(\rho_L - L)} + \frac{(1 - \rho_L)}{2\rho_L r_m} \right) \|x^* - x_m\|^2$$
$$+ \frac{L(1 - \rho_L)}{2(1 - L)\rho_L r_m} \|x^* - x_m\| \|x^* - x_{m-1}\|,$$

which establishes (5.5.10). ∎

The inequality (5.5.10) is not a quadratic convergence law. If $\|x^* - x_m\| \le 10^{-q}$ and $\|x^* - x_{m-1}\| \le 10^{-t}$, then $\|x^* - x_{m+1}\| \le C_2 10^{-(q+t)} + C_1 10^{-2q}$. Assuming that $q \ge t$, the first term dominates. The number of correct digits in x_{m+1} is roughly the sum of the number of correct digits in x_m and x_{m-1}. (One might call this "Fibonacci convergence," in reference to the Fibonacci sequence 1, 1, 2, 3, 5, ..., [5.5].)

The method of Theorem 5.5.3 was developed as a generalization of the regula falsi given by Eq. (5.2.4). The latter is also known as the *secant method*. In the one-dimensional case, it is easy to see that (5.2.4) amounts to approximating the curve $y = f(x)$ by the secant (or chord) joining the points $(x_m, f(x_m))$ and $(x_{m-1}, f(x_{m-1}))$. This line has the equation

$$y = f(x_m) + \frac{f(x_m) - f(x_{m-1})}{x_m - x_{m-1}}(x - x_m).$$

Setting $y = 0$ and solving for x yields (5.2.4). If we consider the inverse function $x = f^{-1}(y)$ (assumed to exist in a neighborhood of the zero x^*), we have $x_m = f^{-1}(y_m)$ and $x_{m-1} = f^{-1}(x_{m-1})$, where $y_m = f(x_m)$, $y_{m-1} = f(x_{m-1})$. The problem is to find $x^* = f^{-1}(0)$. This may now be regarded as an interpolation problem for the function f^{-1}; i.e. we wish to find the unknown value $f^{-1}(0)$ by interpolating the known values $f^{-1}(y_m)$ and $f^{-1}(y_{m-1})$. As we shall see in Chapter 8, (5.2.4) is a linear interpolation formula for f^{-1}.

In the one-dimensional case, the regula falsi can be modified in a way which insures convergence for any continuous function f. Suppose x_m and x_{m-1} are such that $f(x_m)$ and $f(x_{m-1})$ have opposite sign. Then there exists a zero x^* between x_m and x_{m-1}. We determine x_{m+1} by formula (5.2.4). However, before proceeding with the next iteration, we determine which of the two points x_m, x_{m-1} is such that the value of f has the opposite sign to $f(x_{m+1})$. We relabel that point as x'_m and proceed to find x_{m+2} by the iteration step (5.2.4) using x_{m+1} and x'_m. It is clear that

$$\min\{x_{m-1}, x_m\} \leq \min\{x'_m, x_{m+1}\}$$
$$\leq x^* \leq \max\{x'_m, x_{m+1}\} \leq \max\{x_{m-1}, x_m\}.$$

Thus, x^* is enclosed in a sequence of intervals $[a_m, b_m]$ such that $a_m \leq a_{m+1} \leq b_{m+1} \leq b_m$ for all m. Furthermore, either $b_m - a_m \to 0$, or $x^* = \lim a_m$ or $\lim b_m$ for some m. Indeed we have $a = \sup\{a_m\} \leq x^* \leq \inf\{b_m\} = b$, where $f(a) \leq 0$ and $f(b) \geq 0$. If $f(a) < 0 < f(b)$, then for m sufficiently large, there is a pair x_m, x_{m-1} such that the point x_{m+1} determined by (5.2.4) lies between a and b. This contradicts the definition of a and b.

Further results on the convergence of Newton's method without assumptions on the second derivative are to be found in [5.7] and [5.8].

A variation of the modified Newton method in which the derivative is re-evaluated every few iterations is discussed in [5.13]. This leads to a scheme of the type

$$x_{n+1} = x_n - (f'(x_{n_i}))^{-1} f(x_n),$$

where (x_{n_i}) is a subsequence of (x_n) with $n_0 = 0$.

5.6 ZEROS OF POLYNOMIALS IN ONE VARIABLE

In previous sections of this chapter, we have imposed only the weakest restrictions on the function f and these restrictions have been topological (contractive mapping) or analytical (differentiability). No algebraic properties of f were considered. Of course, in Chapter 4 we considered the finite-dimensional linear function $f(x) = Ax - b$ and developed methods of solving $f(x) = 0$ based on algebraic properties. The simplest example of a nonlinear function is the polynomial of degree n ($n \geq 0$),

$$p(x) = a_n x^n + a_{n-1} x^{n-1} + \cdots + a_1 x + a_0, \qquad a_n \neq 0.$$

In Exercise 5.15, an efficient algorithm is suggested for applying Newton's method to the problem of solving $p(x) = 0$. However, this requires initial guesses for the roots and modifications must be made when the roots (or coefficients) are complex. Since polynomials have many algebraic properties, it seems reasonable to consider special methods which take advantage of them. There is a rather extensive literature on numerical solution of polynomial equations and we shall not attempt to cover it all. One of the most frequently used methods (Muller's) will be discussed in Chapter 8 after we have developed the theory of the interpolating polynomial. Another method based on *Sturm sequences* is especially suited to finding eigenvalues of a symmetric matrix. (See Section 6.6.)

Sturm Sequences

A sequence of polynomials $(p_0, p_1, \ldots, p_n)$ is called a *Sturm sequence* on the interval $[a, b]$ if it has the following properties:

i) p_0 has no zeros in $[a, b]$;

ii) p_n has no multiple roots in $[a, b]$;

iii) if $p_r(\alpha) = 0$, then $p_{r-1}(\alpha) p_{r+1}(\alpha) < 0$ for α in $[a, b]$;

iv) if $p_n(\alpha) = 0$ for α in $[a, b]$, then $p_n'(\alpha) p_{n-1}(\alpha) > 0$. *Sturm's theorem* states that if neither a nor b are zeros of p_n, then the number of zeros of p_n in $[a, b]$ is equal to $w(a) - w(b)$, where $w(x)$ is the number of changes of sign in the sequence $(p_0(x), p_1(x), \ldots, p_n(x))$. In proof, observe that the value of $w(x)$ can change only when x passes through a zero, α say, of some p_r. By (i), $r \geq 1$. By (iii), $p_{r-1}(\alpha)$ and $p_{r+1}(\alpha)$ have opposite sign. Therefore, regardless of how the sign of $p_r(x)$ changes ($+$ to $-$ or $-$ to $+$), the value of $w(x)$ does not change unless $r = n$. Suppose $p_n(\alpha) = 0$. If $p_n'(\alpha) > 0$ and $p_{n-1}(\alpha) > 0$, this means that $p_n(x)$ goes from $-$ to $+$ and $w(x)$ decreases by 1 sign change. If $p_n'(\alpha) < 0$ and $p_{n-1}(\alpha) < 0$, there is also a decrease of 1 in $w(x)$. These are the only two possibilities, by condition (iv). Hence, $w(b)$ is less than $w(a)$ by the number of zeros of p_n in $[a, b]$.

For a given polynomial f of degree n a Sturm sequence is generated by the Euclidean algorithm for finding the greatest common divisor of f and its derivative f'. Set $g_0 = f, g_1 = f'$ and define g_r recursively by

$$g_{r-2} = q_r g_{r-1} - g_r, \qquad r \geq 2;$$

i.e., g_r is the remainder after dividing g_{r-2} by g_{r-1}. The process terminates with $g_{m-1} = q_{m+1} g_m$, where g_m is the greatest common divisor. The sequence of polynomials $p_{m-i} = g_i/g_m$, $0 \le i \le m$, is a Sturm sequence (Exercise (5.17)). The application of Sturm's theorem to the computation of polynomial zeros is discussed in Section 6.6.

Bernoulli's Method

Let $p(x) = a_n x^n + \cdots + a_1 x + a_0$, $a_n \ne 0$ have zeros $z_1, \ldots, z_n$. Let us define the sequence (x_m) by

$$x_m = c_1 z_1^m + \cdots + c_n z_n^m, \quad m = 0, 1, 2, \ldots, \qquad (5.6.1)$$

where the c_i are arbitrary constants. It is easy to see that for any $m \ge n$,

$$a_n x_m + a_{n-1} x_{m-1} + \cdots + a_0 x_{m-n} = 0. \qquad (5.6.2)$$

Equation (5.6.2) is an nth-order linear difference equation. Difference equations and their solutions are discussed in Section 10.1, where it is shown that the sequence (x_m) defined by (5.6.1) is a solution of (5.6.2) and that when the z_i are n distinct zeros every solution is of the form (5.6.1). If we consider the ratio

$$\frac{x_{m+1}}{x_m} = z_1 \frac{c_1 + c_2 \left(\frac{z_2}{z_1}\right)^{m+1} + \cdots + c_n \left(\frac{z_n}{z_1}\right)^{m+1}}{c_1 + c_2 \left(\frac{z_2}{z_1}\right)^{m} + \cdots + c_n \left(\frac{z_n}{z_1}\right)^{m}},$$

we see at once that when $|z_1| > |z_i|$, $i = 2, \ldots, n$,

$$\lim_{m \to \infty} \frac{x_{m+1}}{x_m} = z_1,$$

provided that $c_1 \ne 0$. This is the basis of Bernoulli's method for finding z_1, the zero of maximum modulus, when it is unique. Then by dividing $p(x)$ by $(x - z_1)$ we can obtain a polynomial $p_1(x)$ having a different root of $p(x)$ as the root of maximum modulus. In this way, we can obtain approximations to each root z_i. The Bernoulli method converges at a linear rate. Therefore, as a practical computational tool, it should be used to obtain starting values for a more rapidly convergent method, such as Newton's.

To compute the sequence (x_m) we simply solve the difference equation (5.6.2). This is easily done for $m \ge n$ once we have chosen the starting values $x_0, x_1, \ldots, x_{n-1}$; i.e., by solving (5.6.2) for x_m. The choice $x_0 = x_1 = \cdots = x_{n-2} = 0$, $x_{n-1} = 1$ will insure that $c_1 \ne 0$ as required. This follows from a direct solution of the n linear equations in $c_1, \ldots, c_n$ obtained from (5.6.1) by setting $m = 0, \ldots, n - 1$. (See Section 10.1, the Wronskian, etc.)

If the roots z_i, $i \ge 2$, have multiplicities greater than 1, Bernoulli's method still converges to z_1 provided z_1 has multiplicity 1. Instead of (5.6.1), the general solution of (5.6.2) will contain terms of the form $m^k z_i^m$. (See Section 10.1.) These terms give rise to ratios which converge to zero as before. If z_1 has multiplicity > 1, the method still converges but at a slower rate. (See Exercise 5.18.)

5.6 Zeros of Polynomials in One Variable

If z_1 has a nonzero imaginary part, then the conjugate $\bar{z}_1$ is also a root. In this case, a modification is required. Let $z_1 = re^{i\theta}$. A general solution of (5.6.2) is given by

$$x_m = c_1 z_1^m + \bar{c}_1 \bar{z}_1^m + c_3 z_3^m + \cdots$$
$$= 2r_1 r^m \cos(m\theta + \alpha) + c_3 z_3^m + \cdots,$$

where $c_1 = r_1 e_i^\alpha$. Now, it is easy to verify that $2r_1 r^m \cos(m\theta + \alpha)$ is a solution of the second-order difference equation

$$x_m + ax_{m-1} + bx_{m-2} = 0,$$

with $b = r^2$ and $a = -2r\cos\theta$. Since

$$x_{m+1} + ax_m + bx_{m-1} = 0$$

also holds, we can solve for a and b in terms of $x_{m-2}, x_{m-1}, x_m, x_{m+1}$, four successive values of the computed sequence (x_m). Thus,

$$a = \frac{x_{m+1} x_{m-2} - x_m x_{m-1}}{x_{m-1}^2 - x_m x_{m-2}}, \qquad (5.6.3)$$

$$b = \frac{x_m^2 - x_{m+1} x_{m-1}}{x_{m-1}^2 - x_m x_{m-2}}. \qquad (5.6.4)$$

It is easy to show that r^2 is the limit of the right side of (5.6.4) and $r\cos\theta$ is the limit of the right side of (5.6.3).

Bairstow's Method

If z_1 and its conjugate $\bar{z}_1$ are both zero's of p, then $p(x)$ is divisible by the real quadratic factor $(x - z_1)(x - \bar{z}_1)$. If this factor occurs with multiplicity greater than 1 in the factorization of $p(x)$, then the modification of Bernoulli's method does not apply. An alternative procedure for finding z_1 and $\bar{z}_1$ is Bairstow's method for determining quadratic factors of polynomials. This is an iterative procedure based on Newton's method.

Let p be an nth-degree polynomial as above. If we divide p by any quadratic polynomial $x^2 + sx + t$, we obtain

$$p(x) = q(x)(x^2 + sx + t) + Ax + B, \qquad (5.6.5)$$

where the coefficients $A = A(s, t)$ and $B = B(s, t)$ are functions of s and t. We wish to determine values for s and t which make both A and B zero, that is, we wish to solve the system of two equations in two unknowns

$$A(s, t) = 0, \qquad B(s, t) = 0.$$

The solution can be effected by means of Newton's method if we can determine the partial derivatives $\partial A/\partial s$, $\partial A/\partial t$, $\partial B/\partial s$ and $\partial B/\partial t$ for arbitrary (s, t). A recursive scheme for computing the partials is given in Exercise 5.19.

Error Analysis

It is well-known that the problem of computing zeros of a polynomial can be "ill-conditioned" in the sense that small perturbations in the coefficients can produce large changes (errors) in the zeros. If $p(x) = a_n x^n + a_{n-1} x^{n-1} + \cdots + a_0$, we may consider perturbations of the form $\Delta p = \varepsilon q$, where $q(x) = b_n x^n + b_{n-1} x^{n-1} + \cdots + b_0$. In the theory of algebraic functions, it is shown that if z_1 is a simple zero of p, then for $|\varepsilon|$ sufficiently small the polynomial $p + \varepsilon q$ has a zero $z_1(\varepsilon)$ given by a convergent power series,

$$z_1(\varepsilon) = z_1 + \varepsilon \frac{q(z_1)}{p'(z_1)} + c_2 \varepsilon^2 + c_3 \varepsilon^3 + \cdots.$$

If we consider the set of polynomials of degree $\leq n$ as an $(n+1)$-dimensional vector space of functions p given by their coefficients $(a_n, \ldots, a_0)$, we may regard $z_1 = z_1(p)$ as a functional defined for all vectors $p + \varepsilon q$, $|\varepsilon|$ sufficiently small. Thus, $z_1(\varepsilon) = z_1(p + \varepsilon q)$ and $z_1(0) = z_1(p)$. Introducing a norm, we can consider the weak differential,

$$\lim_{\varepsilon \to 0} \frac{z_1(p + \varepsilon q) - z_1(p)}{\varepsilon} = dz_1(p; q) = \frac{q(z_1)}{p'(z_1)}.$$

Thus, if $|p'(z_1)|$ is small, we could have relatively large changes in z_1 for small ε and $\|q\| = \|p\|$. If z_1 is a root of multiplicity k, then $z_1(\varepsilon)$ is generally a power series in $\varepsilon^{1/k}$ and we see that the weak differential is infinite and the problem of determining z_1 would be "ill-conditioned."

In view of the possible ill-conditioning, it is useful to have an a posteriori error estimate. Let $z_1, \ldots, z_n$ be the zeros of $p(x) = a_n x^n + \cdots + a_0$, with $a_n a_0 \neq 0$. Suppose α is a number such that $|p(\alpha)| \leq \varepsilon$. Since the product of the roots is $(-1)^n a_0 / a_n$ and $p(\alpha) = a_n (\alpha - z_1) \cdots (\alpha - z_n)$,

$$\left| \frac{p(\alpha)}{a_0} \right| = \left| 1 - \frac{\alpha}{z_1} \right| \left| 1 - \frac{\alpha}{z_2} \right| \cdots \left| 1 - \frac{\alpha}{z_n} \right| \leq \frac{\varepsilon}{|a_0|}.$$

It follows that a bound on the relative error in α is given by

$$\min_i \left| \frac{z_i - \alpha}{z_i} \right| \leq \left(\frac{\varepsilon}{|a_0|} \right)^{1/n}.$$

If $z_1 = z_2 = \cdots = z_n$, then the above inequality becomes an equality. Thus,

$$\left| \frac{z_1 - \alpha}{z_1} \right| = \left(\frac{\varepsilon}{|a_0|} \right)^{1/n}.$$

This shows that $|p(\alpha)|$ may be quite small while the relative error in α is unacceptably large. For example, if $\varepsilon = 10^{-8}$ and $n = 8$, then (assuming $a_0 = 1$) we can have a relative error in α as large as 10^{-1}; i.e., if a precision of 8 digits is being used, then α is accurate only in the first digit. In this case, conclusions based on $|p(\alpha)|$ can be deceptive. On the other hand, if the roots are distinct and well separated, the estimate given above can be much too conservative, since the relative error will very likely be less than $|p(\alpha)/a_0|$.

5.7 ACCELERATION OF CONVERGENCE OF SEQUENCES

Consider a sequence (x_m) generated by (5.1.1) and suppose it converges to a fixed point, x^*, of g; i.e., $x^* = g(x^*)$. As we have seen, the rate of convergence of x_m to x^* is linear, in general. Suppose that x_m is one-dimensional. Then it is possible to transform (x_m) into a new sequence (x'_m) which converges to x^* faster than x_m. One simple way to do this is to apply linear extrapolation to the function $f(x) = x - g(x)$, that is, we construct the line passing through successive points $(x_m, f(x_m))$ and $(x_{m+1}, f(x_{m+1}))$ and find the intercept with the x-axis. This is done by the same procedure as in the method of false position (5.2.4) and the x-intercept, x'_{m+1}, is given by

$$x'_{m+1} = x_m - f(x_m)\left(\frac{x_{m+1} - x_m}{f(x_{m+1}) - f(x_m)}\right). \tag{5.7.0}$$

Since

$$f(x_m) = x_m - g(x_m) = x_m - x_{m+1} \text{ and } f(x_{m+1}) - f(x_m)$$
$$= g(x_m) - g(x_{m+1}) - g(x_{m-1}) + g(x_m) = -g(x_{m+1}) + 2g(x_m) - g(x_{m-1}),$$

we have also

$$x'_{m+1} = x_m - \frac{(x_{m+1} - x_m)^2}{x_{m+2} - 2x_{m+1} + x_m}, \tag{5.7.1}$$

Formula (5.7.1) is known as *Aitken's Δ^2-method* [5.9]. (As we shall see later (Chapter 8), the denominator can be expressed in terms of the second difference, Δ^2, which accounts for the name.) Actually, (5.7.1) can be applied to real sequences (x_m) which do not arise from (5.1.1) provided that they satisfy certain convergence conditions. To see what these may be, let us re-examine the vector sequence generated by (5.1.1) under the assumption that $g'(x^*)$ exists. Since x_m converges to x^*, the error $e_m = x_m - x^*$ satisfies

$$e_m = g(x_{m-1}) - g(x^*) = g'(x^*)e_{m-1} + \varepsilon_m,$$

where

$$\frac{\|\varepsilon_m\|}{\|e_{m-1}\|} \to 0 \quad \text{as } m \to \infty.$$

In fact, in the one-dimensional case, the mean-value theorem and continuity of the derivative give

$$e_m = g'(x^* + \theta e_{m-1})e_{m-1} = (g'(x^*) + \delta_m)e_{m-1},$$

where $\delta_m \to 0$ as $m \to \infty$. These considerations motivate us to assume that for large m,

$$e_{m+1} = Ae_m, \tag{5.7.2}$$

where A is an unknown linear operator independent of m. If $\|A\| < 1$, then

$$x_{m+1} - x^* = Ax_m - Ax^*,$$
$$x^* = (I - A)^{-1}(x_{m+1} - Ax_m)$$
$$= x_m + (I - A)^{-1}(x_{m+1} - Ax_m - (I - A)x_m)$$
$$= x_m + (I - A)^{-1}(x_{m+1} - x_m). \tag{5.7.3}$$

Now, it also follows from (5.7.3) that
$$x_{m+2} - x_{m+1} = A(x_{m+1} - x_m), \tag{5.7.4}$$
and subtracting $x_{m+1} - x_m$ from both sides,
$$x_{m+2} - 2x_{m+1} + x_m = (A - I)(x_{m+1} - x_m).$$
In the one-dimensional case, A is simply a scalar and we can solve for $(I - A)^{-1}$, obtaining
$$(I - A)^{-1} = -\frac{x_{m+1} - x_m}{x_{m+2} - 2x_{m+1} + x_m}.$$
Substituting in the above expression for x^*, we again derive the right side of Aitken's method in (5.7.1). If (5.7.2) held exactly, then one application of Aitken's method would yield x^*. As we have seen, (5.7.2) is an approximation to the actual condition satisfied by the sequence (e_m). Therefore, (5.7.1) will often accelerate convergence, as the next theorem shows.

Theorem 5.7.1 Let (x_m) be a real sequence converging to x^*. Let $e_m = x_m - x^*$ satisfy the condition
$$e_{m+1} = (A + \delta_m)e_m,$$
where A is a constant, $|A| < 1$, and $\delta_m \to 0$ as $m \to \infty$. Furthermore, suppose $e_m \neq 0$ for all m. Then the sequence (x'_m) given by (5.7.1) is well-defined for m sufficiently large and
$$\lim_{m \to \infty} \frac{x'_m - x^*}{x_m - x^*} = 0;$$
i.e., (x'_m) converges to x^* faster than (x_m).

Proof. By the condition of the theorem, $e_{m+2} = (A + \delta_{m+1})(A + \delta_m)e_m$. Hence,
$$x_{m+2} - 2x_{m+1} + x_m = e_{m+2} - 2e_{m+1} + e_m = [(A - 1)^2 + \alpha_m]e_m,$$
where $\alpha_m = A(\delta_m + \delta_{m+1}) - 2\delta_m + \delta_m\delta_{m+1}$. Thus, $\alpha_m \to 0$ and for m sufficiently large, $(A - 1)^2 + \alpha_m \neq 0$. This implies that the denominator in (5.7.1) is not zero, so that x'_m is defined. A straightforward calculation yields
$$x'_m - x^* = e_m - \frac{(A - 1 + \delta_m)^2 e_m}{(A - 1)^2 + \alpha_m}$$
and the result follows. ∎

The previous theorem shows that (5.7.1) may be applied to sequences which do not arise from successive approximation, as long as they satisfy the conditions of the theorem.

If (x_m) is a real sequence generated by successive approximation, $x_{m+1} = g(x_m)$, then (5.7.1) suggests the iterative scheme,
$$x_{m+1} = x_m - \frac{(g(x_m) - x_m)^2}{g(g(x_m)) - 2g(x_m) + x_m} = \frac{x_m g(g(x_m)) - (g(x_m))^2}{x_m - 2g(x_m) + g(g(x_m))} \tag{5.7.5}$$

The iterative procedure (5.7.5) is referred to as *Steffensen's method* [5.10]. Under certain conditions, it converges quadratically (Exercise 5.22). This method—and Aitken's as well—can be iterated to produce sequences $(x'_m), (x''_m)$, etc. However, we shall not pursue this further.

5.8 RATES OF CONVERGENCE

The concept of rate of convergence of an iterative method can be made precise by identifying the method with the set of convergent sequences produced by the method.

Definition 5.8.1 An *r-step iterative method*, I, on the normed vector space X is a sequence of mappings $f_m : X^r \longrightarrow X$, $m = 0, 1, 2, \ldots$, and a nonempty set $D \subset X^r$ such that for any r initial values $(x_0, x_1, \ldots, x_{r-1}) \in D$ the sequence (x_m), where

$$x_{m+1} = f_m(x_m, x_{m-1}, \ldots, x_{m-r+1}), \qquad m \geq r - 1,$$

is defined. If $f_m = f$ for all m, then I is called a *stationary* method with *iteration function* f. If $\lim_{m \to \infty} x_m = x^*$, then x^* is called a *limit point* of I.

Example 1. Newton's method is a one-step stationary method, as is the usual successive approximation method (5.1.1).

Example 2. The regula falsi and secant methods are two-step stationary methods.

Example 3. The gradient methods of Chapter 12 are generally not stationary.

We shall characterize the rate of convergence of an iterative method I by studying the set, $c(I, x^*)$, of sequences produced by I which have x^* as limit point. Recalling that quadratic convergence connotes that $\|x_{m+1} - x^*\| \leq L \|x_m - x^*\|^2$, we are led to consider the quotients $\|x_{m+1} - x^*\|/\|x_m - x^*\|^p$, where $1 \leq p < \infty$.

Definition 5.8.2 Let $(x_m) \in c(I, x^*)$ and for each $p \in (1, \infty)$ define

$$Q_p((x_m)) = \begin{cases} 0 & \text{if } x_m = x^* \text{ except for finitely many } m; \\ \limsup_{m \to \infty} \dfrac{\|x_{m+1} - x^*\|}{\|x_m - x^*\|^p} & \text{if } x_m \neq x^* \text{ except for finitely many } m; \\ +\infty & \text{otherwise.} \end{cases}$$

The numbers $Q_p((x_m))$ are called the *quotient convergence factors* or *Q-factors* of (x_m).

Observe that if $Q_p < \infty$, then for $\varepsilon > 0$ there exists m_0 such that for $m \geq m_0$

$$\|x_{m+1} - x^*\| \leq (Q_p + \varepsilon)\|x_m - x^*\|^p;$$

i.e., Q_p is asymptotically the constant factor in an inequality expressing the error at the $(m+1)$-st iteration in terms of the pth power of the error at the mth step. To measure the rate of convergence of I to x^* it seems reasonable to consider all sequences in $c(I, x^*)$ and take Q_p of the slowest.

Definition 5.8.3 The quantity $Q_p = Q_p(I, x^*)$ defined by

$$Q_p(I, x^*) = \sup_{c(I, x^*)} \{Q_p((x_m))\}$$

is called a *Q-factor* of I at x^*.

For each (I, x^*) one value of p plays a distinguished role which determines the *order* of I.

Theorem 5.8.1 Let x^* be a limit point of the iterative method I. Precisely one of the following cases holds:

a) $Q_p(I, x^*) = 0$ for all p;

b) $Q_p(I, x^*) = \infty$ for all p;

c) There exists p_0 such that

$$Q_p(I, x^*) = 0 \quad \text{for } p < p_0$$
$$Q_p(I, x^*) = \infty \quad \text{for } p_0 < p.$$

(Q_{p_0} can have any value including 0 or ∞.)

Proof. First, we show that for any $(x_m) \in c(I, x^*)$ exactly one of (a), (b) or (c) holds. We set $e_m = \|x_m - x^*\|$. If $e_m = 0$ except for finitely many m, then (a) holds by Definition 5.8.2. If e_m is both zero and nonzero for infinitely many m, then (b) holds. Therefore, suppose $e_m > 0$ except for finitely many m. If $Q_p((x_m)) = \infty$ for all p, case (b) obtains. If not, $Q_p((x_m)) < \infty$ for some p and for $q < p$,

$$Q_q((x_m)) = \limsup_{m \to \infty} e_{m+1} \frac{e_m^{p-q}}{e_m^p} \leq Q_p((x_m)) \limsup_{m \to \infty} e_m^{p-q} = 0.$$

Hence, if $Q_q((x_m)) \neq 0$, then $Q_p((x_m)) = \infty$ for $p > q$, which establishes (c).

Now assume that neither (a) nor (b) holds for I and set

$$p_0 = \inf \{p \in [1, \infty) : Q_p(I, x^*) = \infty\}.$$

Let $p > p_0$. Then there exists $q \in [p_0, p)$ such that $Q_q = \infty$. This implies $Q_q((x_m)) > 0$ for some $(x_m) \in c(I, x^*)$. By the preceding paragraph, $Q_p((x_m)) = \infty$ and therefore, $Q_p = \infty$. Similarly, if $p < p_0$ and $Q_p > 0$, then $Q_q = \infty$ for all $q \geq p$, which contradicts the definition of p_0. Hence, $Q_p = 0$ for $p < p_0$, establishing (c). ∎

Definition 5.8.4 The *Q-order* of I at x^* is the number

$$O_Q = \begin{cases} \infty & \text{if } Q_p = 0 \text{ for all } p \\ \inf \{p : Q_p = \infty\} & \text{otherwise.} \end{cases}$$

For example, if $O_Q = 2$, then $Q_p = \infty$ for $p > 2$. This implies that $e_{m+1} \leq L e_m^{2+\varepsilon}$ cannot hold for any $\varepsilon > 0$ and $m \geq m_0$.

Definition 5.8.5 If $O_Q = 2$ and $0 < Q_2 < \infty$, then I is said to have *Q-quadratic convergence* at x^*. If $O_Q = 1$ and $0 < Q_1 < 1$, the convergence is said to be *Q-linear*. If $Q_1 = 0$, the convergence is called *Q-superlinear*. If $Q_1 \geq 1$, the convergence is *Q-sublinear*.

Note that superlinear convergence does not specify the order; e.g. quadratic convergence is superlinear.

It seems reasonable to say that method I_1 is *Q-faster* than I_2 if there exists a p such that $Q_p(I_1, x^*) < Q_p(I_2, x^*)$. If this is true, there cannot be a q such that $Q_q(I_1) > Q_q(I_2)$. If $q > p$, this would imply $0 < Q_p(I_2)$ and $Q_q(I_2) = \infty$, by Theorem 5.8.1. This contradicts $Q_q(I_1) > Q_q(I_2)$. Thus, the relation of Q-faster is well-defined.

To compare I_1 and I_2, we first examine their orders. If $O_Q(I_1) > O_Q(I_2)$, then I_1 is Q-faster than I_2, since we take $O_Q(I_2) < p < O_Q(I_1)$. Then $Q_p(I_1) = 0$ and $Q_p(I_2) = \infty$. In V^n all norms are equivalent. It is not difficult to show that the order O_Q is independent of the norm in V^n used to define it. In fact, $Q_p = 0$ and $Q_p = \infty$ hold independently of the norm used to define Q_p. Thus, if $0 = Q_p(I_1) < Q_p(I_2)$ or $Q_p(I_1) < Q_p(I_2) = \infty$, then I_1 is Q-faster than I_2 in any norm in V^n. However, if $0 < Q_p(I_1) < Q_p(I_2) < \infty$ holds for some norm in V^n, I_1 is Q-faster in that norm but I_2 may be Q-faster in some other norm. A different set of factors which are norm-independent in V^n are the following:

5.8 Rates of Convergence

Definition 5.8.6 Let $(x_m) \in c(I, x^*)$. The *root convergence factors* or *R*-factors of (x_m) are the numbers

$$R_p((x_m)) = \begin{cases} \limsup\limits_{m \to \infty} \|x_m - x^*\|^{1/m} & \text{if } p = 1, \\ \limsup\limits_{m \to \infty} \|x_m - x^*\|^{1/p^m} & \text{if } p > 1. \end{cases}$$

The numbers

$$R_p = R_p(I, x^*) = \sup_{c(I, x^*)} \{R_p((x_m))\}$$

are called the *R-factors* of I at x^*.

Since $e_m < 1$ for $m \geq m_0$, we have $0 \leq R_p((x_m)) \leq 1$. For linear convergence, $e_m \leq L^m e_0$ so that $e_m^{1/m} \leq L e_0^{1/m}$ and $L = R_1$ asymptotically. In the case of quadratic convergence,

$$e_m \leq L^{2^m - 1} e_0^{2^m}, \quad \text{so that} \quad \limsup e_m^{1/2^m} \leq L e_0.$$

It is easy to see that equivalent norms yield the same value of $R_p((x_m))$. For if $\|x\|$ and $\|x\|_j$ are equivalent norms, then $\|x\| \leq K \|x\|_j$ for some constant K. Therefore,

$$\limsup \|x_m - x^*\|^{1/p^m} \leq \lim K^{1/p^m} \limsup \|x_m - x^*\|_j^{1/p^m}$$

$$= \limsup \|x_m - x^*\|_j^{1/p^m}.$$

Reversing the roles of $\|x\|$ and $\|x\|_j$, we obtain equality.

Analogous to Theorem 5.8.1 we have

Theorem 5.8.2 Let x^* be a limit point of the iterative method I. Exactly one of the following cases holds:

a) $R_p = 0$ for all p;
b) $R_p = 1$ for all p;
c) there exists p_0 such that $R_p = 0$ for $p < p_0$ and $R_p = 1$ for $p > p_0$.

Proof. Let $x_m \to x^*$. We shall prove that exactly one of (a), (b), (c) holds for (x_m). Suppose $R_p < 1$. For $\varepsilon > 0$ such that $R_p + \varepsilon < 1$ there exists m_0 such that $\|x_m - x^*\|^{1/p^m} \leq R_p + \varepsilon$ for $m \geq m_0$. Therefore, for $q < p$,

$$R_q((x_m)) = \limsup (\|x_m - x^*\|^{1/p^m})^{(p/q)^m} \leq \lim (R_p + \varepsilon)^{(p/q)^m} = 0.$$

To obtain the result we take $p_0 = \inf \{p : R_p = 1\}$. ∎

Definition 5.8.7 Let x^* be a limit point of the iterative method I. The number

$$O_R = O_R(I, x^*) = \begin{cases} \infty & \text{if } R_p(I) = 0 \text{ for all } p; \\ \inf \{p : R_p = 1\} & \text{otherwise.} \end{cases}$$

is called the *R-order* of I at x^*.

As with the Q-factors, we see that if $0 < R_p < 1$, then $O_R = p$. If $0 < R_1 < 1$, the convergence is *R-linear*. It is *R-superlinear* if $R_1 = 0$. If $0 < R_2 < 1$, the convergence is called *R-quadratic* at x^*.

Theorem 5.8.3
$$O_Q(I, x^*) \leq O_R(I, x^*).$$

Proof. First, we prove that for any $x_m \to x^*$ and $p > 1$, if $Q_p((x_m)) < \infty$, then $R_p((x_m)) < 1$. (This is not true for $p = 1$; see Exercise 5.3.) Let $K = Q_p((x_m)) + \varepsilon$, $\varepsilon > 0$. There exists M such that for $m > M$

$$\|x_{m+1} - x^*\| \leq K \|x_m - x^*\|^p \leq \cdots \leq K^{1+p+\cdots+p^{m-M}} \|x_M - x^*\|^\beta,$$

where $\beta = p^{m-M+1}$. Hence,

$$\|x_{m+1} - x^*\|^{1/p^{m+1}} \leq K^\alpha \|x_M - x^*\|^{1/p^M},$$

where

$$\alpha = (1/p^M)(1/p + 1/p^2 + \cdots + 1/p^{m-M+1}) \leq 1/p^M(p - 1).$$

Therefore, $\|x_{m+1} - x^*\|^{1/p^{m+1}} \leq (L \|x_M - x^*\|)^{1/p^M}$, where $L = \max\{1, K^{1/p-1}\}$. Since $x_m \to x^*$, M can be chosen so that $L \|x_M - x^*\| < 1$. It follows that $R_p((x_m)) < 1$.

Now, let $p_0 = O_Q > 1$. Then for $p < p_0$ we have $Q_p = 0$, by Theorem 5.8.1. Hence, $Q_p((x_m)) = 0$ and therefore, $R_p((x_m)) < 1$. This implies $O_R \geq p_0$, by Theorem 5.8.2. If $O_Q = 1$, then obviously $O_R \geq O_Q$. This completes the proof. ∎

A sufficient condition for O_Q and O_R to be equal is given in the next theorem.

Theorem 5.8.4 Let $1 \leq p < \infty$.

i) Suppose there exists a constant $K > 0$ such that for all $(x_m) \in c(I, x^*)$, $\|x_{m+1} - x^*\| \leq K \|x_m - x^*\|^p$ whenever $m \geq M$ where M depends on (x_m). Then $O_R \geq O_Q \geq p$.
ii) Suppose there is a constant $L > 0$ and a sequence $(x_m) \in c(I, x^*)$ such that for m sufficiently large $\|x_{m+1} - x^*\| \geq L \|x_m - x^*\|^p > 0$. Then $O_Q \leq O_R \leq p$.

Hence, if (i) and (ii) hold for some p, then $O_Q = O_R = p$.

Proof. If (i) holds then obviously $Q_p((x_m)) \leq K$ for all $(x_m) \in c(I, x^*)$. Hence, $Q_p \leq K < \infty$, which implies $O_Q \geq p$. By the preceding theorem, $O_R \geq O_Q \geq p$.

If (ii) holds, then for $m \geq M$,

$$\|x_{m+1} - x^*\| \geq L^{1+p+\cdots+p^{m-M}} \|x_M - x^*\|^{p^{m-M+1}}.$$

If $p = 1$, then $R_1((x_m)) \geq \lim_m (L^{m-M+1} \|x_M - x^*\|)^{1/m} = L > 0$.

If $p > 1$, then as in the preceding proof,

$$\|x_{m+1} - x^*\|^{1/p^{m+1}} \geq \min(1, L^{1/(p-1)}) \|x_M - x^*\|^{1/p^M} > 0.$$

This implies $R_p((x_m)) > 0$ and consequently $R_p > 0$. By Theorem 5.8.2, $O^R \leq p$ and by Theorem 5.8.3, $O_Q \leq O_R \leq p$. This completes the proof. ∎

It is easy to see that $R_1 \leq Q_1$. This is certainly true when $Q_1 = \infty$. Suppose $Q_1 < \infty$ and $Q_1((x_m)) < \infty$. Take $K = Q_1((x_m)) + \varepsilon$, $\varepsilon > 0$ arbitrary. Then for $m \geq M$, $\|x_m - x^*\| \leq K^{m-M} \|x_M - x^*\|$. This implies $R_1((x_m)) \leq K \lim \sup (\|x_M - x^*\|/K^M)^{1/m} = K$. Since ε is arbitrary, $R_1((x_m)) \leq Q_1((x_m))$. This holds for all $(x_m) \in c(I, x^*)$. Hence, $R_1 \leq Q_1$. This gives a lower bound for Q_1. (See Exercise 5.3 for further results.)

REFERENCES

5.1. M. M. VAINBERG, *Variational Methods for the Study of Nonlinear Operators*, translated by A. Feinstein, Holden-Day, San Francisco (1964)

5.2. L. A. LIUSTERNIK and V. I. SOBOLEV,, *Elemente der Funktionalanalysis,* Akademie-Verlag, Berlin (1955)

5.3. L. V. KANTOROVICH and G. P. AKILOV, *Functional Analysis in Normed Spaces*, translated by D. E. Brown, Macmillan, New York (1964)

5.4. J. DIEUDONNÉ, *Fundamentals of Modern Analysis*, Academic Press, New York (1960)

5.5. H. A. ANTOSIEWICZ, "Newton's method and boundary value problems," *Journal of Computer and System Sciences* **2**, No. 2, 177–202 (June 1968)

5.6. L. B. RALL, *Computational Solution of Nonlinear Operator Equations*, Wiley, New York (1969)

5.7. H. B. KELLER, "Newton's method under mild differentiability conditions, *Journal of Computer and System Sciences* **4**, No. 1, 15–28 (Feb. 1970)

5.8. A. A. GOLDSTEIN, *Constructive Real Analysis*, Harper and Row, New York (1967)

5.9. A. C. AITKEN, "On Bernoulli's numerical solution of algebraic equations," *Proc. Roy. Soc. Edinburgh* **46**, 289–305 (1926)

5.10. J. F. STEFFENSEN, "Remarks on iteration, *Skand. Aktuar. Tidskr.* **16**, 64–72 (1933)

5.11. A. M. OSTROWSKI, *Solution of Equations and Systems of Equations* (second edition), Academic Press, New York (1966)

5.12. D. C. MULLER, "A method for solving algebraic equations using an automatic computer," *Math. Comp.* **10**, 208–215 (1956)

5.13. J. E. DENNIS, "On the Kantorovich hypothesis for Newton's method," *SIAM J. Numer. Anal.* **6**, No. 3, 493–507 (1969)

5.14. J. M. ORTEGA and W. C. RHEINBOLDT, *Iterative Solution of Nonlinear Equations in Several Variables,* Academic Press, New York (1970)

5.15. M. EDELSTEIN, "On fixed and periodic points under contractive mappings," *J. London Math. Soc.* **37**, 74–79 (1962)

5.16. A. S. HOUSEHOLDER, *The Numerical Treatment of a single Nonlinear Equation*, McGraw-Hill, New York (1970)

5.17. H. RUTISHAUSER, *Der Quotienten-Differenzen-Algorithmus,* Birkhäuser Verlag, Basel (1957)

5.18. P. HENRICI, "Methods of search for polynomial equations, *J.A.C.M.* **17**, (1970)

5.19. D. SHANKS, "Nonlinear transformations of divergent and slowly convergent sequences," *J. Math. Phys.* **34**, 1–42 (1955)

5.20. P. WYNN, "On a device for computing the $e_m(S_n)$ transformation," *Math. Tables Aids Comp.* **10**, 91–96 (1956)

5.21. M. Z. NASHED, "Differentiability and other properties of nonlinear operators," in *Nonlinear Functional Analysis and Applications*, ed. by L. B. Rall, Academic Press, N.Y. (1971)

EXERCISES

5.1 Use formula (5.2.6) to compute the square root of 17 correctly to eight decimal places. Choose different starting values and compare the number of iterations required.

5.2 a) Let $f: R^2 \longrightarrow R$ have partial derivatives $f_x = \partial f/\partial x$ and $\partial f/\partial y = f_y$ in a neighborhood of the point (x_0, y_0). Assume that f_x and f_y are continuous at (x_0, y_0). Let $x(t)$, $y(t)$ be real functions such that $x(t_0) = x_0$ and $y(t_0) = y_0$. Assume further that $x'(t_0)$

220 Solution of Nonlinear Equations

and $y'(t_0)$ exist. Prove that the composite function $g(t) = f(x(t), y(t))$ is differentiable at t_0 and

$$g'(t_0) = f_x(x_0, y_0)x'(t_0) + f_y(x_0, y_0)y'(t_0). \tag{1}$$

[*Hint:* Use the definition of partial derivative to obtain

$$\frac{\Delta y}{\Delta t}\left[\frac{f(x_0 + \Delta x, y_0 + \Delta y) - f(x_0 + \Delta x, y_0)}{\Delta y}\right] = [f_y(x_0 + \Delta x, y_0) + \varepsilon]\frac{\Delta y}{\Delta t}$$

where $\Delta x = x(t) - x_0$, $\Delta y = y(t) - y_0$, and $\varepsilon \to 0$ as $\Delta y \to 0$. Then use the continuity of f_y as Δx, Δy and Δt approach zero.]

b) Let $f(x, y) = xy/\sqrt{x^2 + y^2}$ and $f(0, 0) = 0$. Verify that $f_x(0, 0) = f_y(0, 0) = 0$. Let $x = t$ and $y = mt$, $m \neq 0$. Define $g(t) = f(t, mt) = mt/\sqrt{1 + m^2}$. Verify that $g'(0) = m/\sqrt{1 + m^2}$ and hence, that the composite formula of Exercise 5.2a fails to hold for $g'(0)$. Show that f_x and f_y are not continuous at $(0, 0)$ and that $f(x, y)$ has no weak derivative at $(0, 0)$, although f_x and f_y exist at $(0, 0)$. [*Hint:* If $f_w'(0)$ exists, then $f_w'(0) =$ grad $f(0)$.] However, f has a one-sided directional differential at $(0, 0)$ for every increment h.

c) Prove the result of advanced calculus that if $f : R^n \longrightarrow R$ is such that grad $f = (\partial f/\partial \xi_1, \ldots, \partial f/\partial \xi_n)$ exists in a neighborhood of x and is continuous at x, then

$$f(x + h) = f(x) + \langle \text{grad } f(x), h\rangle + \varepsilon(x, h)$$

where $\varepsilon/\|h\| \to 0$ as $\|h\| \to 0$. Thus, in this case, grad $f(x) = f'(x)$, the Fréchet derivative. Extend this result to the case $f : R^n \to R^n$ by replacing grad f by the Jacobian of f. [*Hint:* Set $h = \sum_1^n a_i e_i$, $h_0 = 0$, $h_j = \sum_1^j a_i e_i$. Then

$$f(x + h) - f(x) = \sum_1^n (f(x + h_j) - f(x + h_{j-1})) = \sum_1^n a_j \frac{\partial f}{\partial \xi_j}(x + h_{j-1} + \theta_j a_j e_j).\ ]$$

d) Let $f : R^2 \longrightarrow R$ be defined by $f(r, \theta) = \cos(r/\theta)$ for $0 < \theta < 2\pi$ and $f(r, 0) \equiv 1$. Verify that the directional differential at $(0, 0)$ is 0 along any ray; i.e., for $\theta \neq 0$, $\partial f/\partial r = (-1/\theta)\sin(r/\theta)$. To see that the weak derivative $f_w'(0) = 0$ exists, observe that $f(r, \theta) - f(0, \theta) = -(r/\theta)\sin(\rho/\theta)$, where $0 < \rho < r$. By continuity, $\sin(\rho/\theta) = \sin(0) + \varepsilon(\rho) = \varepsilon(\rho)$, where $\varepsilon(\rho) \to 0$ as $\rho \to 0$. Hence, $f(r, \theta) - f(0, \theta) = -r\varepsilon/\theta$. Show that the strong (or Fréchet) derivative does not exist at 0. [*Hint:* The strong derivative $f'(0)$ would have to equal the weak derivative. Thus, $f'(0) = 0$. This would imply that $\cos(r/\theta) - 1 = f(r, \theta) - f(0, 0) = \varepsilon$, where $\varepsilon/r \to 0$ as $r \to 0$ (uniformly in θ). But $|\cos r/\theta - 1|/r = 1/r$ for $\theta = 2r/\pi$.]

e) Show that for f in 5.2d the partial derivative $\partial f/\partial y$ is not continuous at $(0, 0)$.

f) Show that $f(r, \theta)$ in 5.2d is not continuous at $(0, 0)$. [*Hint:* In any ball $0 < r < \rho$, find a point (r, θ) with $\theta = 2r/\pi$.]

g) Let $f(x, y) = (x^2 + y^2)^{1/2}$. Show that the one-sided directional differential of f exists at $(0, 0)$ for all increments h but the weak differential does not exist. [*Hint:* For $u = (0, 0)$ and $h = (h_1, h_2)$,

$$\delta f(0; h) = \lim_{t \to 0} \frac{f(u + th) - f(u)}{t} = (h_1^2 + h_2^2)^{1/2} = \|h\|_2. \ ]$$

5.3 a) Given that $x_m \to x^*$ and $x_m \neq x^*$, define the asymptotic error constants,

$$C_p = \lim_{m \to \infty} \frac{\|x_{m+1} - x^*\|}{\|x_m - x^*\|^p}$$

when the limit exists. Prove that $Q_p = 0$ implies $C_p = 0$ and if C_p exists, then $C_p = Q_p$. Further, prove that $0 < C_p < \infty$ implies that $O_Q = O_R = p$.

b) Let $x_m = 1/m$. Verify that $R_1((x_m)) = Q((x_m)) = 1$.

5.4 Let X be a normed vector space. Let $f: X \longrightarrow X$ be defined in a neighborhood of x. Let $A: X \longrightarrow L(X; X)$ be defined in the same neighborhood. Consider $L(X; X)$ as a normed vector space (Chapter 3). Suppose $f'(x)$ and $A'(x)$ exist (Definition 5.3.4). Let $g(x) = A(x) \cdot f(x)$, where $\cdot$ signifies the linear operation. Derive the formula for the "derivative of a product,"

$$g'(x)h = A(x)f'(x)h + A'hf(x).$$

5.5 Prove Lemma 5.3.2 in detail.

5.6 Let $T_s \in L_c(X; Y)$ for all real s and let $f(s) \in X$. Suppose $T_s \longrightarrow T_{s_0}$ as $s \to s_0$, the convergence being in the operator norm, and $f(s) \to f(s_0)$. Prove that $\lim_{s \to s_0} T_s f(s) = T_{s_0} f(s_0)$. [*Hint:* $\|T_s f(s) - T_{s_0} f(s_0)\| \le \|T_s - T_{s_0}\| \|f(s)\| + \|T_{s_0}\| \|f(s) - f(s_0)\|$.]

5.7 a) Let f be continuous on the closed bounded interval $[a, b]$ and have values in a Banach space. Prove that the Riemann integral $\int_a^b f(t)\, dt$ exists. [*Hint:* Since $[a, b]$ is compact, f is uniformly continuous, hence has a modulus of continuity $\omega(\delta)$ such that $\|f(t') - f(t)\| \le \omega(\delta)$ whenever $|t' - t| \le \delta$, where $\omega(\delta) \to 0$ as $\delta \to 0$. Let d be a subdivision $a = t_0 < t_1 < \cdots < t_m = b$ and d' a subdivision $a = t'_0 < t'_1 < \cdots < t'_k = b$ which is a refinement of d; i.e., $\{t_i\} \subset \{t'_j\}$. Prove that

$$\left\|\sum_{i=0}^{m-1} f(t_i)\, \Delta t_i - \sum_{j=0}^{k-1} f(t'_j)\, \Delta t'_j\right\| \le \omega(\delta)(b - a),$$

where

$$\delta = \max_{0 \le j \le k-1} |\Delta t'_j| = \|d'\|.$$

Next, show that for any sequence (d_i) such that $\|d_i\| \to 0$ the sums $S_{d_i} = \sum_{i=0}^{m_i - 1} f(t_i)\, \Delta t_i$ form a Cauchy sequence in Y. Thus, for S_{d_q}, S_{d_p} consider S_d, where d is a subdivision containing all points of d_q and d_p. Since d is a refinement of both d_q and d_p, it follows that $\|S_d - S_{d_q}\| \le \omega(\delta)(b - a)$, where $\delta = \max\{|\Delta t_i|\} = \|d\|$. Similarly, $\|S_d - S_{d_p}\| \le \omega(\delta)(b - a)$. For p, q sufficiently large, $\|d\| \le \min\{\|d_p\|, \|d_q\|\}$ can be made arbitrarily small.]

b) Prove that if $f'(t)$ is continuous in $[a, b]$, then

$$\int_a^b f'(t)\, dt = f(b) - f(a).$$

[*Hint:* Let F be an arbitrary bounded linear functional on Y. Show that

$$F \int_a^b f'(t)\, dt = \int_a^b F(f'(t))\, dt = \int_a^b \frac{d}{dt} F(f(t))\, dt$$

$$= Ff(b) - Ff(a). \ ]$$

5.8 Prove Theorem 5.3.2. [*Hint:* Let $\varepsilon(u, h) = f(u + h) - f(u) - f'_w(u) \cdot h$. Let $g^* \in Y'$. By Lemma 5.3.6,

$$\langle \varepsilon(u, h), g^* \rangle = \langle f(u + h) - f(u), g^* \rangle - \langle f'_w(u) \cdot h, g^* \rangle$$
$$= \langle f'_w(u + \theta h) \cdot h, g^* \rangle - \langle f'_w(u) \cdot h, g^* \rangle$$
$$= \langle [f'_w(u + \theta h) - f'_w(u)] \cdot h, g^* \rangle.$$

By the Hahn-Banach theorem (Appendix II) there exists $g^* \in Y'$ with $\|g^*\| = 1$ such that $|\langle \varepsilon(u, h), g^* \rangle| = \|\varepsilon(u, h)\|$. For this g^*, the above gives

$$\|\varepsilon(u, h)\| \le \|f'_w(u + \theta h) - f'_w(u)\| \|h\|.$$

By the continuity of f'_w, it follows that

$$\lim_{\|h\| \to 0} \frac{\|\varepsilon(u, h)\|}{\|h\|} = 0.$$

Even simpler, use (5.3.11).]

5.9 Let $f: X^n \longrightarrow V$ be a multilinear mapping, X and V real vector spaces.

a) Show that $f(sh)^n = s^n f h^n$ for any real scalar s and $h \in X$.

b) Show that if f is symmetric, then the multinomial expansion

$$f(s_1 h_1 + \cdots + s_k h_k)^n = \sum_{n_1 + \cdots + n_k = n} \frac{n!}{n_1! \cdots n_k!} s_1^{n_1} \cdots s_k^{n_k} f h_1^{n_1} \cdots h_k^{n_k}$$

holds.

c) Show that

$$\frac{\partial^n}{\partial s_1 \cdots \partial s_n} f(s_1 h_1 + \cdots + s_n h_n)^n = n! f h_1 \cdots h_n.$$

d) Using (b), (c) prove that $fh^n = gh^n$ for all $h \in X$ implies that $f h_1 \cdots h_n = g h_1 \cdots h_n$ for all $h_1, \cdots, h_n \in X$; i.e., $f = g$ as multilinear mappings.

e) Let $f: X \longrightarrow V$ and suppose the strong derivative f'' exists in a neighborhood of $u \in X$ and is continuous at u. Show that $f''(u)$ is a symmetric bilinear mapping.

[*Hint:* Let $g(t_1, t_2) = f(u + t_1 h_1 + t_2 h_2)$. By (5.3.1),

$$\frac{\partial^2 g}{\partial t_2 \partial t_1} = f''(u + t_1 h_1 + t_2 h_2) h_2 h_1,$$

$$\frac{\partial^2 g}{\partial t_1 \partial t_2} = f''(u + t_1 h_1 + t_2 h_2) h_1 h_2.$$

Use the result from ordinary calculus to obtain equality of the cross partials. In fact, it can be proved that $f''(u)$ is symmetric even without the assumption that f'' is continuous at u. (See [5.14], page 77.) This is not true of the weak derivative.]

5.10. The system of equations

$$f_1(x, y, z) = x^2 + y^2 + z^2 - 4.12 = 0,$$
$$f_2(x, y, z) = x^2 + y^2 + z^2 - 6.43 = 0,$$
$$f_3(x, y, z) = x^2 + y^2 + z - 6.52 = 0,$$

has a solution in a ball B centered at $(2, 1, 1)$ and of radius 0.2 in the l_∞-norm. Use Newton's method to determine the solution correctly to six digits. [*Hint:* Obtain expressions for $\partial f_i/\partial x$, $\partial f_i/\partial y$ and $\partial f_i/\partial z$. Form the Jacobian J of $f = (f_1, f_2, f_3)$ and solve the linear systems

$$J(u_m) \Delta u_m = J(u_m) \cdot (u_{m+1} - u_m) = -f(u_m), \quad m = 0, 1, 2, \ldots,$$

where $u_0 = (x_0, y_0, z_0,)$ is some initial guess in B, and $u_{m+1} = u_m + \Delta u_m$. (See Exercise 5.25.)]

5.11 Show by a counter-example that if the hypothesis of Theorem 5.1.2 is changed to the weak inequality

$$d(g(x), g(y)) \leq d(x, y), \tag{1}$$

then the conclusion is no longer true. [*Hint:* Consider a real function g which maps the set $D = [0, 1] \cup [2, 3]$ into itself. Since $|g(x) - g(y)| \leq |x - y|$, the graph may look as shown in Fig. 5.5.]

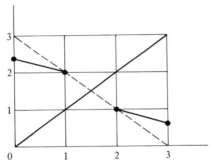

Figure 5.5

5.12 Consider the problem of solving $Ax = b$, A a bijective linear operator on a vector space X to itself. Let $f(x) = Ax - b$. Convert the problem of finding a zero of $f(x)$ to a fixed-point problem for a function $g(x) = x - M^{-1}f(x)$, where M is an arbitrary bijective linear operator. Show that the method of successive approximations applied to g is just the iterative method given in (4.4.4). Show that a sufficient condition that g be contractive is that $\|Q\| = \|I - M^{-1}A\| < 1$.

5.13 Let $f: X \longrightarrow X$, where X is a normed vector space. Suppose $g(x) = (f'(x))^{-1}$ exists in a neighborhood of $x_0 \in X$ and $f''(x_0)$ exists. Prove that

$$g'(x_0)h = -(f'(x_0))^{-1}(f''(x_0)h)(f'(x_0))^{-1}.$$

[*Hint:* Note that $f'(x_0)$, $f''(x_0)h$ and $(f'(x_0))^{-1}$ are bounded linear operators on X to X.

$$\begin{aligned}
g(x_0 + h) - g(x_0) &= (f'(x_0 + h))^{-1}[I - f'(x_0 + h)(f'(x_0))^{-1}] \\
&= (f'(x_0 + h))^{-1}[f'(x_0) - f'(x_0 + h)](f'(x_0))^{-1} \\
&= (f'(x_0 + h))^{-1}[-f''(x_0)h + \varepsilon_1](f'(x_0))^{-1} \\
&= (f'(x_0))^{-1}(f''(x_0)h)f'(x_0)^{-1} + \varepsilon,
\end{aligned}$$

where $\|\varepsilon\|/\|h\| \to 0$ as $\|h\| \to 0$, since $(f'(x))^{-1}$ is continuous in x at x_0.]

5.14 Let g be a continuous mapping of the closed bounded interval $[a, b]$ into itself. Show that g has a fixed point.

5.15 Let $p(x) = a_n x^n + \cdots + a_1 x + a_0$, $a_n \neq 0$, be a polynomial of degree n. Define the quantities

$$p_0 = a_n, \quad p_k = xp_{k-1} + a_{n-k}, \quad k = 1, \ldots, n$$
$$q_0 = p_0, \quad q_k = xq_{k-1} + p_k, \quad k = 1, \ldots, n-1.$$

Show that $p(x) = p_n$ and $p'(x) = q_{n-1}$. (This yields an efficient algorithm for computing $p(x_m)$ and $p'(x_m)$ when applying Newton's method to the computation of the real zero's of a polynomial.)

5.16 Let $f: X \longrightarrow Y$, X and Y normed vector spaces. Let $f'(x)$ exist for all $x \in B(x_0, r)$. Prove that for any two points $x_1, x_2 \in B(x_0, r)$,

$$\|f(x_2) - f(x_1) - f'(x_0)(x_2 - x_1)\| \leq \sup_{x \in B} \|f'(x) - f'(x_0)\| \|x_2 - x_1\|.$$

[*Hint:* Let $g(x) = f(x) - f(x_1) - f'(x_0)(x - x_1)$. Apply the mean-value theorem (formula (5.3.10)) to $g(x_2) - g(x_1)$.]

5.17 Let f be a polynomial of degree n. Show that the sequence of polynomials $p_{m-i} = g_i/g_m$, where the g_i are the remainders in the algorithm for the greatest common divisor g_m of f and f' (see Section 5.6), is a Sturm sequence. [*Hint:* $p_r = q_r p_{r-1} - p_{r-2}$ and $p_0 \equiv 1$. If $p_{r-1}(\alpha) = 0$, then $p_r(\alpha) \neq 0$, since otherwise $p_0(\alpha) = 0$. Thus, $p_{r-2}(\alpha) p_r(\alpha) < 0$.]

5.18 Show that Bernoulli's method converges when the root z_1 of maximum modulus has multiplicity 2. [*Hint:* The general solution of (5.6.2) is with $z_1 = z_2$,

$$x_m = c_1 m z_1^m + c_2 z_1^m + c_3 z_3^m + \cdots$$

$$\frac{x_{m+1}}{x_m} = z_1 \left[\frac{c_1(m+1) + c_2}{mc_1 + c_2} \right] \left[\frac{1 + \frac{c_3}{(m+1)c_1 + c_2} \left(\frac{z_3}{z_1}\right)^{m+1} + \cdots}{1 + \frac{c_3}{mc_1 + c_2} \left(\frac{z_3}{z_1}\right)^{m+1} + \cdots} \right]$$

and both bracketed terms converge to 1.]

5.19 Let

$$x^n + a_1 x^{n-1} + \cdots + a_n = (x^2 + sx + t)(x^{n-2} + b_1 x^{n-3} + \cdots + b_{n-2}) + Ax + B.$$

a) Verify that $a_1 = b_1 + s$, $a_i = b_i + sb_{i-1} + tb_{i-2}$ for $2 \leq i \leq n-2$, $a_{n-1} = A + sb_{n-2} + tb_{n-3}$ and $a_n = B + tb_{n-2}$.

b) Define $b_{-1} = 0$ and show that

$$b_i = a_i - sb_{i-1} - tb_{i-2}, \quad 1 \leq i \leq n-2.$$

Further define

$$b_{n-1} = a_{n-1} - sb_{n-2} - tb_{n-3},$$
$$b_n = a_n - sb_{n-1} - tb_{n-2},$$

and show that $b_{n-1} = A$ and $B = b_n + sb_{n-1}$.

c) Using recursions, determine $\partial A/\partial s$, $\partial A/\partial t$, $\partial B/\partial s$ and $\partial B/\partial t$. [*Hint:* $\partial b_0/\partial s = \partial b_{-1}/\partial s = 0$ and $\partial b_0/\partial t = \partial b_{-1}/\partial t = 0$. Since the b_i are functions of (s, t), differentiate the recursion equation in part (b) to get

$$\frac{\partial b_i}{\partial s} = -b_{i-1} - s\frac{\partial b_{i-1}}{\partial s} - t\frac{\partial b_{i-2}}{\partial s}. \Bigg]$$

5.20 Let $f(x, y)$ be a vector function of two vector variables $x \in X$, $y \in Y$ as in Section 5.3 (Partial Differentiation). Now, let $g: W \longrightarrow X \oplus Y$ be a mapping from the normed vector space W to the direct sum space. Define the composite function $\phi(w) = f(g(w)) = f(g_1(w), g_2(w))$, where $g(w) = (g_1(w), g_2(w))$. Show that, under hypotheses like those in Lemma 5.3.4,

$$\varphi'(w)h = f_x(g(w))\cdot g_1'(w)h + f_y(g(w))\cdot g_2'(w)h.$$

[*Hint:* $\varphi'(w) = f'(g(w))\cdot g'(w)$ by the lemma, and $g'(w)h = (g_1'(w)h, g_2'(w)h)$.]

5.21 Let f be mapping from a subset of a normed space X into a normed space Y. Let $f(z) = 0$. Show that z is an isolated zero of f iff f has a Fréchet derivative at z which is nonsingular; i.e. $f'(z)$ has an inverse. [*Hint:* By Lemma 3.5.1, there exists $m > 0$ such that

$$\|f'(z)h\| \geq m\|h\| \qquad \text{for all } h \in X.$$

Hence,

$$\|f(z + h)\| \geq m\,\|h\| - \|\varepsilon\|,$$

where

$$\|\varepsilon\|/\|h\| \to 0 \text{ as } \|h\| \to 0.]$$

5.22 Steffensen's iteration, as defined by Eq. (5.7.5), may be viewed as successive approximation applied to the function,

$$G(x) = \frac{xg(g(x)) - (g(x))^2}{x - 2g(x) + g(g(x))}.$$

As such, it differs from the Aitken procedure as given by (5.7.1). The latter is a transformation on sequences and, in general, the new sequence is only linearly convergent when the original sequence is linearly convergent. By contrast, Steffensen's iteration, $x_{m+1} = G(x_m)$, can produce a sequence which converges quadratically [5.11]. Assume that $g(x)$ has a fixed point, x^*, and has a continuous third derivative in a ball centered at x^*. Further, assume that $g'(x^*) \neq 1$. Show that these assumptions imply quadratic convergence of the Steffensen iterates for any starting value, x_0, sufficiently close to x^*. [*Hint:* Define $G(x^*) = x^*$. Verify that $G'(x^*) = 0$ and $G''(x)$ is bounded in a neighborhood of x^*. Then apply Corollary 2 of Theorem 5.1.4. Observe that $g(x) = g(x^*) + g'(\bar{x})h$, where $x = x^* + h$ and $\bar{x} = x^* + \theta h$. Use this to evaluate $G'(x^*)$.]

5.23 Suppose that (5.3.7) is used as the definition of the strong derivative; i.e. $f'(x)$ is a linear operator on X to Y satisfying (5.3.7). Show that this defines $f'(x)$ uniquely. Similarly, show that (5.3.25) could have been taken as the definition of $f''(x)$. [*Hint:* Prove that a linear operator A such that $\lim_{\|h\| \to 0} Ah/\|h\| = 0$ must be the zero operator. If not, $Ah_0 \neq 0$ for some h_0 and setting $h_n = h_0/n$, we have $\lim Ah_n/\|h_n\| \neq 0$. Now, let $A = f'(x) - \bar{f}'(x)$, where $\bar{f}'(x)$ satisfies (5.3.7).]

5.24 Prove the following theorems which establish quadratic convergence of Newton's method without the requirement that the second derivative exist. Let f be a mapping

$X \longrightarrow Y$, where X, Y are Banach spaces, and let x_0 be such that $(f'(x_0))^{-1}$ exists. Set

$$\eta_0 = \|(f'(x_0))^{-1} f(x_0)\|.$$

Theorem 5.5.4 For x, x' in the closed ball $B = \bar{B}(x_0, 2\eta_0)$, let

i) $\|f'(x) - f'(x')\| \leq L \|x - x'\|$,
ii) $\eta_0 L \|f'(x_0))^{-1}\| \leq \frac{1}{4}$.

Then f has a zero x^* in B and the Newton iterates (5.2.1) converge to x^* quadratically. [*Hint:* See [5.3] or [5.8].]

Theorem 5.5.5 Let $r > 0$ be such that

$$\|(f'(x_0))^{-1}(f'(x) - f'(x'))\| \leq K \|x - x'\|$$

for x, x' in the closed ball $B = \bar{B}(x_0, r)$. If

i) $Kr \leq \frac{2}{3}$
and
ii) $\eta_0 \leq (1 - \frac{3}{2}Kr)r$,

then f has a zero x^* in B and the Newton iterates converge to x^* quadratically. [*Hint:* See [5.7]. Also see Theorem 5.5.1 and compare conditions (i) and (ii).]

5.25 a) Let $\bar{x}_{m+1} = g(\bar{x}_m) + \delta_m$, with $\|\delta_m\| < \delta$, as in Theorem 5.1.5. If the contractive constant L is not known, then formula (5.1.10) cannot be used to halt the computation. Show that a practical stopping criterion in many cases is

$$\max_{1 \leq i \leq n} \frac{|\xi_i^{(m+1)} - \xi_i^{(m)}|}{|\xi_i^{(m)}|} < 10^{-q},$$

where $\bar{x}_m = (\xi_1^{(m)}, \ldots, \xi_n^{(m)})$, and the desired number of correct digits in $\xi_i^{(m)}$ is $q - 1$. [*Hint:* Compare Exercise 4.23. Note that $x_m = x_0 + \sum_{i=0}^{m-1}(x_{i+1} - x_i)$. Hence,

$$\|e_m\| = \|x_m - x^*\| \leq \sum_{m}^{\infty} \|x_{i+1} - x_i\| \leq \frac{\|x_{m+1} - x_m\|}{1 - L} \leq \frac{10^{-q}\|\bar{x}_m\| + 2\delta}{1 - L}, \quad (1)$$

(using the sup norm). Using (5.1.11), we have

$$\|\bar{e}_m\| = \|\bar{x}_m - x^*\| \leq (10^{-q}\|x_m\| + 3\delta)/(1 - L)$$

Thus, if $L < .9$, say, and $\delta < 10^{-q}\|x_m\|$, the desired accuracy will be achieved.]

b) Under the hypotheses of Theorem 5.4.1, show that the Newton iterations satisfy the inequality,

$$\|x_{m+1} - x_m\| \leq K \|x_m - x_{m-1}\|^2 \quad (2)$$

[*Hint:* As in the proof,

$$\|g'(x_m)\| \leq (L/r) \|(f'(x_m))^{-1} f(x_m)\| = (L/r) \|x_{m+1} - x_m\|.$$

By the mean-value theorem,

$$\|x_{m+1} - x_m\| = \|g(x_m) - g(x_{m-1})\| \leq \sup \|g'(x_{m-1} + \theta h_m)\| \|h_m\|, \quad (3)$$

where $h_m = x_m - x_{m-1}$. Again, by the mean-value theorem,
$$\|g'(x_{m-1} + \theta h_m)\| \leq \|g'(x_{m-1})\| + \sup \|g''(x_{m-1} + \alpha h_m)\| \|h_m\|.$$
From (3), we then have $\|h_{m+1}\| \leq (L/r + \mu) \|h_m\|^2$, where $\mu = \sup \|g''(x)\|$ for $x \in B(x_0, r)$.]

c) Use (2) in part (b) to show that the stopping criterion in (a) is often practical. Assume $\delta_m = 0$. [*Hint:*
$$\|e_{m-1}\| \leq \|h_m\| + K \|h_m\|^2 + K^3 \|h_m\|^4 + \cdots$$
$$\|e_{m-1}\| \leq \|h_m\|/(1 - K\|h_m\|), \tag{4}$$
and $K \|h_m\|$ is small compared to 1.]

d) Next, suppose that $\|\delta_m\| < \delta$ and $K\delta < L_1^2$, where $L_1 \leq \frac{1}{2}$. If $K \|\bar{x}_0 - x_0\| < L_1$, show that $K \|\bar{x}_m - x_m\| < L_1$ for all m and, consequently $\|\bar{x}_{m+1} - x_{m+1}\| \leq L_1 \|\bar{x}_m - x_m\| + \delta$. This implies
$$\|\bar{x}_{m-1} - x^*\| \leq \frac{\|\bar{h}_m\| + \bar{\delta}}{1 - K(\|\bar{h}_m\| + \bar{\delta})}, \tag{5}$$
where $\bar{h}_m = \bar{x}_m - \bar{x}_{m-1}$ and $\bar{\delta} = 2\delta/(1 - L_1)$. [*Hint:*
$$\|\bar{x}_{m+1} - x_{m+1}\| \leq \|g(\bar{x}_m) - g(x_m)\| + \delta \leq K \|\bar{x}_m - x_m\|^2 + \delta.$$
$$K\|\bar{x}_{m+1} - x_{m+1}\| \leq 2L_1^2 \leq L_1.$$
As in (5.1.11), we calculate
$$\|\bar{x}_{m+1} - x_{m+1}\| \leq L_1 \|\bar{x}_m - x_m\| + \delta \leq \delta/(1 - L_1).$$
Using $\bar{h}_m \leq h_m + \|\bar{x}_{m-1} - x_{m-1}\| + \|\bar{x}_m - x_m\|$ together with (4) above, we obtain (5).]

e) Discuss the implications of (5) when the δ_m are the results of rounding errors in the solution of the linear equations $J(x_m) \Delta x_m = -f(x_m)$ at the mth iteration.

5.26 In [5.16], it is shown how Bernoulli's method (Section 5.6) can be improved and leads to a method known as the *q-d algorithm* (or *quotient-difference algorithm*) [5.17]. Let $p(x) = a_n x^n + \cdots + a_1 x + a_0$, $a_n \neq 0$, have simple roots $0 < |z_1| < \cdots < |z_n|$. Then we can write
$$\frac{1}{p(x)} = \frac{b_1}{x - z_1} + \cdots + \frac{b_n}{x - z_n}.$$
Expanding in geometric series, we have
$$\frac{b_i}{x - z_i} = \frac{-b_i}{z_i}\left(1 + \frac{x}{z_i} + \frac{x^2}{z_i^2} + \cdots\right).$$
Now, adding the series and collecting like powers of x, we get
$$\frac{1}{p(x)} = \sum_{j=0}^{\infty} c_j x^j,$$
where
$$c_j = -\sum_{i=1}^{n} b_i/z_i^{j+1}.$$

228   Solution of Nonlinear Equations

Define the quotient,

$$q_{j1} = \frac{c_j}{c_{j-1}} = \frac{1}{z_1}\left(\frac{1 + b_2(z_1/z_2)^{j+1} + \cdots + b_n(z_1/z^n)^{j+1}}{1 + b_2(z_1/z_2)^j + \cdots + b_n(z_1/z_2)^j}\right).$$

Since $|z_1/z_i| < 1$ for $i = 2, \ldots, n$, we have $\lim_{j \to \infty} q_{j1} = 1/z_1$.

a) Show that

$$\lim_{j \to \infty} \frac{(1/z_1) - q_{j1}}{(z_1/z_2)^j} = \frac{b_2}{z_1 b_1}\left(1 - \frac{z_1}{z_2}\right)$$

$$\lim_{j \to \infty} \frac{(1/z_1) - q_{j+1,1}}{(z_1/z_2)^j} = \frac{b_2}{z_2 b_1}\left(1 - \frac{z_1}{z_2}\right).$$

b) Define the difference, $e_{j1} = q_{j+1,1} - q_{j1}$ and verify that

$$\lim_{j \to \infty} \frac{e_{j+1,1}}{e_{j1}} = \frac{z_1}{z_2}, \quad \lim_{j \to \infty} \frac{e_{j+1,1} q_{j+1,1}}{e_{j1}} = \frac{1}{z_2}.$$

c) Define $q_{j2} = (e_{j+1,1}/e_{j1}) q_{j+1,1}$. Show that $\lim_{j \to \infty} q_{j2} = 1/z_2$. Continue by defining $e_{j2} = q_{j+1,2} - q_{j2} + e_{j+1,1}$, etc. Verify that $\lim_{j \to \infty} e_{j+1,2}/e_{j2} = z_2/z_3$, and so on. This leads to the q-d algorithmic formulas,

$$e_{j-1,k} q_{j-1,k+1} = e_{jk} q_{jk}, \quad q_{j-1,k} + e_{j-1,k} = q_{jk} + e_{j,k-1}.$$

Arrange the q_{jk} in columns (with $k = 1, \ldots, n$ as column index) alternating with columns of e_{jk} to form a rhombic pattern. Verify that $z_k = \lim_{j \to \infty} q_{jk}$.
Initial values are as follows:

e_0	q_1	e_1	q_2	e_2	q_3	$\cdots$
	$\dfrac{-a_{n-1}}{a_n}$		0		0	$\cdots$
0		$\dfrac{a_{n-2}}{a_{n-1}}$		$\dfrac{a_{n-3}}{a_{n-2}}$		$\cdots$
	q_{11}		q_{12}		q_{13}	$\cdots$
0		e_{11}		e_{12}		$\cdots$

5.27 a) Show that Aitken's Δ^2-method (5.7.1) can be rewritten in the form

$$x'_{m+1} = \frac{\det\begin{bmatrix} x_m & x_{m+1} \\ \Delta x_m & \Delta x_{m+1} \end{bmatrix}}{\det\begin{bmatrix} 1 & 1 \\ \Delta x_m & \Delta x_{m+1} \end{bmatrix}},$$

where $\Delta x_m = x_{m+1} - x_m$ is the first forward difference at x_m. (See Section 8.2a.)

b) The formula in part (a) is generalized in [5.19] as follows:

$$x_{m+1}^{(p)} = \frac{\det \begin{bmatrix} x_m & x_{m+1} & \cdots & x_{m+p} \\ \Delta x_m & \Delta x_{m+1} & & \Delta x_{m+p} \\ \vdots & & & \\ \Delta x_{m+p-1} & \Delta x_{m+p} & & \Delta x_{m+2p-1} \end{bmatrix}}{\det \begin{bmatrix} 1 & 1 & \cdots & 1 \\ \Delta x_m & \Delta x_{m+1} & & \Delta x_{m+p} \\ \vdots & & & \\ \Delta x_{m+p-1} & \Delta x_{m+1} & & \Delta x_{m+2p-1} \end{bmatrix}}$$

The following convenient algorithm for computing $x_{m+1}^{(p)}$ is given in [5.20]. (See also [5.16]). The ε-*algorithm* is defined by the formula,

$$\varepsilon_{p+1}(x_m) = \varepsilon_{p-1}(x_{m+1}) + 1/(\varepsilon_p(x_{m+1}) - \varepsilon_p(x_m)),$$

where

$$\varepsilon_0(x_m) = x_m \quad \text{and} \quad \varepsilon_1(x_m) = 1/\Delta x_m.$$

Show that $x_{m+1}^{(p)} = \varepsilon_{2p}(x_m)$ and $\varepsilon_{2p+1}(x_m) = 1/\varepsilon_{2p}(\Delta x_m)$.

CHAPTER 6

THE COMPUTATION OF MATRIX EIGENVALUES AND EIGENVECTORS

Before reading this chapter, the student should review Section 3.6.

6.1 INTRODUCTION

Let $A = (a_{ij})$ be an $n \times n$ real matrix. Then λ is an eigenvalue of A if and only if there is a vector $x \neq 0$ such that

$$Ax - \lambda x = 0. \tag{1}$$

As we have seen, (1) holds if and only if

$$\det(\lambda I - A) \equiv p_A(\lambda) = 0, \tag{2}$$

where $p_A(\lambda)$ is the characteristic polynomial of A. Equations (1) and (2) suggest several possible approaches to the problem of computing eigenvalues and eigenvectors.

First, one might attempt to solve (1) directly for the $n + 1$ unknowns $(\xi_1, \ldots, \xi_n, \lambda)$. The ith equation, $1 \leq i \leq n$, of (1) may be written as

$$f_i(\xi_1, \ldots, \xi_n, \lambda) = a_{i1}\xi_1 + \cdots + (a_{ii} - \lambda)\xi_i + \cdots + a_{in}\xi_n = 0.$$

Clearly, f_i is nonlinear with respect to ξ_i and λ. By adjoining the equation

$$f_{n+1}(\xi_1, \ldots, \xi_n) = \xi_1^2 + \cdots + \xi_n^2 - 1 = 0$$

to system (1) we obtain a system of $n + 1$ nonlinear equations in the $n + 1$ unknowns. Newton's method is suggested. However, this requires initial guesses and does not take full advantage of the algebraic structure of the problem. For example, if A is symmetric, there exists an orthogonal matrix R such that $R^{-1}AR$ is a diagonal matrix. (See Section 3.6 and Exercise 3.15.) Several methods are based on this algebraic property. However, even these methods must be iterative, since the roots of a polynomial of degree ≥ 5 cannot be determined by a finite number of algebraic operations (Galois).

Second, one could try to solve for the roots of (2). Again, Newton's method or Bernoulli's method for finding zero's of a polynomial is suggested. (Since λ may be complex, Newton's method would have to be modified to find complex roots and is not recommended.) Having determined a root, we could, in principle, substitute this value for λ in (1) and solve (1) as a singular linear system in x. This approach, if applied directly, would require the computation of the coefficients of $p_A(\lambda)$; i.e., we would have to compute $\det(\lambda I - A)$. This could be carried out by the elimination process of reducing $A - \lambda I$ to upper triangular form, treating λ as an unknown constant. As in Gaussian elimination, rounding errors occur and the resulting polynomial is only an approximation to $p_A(\lambda)$. Small errors in the coefficients of a polynomial can cause large errors in the roots. Therefore, this technique is not recommended, in general. However, if we can find a similar matrix $B = P^{-1}AP$, such that det $(\lambda I - B)$ is easier to evaluate, then since B has the same eigenvalues as A, the technique of finding roots of $p_B(\lambda)$ can be used. Certainly this is the case if B is diagonal. We shall see that likewise if B is tridiagonal, then $p_B(\lambda)$ is easily determined with high accuracy.

Before considering specific methods, let us examine certain general metric aspects of the eigenvalue problem which have a bearing on the accuracy that one can expect to attain.

6.2 ERROR ESTIMATES

Theorem 6.2.1 (Hadamard). Let $A = (a_{ij})$. Let

$$r_i = \sum_{j=1}^{n} |a_{ij}| - |a_{ii}|, \qquad c_j = \sum_{i=1}^{n} |a_{ij}| - |a_{jj}|.$$

Let

$$R_i = \{z : |z - a_{ii}| \le r_i\}, \qquad 1 \le i \le n,$$

be closed balls in the complex plane and $R = \bigcup_{1}^{n} R_i$. Similarly, let

$$C_j = \{z : |z - a_{jj}| \le c_j\}, \qquad 1 \le j \le n,$$

and $C = \bigcup_{j=1}^{n} C_j$. Then for each eigenvalue λ of A we have $\lambda \in R$ and $\lambda \in C$.

Proof. Let $x = (\xi_1, \ldots, \xi_n)$ be an eigenvector belonging to λ. Choose ξ_i such that $|\xi_i| = \|x\|_\infty$. Since $(A - \lambda I)x = 0$, the ith equation is

$$(a_{ii} - \lambda)\xi_i = -\sum_{\substack{k=1 \\ k \ne i}}^{n} a_{ik}\xi_k.$$

Since $|\xi_k|/|\xi_i| \le 1$, $|a_{ii} - \lambda| \le r_i$. Hence, $\lambda \in R_i \subset R$. Since $\sigma(A) = \sigma(A^T)$, (see example following Definition 3.6.1), the above proof applied to A^T yields $\lambda \in C$. ∎

Theorem 6.2.2 (Gerschgorin). Let $R_i(C_j)$ be the closed balls of Theorem 6.2.1. Each component of $R(C)$ contains as many eigenvalues as there are balls in that component, where each eigenvalue and ball is counted a number of times equal to its multiplicity.

Proof. Let $A(\alpha) = D + \alpha(A - D)$, where $D = (a_{ij}\delta_{ij})$; i.e., D is the diagonal matrix whose diagonal elements agree with those of A. Then $A(1) = A$ and $A(0) = D$ and for $0 < \alpha < 1$, the balls $R_i(\alpha)$ have centers a_{ii} and radii αr_i. In particular, for $\alpha = 0$, the radii are 0 and the balls are the points a_{ii}. Since $A(0) = D$, the number of its eigenvalues in $R_i(0)$ equals the multiplicity of a_{ii}. In fact, the a_{ii} are the eigenvalues of $A(0)$. As α varies continuously from 0 to 1, the radii αr_i vary continuously from 0 to r_i and the number of components in $R(\alpha)$ (i.e., the number of maximal connected subsets, Section 1.7) remains constant except for a finite set of values α_q of α, where balls intersect which were previously disjoint. At these values of α, the number of balls in some component increases (and the number of components decreases). Now, the eigenvalues $\{\lambda_i(\alpha)\}$ of $A(\alpha)$ are the zeros of $\det(\lambda I - A(\alpha)) = p_\alpha(\lambda)$. The coefficients of $p_\alpha(\lambda)$ are continuous in α and each $\lambda_i(\alpha)$ is a continuous function of the coefficients, hence of α. Thus, $\lambda_i(\alpha)$, $1 \leq i \leq n$, is a continuous arc in the complex plane starting at a_{ii} and ending at some eigenvalue of A. (If a_{ii} has multiplicity m_i, there will be m_i such arcs.) By the previous theorem, every eigenvalue is in some component. Hence, the number of eigenvalues in each component is a continuous function of α and is, therefore, constant except at the values α_q where the number of components decreases. Again, by continuity of $\lambda_i(\alpha)$ at α_q, the number of eigenvalues and the number of balls in a component must both increase by the same amount at α_q. Since these two numbers are equal for $\alpha = 0$, they must be equal at $\alpha = 1$. ∎

Remark The balls R_i are called the *Gerschgorin disks*.

Theorem 6.2.3 Let A be an $n \times n$ matrix having n linearly independent eigenvectors. Let P be any matrix such that $\|P\| = \|A\|$ and define $A(\varepsilon) = A + \varepsilon P$. For any eigenvalue $\lambda(\varepsilon)$ of $A(\varepsilon)$ there is an eigenvalue λ_i of A such that $|\lambda(\varepsilon) - \lambda_i| = O(\varepsilon)$. If λ_i has multiplicity 1, then

$$\lim_{|\varepsilon| \to 0} \frac{\lambda(\varepsilon) - \lambda_i}{\varepsilon} = \frac{y^*Px_i}{y^*x_i} \qquad (6.2.1)$$

where x_i is any eigenvector of A belonging to λ_i and y is any eigenvector of A^* belonging to $\bar{\lambda}_i$.

Proof. Let U be the matrix having as its columns the eigenvectors of A. Then

$$AU = U\Lambda$$

where Λ is a diagonal matrix of the eigenvalues of A in some order, say $(\lambda_1, \ldots, \lambda_n)$. Let $C = U^{-1}PU$. Since U has n linearly independent columns, U^{-1} exists and $U^{-1}A(\varepsilon)U = U^{-1}AU + \varepsilon U^{-1}PU = \Lambda + \varepsilon C$ has the same eigenvalues as $A(\varepsilon)$. By Hadamard's theorem, for any eigenvalue $\lambda(\varepsilon)$ of $A(\varepsilon)$, we have

$$|\lambda(\varepsilon) - \lambda_i - \varepsilon c_{ii}| \leq |\varepsilon| \sum_{\substack{k=1 \\ k \neq i}}^{n} |c_{ik}| \qquad \text{for some } i. \qquad (1)$$

Hence,

$$|\lambda(\varepsilon) - \lambda_i| \leq |\varepsilon| \sum_{k=1}^{n} |c_{ik}| \qquad \text{and} \qquad |\lambda(\varepsilon) - \lambda_i| = O(\varepsilon).$$

If λ_i has multiplicity 1, then by (1) above, we see that for ε sufficiently small the ball $R_i(\varepsilon)$ does not intersect any other ball $R_j(\varepsilon)$, $j \neq i$. By Gerschgorin's theorem, there is exactly one eigenvalue $\lambda(\varepsilon) \in R_i(\varepsilon)$; i.e., $\lambda(\varepsilon)$ has multiplicity 1 and a unique eigenvector $x(\varepsilon)$. Thus,

$$(A + \varepsilon P)x(\varepsilon) = \lambda(\varepsilon)x(\varepsilon),$$
$$Ax_i = \lambda_i x_i,$$
$$A[x(\varepsilon) - x_i] + \varepsilon Px(\varepsilon) = (\lambda(\varepsilon) - \lambda_i)x(\varepsilon) + \lambda_i(x(\varepsilon) - x_i).$$

Let y be any eigenvector of A^* belonging to $\bar{\lambda}_i$. Then since $y^*A = \lambda_i y^*$, multiplying by y^* gives

$$\varepsilon y^* P x(\varepsilon) = (\lambda(\varepsilon) - \lambda_i) y^* x(\varepsilon).$$

Now, $y^*x = 0$ for all eigenvectors $x \neq x_i$, since the x belong to other eigenvalues. (See Exercise 3.24b.) Hence, $y^*x_i \neq 0$, since otherwise $y = 0$.

The matrix $A - \lambda_i I$ has nullity 1 and rank $n - 1$. By continuity, for ε sufficiently small $A(\varepsilon) - \lambda(\varepsilon)I$ also has rank $n - 1$ and in fact the same $(n-1) \times (n-1)$ submatrix has a nonzero determinant so that we can delete the same column, the jth say, and row from both matrices to obtain nonsingular submatrices $B(\varepsilon)$ and B. Determine $x(\varepsilon)$ by taking $\xi_j(\varepsilon) = 1$ and solving $(A(\varepsilon) - \lambda(\varepsilon)I)x(\varepsilon) = 0$ for the other components. Since $B^{-1}(\varepsilon) \to B^{-1}$, we have $x(\varepsilon) \to x_i$. Hence,

$$\frac{\lambda(\varepsilon) - \lambda_i}{\varepsilon} = \frac{y^* P x(\varepsilon)}{y^* x(\varepsilon)} \to \frac{y^* P x_i}{y^* x_i}. \quad \blacksquare$$

Corollary

$$\min_{\lambda_i} |\lambda(\varepsilon) - \lambda_i| \leq |\varepsilon| \, \|P\|_p \, \|U^{-1}\|_p \, \|U\|_p,$$

for any matrix norm induced by an l_p vector norm, $1 \leq p \leq \infty$.

Proof. For $p = 1$, this follows directly from (1) above. For $p > 1$, see Exercise 6.12. $\blacksquare$

Remark Since εP can be viewed as a perturbation of A, the quantity $\|U^{-1}\| \, \|U\|$ is sometimes regarded as a condition number for the eigenvalue problem. (See [6.9].) If it is near unity, then perturbations of relative magnitude ε in the data (A) produce errors of the same order. The eigenvalue problem for λ_i is said to be *ill-conditioned* if such perturbations produce errors in λ_i several orders of magnitude larger than ε.

If we take $\|y\|_2 = \|x_i\|_2 = 1$ in (6.2.1) above, then

$$\lim_{\varepsilon \to 0} \frac{\lambda(\varepsilon) - \lambda_i}{\varepsilon} = \frac{y^* P x_i}{y^* x_i} \leq \frac{\|P\|_s}{|y^* x_i|} = \frac{\|A\|_s}{|y^* x_i|}.$$

This inequality suggests that a sufficient condition for the eigenvalue problem for a simple eigenvalue λ_i to be well-conditioned is that the number $|y^*x_i|$ not be small compared to 1; e.g., $|y^*x_i| > 0.1$ would suffice. If λ_i has multiplicity > 1, then it can happen that $y^*x_i = 0$ for all eigenvectors y of A^* belonging to $\bar{\lambda}_i$. For example, this is the case for a Jordan matrix, as shown in Exercise 3.21. When $y^*x_i = 0$ for all such eigenvectors y, the result (6.2.1) is no longer valid.

If $|y^*x_i| > 0.1$, one would expect $|\Delta \lambda_i| \leq 10 \, \|\Delta A\|$, where $\Delta \lambda_i$ is the error in the eigenvalue λ_i caused by a small error (or perturbation) ΔA in the matrix A. In this case, we would say that the eigenvalue problem for λ_i is not ill-conditioned. It is quite possible that for some eigenvalues of A the problem is ill-conditioned while for others it is not. However, when A is Hermitian, the problem is well-conditioned for every eigenvalue. We state this as the next theorem.

Theorem 6.2.4 Let $A = A^*$ and suppose λ_i are the eigenvalues of A. For any perturbation εP where $\|P\|_s = \|A\|_s$, the matrix $A + \varepsilon P$ has eigenvalues $\lambda_j(\varepsilon)$ such that

$$\min_i |\lambda_j(\varepsilon) - \lambda_i| \leq |\varepsilon| \|A\|_s.$$

Proof. Since A is Hermitian, there exists a unitary matrix U of eigenvectors of A. Since $\|U\|_s = \|U^{-1}\|_s = 1$, Theorem 6.2.3, Corollary, yields

$$\min_i |\lambda(\varepsilon) - \lambda_i| \leq |\varepsilon| \|P\|_s \|U^{-1}\|_s \|U\|_s = |\varepsilon| \|A\|_s. \blacksquare$$

If we interpret $\Delta \lambda_j = \min_i |\lambda_j(\varepsilon) - \lambda_i|$ as the error in λ_j caused by a perturbation $\Delta A = \varepsilon P$, then the preceding theorem allows us to assert that $|\Delta \lambda_j| \leq \|\Delta A\|_s$ for all eigenvalues λ_j, which shows that the problem is not ill-conditioned. If we assume that $\|A\|_s = 1$ (as we may by multiplying A by a scalar, the eigenvalues of αA being $\alpha \lambda_i$), then $|\Delta \lambda_j| \leq \varepsilon$ and the error in any λ_j is of the same order of magnitude as the relative error in A. Observe, however, that for small $|\lambda_j|$ the relative error $|\Delta \lambda_j|/|\lambda_j|$ may be much larger than ε.

The preceding results, and (6.2.1) in particular, suggest a more precise definition of "condition number" and "ill-conditioned." With a simple eigenvalue λ_i of A we may associate a real-valued function $f(X)$ defined for all matrices X in a ball $B(A, r)$ and such that $f(A) = \lambda_i$. (We use the topology of the spectral norm in the Banach algebra $A^{(n)}$ of nth order matrices.) For any matrix X, $f(X)$ is an eigenvalue of X. If $X = A_\varepsilon = A + \varepsilon P$, the preceding corollary shows that

$$|\lambda(\varepsilon) - \lambda_i| = |f(A + \varepsilon P) - f(A)| \leq |\varepsilon| \|U\| \|U^{-1}\| \|P\|.$$

Letting $\Delta A = \varepsilon P$, we may also write

$$|f(A + \Delta A) - f(A)| \leq \|U\| \|U^{-1}\| \|\Delta A\|.$$

This is a continuity condition on f at A which is similar to a Lipschitz condition with constant $\|U\| \|U^{-1}\|$. Thus, we may assert the existence of a constant $L > 0$ such that

$$|f(A + \Delta A) - f(A)| \leq L \|\Delta A\| \qquad (6.2.2)$$

for all ΔA with $\|\Delta A\| < r$. The *condition number* of λ_i could then be defined as the smallest L which satisfies (6.2.2). If the condition number is <10, the problem of computing λ_i would be called *well*-conditioned. If the condition number is $\geq 10^n$, $n \geq 1$, the problem would be called *ill-conditioned* of degree n.

From (6.2.1) it follows that the weak differential (Definition (5.3.2)) of f at A exists and

$$df(A; P) = \lim_{\varepsilon \to 0} \frac{f(A + \varepsilon P) - f(A)}{\varepsilon} = \frac{y^* P x_i}{y^* x_i}. \qquad (6.2.3)$$

In fact, the weak derivative $f'_w(A)$ exists, since $y^* P x_i / y^* x_i$ is linear in P. It is clear that $f'_w(A_\varepsilon)$ exists for all A_ε with ε sufficiently small. Applying the mean-value theorem as given by formula (5.3.10) we obtain

$$|f(A + \Delta A) - f(A)| \leq \sup_{0 < \theta < 1} \|f'_w(A + \theta \Delta A)\| \|\Delta A\|$$

for $\|\Delta A\|$ sufficiently small. Now, from (6.2.3) it follows that for $\|P\| = 1$,

$$\|f'_w(A)\cdot P\| \le \frac{\|y^*\|\,\|x_i\|}{|y^*x_i|} \le \frac{1}{|y^*x_i|}.$$

Indeed, choosing P to be such that $Px_i = y$, we obtain

$$|f'_w(A)\cdot P| = \frac{|y^*Px_i|}{|y^*x_i|} = \frac{1}{|y^*x_i|}.$$

Hence, $\|f'_w(A)\| = 1/|y^*x_i|$. In the preceding, y^* and x_i were unit left and right eigenvectors of A belonging to λ_i. Proceeding similarly for A_ε, we have for ε sufficiently small, $\|f'_w(A_\varepsilon)\| = 1/|y^*(\varepsilon)x_i(\varepsilon)|$. Since $y^*(\varepsilon) \to y^*$ and $x_i(\varepsilon) \to x_i$ as $\varepsilon \to 0$, we see that $\sup_{0<\theta<1}\|f'_w(A + \theta\cdot\Delta A)\|$ is close to $1/|y^*x_i|$ for $\|\Delta A\|$ sufficiently small. Therefore, $1/|y^*x_i|$ is a good approximation to the condition number of λ_i for small $\|\Delta A\|$. How small $\|\Delta A\|$ must be depends on the rate of convergence of $x_i(\varepsilon)$ to x_i. We shall now consider the effect of perturbations in A on the eigenvectors of A. We shall restrict our discussion to matrices which have a basis of eigenvectors.

Error in Eigenvectors of a Diagonalizable Matrix

Let A be an arbitrary matrix and again P a perturbation matrix such that $\|P\| = \|A\|$. We may assume that $|a_{ij}| < 1$ and $|p_{ij}| < 1$. (The eigenvalues of αA are $\alpha\lambda_i$, where λ_i are the eigenvalues of A.) As before, let $A(\varepsilon) = A + \varepsilon P$.

We have

$$p(\lambda) = \det(\lambda I - A) = \lambda^n + b_{n-1}\lambda^{n-1} + \cdots + b_0,$$

$$p_\varepsilon(\lambda) = \det(\lambda I - A(\varepsilon)) = \lambda^n + b_{n-1}(\varepsilon)\lambda^{n-1} + \cdots + b_0(\varepsilon),$$

where $b_i(\varepsilon)$ is a polynomial of degree $n - i$ in ε. Suppose λ_1 is a simple zero of $p(\lambda)$. From the theory of algebraic functions, it is known that there is a zero, $\lambda_1(\varepsilon)$, of $p_\varepsilon(\lambda)$ given by a convergent power series in ε for ε sufficiently small and $|\lambda_1(\varepsilon) - \lambda_1| = O(\varepsilon)$. Thus,

$$\lambda_1(\varepsilon) = \lambda_1 + \alpha_1\varepsilon + \cdots.$$

Suppose $x_1(\varepsilon)$ is an eigenvector such that

$$(A + \varepsilon P)x_1(\varepsilon) = \lambda_1(\varepsilon)x_1(\varepsilon). \tag{1}$$

The components of the eigenvector $x_1(\varepsilon)$ are polynomials in $\lambda_1(\varepsilon)$ and ε; (i.e., solve for $n - 1$ components in terms of the other and multiply by a suitable polynomial in $\lambda_1(\varepsilon)$ and ε to eliminate fractions). Since $\lambda_1(\varepsilon)$ is given by a convergent power series, so is $x_1(\varepsilon)$ and $x_1(\varepsilon) = x_1 + \varepsilon v_1 + \varepsilon^2 v_2 + \cdots$.

Now, we assume that A is diagonalizable. This implies the existence of a basis of eigenvectors $\{x_1, \ldots, x_n\}$ and a basis of left eigenvectors $\{y_1, \ldots, y_n\}$ such that $y_i^T x_j = 0$ for $i \ne j$ (Exercise 6.15). We choose the x_i and y_i to have unit l_2-norm. Each vector v_i in the expansion of $x_1(\varepsilon)$ can be expressed in terms of the basis vectors as $v_i = \sum_{j=1}^n \omega_{ij}x_j$. Therefore, rearranging the series, we have

$$x_1(\varepsilon) = (1 + \varepsilon\omega_{11} + \varepsilon^2\omega_{21} + \cdots)x_1 + (\varepsilon\omega_{12} + \varepsilon^2\omega_{22} + \cdots)x_2 + \cdots$$

$$+ (\varepsilon\omega_{1n} + \varepsilon^2\omega_{2n} + \cdots)x_n.$$

Normalizing by dividing by the coefficient of x_1, we obtain an eigenvector which we again designate by $x_1(\varepsilon)$. Thus, for ε sufficiently small,

$$x_1(\varepsilon) = x_1 + \left(\sum_{i=1}^{\infty} \varepsilon^i t_{i2}\right) x_2 + \cdots + \left(\sum_{i=1}^{\infty} \varepsilon^i t_{in}\right) x_n.$$

Substituting for $x_1(\varepsilon)$ and $\lambda_1(\varepsilon)$ in Eq. (1) above and equating the coefficients of ε in the left and right members we obtain

$$A\left(\sum_{j=2}^{n} t_{1j} x_j\right) + Px_1 = \lambda_1 \sum_{j=2}^{n} t_{1j} x_j + \alpha_1 x_1,$$

$$Px_1 + \sum_{j=2}^{n} (\lambda_j - \lambda_1) t_{1j} x_j = \alpha_1 x_1.$$

Multiplying by y_1^T, we have $|\alpha_1| = |y_1^T Px_1|/|y_1^T x_1|$, which gives an alternative proof of (6.2.1). Multiplying by y_i^T, we get for $i = 2, \ldots, n$,

$$(\lambda_i - \lambda_1) t_{1i} y_i^T x_i + y_i^T Px_1 = 0.$$

Therefore, the coefficient of ε in $x_1(\varepsilon)$ is

$$\sum_{j=2}^{n} t_{1j} x_j = \sum_{j=2}^{n} \frac{y_j^T Px_1}{y_j^T x_j (\lambda_1 - \lambda_j)} x_j.$$

For a symmetric matrix, we can choose $y_j = x_j$, whence $y_j^T x_j = 1, j = 2, \ldots, n$. However, if λ_1 is close to λ_j for some $\lambda_j, j = 2, \ldots, n$, then the coefficient of ε can be large. Thus, even when the eigenvalue problem is well-conditioned, the eigenvector problem may be ill-conditioned; (i.e., choose P so that $y_j^T Px_1 \neq 0$).

The preceding error analysis gives a priori error estimates. As in Chapter 4, we shall now consider a posteriori error estimates.

A Posteriori Error Estimates in Eigenvalues and Eigenvectors

Theorem 6.2.5 Let A be a Hermitian matrix of order n with eigenvalues $\{\lambda_i\}$. Let $r = Ax - \lambda x, x \neq 0$. Then

$$\min_{i} |\lambda - \lambda_i| \leq \frac{\|r\|_2}{\|x\|_2}.$$

Proof. Since A is Hermitian, there exists an orthonormal basis of eigenvectors $\{u_i\}$ such that $Au_i = \lambda_i u_i$, $1 \leq i \leq n$. Let $x = \sum_{i=1}^{n} \xi_i u_i$. Then $\|x\|_2^2 = \sum_{i=1}^{n} \xi_i^2$. Since $Ax = \sum_{i=1}^{n} \xi_i \lambda_i u_i$, it follows that $r = Ax - \lambda x = \sum_{i=1}^{n} \xi_i (\lambda_i - \lambda) u_i$,

$$\|r\|_2^2 = \langle r, r \rangle = \sum_{i=1}^{n} |\lambda_i - \lambda|^2 |\xi_i|^2 \geq \min_{i} |\lambda_i - \lambda|^2 \sum_{j=1}^{n} |\xi_j|^2.$$

Therefore,

$$\frac{\|r\|_2^2}{\|x\|_2^2} \geq \min_{i} |\lambda - \lambda_i|^2,$$

which proves the theorem. ∎

If λ is a computed approximation to an eigenvalue λ_j of a Hermitian matrix and x is an approximation to an eigenvector belonging to λ_j, then we may assume that $|\lambda - \lambda_j| = \min_i |\lambda - \lambda_i|$. We can bound the error $|\lambda - \lambda_j|$ by computing $\|r\|_2$ and $\|x\|_2$ and using Theorem 6.2.5.

An estimate of the error in x can be obtained if we have information on the spacing of all the eigenvalues, as the next theorem shows.

Theorem 6.2.6 Let A be Hermitian with eigenvalues $\{\lambda_i\}$. Let $\{u_i\}$ be an orthonormal basis of eigenvectors, with $Au_i = \lambda_i u_i$, $1 \leq i \leq n$. Suppose λ is such that $|\lambda_i - \lambda| \geq d > 0$ for $q + 1 \leq i \leq n$. If M_q is the subspace spanned by $\{u_1, \ldots, u_q\}$ and $r = Ax - \lambda x$, then

$$\|x - M_q\|_2 = \inf_{y \in M_q} \|x - y\|_2 \leq \frac{\|r\|_2}{d}.$$

Proof. (We write $\|x - M_q\|_2 = \inf_{y \in M_q} \|x - y\|_2$, since this is the l_2-distance from x to the subspace M_q.) Since $\{u_i\}$ is a basis, $x = \sum_{i=1}^n \xi_i u_i$ and by Theorem 3.3.11 (see also Theorem 7.4.1),

$$\|x - M_q\|_2^2 = \|x - \sum_{i=1}^q \xi_i u_i\|_2^2 = \sum_{i=q+1}^n |\xi_i|^2.$$

As in Theorem 6.2.5, we calculate

$$\|r\|_2^2 = \sum_{i=1}^n |\xi_i|^2 |\lambda_i - \lambda|^2 \geq \sum_{q+1}^n |\xi_i|^2 |\lambda_i - \lambda|^2 \geq d^2 \sum_{i=q+1}^n |\xi_i|^2.$$

Therefore,

$$\|x - M_q\|_2^2 \leq \frac{\|r\|_2^2}{d^2}. \blacksquare$$

Remark The preceding theorem is consistent with the fact that the eigenvector problem can be ill-conditioned if there are close or coincident (multiplicity > 1) eigenvalues. In application of Theorem 6.2.6, λ would be a computed approximation to some eigenvalue, say λ_1, and x a computed approximation to an eigenvector belonging to λ_1. We can then compute $r = Ax - \lambda x$. Suppose, furthermore, we know that λ_1 has multiplicity q and that the nearest eigenvalue to λ_1 is at least at a distance $2d$ from λ_1. Using Theorem 6.2.5, if $\|r\|_2/\|x\|_2 < d$, we assume that $|\lambda - \lambda_1| \leq d$ and therefore that $|\lambda - \lambda_i| \geq d$ for $\lambda_i \neq \lambda_1$. By Theorem 6.2.6, we would then conclude that $\|x - M_1\|_2 \leq \|r\|_2/d$, which would be an upper bound on the distance from x to the nearest eigenvector belonging to λ_1.

We turn now to a consideration of several methods of computing eigenvalues and eigenvectors.

6.3 THE POWER METHOD

Theorem 6.3.1 Let A be a real $n \times n$ matrix having n linearly independent eigenvectors $\{u_1, \ldots, u_n\}$ belonging respectively to eigenvalues $\{\lambda_i\}$ satisfying

$$|\lambda_1| > |\lambda_2| \geq |\lambda_3| \geq \cdots \geq |\lambda_n|.$$

Let $x_0 = \sum_{i=1}^{n} a_i u_i$ be such that $a_1 \neq 0$ and define vectors $x_q = (\xi_1^{(q)}, \ldots, \xi_n^{(q)})$ and scalars r_q by

$$x_{q+1} = A x_q, \quad q = 0, 1, \ldots, \tag{1}$$

$$r_{q+1} = \frac{\xi_k^{(q+1)}}{\xi_k^{(q)}}, \quad \text{where } k \text{ is such that } u_{1k} \neq 0. \tag{2}$$

Then

$$\lim_{q \to \infty} r_q = \lambda_1 \quad \text{and} \quad \lim_{q \to \infty} \frac{x_q}{\lambda_1^q} = a_1 u_1. \tag{3}$$

Proof.

$$x_q = A^q x_0 = \sum_{1}^{n} \lambda_i^q a_i u_i$$

$$x_q = \lambda_1^q \left(a_1 u_1 + \sum_{2}^{n} \left(\frac{\lambda_i}{\lambda_1}\right)^q a_i u_i \right).$$

Choosing k so that $\xi_k^{(q)} \neq 0$, we obtain

$$r_{q+1} = \lambda_1 \frac{a_1 u_{1k} + \sum_{i=2}^{n} \left(\frac{\lambda_i}{\lambda_1}\right)^{q+1} a_i u_{ik}}{a_1 u_{1k} + \sum_{2}^{n} \left(\frac{\lambda_i}{\lambda_1}\right)^{q} a_i u_{ik}}, \tag{4}$$

where u_{ik} is the kth component of u_i. Since $|\lambda_i/\lambda_1| < 1$ for $2 \leq i \leq n$, (3) follows. Furthermore, for q sufficiently large, we have

$$r_{q+1} = \lambda_1 + O\left(\left|\frac{\lambda_2}{\lambda_1}\right|^q\right) \quad \text{as } q \to \infty. \quad \blacksquare$$

Remark If $|\lambda_1| > 1$, $\|x_q\| \to \infty$. If $|\lambda_1| < 1$, $\|x_q\| \to 0$. Hence, in actual computation, scaling is usually necessary. Thus, we define

$$y_q^T = (\eta_1^{(q)}, \ldots, \eta_n^{(q)}),$$
$$y_{q+1} = A x_q,$$
$$x_{q+1} = \frac{1}{\|y_{q+1}\|_\infty} y_{q+1},$$
$$r_{q+1} = \frac{\eta_k^{(q+1)}}{\xi_k^{(q)}}.$$

r_{q+1} is again given by (4) above. A component u_{1k} can be determined by taking the

6.3 The Power Method

largest component of y_q after a few iterations. This method is known as the *power method*.

Corollary Let A have a unique largest (in absolute value) eigenvalue which is real and has multiplicity m; i.e.,

$$|\lambda_1| = \cdots = |\lambda_m| > |\lambda_{m+1}| \geq \cdots \geq |\lambda_n|$$

and

$$\lambda_1 = \lambda_2 = \cdots = \lambda_m.$$

If A has n independent eigenvectors, the power method converges provided that x_0 has a nonzero component in the subspace spanned by the eigenvectors belonging to λ_1.

Proof. Let $(u_1, \ldots, u_m)$ belong to λ_1. Then

$$x_q = A^q x_0 = \lambda_1^q \left(\sum_{j=1}^m a_j u_j\right) + \sum_{i=m+1}^n \lambda_i^q a_i u_i$$

$$r_{q+1} = \lambda_1 \frac{\left(\sum_1^m a_j u_j\right)_k + \sum_{i=m+1}^n \left(\frac{\lambda_i}{\lambda_1}\right)^{q+1} a_i u_{ik}}{\left(\sum_1^m a_j u_j\right)_k + \sum_{i=m+1}^n \left(\frac{\lambda_i}{\lambda_1}\right)^q a_i u_{ik}} \rightarrow \lambda_1. \blacksquare$$

Having found the largest eigenvalue λ_1 and its eigenvector u_1 by the power method, we can apply the method to find λ_2 if $|\lambda_2| > |\lambda_3| \geq \cdots \geq |\lambda_n|$. To do this, we must transform the given matrix A into a matrix B which has λ_2 as its largest eigenvalue. The next theorem indicates how this can be done by a process known as *deflation*.

Theorem 6.3.2 As above, let λ_1 be the largest eigenvalue of A and u_1 its eigenvector. Let v be any (column) vector such that $v^T u_1 = 1$. The eigenvalues of $B = A - \lambda_1 u_1 v^T$ are $\{0, \lambda_2, \lambda_3, \ldots, \lambda_n\}$.

Proof. We have $Bu_1 = Au_1 - \lambda_1 u_1 v^T u_1 = Au_1 - \lambda u_1 = 0$. Thus, u_1 is an eigenvector of B belonging to 0.

Now, we may assume that the first component u_{11} of u_1 is not zero. (Otherwise, we consider $U^{-1}AU$, where U is an orthogonal matrix which rotates the coordinate axes.). Thus, we may take $u_{11} = 1$. Let $x = u_1 - e_1$, where $e_1^T = (1, 0, \ldots, 0)$ and define $T = I + xe_1^T$. Then $T^{-1} = I - xe_1^T$, since $xe_1^T xe_1^T = 0$. Letting $A_1^{(r)}$ and $A_1^{(c)}$ be the first row and column of A respectively, we have

$$C = T^{-1}AT = (I - xe_1^T)A(I + xe_1^T) = A - xA_1^{(r)} + (A - xA_1^{(r)})xe_1^T.$$

But

$$(A - xA_1^{(r)})x = Au_1 - Ae_1 - (A_1^{(r)}x)x = \lambda_1 u_1 - A_1^{(c)} - (\lambda_1 u_{11} - a_{11})x$$

$$= a_{11}x - A_1^{(c)} + \lambda_1 e_1.$$

The first column of C is

$$Ce_1 = A_1^{(c)} - xA_1^{(r)}e_1 + a_{11}x - A_1^{(c)} + \lambda_1 e_1 = \lambda_1 e_1.$$

Now consider $D = T^{-1}u_1v^{\mathrm{T}}T$. We have
$$D = (I - xe_1^1)u_1v^{\mathrm{T}}(I + xe_1^{\mathrm{T}}) = e_1v^{\mathrm{T}} + (1 - v_1)e_1e_1^{\mathrm{T}}.$$
Thus, $D_i^{(r)} = 0$ for $i \geq 2$ and $D_1^{(r)} = v^{\mathrm{T}} + (1 - v_1)e_1^{\mathrm{T}}$. $D_1^{(r)} = (1, v_2, \ldots, v_n)$. Hence, $C - \lambda_1 D$ has the same eigenvalues as C except for λ_1 which is replaced by zero. Since $C - \lambda_1 D = T^{-1}BT$, the same is true of B. Since C and A have the same eigenvalues, the theorem is proved. ∎

Corollary If $(w_1, \ldots, w_n)$ are the eigenvectors of B belonging to $(0, \lambda_2, \ldots, \lambda_n)$ respectively, then for λ_1 simple, $u_1 = w_1$ and $u_i = (\lambda_i - \lambda_1)w_i + \lambda_1(v^{\mathrm{T}}w_i)u_1, i > 1$.

Remark If A has a nonlinear elementary divisor $(\lambda - \lambda_1)^m$, $m > 1$, then not all w_i corresponding to λ_1 are equal to u_1.

For $\lambda_i \neq 0$, $\lambda_i \neq \lambda_1$ we find
$$w_i = \frac{1}{(\lambda_i - \lambda_1)}\left(u_i - \frac{\lambda_1}{\lambda_i}(v^{\mathrm{T}}u_i)u_1\right).$$

Wielandt's Deflation. Taking
$$v^{\mathrm{T}} = \frac{1}{\lambda_1 u_{1j}} A_j^{(r)}$$
we have
$$v^{\mathrm{T}}u_1 = \frac{1}{\lambda_1 u_{1j}} \lambda_1 u_{1j} = 1$$
and
$$B = A - \lambda_1 u_1 v^{\mathrm{T}} = A - \frac{1}{u_{1j}} u_1 A_j^{(r)}.$$
Thus,
$$B_j^{(r)} = A_j^{(r)} - \frac{u_{1j}}{u_{1j}} A_j^{(r)} = 0.$$

Since $Bw_i = \lambda_i w_i$, we see that for $\lambda_i \neq 0$, the jth component of w_i must be zero and the jth column of B does not enter into this equation. Therefore, $|\lambda_2| > |\lambda_3| \geq \cdots \geq |\lambda_n|$, and we may apply the power method to the matrix of order $n - 1$ obtained by deleting the jth row and column of B.

The power method converges even when A does not have n linearly independent eigenvectors; i.e. is not diagonalizable. (See Exercise 6.18.) However, in this case, the iteration converges at a slower rate.

6.4 THE JACOBI METHOD (FOR A SYMMETRIC MATRIX)

Let A be a real symmetric $n \times n$ matrix. In Exercise 3.25, it is shown how to choose a two-dimensional rotation matrix U so that for $B = UAU^*$, a specified row r and column s, $r \neq s$, $b_{rs} = 0$. In this case, we say that the rotation U *annihilates* the (r, s)-

element of A. This suggests the possibility of annihilating all the off-diagonal elements of A by a sequence $\{U_n\}$ of such two-dimensional rotations. Since each similarity transformation preserves the eigenvalues of A, the resulting diagonal matrix would have these eigenvalues as its diagonal elements. Of course, a symmetric matrix can be reduced to diagonal form by a single rotation (Exercise 3.15). However, this rotation is unknown. The Jacobi method seeks to approximate it by a sequence (U_n) of two-dimensional rotations, where U_n is chosen to annihilate the maximum modulus off-diagonal element at the nth step. Unfortunately, if the (r, s)-element is annihilated by rotation U_n and the next rotation U_{n+1} is selected to annihilate the (r', s')-element, then the new (r, s)-element will not in general be zero. Thus, it is to be expected that an infinite sequence of two-dimensional rotations will be required. For numerical computations, the procedure is carried out until the matrix $U_n U_{n-1} \cdots U_1 A U_1^* U_2^* \cdots U_n^*$ has all of its off-diagonal elements of absolute value less than some prescribed error bound $\varepsilon > 0$. That this is always possible for arbitrary $\varepsilon > 0$ is proved in the next theorem.

Theorem 6.4.1 (*Jacobi*). Let A be a real symmetric $n \times n$ matrix. Let $\{U_m\}$ be a sequence of two-dimensional rotation matrices, with U_m chosen to annihilate the maximum modulus off-diagonal element of $A^{(m-1)} = U_{m-1} \cdots U_1 A U_1^* \cdots U_{m-1}^*$, $m = 1, 2, \ldots$, with $A^{(0)} = A$. Then $\lim_{m \to \infty} A^{(m)} = D$, where D is a diagonal matrix having the eigenvalues of A, with the correct multiplicities, as its diagonal elements.

Proof. Let $A^{(m)} = (a_{ij}^{(m)})$. Then by Exercise 3.25,

$$\sum_{i \neq j} |a_{ij}^{(m)}|^2 = \sum_{i \neq j} |a_{ij}^{(m-1)}|^2 - 2|a_{rs}^{(m-1)}|^2, \tag{6.4.1}$$

where $a_{rs}^{(m-1)}$ is an element of maximum modulus in the matrix $A^{(m-1)}$ and $A^{(m)} = U_m A^{(m-1)} U_m^*$, with U_m chosen to annihilate the (r, s)-element of $A^{(m-1)}$. By the maximality of $a_{rs}^{(m-1)}$ we have

$$(n^2 - n) |a_{rs}^{(m-1)}|^2 \geq \sum_{i \neq j} |a_{ij}^{(m-1)}|^2.$$

Therefore,

$$\sum_{i \neq j} |a_{ij}^{(m)}|^2 \leq \left(1 - \left(\frac{2}{n^2 - n}\right)\right) \sum_{i \neq j} |a_{ij}^{(m-1)}|^2.$$

The factor $K = 1 - 2/(n^2 - n)$ is less than 1 for $n > 1$. Hence,

$$\sum_{i \neq j} |a_{ij}^{(m)}|^2 \leq K^m \sum_{i \neq j} |a_{ij}|^2 \to 0 \quad \text{as } m \to \infty.$$

This shows that $a_{ij}^{(m)} \to 0$ for any off-diagonal element, $i \neq j$.

Now, the eigenvalues of $A^{(m)}$ are the same as those of A, since $A^{(m)}$ is obtained from A by a sequence of similarity transformations. Since the Gerschgorin disks of $A^{(m)}$ have radii which approach 0 as $m \to \infty$, each diagonal element $a_{ii}^{(m)}$ must converge to an eigenvalue λ_i of A. For suppose $\{\lambda_j\}$ are the distinct eigenvalues of A. Let λ_j have multiplicity n_j and let the minimum distance between distinct eigenvalues be $2d$. Then for m sufficiently large so that each Gerschgorin disk has radius $< d/8n$ there must be precisely n_j disks whose union form a connected component containing λ_j and no other eigenvalue. (See Theorem 6.2.2.) Let $a_{i_1 i_1}^{(m)}, \ldots, a_{i_{n_1} i_{n_1}}^{(m)}$ be the centers of the disks containing λ_1. We say that these elements are *linked to* λ_1. Thus, $|a_{ii}^{(m)} - \lambda_1| < d/4$ for $i = i_1, \ldots, i_{n_1}$, and similarly for the other eigenvalues. Now suppose U_{m+1} annihilates the (r, s)-element. Then $a_{rr}^{(m)}$ and $a_{ss}^{(m)}$ are the only diagonal elements which change. If both $a_{rr}^{(m)}$ and $a_{ss}^{(m)}$ are linked to the same eigenvalue, then so are $a_{rr}^{(m+1)}$ and $a_{ss}^{(m+1)}$, since the other linkings are unchanged and multiplicities must conform to Gerschgorin's theorem. If $a_{rr}^{(m)}$ is linked to λ_i and $a_{ss}^{(m)}$ to λ_j, $\lambda_i \neq \lambda_j$, then

$$a_{rr}^{(m+1)} - \lambda_j = (a_{ss}^{(m)} - \lambda_j) \sin^2 \theta - 2a_{rs}^{(m)} \cos \theta \sin \theta + (a_{rr}^{(m)} - \lambda_i + \lambda_i - \lambda_j) \cos^2 \theta$$

$$|a_{rr}^{(m+1)} - \lambda_j| \geq |\lambda_i - \lambda_j| \cos^2 \theta - |a_{ss}^{(m)} - \lambda_j| \sin^2 \theta - |a_{rr}^{(m)} - \lambda_i| \cos^2 \theta - |a_{rs}^{(m)}|$$

$$|a_{rr}^{(m+1)} - \lambda_j| \geq d \cos^2 \theta - \frac{d}{4}(\sin^2 \theta + \cos^2 \theta) - \frac{d}{4} = \frac{d}{2} \cos 2\theta.$$

Since $|a_{rr}^{(m)} - a_{ss}^{(m)}| \geq d/2$, we have for m sufficiently large so that

$$|a_{rs}^{(m)}| \leq d/4, \qquad |\tan 2\theta| = |2a_{rs}^{(m)}/(a_{rr}^{(m)} - a_{ss}^{(m)})| \leq \frac{d/2}{d/2} = 1.$$

Hence,

$$|2\theta| \leq \pi/4 \quad \text{and} \quad |a_{rr}^{(m+1)} - \lambda_j| \geq \frac{d}{4}\sqrt{2},$$

which shows that $a_{rr}^{(m+1)}$ cannot be linked to λ_j. It must, therefore, remain linked to λ_i. Thus, for all m sufficiently large, the linkings do not change. Since the radii of the disks converge to 0, we must have $a_{ii}^{(m)} \to \lambda_1$ for all $a_{ii}^{(m)}$ linked to λ_1 and similarly for the other eigenvalues. ∎

Corollary Let $V_m = U_m \cdots U_1$, where the U_i are the rotations of the preceding theorem. Let $v_j^{(m)}$ be the jth column of V_m^* and let $a_{jj}^{(m)} \to \lambda_j$, where λ_j is an eigenvalue of A of multiplicity n_j. Let M_j be the subspace of eigenvectors belonging to λ_j. Then $\lim_{m \to \infty} \|v_j^{(m)} - M_j\| = 0$.

Proof. Consider the first column $v_1^{(m)}$ of V_m. (The proof for the other columns is identical.) Since $AV_m^* = V_m^* A^{(m)}$ we have

$$Av_1^{(m)} = a_{11}^{(m)} v_1^{(m)} + r^{(m)},$$

where

$$r_i^{(m)} = \sum_{j=2}^{n} a_{j1} v_{ij}^{(m)}, \qquad 1 \leq i \leq n.$$

Thus, $r^{(m)} \to 0$, and $a_{11}^{(m)} \to \lambda_1$. Referring to Theorem 6.2.6, we see that if m is sufficiently large so that $|a_{11}^{(m)} - \lambda_i| \geq d$, $\lambda_i \neq \lambda_1$, then

$$\|v_1^{(m)} - M_1\|_2 \leq \|r^{(m)}\|_2/d.$$

Since $r^{(m)} \to 0$, the result follows. ∎

Remark If the multiplicity n_1 of λ_1 is 1, then the proof of the corollary shows that $v_1^{(m)}$ converges to a unit eigenvector belonging to λ_1 ($v_1^{(m)}$ is a unit vector, since $V^{(m)}$ is an orthogonal matrix). As $m \to \infty$, $a_{ii}^{(m)} \to \lambda_i$ and $a_{1i}^{(m)} \to 0$. Since $\lambda_1 \neq \lambda_i$, $i \geq 2$, the angles θ_m for rotations applied to the first row (and column) must approach 0; i.e., $\tan 2\theta_m = 2a_{1i}^{(m)}/(a_{ii}^{(m)} - a_{11}^{(m)}) \to 0$. Hence, $v_1^{(m)}$ converges to a particular unit eigenvector. For multiplicity $n_1 > 1$, the proof shows that the subspace spanned by the orthonormal set of column vectors corresponding to λ_1 converges to the subspace of eigenvectors belonging to λ_1. However, we cannot conclude that the individual eigenvectors converge. For computational purposes this poses no difficulties, since if $\{v_1^{(m)}, \ldots, v_{n_1}^{(m)}\}$ is an orthonormal set such that $\|v_i^{(m)} - M_1\|_2 < \varepsilon$ for $1 \leq i \leq n_1$, then any unit vector in M_1 may be approximated by a linear combination of the $v_i^{(m)}$ within ε-accuracy.

Computational Aspects

The search for the largest off-diagonal element can be simplified by recording a vector $w^{(m)}$ containing for each row of $A^{(m)}$ the column index and modulus of the largest element in that row. Since $A^{(m+1)}$ will differ from $A^{(m)}$ only in the r-row and s-column, the vector $w^{(m+1)}$ is obtained with fewer than the $n(n-1)/2$ comparisons required in a complete search.

A variation of the Jacobi method is the *cyclic* Jacobi method in which the elements are annihilated in the cyclic order $(1, 2), (1, 3), \ldots, (1, n), (2, 3)(2, 4), \ldots, (n-1, n)$, regardless of their magnitude. If the angles of rotation are suitably restricted, the cyclic variation also converges. (See [6.3].) In still another variation of the cyclic method), if the (r, s)-element is already small, say below some specified *threshold* value, the annihilation of that element can be skipped over in that cycle and in subsequent cycles until all off-diagonal elements are below the threshold. The threshold is then lowered and cyclic annihilation is resumed with the new threshold. The threshold is lowered repeatedly until all off-diagonal elements are acceptably small. Since only a finite number of cycles are required to reduce all off-diagonal elements below a given threshold, the *threshold cyclic* variation also converges.

To perform a rotation through the angle θ we must calculate $\sin \theta$ and $\cos \theta$. It is not necessary to compute θ itself. To minimize rounding errors, $\cos \theta$ and $\sin \theta$ should be computed by the formulas

$$z = (4a_{rs}^2 + |a_{ss} - a_{rr}|^2)^{1/2},$$

$$\cos \theta = \left(\frac{z + a_{rr} - a_{ss}}{2z}\right)^{1/2},$$

$$\sin \theta = a_{rs}/z \cos \theta.$$

Rate of Convergence

The rate of convergence of the Jacobi method is linear, in general, by Theorem 6.4.1. However, suppose A has n distinct eigenvalues $\{\lambda_i\}$ and that $|\lambda_i - \lambda_j| \geq 2d$, $i \neq j$. Suppose further that m is sufficiently large so that $\sum_{i \neq j} |a_{ij}^{(m)}|^2 < d/2$. It can be proved

[6.4] that after a complete cycle of $N = \frac{1}{2}(n^2 - n)$ rotations the sum of the squares of the off-diagonal elements satisfies the inequality,

$$\left(\sum_{i \neq j} |a_{ij}^{(m+N)}|^2\right)^{1/2} \leq \frac{1}{d}\left(\frac{n}{2} - 1\right)^{1/2} \sum_{i \neq j} |a_{ij}^{(m)}|^2.$$

In this case, we can say that the rate of convergence of the Jacobi method eventually becomes quadratic (counting a complete cycle as one iteration). For the cyclic variation, convergence again becomes quadratic (every N steps) but with a factor $1/\sqrt{2d}$. (See [6.5].) If there are p pairs of identical eigenvalues, then a quadratic error reduction takes place every $N + p(n - 2)$ steps, with a factor $n/2d$. (See [6.4].) The factor $1/d$ in these results suggests that quadratic convergence no longer obtains for eigenvalues of multiplicity > 2. However, Wilkinson [6.1] indicates that even this case has, in practice, involved no decrease in the rate of convergence. About 5 or 6 cycles are generally sufficient for matrices of orders up to 50, with precisions of 32 to 48 binary digits.

Error Analysis

In applying the Jacobi method, rounding errors occur both in the computation of the rotation matrices U_m and in the multiplication of $A^{(m-1)}$ by U_m and U_m^*. An analysis of these errors can be found in [6.1], Chapter 3. The end result of this analysis can be stated as follows. Suppose r complete cycles (i.e., $k = \frac{1}{2}n(n - 1)r$ annihilations) have been completed. Let $U = U_k U_{k-1} \cdots U_1$ be the exact product of the $k = rN$ exact rotations U_m determined from the actual computed matrices $\bar{A}^{(m)}$ corresponding to the exact $A^{(m)}$, $m = 1, \ldots, k$. Thus, each U_m is the exact plane rotation determined by $\bar{A}^{(m-1)}$, where $\sin \theta_m$, $\cos \theta_m$, are computed with no rounding error, and $U_m \bar{A}^{(m-1)} U_m^*$ is an exact similarity transformation which preserves the eigenvalues of $\bar{A}^{(m-1)}$. Hence UAU^* also has the same eigenvalues as A. Now, suppose that D is the actual computed diagonal matrix after $k = rN$ annihilations. It is shown in [6.1] that

$$\|\bar{A}^{(k)} - UAU^*\|_2 \leq 18 \cdot 2^{-s} r n^{3/2} (1 + 9 \cdot 2^{-s})^{r(4n-7)} \|A\|_2,$$

where s is the number of binary digits of floating-point precision used. Now, UAU^* is symmetric. Hence, by the Hoffman–Wielandt theorem (Exercise 6.16), if the eigenvalues λ_i of A and the computed diagonal elements $\bar{\lambda}_i$ are arranged in order of magnitude, we have for $r = 6$ cycles,

$$\left(\frac{\sum_{i=1}^n (\bar{\lambda}_i - \lambda_i)^2}{\sum_{i=1}^n \lambda_i^2}\right)^{1/2} \leq 108 \cdot 2^{-s} n^{3/2} (1 + 9 \cdot 2^{-s})^{6(4n-7)}.$$

The factor $(1 + 9 \cdot 2^{-s})^{24n-42} = 1 + K 2^{-s} n + O(2^{-s} n)^2$, so that the dominant term in the bound is $108(2^{-s} n^{3/2})$. For $s = 40$ and $n = 10^2$, this dominant term is about 10^{-7}, indicating an accuracy of 7 digits. Statistical bounds on the rounding error would be considerably better, about $11(2^{-s}n^{3/4})$. Note that these are a priori bounds. They indicate a rather high accuracy with respect to rounding error.

6.5 HOUSEHOLDER'S METHOD

A matrix $B = (b_{ij})$ is said to be an *upper Hessenberg* matrix if all entries below the subdiagonal are zero; i.e., $b_{ij} = 0$ for $1 \leq j \leq i - 2$. In Householder's method, an arbitrary matrix A is reduced to upper Hessenberg form by applying $n - 2$ similarity transformations. If A is symmetric, it is reduced to tridiagonal form. The eigenvalues of a tridiagonal matrix can be obtained by solving the characteristic equation. This will be treated in the next section.

The similarity transformations in Householder's method are of a special type in which the matrices are *elementary Hermitian unitary*. An elementary Hermitian unitary matrix P is one having the form

$$P = I - 2vv^*$$

where v is an n-dimensional column vector of unit l_2-norm, that is, $v^*v = 1$. (Note that $v^* = v^T$ if v is real.) It is easy to verify that P is Hermitian and unitary (Exercise 6.1). Therefore, $P = P^{-1}$ and PAP is a similarity transformation.

Theorem 6.5.1 Let A be an arbitrary $n \times n$ matrix. There exist $n - 2$ elementary Hermitian unitary matrices, $P_1, \ldots, P_{n-2}$, such that $A_{n-2} = P_{n-2} \cdots P_1 A P_1 \cdots P_{n-2}$ is upper Hessenberg. If A is symmetric, then A_{n-2} is tridiagonal.

Proof. Suppose that the first $r - 1$ columns of A have been reduced to upper Hessenberg form by $P_1, \ldots, P_{r-1}, r \geq 1$. Then $A_{r-1} = P_{r-1} \cdots P_1 A P_1 \cdots P_{r-1}$ has the following form:

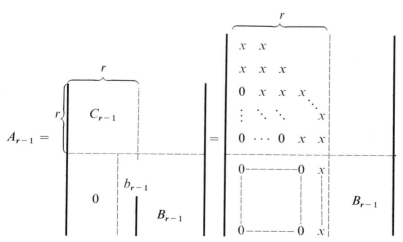

If A is symmetric, then C_{r-1} is a tridiagonal $r \times r$ matrix. The vector b_{r-1} is $(n - r)$-dimensional. To reduce the rth column to upper Hessenberg form, we seek a transformation $P_r A_{r-1} P_r$, where $P_r = I - 2v_r v_r^*$ and the first r components of v_r are 0. Denoting the $r \times r$ identity by I_r, we have

$$P_r = \begin{array}{c} r\{ \\ \\ n-r\{ \end{array} \overbrace{\begin{pmatrix} I_r & 0 \\ \hline 0 & Q_r \end{pmatrix}}^{r},$$

where $Q_r = I_{n-r} - 2w_r w_r^*$ and w_r consists of the last $n - r$ components of v_r. The matrix $A_r = P_r A_{r-1} P_r$ is easily seen to be of the form

$$A_r = \begin{array}{c} r\{ \\ \\ n-r\{ \end{array} \begin{pmatrix} C_{r-1} & & \\ \hline 0 & c_r & Q_r B_{r-1} Q_r \end{pmatrix},$$

where $c_r = Q_r b_{r-1}$. We must determine w_r so that all components of c_r except the first are zero, for then the first r columns of A_r will be in lower Hessenberg form. Dropping subscripts, we can write the required condition as

$$Qb = (I - 2ww^*)b = ke_1,$$

where $e_1 = (1, 0, \ldots, 0)$. Thus, we have $b - 2ww^*b = ke_1$, which yields

$$2w_1(w^*b) = b_1 - k, \tag{1}$$

$$2w_i(w^*b) = b_i, \quad i \geq 2. \tag{2}$$

Squaring (1) and (2) and summing, we obtain

$$4(w^*b)^2 = \|b\|_2^2 - 2b_1 k + k^2,$$

since $\|w\|_2 = 1$ by definition. Since Q is unitary, $|k| = \|Qb\|_2 = \|b\|_2$. Letting $\alpha = 4(w^*b)^2$, we have

$$\alpha = 2\|b\|_2(\|b\|_2 \pm b_1). \tag{3}$$

From (1) and (2) we obtain the desired values of w_r as

$$w_1 = (b_1 \pm \|b\|_2)/\sqrt{\alpha},$$
$$w_i = b_i/\sqrt{\alpha}, \quad 2 \leq i \leq n - r. \blacksquare$$

Computational Aspects

Setting $u_1 = b_1 \pm \|b\|_2$, $u_i = b_i$, $i \geq 2$, we see that $Q = I - (2/\alpha)uu^*$. The sign in u_1 and α should be chosen to be the same as the sign of b_1 in order to reduce rounding errors. In summary, the Householder reduction to Hessenberg form can be stated in terms of the matrices $A^{(r)} = (a_{ij}^{(r)})$ as follows:

$$A^{(0)} = A$$
$$A^{(r)} = P_r A^{(r-1)} P_r, \quad r = 1, \ldots, n-2,$$
$$P_r = I - \left(\frac{1}{\beta_r}\right) u_r u_r^*, \tag{4}$$

$$\begin{aligned}(u_r)_i &= 0, & 1 \le i \le r, \\ (u_r)_{r+1} &= a_{r+1,r}^{(r-1)} \mp S_r, \\ (u_r)_i &= a_{ir}^{(r-1)} & r+2 \le i \le n,\end{aligned} \tag{5}$$

$$S_r = \left(\sum_{i=r+1}^{n} (a_{ir}^{(r-1)})^2\right)^{1/2} \tag{6}$$

$$\beta_r = S_r(S_r \mp a_{r+1,r}^{(r-1)}) = \mp S_r(u_r)_{r+1}. \tag{7}$$

To simplify the calculation of $A^{(r)}$ in the symmetric case, we compute

$$y_r = (1/\beta_r) A^{(r-1)} u_r, \tag{8}$$
$$z_r = y_r - (1/2\beta_r) u_r (u_r^* y_r), \tag{9}$$
$$A^{(r)} = A^{(r-1)} - u_r z_r^* - z_r u_r^*. \tag{10}$$

Taking advantage of the symmetric form of $A^{(r-1)}$, we need only be concerned with the last $n - r$ rows and columns. (The first $r - 1$ rows and columns will not be changed and $a_{rj}^{(r-1)}$, $j \ge r + 2$, will become 0, while $a_{r,r+1}^{(r)} = \pm S_r$.) Also note that the first $r - 1$ components of y_r are 0 and the rth component is not used. An operation count reveals that a total of $\frac{2}{3}n^3 + O(n^2)$ multiplications are required to reduce an nth order matrix to tridiagonal form.

Wilkinson's [6.1] analysis of the rounding error in floating-point computations using a precision of s binary digits with double-precision accumulation of products shows that

$$\left(\frac{\sum_{i=1}^{n}(\bar{\lambda}_i - \lambda_i)^2}{\sum_{i=1}^{n}\lambda_i^2}\right)^{1/2} \le 2Kn \cdot 2^{-s}(1 + K \cdot 2^{-s})^{2n},$$

where $\bar{\lambda}_i$ are the eigenvalues of the computed tridiagonal matrix, $\bar{A}^{(n)}$, and λ_i are the eigenvalues of A. In practice, a value of 20 for K is reasonable. (The value of K depends on the specific details of the rounding procedure in the arithmetic operations. See, for example, the rounding error analysis in Gauss elimination.)

6.6 GIVENS' METHOD
(EIGENVALUES OF A SYMMETRIC TRIDIAGONAL MATRIX)

Given a symmetric tridiagonal matrix $B = (b_{ij})$, we define $b_i = b_{ii}$ and $c_{i+1} = b_{i+1,i} = b_{i,i+1}$, $1 \le i \le n$. All other elements are 0. If some $c_i = 0$, then B is the direct sum of two tridiagonal matrices B_1 and B_2. The eigenvalues and eigenvectors of A may be found directly from those of B_1 and B_2. Henceforth, we assume $c_i \neq 0$, $1 \leq i \leq n$, and

$$B = \begin{pmatrix} b_1 & c_2 & 0 & \cdots & & 0 \\ c_2 & b_2 & c_3 & & & \\ 0 & & \ddots & \ddots & & \vdots \\ & \ddots & \ddots & \ddots & & 0 \\ \vdots & & & c_{n-1} & b_{n-1} & c_n \\ 0 & \cdots & & 0 & c_n & b_n \end{pmatrix}$$

Consider the leading principal minors of $\lambda I - B$ of order $r = 1, \ldots, n$. It is easy to verify that they are polynomials $p_r(\lambda)$ satisfying the recursion

$$p_r(\lambda) = (\lambda - b_r)p_{r-1}(\lambda) - c_r^2 p_{r-2}(\lambda), \qquad r = 2, \ldots, n, \tag{1}$$

with $p_0(\lambda) = 1$ and $p_1(\lambda) = \lambda - b_1$. It is not hard to show that the r zeros of $p_r(\lambda)$ separate the $r + 1$ zeros of $p_{r+1}(\lambda)$. Recall, first, that the zeros of $p_r(\lambda)$ are the eigenvalues of a symmetric $r \times r$ submatrix of A, and therefore are all real. Equation (1) implies that two successive polynomials p_{r-1} and p_r cannot have a common zero, α, for then α would be zero of p_{r-2} also. By repeated application of (1), this would imply that $p_0(\alpha) = 0$, which is impossible. Now,

$$p_2(\lambda) = (\lambda - b_2)p_1(\lambda) - c_2^2 p_0(\lambda) = (\lambda - b_2)(\lambda - b_1) - c_2^2,$$

has zeros $\frac{1}{2}(b_1 + b_2) \pm \frac{1}{2}((b_1 - b_2)^2 + 4c_2^2)^{1/2}$. Since $c_2 \neq 0$, these are distinct and are separated by b_1, the zero of $p_1(\lambda)$. Now assume (induction hypothesis) that the $r - 2$ zeros of p_{r-2} separate the $r - 1$ zeros of p_{r-1}. By (1) above, this implies that at each zero α of p_{r-1}, $p_r(\alpha)$ and $p_{r-2}(\alpha)$ have opposite signs. But p_{r-2} must change sign exactly once between adjacent zeros of p_{r-1}, since it has one zero between two successive zeros of p_{r-1} and is of degree $r - 2$. Therefore, p_r also changes sign at least once and has one zero between adjacent zeros of p_{r-1}. This accounts for $r - 2$ zeros of p_r. Further, $p_i(+\infty) = +\infty$ and $p_i(-\infty) = (-1)^i \infty$, $1 \leq i \leq n$, so that $p_{r-2}(\lambda)$ must change from $-$ to $+$ as λ increases through the largest zero α_0 of p_{r-2} and p_r must change from $+$ to $-$. But $p_r(+\infty) = +\infty$ and $p_r(\alpha_1) < 0$ at the largest zero α_1 of p_{r-1}, since $\alpha_1 > \alpha_0$ and $p_{r-2}(\alpha_1) > 0$. Therefore, p_r must have another zero greater than α_1. Similar reasoning with signs shows that p_r must have another zero less than the smallest zero of p_{r-1}. This accounts for all the zeros of p_r and shows that they are distributed so that the $r - 1$ zeros of p_{r-1} separate the r zeros of p_r. By induction, the result holds for all p_r. Note that this implies that p_r has r distinct roots.

Givens' method is based on the following lemma, which is similar to Sturm's theorem (Section 5.6).

Lemma For any α, let $m_r(\alpha)$ be the number of places i in the sequence $(p_0(\alpha), p_1(\alpha), \ldots, p_r(\alpha))$ at which there is a change of sign in going from $p_i(\alpha)$ to $p_{i+1}(\alpha)$. Then there are $m_n(\alpha)$ eigenvalues of A ($=$ zeros of $p_n(\alpha)$) which are greater than α. (If some $p_i(\alpha) = 0$, then $p_i(\alpha)$ is considered to have the same sign as $p_{i-1}(\alpha)$.)

Proof. By induction on r. The lemma is trivially true for $r = 1$. Assume it holds for r. Let $q = m_r(\alpha)$. Then there are q zeros of $p_r(\lambda)$ greater than α. Arranging the zeros, z_i of p_r in descending order of magnitude, we have $z_r < \cdots < z_{q+1} \leq \alpha < z_q < \cdots < z_2 < z_1$. Since the zeros z_i of p_r separate the zeros y_i of p_{r+1}, we have $(z_{q+2} < y_{q+2} \leq z_{q+1} \leq \alpha < z_q < y_q < \cdots < y_2 < z_1 < y_1$. Either $\alpha < y_{q+1}$, in which case we must show that $m_{r+1}(\alpha) = 1 + m_r(\alpha)$, or $y_{q+1} \leq \alpha$ and we must show $m_{r+1}(\alpha) = m_r(\alpha)$. Suppose first that $p_r(\alpha) \neq 0$ and $p_{r+1}(\alpha) \neq 0$. Then $z_{q+1} < \alpha < y_{q+1}$ implies that $p_r(\alpha)$ and $p_{r+1}(\alpha)$ have opposite signs, since $p_{r+1}(\alpha) = \Pi_{i=1}^{r+1}(\alpha - y_i)$ has one more negative factor than $p_r(\alpha) = \Pi_{i=1}^r(\alpha - z_i)$. Hence, $m_{r+1}(\alpha) = 1 + m_r(\alpha)$. Similarly, if $y_{q+1} < \alpha < z_q$, then $p_r(\alpha)$ and $p_{r+1}(\alpha)$ have the same sign and $m_r(\alpha) = m_{r+1}(\alpha)$.

Now, suppose $\alpha = y_{q+1}$. Then $p_{r+1}(\alpha) = 0$ and $p_r(\alpha) \neq 0$. By our agreement, $p_{r+1}(\alpha)$ has the same sign as $p_r(\alpha)$ and $m_r(\alpha) = m_{r+1}(\alpha)$ in accordance with both p_r and p_{r+1} having q zeros greater than α.

Finally, if $\alpha = z_{q+1}$, then $p_r(\alpha) = 0$ and is assigned the same sign as $p_{r-1}(\alpha) \neq 0$. By our preceding reasoning, $p_{r-1}(\alpha)$ and $p_{r+1}(\alpha)$ must have opposite signs. Hence, $m_{r+1}(\alpha) = 1 + m_r(\alpha)$. This corresponds to the fact that $\alpha = z_{q+1} < y_{q+1} < z_q < y_q$; i.e., p_{r+1} has one more zero greater than α. This completes the induction. ∎

Computational Aspects

Again, assume that the matrix elements $c_i \neq 0$. We have seen that this implies that all eigenvalues have multiplicity one. Index them so that $\lambda_n < \lambda_{n-1} < \cdots < \lambda_1$. Suppose that we have found an interval (a_0, b_0) such that $m_n(a_0) \geq k$ and $m_n(b_0) < k$. Then $\lambda_k \in (a_0, b_0)$. Choosing the midpoint $\alpha_0 = (a_0 + b_0)/2$, compute $m_n(\alpha_0)$. If $m_n(\alpha_0) \geq k$, take $a_1 = \alpha_0$ and $b_1 = b_0$. If $m_n(\alpha_0) < k$, take $a_1 = a_0$ and $b_1 = \alpha_0$. Repeating the procedure with (a_1, b_1) and the midpoint α_1, and so on, λ_k can be located within an arbitrarily small interval (assuming no rounding errors).

As starting values, it is reasonable to take $a_0 = -\|A\|_\infty$ and $b_0 = \|A\|_\infty$, since $|\lambda_i| \leq \|A\|_\infty$ for every eigenvalue λ_i (Theorem 3.6.1, Corollary.).

The *bisection* procedure just described is, of course, subject to rounding error. Therefore, there is a lower bound to the accuracy with which λ_k can be determined. Wilkinson shows that if the bisection procedure is carried out with s-digit floating-point precision, then

$$|\lambda_k - \alpha_q| < (15.06)2^{-s} + 3(2^{-q}),$$

where α_q is the center of the qth interval (a_q, b_q); i.e., after q bisections. (Remember, of course, that the λ_k are the true eigenvalues of a tridiagonal matrix, A. If A has been obtained from a matrix A_0 by Householder's method, then error estimates of the eigenvalues of A_0 are subject to the additional bounds of the preceding section.)

6.7 EIGENVECTORS OF A SYMMETRIC TRIDIAGONAL MATRIX

Suppose $\bar{\lambda}_k$ is an approximation to the eigenvalue λ_k of the symmetric matrix A. If u_k is an eigenvector of the tridiagonal matrix $B = P_{n-2} \cdots P_1 A P_1^* \cdots P_{n-2}^*$ belonging to λ_k, then $P_1^* \cdots P_{n-2}^* u_k$ is an eigenvalue of A belonging to λ_k. Therefore, we consider the general problem of finding an eigenvector of a symmetric tridiagonal matrix B.

The obvious procedure of solving $(B - \bar{\lambda}_k I)y = 0$ is unsatisfactory, since even if the error $|\lambda_k - \bar{\lambda}_k|$ is small, the eigenvector y can be a poor approximation to the exact eigenvector u_k belonging to λ_k. For a discussion of this point, see Wilkinson [6.1]. A version of the power method is recommended. The power method is applied to $(B - \bar{\lambda}_k I)^{-1}$ as follows. Let $b = \sum_{i=1}^{n} \beta_i u_i$, where $\{u_1, \ldots, u_n\}$ is an orthonormal basis of eigenvectors of B. (We retain the hypothesis that B has n distinct eigenvalues $\lambda_1, \ldots, \lambda_n$.) If we solve for x_1 in $(B - \bar{\lambda}_k I) x_1 = b$, we obtain

$$x_1 = \sum_{i=1}^{n} \frac{\beta_i}{(\lambda_i - \bar{\lambda}_k)} u_i.$$

Since $\bar{\lambda}_k$ is presumably closer to λ_k than to λ_i, $i \neq k$, the kth component of x should be relatively large compared to the others, provided that $\beta_k \neq 0$. Repeating the process with x_1 substituted for b, we obtain $(B - \bar{\lambda}_k I)x_2 = x_1$ or $(B - \bar{\lambda}_k I)^2 x_2 = b$. Hence,

$$x_2 = \sum_{i=1}^{n} \frac{\beta_i}{(\lambda_i - \bar{\lambda}_k)^2} u_i.$$

After q iterations, the vector x_q should be a good approximation to u_k. To determine b, we first reduce $B - \bar{\lambda}_k I$ to an upper triangular matrix U by Gauss elimination and then solve $U x_1 = e$, where $e^T = (1, \ldots, 1)$. (Thus, b itself does not have to be determined.) The method just described is sometimes called *inverse iteration*.

6.8 EIGENVALUES OF AN ARBITRARY MATRIX BY REDUCTION TO UPPER HESSENBERG FORM

If A is not symmetric, then Householder's reduction will reduce A to upper Hessenberg form. The relevant calculations are given in Eqs. (5), (6), (7) of Section 6.5 and the following equations (to replace (8), (9), (10) which are valid only for the symmetric case):

$$x_r^* = u_r^* A^{(r-1)}, \qquad y_r = A^{(r-1)} u_r,$$
$$z_r = (1/\beta) u_r, \qquad x_r^* z_r = d_r,$$
$$A^{(r)} = A^{(r-1)} - z_r x_r^* - (y_r - d_r u_r) z_r^*.$$

Let $B = A^{(n-2)}$. To evaluate the determinant of the upper Hessenberg matrix $B - \lambda I$ the following procedure ([6.6]) may be used. Suppose $B_1, \ldots, B_n$ are the columns of $B - \lambda I$. Polynomials $q_i(\lambda)$, $0 \leq i \leq n - 1$, such that

$$q_1 B_1 + \cdots + q_{n-1} B_{n-1} + B_n = q_0 e_1, \qquad (1)$$

where $e_1^T = (1, 0, \ldots, 0)$ are given by the recursion

$$-b_{n,n-1}q_{n-1} = (b_{nn} - \lambda)$$
$$-b_{i,i-1}q_{i-1} = (b_{ii} - \lambda)q_i + b_{i,i+1}q_{i+1} + \cdots + b_{in}, \quad 2 \leq i \leq n-1, \quad (2)$$
$$q_0 = (b_{11} - \lambda)q_1 + b_{12}q_2 + \cdots + b_{1,n-1}q_{n-1} + b_{1n}.$$

If any $b_{i,i-1} = 0$, then the determinant of $B - \lambda I$ splits into the product of two smaller upper Hessenberg determinants (Exercise 6.4). Hence, we may assume $b_{i,i-1} \neq 0$ for all i. With this assumption, the $q_i(\lambda)$ determined by (2) above satisfy (1). We claim that

$$\det(B - \lambda I) = \pm q_0(\lambda)b_{21}b_{32}\cdots b_{n,n-1}. \quad (3)$$

To prove this, let C be the matrix having columns $B_1, \ldots, B_{n-1}, q_0 e_1$. As a consequence of (1), we see that $\det C = \det(B - \lambda I)$. But, expanding $\det C$ by minors of the last column, we obtain

$$\det C = \pm q_0 \det C',$$

where $\det C'$ is the cofactor of the $(1, n)$-element of C. Since C' is an upper triangular matrix, its determinant is the product of its diagonal elements, $b_{24}, \ldots, b_{n,n-1}$, which establishes (3). Since we wish to find the zeros of $\det(B - \lambda I)$, it suffices to find the zeros of $q_0(\lambda)$. This can be done by any of the methods for finding zeros of polynomials. Note that the coefficients of $q_0(\lambda)$ need not be determined explicitly. It is only required that we be able to evaluate $q_0(\lambda)$ for arbitrary λ and this can be done by the recursion (2). If the derivative $q_0'(\lambda)$ must be evaluated as well (e.g., if Newton's method is used to find zeros), then this can also be done by recursion. Indeed, differentiation of (2) with respect to λ yields a recursion formula for the derivatives $q_i'(\lambda)$.

6.9 THE QR ALGORITHM (FOR REDUCTION OF A MATRIX TO TRIANGULAR FORM BY SIMILARITY TRANSFORMATIONS)

For an arbitrary matrix A there exists a nonsingular matrix Q such that $U = QAQ^{-1}$ is upper triangular (Exercise 6.5). In fact, there is a unitary matrix Q such that $U = QAQ^*$ is upper triangular (Exercise 6.5). The eigenvalues of A are the diagonal elements of U. Q cannot be obtained from A in a finite number of arithmetic operations, since this would imply that the zeros of any nth-degree polynomial can be obtained in a finite number of such operations (Exercise 6.7). However, the existence of Q suggests the possibility of obtaining an infinite sequence of unitary matrices (Q_k) such that $Q_k A Q_k^*$ approaches an upper triangular form which has the eigenvalues of A on its diagonal. One such sequence is given by the QR algorithm of Francis [6.7]. (A related method which does not use unitary transformations is the LR algorithm [6.8].)

In the QR algorithm, we obtain a sequence of unitary matrices Q_k, upper triangular matrices R_k and matrices $A_k = (a_{ij}^{(k)}), k = 0, 1, 2, \ldots$, defined by the relations.

$$A_0 = A,$$
$$A_k = Q_k R_k \quad (1)$$
$$A_{k+1} = R_k Q_k, \quad k = 0, 1, 2, \ldots.$$

Since $A_{k+1} = Q_k^* Q_k R_k Q_k = Q_k^{-1} A_k Q_k$, each matrix A_k is similar to $A_0 = A$ and therefore has the same eigenvalues.

Also, note that $\|A_{k+1}\|_s \leq \|Q_k^{-1}\|_s \|A_k\|_s \|Q_k\|_s = \|A_k\|_s$ for the spectral norm.

Under conditions to be stated below (Theorem 6.9.1) it can be shown that A_k becomes upper triangular in the limit as $k \to \infty$, so that, $\lim_{k \to \infty} a_{ij}^{(k)} = 0$ for $i > j$. Further, the diagonal elements $a_{ii}^{(k)}$ approach the eigenvalues of A. Unfortunately, for arbitrary A the amount of computation makes the QR algorithm impractical. However, if A is of upper Hessenberg form, tridiagonal or a *band matrix* (i.e., a matrix having zero entries except in a narrow band along the diagonal) then the QR algorithm is practical. Therefore, a reduction to upper Hessenberg form should be carried out first for the general matrix A.

The basis of the QR algorithm is the fact that a factorization $B = QR$ with Q unitary and R upper triangular exists for any matrix B. (If B is nonsingular, the factorization is essentially unique.) We can calculate Q by Householder's technique of applying a finite sequence of elementary Hermitian unitary matrices, $P_r = I - 2v_r v_r^*$. Suppose that we have obtained $B_r = P_{r-1} \cdots P_0 B$ and

$$B_r = \left[\begin{array}{c|c} U & V \\ \hline O & W \end{array} \right]$$

where U is an $r \times r$ upper triangular matrix. We define

$$P_r = \left[\begin{array}{c|c} I_r & O \\ \hline O & I_{n-r} - 2v_r v_r^* \end{array} \right],$$

where I_r is the $r \times r$ identity and $\|v_r\|_2 = 1$. Then

$$B_{r+1} = P_r B_r = \left[\begin{array}{c|c} U & V \\ \hline O & (I_{n-r} - 2v_r v_r^*)W \end{array} \right].$$

The vector v_r must be determined so that the first column of $(I_{n-r} - 2v_r v_r^*)W$ has all zero entries except the first. This is like the situation in the reduction to lower Hessenberg form (Section 6.5) and we obtain in similar fashion,

$$(u_r)_1 = b_{r+1, r+1}^{(r)} \mp S_r, \quad \text{(Sign of } S_r \text{ same as that of } b_{r+1, r+1}^{(r)}\text{)}$$
$$(u_r)_i = b_{r+i, r+1}^{(r)}, \quad i = 2, \ldots, n - r,$$
$$S_r^2 = \sum_{i=1}^{n-r} (b_{r+i, r+1}^{(r)})^2, \quad (2)$$
$$\beta_r = S_r(S_r \mp b_{r+1, r+1}^{(r)}) = \mp S_r(u_r)_1,$$
$$I_{n-r} - 2v_r v_r^* = I_{n-r} - u_r u_r^*/\beta_r, \quad r = 0, \ldots, n - 2,$$
$$Q = P_{n-2} \cdots P_0, R = QB.$$

6.9 The QR Algorithm (Reduction of a Matrix to Triangular Form by Similarity Transformations)

Thus, Q is a unitary matrix and $B_{n-1} = QB$ is upper triangular; i.e., $B = Q^*B_{n-1}$ is the required factorization. In the QR algorithm, this triangularization process must be performed for each A_k. If $b_{r+i,r+1} = 0$ for most i, many computations can be eliminated. Otherwise, the amount of work becomes prohibitive.

We now give sufficient conditions on A to insure that A_k approaches triangular form with the eigenvalues of A on the diagonal. Since all computations in Eqs. (1) and (2) above are carried out with real numbers, this implies that the eigenvalues of A must necessarily be real. Thus, when A has some complex eigenvalues, we cannot expect A_k to become upper triangular.

Theorem 6.9.1 Let A have n distinct eigenvalues such that $|\lambda_1| > |\lambda_2| > \cdots > |\lambda_n| > 0$. Choose A_k, Q_k and R_k to be the matrices defined in (1) above, supposing further that R_k has positive diagonal elements. Let $AX = XD$, where $D = (\lambda_i \delta_{ij})$ and assume that $X^{-1} = LU$, where L is lower triangular with 1's on the diagonal and U is upper triangular. Then $\lim_{k \to \infty} a_{ij}^{(k)} = 0$ for $i > j$ and $\lim_{k \to \infty} a_{ii}^{(k)} = \lambda_i$, $1 \leq i \leq n$.

Proof. We observe that all eigenvalues of A are distinct and real (since no two have the same modulus). Also, A is nonsingular, since 0 is not an eigenvalue. By Exercise 6.6, the hypothesis on R_k insures that the Q_k and R_k are unique. The condition that L have 1's on the diagonal implies that the leading principal minors of X^{-1} are nonzero. (See Chapter 4 on Gauss elimination.) Of course, X^{-1} exists, since X is a matrix of linearly independent eigenvectors.

From relations (1) above, we deduce

$$A_{k+1} = Q_k^* A_k Q_k = Q_k^* \cdots Q_0^* A Q_0 \cdots Q_k.$$

If we set $T_k = Q_0 \cdots Q_k$, this yields

$$T_k A_{k+1} = A T_k.$$

Setting $U_k = R_k \cdots R_0$, we obtain from (1)

$$T_k U_k = Q_0 \cdots Q_{k-1}(Q_k R_k) R_{k-1} \cdots R_0 = T_{k-1} A_k U_{k-1} = A T_{k-1} U_{k-1}.$$

From this recursion, we have

$$T_k U_k = A^{k+1}, \tag{3}$$

which is a QR type factorization of A^{k+1} in which the upper triangular matrix U_k has positive diagonal elements.

Now, the eigenvalues of A^k are $\{\lambda_i^k\}$ and $A^k X = XD^k$, or $A^k = XD^k X^{-1}$. Since X is nonsingular, there exists a unitary matrix Q' and a nonsingular upper triangular matrix R' such that $X = Q'R'$ and R' has positive diagonal elements (Exercise 6.6.b). Since $X^{-1} = LU$, we have

$$A^k = Q'R'D^k LU = Q'R'(D^k L D^{-k})D^k U.$$

The matrix $D^k LD^{-k}$ is lower triangular with elements $l_{ij}(\lambda_i/\lambda_j)^k$ for $i > j$ and 1's on the diagonal. Since $|\lambda_i/\lambda_j|^k \to 0$ for $i > j$, we can write $D^k L D^{-k} = I + E_k$, where $E_k \to 0$ as $k \to \infty$. Therefore,

$$A^k = Q'(I + R'E_k R'^{-1})R'D^k U.$$

But

$$I + R'E_k R'^{-1} = \bar{Q}_k \bar{R}_k, \tag{4}$$

where $\bar{Q}_k$ is unitary and $\bar{R}_k$ upper triangular with positive diagonal elements. Hence,

$$\bar{R}_k - \bar{Q}_k^* = \bar{Q}_k^* R' E_k R'^{-1} \to 0,$$

showing that $\bar{Q}_k^*$ approaches upper triangular form. Since $\bar{Q}_k$ is unitary, it must approach a diagonal form. Further, since $\bar{R}_k$ has positive diagonal elements, $\bar{Q}_k \to I$. Now,

$$A^k = (Q'\bar{Q}_k)(\bar{R}_k R' D^k U)$$

is a QR factorization of A^k. But (3) is also such a factorization. Since A^k is nonsingular, T_{k-1} and $Q'\bar{Q}_k$ must agree up to a unitary diagonal factor. To determine this factor, let D_1 be the unitary diagonal matrix with entries $\lambda_i/|\lambda_i|$. We write $D = |D|D_1$. Further, let D_2 be a unitary diagonal matrix such that $D_2^{-1}U$ has positive diagonal elements. Then

$$A^k = Q'\bar{Q}_k D_2 D_1^k [(D_2 D_1^k)^{-1} \bar{R}_k R'(D_2 D_1^k) |D|^k D_2^{-1} U]$$

is a QR factorization of A^k in which the bracketed quantity is an upper triangular matrix with positive diagonal elements. Since U_{k-1} in (3) above is also such a matrix, then by Exercise 6.6, we have

$$T_{k-1} = Q'\bar{Q}_k D_2 D_1^k.$$

Therefore,

$$Q_k = T_{k-1}^{-1} T_k = D_1^{-k} D_2^{-1} \bar{Q}_k^{-1} \bar{Q}_{k+1} D_2 D_1^{k+1}.$$

Since $\bar{Q}_k = I + F_k$, where $F_k \to 0$, we find that

$$Q_k = D_1 + O(F_k),$$

where $O(F_k) \to 0$ as $k \to \infty$. Hence, from (1), $A_k = Q_k R_k$ must approach upper triangular form. Furthermore,

$$A_{k+1} = Q_k^{-1} A_k Q_k = D_1 A_k D_1 + O_1(F_k),$$

where again $O_1(F_k)$ is a term which approaches 0 as $k \to \infty$. (This shows that the off-diagonal elements of the A_k can change sign repeatedly.) Finally, since

$$A_k = T_{k-1}^{-1} A T_{k-1} = T_{k-1}^{-1} X D X^{-1} T_{k-1} \quad \text{and} \quad X^{-1} T_k = R'^{-1} Q'^{-1} Q' \bar{Q}_k D_2 D_1^k,$$

we obtain

$$A_k = D_1^{-k} D_2^{-1} \bar{Q}_k^{-1} R' D R'^{-1} \bar{Q}_k D_2 D_1^k = [D_1^{-k} D_2^{-1} R' D R'^{-1} D_2 D_1^k] + G_k, \tag{5}$$

where $G_k \to 0$ as $k \to \infty$. Since the diagonal elements of the bracketed matrix are the elements of D, we see that $\lim_{k \to \infty} a_{ii}^{(k)} = \lambda_i$, $1 \leq i \leq n$, as was to be proved. ∎

Analysis of the preceding proof shows that the rate of convergence of the A_k depends on the rate of convergence of $\bar{Q}_k$ (see (5) above) and this in turn depends on the rate of convergence of E_k to 0 (see (4) above). The calculations preceding Eq. (4) show that $E_k \to 0$ at a rate determined by the ratios $|\lambda_i/\lambda_j|$, $i > j$. Therefore, the rate is governed by $\min_{2 \leq i \leq n}\{|\lambda_i/\lambda_{i-1}|\}$. As in most iterative methods, convergence will be slow if two eigenvalues of A are close together.

It can be proved that the QR algorithm converges even when some leading principal minors of X^{-1} are zero. (See Exercise 6.8.) Also, if A is diagonalizable and has real eigenvalues of multiplicity >1, convergence still obtains (Exercise 6.9).

6.9 The QR Algorithm (Reduction of a Matrix to Triangular Form by Similarity Transformations)

If $|\lambda_q| = |\lambda_{q+1}| = \cdots = |\lambda_{q+r}|$ are of equal modulus, then for $q + r \geq i > j \geq q$, $a_{ij}^{(k)}$ does not converge to 0. All other subdiagonal elements do converge to zero (Exercise 6.9). In particular, if A has a complex eigenvalue λ_q, then $\lambda_{q+1} = \bar{\lambda}_q$ and $a_{ij}^{(k)} \to 0$ $(i > j)$ except for $a_{q+1,q}^{(k)}$. However, in this case, the eigenvalues of the 2×2 submatrix

$$\begin{bmatrix} a_{qq}^{(k)} & a_{q,q+1}^{(k)} \\ a_{q+1,q}^{(k)} & a_{q+1,q+1}^{(k)} \end{bmatrix}$$

converge to λ_q and $\bar{\lambda}_q$.

Computational Aspects

As we have remarked above, the first step in the application of the QR algorithm should be a reduction of A to upper Hessenberg form by means of the elementary Hermitian unitary matrices P_r as in Householder's method. It is not difficult to prove that if A_k is upper Hessenberg, then so is A_{k+1}. Thus, in each iteration of the QR algorithm, the matrix A_k is upper Hessenberg.

The factorization $A_k = Q_k R_k$ can be carried out by using elementary Hermitian unitary matrices P_r as given in (2) above. In carrying out this process, the annihilation is accomplished by multiplying A_k on the left by $P_{n-1} \cdots P_1$. To form A_{k+1}, the matrix R_k must be multiplied on the right by $P_1 \cdots P_{n-1}$. The left and right multiplications can be combined in such a way that A_{k+1} is formed one row at a time without the need to accumulate the product $Q_k = P_{n-1} \cdots P_1$ or to form R_k. (See [6.11].) Note that the upper Hessenberg form implies that only 4 elements of P_r need be computed and only two rows and columns of A_k are involved for each r.

Since A_k approaches upper triangular form, at some stage one of the subdiagonal elements may become sufficiently small (e.g., $< 10^{-s}\|A\|_\infty$) to be treated as zero. In that event, A_k can be treated as the direct sum of two upper Hessenberg matrices B_k and D_k as shown below,

$$A_k = \left[\begin{array}{c|c} B_k & C_k \\ \hline 0 & D_k \end{array} \right],$$

since the matrix C_k does not enter into the determinant of $A_k - \lambda I$.

The proof of Theorem 6.9.1 shows that $a_{ij}^{(k)}$ for $i > j$ tends to zero approximately like $(\lambda_i/\lambda_j)^j$. For Hessenberg matrices we are concerned only with the elements $a_{i,i-1}^{(k)}$. To improve the rate of convergence we can consider the matrix $A - \alpha I$ for some real α. This matrix has eigenvalues $\lambda_i - \alpha$ and the rate of convergence to zero of the element in the $(n, n-1)$ position depends on $(\lambda_n - \alpha)/(\lambda_{n-1} - \alpha)$. If α is close to λ_n, the convergence will be quite rapid. Now, the proof shows that (when the $|\lambda_i|$ are distinct) $a_{nn}^{(k)} \to \lambda_n$ and $a_{n,n-1}^{(k)} \to 0$. When $a_{n,n-1}^{(k)}$ becomes small, $a_{nn}^{(k)}$ should be close to λ_n and can therefore be taken as the α in "shifting" A_k to $A_k - \alpha I$. At this point, the QR iteration is applied to $A_k - \alpha I$.

The eigenvectors of A can be computed by *inverse iteration* described in Section 6.7 for the symmetric tridiagonal matrix B. However, no use is made of the special properties of B and, in fact, the method applies to any matrix A which has n linearly independent eigenvectors. If λ_k is an eigenvalue of multiplicity $q > 1$, then the sequence of vectors (x_m) generated by this method converges to some vector in the invariant subspace belonging to λ_k, just as in the power method (Section 6.3). To obtain a set of q linearly independent eigenvectors for λ_k the initial vector b must be given q different values. Let $\bar{\lambda}_k$ be an approximation to the eigenvalue λ_k of A, computed by the QR algorithm, say. An eigenvector belonging to λ_k may be determined by the method of Section 6.7, where $(A - \bar{\lambda}_k I)^{-1}$ is being applied iteratively, in effect, as in the power method.

A rereading of Section 6.7 will show that inverse iteration does not depend on any special properties of A (other than that A have a basis of eigenvectors). As in the power method, the rate of convergence depends on the distance from λ_k to the nearest eigenvalue. If λ_k has multiplicity greater than 1, the vectors x_q generated by the iterations will converge to a vector in the invariant subspace of eigenvectors belonging to λ_k. To obtain a linearly independent set of eigenvectors in this subspace, one may consider changing the initial vector $e^T = (1, \ldots, 1)$ to some vector which is linearly independent of e^T. In principle, this should produce a new sequence (x_q) which converges to another eigenvector, unless the initial vector happens to be the same linear combination of eigenvectors as is e^T. The latter is unlikely. A more likely difficulty arises from rounding errors. Suppose, for example, that λ_k has multiplicity 2. As a result of rounding errors, the matrix A will behave as if it has two very close eigenvalues λ_k and λ_k'. The approximation $\bar{\lambda}_k$ will be closer to one of these in general, say λ_k'. In that case, the inverse powers of $(\lambda_k' - \bar{\lambda}_k)$ will dominate and the vectors x_q will converge to an eigenvector belonging to λ_k' regardless of the starting vector. To obtain other eigenvectors of a multiple root λ_k we must choose a starting vector which is orthogonal to the eigenvectors already computed. Furthermore, since components along these eigenvectors are introduced by rounding errors in each iteration, the orthogonalization procedure must be repeated after each iteration. Thus, after computing x_q, the components of x_q along the eigenvectors already computed should be subtracted from x_q. This yields a new vector x_q' which can then be used to obtain x_{q+1} by one step of inverse iteration. In Exercise 6.11, the eigenvectors belonging to an eigenvalue of multiplicity 2 must be computed. If Householder's method is used to obtain the eigenvalues (A is symmetric), then inverse iteration with orthogonalization will yield two linearly independent eigenvectors.

6.10 EXTREMAL PROPERTIES OF EIGENVALUES

In this chapter, we have concentrated on algebraic methods of determining matrix eigenvalues and eigenvectors. The eigenvectors have important nonalgebraic properties which can also be used as the basis of computational methods. For a symmetric matrix A, it is not difficult to show that the eigenvectors are stationary points of the functional $\langle Ax, x \rangle$, where x is constrained to lie on the unit l_2-sphere. Thus, for $\|x\|_2 = 1$, the maximum value of $\langle Ax, x \rangle$ occurs at an eigenvector u belonging to the largest eigenvalue λ and $\langle Au, u \rangle = \lambda$. We shall consider these properties in

6.11 SINGULAR VALUES AND THE PSEUDOINVERSE

In Section 4.6, we discussed the solution of singular linear systems $Ax = b$, where A is an $m \times n$ matrix of rank r. We made use of the pseudoinverse $A^\dagger$ and gave a method of computing it. An alternative approach involves the singular values of A. It will be recalled from Section 3.6 that $\lambda > 0$ is a singular value of A if and only if λ^2 is an eigenvalue of A^*A. Since A^*A has rank r also, the nullity is $n - r$. Hence, there are $n - r$ linearly independent eigenvectors $u_{r+1}, \ldots, u_n$ belonging to the eigenvalue of 0. Since A^*A is symmetric and positive semi-definite, there are exactly r nonzero eigenvalues $\lambda_1^2, \ldots, \lambda_r^2$; i.e., A has r singular values. Let $u_1, \ldots, u_r$ be the corresponding orthonormal set of eigenvectors. If we let $v_i = (1/\lambda_i) A u_i$, then we have

$$Au_i = \lambda_i v_i \quad \text{and} \quad A^*v_i = \lambda_i u_i, \; 1 \leq i \leq r; \tag{6.11.1}$$

u_i, v_i are called a pair of *singular vectors* of A belonging to λ_i. Note that $AA^*v_i = \lambda_i^2 v_i$, so that the v_i are orthonormal eigenvectors of AA^* belonging to the eigenvalues $\lambda_1^2, \ldots, \lambda_r^2$. The nullity of AA^* is $m - r$. Let $v_{r+1}, \ldots, v_m$ be an orthonormal set spanning the null space of AA^*. Let U be the $n \times n$ matrix having $u_1, \ldots, u_n$ as its columns and V the $m \times m$ matrix having $v_1, \ldots, v_m$ as columns. Then by (6.11.1),

$$V^*AU = \begin{bmatrix} \Lambda & 0 \\ 0 & 0 \end{bmatrix} = D,$$

where Λ is the $r \times r$ diagonal matrix having $\lambda_1, \ldots, \lambda_r$ as its diagonal elements. Note that U and V are orthogonal matrices.

Now, suppose there exists x such that $Ax = b$. Then $DU^*x = V^*AUU^*x = V^*Ax = V^*b$. Hence,

$$\langle v_i, b \rangle = \lambda_i \langle u_i, x \rangle, \quad i = 1, \ldots, r,$$
$$\langle v_i, b \rangle = 0, \quad i = r+1, \ldots, m. \tag{6.11.2}$$

Conversely, (6.11.2) implies that $DU^*x = V^*b$ has a solution, from which it follows that $Ax = VDU^*x = b$. Thus, (6.11.2) is a necessary and sufficient condition for existence of a solution of $Ax = b$. The general solution x is easily obtained. Since $x = \sum_1^n a_i u_i$, by linear independence of the u_i, we see that $a_i = \langle x, u_i \rangle = (1/\lambda_i)\langle v_i, b \rangle$, $i = 1, \ldots, r$, and $a_{r+1}, \ldots, a_n$ may be arbitrary. Therefore,

$$x = \sum_{i=1}^{r} \frac{1}{\lambda_i} \langle v_i, b \rangle u_i + \sum_{r+1}^{n} a_i u_i. \tag{6.11.3}$$

Thus, x may be computed by determining the eigenvalues of A^*A by one of the methods of this chapter and finding the corresponding eigenvectors of A^*A and AA^*.

In matrix form, (6.11.3) with $a_i = 0$ can be written as $x = A^\dagger b$, where

$$A^\dagger = U \begin{bmatrix} \Lambda^{-1} & 0 \\ 0 & 0 \end{bmatrix} V^*. \qquad (6.11.4)$$

It is easily verified that (6.11.4) does, in fact, give the pseudoinverse and $x = A^\dagger b$ is the least-squares solution of minimum l_2-norm.

As with the method of Section 4.6, the delicate computational aspects arise in deciding which computed eigenvalues are actually zero. Since rounding errors are present, this may lead to difficulties when A^*A has an ill-conditioned eigenvalue problem. Note that if some λ_i is actually zero, but is computed as some small value ε_i, then the term $(1/\varepsilon_i) \langle v_i, b \rangle$ must be evaluated with high precision to avoid large errors in x.

REFERENCES

6.1. J. H. Wilkinson, *The Algebraic Eigenvalue Problem*, Clarendon Press, Oxford (1965)
6.2. C. G. J. Jacobi, "Über ein leichtes Verfahren die in der Theorie der Säculärstörungen vorkommenden Gleichungen numerisch aufzulösen," *Crelle's J.* **30**, 51–94 (1846)
6.3. P. Henrici, "On the speed of convergence of cyclic and quasicyclic Jabobi methods for computing eigenvalues of Hermitian matrices," *J. SIAM* **6**, 144–162 (1958)
6.4. A. Schönhage, "Zur Konvergenz des Jacobi-Verfahrens," *Numerische Mathematik* **3**, 374–380 (1961)
6.5. J. H. Wilkinson, "Note on the quadratic convergence of the cyclic Jacobi process," *Numerische Mathematik* **4**, 296–300 (1962)
6.6 M. A. Hyman, "Eigenvalues and eigenvectors of general matrices," *Proc. Twelfth National Meeting, ACM* (1957)
6.7. J. G. F. Francis, "The QR transformation," *Computer Journal* **4** (Part I), 265–271 (Part II), 332–345 (1961)
6.8. H. Rutishauser, "Solution of eigenvalue problems with the LR transformation," *Applied Math. Ser. Nat. Bureau of Standards* **49**, 47–81 (1958)
6.9. F. L. Bauer and C. T. Fike, "Norms and exclusion theorems," *Numerische Math.* **2**, 42–53 (1960)
6.10. A. J. Hoffman and H. W. Wielandt, "The variation of the spectrum of a normal matrix," *Duke Math. Journal* **20**, 37–39 (1953)
6.11. J. Grad and M. A. Brebner, "Eigenvalues and eigenvectors of a real general matrix," *Comm. A.C.M.* **11**, 820–825, No. 12 (December 1968)
6.12. B. Parlett, "The development and use of methods of LR type," *SIAM Review* **6**, 275–295 (1964)

EXERCISES

6.1 Let $P = I - 2vv^*$, where v is an n-dimensional column vector such that $v^*v = 1$. Verify that $PP^* = I$. If $v = e_{j+1}$ (having 1 as its $(j+1)$-st component and 0 elsewhere), show that P has the form

$$\begin{array}{c} j \\ {}_{n-j}\end{array}\!\left\{\left(\begin{array}{c|c} \overset{j}{I} & 0 \\ \hline 0 & I - 2v_j v_j^* \end{array}\right)\right.$$

where v_j is an $(n-j)$-dimensional vector.

6.2 Verify that Eqs. (8), (9), (10) in Section (6.5) do indeed yield $A^{(r)} = P_r A^{(r-1)} P_r$, where P_r is given by Eqs. (4)–(7).

6.3 Referring to Section (6.5), let A be an arbitrary matrix. Assuming that A_{r-1} is such that the $r \times r$ submatrix C_{r-1} is upper Hessenberg, show that $A_r = P_r A_{r-1} P_r$ has the corresponding submatrix C_r in upper Hessenberg form.

6.4 Let $B = (b_{ij})$ be upper Hessenberg. Suppose $b_{i,i-1} = 0$ for some i. Show that $\det B = \det C \det D$, where C is the submatrix of the first $i - 1$ rows and columns and D is the submatrix of the last $n - i + 1$ rows and columns.

6.5 Prove that any $n \times n$ matrix A can be reduced to upper triangular form by a similarity transformation $B^{-1}AB$. Show that B can be taken to be unitary (Schur's Theorem). [*Hint:* Use induction on n. Let $Av_1 = \lambda v_1$. Extend $\{v_1\}$ to a basis $\{v_1, v_2, \ldots, v_n\}$ (by Theorem 2.1.2). Let B be the matrix having $v_1, \ldots, v_n$ as its columns. Then by Theorem 2.1.7, $C = B^{-1}AB$ is the matrix of the transformation A relative to $\{v_i\}$ and therefore is of the form

$$\begin{pmatrix} \lambda & c_{12} & \cdots & c_{1n} \\ 0 & & & \\ \vdots & & C_1 & \\ 0 & & & \end{pmatrix}.$$

By induction, there exists B_1 such that $B_1^{-1} C_1 B_1$ is upper triangular. Take the direct sum of B_1 and the 1×1 identity. For the rest, take $\{v_i\}$ orthonormal.]

6.6 Let A be nonsingular.

a) Prove that A has a QR factorization unique up to a diagonal unitary factor.

b) Show that the diagonal elements of R can be taken to be positive if A is real.

[*Hint:* a) $A = Q_1 R_1 = Q_2 R_2$ implies $Q_2^{-1} Q_1 = R_2 R_1^{-1}$. Since $Q_2^{-1} Q_1$ is unitary and upper triangular, it must be a diagonal matrix with entries of unit modulus. Note that if R_1 and R_2 are real and have positive diagonal elements, then $R_2 R_1^{-1} = I$ and so $Q_2 = Q_1$.
b) If $A = Q_1 R_1$ and $D = (\bar{r}_{ii}/|r_{ii}|)$ is diagonal, then $A = (Q_1 D^*)(DR_1)$ is a QR factorization.]

6.7 Let $p(\lambda) = \lambda^n + a_1 \lambda^{n-1} + \cdots + a_n$. Show that $p(\lambda)$ is the characteristic polynomial of the nth order matrix

$$A = \begin{bmatrix} 0 & 1 & 0 & \cdots & & 0 \\ & \ddots & 1 & 0 & & \vdots \\ & & \ddots & & & \\ & & & & & 0 \\ 0 & \cdots & & & 0 & 1 \\ -a_n & -a_{n-1} & & \cdots & & -a_1 \end{bmatrix}$$

having 1's on the superdiagonal, the negative coefficients of p as the last row and zero's everywhere else. Thus, every polynomial is the characteristic polynomial of some matrix.

6.8 In Theorem 6.9.1, suppose that X^{-1} does not have a triangular decomposition.

a) Show that there is a permutation matrix P such that $PX^{-1} = LU$. (Refer to column pivoting in Gauss elimination, Chapter 4.)

b) Prove Theorem 6.9.1 without the hypothesis that $X^{-1} = LU$. [Hint: $A^k = XD^kP^*LU = XP^*(PD^kP^*)LU$, where $PD^kP^* = D_P^k$ is a diagonal matrix with elements $\lambda_{r_i}^k$ which are the λ_i^k permuted. Let $XP^* = Q'R'$. Then $A^k = Q'R'(D_P^kLD_P^{-k})D^kU$. The (i,j)-element, $i > j$, of $D_P^kLD_P^{-k}$ is $(\lambda_{r_i}/\lambda_{r_j})^k l_{ij}$ and if pivoting is done by choosing the first nonzero element below the diagonal, then $l_{ij} = 0$ whenever $r_i < r_j$. Hence, we have, as in the proof of Theorem 6.9.1, that $D_P^kLD_P^{-k} \to I$.]

6.9 In the statement of Theorem 6.9.1, replace the hypothesis of unequal moduli by the condition
$$|\lambda_1| > |\lambda_2| > \cdots > |\lambda_q| = \cdots = |\lambda_{q+r}| > \cdots > |\lambda_n| > 0.$$

a) Show that $a_{ij}^{(k)} \to 0$ for $i > j$ except if $q + r \geq i > j \geq q$. [Hint: $A^k = XD^kLU$ as before and D^kLD^{-k} behaves as before except that $|(\lambda_i/\lambda_j)^k| |l_{ij}| = |l_{ij}|$ for the exceptional i, j.]

b) If in (a) we have $\lambda_q = \cdots = \lambda_{q+r}$ then the exceptional elements remain fixed and the conclusions of Theorem 6.9.1 hold. [Hint: $D^kLD^{-k} = L_1 + E_k$, $E_k \to 0$. Let $XL_1 = Q'R'$. Then $A^k = Q'R'(I + L_1^{-1}E_k)D^kU = Q'Q_k\bar{R}_kR'D^kU$ as before.]

c) If $\lambda_{q+m} = |\lambda_q| \exp(i\theta_m)$, $m = 0, \ldots, r$ in part (a), then the matrix L_1 in part (b) has elements $l_{ij} \exp[ik(\theta_i - \theta_j)]$.

6.10 Show that the iterations in the QR algorithm preserve upper Hessenberg form, i.e., if A satisfies the hypotheses of Theorem 6.9.1 and is upper Hessenberg, so is A_k for all k.

6.11 Write a computer program to determine eigenvalues and eigenvectors by the threshold Jacobi method. Use the following orthogonal symmetric matrix as a test case. Also program Householder's method.

$$\left\| \begin{array}{cccc|cccc} 611 & 196 & -192 & 407 & -8 & -52 & -49 & 29 \\ 196 & 899 & 113 & -192 & -71 & -43 & -8 & -44 \\ -192 & 113 & 899 & 196 & 61 & 49 & 8 & 52 \\ 407 & -192 & 196 & 611 & 8 & 44 & 59 & -23 \\ \hline -8 & -71 & 61 & 8 & 411 & -599 & 208 & 208 \\ -52 & -43 & 49 & 44 & -599 & 411 & 208 & 208 \\ -49 & -8 & 8 & 59 & 208 & 208 & 99 & -911 \\ 29 & -44 & 52 & -23 & 208 & 208 & -911 & 99 \end{array} \right\|$$

[Hint: The answers are as follows.]

Exact   Approximate

$\lambda_1 = 10\sqrt{10405} = 1020.04901843$ $\lambda_2 = 1020$

$$v_1 = \left\| \begin{array}{c} 2 \\ 1 \\ 1 \\ 2 \\ 102 - \sqrt{10405} \\ 102 - \sqrt{10405} \\ -204 + 2\sqrt{10405} \\ -204 + 2\sqrt{10405} \end{array} \right\| = \left\| \begin{array}{c} 2 \\ 1 \\ 1 \\ 2 \\ -0.004901843 \\ -0.004901843 \\ 0.009803686 \\ 0.009803686 \end{array} \right\| \quad v_2 = \left\| \begin{array}{c} 1 \\ -2 \\ -2 \\ 1 \\ 2 \\ -2 \\ 1 \\ -1 \end{array} \right\|$$

$\lambda_3 = \dfrac{510 + 100\sqrt{26}}{2} = 1019.90195136 \quad \lambda_4 = 1000 \quad \lambda_5 = 1000$

$$v_3 = \begin{Vmatrix} 2 \\ -1 \\ 1 \\ -2 \\ 5+\sqrt{26} \\ -5-\sqrt{26} \\ -10-2\sqrt{26} \\ 10+2\sqrt{26} \end{Vmatrix} = \begin{Vmatrix} 2 \\ -1 \\ 1 \\ -2 \\ 10.09901951 \\ -10.09901951 \\ -20.19803903 \\ 20.19803903 \end{Vmatrix} \quad v_4 = \begin{Vmatrix} 1 \\ -2 \\ -2 \\ 1 \\ -2 \\ 2 \\ -1 \\ 1 \end{Vmatrix} \quad v_5 = \begin{Vmatrix} 7 \\ 14 \\ -14 \\ -7 \\ -2 \\ -2 \\ -1 \\ -1 \end{Vmatrix}$$

$\lambda_6 = \dfrac{510 - 100\sqrt{26}}{2} = 0.09804864072 \quad \lambda_7 = 0$

$$v_6 = \begin{Vmatrix} 2 \\ -1 \\ 1 \\ -2 \\ 5-\sqrt{26} \\ -5+\sqrt{26} \\ -10+2\sqrt{26} \\ 10-2\sqrt{26} \end{Vmatrix} = \begin{Vmatrix} 2 \\ -1 \\ 1 \\ -2 \\ -0.099019514 \\ 0.099019514 \\ 0.198039027 \\ -0.198039027 \end{Vmatrix} \quad v_7 = \begin{Vmatrix} 1 \\ 2 \\ -2 \\ -1 \\ 14 \\ 14 \\ 7 \\ 7 \end{Vmatrix}$$

$\lambda_8 = -10\sqrt{10405} = -1020.04901843$

$$v_8 = \begin{Vmatrix} 2 \\ 1 \\ 1 \\ 2 \\ 102+\sqrt{10405} \\ 102+\sqrt{10405} \\ -204-2\sqrt{10405} \\ -204-2\sqrt{10405} \end{Vmatrix} = \begin{Vmatrix} 2 \\ 1 \\ 1 \\ 2 \\ 204.0049018 \\ 204.0049018 \\ -408.0098037 \\ -408.0098037 \end{Vmatrix}$$

6.12 Let A be an $n \times n$ diagonalizable matrix as in Theorem 6.2.3. Prove the Corollary for the case $p > 1$. [*Hint:* Let $\lambda(\varepsilon)$ be an eigenvalue of $A(\varepsilon) = A + \varepsilon P$. Let $U^{-1}AU = \Lambda$. Then $B = U^{-1}(A(\varepsilon) - \lambda(\varepsilon)I)U = \Lambda - \lambda(\varepsilon)I + \varepsilon U^{-1}PU$ is a singular operator. If $\lambda(\varepsilon) = \lambda_i$ for some i, we are finished. If $\lambda(\varepsilon) \neq \lambda_i$ for all i, then $\Lambda - \lambda(\varepsilon)I$ is nonsingular. Since B is singular, we cannot have

$$\|\varepsilon U^{-1}PU\| < \dfrac{1}{\|(\Lambda - \lambda(\varepsilon)I)^{-1}\|}$$

by Theorem 3.5.2. Hence, for any operator norm,

$$\|\varepsilon U^{-1}PU\| \geq \dfrac{1}{\|(\Lambda - \lambda(\varepsilon)I)^{-1}\|}$$

For any p norm,

$$\|(\Lambda - \lambda(\varepsilon)I)^{-1}x\|_p \leq \dfrac{1}{\min_i |\lambda_i - \lambda(\varepsilon)|} \|x\|_p.$$

Hence, $1/\|(\Lambda - \lambda(\varepsilon)I)^{-1}\|_p \geq \min_i |\lambda_i - \lambda(\varepsilon)|$.]

262   The Computation of Matrix Eigenvalues and Eigenvectors

6.13 Give an alternative proof of Theorem 6.2.1 by using Theorem 4.4.6, Corollary 2. (See Exercise 4.15.) [*Hint:* Note that $|a_{ii} - \lambda| > r_i$ for all $1 \le i \le n$ implies that $A - \lambda I$ is strictly diagonally dominant.]

6.14 Let M_1 and M_2 be two finite-dimensional vector subspaces of dimension r of a pre-Hilbert space V. Show that there exists a basis $\{u'_1, \ldots, u'_r\}$ for M_1 and a basis $\{v'_1, \ldots, v'_r\}$ for M_2 such that $\langle u'_i, v'_j \rangle = 0$ for $i \ne j$. Two such bases are said to be *biorthogonal*. [*Hint:* Let $\{u_1, \ldots, u_r\}$, $\{v_1, \ldots, v_r\}$ be bases of M_1 and M_2 respectively. If $\langle u_i, v_j \rangle = 0$ for every pair $1 \le i, j \le r$, these u_i and v_j are biorthogonal. If not, reindex so that $\langle u_1, v_1 \rangle \ne 0$ and define $u'_1 = u_1, v'_1 = v_1$. For $u_i, v_j, 2 \le i, j \le r$, define $u'_i = \alpha_i u'_1 + u_i, v'_j = \beta_j v'_1 + v_j$, with $\alpha_i = -\langle u_i, v'_1 \rangle / \langle u'_1, v'_1 \rangle, \beta_j = -\langle v_j, u'_1 \rangle / \langle u'_1, v'_1 \rangle$. If some $\langle u'_i, v'_j \rangle \ne 0, i, j \ge 2$, reindex with this $i = 2$ and $j = 2$. Then $\{u'_1, u'_2\}, \{v'_1, v'_2\}$ are biorthogonal, linearly independent and $\langle u'_i, v'_i \rangle \ne 0, 1 \le i \le 2$. If $\langle u'_i, v'_j \rangle = 0$, $2 \le i, j \le r$, then $\{u'_1, \ldots, u'_r\}$ is biorthogonal to $\{v'_1, \ldots, v'_r\}$. Continuing in this way, we obtain $\{u'_1, \ldots, u'_k\}$ biorthogonal to $\{v'_1, \ldots, v'_k\}$ for some $k \le r$, and $\langle u'_i, v'_i \rangle \ne 0$, $1 \le i \le k$. Define

$$u'_{k+q} = \sum_{i=1}^{k} \alpha_{qi} u'_i + u_{k+q}$$

and

$$v'_{k+q} = \sum_{i=1}^{k} \beta_{qi} v'_i + v_{k+q},$$

where

$$\alpha_{qi} = -\langle u_{k+q}, v'_i \rangle / \langle u'_i, v'_i \rangle$$

and similarly for the β's. Again, if some $\langle u'_{k+i}, v'_{k+j} \rangle \ne 0$, reindex with $k + i = k + 1$ and $k + j = k + 1$ and continue. If $\langle u'_{k+1}, v'_{k+j} \rangle = 0$ for all $i, j = 1, \ldots, r - k$, then $u'_1, \ldots, u'_r$ is biorthogonal to $\{v'_1, \ldots, v'_r\}$ and both sets are linearly independent.]

6.15 Let A be a real diagonalizable $n \times n$ matrix. Show that there exists a basis $\{u_1, \ldots, u_n\}$ of eigenvectors of A which is *biorthogonal* to a basis $\{v_1, \ldots, v_n\}$ of eigenvectors of A^*; i.e., $\langle u_i, v_j \rangle = 0$ for $i \ne j$. In fact, $\langle u_i, v_j \rangle = \delta_{ij}$. [*Hint:* There exists an $n \times n$ matrix U such that $U^{-1}AU = \Lambda$, where Λ is a diagonal matrix having the eigenvalues of A on its diagonal. Since $U^{-1}A = \Lambda U^{-1}$, the rows v_j of U^{-1} are a basis of left eigenvectors of A or $A^*U^{-1*} = U^{-1*}\Lambda^*$. The columns u_i of U are a basis of eigenvectors of A. Hence, $\lambda_i \langle u_i, v_j \rangle = \langle Au_i, v_j \rangle = \langle u_i, A^*v_j \rangle = \lambda_j \langle u_i, v_j \rangle$ and $\langle u_i, v_j \rangle = 0$ for $\lambda_i \ne \lambda_j$. For $\lambda_i = \lambda_j$, consider the subspaces M_1, M_2 of eigenvectors of A and A^* respectively and apply Exercise 6.14]

6.16 (The Hoffman–Wielandt theorem [6.10].) Let $C = A + B$, where A, B and C are symmetric matrices. Let $\alpha_i, \beta_i, \gamma_i$ be the eigenvalues of A, B, C respectively and indexed in nonincreasing order. Prove that

$$\sum_{i=1}^{n} (\gamma_i - \alpha_i)^2 \le \sum_{i=1}^{n} \beta_i^2.$$

[*Hint:* (Givens) For any symmetric matrix B, $\|B\|_2 = \sum_{i=1}^{n} \beta_i^2$ by Exercise 3.17. The set $\mathcal{U}$ of unitary matrices U is a compact subset of n^2-dimensional vector space with the l_2-norm, since $\|U\|_2 = 1$ and $\mathcal{U}$ is closed. There exist unitary matrices U_1, U_2 such that $U_1^*BU_1 = \beta, U_2^*CU_2 = \Gamma$, where β and Γ are diagonal matrices. Hence,

$$\beta = U_1^*(C - A)U_1 = U_1^*U_2(\Gamma - U_2^*AU_2)U_2^*U_1.$$

Therefore,
$$\sum_1^n \beta_i^2 = \|\Gamma - U_2^* A U_2\|_2^2$$

Let $f(U) = \|\Gamma - U^*AU\|_2^2$. The function f is continuous on the compact set $\mathcal{U}$, hence has an infimum at some $U_3 \in U$. Now, show that $U_3^* A U_3$ must be a diagonal matrix. It follows that $f(U_3) = \sum_{i=1}^n (\gamma_i - \alpha_{j_i})^2$, where $j_1, \ldots j_n$ is a permutation of $1, \ldots, n$. If $j_1 \neq 1$ and $j_k = 1$ for $k \neq 1$, we could interchange j_1 and j_k by a unitary similarity transformation and the change in the value of f would be

$$(\gamma_1 - \alpha_1)^2 + (\gamma^k - \alpha_{j_1})^2 - (\gamma_1 - \alpha_{j_1})^2 - (\gamma_k - \alpha_1)^2 = -2(\gamma_k - \gamma_1)(\alpha_{j_1} - \alpha_1) \leq 0.$$

Thus, f would be decreased unless $\gamma_k = \gamma_1$ or $\alpha_{j_1} = \alpha_1$. Hence $f(U_3) = \sum_1^n (\gamma_i - \alpha_i)^2$].

6.17 A more general type of *algebraic eigenvalue problem* is to find λ and $x \neq 0$ such that
$$(A - \lambda B)x = 0, \tag{1}$$
where A and B are arbitrary nth order matrices. Suppose that A and B are real symmetric and that B is positive definite.

a) Show that there exists an upper triangular matrix S such that $B = S^T S$. (This is called the *Cholesky decomposition* of B.) The proof is by induction on n. Express B as

$$B = \left[\begin{array}{c|c} B_{n-1} & b \\ \hline b^T & b_{nn} \end{array}\right]$$

where B_{n-1} is the leading principal submatrix of B of order $n - 1$. B_{n-1} is positive definite and has a Cholesky decomposition by the induction hypothesis, say $B_{n-1} = U^T U$. Since B_{n-1} is nonsingular, so is U. Hence, there exists c such that $U^T c = b$. Now, find an element d such that

$$\left[\begin{array}{c|c} U^T & 0 \\ \hline c^T & d \end{array}\right] \left[\begin{array}{c|c} U & c \\ \hline 0 & d \end{array}\right] = \left[\begin{array}{c|c} B_{n-1} & b \\ \hline b^T & b_{nn} \end{array}\right].$$

b) Show that (λ, x) is a solution of Eq. (1) above if and only if λ is an eigenvalue of $(S^T)^{-1} A S^{-1}$ and Sx is an eigenvector belonging to λ.

6.18 Let A be an $n \times n$ matrix having a Jordan canonical form J (Exercise 3.20) in which

$$J_1 = \begin{bmatrix} \lambda_1 & 1 \\ 0 & \lambda_1 \end{bmatrix}$$

and $J_i = [\lambda_i]$, $3 \leq i \leq n$. Thus, J has exactly one nonlinear elementary divisor, J_1, and only $n - 1$ linearly independent eigenvectors. Let λ_1 be the eigenvalue of maximum modulus. Prove that the power method converges to λ_1. [*Hint:* Let $\{u_1, u_2, \ldots, u_n\}$ be the basis relative to which the operator A has matrix J. Then
$$Au_1 = \lambda_1 u_1,$$
$$Au_2 = u_1 + \lambda_1 u_2,$$
$$Au_i = \lambda_i u_i, \quad 3 \leq i \leq n,$$
and
$$A^q u_1 = \lambda_1^q u_1, \quad A^q u_i = \lambda_i^q u_i, \quad 3 \leq i \leq n$$
$$A^q u_2 = \lambda_1^q u_2 + q\lambda_1^{q-1} u_1.$$

For any $x_0 = \sum_1^n \xi_i u_i$,

$$A^q x_0 = \sum_1^n \xi_i A^q u_i = \xi_1 \lambda_1^q u_1 + \xi_2(\lambda_1^q u_2 + q\lambda_1^{q-1} u_1) + \sum_{i=3}^n \xi_i \lambda_i^q u_i$$

$$= \lambda_1^q \left(\left(\xi_1 + \frac{q\xi_2}{\lambda_1}\right) u_1 + \xi_2 u_2\right) + \sum_3^n \xi_i \left(\frac{\lambda_i}{\lambda_1}\right)^q u_i.$$

Thus, $r_{q+1} = \lambda_1(1 + O(1/q))$ as $q \to \infty$.]

6.19 a) Prove that any nth-order matrix A is similar to a matrix J_ε which consists of the Jordan normal form, J, of A with each off-diagonal 1 replaced by ε. [*Hint:* $J = B^{-1}AB$ by Exercise 3.20. Let D be the diagonal matrix having 1, ε, ε^2, ..., ε^{n-1} as its diagonal elements $d_{11}, d_{22}, \ldots, d_{nn}$ respectively. Verify that $J_\varepsilon = D^{-1}JD$.]

b) Referring to part (a), show that $\|J_\varepsilon\|_1 \le r_\sigma(A) + \varepsilon$, where $r_\sigma(A)$ is the spectral radius.

c) Prove that for any $\varepsilon > 0$ there exists an operator norm $\|A\|$ such that

$$\|A\| \le r_\sigma(A) + \varepsilon.$$

[*Hint:* Let $C = BD$, where B and D are given in part (a). Define the vector norm $\|x\| = \|C^{-1}x\|_1$; i.e. for $x = (\xi_1, \ldots, \xi_n)$ define $\|x\| = \sum_1^n |\eta_i|$, where $\eta_i = \sum_{j=1}^n c'_{ij}\xi_j$ and $C^{-1} = (c'_{ij})$. Then

$$\|A\| = \sup_{\|x\|=1} \|Ax\| = \sup_{\|C^{-1}x\|_1=1} \|C^{-1}Ax\|_1 = \sup_{\|y\|_1=1} \|C^{-1}ACy\|_1.]$$

6.20 a) Show that similar matrices have the same minimum polynomial. [*Hint:* See Exercise 3.41 and note that $B = C^{-1}AC$ implies $B^i = C^{-1}A^iC$, hence $m(B) = C^{-1}m(A)C$.]

b) Let J be a matrix in Jordan canonical form, as given in Exercise 3.20. Suppose that each Jordan submatrix J_i has the same eigenvalue λ_1. Let q be the order of the largest submatrix J_i. Prove that the minimum polynomial $m(\lambda)$ of J is $m(\lambda) = (\lambda - \lambda_1)^q$. [*Hint:* $J - \lambda_1 I$ has zero's on the diagonal and $(J - \lambda_1 I)^q = 0$, while $(J - \lambda_1 I)^{q-1} \ne 0$.]

c) Now, let J be in Jordan canonical form and let $\lambda_1, \ldots, \lambda_k$ be its distinct eigenvalues. The *index* (or grade) q_i of λ_i is the order of the largest Jordan submatrix having λ_i on its diagonal. Prove that the minimum polynomial $m(\lambda)$ of J is given by

$$m(\lambda) = (\lambda - \lambda_1)^{q_1} \cdots (\lambda - \lambda_k)^{q_k}.$$

Hence, λ_1 is a simple zero of $m(\lambda)$ if and only if the elementary divisors for λ_1 are linear. [*Hint:* See Exercise 3.22(a). Also, note that

$$m(J) = \begin{bmatrix} m(J_1) & & 0 \\ & \ddots & \\ 0 & & m(J_p) \end{bmatrix}.$$

Hence, $m(J_i) = 0$, whence $(\lambda - \lambda_i)^{q_i}$ divides $m(\lambda)$.]

CHAPTER 7

APPROXIMATION THEORY

7.1 INTRODUCTION

In Chapters 4, 5 and 6, we have considered instances of the problem of finding zeros of functions, that is, of solving the equation $f(x) = 0$. In general, f was a function from one Banach space to another. In Chapter 4, we had $f(x) = Ax - b$, where A was an $m \times n$ matrix and b an m-dimensional vector. In Chapter 5, f was nonlinear and in Chapter 6 we considered $f(x, \lambda) = Ax - \lambda x$. The problem of solving $f(x) = 0$ constructively may be viewed as a main concern of numerical analysis. Later, we shall consider the case $f(x) = dx/dt - g(x)$ where $x = x(t)$ is a function satisfying a differential equation. To obtain a numerical solution of such a differential equation, we shall approximate the function g by a function chosen from a suitable family of functions which are in some sense "simpler" than g and "finitely representable." For example, in the one-dimensional case, $g(x)$ might be approximated by a polynomial

$$p(x) = a_0 x^n + a_1 x^{n-1} + \cdots + a_{n-1} x + a_n$$

of degree $\leq n$ with real coefficients a_i, $0 \leq i \leq n$. Polynomials as functions have been studied thoroughly in analysis and algebra. Besides, they are easy to compute with and can be finitely represented in a computer by the sequence $(a_0, a_1, \ldots, a_n)$ of coefficients (suitably rounded). Thus, our central problem, finding a solution of $f(x) = 0$, gives rise to an ancillary problem, that of approximating f by polynomials.

Approximation also arises in a variety of other contexts. For example, if we have to evaluate $\sin x$ or e^x in a computer, we cannot use their infinite-series definitions, since this would require an infinite computation. Clearly, some approximation must be made by means of functions which can be evaluated in a finite number of arithmetic operations, such as polynomials or rational functions. The theory of such approximations is an important part of the bridge from the continuous to the discrete. It is also of theoretical importance in any study of real and complex functions. Indeed, the theory of approximation of functions stands as a theory in its own right and encompasses much of classical and modern real analysis. In this chapter, we present some aspects of the theory which are of particular importance in numerical analysis. Also, we consider applications to practical computing problems. For the most part, we are concerned with real functions of one real variable.

In any approximation problem, say of $f(x)$ by a polynomial $p(x)$, we must be concerned with a measure of the closeness of $p(x)$ to $f(x)$. This means that we must introduce a topology in the set of functions being considered. In numerical computations, we prefer a numerical measure of closeness. This suggests that we introduce a metric. In fact, it turns out to be more convenient to introduce norms or seminorms.

In Chapter 3, we have shown how norms can be defined in various function spaces. These function spaces provide the proper setting for a general theory of approximation. However, before we consider the general theory, it will be helpful to begin with some facets of the special case of polynomial approximation of a real function of a single real variable. Indeed, since polynomial approximation is one of the most important parts of the entire theory of approximation, this special case deserves special attention.

It should be borne in mind that the numerical framework for approximation theory is the real-number system. We shall assume that all real numbers are represented by infinite-precision decimal expansions and all operations are exact operations on real numbers. For the most part, we do not consider rounding errors, since they are comparatively simple to estimate and are usually small compared to the error of approximation. However, in a few instances, we do treat rounding errors to illustrate their effect in actual computation. The approximation of real functions by finitely representable functions such as polynomials can be viewed as one step in the process of replacing problems formulated in terms of the infinite real continuum by problems formulated in finite terms which can be translated into a computer algorithm.

7.2 THE WEIERSTRASS THEOREM (1885)

Suppose we are given a real function f defined for all x in the closed bounded interval $[a, b]$, $a < b$. In general, we cannot hope to specify a single finite algorithm for computing $f(x)$ for arbitrary $x \in [a, b]$. For example, $e^x = \sum_{n=0}^{\infty} x^n/n!$ can be specified by a single algorithm which defines the coefficients $1/n!$ recursively, but the algorithm requires an infinite number of steps. This is true also for the function $\sqrt{x}$, $x \in [a, b]$, and many functions which arise in common practical computation. Fortunately, common practical computation does not require that $f(x)$ be computed exactly. If $f(x)$ can be determined to within some specified accuracy $\varepsilon = \varepsilon(x)$, this suffices. If there is a finite algorithm to do this in such a way that ε is independent of x, this is even more satisfactory, since the same algorithm can be used for all x in $[a, b]$. We then have a *uniform approximation* to $f(x)$ in $[a, b]$. Can we hope to obtain such an approximation in the form of a polynomial? Since a polynomial is a continuous function in $[a, b]$, it would seem reasonable to require at least this much of $f(x)$. In fact, since ε may be arbitrarily small (depending on varying practical requirements say), we really would like to approximate $f(x)$ uniformly for a sequence (ε_n), where $\varepsilon_n \longrightarrow 0$. Of course, for each ε_n we would use a different polynomial. The result would be a sequence of polynomials $p_n(x)$ converging uniformly to $f(x)$ on $[a, b]$. This would imply that f is continuous on $[a, b]$. (See Exercise 1.3; also 7.2.) Thus, if a function f is uniformly approximable by polynomials on $[a, b]$ with arbitrarily small error, then f is necessarily continuous. Weierstrass' theorem states that the continuity of f is also sufficient. We give a proof due to Bernstein. We recall from Chapter 3 that for any two real functions f and g defined on $[a, b]$

$$\|f - g\|_\infty = \sup_{a \le x \le b} |f(x) - g(x)|.$$

7.2 The Weierstrass Theorem (1885)

Theorem 7.2.1 If f is continuous in the closed bounded interval $[a, b]$, then for every $\varepsilon > 0$ there exists a polynomial $p(x)$ such that
$$\|f - p\|_\infty < \varepsilon.$$

Proof. The translation $x' = (x - a)/(b - a)$ allows us to take $a = 0$ and $b = 1$ without loss of generality (since this translation maps $p(x')$ into a polynomial $q(x)$ on $[a, b]$).

Consider the *Bernstein polynomials* $B_n(x) = B_n(f; x)$ of degree $\leq n$ given by
$$B_n(x) = \sum_{j=0}^{n} \binom{n}{j} x^j (1 - x)^{n-j} f(j/n),$$
where the $\binom{n}{j}$ are the binomial coefficients. We shall prove that $B_n(x) \to f(x)$ uniformly. Consider the binomial expansion
$$(x + y)^n = \sum_{j=0}^{n} \binom{n}{j} x^j y^{n-j}. \tag{1}$$

We obtain
$$\sum_{j=0}^{n} \binom{n}{j} x^j (1 - x)^{n-j} = (x + (1 - x))^n = 1. \tag{2}$$

Differentiating (1) with respect to x and setting $y = 1 - x$, we have
$$\sum_{j=0}^{n} (j/n) \binom{n}{j} x^j (1 - x)^{n-j} = x. \tag{3}$$

Similarly, differentiating (1) twice with respect to x, and using (3), we have
$$\sum_{j=0}^{n} (j^2/n^2) \binom{n}{j} x^j (1 - x)^{n-j} = (1 - 1/n)x^2 + x/n. \tag{4}$$

Multiplying (3) by $-2x$ and (2) by x^2 and adding the results to (4) we get
$$\sum_{j=0}^{n} (j/n - x)^2 \binom{n}{j} x^j (1 - x)^{n-j} = x(1 - x)/n. \tag{5}$$

Multiplying (2) by $f(x)$ and subtracting $B_n(x)$, we obtain
$$f(x) - B_n(x) = \sum_{j=0}^{n} (f(x) - f(j/n)) \binom{n}{j} x^j (1 - x)^{n-j}. \tag{6}$$

Let $\sum_{j \in S_1}$ be the sum of the terms in (6) over the set of indices S_1 such that $|j/n - x| \leq 1/n^{1/4}$ and let $\sum_{j \in S_2}$ be the sum of the remaining terms ($|j - nx|^2 > n^{3/2}$). Since f is continuous on $[0, 1]$, it is bounded (Theorem 1.4.1, Corollary), say $|f(x)| \leq M$. Hence, using (5),

$$\left| \sum_{j \in S_2} \right| \leq 2M \sum_{j \in S_2} \binom{n}{j} x^j (1 - x)^{n-j}$$
$$= 2M \sum_{j \in S_2} \frac{(j - nx)^2}{(j - nx)^2} \binom{n}{j} x^j (1 - x)^{n-j}$$
$$\leq 2M \sum_{j \in S_2} \frac{(j - nx)^2}{n^{3/2}} \binom{n}{j} x^j (1 - x)^{n-j} \leq \frac{2M}{n^{3/2}} nx(1 - x).$$

In the interval $[0, 1]$, $0 \leq x(1 - x) \leq \frac{1}{4}$. Hence, $|\sum_{j \in S_2}| \leq M/2\sqrt{n}$.

Now, let $\omega(n^{-1/4}) = \sup_{|x'-x|\leq n^{-1/4}} \{|f(x) - f(x')|\}$, where ω is a modulus of continuity as in Definition 1.3.3 and $\omega(n^{-1/4}) \to 0$ as $n \to \infty$. We have, therefore,

$$\left|\sum_{j \in S_1}\right| \leq \omega(n^{-1/4}) \sum_{j=0}^{n} \binom{n}{j} x^j (1-x)^{n-j} = \omega(n^{-1/4});$$

and finally,

$$|f(x) - B_n(x)| \leq \omega(n^{-1/4}) + M/2\sqrt{n} \to 0. \blacksquare$$

Corollary Let f satisfy a Lipschitz condition with constant L on $[a, b]$. Then

$$|f(x) - B_n(x)| \leq Ln^{-1/4} + M/2\sqrt{n}. \qquad (7)$$

Proof. $\omega(\Delta x) \leq L \cdot \Delta x. \blacksquare$

The preceding proof is constructive if $\omega(\Delta x)$ can be constructed, as for example, when a Lipschitz constant L is known. Knowing L and the bound M, we can use (7) to determine the degree n for any specified ε and then explicitly obtain $B_n(x)$. (We assume that $f(x)$ can be evaluated at the $n+1$ points, $x_j = j/n$, $0 \leq j \leq n$. For example, if $f(x) = e^x$, we can compute $e^{j/n}$ to any desired accuracy by calculating enough terms of the series expansion. Having done this once for the points x_j, we can determine $B_n(x)$ and use it as a uniform approximation to $f(x)$). $B_n(x)$ provides a single finite procedure for computing $f(x)$ to within the specified accuracy ε for any x in $[a, b]$. We shall not pursue the Bernstein polynomials further, since there are better methods of obtaining polynomial approximations. This raises the question of whether for a given degree n there is a "best" uniform approximating polynomial π_n of degree $\leq n$ in the sense that $\|f - \pi_n\|_\infty \leq \|f - p_n\|_\infty$ for all polynomials p_n of degree $\leq n$. In sections to follow, we shall prove that such a polynomial π_n exists and show how to construct a polynomial close to π_n.

A second theorem of Weierstrass shows that a function $f(\theta)$ which is continuous and of period 2π can be uniformly approximated to within arbitrary $\varepsilon > 0$ by a *trigonometric sum*

$$S_n = \sum_{k=0}^{n} (a_k \cos k\theta + b_k \sin k\theta).$$

We sketch a proof. Let $x = \cos \theta$, $0 \leq \theta \leq \pi$. Define $f_1(\theta) = (\frac{1}{2})(f(\theta) + f(-\theta))$, $f_2(\theta) = \frac{1}{2}(f(\theta) - f(-\theta)) \sin \theta$. Then f_1 and f_2 are even functions of period 2π. Let $g_1(x) = f_1(\theta)$ and $g_2(x) = f_2(\theta)$. The functions g_1 and g_2 are continuous on $[-1, 1]$. Hence, there exist polynomials $p_1(x), p_2(x)$ such that $|g_i(x) - p_i(x)| \leq \varepsilon/4$. Then, $|f_i(\theta) - p_i(\cos \theta)| \leq \varepsilon/4$. Since $f(\theta) \sin \theta = f_1(\theta) \sin \theta + f_2(\theta)$, $|f(\theta) \sin \theta - p_1(\cos \theta) \sin \theta - p_2(\cos \theta)| \leq \varepsilon/2$. Since integral powers of $\cos \theta$ and $\sin \theta$ can be expressed as trigonometric sums, we have approximated $f(\theta) \sin \theta$ by such a sum. By similar reasoning we obtain a trigonometric sum approximating $f(\theta) \cos \theta$. This yields a trigonometric sum approximating $f(\theta) = f(\theta)(\sin^2 \theta + \cos^2 \theta)$.

The Weierstrass theorem deals with approximation in the L^∞-norm. We are also interested in other measures of the accuracy of an approximation, for example the L^2-norm which is used in *least-squares* approximation. In order to treat a variety of measures of approximation, it is convenient to turn now to the general theory.

7.3 CLOSEST APPROXIMATION IN NORMED AND SEMINORMED SPACES

Many of the problems in approximation theory are instances of the following abstract approximation problem.

The Linear Approximation Problem

Let V be a seminormed vector space and let $f_1, \ldots, f_n$ be n vectors in V. Given $f \in V$, determine values of the scalars $\alpha_1, \ldots, \alpha_n$ such that the function

$$g(\alpha_1, \ldots, \alpha_n) = \left\| f - \sum_{j=1}^{n} \alpha_j f_j \right\| \qquad (7.3.1)$$

attains its minimum value.

If $(\alpha_1^*, \ldots, \alpha_n^*)$ minimizes g, then we say that $\sum_{j=1}^{n} \alpha_j^* f_j$ is the *closest* or *best approximation* to f by linear combinations of the f_j.

For example, if $V = C[a, b]$, the space of functions continuous on the closed bounded interval $[a, b]$ with norm $\|f\|_\infty = \sup_{a \le x \le b} \{|f(x)|\}$, and $f_j = x^{j-1}$, then the closest approximation $\sum_{j=0}^{n} \alpha_j^* x^j$ is called the *best approximating polynomial* (B.A.P.) *of degree* $\le n$ *for f on* $[a, b]$.

As another important example, consider the Hilbert space $L^2[a, b]$ described in Chapter 3. (Again, $[a, b]$ is a bounded interval.) The power functions x^j all lie in $L^2[a, b]$. For any function $f \in L^2[a, b]$ we may consider approximations by polynomials of degree $\le n$. The polynomial $\sum_0^n \alpha_j^* x^j$ such that

$$\int_a^b \left| f(x) - \sum_0^n \alpha_j^* x^j \right|^2 dx \le \int_a^b |f(x) - p(x)|^2 dx \qquad (7.3.2)$$

for all polynomials $p(x)$ of degree $\le n$ is called the *least-squares polynomial* of degree $\le n$ for f. Note that $\int_a^b |f(x)|^2 dx$ is a seminorm which becomes a norm if we identify two functions which differ only on a set of measure zero; i.e., we consider equivalence classes of such functions. This is a specific case of a general procedure which can be applied to arbitrary seminormed spaces.

Let V be a seminormed space. It is a simple matter to verify that the set $K = \{f \in V : \|f\| = 0\}$ is a subspace (Exercise 7.23). The set K is called the *kernel* of the seminorm. We construct a normed vector space V/K as follows. Define two vectors $f, g \in V$ to be equivalent (written $f \equiv g$) if $\|f - g\| = 0$; i.e., $f - g \in K$. It is obvious that $\equiv$ is an equivalence relation and, therefore, partitions V into equivalence classes. Let $\hat{f}$ denote the equivalence class which contains f, and let V/K consist of all these equivalence classes. It becomes a vector space by defining $\hat{f} + \hat{g}$ to be the equivalence class containing $f + g$ and $\alpha \hat{f}$ to be the equivalence class containing αf (Exercise 7.23b). (This is the standard construction of forming the *factor space* V/W for any subspace $W \subset V$.) Now, we introduce a norm into V/K by defining $\|\hat{f}\| = \|f\|$, where f is any vector in the class $\hat{f}$. Then $\|\hat{f}\|$ is well-defined (Exercise 7.23c). It is obviously a norm.

The mapping $\varphi : V \longrightarrow V/K$ defined by $f \longmapsto \hat{f}$ is called the *natural morphism* of V onto V/K. It is a linear mapping. Furthermore, it is continuous and open (i.e., it maps open sets onto open sets). This is obvious from the equality $\|f\| = \|\hat{f}\|$.

Lemma 7.3.0 Let V be a seminormed vector space with kernel K. Let E be a compact subset of V/K. Then the preimage $\varphi^{-1}(E)$ is a compact subset of V. A closed bounded subset of a finite-dimensional subspace $W \subset V$ is compact.

Proof. Let $\{O_i\}$ be an open covering of $\varphi^{-1}(E)$. Then the sets $\{\varphi(O_i)\}$ form an open covering of E. Since E is compact, there is a finite subcovering $\{\varphi(O_1), \ldots, \varphi(O_n)\}$. $\{O_1, \ldots, O_n\}$ is a finite subcovering of $\varphi^{-1}(E)$, since $f \in O_i$ implies $\hat{f} \subset O_i$.

Now, suppose W is a finite-dimensional subspace of V. Since φ is a linear mapping, $\varphi(W)$ is a finite-dimensional subspace of V/K. Let $F \subset W$ be a closed bounded set. Then $\varphi(F)$ is a closed bounded subset of $\varphi(W)$. Since $\varphi(W)$ is a finite-dimensional normed space, $\varphi(F)$ is compact. By the preceding paragraph, F is also compact. This completes the proof. ∎

Returning to the linear approximation problem, let W be the subspace generated by the elements $f_1, \ldots, f_n \in V$. Then W consists of all the linear combinations $x = \sum_{j=1}^{n} \alpha_j f_j$, hence is finite-dimensional. In the linear approximation problem, we wish to determine $x^* \in W$ such that $\|f - x^*\| \leq \|f - x\|$ for all $x \in W$. The existence of x^* is established in the next theorem.

Theorem 7.3.1 Let V be a seminormed vector space. Let $W \subset V$ be a finite-dimensional subspace. For any $f \in V$ there exists a closest point $x^* \in W$.

Proof. Consider the closed ball $\bar{B} = \bar{B}(f, \|f\|)$, centered at f with radius $\|f\|$. Since $0 \in \bar{B}$, the intersection of $\bar{B}$ and W is nonempty. This also shows that $\inf_{x \in W} \|f - x\| \leq \|f - 0\| = \|f\|$. Thus, we may restrict our search for a closest point to the set $E = \bar{B} \cap W$. Since E is a closed bounded subset of the finite-dimensional subspace W, E is compact. Now, $\|f - x\|$ is a continuous function of $x \in W$, since

$$|\|f - (x+h)\| - \|f - x\|| \leq \|h\|.$$

Therefore, $\|f - x\|$ assumes its infimum on the compact subset E, that is, there exists $x^* \in E$ such that $\|f - x^*\| = \inf_{x \in E} \|f - x\|$. As we have already observed, this is a closest point to f in all of W; i.e., $\|f - x^*\| = \inf_{x \in W} \|f - x\|$, which completes the proof. ∎

As an immediate application of Theorem 7.3.1, which can be regarded as the *fundamental theorem on the linear approximation problem*, we have several important results which we state as corollaries.

Corollary 1 Let $f \in C[a, b]$, where $-\infty < a < b < \infty$. For any $n \geq 0$ there exists a best approximating polynomial of degree $\leq n$ (in the uniform norm).

Proof. The set of all polynomials of degree $\leq n$ is a finite-dimensional subspace P_n of the normed space $C[a, b]$, since it is spanned by the functions $\{1, x, \ldots, x^n\}$. By the theorem, there must exist a polynomial in P_n which is closest to f. ∎

Corollary 2 Let $f \in L^2[a, b]$, where $-\infty < a < b < \infty$. For any $n \geq 0$, there exists a least-squares polynomial approximation to f of degree $\leq n$.

7.3 Closest Approximation in Normed and Seminormed Spaces

Proof. The set P_n of polynomials of degree $\leq n$ is a finite-dimensional subspace of $L^2[a, b]$, since each power x^j, $0 \leq j \leq n$, is a square integrable function on any bounded interval. The result then follows directly from the theorem. ∎

Let f be a real-valued function defined on some subset S of the real line. Let $E = \{x_1, \ldots, x_k\}$ be a finite subset of S. By the least-squares polynomial of degree $\leq n$ for f on the set E we mean a polynomial p^* of degree $\leq n$ such that

$$\sum_{i=1}^{k} |f(x_i) - p^*(x_i)|^2 \leq \sum_{i=1}^{k} |f(x_i) - p(x_i)|^2$$

for all polynomials p of degree $\leq n$.

Corollary 3 For any real function f defined on some subset S of the real line containing the set $E = \{x_1, \ldots, x_k\}$ there exists a least squares polynomial of degree $\leq n$ on the set E.

Proof. The set of all real functions f defined on S is a vector space. The function $\|f\| = (\sum_{i=1}^{k} |f(x_i)|^2)^{1/2}$ is a seminorm on this space. The set P_n of polynomials of degree $\leq n$ is a subspace of finite dimension. The result then follows directly from the theorem. ∎

Corollary 4 Let f be as in Corollary 3. There exists a best approximating polynomial π_{nE} of degree $\leq n$ for f on the set E; i.e.,

$$\max_{1 \leq i \leq k} |f(x_i) - \pi_{nE}(x_i)| \leq \max_{1 \leq i \leq k} |f(x_i) - p(x_i)|$$

for any polynomial $p \in P_n$.

Proof. $\|f\|_\infty = \max_{1 \leq i \leq k} |f(x_i)|$ is a seminorm on the space of functions f defined on the set S. Again, P_n is a finite-dimensional subspace and the result follows. ∎

Theorem 7.3.1 asserts the existence of closest points in finite-dimensional subspaces of a seminormed space.

This is not always true for infinite-dimensional subspaces W, as Exercise 7.4 shows. However, if V is a Hilbert space, and W is a *closed* infinite-dimensional subspace, then there is a point in W which is closest to any $f \in V$. In fact, a more general result can be proved (Theorem 7.3.2 below) by introducing the notion of uniform convexity. A normed space is called *uniformly convex* if for any $\varepsilon > 0$ there exists a $\delta > 0$ such that

$$\|f\| = \|g\| = 1 \quad \text{and} \quad \|\tfrac{1}{2}(f + g)\| > 1 - \delta \text{ implies } \|f - g\| < \varepsilon. \tag{7.3.3}$$

Geometrically, (7.3.3) means that if the midpoint of a line segment with end points on the unit sphere approaches the unit sphere, then the end points must approach each other. A pre-Hilbert (inner-product) space is uniformly convex (Exercise 7.5).

Theorem 7.3.2 Let E be a closed convex set in a uniformly convex Banach space V. For any $f \in V$ there is a closest point x^* in E; i.e., $\|x^* - f\| = \inf_{x \in E} \|x - f\|$.

Proof. We must find x^* in E which minimizes $\|x - f\|$ for $x \in E$. If we consider the translated set $E - f$, which is also closed convex, we must find a point in $E - f$ closest to the origin. Let $E' = E - f$. Let $d = \inf_{x \in E'} \|x\|$. If $d = 0$, then there is a sequence in E' which converges to the zero vector. Since E' is closed, $0 \in E'$ is the closest point to 0. If $d > 0$, we may take $d = 1$ by considering the set $E'' = (1/d)E'$. Since $d = 1$, there is a sequence $x_n \in E''$ with $\lim_{n \to \infty} \|x_n\| = 1$. Let δ be such that (7.3.3) above holds.

Let $\bar{x}_n = x_n/\|x_n\|$. Note that $x_n - \bar{x}_n \to 0$. Hence, for n sufficiently large, $\|x_n - \bar{x}_n\| < \delta$. Also, since E'' is convex, $\frac{1}{2}(x_p + x_q) \in E''$. Since $d = 1$, $\|\frac{1}{2}(x_p + x_q)\| \geq 1$. For p, q sufficiently large,

$$\tfrac{1}{2}\|\bar{x}_p + \bar{x}_q\| = \tfrac{1}{2}\|x_p + x_q - (x_p - \bar{x}_p) - (x_q - \bar{x}_q)\|$$

$$\geq \tfrac{1}{2}\|x_p + x_q\| - \tfrac{1}{2}\|x_p - \bar{x}_p\| - \tfrac{1}{2}\|x_q - \bar{x}_q\| > 1 - \delta.$$

By the uniform convexity, $\|\bar{x}_p - \bar{x}_q\| < \varepsilon$. Therefore, $(\bar{x}_n)$ is a Cauchy sequence. Since V is complete, there is a limit $\bar{x}$ and $\|\bar{x}\| = 1$. But $\|x_n - \bar{x}\| \leq \|x_n - \bar{x}_n\| + \|\bar{x}_n - \bar{x}\| \to 0$. Hence, $\lim x_n = \bar{x}$. Since E is closed, $\bar{x} \in E''$. ∎

Corollary Let E be an infinite-dimensional closed subspace of a Hilbert space V. For any $f \in V$ there is a closest point $x \in E$.

Proof. E is closed and convex. V is uniformly convex. ∎

We note that in a finite-dimensional space V^n it is enough for a set E to be closed to ensure the existence of a point $x^* \in E$ closest to a given point $f \in E$. In proof, let $d = \inf_{x \in E} \|x - f\| \geq 0$. Then there is a sequence $x_n \in E$ such that $\|x_n - f\| \to d$. Hence, for $n \geq N$, N sufficiently large, $\|x_n - f\| \leq d + \varepsilon$ for some $\varepsilon > 0$, that is, $x_n \in \bar{B}(f, d + \varepsilon)$. The intersection $E \cap \bar{B}(f, d + \varepsilon)$ is a closed bounded set in V^n and is therefore compact. Hence, the sequence $(x_n : n \geq N)$ has a convergent subsequence whose limit x^* is in E. By the continuity of $\|x - f\|$ in x, we have $\|x^* - f\| = d$.

Having established the existence of the best approximation, it is natural to ask whether the best approximation is unique. This is of more than theoretical interest. We wish to obtain constructive methods for computing the best approximation. In such a computation it is important to know whether there is more than one solution to the problem. In Theorem 7.3.2, the point $\bar{x}$ is unique (Exercise 7.6). A simple example of non-uniqueness occurs in the 2-dimensional space V^2 with the l_∞-norm. Consider the 1-dimensional subspace $\{(\xi_1, 0)\}$ and the vector $(0, 1)$. Then $\|(0, 1) - (\xi_1, 0)\|_\infty = \|(-\xi_1, 1)\|_\infty = \max(|\xi_1|, 1)$. Hence, the minimum distance from $(0, 1)$ to the subspace is 1 and is achieved for any point $(\xi_1, 0)$ such that $|\xi_1| \leq 1$. The l_∞-norm is not strict. (See Exercise 7.7 and Definition 3.3.3.) The next theorem shows that strictness is a sufficient condition for uniqueness. However, it is not a necessary condition as we shall see when we prove the uniqueness of the best approximating polynomial in $C[a, b]$.

Theorem 7.3.3 Let V be a strictly normed vector space and $\{f_1, \ldots, f_n\}$ a linearly independent set. The best approximation $\sum_{i=1}^{n} \alpha_i^* f_i$ to any vector f is unique.

7.3 Closest Approximation in Normed and Seminormed Spaces 273

Proof. Suppose $\sum_{i=1}^{n} \beta_i f_i$ is another best approximation. Using the fact that $\sum \alpha_i^* f_i$ and $\sum \beta_i f_i$ both minimize $\|f - \sum \alpha_i f_i\|$, we have

$$\|f - \sum \alpha_i^* f_i\| \leq \|f - \sum \frac{\alpha_i^* + \beta_i}{2} f_i\| \leq \tfrac{1}{2}\|f - \sum \alpha_i^* f_i\| + \tfrac{1}{2}\|f - \sum \beta_i f_i\|$$
$$= \|f - \sum \alpha_i^* f_i\|.$$

Therefore,

$$\|\tfrac{1}{2}(f - \sum \alpha_i^* f_i) + \tfrac{1}{2}(f - \sum \beta_i f_i)\| = \tfrac{1}{2}\|f - \sum \alpha_i^* f_i\| + \tfrac{1}{2}\|f - \sum \beta_i f_i\|.$$

Since the norm is strict, there exists λ such that

$$f - \sum \alpha_i^* f_i = \lambda(f - \sum \beta_i f_i).$$

If $\lambda \neq 1$, then f is a linear combination of the f_i and the best approximation is certainly unique. If $\lambda = 1$, then $\sum (\alpha_i^* - \beta_i) f_i = 0$, which implies $\alpha_i^* = \beta_i$ by the linear independence of the f_i. ∎

The geometric significance of a strict norm lies in the fact that in a strictly normed space every closed ball is *strictly convex*. A set E is called *strictly convex* if for any x, $y \in E$ the points $\alpha x + \beta y$ with $\alpha, \beta > 0$ and $\alpha + \beta = 1$ all lie in the interior of E. A space V is strictly normed if and only if the unit closed ball is strictly convex (Exercise 7.7). The strict convexity of the unit closed ball implies that the unit sphere cannot contain a line segment. Since this property of the unit sphere obviously holds also in a uniformly convex space, every uniformly convex space is strictly normed (Exercise 7.11). Every finite-dimensional strictly normed space is uniformly convex (Exercise 7.11).

The function $g(\alpha_1, \ldots, \alpha_n)$ in (7.3.1) above is continuous and has a minimum over the entire space R^n; i.e., it has a *global* or *absolute* minimum. This global character of the minimum is a property of *convex* functions. A real function f defined on a convex set is said to be a *convex* function if it satisfies the inequality

$$f(\alpha x + \beta y) \leq \alpha f(x) + \beta f(y)$$

whenever $\alpha \geq 0$, $\beta \geq 0$ and $\alpha + \beta = 1$; i.e., in one dimension this means that the graph of f lies below the chord joining any two points.

Theorem 7.3.4 *A relative minimum of a convex function f is a global minimum.*

Proof. Let f have a relative minimum at x_0, that is, suppose $f(x_0) \leq f(x)$ for $\|x - x_0\| \leq \delta$. Suppose there exists x_1 such that $\|x_1 - x_0\| < 2\delta$ and $f(x_1) < f(x_0)$. We must have $\tfrac{1}{2} < \delta/\|x_1 - x_0\| = \alpha < 1$. Let $\beta = 1 - \alpha$. By the convexity of f, $f(\alpha x_0 + \beta x_1) \leq \alpha f(x_0) + \beta f(x_1) < (\alpha + \beta) f(x_0) = f(x_0)$. But $\|\alpha x_0 + \beta x_1 - x_0\| = \beta\|x_1 - x_0\| < \delta$, which means $f(x_0) \leq f(\alpha x_0 + \beta x_1)$. This contradiction implies $f(x_1) \geq f(x_0)$ for all x_1. ∎

The function g in (7.3.1) above is convex (Exercise 7.8).

Remark. It might be argued that for practical computation purposes we should not be concerned with the existence of an $x^* \in E$ closest to f. Since $\|f - E\| = \inf_{x \in E} \|f - x\|$, there always exists a *minimizing sequence* (x_n) of points $x_n \in E$ such that $\lim_{n \to \infty} \|f - x_n\| = \|f - E\|$. We simply choose balls B_n centered at f and having radii $\|f - E\| + \varepsilon_n$, where $\varepsilon_n > 0$ and $\varepsilon_n \to 0$. In each B_n there is a point $x_n \in E$. From a practical standpoint, we should not care whether the sequence (x_n) converges. It is enough that $\|f - x_n\| \to \|f - E\|$, since for n sufficiently large, x_n yields an approximation to f which is "best" within the accuracy limits of computation (i.e. rounding error). Although this kind of argument has some merit, it overlooks one very important and very practical point (as is often the case with many so-called practical arguments). A point x^* which actually yields the minimum of $\|f - x\|$ may possess certain special properties which can be used to compute x^*. On the other hand, there may be nothing special about the points x_n of a minimizing sequence and unless one can show how to construct the sequence (x_n), the mere existence of (x_n) is of no practical value. For example, if $g(x)$ is a differentiable function for $x \in [a, b]$, then there exists $x^* \in [a, b]$ which minimizes $g(x)$. A necessary condition on x^* is that either $x^* = a$ or $x^* = b$ or $g'(x^*) = 0$. In ordinary calculus, minima can be found by finding the points where the derivative is zero and comparing the values of g at these points and at a and b. In the next section, we shall see that in a pre-Hilbert space, the best approximation satisfies a condition which provides us with the basis of a computational method.

In connection with a minimizing sequence, therefore, we raise the question of whether it converges. We shall prove that in the linear approximation problem, there is always a convergent subsequence and if the closest point is unique, every minimizing sequence itself converges. In the next four lemmas, we assume that $\{f_1, \ldots, f_n\}$ is a linearly independent set of vectors in a normed vector space V and E is the subspace spanned by the f_i. The vector $f \in V$ is arbitrary and $\|f - E\| = \inf_{x \in E} \|f - x\|$ is the distance from f to E.

Lemma 7.3.1 Let $S \subset E$ be a bounded set contained in a ball $\bar{B}(0, r)$. There exists a constant $\mu > 0$ such that for any vector $\sum_1^n \alpha_i f_i \in S$,

$$\sum_1^n |\alpha_i| \leq \mu r.$$

Proof. The function $\|\sum_1^n \alpha_i f_i\|$ is a continuous function of $(\alpha_1, \ldots, \alpha_n)$. On the compact unit l_1-sphere ($\sum_1^n |\alpha_i| = 1$) in R^n, it assumes its minimum value. Since the f_i are linearly independent, this minimum is positive. Denote it by $1/\mu$. Then

$$\left\| \sum_1^n \left(\alpha_i / \sum_1^n |\alpha_i| \right) f_i \right\| \geq (1/\mu).$$

If $\sum_1^n \alpha_i f_i \in S$, then

$$\sum_1^n |\alpha_i| \leq \mu \left\| \sum_1^n \alpha_i f_i \right\| \leq \mu r. \blacksquare$$

7.3 Closest Approximation in Normed and Seminormed Spaces

Corollary If the sequence $x_q = \sum_1^n \alpha_{iq} f_i$, $q = 1, 2, \ldots$, converges to $x = \sum_1^n \alpha_i f_i$, then $\alpha_{iq} \to \alpha_i$ for $i = 1, \ldots, n$. ∎

Proof. If $y_q = x - x_q = \sum_{i=1}^n (\alpha_i - \alpha_{iq}) f_i$, then y_q is such that $\|y_q\| \to 0$. Hence, for q sufficiently large and $\varepsilon > 0$, $\|y_q\| \le \varepsilon$. By the lemma $\sum_{i=1}^n |\alpha_i - \alpha_{iq}| \le \mu\varepsilon$. Hence, $|\alpha_i - \alpha_{iq}| \le \mu\varepsilon$. ∎

Lemma 7.3.2 $\|f - \sum_1^n \alpha_i f_i\|$ becomes infinite if $\sum_1^n |\alpha_i| \to \infty$ and conversely.

Proof

$$\left\| f - \sum_1^n \alpha_i f_i \right\| \ge \left\| \sum \alpha_i f_i \right\| - \|f\| \ge \left(\frac{1}{\mu}\right) \sum_1^n |\alpha_i| - \|f\|$$

by the proof of Lemma 7.3.1. Conversely, if $\sum_1^n |\alpha_i| < r$, then $\|f - \sum \alpha_i f_i\| \le \|f\| + r \max \|f_i\|$. ∎

Lemma 7.3.3 Let (x_q) be a minimizing sequence of vectors $x_q \in E$ such that $\|f - x_q\| \to \|f - E\|$. There exists a subsequence (x_{q_j}) which converges to a point $x^* \in E$ such that $\|f - x^*\| = \|f - E\|$.

Proof. (Since a minimizing sequence always exists, this is an alternative proof of Theorem 7.3.1.) For q sufficiently large, $\|f - x_q\| < \|f - E\| + 1$. Hence,

$$\|x_q\| \le \|f - x_q\| + \|f\| < \|f - E\| + 1 + \|f\| = r.$$

Writing $x_q = \sum_{i=1}^n \alpha_{iq} f_i$, Lemma 7.3.1 yields $\sum_1^n |\alpha_{iq}| \le \mu r$. Since a closed l_1-ball in R^n is compact, the sequence of points $\alpha_q = (\alpha_{1q}, \ldots, \alpha_{nq})$ contains a subsequence (α_{q_j}) converging to some point $\alpha^* = (\alpha_1^*, \ldots, \alpha_n^*)$. For the corresponding vectors $x_{q_j} = \sum_1^n \alpha_{iq_j} f_i$ and $x^* = \sum \alpha_i^* f_i$, we have

$$\|x_{q_j} - x^*\| \le \left(\sum_1^n |\alpha_{iq_j} - \alpha_i^*| \right) \max_{1 \le i \le n} \|f_i\| \to 0.$$

Finally,

$$\|f - x^*\| \le \|f - x_{q_j}\| + \|x_{q_j} - x^*\| \to \|f - E\|. \quad \blacksquare$$

For the next result, we require uniqueness of the closest point.

Lemma 7.3.4 Let x^* be the unique closest point to f in E. If (x_q) is a minimizing sequence such that $\|f - x_q\| \to \|f - E\|$, then $x_q \to x^*$ as $q \to \infty$.

Proof. Suppose (x_q) does not converge to x^*. Then there is an $\varepsilon > 0$ and a subsequence (x_{q_i}) such that

$$\|x_{q_i} - x^*\| > \varepsilon, \quad i = 1, 2, \ldots. \tag{1}$$

Since $\|f - x_{q_i}\| \to \|f - E\|$, (x_{q_i}) is also a minimizing sequence. By Lemma 7.3.3, (x_{q_i}) contains a subsequence converging to a closest point in E. Since x^* is the only such closest point, the subsequence of (x_{q_i}) must converge to x^*. This contradicts (1) above. Hence, $x_q \to x^*$. ∎

7.4 CLOSEST (LEAST-SQUARES) APPROXIMATION IN PRE-HILBERT SPACES

When the norm is induced by an inner product, we can draw upon geometric concepts to a much greater extent than heretofore. We can obtain a necessary and sufficient condition on the solution of the linear approximation problem which provides a basis for constructing the solution. (In Theorem 7.3.1, we proved the existence of such a solution, but the proof is not constructive.)

Theorem 7.4.1 Let E be a subspace of a pre-Hilbert space V. Let $f \in V$. The point $x^* \in E$ is the closest point in E to f if and only $f - x^*$ is orthogonal to every $x \in E$.

Proof. (We may assume $f \neq 0$, since otherwise 0 is the closest point to f and the result follows trivially.) Suppose $x_0 \in E$ is such that $f - x_0$ is not orthogonal to every vector in E. We shall prove that x_0 is not the closest point to f. Let $x \in E$ be such that $\langle f - x_0, x \rangle = \alpha \neq 0$. Take $x' = x_0 + (\alpha/\langle x, x \rangle)x$. Then

$$\|f - x'\|^2 = \langle f - x', f - x' \rangle = \left\langle f - x_0 - \frac{\alpha}{\langle x, x \rangle} x, f - x_0 - \frac{\alpha x}{\langle x, x \rangle} \right\rangle$$

$$= \|f - x_0\|^2 - \frac{\bar{\alpha}}{\langle x, x \rangle} \langle f - x_0, x \rangle - \frac{\alpha}{\langle x, x \rangle} \langle x, f - x_0 \rangle + \frac{|\alpha|^2}{\langle x, x \rangle}$$

$$= \|f - x_0\|^2 - \frac{|\alpha|^2}{\langle x, x \rangle} < \|f - x_0\|^2;$$

i.e., x' is closer to f than is x_0. Thus, if x^* is the closest point, $f - x^*$ is orthogonal to every x in E.

Conversely, suppose $\langle f - x^*, x \rangle = 0$ for all $x \in E$. For any $y \in E$, we may write $y = x^* + (y - x^*) = x^* + h$, where $h = y - x^* \in E$. We have $\langle f - x^*, h \rangle = 0$. Hence,

$$\|f - (x^* + h)\|^2 = \langle f - x^* - h, f - x^* - h \rangle = \|f - x^*\|^2 + \|h\|^2 \geq \|f - x^*\|^2.$$

Therefore, x^* is the closest point in E to f. ∎

The geometric interpretation of Theorem 7.4.1 is immediate. Recalling Chapter 3, Theorem 3.3.11, we see that $f - x^* \in \bar{E}^\perp$, the orthogonal complement of the closure $\bar{E}$. Therefore, $f = (f - x^*) + x^*$ and x^* is just the orthogonal projection of f on $\bar{E}$. The orthogonal projection of f on any finite-dimensional subspace E may be determined as follows. Let E be spanned by $\{f_1, \ldots, f_n\}$. Then $\langle f - x^*, f_i \rangle = 0$, $i = 1, \ldots, n$, so that

$$\langle x^*, f_i \rangle = \langle f, f_i \rangle, \quad 1 \leq i \leq n. \tag{1}$$

Let us write $x^* = \sum_1^n \alpha_i f_i$, since $x^* \in E$.

Applying (1), we obtain a system of n linear equations for the α_i,

$$\begin{aligned} \alpha_1 \langle f_1, f_1 \rangle + \cdots + \alpha_n \langle f_n, f_1 \rangle &= \langle f, f_1 \rangle, \\ &\vdots \\ \alpha_1 \langle f_1, f_n \rangle + \cdots + \alpha_n \langle f_n, f_n \rangle &= \langle f, f_n \rangle. \end{aligned} \tag{7.4.1}$$

7.4 Closest (Least-Squares) Approximation in pre-Hilbert Spaces

Equations (7.4.1) are usually known as the *normal equations*. The $n \times n$ matrix $(\langle f_i, f_j \rangle)$ is called the *Gram* matrix of the set $\{f_1, \ldots, f_n\}$. If this set is linearly independent, the Gram determinant must be nonzero (Exercise 7.12). This implies that system 7.4.1 has a unique solution $(\alpha_1, \ldots, \alpha_n)$. Conversely, for such a solution, the vector $x^* = \sum_1^n \alpha_i f_i$ satisfies (1). Hence, $\langle f - x^*, f_i \rangle = 0$, and $f - x^* \in E^\perp$. Therefore, x^* is the orthogonal projection of f on E.

Since the closest point in E to f is the orthogonal projection of f on $\bar{E}$ by Theorem 7.4.1, we have an independent proof of the existence and uniqueness of a solution of the linear approximation problem in Hilbert spaces when E is closed. Furthermore, when E is finite-dimensional the proof is, in principle, constructive; i.e., given $f_1, \ldots, f_n, f$, we can determine $x^* = \sum \alpha_i f_i$ by solving Eqs. (7.4.1) for the α_i. Of course, in practice we must be able to compute $\langle f_i, f_j \rangle$ and $\langle f, f_i \rangle$. Then the solution of the linear system (7.4.1) can be computed by the methods of Chapter 4. Sometimes, the latter computation can be avoided by orthonormalization of the f_i by the Gram-Schmidt process (Theorem 3.3.4). Suppose $\{e_1, \ldots, e_n\}$ is the resulting orthonormal set. Since the $\{e_i\}$ span E, we have $x^* = \sum_1^n \beta_i e_i$. Again applying the necessary conditions, $\langle x^*, e_i \rangle = \langle f, e_i \rangle$, $1 \leq i \leq n$, we now obtain immediately that $\beta_i = \langle f, e_i \rangle$. Hence,

$$x^* = \sum_{i=1}^n \langle f, e_i \rangle e_i. \tag{7.4.2}$$

Furthermore, $\|f - x^*\|^2 = \|f\|^2 - \sum_{i=1}^n |\langle f, e_i \rangle|^2$.

For an infinite subspace of a pre-Hilbert space, we must consider a denumerably infinite orthonormal set $\{e_i\}$ and Bessel's inequality (Theorem 3.3.5) holds. We now have an infinite number of necessary conditions $\langle x^*, e_i \rangle = \langle f, e_i \rangle$, $i = 1, 2, \ldots$. If V is complete, $x^* = \sum_1^\infty \langle x^*, e_i \rangle e_i$, by Theorem 3.3.5. But then $x^* = \sum_1^\infty \langle f, e_i \rangle e_i$ and

$$\|f - x^*\|^2 = \|f\|^2 - \sum_{i=1}^\infty |\langle f, e_i \rangle|^2 \geq 0.$$

Hence, to obtain arbitrarily close approximations to an arbitrary f by linear combinations of the e_i, it is necessary and sufficient that $\{e_i\}$ be a complete orthonormal set in V.

An alternative approach to the (real) least-squares approximation problem is the following. We must choose $(\alpha_1, \ldots, \alpha_n)$ to minimize

$$\varphi(\alpha_1, \ldots, \alpha_n) = \|f - \sum_1^n \alpha_j f_j\|^2 = \left\langle f - \sum_1^n \alpha_j f_j, f - \sum_1^n \alpha_j f_j \right\rangle$$

$$= \|f\|^2 - 2\sum_1^n \langle f, f_j \rangle \alpha_j + 2 \sum_{1 \leq i < j \leq n} \alpha_i \alpha_j \langle f_i, f_j \rangle$$

$$+ \sum_{i=1}^n \alpha_i^2 \langle f_i, f_i \rangle.$$

A necessary condition that φ have a minimum at $a^* = (\alpha_1^*, \ldots, \alpha_n^*)$ is that $(\partial \varphi / \partial \alpha_j)_* = 0$, $1 \leq j \leq n$, where the * subscript indicates that the partial derivatives are evaluated at a^*. Differentiating φ, we obtain

$$\partial \varphi / \partial \alpha_j = -2 \langle f, f_j \rangle + 2 \sum_{i=1}^{n} \langle f_i, f_j \rangle \alpha_i, \quad 1 \leq j \leq n.$$

Setting $\partial \varphi / \partial \alpha_j = 0$, we again obtain the system (7.4.1) above.

Finally, let us consider the case where $\{f_1, \ldots, f_n\}$ is not linearly independent. Again, let E be the subspace spanned by the f_i. Theorem 7.4.1 is valid so that $f - x^* \in E^\perp$ is still a necessary and sufficient condition that x^* be the closest point in E to f. Writing $x^* = \sum_{i=1}^{n} \alpha_i f_i$ as before, we see that again a necessary and sufficient condition on x^* is that the α_i satisfy the linear system (7.4.1) of normal equations. However, the Gram determinant will now be zero. Hence, there will not be a unique solution $(\alpha_1, \ldots, \alpha_n)$ of system (7.4.1). This simply reflects the fact that the unique point x^* is not uniquely represented as a linear combination of the f_i when the f_i are linearly dependent.

7.4(a) The Linear Least-Squares Problem

In Section 7.7, we shall consider polynomial least-squares approximation as a specific example of closest approximation in Hilbert spaces. (See also Exercise 7.19.) Another application is to be found in the familiar problem of Chapter 4, solving $Ax = b$, where A is an $m \times n$ matrix and b is an m-dimensional vector. If b is not in the range of A, then this equation has no solution. However, we may then be willing to settle for the *least-squares solution* x^* which is such that

$$\|Ax^* - b\|_2 \leq \|Ax - b\|_2$$

for all x in V^n. Thus, Ax^* is the closest vector to b in the range of A. Now, the range of A is a subspace of V^m spanned by the column vectors $A_1, \ldots, A_n$ of A. Letting $x^* = (\alpha_1, \ldots, \alpha_n)$, we see that $Ax^* = \sum_{j=1}^{n} \alpha_j A_j$. From the preceding discussion it follows that the α_j must satisfy the normal equations (7.4.1) which in this instance becomes

$$\begin{aligned} \alpha_1 \langle A_1, A_1 \rangle + \cdots + \alpha_n \langle A_n, A_1 \rangle &= \langle b, A_1 \rangle \\ &\vdots \\ \alpha_1 \langle A_1, A_n \rangle + \cdots + \alpha_n \langle A_n, A_n \rangle &= \langle b, A_n \rangle. \end{aligned} \quad (7.4.3)$$

It is easily seen that the matrix form of these equations is

$$A^T A x = A^T b. \quad (7.4.4)$$

The rank of $A^T A$ is the maximum number of linearly independent vectors among the A_j, that is, it is equal to the rank of A. If rank $A < n$, there will be $v + 1$ linearly independent solutions x, where v is the nullity of A. However, by the uniqueness of the closest point, all solutions x^* will give rise to the same vector Ax^* which is the closest vector to b in the range of A.

7.4 Closest (Least-Squares) Approximation in pre-Hilbert Spaces

In practice, it may be advantageous to compute that least-squares solution $x^*_{\min}$ which has minimum $\sum_1^n |\alpha_j|^2$. A procedure for computing $x^*_{\min}$ was given in Section 4.6. Further results on the computation of solutions to linear least-squares problems and their relation to singular linear systems can be found in references [7.15]–[7.21]. For example, it can be shown [7.21] that $x^*_{\min}$ (called the *best approximate solution*) is unique and is given by $x^*_{\min} = A^\dagger b$, where $A^\dagger$ is the pseudoinverse of A (Exercises 2.13, 7.31). The pseudoinverse is itself the best approximate solution of the matrix equation $AX = I$.

Let us prove the result concerning $x^*_{\min}$ and the pseudoinverse for Hilbert spaces [7.31].

Theorem 7.4.2 Let A be a bounded linear operator mapping a Hilbert space X into a Hilbert space Y. Let the range of A, $R(A)$, be closed. Let $x^*_{\min} = A^\dagger b$, where $A^\dagger$ is the pseudoinverse of A and $b \in Y$. Then

$$\|Ax^*_{\min} - b\| \leq \|Ax - b\| \quad \text{for all } x \in X;$$

i.e. $x^*_{\min}$ is a least-squares solution of $Ax = b$. Furthermore, $\|x^*_{\min}\| \leq \|x_0\|$ for any least squares solution x_0.

Proof. We refer to Exercise 7.31 for the properties of A, $R(A)$ and the null space, $N(A)$, of A. Writing $b = b_1 + b_2$, where $b_1 \in R(A)$ and $b_2 \in R(A)^\perp$, we have $Ax^*_{\min} = AA^\dagger b = b_1$, since $AA^\dagger = P_{R(A)}$. Hence,

$$\|Ax^*_{\min} - b\| = \|b_1 - b\| = \|b_2\|.$$

For arbitrary $x \in X$, $Ax \in R(A)$ so that

$$\|Ax - b\|^2 = \|Ax - b_1\|^2 + \|b_2\|^2 \geq \|Ax^*_{\min} - b\|^2,$$

proving that $x^*_{\min}$ is a least-squares solution.

Now, let x_0 be any least-squares solution. Since $A^*Ax_0 = A^*b$, the vector $x_1 = x_0 - x^*_{\min}$ must be in the null space of A^*A. By Exercise 7.31 (a) and (b), $N(A^*A) = N(A)$; (i.e., $x \in N_A^\perp$ implies $Ax \neq 0$ is in $N_{A^*}^\perp$). By Exercise 7.31(e), $A^\dagger b \in R(A^*) = N_A^\perp$. Hence

$$\|x_0\|^2 = \|x_1\|^2 + \|x^*_{\min}\|^2 \geq \|x^*_{\min}\|^2,$$

which completes the proof. ∎

The preceding theorem shows that a least-squares solution x^* of $Ax = b$ exists when A is a bounded linear operator on one Hilbert space, X, to another, Y, provided that the range of A is closed. (This also follows from Theorem 7.4.1.) We have shown that in the finite-dimensional case x^* must satisfy the normal equations (7.4.4). The same is true when X and Y are infinite-dimensional; i.e. x^* is a solution of $A^*Ax = A^*b$. This is proved in Section 12.3, where another method of solving the linear least-squares problem is given in Example 1 following Theorem 12.3.3.

An especially important application of linear least-squares approximation is to be found in the statistical analysis of empirical data. Suppose it is known that a measurable quantity y depends linearly on other measurable quantities $a_1, \ldots, a_n$. Thus,

$$y = x_1 a_1 + \cdots + x_n a_n \tag{1}$$

and it is desired to determine the parameters x_i. Experiments are performed and repeated measurements yield a set of corresponding values $(a_{i1}, \ldots, a_{in}, y_i)$, $1 \leq i \leq m$, where usually m is much larger than n. Since the measurements are subject to random errors, the quantities $a_1, \ldots, a_n, y$ may be viewed as random variables related by Eq. (1). A reasonable estimate of the x_i would be obtained by determining values of the x_i which minimize the *mean square error*

$$\frac{1}{m} \sum_{i=1}^{m} (y_i - x_1 a_{i1} - \cdots - x_n a_{in})^2 \rho_i,$$

where the ρ_i are positive weights determined by the accuracy of measurement i, $1 \leq i \leq m$. If we let A be the $m \times n$ matrix (a_{ij}) and define the inner product in V^m to be $\sum_{i=1}^{m} \rho_i y_i z_i$ for any vectors y, z in V^m, then minimizing the mean square error is equivalent to minimizing $\|Ax - b\|_2$, where $b = (y_1, \ldots, y_m)$. The solution is obtained by solving the normal equations. We shall continue this discussion in Section 8.8.

7.5 POLYNOMIAL APPROXIMATION IN THE UNIFORM NORM

Definition 7.5.1 Let f be a bounded real-valued function defined on a bounded subset E of the real line. By the *best approximating polynomial* (B.A.P.) of degree $\leq n$ for f on E we mean a polynomial π_n of degree $\leq n$ such that

$$\sup_{x \in E} |f(x) - \pi_n(x)| \leq \sup_{x \in E} |f(x) - p(x)| \tag{1}$$

holds for all polynomials p of degree $\leq n$.

We shall call the function

$$r(x) = r_p(x) = f(x) - p(x)$$

the *residual* of f relative to p. The set m_E of all functions f bounded on E is a vector space with norm $\|f\|_\infty = \sup_{x \in E} |f(x)|$. Inequality (1) above may be rewritten as

$$\|f - \pi_n\|_\infty \leq \|f - p\|_\infty.$$

If E is a closed bounded interval $[a, b]$ and f is continuous on $[a, b]$, then it attains its supremum and infimum values. This is also true for any function f if E is a finite set. In these cases, we have $\|f\|_\infty = \max_{x \in E} |f(x)|$ and the B.A.P. for f on $[a, b]$ minimizes the maximum residual obtained with polynomials of degree $\leq n$; i.e., $\|r_{\pi_n}\|_\infty \leq \|r_p\|_\infty$ for all polynomials p of degree $\leq n$. Therefore, we shall call the quantity

$$M_n = M_n(E) = \|f - \pi_n\|_\infty$$

the *minimax of degree n* for f on E.

7.5 Polynomial Approximation in the Uniform Norm

By the fundamental theorem (Theorem 7.3.1), there is a B.A.P. for any f in m_E. Although B.A.P.'s over finite sets are of some interest in themselves, our main reason for considering them is that they provide us with a tool for obtaining the B.A.P. of a continuous function over an interval $[a, b]$. It is one of the basic tenets of numerical analysis that problems stated in terms of the real continuum may be solved by *discretizing*, that is, by choosing a finite set of N points and solving the problem with respect to this finite set. Then by allowing $N \to \infty$, we hope to obtain a sequence of solutions which converge to the solution of the continuous problem. The problem of determining the B.A.P. of a function $f \in C[a, b]$ could be approached in this way. We could select a set E_N of N points in the interval $[a, b]$ and obtain the B.A.P. π_{nN} of f over E_N. If $N \to \infty$ in a suitable manner, we would expect that π_{nN} converge to π_n, the B.A.P. on $[a, b]$. Since we are using more of the values of f as we increase N, it is intuitively reasonable to expect that $\pi_{nN} \to \pi_n$. In fact, we shall prove this as a theoretical result. In the process, we shall obtain results which are the basis of a practical procedure for computing π_n for $f \in C[a, b]$.

The Interpolating Polynomial

Let us begin with the case of E containing $n + 1$ distinct points $x_0, x_1, \ldots, x_n$. We wish to determine a polynomial of degree $\leq n$ which minimizes $\max_{0 \leq i \leq n} |f(x_i) - p(x_i)|$. It is a well-known result that the minimax is zero, that is, there is a unique polynomial $L_n(x)$ of degree $\leq n$, called the *interpolating polynomial* of f on E, such that $f(x_i) - L_n(x_i) = 0$ for $0 \leq i \leq n$. The proof consists of a construction of $L_n(x)$. There are several ways to carry out this construction. At this point, we give the *Lagrange interpolation formula*. (In Chapter 8, we shall give alternative forms.) Consider the polynomials, called the *Lagrange coefficients*,

$$l_i(x) = \prod_{\substack{j=0 \\ j \neq i}}^{n} \frac{x - x_j}{x_i - x_j}, \qquad 0 \leq i \leq n.$$

Each l_i is of degree n and $l_i(x_j) = \delta_{ij}$ (Kronecker delta). Therefore, the polynomial

$$L_n(x) = \sum_{i=0}^{n} f(x_i) l_i(x)$$

is such that $L_n(x_i) = f(x_i), 0 \leq i \leq n$. Furthermore, if $Q(x)$ were another polynomial of degree $\leq n$ with this property, then $Q(x_i) - L_n(x_i) = 0$. Since $Q(x) - L_n(x)$ is a polynomial of degree $\leq n$ which vanishes at $n + 1$ distinct points, it must be identically zero; i.e., $Q = L_n$.

Another method of obtaining the interpolating polynomial is to set $L_n(x) = a_n x^n + \cdots + a_0$ and simply write down the necessary and sufficient conditions on the coefficients, namely,

$$L_n(x_i) = a_n x_i^n + a_{n-1} x_i^{n-1} + \cdots + a_1 x_i + a_0 = f(x_i), \qquad i = 0, \ldots, n.$$

This is a system of $n+1$ linear equations in the $n+1$ unknowns $a_0, a_1, \ldots, a_n$. Its matrix is

$$\begin{bmatrix} 1 & x_0 & x_0^2 & \cdots & x_0^n \\ \vdots & & & & \vdots \\ 1 & x_n & x_n^2 & \cdots & x_n^n \end{bmatrix}.$$

The determinant D of such a matrix is known as *Vandermonde's determinant*. It can be shown (Exercise 7.13) that

$$D = \prod_{0 \le j < i \le n} (x_i - x_j).$$

Since $D \ne 0$ when the x_i are distinct, this again shows that L_n exists and is unique.

Since a polynomial of degree $\le n$ is completely determined by its values on a set of $n+1$ points $\{x_0, x_1, \ldots, x_n\}$ its value at a point x_{n+1} distinct from the points in the given set cannot in general be equal to $f(x_{n+1})$. Hence, the minimax of f on a set of $n+2$ points will not be zero in general. We shall see that the behavior of the B.A.P. on such a set exhibits a property characteristic of B.A.P.'s on larger sets.

Theorem 7.5.1 Let E be a set of $n+2$ points x_i in $[a, b]$, where

$$-\infty < a \le x_0 < x_1 < \cdots < x_{n+1} \le b < \infty.$$

Let f be a function defined on E. There exists a unique B.A.P. π_n of degree $\le n$ for f on E. The residual $r(x) = f(x) - \pi_n(x)$ assumes the minimax value, M_n, on all points of E with successively alternating sign.

Proof. The existence of π_n is assured by Theorem 7.3.1. However, we shall give an independent proof (due to de la Vallée Poussin [7.12]) for this case, since it is constructive and demonstrates uniqueness at the same time. (Note that the L_∞-norm is not strict, so that the general uniqueness theorem (Theorem 7.3.3) does not apply.)

Let $p(x) = a_n x^n + \cdots + a_1 x + a_0$ be an arbitrary polynomial of degree $\le n$. Let $m = \max_{x \in E} |f(x) - p(x)|$ and define quantities u_i by the equations

$$f(x_i) - p(x_i) = (-1)^i u_i m, \qquad 0 \le i \le n+1.$$

Note that $|u_i| \le 1$. Assume for the moment that the u_i are known and consider the system of $n+2$ equations

$$\begin{aligned} u_0 m + a_0 + a_1 x_0 + \cdots + a_n x_0^n &= f(x_0) \\ &\vdots \\ (-1)^{n+1} u_{n+1} m + a_0 + a_1 x_{n+1} + \cdots + a_n x_{n+1}^n &= f(x_{n+1}) \end{aligned} \qquad (7.5.1)$$

as a linear system in the unknowns $m, a_0, \ldots, a_n$. Its matrix is

$$\begin{bmatrix} u_0 & 1 & x_0 & \cdots & x_0^n \\ -u_1 & 1 & x_1 & & x_1^n \\ \vdots & & & & \\ \pm u_{n+1} & 1 & x_{n+1} & & x_{n+1}^n \end{bmatrix}.$$

Let $D_0, D_1, \ldots, D_{n+1}$ be the minors of the first column. The D_i are positive Vandermonde determinants, since $x_i - x_j > 0$ for $i > j$. The determinant D of the system is given by (Exercise 2.12) $D = u_0 D_0 + u_1 D_1 + \cdots + u_{n+1} D_{n+1}$ and using Cramer's rule, we obtain for the solution m in terms of the u_i,

$$m = \frac{D_0 f(x_0) - D_1 f(x_1) + \cdots + (-1)^{n+1} D_{n+1} f(x_{n+1})}{u_0 D_0 + u_1 D_1 + \cdots + u_{n+1} D_{n+1}}.$$

The numerator is independent of $p(x)$. Suppose the numerator is zero. By taking all $u_i = 1$, we obtain $m = 0$ (since $D \neq 0$) and a unique polynomial p with $p(x_i) = f(x_i)$ for all $x_i \in E$. Hence, $M_n(E) = 0$; i.e., the B.A.P. is the (unique) interpolating polynomial on any subset of $n + 1$ points in E. If the numerator is not zero, then to find π_n it is necessary and sufficient to determine the u_i so that m is a minimum. This will yield the minimax M_n. To minimize m it is necessary and sufficient to maximize $|D|$. Since $|u_i| \leq 1$ and $D_i > 0$, we see at once that $|D| \leq \sum_0^{n+1} D_i$. Hence, $|D|$ is maximized either by taking all $u_i = 1$ or all $u_i = -1$. In either choice, we can solve system (7.5.1) uniquely, obtaining as the minimax value.

$$M_n = \frac{|D_0 f(x_0) - D_1 f(x_1) + \cdots + (-1)^{n+1} D_{n+1} f(x_{n+1})|}{D_0 + D_1 + \cdots + D_{n+1}}. \quad (7.5.2)$$

(In one choice of the u_i, the solution of (7.5.1) will actually yield $m = M_n$. The opposite choice of u_i would yield $m = -M_n$.) The choice is determined by the sign of the numerator in the expression for m above. Therefore, there will be a unique solution for the coefficients $a_0, a_1, \ldots, a_n$. Hence, this polynomial must be the unique B.A.P. for f on E.

Finally, $r(x_i) = (-1)^i u_i m = (-1)^i M_n$ for the B.A.P.

Remark To compute the B.A.P. π_n of degree $\leq n$ of f on a set E of $n + 2$ points, solve the linear system (7.5.1) with all $u_i = 1$. Use Gaussian elimination. The solutions a_j, $0 \leq j \leq n$, are the coefficients of π_n and $M_n(E) = |m|$. Thus, formula (7.5.2) would not be used in a practical computation.

The property of the residual of alternating in sign at its extreme values is characteristic of best approximations in the uniform norm. If we drop the requirement that the values be the same but retain the requirement of alternating sign, we obtain a useful theorem.

Theorem 7.5.2 Let f be defined on $E = \{x_0 < x_1 < \cdots < x_{n+1}\}$. Let q be a polynomial of degree $\leq n$ such that $r(x) = f(x) - q(x)$ alternates in sign on E. If not all $|r(x_i)|$ are equal, then

$$\min_{0 \leq i \leq n} |r(x_i)| < M_n(E) < \max_{0 \leq i \leq n} |r(x_i)|.$$

Proof. The minimax of the function $r(x)$ is equal to the minimax of $f(x)$. (If π_n^* is the B.A.P. of $r(x)$, then $\pi_n(x) = q(x) + \pi_n^*(x)$ must be the B.A.P. of $f(x)$. Hence,

$$\max |f(x) - \pi_n(x)| = \max |f(x) - q(x) - \pi_n^*(x)|.)$$

Setting $r_i = |r(x_i)|$ and using the hypothesis that $r(x_i)$ alternates in sign, we obtain from (7.5.2)

$$M_n = \frac{D_0 r_0 + D_1 r_1 + \cdots + D_{n+1} r_{n+1}}{D_0 + D_1 + \cdots + D_{n+1}}.$$

Since not all the r_i are equal, the result follows directly. ∎

Using this result, we can obtain the B.A.P. over any finite set of $N \geq n + 2$ points as follows.

Theorem 7.5.3 Let f be defined on a finite set G of $N \geq n + 2$ points. Let $E \subset G$ consist of the $n + 2$ points $x_0 < x_1 < \cdots < x_{n+1}$ and be such that $M_n(E) \geq M_n(E')$ for any subset $E' \subset G$ of $n + 2$ points. Then $M_n(E) = M_n(G)$ and the B.A.P. π_n for f on E is also a B.A.P. for f on G. Further, the B.A.P. on G is unique.

Proof. Since E is a subset of G, $\max_{x \in E} |f(x) - p(x)| \leq \max_{x \in G} |f(x) - p(x)|$ for any polynomial p of degree $\leq n$. Hence, $M_n(E) \leq M_n(G)$. Let π_n be the B.A.P. of f on E. We shall show that $r(x) = f(x) - \pi_n(x)$ satisfies the inequality

$$|r(x)| \leq M_n(E) \qquad \text{for all } x \text{ in } G. \tag{1}$$

We establish this inequality by contradiction. Suppose there is a y in G but not in E such that $|r(y)| > M_n(E)$.

CASE 1. Suppose there exists $x_i \in E$ such that $x_i < y < x_{i+1}$. By Theorem 7.5.1, $r(x_i)$ and $r(x_i + 1)$ are equal in magnitude and opposite in sign. Hence, one of them, say $r(x_i)$, has the same sign as $r(y)$. Consider the set

$$E' = \{x_0 < x_1 < \cdots < x_{i-1} < y < x_{i+1} < \cdots < x_{n+1}\}.$$

The function $r(x)$ alternates in sign on E'. By Theorem 7.5.2, $M_n(E') > \min_{x \in E'} |r(x)|$. By the assumption on y, $\min_{x \in E'} |r(x)| = M_n(E)$. Hence, $M_n(E') > M_n(E)$, contradicting the hypothesis that $M_n(E) \geq M_n(E')$.

CASE 2. Suppose $y < x_0$. If $r(y)$ and $r(x_0)$ have the same sign, let

$$E' = \{y < x_1 < \cdots < x_{n+1}\}.$$

If $r(y)$ and $r(x_0)$ have opposite sign, let $E' = \{y < x_0 < x_1 < \cdots < x_n\}$. Then apply the argument of Case 1.

CASE 3. Suppose $x_{n+1} < y$. Similar to Case 2. Thus $|r(x)| = |f(x) - \pi_n(x)| \leq M_n(E)$ for all $x \in G$. Using the inequality (1) above and the definition of the minimax $M_n(G)$ on G, we have

$$M_n(G) \leq \max_{x \in G} |f(x) - \pi_n(x)| \leq M_n(E). \tag{2}$$

Since $M_n(E) \leq M_n(G)$, it follows that $M_n(E) = M_n(G)$. Inequality (2) above then yields the equality, $\max_{x \in G} |f(x) - \pi_n(x)| = M_n(G)$. This means that π_n is a B.A.P.

on G. To show that the B.A.P. on G is unique, suppose π_n^* is such a B.A.P. Then $\max_{x \in G} |f(x) - \pi_n^*(x)| = M_n(G)$ and we have

$$M_n(E) \leq \max_{x \in E} |f(x) - \pi_n^*(x)| \leq \max_{x \in G} |f(x) - \pi_n^*(x)| = M_n(G) = M_n(E).$$

Hence, $\pi_n^*(x)$ is also a B.A.P. on E. Since the latter is unique, $\pi_n^* = \pi_n$. ∎

Remark In principle, the existence proof of the previous theorem is constructive. To compute the B.A.P. of degree $\leq n$ on any finite set of $N \geq n + 2$, we may consider all $\binom{N}{n+2}$ subsets of $n + 2$ points. For each subset E, we compute $M_n(E)$ by formula (7.5.2) and then select a set E for which M_n is a maximum. For this E, $M_n(E) = M_n(G)$ and the B.A.P. is the B.A.P. on G. Note that there may be more than one such subset E on which M_n assumes its maximum. However, the B.A.P. on each of them is the same, by the uniqueness result. Clearly, an exhaustive search procedure through all $\binom{N}{n+2}$ sets would be prohibitive for N large. We shall present a convenient algorithm for finding E after extending this result to a continuous interval $[a, b]$.

Theorem 7.5.4 Let $f \in C[a, b]$, where $-\infty < a < b < \infty$. There exists a unique B.A.P. of degree $\leq n$ for f on $[a, b]$. It coincides with the B.A.P. π_n for f on a set E of $n + 2$ points in $[a, b]$ so chosen that $M_n(E) \geq M_n(E')$ for any set E' of $n + 2$ points in $[a, b]$. Further, $M_n(E) = M_n([a, b])$.

Proof. Consider R^{n+2}. Let G be the subset of R^{n+2} consisting of all points $(x_0, x_1, \ldots, x_{n+1})$ such that $a \leq x_0 \leq x_1 \cdots \leq x_{n+1} \leq b$. Since G is closed and bounded, it is compact. A point in G such that $a \leq x_0 < x_1 < \cdots < x_{n+1} \leq b$ will be called a *pseudointerior* point of G. Each such point corresponds to a set $E \subset [a, b]$ having $n + 2$ distinct points. The function $M_n = M_n(x_0, x_1, \ldots, x_{n+1})$, given by formula (7.5.2) above for the minimax on a set $x_0 < x_1 < \cdots < x_{n+1}$, is defined and continuous at all pseudointerior points of G. We define M_n to be 0 on all boundary points of G except pseudointerior points. This extension of M_n is continuous on G. (See Exercise 7.25.) Since M_n is continuous on the compact set G, there is a point at which it attains its maximum value. Since $M_n \geq 0$ and is zero except on pseudointerior points, the maximum occurs at a pseudointerior point having coordinates $a \leq x_0 < x_1 < \cdots < x_n \leq b$. Thus, on the set $E = \{x_i\}$, $M_n(E) \geq M_n(E')$ for any other set E' of $n + 2$ distinct points in $[a, b]$. Let π_n be the B.A.P. on E. By the argument used in proving Theorem 7.5.3, we can show that $|r(x)| \leq M_n(E)$ for all $x \in [a, b]$. As before, this implies that $M_n(E) = M_n([a, b])$ and π_n is the unique B.A.P. on $[a, b]$. ∎

Corollary A necessary condition that a polynomial p of degree $\leq n$ be the B.A.P. of a continuous function f on the closed bounded interval $[a, b]$ is that the residual $r(x) = f(x) - p(x)$ assumes its extreme values $\pm \max_{a \leq x \leq b} |r(x)|$ with successively alternating sign on at least $n + 2$ points in $[a, b]$.

The next theorem shows that this necessary condition is also sufficient.

Theorem 7.5.5 Let $f \in C[a, b]$. If $q(x)$ is a polynomial of degree $\leq n$ such that the residual $r(x) = f(x) - q(x)$ assumes its extreme values $\pm \max_{a \leq x \leq b} |r(x)|$ with successively alternating sign on at least $n + 2$ points $a \leq x_0 < x_1 < \cdots < x_{n+1} \leq b$, then $q(x) = \pi_n(x)$, the B.A.P. for f on $[a, b]$.

Proof. Let $\max_{a \leq x \leq b} |f(x) - q(x)| = m$. By definition, $\pi_n(x)$ satisfies the inequality $\max_{a \leq x \leq b} |f(x) - \pi_n(x)| \leq m$. If the inequality is in fact an inequality, then $m = M_n$ and by the uniqueness of the B.A.P. $q(x) = \pi_n(x)$. If $M_n < m$, then

$$\pi_n(x_i) - q(x_i) = (f(x_i) - q(x_i)) - (f(x_i) - \pi_n(x_i)) \neq 0$$

and its sign is opposite to the sign of $\pi_n(x_{i+1}) - q(x_{i+1})$. Thus, $\pi_n(x) - q(x)$ alternates in sign on the $n + 2$ points x_i. Therefore, it has at least $n + 1$ zeros in $[a, b]$. Since $\pi_n - q$ is a polynomial of degree $\leq n$, it must be identically zero; i.e., $\pi_n = q$. But then $M_n = m$. This contradiction shows that $M_n \geq m$. Hence, $M_n = m$ and q is the unique B.A.P. ∎

As a corollary to this theorem, we obtain a result which will be useful in constructing the B.A.P.

Corollary Let $f \in C[a, b]$. Let q be a polynomial of degree $\leq n$ such that the residual $r(x) = f(x) - q(x)$ alternates in sign on a set $a \leq x_0 < x_1 < \cdots < x_{n+1} \leq b$ of $n + 2$ points. Then

$$\min_{0 \leq i \leq n+1} \{|r(x_i)|\} \leq M_n[a, b].$$

Proof. We shall show that the contrary assumption,

$$\min_{0 \leq i \leq n+1} \{|r(x_i)|\} > M_n[a, b],$$

leads to a contradiction. Indeed, it implies that $\pi_n(x) - q(x) = (f(x) - q(x)) - (f(x) - \pi_n(x))$ alternates in sign on the points x_i. By the same argument used in the proof of the theorem, this implies $q = \pi_n$, the B.A.P. for f on $[a, b]$. But then $r(x) = f(x) - \pi_n(x)$ and

$$\min_{0 \leq i \leq n+1} |r(x_i)| \leq \max_{0 \leq i \leq n+1} |r(x_i)| \leq \max_{a \leq x \leq b} |r(x)| = M_n[a, b],$$

which is a contradiction. ∎

Computation of the B.A.P. of Degree $\leq n$

The remark following Theorem 7.5.3 provides us with a method of constructing the B.A.P. for a function f on any finite set. Let us suppose that we have an efficient algorithm for doing this. Can this be used to compute the B.A.P. for f on an interval $[a, b]$; that is, if we compute the B.A.P. π_{nN} for f on a set of N points in $[a, b]$, does the sequence (π_{nN}) converge to the B.A.P. π_n for f on $[a, b]$ as $N \to \infty$?

Theorem 7.5.6 Let $f \in C[a, b]$. Let E_N, $N = 1, 2, \ldots$, be a sequence of subsets of N points in $[a, b]$ such that $E_N \subset E_{N+1}$ for every N. Let d_N be the maximum distance

7.5 Polynomial Approximation in the Uniform Norm

between adjacent points of E_N and π_{nN} the B.A.P. for f on E_N. If $d_N \to 0$ as $N \to \infty$, then

$$\max_{a \le x \le b} |\pi_{nN}(x) - \pi_n(x)| \to 0;$$

that is, π_{nN} converges uniformly to π_n in $[a, b]$.

Proof. Let M_{nN} be the minimax on E_N. Since $E_N \subset E_{N+1} \subset \cdots \subset [a, b]$, it follows that $M_{nN} \le M_{n,N+1} \le \cdots \le M_n = M_n[a, b]$. Hence, $\lim_{N \to \infty} M_{nN}$ exists. By Theorem 7.5.4, there exists a set E of $n + 2$ points in $[a, b]$ such that π_n is the B.A.P. on E as well as on $[a, b]$; i.e., $M_n(E) = M_n$. For N sufficiently large, d_N is so small that each of the subintervals formed by the N points of E_N contains at most one point of E. Hence, there is a subset $G_N \subset E_N$ consisting of $n + 2$ points $x_0, \ldots, x_{n+1}$ each of which is within distance d_N of a point of E. Clearly, $M_n(G_N) \le M_{nN} \le M_n$. Since M_n is a continuous function of $(x_0, \ldots, x_{n+1})$, $|M_n(G_N) - M_n(E)| \to 0$ as $N \to \infty$, (since $G_N \to E$). This implies that $M_{nN} \to M_n$. Hence, $\|\pi_{nN} - f\|_\infty \to M_n$; i.e., the sequence π_{nN} is a minimizing sequence in the subspace of polynomials of degree $\le n$. Since π_n is the unique closest point to f in this subspace, Lemma 7.3.4 applies and yields $\pi_{nN} \to \pi_n$ in the uniform norm. ∎

For computational purposes, we require a more efficient algorithm than the one suggested by the preceding theorem. Clearly, as the number of points N becomes large, the determination of π_{nN} requires more computation. To avoid this, Remes devised an algorithm which involves only sets of $n + 2$ points. However, it also requires the determination of extremal points, which is often difficult computationally. We present two versions of this technique.

The Single-Exchange Algorithm

Let $f \in C[a, b]$. The algorithm described below produces a sequence of polynomials $\pi^{(k)}$ of degree $\le n$ for $k = 1, 2, \ldots$, such that $\|\pi^{(k)} - \pi_n\|_\infty \to 0$ as $k \to \infty$. Here, the ∞-norm refers to uniform convergence on $[a, b]$. In practice, we would specify a desired accuracy $\varepsilon > 0$ and stop when

$$M_n \le \max_{a \le x \le b} |f(x) - \pi^{(k)}(x)| \le M_n + \varepsilon, \tag{1}$$

where $M_n = M_n[a, b]$ is the minimax for f on $[a, b]$. The algorithm for determining the $\pi^{(k)}$ is as follows.

Step 1. Let $E_k = \{a \le x_0 < x_1 < \cdots < x_{n+1} \le b\}$ be a set of $n + 2$ points in $[a, b]$.
Step 2. Determine the B.A.P. $\pi^{(k)}$ for f on E_k. This may be done by solving the system of equations (7.5.1) with all $u_i = 1$.
Step 3. Determine a point $y \in [a, b]$ which maximizes the residual $|r_k(x)|$ for x in $[a, b]$. If $x_j \le y \le x_{j+1}$ for some $j = 0, \ldots, n$, determine the signs of $r_k(x_j)$ and $r_k(x_{j+1})$. Since they are opposite, one of them agrees with the sign of $r_k(y)$. Replace the point where the signs agree by y. If $y \le x_0$ and the sign of $r_k(y)$ agrees with the sign of $r_k(x_0)$, replace x_0 by y. If the signs are opposite, adjoin y to the set E in place of x_{n+1}.

Similarly, if $x_{n+1} \le y$ and the signs of $r_k(x_{n+1})$ and $r_k(y)$ agree, replace x_{n+1} by y and otherwise replace x_0 by y. The new set with exactly one of the x_i exchanged for y is called E_{k+1}.

Step 4. Compute $|r_k(y)| - M_n(E_k) = d_k$. If $d_k > \varepsilon$, repeat the procedure with E_{k+1}. If $d_k \le \varepsilon$, $\pi^{(k)}$ is a polynomial satisfying inequality (1) above.

Note. $|r_k(x_i)| = M_n(E_k)$ and $r_k(x_i)$ alternates in sign on $x_i \in E_k$. If $|r_k(y)| = M_n(E_k)$, then $\pi^{(k)}$ satisfies the hypotheses of Theorem 7.5.5. Hence, $\pi^{(k)}$ is the B.A.P. on $[a, b]$. On the set E_{k+1}, $\pi^{(k)}$ satisfies the hypotheses of the corollary of Theorem 7.5.5. Hence,

$$M_n(E_k) = \min_{x \in E_{k+1}} |r_k(x)| \le M_n[a, b] = M_n.$$

Since M_n is the minimax of f, we also have

$$M_n \le \max_{a \le x \le b} |r_k(x)| = |r_k(y)|.$$

Therefore,

$$M_n(E_k) \le M_n \le |r_k(y)|. \tag{2}$$

If $|r_k(y)| - M_n(E_k) \le \varepsilon$, then $|r_k(y)| - M_n \le \varepsilon$, which establishes inequality (1) above.

Theorem 7.5.7 Let $f \in C[a, b]$. Let $(\pi^{(k)})$ be a sequence of polynomials obtained by the single-exchange algorithm and $r_k = f - \pi^{(k)}$. Let π_n be the B.A.P. for f on $[a, b]$ of degree $\le n$. There exist $C > 0$ and $0 < \theta < 1$ such that

$$0 \le \max_{a \le x \le b} |r_k(x)| - M_n \le C\theta^k.$$

Hence, $\|f - \pi^{(k)}\|_\infty \to \|f - \pi_n\|_\infty$ and $\|\pi^{(k)} - \pi_n\|_\infty \to 0$.

Proof. Let the points in the set E_{k+1} of the algorithm be denoted by $\{y_0 < y_1 < \cdots < y_{n+1}\}$. Since the minimax for f is also the minimax for $r_k(x)$, we have, by formula 7.5.2,

$$M_n(E_{k+1}) = \frac{\sum_{i=0}^{n+1} D_i |r_k(y_i)|}{\sum_0^{n+1} D_i} = \sum_{i=0}^{n+1} \theta_i^{(k)} |r_k(y_i)|,$$

where $0 < \theta_i^{(k)} = D_i / \sum_0^{n+1} D_i < 1$. Since $\sum_{i=0}^{n+1} \theta_i^{(k)} = 1$,

$$M_n(E_{k+1}) \ge \min_{0 \le i \le n+1} |r_k(y_i)| = M_n(E_k). \tag{3}$$

Suppose that there exists $0 < \theta < 1$ such that $\theta_i^{(k)} > 1 - \theta$ for all k. Then (with y being the exchanged point),

$$M_n(E_{k+1}) - M_n(E_k) = \sum_{i=0}^{n+1} \theta_i^{(k)} (|r_k(y_i)| - M_n(E_k)) \ge (1 - \theta)(|r_k(y)| - M_n(E_k)), \tag{4}$$

since $|r_k(y_i)| = M_n(E_k)$ unless $y_i = y$. As observed in the note above, $M_n \le |r_k(y)|$. Hence

$$M_n(E_{k+1}) - M_n(E_k) \ge (1 - \theta)(M_n - M_n(E_k)).$$

7.5 Polynomial Approximation in the Uniform Norm

Therefore,

$$M_n - M_n(E_{k+1}) = (M_n - M_n(E_k)) - (M_n(E_{k+1}) - M_n(E_k))$$
$$\leq \theta(M_n - M_n(E_k)).$$

Applying this inequality k times we obtain

$$M_n - M_n(E_{k+1}) \leq \theta^{k+1}(M_n - M_n(E_0)). \tag{5}$$

Since $M_n(E_k) \leq M_n$ for all k, (5) implies that $M_n(E_k) \to M_n$ as $k \to \infty$. Using (4), we obtain

$$|r_k(y)| - M_n \leq |r_k(y)| - M_n(E_k) \leq (1 - \theta)^{-1}(M_n(E_{k+1}) - M_n(E_k))$$
$$\leq (1 - \theta)^{-1}(M_n - M_n(E_k)) \leq (1 - \theta)^{-1}\theta^k(M_n - M_n(E_0)).$$

Since $|r_k(y)| = \max_{a \leq x \leq b} |r_k(x)|$, taking $C = (1 - \theta)^{-1}(M_n - M_n(E_0))$ we obtain the inequality of the theorem.

It remains to prove that θ exists. For this it suffices to show that there exists $\varepsilon > 0$ such that $y_{i+1} - y_i \geq \varepsilon$, $0 \leq i \leq n$, holds for all sets E_k. Since $\theta_i^{(k)} = D_i/D$, this will show that $\theta_i^{(k)}$ remains bounded away from 0. Assume the contrary. Then the set $\{E_k\}$ of $(n + 1)$-tuples $(y_0, y_1, \ldots, y_{n+1})$ has an accumulation point $E' = (y'_0, \ldots, y'_{n+1})$ with at least two y'_i equal. Since E' has at most $n + 1$ points, the B.A.P. is the interpolating polynomial $L_n(x)$ on E'. Let $r'(x) = f(x) - L_n(x)$. Choose δ such that for $|h| \leq \delta$, $|r'(x + h) - r'(x)| < \omega < M_n(E_0)$, by the uniform continuity of r'. Since E' is an accumulation point, there exists $E_k = (y_0, \ldots, y_{n+1})$ such that $\max_{0 \leq i \leq n+1} |y'_i - y_i| < \delta$. Then $|r'(y_i)| \leq |r'(y'_i)| + \omega$ for all i. Hence,

$$M_n(E_k) \leq \max_i |L_n(y_i) - f(y_i)| \leq \max_i |L_n(y'_i) - f(y'_i)| + \omega = \omega < M_n(E_0).$$

This contradicts (3) above.

Since $(\pi^{(k)})$ is a minimizing sequence and π_n is the unique closest point to f, Lemma 7.3.4 yields the uniform convergence of $\pi^{(k)}(x)$ to $\pi_n(x)$ for x in $[a, b]$. By the corollary to Lemma 7.3.1, we see that the coefficients of $\pi^{(k)}$ converge to corresponding coefficients of π_n. ∎

Second Algorithm of Remes (Multiple Exchanges)

A more complex version of the preceding algorithm is the following. Again, perform steps 1 and 2 of the single-exchange algorithm and again find a point y such that $|r_k(y)| = \max_{a \leq x \leq b} |r_k(x)|$. Now, let z_i be a zero of $r_k(x)$ in the interval (x_{i-1}, x_i), $1 \leq i \leq n + 1$. (z_i exists since $r_k(x_{i-1})$ and $r_k(x_i)$ have opposite sign.) Also define $z_0 = a$ and $z_{n+2} = b$. In each interval $[z_i, z_{i+1}]$, $0 \leq i \leq n + 1$, determine a point y_i which maximizes $(\text{sgn } r_k(x_i)) \cdot r_k(x)$ in that interval. Consider the set $\{y_0, y_1, \ldots, y_{n+1}\}$. If $\max_i |r_k(y_i)| < |r_k(y)|$, exchange a y_i for y in the same manner as in the single-exchange algorithm. This yields a set E_{k+1} of $n + 2$ points y_i such that r_k alternates in sign on E_{k+1} and such that $M_n(E_k) \leq \min_i |r_k(y_i)|$ instead of $M_n(E_k) = \min |r_k(y_i)|$ as in the single-exchange algorithm. All other inequalities (2)–(5) above hold and the proof of Theorem 7.5.7 can be used to obtain the same result for the multiple-exchange algorithm. However, for many functions f, it can be proved that the multiple-exchange method converges quadratically. (See [7.7].)

Computational Aspects

In both algorithms, steps 1 and 2 can be carried out by solving linear systems of equations. Step 3, however, requires the determination of an extremum y of $r_k(x)$.

This, in itself, is often a difficult computational problem. If f is differentiable in $[a, b]$ and we can compute the derivative $r'(x) = f'(x) - p'(x)$, then we can determine the zero's of $r'(x)$. These points and the end-points a, b are then possible extrema. By computing the values of r at these points we can determine y. Otherwise, we simply compute values $r(x)$ for a *mesh* of points x equally spaced in $[a, b]$, with the mesh spacing h chosen sufficiently small to yield a good estimate of the extrema. (See Exercise 7.26.) Thus, h will depend on f and n. In Chapter 12, we shall discuss several computational schemes for finding extrema. For a compilation of approximations to the elementary functions (e^x, log x, sin x, tan x etc.) and other important functions, see [7.10]. These approximations are obtained by the second Remes algorithm.

7.6 GENERALIZED POLYNOMIAL APPROXIMATION IN THE UNIFORM NORM, CHEBYSHEV SYSTEMS

A careful review of Section 7.5 will reveal that only two properties of the system of functions $\{1, x, \ldots, x^n\}$ were used in developing the theory of the best approximating polynomial of degree $\leq n$. First, we used the fact that any nontrivial (i.e., not all coefficients 0) polynomial of degree $\leq n$, $\sum_0^n a_i x^i = a_0 + a_1 x + \cdots + a_n x^n$, has at most n distinct real roots. It is clear that a necessary and sufficient condition for this is that the Vandermonde determinant

$$D(x_0, \ldots, x_n) = \begin{vmatrix} 1 & x_0 & x_0^2 & \cdots & x_0^n \\ \vdots & & & & \\ 1 & x_n & x_n^2 & \cdots & x_n^n \end{vmatrix}$$

be nonzero for any set of $n + 1$ distinct points. This, in turn, is equivalent to the condition that a polynomial of degree $\leq n$ is uniquely determined by specifying its values at $n + 1$ distinct points. Hence, there is an interpolating polynomial of degree $\leq n$ for any function f on a set of $n + 1$ points. This property of polynomials, or of the functions $\{1, x, \ldots, x^n\}$ can be generalized. We also used the property that the x^i are continuous functions on a set $X = [a, b] \subset R$. Putting these two together, we shall say that $n + 1$ functions $\{f_0, \ldots, f_n\}$ satisfy the *Haar condition* on a compact set X if (i) each f_i is continuous on X and (ii) the generalized Vandermonde determinant

$$D(x_0, \ldots, x_n) = \begin{vmatrix} f_0(x_0) & f_1(x_0) & \cdots & f_n(x_0) \\ \vdots & & & \\ f_0(x_n) & f_1(x_n) & \cdots & f_n(x_n) \end{vmatrix}$$

is nonzero for any set of $n + 1$ distinct points $x_i \in X$. A system of functions satisfying the Haar condition is often called a *Chebyshev system*. A linear combination $\sum_{i=0}^n a_i f_i(x)$ will be called a *generalized polynomial*. As for ordinary polynomials, it is a simple matter to verify that $\{f_0, \ldots, f_n\}$ satisfies the Haar condition if and only if any nontrivial generalized polynomial $\sum_0^n a_i f_i$ has at most n distinct zeros in X. Also,

7.6 Generalized Polynomial Approximation in the Uniform Norm, Chebyshev Systems

the Haar condition holds if and only if any generalized polynomial is determined by its values at $n + 1$ points. Thus, a unique interpolating generalized polynomial exists for any function f on $n + 1$ points.

Inspection of the proof of Theorem 7.5.1 shows that the proof is valid if the functions $\{1, x, \ldots, x^n\}$ are replaced by any Chebyshev system and the polynomials $p(x)$ and $\pi_n(x)$ are replaced by generalized polynomials. The only property that needs verification is the following: for all vectors $(x_0, x_1, \ldots, x_n)$ such that $x_0 < x_1 < \cdots < x_n$, $D(x_0, \ldots, x_n)$ has the same sign. (It does not matter whether this is positive or not.) If this can be established for the generalized Vandermonde determinant, then the proof of Theorem 7.5.1 is valid, mutatis mutandis, for any Chebyshev system. (We leave this to the reader as an exercise.) To verify the property of the generalized Vandermonde for point sets $x_0 < x_1 < \cdots < x_n$, we suppose that $D(x_0, \ldots, x_n) < 0 < D(y_0, \ldots, y_n)$, where $y_0 < y_1 < \cdots < y_n$. By the continuity of D (since the f_i are continuous), there exists $0 < \alpha < 1$ such that

$$D(\alpha x_0 + (1 - \alpha)y_0, \alpha x_1 + (1 - \alpha)y_1, \ldots, \alpha x_n + (1 - \alpha)y_n) = 0.$$

The Haar condition implies that this can only happen if

$$\alpha x_i + (1 - \alpha)y_i = \alpha x_j + (1 - \alpha)y_j$$

for some $i < j$, whence $\alpha(x_i - x_j) = -(1 - \alpha)(y_i - y_j)$. Since $x_i < x_j$ and $y_i < y_j$, this is a contradiction. Thus, all the generalized determinants $D_0, \ldots, D_{n+1}$ in the proof of Theorem 7.5.1 have the same sign and so formula 7.5.2 holds. The remainder of the proof carries through. In similar fashion, one verifies that Theorem 7.5.2 generalizes. Theorems 7.5.3 and 7.5.4 use only inequalities depending on the notion of best approximation rather than on any properties of polynomials other than their continuity. Hence, these results also apply to generalized polynomials over arbitrary Chebyshev systems. Finally, we observe that the exchange algorithm and Theorem 7.5.7 are likewise valid for arbitrary Chebyshev systems.

It can be shown that the set of trigonometric functions

$$\{1, \cos \theta, \cos 2\theta, \ldots, \cos n\theta, \sin \theta, \ldots, \sin n\theta\}$$

is a Chebyshev system on the interval $[0, 2\pi)$ or $[-\pi, \pi)$. (See Exercise 7.14.) Therefore, for a function f continuous on $[-\pi, \pi - \varepsilon]$ there is a *best trigonometric sum*, $\sum_{k=0}^{n} (a_k^* \cos k\theta + b_k^* \sin k\theta)$, of order $\leq n$ which is closer to f in the uniform norm than any other such trigonometric sum. For even functions, we obtain trigonometric sums of the form $\sum_0^n a_k \cos k\theta$. These are polynomials in x, where $x = \cos \theta$. This follows from the recursion,

$$\cos k\theta + \cos (k - 2)\theta = 2 \cos (k - 1)\theta \cdot \cos \theta, k \geq 2,$$

by induction on k.

Definition 7.6.1 Let $x = \cos \theta$ so that $\theta = \arccos x$. The polynomial in x of degree n

$$T_n(x) = \cos (n \arccos x)$$

is called the *Chebyshev polynomial of the first kind* of degree n.

Approximation Theory

The Chebyshev polynomials have certain interesting properties with respect to uniform approximation.

Theorem 7.6.1 Of all nth degree monic polynomials (i.e., leading coefficient equal to 1), the polynomial $2^{1-n} T_n(x)$ has the smallest maximum absolute value 2^{1-n} in the interval $[-1, 1]$.

Proof. Since $\cos n\theta = \frac{1}{2}(e^{-in\theta} + e^{in\theta})$,

$$T_n(x) = \frac{1}{2}[(x + (x^2 - 1)^{1/2})^n + (x - (x^2 - 1)^{1/2})^n].$$

(We have used $e^{i\theta} = \cos \theta + i \sin \theta = x + \sqrt{x^2 - 1}$, where $x = \cos \theta$.) Expanding the binomials, we obtain for the coefficient of x^n

$$\sum_{r=0}^{[n/2]} \binom{n}{2r} = \sum_{r=0}^{n-1} \binom{n-1}{r} = 2^{n-1}.$$

Hence, $2^{1-n} T_n(x)$ has leading coefficient equal to 1.

Since $\cos n\theta$ oscillates between -1 and 1, $2^{1-n} T_n(x)$ oscillates between $\pm 2^{1-n}$ for x in $[-1, 1]$. Further, $\cos n\theta_k = \pm 1$ for $\theta_k = \pi k/n$, $k = 0, 1, \ldots, n$, so that $2^{1-n} T_n(x_k) = \pm 2^{1-n}$ for $x_k = \cos \theta_k$. Therefore, $2^{1-n} T_n(x)$ assumes its extreme value with successively alternating sign on $n + 1$ points in $[-1, 1]$. Writing

$$2^{1-n} T_n(x) = x^n - t_{n-1} x^{n-1} - \cdots - t_1 x - t_0 = x^n - q(x),$$

we may regard $2^{1-n} T_n(x)$ as a residual for the function x^n. Since $q(x)$ is of degree $\leq n - 1$, it follows from Theorem 7.5.5 that q is the B.A.P. of degree $\leq n - 1$ for x^n on $[-1, 1]$. Therefore

$$\max_{-1 \leq x \leq 1} |2^{1-n} T_n(x)| \leq \max_{-1 \leq x \leq 1} |x^n - p(x)|$$

for any polynomial p of degree $\leq n$, which proves the theorem. ∎

Corollary Of all nth-degree polynomials p such that $\max_{-1 \leq x \leq 1} |p(x)| \leq 1$, $T_n(x)$ has the largest leading coefficient, 2^{n-1}.

Proof. Let $p(x) = a_n x^n + \cdots + a_1 x + a_0$, $a_n \neq 0$. For $-1 \leq x \leq 1$,

$$\max_{-1 \leq x \leq 1} |p(x)| = |a_n| \cdot \max \left| x^n + \frac{a_{n-1}}{a_n} x^{n-1} + \cdots + \frac{a_0}{a_n} \right| = \mu |a_n| \leq 1.$$

By Theorem 7.6.1, $\mu \geq 2^{1-n}$. Hence, $|a_n| \leq 2^{n-1}$. ∎

The Chebyshev polynomials $\{T_n(x), n = 0, 1, \ldots,\}$ are also useful in obtaining a polynomial approximation to a function f in the form of a sum $\sum_{k=0}^{n} A_k T_k(x)$. The determination of the *Chebyshev coefficients* A_k leads to a problem in least-squares approximation, as we shall show in the next section. If we have a polynomial approximation in the form $\sum_{k=0}^{n} A_k T_k(x)$, then $\max_{a \leq x \leq b} |f(x) - \sum_{0}^{n} A_k T_k(x)|$ is an upper bound for the minimax $M_n[a, b]$. In fact, it is quite a good upper bound, as we shall see (Theorem 7.7.8). This can be of practical value in determining the degree n required to achieve a polynomial approximation to f with a specified accuracy. For example, suppose it is required to obtain a polynomial which approximates e^x on $[a, b]$ with an accuracy of 10^{-8} over the entire interval. Such a polynomial can be used as the basis of a finite computer algorithm to compute values of e^x with the prescribed accuracy for x in $[a, b]$. By determining the Chebyshev coefficients $A_0, A_1, \ldots$, we can obtain estimates of $M_n[a, b]$ for $n = 0, 1, 2, \ldots$. These estimates can be used to estimate the degree n of the B.A.P. required to approximate f with an accuracy of 10^{-8}. Then using an algorithm such as the single-exchange algorithm, we can compute a polynomial $\pi_n^{(k)}$ which is within 10^{-9} of π_n. This polynomial should approximate e^x with an accuracy of 10^{-8}. If it yields even better accuracy, we can then compute the B.A.P. of degree $\leq n - 1$ and see if it suffices. In this way, we can find a polynomial of nearly minimal degree which can be used to approximate e^x with the desired accuracy. Such a polynomial will provide a more efficient algorithm (i.e., fewer computations) than a higher-degree polynomial and yet will yield the desired accuracy. This is important if the algorithm is to be used frequently, as in a *computer subroutine*.

7.7 LEAST-SQUARES APPROXIMATION OF FUNCTIONS OF A REAL VARIABLE; ORTHOGONAL FAMILIES OF POLYNOMIALS

We shall now apply the results of Sections 7.4 and 3.3 on abstract pre-Hilbert spaces to some specific spaces of functions which arise from different definitions of the inner product. We have seen that for any positive *weight* function $\rho(x)$ integrable on $[a, b]$ (i.e., $\rho(x) > 0$ and $\int_a^b \rho(x)\, dx < \infty$), the set of functions f such that

$$\int_a^b |f(x)|^2 \rho(x)\, dx < \infty$$

is a Hilbert space $L_2^{(\rho)}[a, b]$ (Exercise 3.32). For any $f, g \in L_2^{(\rho)}[a, b]$, the inner product is given by

$$\langle f, g \rangle = \int_a^b f(x) g(x) \rho(x)\, dx. \tag{7.7.1}$$

(Recall that results remain valid if ρ has a finite number of zero's.) Choosing different weight functions over different intervals, we obtain different spaces and different orthonormal families of functions.

Example (a) Let $\rho(\theta) = 1$ for all θ in $[0, 2\pi]$. $L^2[0, 2\pi]$ is the Hilbert space of all Lebesgue square-integrable functions on $[0, 2\pi]$. The familiar relations

$$\int_0^{2\pi} \cos k\theta \cos j\theta\, d\theta = \pi \delta_{jk}, \qquad j, k \neq 0,$$

$$\int_0^{2\pi} \sin k\theta \sin j\theta\, d\theta = \pi \delta_{jk},$$

$$\int_0^{2\pi} \cos k\theta \sin j\theta\, d\theta = 0,$$

show that

$$\left\{\frac{1}{\sqrt{2\pi}}, \frac{1}{\sqrt{\pi}}\cos\theta, \frac{1}{\sqrt{\pi}}\sin\theta, \frac{1}{\sqrt{\pi}}\cos 2\theta, \frac{1}{\sqrt{\pi}}\sin 2\theta, \ldots,\right\}$$

is an orthonormal set in $L^2[0, 2\pi]$. It is complete (by Exercise 3.33).

Example (b) Let $\rho(x) = (2/\pi)(1 - x^2)^{-1/2}$ on the interval $[-1, 1]$. Then $L_2^{(\rho)}$ is a Hilbert space and the *Chebyshev polynomials* $\{(1/\sqrt{2})T_0(x), T_1(x), T_2(x), \ldots,\}$ form a complete orthonormal set. To show this, we simply make the substitution $x = \cos\theta$, obtaining $\langle f, g \rangle = \int_{-1}^{1} f(x)g(x)(2/\pi)(1 - x^2)^{-1/2}\,dx = (2/\pi)\int_0^\pi f(\cos\theta)g(\cos\theta)\,d\theta$, and

$$\langle T_k, T_j \rangle = \int_{-1}^{1} T_k(x)T_j(x)\cdot\frac{2}{\pi}(1 - x^2)^{-1/2}\,dx = \frac{1}{\pi}\int_{-\pi}^{\pi}\cos k\theta \cos j\theta\,d\theta.$$

Example (c) Let $\rho(x) = x^p(1 - x)^q$ on the interval $[0, 1]$, $p > -1$, $q > -1$. A family of polynomials in x which are orthogonal with respect to this weight function are the *Jacobi* polynomials, $J_n^{(p,q)}$. (See Section 7.8.) For $p = q$, they are called *ultraspherical* polynomials. For $p = q = 0$ and the interval $[-1, 1]$, they are the *Legendre* polynomials. For $p = q = -\frac{1}{2}$, $\rho(x) = [x(1 - x)]^{-1/2}$. Making the transformation, $y = 2x - 1$, we obtain the weight function $(1 - y^2)^{-1/2}$ on the interval $[-1, 1]$, which is recognized as the Chebyshev polynomial case.

From Section 7.4 we see at once that if $\{q_0(x), \ldots, q_n(x)\}$ is an orthonormal set of functions in $L_2^{(\rho)}$, then the least-squares approximation $\sum_0^n a_i q_i(x)$ to any $f \in L_2^{(\rho)}$ is obtained by taking

$$a_i = \langle f, q_i \rangle = \int f(x) q_i(x) \rho(x)\,dx.$$

These a_i are called the *generalized Fourier coefficients* of f with respect to the q_i. In Example (a) above, we obtain the ordinary Fourier coefficients. The *Fourier sum* $\sum_0^n (A_k \cos k\theta + B_k \sin k\theta)$ of order n of the function f is the trigonometric sum in which

$$A_k = \frac{1}{\sqrt{\pi}}\int_0^{2\pi} f(\theta)\cos k\theta\,d\theta,$$

$$B_k = \frac{1}{\sqrt{\pi}}\int_0^{2\pi} f(\theta)\sin k\theta\,d\theta, \quad k \geq 1,$$

$$A_0 = \frac{1}{\sqrt{2\pi}}\int_0^{2\pi} f(\theta)\,d\theta.$$

The Fourier sum is the least-squares trigonometric sum for f, i.e., it minimizes

$$\int_0^{2\pi}\left|f(x) - \sum_0^n (a_k \cos k\theta + b_k \sin k\theta)\right|^2 dx,$$

the *mean-square error*.

Note that to obtain the next higher-order least-squares trigonometric sum, we need only calculate A_{n+1} and B_{n+1}. This is true of any least-squares approximation by orthonormal functions.

Since the Chebyshev polynomials form an orthonormal set on $[-1, 1]$, with respect to the weight function $\rho = (2/\pi)(1 - x^2)^{-1/2}$, we can obtain the least-squares polynomial approximation to any function $f \in L_2^{(\rho)}[-1, 1]$ by computing its *Fourier–Chebyshev coefficients*,

$$A_k = \frac{2}{\pi} \int_{-1}^{1} f(x) T_k(x) \cdot (1 - x^2)^{-1/2} \, dx = \langle f, T_k \rangle.$$

The nth-degree polynomial $\sum_{k=0}^{n} A_k T_k(x)$ minimizes the *weighted mean-square error*

$$\|f - p\|_2^2 = \frac{2}{\pi} \int_{-1}^{1} |f(x) - p(x)|^2 (1 - x^2)^{-1/2} \, dx$$

over all polynomials p of degree $\leq n$. This follows from the fact that the set of polynomials $\{T_k(x)\}$ is a linearly independent set (being orthonormal) of polynomials. Also, each $T_k(x)$ is of degree k (Theorem 7.6.1, Corollary). Hence, any polynomial p of degree $\leq n$ can be expressed uniquely as a linear combination of $T_0, T_1, \ldots, T_n$. (See Exercise 7.15.) Writing $p = \sum_{k=0}^{n} a_k T_k$, we can apply the general theory of Section 7.4 to prove that the weighted least-squares polynomial is given by taking $a_k = A_k = \langle f, T_k \rangle$. (See Eq. (7.4.2).) We call $\sum_{k=0}^{n} A_k T_k(x)$ a *Fourier–Chebyshev expansion* of f.

7.7.1 COMPUTATION OF ORTHOGONAL POLYNOMIALS

We have given examples of orthogonal families of polynomials, e.g., the Chebyshev, Legendre and more generally, the Jacobi polynomials. A general method can be given for constructing an orthogonal family of polynomials with respect to any positive weight function on a finite interval $[a, b]$ or on any finite set of points. (In the case of a finite set, the family will also be finite.) One can construct such a family on an interval $[a, b]$ by applying the general Gram–Schmidt procedure to the set $\{1, x, x^2, \ldots,\}$ and using Eq. (7.7.1) as the definition of the inner product. The Gram–Schmidt process makes no use of the algebraic properties of polynomials. Furthermore, it is very sensitive to rounding errors. (See below.) A simpler scheme which makes use of the algebraic polynomial structure is the following:

Suppose $\{Q_0, Q_1, \ldots, Q_{n-1}\}$ is an orthonormal set of polynomials such that each Q_i is of degree i. Let $\bar{Q}_n$ be a polynomial of degree n which is orthogonal to all Q_i. Consider $\bar{Q}_n(x) - \alpha x Q_{n-1}(x)$. For a suitable choice of $\alpha \neq 0$, this is a polynomial of degree $\leq n - 1$. Hence, by Exercise 7.15,

$$\bar{Q}_n - \alpha x Q_{n-1} = \sum_{i=0}^{n-1} \alpha_i Q_i.$$

If $\langle \bar{Q}_n, Q_i \rangle = 0$ for all $0 \le i \le n - 1$, we must have

$$0 = \langle \bar{Q}_n, Q_{n-1} \rangle = \alpha \langle xQ_{n-1}, Q_{n-1} \rangle + \alpha_{n-1},$$

$$0 = \langle \bar{Q}_n, Q_{n-2} \rangle = \alpha \langle xQ_{n-1}, Q_{n-2} \rangle + \alpha_{n-2}.$$

Now, we may set $\alpha = 1$, as this amounts to multiplying $\bar{Q}_n$ by a constant and does not change the orthogonality relations. Thus, α_{n-1} and α_{n-2} are determined by the above equations. Applying a similar argument to the Q_i for $i < n - 2$, we would find that $\alpha_i = 0$ for $i < n - 2$. This suggests the following recursive formula for computing $\bar{Q}_n$:

$$\bar{Q}_n(x) = (x + a_n)Q_{n-1}(x) + b_n Q_{n-2}(x), \quad n \ge 2. \tag{7.7.2}$$

Then Q_n is $(1/\|\bar{Q}_n\|)\bar{Q}_n$ and

$$a_n = -\langle xQ_{n-1}, Q_{n-1} \rangle, \tag{7.7.3}$$

$$b_n = -\langle xQ_{n-1}, Q_{n-2} \rangle. \tag{7.7.4}$$

These values of a_n and b_n make $\bar{Q}_n$ orthogonal to Q_{n-1} and Q_{n-2}. We shall show that $\langle \bar{Q}_n, Q_i \rangle = 0$ for $0 \le i < n - 2$ as well. From (7.7.2) we have

$$xQ_i = \bar{Q}_{i+1} - a_{i+1}Q_i - b_{i+1}Q_{i-1}.$$

For $i < n - 2$, (7.7.2) yields

$$\langle \bar{Q}_n, Q_i \rangle = \langle xQ_{n-1}, Q_i \rangle = \langle Q_{n-1}, xQ_i \rangle$$
$$= \langle Q_{n-1}, \bar{Q}_{i+1} \rangle - a_{i+1} \langle Q_{n-1}, Q_i \rangle - b_{i+1} \langle Q_{n-1}, Q_{i-1} \rangle = 0,$$

establishing the required orthogonality. This also shows that $\bar{Q}_n$ is uniquely determined by (7.7.2). For suppose $\langle \bar{Q}_n, Q_i \rangle = 0$, $i \le n - 1$, and all Q_i, $i \le n - 1$, are given by (7.7.2). Then $\bar{Q}_n = xQ_{n-1} + \sum_0^{n-1} \alpha_i Q_i$ and as we have seen, α_{n-1} and α_{n-2} must satisfy (7.7.3) and (7.7.4) respectively. Further, for $i < n - 2$,

$$0 = \langle \bar{Q}_n, Q_i \rangle = \langle xQ_{n-1}, Q_i \rangle + \alpha_i$$
$$= \langle Q_{n-1}, \bar{Q}_{i+1} \rangle - a_{i+1} \langle Q_{n-1}, Q_i \rangle - b_{i+1} \langle Q_{n-1}, Q_{i-1} \rangle + \alpha_i = \alpha_i.$$

Thus, (7.7.2) provides a recursive scheme for computing the unique orthonormal sequence of polynomials in $L_2^{(\rho)}[a, b]$. To start the recursion we set $Q_0 = b_0$ where b_0 is a constant such that $\|Q_0\| = 1$, and set $\bar{Q}_1 = (x + a_1)Q_0$. Then $\langle \bar{Q}_1, Q_0 \rangle = \langle xQ_0, Q_0 \rangle + a_1 = 0$ defines $a_1 = -\langle xQ_0, Q_0 \rangle$.

Let $T_n(x) = \cos(n \arccos x)$. Using the orthogonality and (7.7.2)–(7.7.4), we would find that

$$T_n(x) = 2xT_{n-1}(x) - T_{n-2}(x). \tag{7.7.5}$$

However, if we observe that $\cos(n + 1)\theta = 2 \cos \theta \cos n\theta - \cos(n - 1)\theta$, then letting $x = \cos \theta$, (7.7.5) is immediate.

7.7.1 Computation of Orthogonal Polynomials

Least-Squares Polynomials on Finite Sets

For a finite set $\{x_1, \ldots, x_N\}$ with $N \geq n + 1$, we may determine an orthonormal set $\{Q_0, \ldots, Q_n\}$ using as the definition of inner product $\langle f, g \rangle = \sum_{i=1}^{N} f(x_i)g(x_i)\rho(x_i)$. Thus, if we are given the problem of determining the least-squares polynomial $p(x) = a_0 + a_1 x + \cdots + a_n x^n$ of degree $\leq n$ which minimizes

$$\|f - p\|_2^2 = \sum_{i=1}^{N} |f(x_i) - p(x_i)|^2 \rho(x_i),$$

we can either determine the a_i directly by solving the normal equations (7.4.1) or we can determine the orthonormal family $\{Q_0, \ldots, Q_n\}$ by formula (7.7.2) and compute $p = \sum_{i=0}^{n} A_k Q_k$, where

$$A_k = \sum_{i=1}^{N} f(x_i) Q_k(x_i) \rho(x_i).$$

Again, the advantage of the second method is that A_k is independent of $A_0, \ldots, A_{k-1}$. If we wish to make $\|f - p\|_2 < \varepsilon$, where $\varepsilon > 0$ is some prescribed accuracy, and we do not know the degree n necessary to achieve this accuracy, then we can successively determine polynomials $p_k = \sum_{i=0}^{k} A_i Q_i(x)$, $k = 0, 1, 2, \ldots$, and compute $\|f - p\|_2$ until the accuracy is achieved. Each step requires only the computation of another generalized Fourier coefficient A_k and the formation of $P_k = P_{k-1} + A_k Q_k$. If the normal equations (7.4.1) are used to determine $p_k = a_{0k} + a_{1k} + \cdots + a_{kk} x^k$ directly, the entire computation must be repeated for each k, since the coefficients a_{ik} all differ, in general, from the $a_{i,k-1}$.

If the finite set of N points $\{x_i\}$ is such that the x_i are equally spaced, we may (by a translation) without loss of generality assume that $x_i = \pm i/m$, $i = 0, \ldots, m$. (We shall assume that $N = 2m + 1$ is odd. In practice, this can usually be arranged.) The orthogonal polynomials on such a set are called *Gram polynomials* (sometimes also Chebyshev polynomials, but not to be confused with the $T_n(x)$, the Chebyshev polynomials of the first kind).

In the case of uniform polynomial approximation of $f \in C[a, b]$, we saw (Theorem 7.5.6) that the sequence π_{nN} of B.A.P.'s on finite sets of N points converges uniformly to the B.A.P. π_n. It is of some theoretical and practical interest to establish an analogous result for least-squares polynomial approximation. Thus, if we wish to compute the least-squares polynomial, $p^*(x)$, of degree n for f on $[a, b]$ with weight function $\rho(x)$, we could proceed by discretizing the entire problem immediately; i.e., we choose a set E_N of N points in $[a, b]$, determine an orthonormal family of polynomials $Q_i^{(N)}$ on E_N using ρ as the weight function, and compute

$$p^{(N)}(x) = \sum_{0}^{n} A_i^{(N)} Q_i^{(N)}(x),$$

where

$$A_i^{(N)} = \langle f, Q_i^{(N)} \rangle = \sum_{j=1}^{N} f(x_j) Q_i^{(N)}(x_j) \rho(x_j).$$

The $p^{(N)}(x) \to p^*(x)$ as $N \to \infty$. This discretization is to be compared with the alternative of first determining an orthonormal family of polynomials Q_i on $[a, b]$ and then determining $p^*(x) = \sum_0^n A_i Q_i(x)$ by evaluating

$$A_i = \langle f, Q_i \rangle = \int_a^b f(x) Q_i(x) \rho(x)\, dx$$

by one of the quadrature formulas of Chapter 9; that is,

$$A_i^{(K)} = \sum_{j=0}^K a_j f(x_j) Q_i(x_j) \rho(x_j).$$

The choice of method must depend on an analysis of the two rates of convergence of $A_i^{(K)}$ to A_i on the one hand and on $p^{(N)}$ to p^* on the other. The behavior of the $A_i^{(K)}$ can be analyzed with the help of the formulas for the error in numerical quadrature given in Chapter 9. For the convergence of $p^{(N)}$, where the sets E_N are arbitrary and ρ is an arbitrary positive weight function, the situation is more complicated. We limit our attention to sets of equally spaced points, although the following result remains valid for arbitrary E_N provided that the maximum distance between adjacent points approaches zero. (See [7.11].)

Theorem 7.7.0 Let $f \in C[0, 1]$. Let $E_N = \{i/N : i = 0, \ldots, N\}$ be a sequence of sets of equally spaced points. Let $\{Q_i^{(N)} : i = 0, \ldots, n\}$ be the orthonormal family of polynomials on E_N with respect to the weight function $(1/N)\rho(x)$ and let $\{Q_i : i = 0, \ldots, n\}$ be the orthonormal family of polynomials (of degrees $i = 0, \ldots, n$) on $[a, b]$ with respect to the nonnegative weight function $\rho(x)$. Then $Q_i^{(N)}$ converges uniformly to Q_i on $[a, b]$ as $N \to \infty$ and a subsequence of the sequence of least-squares polynomials $p^{(N)}$ on E_N of degree $\leq n$ converges uniformly to the least-squares polynomial p^* on $[a, b]$.

Proof. If we replace the weight function $\rho(x)$ by $(1/N)\rho(x)$ on E_N, we obtain the same least-squares polynomial $p^{(N)}$. For the norm induced by the inner product

$$\langle f, g \rangle = \sum_0^N (1/N)\rho(x_i) f(x_i) g(x_i),$$

we have

$$\|f - p^{(N)}\|^2 \leq \|f - p^*\|^2,$$

since $p^{(N)}$ is the least-squares polynomial for this norm. By the definition of the Riemann integral (see Chapter 9), for any $\varepsilon > 0$ there exists N_0 such that for all $N > N_0$,

$$\|f - p^*\|^2 = \sum_{i=0}^N (1/N)\rho(x_i)(f(x_i) - p^*(x_i))^2 \tag{1}$$

$$\leq \varepsilon + \int_0^1 \rho(x)(f(x) - p^*(x))^2\, dx = K_\varepsilon.$$

Letting $Q_i^{(N)}$ be the orthonormal polynomials on E_N with weight function $(1/N)\rho(x)$, we may write $p^{(N)} = \sum_0^n A_i^{(N)} Q_i^{(N)}$ and it follows that

$$\|f - p^{(N)}\|^2 = \|f\|^2 - \sum_0^n (A_i^{(N)})^2 \leq \|f - p^*\|^2 \leq K_\varepsilon, \tag{2}$$

where K_ε is the constant in (1). Hence, the sequence of vectors $A^{(N)} = (A_0^{(N)}, \ldots, A_n^{(N)})$ is bounded and contains a convergent subsequence. We reindex and call the convergent subsequence $A^{(N)}$ again. Suppose that we have shown that $\|Q_i^{(N)} - Q_i\|_\infty \to 0$. It follows easily that $p^{(N)}$ is a uniformly convergent sequence of polynomials, since

$$\|p^{(r)} - p^{(s)}\|_\infty \leq \sum_{i=0}^n |A_i^{(r)} - A_i^{(s)}| \|Q_i^{(r)}\|_\infty + \sum_{i=0}^n |A_i^{(s)}| \|Q_i^{(r)} - Q_i^{(s)}\|_\infty.$$

Let $\lim p^{(N)} = p$. For the polynomial p,

$$\int_0^1 \rho(x)(f(x) - p(x))^2 \, dx = \lim_{N \to \infty} \sum_{i=0}^N (1/N)\rho(x_i)(f(x_i) - p(x_i))^2$$

$$= \lim_{N \to \infty} \sum_{i=0}^N (1/N)\rho(x_i)(f(x_i) - p^{(N)}(x_i))^2$$

$$= \lim_{N \to \infty} \|f - p^{(N)}\|^2 < K_\varepsilon,$$

by inequality (2) above. Allowing $\varepsilon \to 0$, we obtain

$$\int_0^1 \rho(x)(f(x) - p(x))^2 \, dx \leq \int_0^1 \rho(x)(f(x) - p^*(x))^2 \, dx.$$

This implies that $p = p^*$, since the least-squares polynomial is unique.

To complete the proof, we must establish that $Q_i^{(N)}$ converges uniformly to Q_i. This is proved by induction on i, using the recurrence relation (7.7.2) and (7.7.3), (7.7.4). The result is clearly valid for $i = 0$, since $Q_0^{(N)} = b_0$ for all N. Now, $Q_1^{(N)} = (x + a_1^{(N)})Q_0^{(N)}$, where

$$a_1^{(N)} = -\sum_{i=0}^N (1/N)x_i Q_0^2 \rho(x_i)$$

by (7.7.3). We see that $\lim_{N \to \infty} a_1^{(N)} = a_1$. Hence, $Q_1^{(N)} \to Q_1$. Now, assuming that $Q_i^{(N)} \to Q_i$ for $i \leq n - 1$, we again use (7.7.3) to obtain

$$a_n^{(N)} = -\sum_{i=0}^N (1/N)x_i (Q_{n-1}^{(N)}(x_i))^2 \rho(x_i)$$

and the latter sum converges to

$$a_n = -\int_0^1 x(Q_{n-1}(x))^2 \rho(x) \, dx.$$

Similarly, we prove that $b_n^{(N)} \to b_n$ as $N \to \infty$, using (7.7.4) and the induction hypothesis. It then follows from (7.7.2) that $Q_n^{(N)} \to Q_n$ uniformly, which completes the proof. ∎

Computational Aspects

From Exercise 7.21c, d, it will be clear that the computation of the least-squares polynomial on a finite set by means of orthonormal polynomials requires less work than the method of solving the normal equations. In practice, we must also consider the effect of rounding errors. As the degree of the polynomial becomes large, the normal matrix becomes more ill-conditioned (Exercise 7.19b). In the orthonormal polynomial method, the effect of rounding errors is evidenced as a deterioration of the orthogonality relations. This is sometimes referred to as numerical instability. It may be partly corrected by additional computations as follows. Suppose that $\{f_1, f_2, \ldots, f_{n-1}\}$ is an orthonormal set of n-dimensional vectors. Let $\bar{f}_n$ be a computed approximation to the exact orthonormal vector f_n. Then $\langle f_n, f_i \rangle = 0$ for $1 \le i \le n - 1$, but $\langle \bar{f}_n, f_i \rangle \ne 0$ in general. Since $f_n - \bar{f}_n$ is an n-dimensional vector, we have

$$f_n - \bar{f}_n = \sum_{i=1}^{n} \langle f_n - \bar{f}_n, f_i \rangle f_i$$

$$= \langle f_n - \bar{f}_n, f_n \rangle f_n - \sum_{i=1}^{n-1} \langle \bar{f}_n, f_i \rangle f_i.$$

Since normalization produces a small rounding error, we may assume $\langle \bar{f}_n, \bar{f}_n \rangle = 1$. Letting $h = f_n - \bar{f}_n$, we see that $1 = \langle f_n - h, f_n - h \rangle = 1 - 2\langle f_n, h \rangle + \|h\|^2$, or $\langle f_n - \bar{f}_n, f_n \rangle = \|h\|^2/2$. On the other hand $\langle \bar{f}_n, f_i \rangle = \langle f_n - h, f_i \rangle = -\langle h, f_i \rangle$, so that some of these quantities should be of order $\|h\|$ and we may neglect the term of order $\|h\|^2$. Therefore, an improved approximation to f_n should be given by the corrected vector $f_n^*/\|f_n^*\|$, where

$$f_n^* = \bar{f}_n - \sum_{i=1}^{n-1} \langle \bar{f}_n, f_i \rangle f_i.$$

This correction should be made for each computation of a vector in the orthonormalization procedure. Thus, suppose $\{v_1, \ldots, v_n\}$ is a linearly independent set to be orthonormalized. Take $f_1 = v_1/\|v_1\|$. Next, compute $\bar{f}_2$ by the Gram-Schmidt procedure; i.e.

$$g_2 = v_2 - \langle v_2, f_1 \rangle f_1$$

and $\bar{f}_2 = g_2/\|g_2\|$. Then "reorthogonalize" by computing

$$f_2^* = \bar{f}_2 - \langle \bar{f}_2, f_1 \rangle f_1$$

and take $f_2 = f_2^*/\|f_2^*\|$. Continue in this way with Gram-Schmidt applied to $v_3, \ldots, v_n$ followed by reorthonalization at each step. For obvious reasons, this process is sometimes called *reorthogonalization*. An alternative procedure which appears to produce good results without reorthogonalization is obtained by a rearrangement or modification of the Gram-Schmidt process. It is referred to as the *modified Gram-Schmidt algorithm* [4.19] and goes as follows.

Again, let $\{v_1, \ldots, v_n\}$ be a linearly independent set. Take $f_1 = v_1/\|v_1\|$. Then subtract from each vector v_j its component along f_1, calling the result $v_j^{(1)}$. Thus,

$$v_j^{(1)} = v_j - \langle f_1, v_j \rangle f_1, \quad 2 \le j \le n.$$

Clearly, $\langle f_1, v_j^{(1)} \rangle = 0$. Next, take $f_2 = v_2^{(1)}/\|v_2^{(1)}\|$ and compute $v_j^{(2)} = v_j^{(1)} - \langle f_2, v_j^{(1)} \rangle f_2$, $3 \le j \le n$. The vectors $v_j^{(2)}$ are orthogonal to both f_1 and f_2. The procedure is continued until we reach $f_n = v_n^{(n-1)}/\|v_n^{(n-1)}\|$. Rounding errors for this algorithm are analyzed in [7.32] and experimental results given in [7.33].

7.7.1 Computation of Orthogonal Polynomials

Recall (Section 7.4) that if $N < n + 1$, then the least-squares polynomial of degree $\leq n$ need not be unique. This follows at once from the normal equations and the fact that $\{1, x, \ldots, x^n\}$ is not a linearly independent set of functions on $\{x_1, \ldots, x_N\}$. It is also a consequence of the fact that the orthonormal family given by (7.7.2) is not unique. To see this note that a polynomial of degree $n > N - 1$ can be zero on all x_i, $1 \leq i \leq N$ without being the zero polynomial. Hence, such a polynomial is orthogonal to all polynomials with respect to the inner product $\sum_{i=1}^{N} f(x_i)g(x_i)\rho(x_i)$ on the set $E = \{x_1, \ldots, x_N\}$. In fact, the functions f with domain restricted to the finite set E can be viewed as N-dimensional vectors $(f(x_1), \ldots, f(x_N))$ and there are precisely N linearly independent ones, including those obtained as the restriction of a polynomial to the set E. Hence, an orthonormal set can have at most N elements. For further remarks on the finite case, see Exercise 7.21.

The question of convergence in $L_2^{(\rho)}[a, b]$ can be referred to the results in Section 3.3. $L_2^{(\rho)}$ is a separable Hilbert space. Hence, the generalized Fourier expansion

$$\sum_0^\infty A_k \varphi_k(x), \qquad A_k = \int_a^b f(x)\varphi_k(x)\rho(x)\,dx,$$

converges in the mean to $f \in L_2^{(\rho)}$ if and only if $\{\varphi_k(x)\}$ is a complete orthonormal family.

Let $[a, b]$ be a finite interval and let $\rho(x)$ be such that $x^n \in L_2^{(\rho)}[a, b]$ for $n = 0, 1, 2, \ldots$. Let $\{Q_n(x)\}$ be the orthonormal family of polynomials determined by (7.7.2). We shall prove that $\{Q_n(x)\}$ is complete.

Theorem 7.7.1 Let f be continuous on $[a, b]$ and $\{Q_n(x)\}$ an infinite orthonormal family of polynomials determined by (7.7.2)–(7.7.4). Then $\|f - \sum_0^n A_k Q_k\|_2 \to 0$ as $n \to \infty$, where $A_k = \langle f, Q_k \rangle$.

Proof. Since $\sum_0^n A_k Q_k(x)$ is the least-squares polynomial,

$$\left\| f - \sum_0^n A_k Q_k \right\|_2 \leq \|f - \pi_n\|_2,$$

where π_n is the B.A.P. of degree $\leq n$ for f on $[a, b]$. But

$$\|f - \pi_n\|_2^2 = \int_a^b |f(x) - \pi_n(x)|^2 \rho(x)\,dx \leq \|f - \pi_n\|_\infty^2 \int_a^b \rho(x)\,dx \to 0,$$

since $\|f - \pi_n\|_\infty \to 0$ by the Weierstrass Theorem. ∎

Corollary In the pre-Hilbert space $C[a, b] \subset L_2^{(\rho)}[a, b]$, the polynomials $\{Q_k(x)\}$ are complete.

Proof. $C[a, b]$ is a subspace of $L_2^{(\rho)}[a, b]$ and is therefore a pre-Hilbert space. Since $f = \lim_{n \to \infty} \sum_0^n A_k Q_k$ in the L_2-norm for any $f \in C[a, b]$ according to the theorem just proved, $\{Q_k\}$ must be complete in $C[a, b]$ by Theorem 3.3.7, Corollary.

Theorem 7.7.2 Let $\{Q_k\}$ be determined by (7.7.2)–(7.7.4) with inner product given by (7.7.1). The $\{Q_k\}$ is a complete orthonormal set in $L_2^{(\rho)}[a, b]$.

Proof. Let $f \in L_2^{(\rho)}[a, b]$ and suppose that $\langle f, Q_k \rangle = \int_a^b f(x) Q_k(x) \rho(x)\,dx = 0$ for all $k = 0, 1, 2, \ldots$. Then since x^n can be expressed as a linear combination of the $Q_k(x)$, it follows that $\int_a^b x^n f(x)\rho(x)\,dx = 0$ for $n = 0, 1, 2, \ldots$. Let $F(x) = \int_a^x f(s)\rho(s)\,ds$. Then F is continuous on $[a, b]$ and $F(a) = 0$. Also, $F(b) = \int_a^b f(s)\rho(s)\,ds = \langle f, x^0 \rangle = 0$. Hence, integration by parts gives

$$0 = \int_a^b x^n f(x)\rho(x)\,dx = x^n F(x)\Big|_a^b - n\int_a^b x^{n-1} F(x)\,dx.$$

This implies that F is orthogonal to x^n for all n, which implies that $\langle F, Q_n \rangle = 0$, all n. By the Corollary above, since F is continuous, this implies $F = 0$. Hence, $f(x) = 0$ almost everywhere (otherwise, $\int_a^x f(s)\rho(s)\,ds \neq 0$ for some x); i.e., $f = 0$ in $L_2^{(\rho)}[a, b]$. By Definition 3.3.5, $\{Q_k\}$ is complete. ∎

We have shown that a Fourier expansion $\sum_0^\infty A_k Q_k(x)$ of a function $f \in C[a, b]$ converges in the mean to f. What can be said about uniform convergence? Does the series converge uniformly and if so, does it converge to f? It is known that the Fourier series $\sum_0^\infty (A_k \cos kx + B_k \sin kx)$ of a continuous function need not converge at every point x. (See [7.8], p. 360.) Therefore, for a continuous function f we cannot expect its Fourier expansion to converge uniformly. However, if we require f to have certain differentiability properties, then convergence in the uniform norm is assured.

Remark Observe that if $f \in C[0, 2\pi]$ and $f(0) = f(2\pi)$ and if the Fourier series of f converges uniformly, then it converges to f. Let $g(x) = \sum_0^\infty (A_k \cos kx + B_k \sin kx)$. By virtue of the uniform convergence, we may integrate term by term to obtain $\langle f - g, \cos kx \rangle = \langle f - g, \sin kx \rangle = 0$ for all k. Since $\{(1/\sqrt{\pi})\sin kx, (1/\sqrt{\pi})\cos kx\}$ is a complete orthonormal set, $f = g$. If $\sum_0^\infty (|A_k| + |B_k|) < \infty$, the Fourier series converges uniformly and absolutely, since it is majorized by $\sum(|A_k| + |B_k|)$.

Theorem 7.7.3 Let f have a continuous second derivative in $[0, 2\pi]$; i.e., $f \in C^2[0, 2\pi]$. Let $f(0) = f(2\pi)$ and $f'(0) = f'(2\pi)$.
Then the Fourier series of f converges uniformly and absolutely to f.

Proof.

$$A_k = \frac{1}{\pi}\int_0^{2\pi} f(x)\cos kx\,dx.$$

Integrating twice by parts, we obtain

$$A_k = -(1/\pi k^2)\int_0^{2\pi} f''(x)\cos kx\,dx.$$

Since f'' is continuous, $|f''(x)| \leq m$ for some constant m. Hence, $|A_k| \leq 2m/k^2$ and similarly, $B_k \leq 2m/k^2$. Therefore, $\sum_0^\infty (|A_k| + |B_k|)$ converges (since $\sum_1^\infty 1/k^2 < \infty$). By the above remark, the theorem follows. ∎

A somewhat better result is the following.

Theorem 7.7.4 Let f be a periodic function of period 2π having a continuous derivative of order $r \geq 1$. Suppose that $f^{(r)}$ satisfies a Lipschitz condition of order α, that is, for all x

$$|f^{(r)}(x + h) - f^{(r)}(x)| \leq L_r|h|^\alpha, \qquad 0 < \alpha \leq 1.$$

Then the Fourier coefficients of f satisfy the inequality

$$|A_k|, |B_k| \leq c/k^{r+\alpha}, \qquad k = 1, 2, \ldots, c = 2L_r,$$

and the Fourier series converges uniformly to $f(x)$ in $[0, 2\pi]$.

7.7.1 Computation of Orthogonal Polynomials

Proof. We give the proof for the case $r = 1$, the cases $r > 1$ being similar after applying r successive integrations by parts. Integration by parts gives

$$A_k = -\frac{1}{\pi k} \int_0^{2\pi} f'(x) \sin kx \, dx.$$

Changing the variable of integration to $x' = x - \pi/k$, we get

$$A_k = (1/\pi k) \int_0^{2\pi} f'(x + \pi/k) \sin kx \, dx.$$

Therefore,

$$A_k = (1/2\pi k) \int_0^{2\pi} (f'(x + \pi/k) - f'(x)) \sin kx \, dx,$$

$$|A_k| \leq (2/\pi k) L_1 (\pi/k)^\alpha \leq c/k^{1+\alpha}.$$

The same inequality holds for $|B_k|$ and the theorem follows.

A function is *piecewise continuous* on an interval $[a, b]$ if it has only a finite number of discontinuities in $[a, b]$.

Theorem 7.7.5 Let f be continuous and periodic of period 2π. Let the derivative f' be piecewise continuous. Then the Fourier series of f converges uniformly and absolutely to f.

Proof. Let

$$\pi a_k = \int_0^{2\pi} f'(x) \sin kx \, dx.$$

Since $f' \in L^2[0, 2\pi]$, Bessel's inequality yields $\sum_1^\infty a_n^2 < \infty$. By integration by parts, we have $|A_k| = |a_k|/k$, where A_k is the Fourier coefficient of f. Therefore, by the Schwarz inequality

$$\sum_1^\infty |A_k| = \sum_1^\infty |a_k|/k \leq \left(\sum_1^\infty (1/k^2) \right)^{1/2} \left(\sum_1^\infty |a_k|^2 \right)^{1/2} < \infty.$$

Similarly, $\sum_1^\infty |B_k| < \infty$, and the result follows. ∎

Let $f(x)$ be defined on $[-1, 1]$. Define $g(\theta) = f(\cos \theta)$. g is an even function of θ and has period 2π. If f is differentiable at a point x in $[-1, 1]$, then g is differentiable at all θ such that $x = \cos \theta$. Since $g(\theta) = g(-\theta)$, the Fourier coefficients $1/\pi \int_{-\pi}^{\pi} g(\theta) \sin k\theta \, d\theta$ are all zero. Thus, the Fourier series of g has the form

$$A_0/2 + \sum_{k=1}^\infty A_k \cos k\theta, \qquad A_k = (2/\pi) \int_0^\pi g(\theta) \cos k\theta \, d\theta.$$

By Example (b) at the beginning of this section, we see that

$$A_k = (2/\pi) \int_{-1}^1 f(x) T_k(x) (1 - x^2)^{-1/2} \, dx.$$

are the Fourier-Chebyshev coefficients of f on $[-1, 1]$. If the Fourier series of g converges to $g(\theta)$, then we also have $f(x) = A_0/2 + \sum_1^\infty A_k T_k(x)$, that is, the Chebyshev series converges to $f(x)$ at $x = \cos \theta$. In this way, we can use the preceding results on Fourier series to establish properties of Chebyshev expansions. As a direct consequence of Theorem 7.7.5, we have

Theorem 7.7.6 Let f have a piecewise-continuous derivative on $[-1, 1]$. The Chebyshev series expansion of f converges uniformly to f on $[-1, 1]$.

We are now in a position to use Chebyshev expansions to obtain estimates of M_n, the minimax, as a function of n.

For $f \in C[-1, 1]$, let π_n be the best approximating polynomial of degree $\leq n$. Letting $x = \cos \theta$ and $g(\theta) = f(\cos \theta)$, we have $|g(\theta) - \pi_n(\cos \theta)| \leq M_n, 0 \leq \theta \leq 2\pi$. Since $(\cos \theta)^k$ can be expressed as a trigonometric cosine sum, we can write $\pi_n(\cos \theta) = \sum_0^n a_k \cos k\theta$. By the least-squares property of the Fourier expansion and Parseval's identity,

$$\pi \sum_{n+1}^\infty A_k^2 = \int_0^{2\pi} \left| g(\theta) - A_0/2 - \sum_1^n A_k \cos k\theta \right|^2 d\theta$$

$$\leq \int_0^{2\pi} |g(\theta) - \pi_n(\cos \theta)|^2 d\theta \leq 2\pi M_n^2.$$

This yields the inequality

$$M_n^2 \geq \tfrac{1}{2} \sum_{n+1}^\infty A_k^2 \geq \tfrac{1}{2} A_{n+1}^2 \tag{7.7.6}$$

for any function $f \in C[-1, 1]$. This provides a lower bound for M_n. The next theorem gives an upper bound.

Theorem 7.7.7 Let the derivative f' satisfy a Lipschitz condition on $[-1, 1]$, with Lipschitz constant L. Let $C = \max_{-1 \leq x \leq 1} |f'(x)|$. Then

$$M_n < 2(L + C)/n.$$

Proof. $g'(\theta) = -f'(\cos \theta) \sin \theta$ satisfies a Lipschitz condition on $[0, 2\pi]$ with constant $L + C$. By Theorem 7.7.4, $|A_k| \leq 2(L + C)/k^2$, $k = 1, 2, \ldots$. By Theorem 7.7.6, $\sum_0^\infty A_k T_k(x)$ converges uniformly to $f(x)$ on $[-1, 1]$. Hence,

$$M_n \leq \max_{-1 \leq x \leq 1} \left| f(x) - \sum_0^n A_k T_k(x) \right| \leq \max \sum_{n+1}^\infty |A_k T_k(x)| \leq \sum_{n+1}^\infty |A_k|.$$

Since

$$\sum_{n+1}^\infty |A_k| \leq 2(L + C) \sum_{n+1}^\infty (1/k^2) < \frac{2(L + C)}{n},$$

the proof is complete. ∎

7.7.1 Computation of Orthogonal Polynomials

Corollary If f has a derivative of order r ($r \geq 1$) which satisfies a Lipschitz condition on $[-1, 1]$, then there is a constant c_r such that

$$M_n < c_r/n^r r.$$

Proof. By Theorem 7.7.4, $|A_k| \leq c/k^{r+1}$, where c depends on the Lipschitz constant of $g^{(r)}(\theta)$. We have again, $M_n \leq \sum_{n+1}^{\infty} |A_k|$ and the result follows from $\sum_{n+1}^{\infty} (1/k^{r+1}) < 1/rn^r$. ∎

Lemma 7.7.1 Let $g(\theta)$ be continuous and of period 2π. Then the Fourier sum

$$S_n(\theta) = \sum_0^n A_k \cos k\theta + B_k \sin k\theta$$

can be represented in the *Dirichlet* integral form

$$S_n(\theta) = (1/\pi) \int_0^{2\pi} g(t + \theta) \frac{\sin (n + 1/2)t}{2 \sin (t/2)} dt.$$

Proof.

$$S_n(\theta) = (1/\pi) \int_0^{2\pi} g(t)(\tfrac{1}{2} + \sum_{k=1}^n \cos k\theta \cos kt + \sin k\theta \sin kt) \, dt$$

$$= (1/\pi) \int_0^{2\pi} g(t) \left(\tfrac{1}{2} + \sum_1^n \cos k(\theta - t) \right) dt.$$

Changing the variable of integration, we get

$$S_n(\theta) = 1/\pi \int_0^{2\pi} g(t + \theta) \left(\tfrac{1}{2} + \sum_1^n \cos kt \right) dt.$$

Using the trigonometric identity,

$$2 \cos kt \sin t/2 = \sin (k + \tfrac{1}{2})t - \sin (k - \tfrac{1}{2})t,$$

the result follows. ∎

Lemma 7.7.2 Let g be continuous of period 2π. Let $S_n(\theta)$ be the nth-order Fourier sum of g. Then

$$\|S_n\|_\infty = \max_{0 \leq \theta \leq 2\pi} |S_n(\theta)| \leq (3 + \log n) \max_{0 \leq \theta \leq 2\pi} |g(\theta)| = (3 + \log n) \|g\|_\infty.$$

Proof. From Lemma 7.7.1,

$$\|S_n\|_\infty \leq \|g\|_\infty \int_0^\pi \left| \frac{\sin (n + \tfrac{1}{2})t}{\pi \sin (t/2)} \right| dt.$$

Now,

$$\frac{2}{\pi} \int_0^{1/n} \frac{\sin (n + \tfrac{1}{2})t}{2 \sin (t/2)} dt = \frac{2}{\pi} \int_0^{1/n} |\tfrac{1}{2} + \cos t + \cdots + \cos nt| \, dt$$

$$\leq \frac{2}{\pi n} (\tfrac{1}{2} + n) < 1.$$

On $[1/n, \pi]$, $\sin(t/2) \geq t/\pi$. Hence,

$$\frac{1}{\pi}\int_{1/n}^{\pi}\left|\frac{\sin(n+\tfrac{1}{2})t}{\sin(t/2)}\right|dt \leq \frac{1}{\pi}\int_{1/n}^{\pi}\frac{\pi}{t}dt = \log\pi - \log\frac{1}{n}$$
$$< 2 + \log n.$$

This completes the proof. ∎

Theorem 7.7.8 Let f be continuous on $[-1, 1]$ and M_n the minimax of degree $\leq n$. Then

$$\max_{-1 \leq x \leq 1}\left|f(x) - \sum_{0}^{n}A_k T_k(x)\right| \leq (4 + \log n)M_n.$$

Proof. Let $g(\theta) = f(\cos\theta)$. Let $\pi_n(x)$ be the best approximating polynomial for f in the uniform norm. Since $\pi_n(x)$ is a polynomial, $\pi_n(\cos\theta)$ is a trigonometric sum of order n. Hence, it coincides with its Fourier sum $F(\theta)$ and we have

$$\max_{0 \leq \theta \leq 2\pi}\left|g(\theta) - \sum_{0}^{n}A_k \cos k\theta\right|$$

$$= \max_{0 \leq \theta \leq 2\pi}\left|\left(\sum_{0}^{n}A_k \cos k\theta - F(\theta)\right) - (g(\theta) - \pi_n(\cos\theta))\right|$$

$$\leq \max_{\theta}\left|\sum_{0}^{n}A_k \cos k\theta - F(\theta)\right| + M_n.$$

Now, $\sum_{0}^{n}A_k \cos k\theta - F(\theta)$ is the Fourier sum of the function $g(\theta) - \pi_n(\cos\theta)$. By Lemma 7.7.2, its maximum is $\leq (3 + \log n)\|g - \pi_n\|_\infty = (3 + \log n)M_n$. ∎

Remark Since $\log 400 < 6$, $4 + \log n < 10$ for $n \leq 400$. Thus, the Chebyshev expansion, $\sum_{0}^{n}A_k T_k(x)$, yields a maximum residual of at most $10M_n$; i.e., one digit of accuracy is lost by using the polynomial approximation $\sum_{0}^{n}A_k T_k(x)$ instead of the best polynomial $\pi_n(x)$.

We conclude this section with a well-known theorem of Jackson which gives a bound on M_n in the case where the function is merely continuous. We first consider approximation by trigonometric sums.

Theorem 7.7.9 Let f be continuous on the interval $[0, 2\pi]$ and $f(0) = f(2\pi)$. Then $M_n^* \leq 12\omega(1/n)$, where $\omega(\delta)$ is the modulus of continuity of f and M_n^* is the minimax of f with respect to trigonometric sums of order $\leq n$.

Proof. We extend the domain of f to the entire real line by defining f to be periodic of period 2π. Let

$$S_n(x) = \frac{3}{2\pi n(2n^2 + 1)}\int_{-\pi}^{\pi}f(\theta)\left(\frac{\sin n(\theta - x)/2}{\sin(\theta - x)/2}\right)^4 d\theta. \tag{1}$$

First, we shall show that S_n is a trigonometric sum of order $2n - 2$. Then we shall prove that $\|f(x) - S_n(x)\| \leq 6\omega(1/n)$. For $n = 2k$, it then follows that

$$M_n^* \leq M_{2k-2}^* \leq 6\omega(1/k) = 6\omega(2/n) \leq 12\omega(1/n).$$

For $n = 2k - 1$, it follows that

$$M_n^* \leq M_{2k-2}^* \leq 6\omega(1/k) = 6\omega(2/(n+1)) \leq 12\omega(1/(n+1)) \leq 12\omega(1/n),$$

which proves the result for all n.

7.7.1 Computation of Orthogonal Polynomials

To see that $S_n(x)$ is a trigonometric sum, we use the trigonometric identity

$$\left(\frac{\sin(n\theta/2)}{\sin(\theta/2)}\right)^4 = a_0 + \sum_{j=1}^{2n-2} a_j \cos j\theta.$$

(The left member is the square of a polynomial in $\cos\theta$ of degree $n-1$. See Exercise 7.22.) Hence,

$$S_n(x) = \frac{3}{2\pi n(2n^2+1)} \int_{-\pi}^{\pi} f(\theta)\left(a_0 + \sum_{j=1}^{2n-2} a_j(\cos j\theta \cos jx + \sin j\theta \sin jx)\right) d\theta.$$

Integration yields the trigonometric sum for $S_n(x)$.

Now, let $\theta = t + x$ in the integral in (1) above. With this change of the variable of integration, we obtain, as a consequence of the periodicity of f,

$$S_n(x) = \frac{3}{2\pi n(2n^2+1)} \int_{-\pi}^{\pi} f(t+x)\left(\frac{\sin(nt/2)}{\sin(t/2)}\right)^4 dt$$

$$= \frac{3}{2\pi n(2n^2+1)} \int_{0}^{\pi} [f(x+t) + f(x-t)]\left(\frac{\sin(nt/2)}{\sin(t/2)}\right)^4 dt$$

$$= \frac{3}{\pi n(2n^2+1)} \int_{0}^{\pi/2} [f(x+2t) + f(x-2t)]\left(\frac{\sin nt}{\sin t}\right)^4 dt.$$

Now, we use the formula (see Exercise 7.22)

$$1 = \frac{6}{\pi n(2n^2+1)} \int_{0}^{\pi/2} \left(\frac{\sin nt}{\sin t}\right)^4 dt, \tag{2}$$

multiplying it by $f(x)$ and subtracting from it the formula for $S_n(x)$ to get

$$f(x) - S_n(x) = \frac{3}{\pi n(2n^2+1)} \int_{0}^{\pi/2} [f(x+2t) + f(x-2t) - 2f(x)]\left(\frac{\sin nt}{\sin t}\right)^4 dt.$$

Clearly, from the definition of $\omega(\delta)$, it follows that

$$|f(x+2t) + f(x-2t) - 2f(x)| \le 2\omega(2t) = 2\omega\left(2nt\cdot\frac{1}{n}\right) \le 2(2nt+1)\omega(1/n).$$

Hence, using (2) again, we have

$$|f(x) - S_n(x)| \le \omega(1/n)\left(1 + \frac{12}{\pi(2n^2+1)} \int_{0}^{\pi/2} t\left(\frac{\sin nt}{\sin t}\right)^4 dt\right).$$

Since

$$\int_{0}^{\pi/2} t\left(\frac{\sin nt}{\sin t}\right)^4 dt < \frac{\pi^2 n^2}{4}$$

(see Exercise 7.22), we obtain the desired result,

$$|f(x) - S_n(x)| \le (1 + 3\pi/2)\omega(1/n) < 6\omega(1/n). \blacksquare$$

Corollary Let M_n be the minimax of degree n for a function $f \in C[-1, 1]$. If $\omega(\delta)$ is the modulus of continuity of f

$$M_n \le 12\omega\left(\frac{\pi}{n}\right).$$

Proof. Let $g(\theta) = f(\cos \theta)$, as in preceding proofs. Then g is even and periodic of period 2π. Its best approximation by a trigonometric sum, S^*, must be even, hence contains only cosine terms, since

$$\|g - S^*\|_\infty = \max_{-\pi \le \theta \le \pi} |g(\theta) - S^*(\theta)| = \max_{-\pi \le \theta \le \pi} |g(-\theta) - S^*(\theta)|$$

$$= \max_{-\pi \le \theta \le \pi} |g(\theta) - S^*(-\theta)|.$$

Thus, $S^*(-\theta)$ is also a best approximating sum. By the uniqueness, $S^*(-\theta) = S^*(\theta)$. Converting the cosine sum to a polynomial $p(x)$ in $x = \cos \theta$, we have

$$M_n \le \max_{-1 \le x \le 1} |f(x) - p(x)| = \|g - S^*\|_\infty \le 12\omega_g(\pi/n),$$

where $\omega_g(\delta)$ is the modulus of continuity of g. But

$$\omega_g(\delta) = \max_{|\theta_1 - \theta_2| \le \delta} |g(\theta_1) - g(\theta)_2| \le \max_{|\cos \theta_1 - \cos \theta_2| \le \delta} |f(\cos \theta_1) - f(\cos \theta_2)| = \omega(\delta),$$

since $|\cos \theta_1 - \cos \theta_2| \le |\theta_1 - \theta_2|$. Hence, $M_n \le 12\omega(\pi/n)$. ∎

Computational Aspects

We have seen that the theory of least-squares approximation yields results in the theory of uniform approximation, namely, the various upper bounds for the minimax M_n given above and the result of Theorem 7.7.8 which suggests that the Chebyshev expansion is a good initial approximation to the B.A.P. As always, theory and practice diverge, the theory being a guide good only up to a point at which ingenuity must prevail. In Exercise 7.16, it is required to compute the B.A.P. for e^x giving accuracy of 10^{-8} on $[-1, 1]$. To begin, one must first estimate the degree n of the polynomial. This could be done by using one of the upper bounds for M_n. For example, Theorem 7.7.7 yields in this case

$$M_n < 4e/n,$$

since $|e^x| \le e$ on the interval $[-1, 1]$. Setting $e/n = 10^{-9}$, we obtain $n = e \cdot 10^9$ as a degree which will insure that $M_n < 10^{-9}$. Of course, this bound is too large to be of practical use. The other bounds are also too large or too difficult to compute. (This is to be expected, since these bounds must apply to a wide class of functions, most of which do not behave as well as e^x.) In Exercise 7.16, the Chebyshev expansion of e^x is suggested as a means of estimating n. However, a much easier way to estimate n is to use the Taylor series expansion of e^x. The series $\sum_{j=0}^{\infty} x^j/j!$ converges fairly rapidly, although not as rapidly as the Chebyshev expansion. Also, the error $e^x - \sum_{j=0}^{n} x^j/j!$ is given by Taylor's formula and is bounded by $|e^y x^{n+1}/(n + 1)!| < e/(n + 1)!$. This can be used to obtain an initial estimate for n. Another estimate can be obtained for functions which are entire (holomorphic in the entire complex plane, that is, single-valued and analytic everywhere in the complex plane) by using a result in [7.14] which states that M_n approaches the Chebyshev coefficient A_{n+1} asymptotically as $n \to \infty$. The Chebyshev coefficients can be estimated by quadrature (Chapter 9).

Unfortunately, the rate of convergence of $M_n(f)$ to zero for a continuous function

f can be so slow that polynomial approximation is not practical when high accuracy is needed. For example, consider arcsin x. To obtain its Chebyshev series, we recall first that $|\theta|$ has a Fourier series

$$|\theta| = \frac{\pi}{2} - \frac{4}{\pi} \sum_{n=1}^{\infty} \frac{\cos(2n-1)\theta}{(2n-1)^2},$$

convergent for $-\pi \le \theta \le \pi$. Setting $\theta = \arccos x$, we obtain

$$\arcsin x = \frac{\pi}{2} - \arccos x = \frac{4}{\pi} \sum_{n=1}^{\infty} \frac{T_{2n-1}(x)}{(2n-1)^2}.$$

Applying inequality (7.7.6), we find that

$$M_{2n} \ge \frac{\sqrt{2/\pi}}{(2n+1)}.$$

Therefore, to obtain a polynomial which approximates arcsin x on $[-1, 1]$ with a uniform error less than 10^{-8}, we must take the degree to be at least 10^8. Finally, we cite the following theorem which shows that polynomial approximation can be "arbitrarily slow".

Bernstein's Theorem Let (ε_n) be an arbitrary sequence such that $\varepsilon_n \ge \varepsilon_{n+1}$ and $\varepsilon_n \to 0$. There exists $f \in C[a, b]$ such that $M_n(f) = \varepsilon_n$ for all n.

For functions like arcsin x another mode of approximation is clearly indicated. In Section 7.9, we consider another important and useful family of approximating functions, the rational functions.

7.8 ORTHOGONAL POLYNOMIALS AS EIGENFUNCTIONS

One of the most important means of generating orthogonal functions is by the solution of Sturm–Liouville boundary-value problems. We shall discuss boundary-value problems in Chapter 11. For the present chapter, it suffices to define the special *Sturm–Liouville problem* as that of finding a solution of a differential equation of the form

$$\frac{d}{dx}\left(\psi \frac{dy}{dx}\right) + (\lambda \rho - \sigma)y = 0,$$

where $\rho = \rho(x)$ and $\sigma = \sigma(x)$ are given continuous functions on $[a, b]$, ψ is continuously differentiable on $[a, b]$ and λ is an arbitrary parameter. The solution must satisfy boundary conditions of the form

$$y(a) + cy'(a) = 0, \qquad y(b) + dy'(b) = 0,$$

where c and d are given constants. In fact, we shall limit our discussion to the special case of the *hypergeometric equation*

$$x(1-x)y'' + [\gamma - (\alpha + \beta + 1)x]y' - \alpha\beta y = 0. \tag{H}$$

To transform (H) into the Sturm–Liouville form, we put

$$\rho(x) = x^{\gamma-1}(1-x)^{\alpha+\beta-\gamma} = x^p(1-x)^q, \quad \text{where } p = \gamma - 1, q = \alpha + \beta - \gamma.$$

We assume that $p > -1$ and $q > -1$. This gives for (H)

$$\frac{1}{\rho(x)} \frac{d}{dx}\left[x(1-x)\rho(x)\frac{dy}{dx}\right] + \lambda y = 0,$$

where $\lambda = -\alpha\beta$. If we define the differential operator L by

$$Ly = \frac{1}{\rho}\frac{d}{dx}\left[x(1-x)\rho(x)\frac{dy}{dx}\right], \tag{7.8.1}$$

then (H) can be written simply as

$$Ly + \lambda y = 0. \tag{7.8.2}$$

We take the end-points $a = 0$ and $b = 1$. As boundary conditions we require only that a solution $u(x)$ and its derivative $u'(x)$ be finite at $x = 0$ and $x = 1$. The operator L is a linear operator on the space $C^2[0, 1]$. Therefore, the values of λ for which the Sturm–Liouville problem has nonzero solutions are called *eigenvalues* and the corresponding solutions are called *eigenfunctions*. With respect to the inner product

$$\langle u, v \rangle = \int_0^1 u(x)v(x)\rho(x)\,dx \tag{7.8.3}$$

the operator L behaves like a self-adjoint operator, that is, $\langle Lu, v \rangle = \langle u, Lv \rangle$ as we shall see in the next theorem.

Theorem 7.8.1 If u_1 and u_2 are eigenfunctions of (H) belonging to the distinct eigenvalues λ_1 and λ_2 respectively, then u_1 and u_2 are orthogonal with respect to the inner product (7.8.3).

Proof. Since $Lu_i = \lambda_i u_i$, $i = 1, 2$, we have $\langle u_1, Lu_2\rangle - \langle u_2, Lu_1\rangle = (\lambda_2 - \lambda_1)\langle u_1, u_2\rangle$. Now, from (7.8.1) it follows, by integration by parts, that

$$\langle u_1, Lu_2\rangle = \int_0^1 u_1 \frac{d}{dx}\left[x(1-x)\rho(x)\frac{du_2}{dx}\right]dx$$

$$= x^{p+1}(1-x)^{q+1}u_1 u_2'\big|_0^1 - \int_0^1 x^{p+1}(1-x)^{q+1}u_1' u_2'\,dx,$$

$$\langle u_2, Lu_1\rangle = x^{p+1}(1-x)^{q+1}u_2 u_1'\big|_0^1 - \int_0^1 x^{p+1}(1-x)^{q+1}u_1' u_2'\,dx.$$

Therefore,

$$\langle u_1, Lu_2\rangle - \langle u_2, Lu_1\rangle = x^{p+1}(1-x)^{q+1}(u_1 u_2' - u_2 u_1')\big|_0^1 = 0,$$

since $p + 1 > 0$ and $q + 1 > 0$. Hence, $\langle u_1, u_2\rangle = 0$, as was to be proved. ∎

Now, from the theory of ordinary differential equations we recall that the general solution of Eq. (H) is the *hypergeometric function*,

$$F(\alpha, \beta, \gamma; x) = 1 + \frac{\alpha \cdot \beta}{1 \cdot \gamma}x + \frac{\alpha(\alpha+1)\beta(\beta+1)}{2!\,\gamma(\gamma+1)}x^2 + \cdots.$$

Since our boundary conditions require $F(x)$ to be finite at $x = 1$, the series must terminate after a finite number of terms. This implies that α (or β) is a negative integer, say $\alpha = -n$, $n = 0, 1, 2, \ldots$, and F is an nth-degree polynomial in x. Then $\beta = n + p + q + 1$ and

$\lambda = n(n + p + q + 1)$. Thus, the eigenfunctions of the Sturm–Liouville problem for Eq (H) with a given p, q are the orthogonal polynomials with respect to the weight function $x^p(1 - x)^q$ on the interval $[0, 1]$; i.e., they are precisely the Jacobi polynomials $J_n^{(p,q)}$ of Section 7.7, Example (c). If we transform the interval to $[-1, 1]$ by letting $x = (1 + \xi)/2$ we obtain the weight function $(1 + x)^p(1 - x)^q$ and the differential equation (H) becomes

$$(1 - x^2)y'' + [p - q - (p + q + 2)x]y' + n(n + p + q + 1)y = 0. \quad (7.8.4)$$

For $p = q = 0$, we obtain the *Legendre equation*,

$$(1 - x^2)y'' - 2xy' + n(n + 1)y = 0,$$

which is satisfied by the Legendre polynomial of degree n. (Section 7.7, Example (c).)

7.9 RATIONAL APPROXIMATION

In a computational structure sense, polynomials are the simplest "finitely representable" functions that can be used in approximating continuous functions. It seems reasonable to consider the next simplest class to be the rational functions. A rational function, $R(x)$, is one which can be evaluated as the quotient of two polynomials. Thus,

$$R(x) = \frac{P(x)}{Q(x)} = \frac{a_0 + a_1 x + \cdots + a_n x^n}{b_0 + b_1 x + \cdots + b_m x^m} \quad (7.9.1)$$

where $a_n \neq 0$, $b_m \neq 0$. When the degree $m = 0$ and Q is a constant function $b_0 \neq 0$, $R(x)$ reduces to a polynomial. In the computer, $R(x)$ is finitely representable by the coefficients $(a_0, \ldots, a_n)$ and $(b_0, \ldots, b_m)$ suitably rounded. (We restrict our attention to real coefficients.) Clearly, $R(x)$ is computable by a finite number of arithmetic operations. The polynomial P can be evaluated by the well-known "nesting" algorithm described by the parenthesized expression,

$$(\cdots((a_n x + a_{n-1})x + a_{n-2}) + \cdots + a_1)x + a_0. \quad (7.9.2)$$

This algorithm is programmed easily as a recursive scheme requiring n additions (of the a_i, $0 \leq i \leq n - 1$) and n multiplications (by x). Similarly, $Q(x)$ can be evaluated by m additions and multiplications. If $Q(x) \neq 0$, then a single division yields the final result. Actually, a rather different algorithm may be less time consuming depending on the time required for the multiplication and division operations in the particular computer being used. The alternative algorithm is derived by transforming $R(x)$ into a *continued fraction*, represented by the following scheme,

$$R(x) = Q_1(x) + \cfrac{c_2}{Q_2(x) + \cfrac{c_3}{Q_3(x) + \cfrac{\ddots}{\ddots + \cfrac{c_k}{Q_k(x)}}}}. \quad (7.9.3)$$

Here, the Q_i are monic polynomials and the c_i are constants determined by successive division as follows.

Let $P_0 = P$ and $P_1 = Q$ and assume that deg $P_0 \geq$ deg P_1, where deg P denotes the degree of polynomial P. Divide P_0 by P_1, obtaining quotient Q_1 and remainder P_2, where deg $P_2 <$ deg P_1. Thus,

$$P_0 = P_1 Q_1 + P_2$$

and

$$R = \frac{P_0}{P_1} = Q_1 + \frac{1}{P_1/P_2}.$$

Apply the same procedure to P_1/P_2 and, generally, to P_i/P_{i+1}, where

$$P_i = P_{i+1} Q_{i+1} + P_{i+2}, \qquad 0 \leq i \leq k - 2. \tag{7.9.4}$$

Since deg $P_{i+2} <$ deg P_{i+1}, after a finite number of steps, we obtain deg $P_k = 0$ and

$$P_{k-1} = P_k Q_k.$$

The result is

$$R = Q_1 + \cfrac{1}{Q_2 + \cfrac{1}{Q_3 + \phantom{\cfrac{1}{1}}}}$$

$$+ \cfrac{1}{Q_{k-1} + \cfrac{1}{Q_k}}.$$

This becomes (7.9.3) when the leading coefficient of each Q_i ($i \geq 2$) is factored out, leaving a monic polynomial (leading coefficient unity). Using the nesting algorithm, we can evaluate a monic polynomial of degree p with only $p - 1$ multiplications. From (7.9.4) we see that

$$\deg P_i = \deg P_{i+1} + \deg Q_{i+1},$$

since deg $P_{i+2} <$ deg P_{i+1}. Therefore,

$$\sum_{i=1}^{k} \deg Q_i = \sum_{i=0}^{k-1} (\deg P_i - \deg P_{i+1}) = \deg P_0 = n.$$

To evaluate $R(x)$ by (7.9.3) we must compute each $Q_i(x)$ and do one division, starting with $i = k$. Since Q_i is monic for $i \geq 2$, the computation of $Q_i(x)$ requires deg $Q_i - 1$ multiplications. Therefore, the computation of $R(x)$ by (7.9.3) requires a total of $n - k + 1$ multiplications, $k - 1$ divisions and $n + k - 1$ additions. If the division and multiplication operations take about the same time in the computer (as is roughly the case in many computers), then there is a decided advantage in using a continued-fraction expansion. Note that in the case $n < m$, we simply write $R = c_1/(c_1 Q/P)$ and proceed to convert Q/P to a continued fraction.

In any case, since the rational functions include the polynomials as a special case, it seems possible that more efficient (i.e., involving fewer computations) approximations can be found by searching in this wider class. In a sense, (7.9.1) allows us to select $n + m + 2$ parameters (the coefficients a_i, b_j) in determining a best approximation and to evaluate the resulting function in at most max (n, m) multiplications plus divisions. Therefore, we appear to be gaining the benefits of polynomial approximation of degree $n + m + 1$ with less computation. Furthermore, for a fixed n and m a wider choice of functions is determined by the space of coefficients $(a_0, \ldots, a_n, b_0, \ldots, b_m)$, since the b_j are permitted to enter in a nonlinear manner, i.e., in the denominator. This is, at the same time, something of a disadvantage because the previous theory of linear best approximation does not apply. Therefore, we must consider anew the question of existence of best approximations. As before, we shall prove the existence of best rational approximations in the uniform norm. Unfortunately the existence theorem given below is nonconstructive, as was the case with the fundamental theorem of linear approximation, Theorem 7.3.1.

We shall seek the best rational approximation of the form (7.9.1), that is, we consider the class of rational functions, $R_m^n[a, b]$, defined as follows:

$$R_m^n[a, b] = \left\{ \frac{P}{Q} : \deg P \leq n, \deg Q \leq m, Q(x) > 0 \text{ on } [a, b] \right\}.$$

Since we shall confine our attention to approximation of a function $f \in C[a, b]$, we must insist that $Q(x) \neq 0$ for x in $[a, b]$. Therefore, either $Q(x) < 0$ or $Q(x) > 0$ for all x in $[a, b]$. Since we can multiply both Q and P by an arbitrary constant without changing the value of $R = P/Q$, there is no loss in generality in taking $Q(x) > 0$. (By continuity, $Q(x) > 0$ in a slightly larger interval containing $[a, b]$.) We may also assume that R is *irreducible*, that is, P and Q have no common factors (other than constants). Note that the set of all rational functions is a linear vector space. However, the class $R_m^n[a, b]$ is not a subspace, since the sum of two rational functions in this class may have a denominator of degree $>m$. Therefore, the linear theory of Section 7.3 does not apply. Furthermore, regarding $R_m^n[a, b]$ as a subset of $C[a, b]$ with the uniform norm topology, we see that a subset of $R_m^n[a, b]$ which is bounded in the uniform norm need not have bounded coefficients a_i, b_j. (E.g., the family $\{R_s = 1/(sx^2 + 1) : s > 1\}$ is bounded, with $\|R_s\|_\infty \leq 1$ on the interval $[-1, 1]$.) This rules out compactness arguments like those used in the proof of Theorem 7.3.1. However, we shall be able to circumvent this difficulty and still use a compactness type of proof—which, of course, is then nonconstructive.

Theorem 7.9.1 For every function $f \in C[a, b]$ there is a best uniform rational approximation in the class $R_m^n[a, b]$.

Proof. Consider $R_m^n[a, b]$ to be a subset of the metric space $C[a, b]$, with the uniform norm as metric. Let d be the distance from f to $R_m^n[a, b]$. Then there exists a sequence (R_i) of elements of $R_m^n[a, b]$ such that $\|f - R_i\|_\infty \to d$. Hence, $\|R_i\|_\infty \leq \|f - R_i\|_\infty + \|f\|_\infty \leq 1 + d + \|f\|_\infty = K$ for all i sufficiently large. Now, $R_i = P_i/Q_i$, where $\deg P_i \leq n$ and $\deg Q_i \leq m$. Furthermore, by dividing numerator and denominator by $\|Q_i\|_\infty$, which is nonzero by virtue of $Q_i(x) > 0$, we may take $\|Q_i\|_\infty = 1$. Therefore, the coefficient vectors $b_i = (b_{0i}, \ldots, b_{mi})$ of the Q_i lie in some closed bounded, hence compact, set in $(m + 1)$-dimensional space. Also,

$$|P_i(x)| = |Q_i(x)| |R_i(x)| \leq \|Q_i\|_\infty \|R_i\|_\infty \leq K.$$

Therefore, $\{\|P_i\|_\infty\}$ is a bounded set and the same holds for the coefficient vectors $a_i = (a_{0i}, \ldots, a_{ni})$ of the P_i. (See Lemma 7.3.1.) By sequential compactness, there exist subsequences of (b_i) and (a_i) which converge. Taking a common subsequence and reindexing, we may write $P_i \to P$ and $Q_i \to Q$. Since all $\|Q_i\|_\infty = 1$, it follows that $\|Q\|_\infty = 1$. If $Q(x) \neq 0$, then $R(x) = P(x)/Q(x)$ is the limit of $R_i(x)$. Suppose $Q(x_j) = 0$ for some x_j in $[a, b]$. (There are at most m such points because $\|Q\|_\infty = 1$.) Since $R_i(x) \to R(x)$ for any x such that $Q(x) \neq 0$, it follows that for such x, $|P(x)| \leq K|Q(x)|$. Now, by the continuity of P and Q at x_j, this inequality must remain valid at x_j. This implies $P(x_j) = 0$. Thus, the linear factors $(x - x_j)$ may be canceled from both P and Q, yielding a rational function P^*/Q^* which is equal to R and such that $Q^*(x) > 0$ for all x in $[a, b]$. Therefore, P^*/Q^* is in $R_m^n[a, b]$ and $\|f - P^*/Q^*\|_\infty = \lim \|f - R_i\|_\infty = d$; i.e., P^*/Q^* is the best approximation, which proves the theorem. ∎

As observed previously, it is important to know whether a best approximation is unique when attempting to compute one. (At the very least, a uniqueness proof settles any argument which might arise in case two different approximations are offered as the "best".) The general uniqueness Theorem 7.3.3 for the linear approximation problem depends on the norm being strict. The uniform norm is not strict (Exercise 7.7b). Therefore, we had to use the Haar condition (Section 7.6) to establish uniqueness (Theorem 7.5.4). Similar ideas will serve for rational approximation.

Let R^* be a fixed rational function in $R_m^n[a, b]$. Let V_P be the finite-dimensional subspace of $C[a, b]$ consisting of all polynomials P of degree $\leq n$. Similarly, let V_Q be the subspace of polynomials Q of degree $\leq m$. We define the new subspace of functions,

$$V_{R^*} = V_P + R^* V_Q = \{P + R^*Q : P \in V_P, Q \in V_Q\}.$$

It is clear that V_{R^*} is a subspace. Such subspaces will play a central role in our analysis of approximation by functions in $R_m^n[a, b]$. To see why, observe that for any $R = P/Q$ in $R_m^n[a, b]$ we have $Q(R - R^*) = P - R^*Q \in V_{R^*}$. Since $Q(x) > 0$, the sign of $R - R^*$ is the same as that of an element in V_{R^*}. Thus, we shall need to know the changing sign behavior of functions in V_{R^*} to be able to compare two rational approximations R and R^*. We shall now show that this behavior depends on the dimension d, of V_{R^*}.

By Exercise 2.14,

$$d = \dim(V_P + R^*V_Q) = \dim V_P + \dim R^*V_Q - \dim(V_P \cap R^*V_Q).$$

Since R^* is fixed, $\dim R^*V_Q = \dim V_Q = m + 1$; i.e., $R^* = P^*/Q^*$, say, and $R^*V_Q = \{(P^*/Q^*)Q : Q \in V_Q\}$. Of course, $\dim V_P = n + 1$. It remains to find $\dim(V_P \cap R^*V_Q)$. Now, since we assume that P^* and Q^* have no common factors, an element $(P^*/Q^*)Q$ of R^*V_Q is in V_P if and only if Q^* divides Q and the quotient Q/Q^* is a polynomial Q_1, of degree $\leq n - \deg P^*$; i.e., $Q = Q^*Q_1$. Since $\deg Q \leq m$, this implies $\deg Q_1 \leq m - \deg Q^*$. Hence,

$$\deg Q_1 \leq \min\{n - \deg P^*, m - \deg Q^*\} = c,$$

which implies $\dim(V_P \cap R^*V_Q) = 1 + c$. But $n - \deg P^* \leq m - \deg Q^*$ implies that $n + \deg Q^* \leq m + \deg P^*$. It follows that $c = m + n - \max\{n + \deg Q^*, m + \deg P^*\}$. Therefore,

$$d = \dim(V_P + R^*V_Q) = 1 + \max\{n + \deg Q^*, m + \deg P^*\}. \quad (7.9.5)$$

(Note that if $\deg Q^* = m$ or $\deg P^* = n$, this is simply $1 + m + n$, the dimension of the space of polynomials of degree $m + n$.)

It is quite easy to see how d is related to the number of sign changes of an arbitrary nonzero function $g = P + R^*Q$ in $V_P + R^*V_Q$. Every such g must have at most $d - 1$ zero's, counting multiplicities. (x_0 is a zero of multiplicity k if g and its first $k - 1$ derivatives vanish at x_0.) To see this, note that on $[a, b]$ Q^*g has precisely the same zero's and sign behavior as g, since $Q^*(x) > 0$ on $[a, b]$. But $Q^*g = Q^*P + P^*Q$ is a polynomial of degree at most $d - 1$, hence has at most $d - 1$ zero's, counting multiplicities. Observe that this holds for the entire real line and not just on $[a, b]$.

Now, recall (Theorem 7.5.5) that a sufficient condition for a polynomial of degree n to be a B.A.P. of $f \in C[a, b]$ is that the residual function assume its extreme values with alternating sign on at least $n + 2$ points. This condition is also necessary (Theorem 7.5.4, Corollary). An analogous condition characterizes the best rational approximation $R^* = P^*/Q^*$ of $f \in C[a, b]$, with the exception that $n + 2$ is replaced by $2 + \max\{n + \deg Q^*, m + \deg P^*\}$.

Theorem 7.9.2 Let $R^* = P^*/Q^* \in R_m^n[a, b]$ and let $f \in C[a, b]$. The residual $r^* = f - R^*$ assumes its extreme values $\pm \|f - R^*\|_\infty$ with successively alternating sign on at least $2 + \max\{n + \deg Q^*, m + \deg P^*\}$ points in $[a, b]$ if and only if R^* is the best approximation to f in $R_m^n[a, b]$.

Proof. We prove sufficiency first. Suppose R^* is not the best approximation. Then there exists $P/Q = R \in R_m^n[a, b]$ such that $\|f - R\|_\infty < \|f - R^*\|_\infty$. Let $E = \{x : |r^*(x)| = \|f - R^*\|_\infty\}$. For $x \in E$, $(\operatorname{sgn} r^*(x))r^*(x) = \|f - R^*\|_\infty$ and

$$\operatorname{sgn} r^*(x)(f(x) - R(x)) \leq \|f - R\|_\infty < \operatorname{sgn} r^*(x)(f(x) - R^*(x)).$$

Hence, $\operatorname{sgn} r^*(x)(R(x) - R^*(x)) > 0$ on E, so that $r^*(x)$ and $Q(x)(R(x) - R^*(x))$ have the same sign on E. ($Q(x) > 0$.) But we have seen that

$$QR - QR^* = P - R^*Q \in V_P + R^*V_Q$$

has at most max $\{n + \deg Q^*, m + \deg P^*\}$ changes of sign in $[a, b]$. Therefore, $r^*(x)$ changes sign on E at most that many times; i.e., has less than

$$1 + \max\{n + \deg Q^*, m + \deg P^*\}$$

successive changes in sign on E. This completes the proof of sufficiency.

To prove the alternating sign condition is necessary, let $R^* = P^*/Q^* \in R_m^n[a, b]$ be such that the residual $r^* = f - R^*$ does not satisfy the condition. We shall show that this implies the existence of a better rational approximation in $R_m^n[a, b]$. In fact, such an improved approximation will be found in the family

$$R_\alpha = \frac{P^* + \alpha P}{Q^* - \alpha Q},$$

where α is a positive real parameter and $\deg P \leq n$, $\deg Q \leq m$. To determine the value of α, we compute the residual

$$r_\alpha = f - R_\alpha = f - R^* + R^* - R_\alpha = r^* + (R^* - R_\alpha).$$

Now,

$$R^* - R_\alpha = \frac{-\alpha(P + R^*Q)}{Q^* - \alpha Q} = \frac{-\alpha g}{Q^* - \alpha Q}, \qquad (7.9.5)$$

where the function $g = P + R^*Q$ is an arbitrary element of $V_{R^*} = V_P + R^*V_Q$. For any Q we can choose α sufficiently small so that $Q^*(x) - \alpha Q(x) > 0$ on $[a, b]$. For such α the sign of $R^*(x) - R_\alpha(x)$ is opposite to the sign of $g(x)$. Therefore, if we could find P and Q such that g has the same sign as r^*, then $|r_\alpha| < |r^*|$. Actually, something less suffices.

Since r^* does not satisfy the necessary condition, the maximum number, k, of points in E on which r^* assumes its extreme values with alternating sign must satisfy the inequality

$$1 \leq k < 2 + \max\{n + \deg Q^*, m + \deg P^*\} = 1 + d. \qquad (7.9.6)$$

Hence, there exists $x_1 > a$ such that $r^*(x)$ has the same sign for all $x \in E \cap [a, x_1]$. (Otherwise, $r^*(x)$ alternates infinitely many times.) If $r^*(x)$ has the same sign for all $x \in E$, take $x_1 = b$. Otherwise, let

$$x^* = \sup\{x_1 : r^*(x) \text{ has constant sign in } E \cap [a, x_1]\}.$$

Let

$$y_1 = \sup\{x < x^* : r^*(x) = 0\}.$$

Then $r^*(y_1) = 0$ and $r^*(x)$ has constant sign for all $x \in E \cap [a, y_1]$. Continuing in the same way with the interval $[y_1, b]$, we find y_2 such that $r^*(y_2) = 0$ and $r^*(x)$ has constant sign in $[y_1, y_2] \cap E$, this sign being opposite to the sign in $[a, y_1]$. Since there are k points in E on which $r^*(x)$ alternates in sign, the above process yields a partition $a < y_1 < \cdots < y_{k-1} < b$ such that $r^*(y_i) = 0$ $(1 \leq i \leq k - 1)$ and $r^*(x)$

$V_P + R_0 V_Q$. This can be done by choosing a basis in V_P (say $1, x, \ldots, x^n$) and then testing successively $R_0, xR_0, x^2 R_0, \ldots$, to see if each $x^i R_0$ is a linear combination of the functions already chosen. Next, determine $g = \sum_1^d a_i g_i$ such that $g(y_j) = 0$, $j = 1, \ldots, d - 1$ and $g(y_d) = \text{sgn } r_0(y_d)$, where y_d is an extremum point of r_0. Note that $g = P + R_0 Q$. This determines the polynomials P and Q. The function $g(x)$ agrees in sign with $r_0(x)$ at all its extreme points $x \in E$. There exists $\alpha > 0$ such that

$$\|f - R_\alpha\|_\infty < \|f - R_0\|_\infty.$$

This α can be determined either by trial and error or by examining the sets F and G of Theorem 7.9.2.

Prior to the question of which method to use in computing a best rational approximation is the question of whether to prefer rational to polynomial approximation—or possibly other forms such as trigonometric series, piecewise-polynomial or piecewise-rational functions [7.24] or spline functions (Section 8.5 below). This is a difficult question to which there is no clearcut general mathematical answer. The answer depends to some extent on the intended use of the approximation. If it is to be the basis of a computer "library" subroutine which will be used many times, the maximum efficiency (i.e., minimum computer time for a prescribed accuracy) might be the criterion for choosing a particular approximation. However, in some computers, the amount of storage required might also be an important consideration. Finally, the effect of rounding errors must be taken into account. Here, as in the solution of linear systems (Chapter 4), the particular algorithm (i.e., sequence of computational steps) used to evaluate the approximation must be analyzed. For example, there are many ways to compute the value, $P(x)$, of a polynomial. We have mentioned the nested form (7.9.2), also known as *Horner's form*. Another possible method is to express P as a linear combination of orthogonal polynomials, for example, the Chebyshev polynomials. Thus, $P(x) = \sum_0^n A_k T_k(x)$. We can then use the recursion formula (7.7.5) to evaluate each $T_k(x)$. This technique requires $n + 1$ multiplications and $2n$ additions. Although there are more operations than in the nested form, the Chebyshev form is "better-conditioned"; i.e., rounding errors do not grow as fast as in the nested form. (See [7.25].) There are methods which use fewer operations than the nested form. In fact, it can be proved ([7.27]) that any polynomial of degree n can be evaluated with $[(n + 1)/2] + 1$ multiplications and $n + 1$ additions. (The symbol $[x]$ denotes the nearest integer $\leq x$.) (Also see [7.26].) However, it appears that these more efficient methods may be ill-conditioned. See [7.10] for further discussion of these matters and for similar considerations on the evaluation of rational functions.

7.10 MULTIVARIATE APPROXIMATION

Much of the application of approximation theory has dealt with (real) functions of a single (real) variable. For (real) functions of several (real) variables, less is known. Of course, the abstract results of Sections 7.3 and 7.4 can be applied to the multivariate case of $m \geq 2$ variables wherever the particular functions satisfy the abstract conditions. For example, if X is a compact subset of R^m, we may consider the normed space $C(X)$ of continuous m-ary functions on X with the uniform norm. The polynomials in m variables of degree $\leq n$ constitute a finite-dimensional subspace of $C(X)$; e.g. for $m = 2$, this subspace is generated by the monomials $x^i y^j$ with $i + j \leq n$. Therefore, the fundamental theorem of the linear approximation problem (Theorem 7.3.1) applies and establishes the existence of a best approximating polynomial of degree $\leq n$ for any $f \in C[X]$. However, generalizations of specific results such as, for instance, Jackson's theorem (Theorem 7.7.9), must be obtained by analyzing specific cases and have been obtained quite recently [7.30]. One result which can be generalized without considering the specific structure of multivariate polynomials is the Weierstrass theorem.

In Section 7.2, we proved the Weierstrass theorem, which asserts that every real continuous function f on a closed bounded interval $[a, b]$ can be uniformly approximated with arbitrary accuracy by a polynomial. Taking $C[a, b]$ to be the space of real continuous functions with uniform norm, we can restate the theorem as follows: the set A of polynomial functions is dense in $C[a, b]$. Thus, $C[a, b] = \bar{A}$, the closure of A. How can this result be generalized? One theorem due to Stone [7.29] is based on the observation that the set of polynomials has two basic properties which enter into the generalization:

i) The space $C[a, b]$ becomes a linear algebra (Definition 2.1.9) if we define multiplication of two functions f, g by $(fg)(x) = f(x)g(x)$. The polynomials are a subalgebra, A, in $C[a, b]$, since the product of any two polynomials is a polynomial. The identity element, I, in $C[a, b]$ is the function which is identically 1 for all x. Thus, I is a polynomial.

ii) For any two distinct points x, y in $[a, b]$, there is a polynomial $p \in A$ such that $p(x) \neq p(y)$. Any family A having this property is said to *separate points* of $[a, b]$.

To generalize the Weierstrass theorem, we consider a compact metric space X instead of $[a, b]$. Let $C(X)$ be the space of real continuous functions on X with uniform norm. $C(X)$ is an algebra under the multiplication in (i) above.

Stone–Weierstrass Theorem Let X be a compact metric space. If A is a subalgebra of $C[X]$ which contains the identity and separates points of X, then A is dense in $C[X]$; i.e. $C[X] = \bar{A}$.

Proof. See Appendix III. ∎

The application of the Stone–Weierstrass theorem to multivariate polynomials is immediate. Consider the case of $m = 2$ variables. Let X be any closed bounded subset of R^2. Then X is compact. The polynomials in x and y form a subalgebra of $C[X]$, since the product $p(x, y) q(x, y)$ of two such polynomials is again a polynomial in x and y. For any two distinct points $(x_0, y_0), (x_1, y_1)$ in X there exists a polynomial p such that $p(x_1, y_1) \neq p(x_2, y_2)$. This follows easily by using the bivariate interpolating polynomial of degree 2 as given in formulas (8.1.2), (8.1.3) of the next in [7.42]. In [7.42], computational algorithms are given in the form of ALGOL programs for both L_1 and L_∞ approximations. In practice, the dual problem (see Ex. 12.14) is often solved, since it requires less computation when m is much larger than n.

Linear programming has also been applied to the rational approximation problem of Section 7.9. In one method [7.39], the problem is formulated as follows. We minimize

$$\max_{a \le x \le b} (1/|Q(x)|) |f(x)Q(x) - P(x)|$$

by an iterative procedure. Having found P_{k-1}, Q_{k-1}, we determine P_k, Q_k to minimize $\max (1/|Q_{k-1}(x)|) |f(x)Q_k(x) - P_k(x)|$. This latter quantity is linear in the coefficients of P_k and Q_k. If we replace the interval $[a, b]$ by a finite set of points $E = \{x_1, \ldots, x_m\}$, we obtain a weighted L_∞ linear approximation problem which can be solved as a linear programming problem. The iteration is continued until the residual function assumes its extreme value with alternating sign on $n + m + 2$ consecutive points.

Another method [7.41] for finding rational approximations uses the Remes algorithm with each step of the algorithm being effected by a sequence of linear programming solutions. Other applications of linear programming to rational approximation are given in [7.43].

REFERENCES

7.1. N. I. ACHIESER, *Theory of Approximation*, translated by C. J. Hyman, Ungar, New York (1956)
7.2. S. N. BERNSTEIN, "Démonstration du théorème de Weierstrass fondée sur le calcul de probabilité," *Proc. Math. Soc. Kharkov* **13**, 1–2 (1912)
7.3. W. RUDIN, *Real and Complex Analysis*, McGraw-Hill, New York (1966)
7.4. E. REMES, "Sur le calcul effectif des polynomes d'approximation de Tchebichef," *C. R. Acad. Sci.* (Paris) **199**, 337–340 (1935)
7.5. E. N. NOVODWORSKII and I. SH. PINSKER, "On a process of equalization of maxima," *Usp. Mat. Nauk* **6**, 174–181 (1951)
7.6. E. W. CHENEY, *Introduction to Approximation Theory*, McGraw-Hill, New York (1966)
7.7. L. VEIDINGER, "On the numerical determination of the best approximations in the Chebyshev sense," *Numerische Mathematik* **2**, 99–105 (1960)
7.8. P. J. DAVIS, *Interpolation and Approximation*, Blaisdell (Ginn & Co.), New York (1963)
7.9. A. J. FLETCHER, C. P. MILLER, and L. ROSENHEAD, *An Index of Mathematical Tables*, Second edition, Blackwell, London (1962)
7.10. J. F. HART, E. W. CHENEY, *et al.*, *Computer Approximations*, Wiley, New York (1968)
7.11. J. R. RICE, *The Approximation of Functions*, Vol. I, Addison-Wesley Reading, Mass. (1964)
7.12. C. J. DE LA VALLÉE POUSSIN, "Leçons sur l'approximation des fonctions d'une variable réelle," Gauthier-Villars, Paris (1919)
7.13. J. TODD, *Introduction to the Constructive Theory of Functions*, Birkhauser Verlag, Basel (1963)

7.14. E. K. Blum and P. C. Curtis, "Asymptotic behavior of the best polynomial approximation," *J. ACM* **8**, 645–647 (1961)
7.15. E. E. Osborne, *Smallest Least Squares Solutions of Linear Equations, TRW Systems Report* 9851-195 (December 1964)
7.16. E. E. Osborne, "On least squares solutions of linear equations," *J. ACM* **8**, 628–636 (1961)
7.17. G. W. Golub, "Numerical methods for solving linear least squares problems," *Num. Math.* **7**, 206–216 (1965)
7.18. A. Bjorck and G. Golub, "Iterative refinements of linear least squares solutions by Householder transformations," *BIT* **7**, 322–337 (1967)
7.19. J. B. Rosen, "Minimum and basic solutions to singular linear systems," *J. SIAM* **12**, 156–162 (1964)
7.20. T. N. E. Greville, "*Some applications of the pseudo-inverse of a matrix*," *SIAM Review* **2**, 15–22 (1960)
7.21. R. Penrose, "On best approximate solutions of linear matrix equations," *Proc. Cambridge Phil. Soc.* **52**, 17–19 (1956)
7.22. J. R. Rice, "A theory of condition," *J. SIAM, Numer. Anal.* **3**, No. 2, 87–310.
7.23. H. Werner, "Rationale Tschebyscheff-Approximation, Eigenwert-theorie und Differenzenrechnung," *Arch. Rational Mech. Anal.* **13**, 330–347 (1963)
7.24. C. L. Lawson, "Characteristic properties of the segmented rational minimax approximation problem," *Numer. Math.* **6**, 293–301 (1964)
7.25. J. R. Rice, "On the conditioning of polynomial and rational forms," *Numer. Math.* **7**, 426–435 (1965)
7.26. T. S. Motzkin, "Evaluation of polynomials," *Bull. Amer. Math. Soc.* **61**, 163 (1955)
7.27. E. G. Belaga, "Some problems involved in the computation of polynomials," *Dokl. Akad. Nauk. SSSR* **123**, 775–777 (1958)
7.28. D. C. Handscomb (ed.), *Methods of Numerical Approximation*, Pergamon, Oxford (1966)
7.29. M. H. Stone, "The generalized Weierstrass approximation theorem," *Mathematics Magazine* **21**, 167–183, 237–254 (1948)
7.30. M. H. Schultz, "L^∞-multivariate approximation theory," *SIAM J. Numer. Anal.* **6**, No. 2, 161–183 (June 1969)
7.31. C. A. Desoer and B. H. Whalen, "A note on pseudoinverses," *SIAM Journal*, **11**, No. 2 (June 1963)
7.32. A. Bjorck, "Solving linear least squares problems by Gram-Schmidt orthogonalization," *BIT* **7**, 1–21 (1967)
7.33. J. R. Rice, "Experiments on Gram-Schmidt orthogogonalization," *Math. Comp.* **20**, 325–328 (1966)
7.34. P. Rabinowitz, "Application of linear programming to numerical analysis," *SIAM Rev.* **10**, 121–159 (1968)
7.35. P. Rabinowitz, "Mathematical programming and approximation," in *Approximation Theory*, edited by A. Talbot, Academic Press, New York (1970)
7.36. W. J. Cody, "A survey of practical rational and polynomial approximation of functions," *S.I.A.M. Rev.* **12**, 400–423 (July 1970)
7.37. C. L. Lawson, Bibliography 18, "Recent publications in approximation theory, with emphasis on computer applications," *Comput. Rev.* **11**, 691–699 (1968)
7.38. P. Fox, A. A. Goldstein, and G. Lastman, "Rational approximation on finite point sets," in *Approximation of Functions*, edited by H. L. Garabedian, American Elsevier, New York, 57–67 (1965)

7.39. H. LOEB, "Algorithms for Chebyshev approximations using the ratio of linear forms," *SIAM Journal* **8**, 458–465 (1960)
7.40. E. L. STIEFEL, "Note on Jordan elimination, linear programming and Tchebycheff approximation," *Numer. Math.* **2**, 1–17 (1960)
7.41. E. L. STIEFEL, "Methods old and new for solving the Tchebycheff approximation problem," *S.I.A.M. J. Num. Anal.* 164–176 (1964)
7.42. I. BARRODALE and A. YOUNG, "Algorithms for the best L_1 and L_∞ linear approximations on a discrete set," *Numer. Math.* **8**, 295–306 (1966)
7.43. S. I. ZUHOVICKII and R. A. POLJAK, "An algorithm for solving the problem of rational Chebyshev approximation," *Soviet Math.* **5**, 1574–1577 (1964)
7.44. J. B. ROSEN, "Minimum error bound for multidimensional spline approximation," *J. Comp. and Syst. Sciences* **5**, No. 4 (1971)

EXERCISES

7.1 a) Extend Weierstrass' theorem to the case of a function of two variables $f(x, y)$ continuous on the unit square $0 \le x \le 1$, $0 \le y \le 1$ by using the Bernstein polynomials

$$B_{mn}(x, y) = \sum_{j=0}^{m} \sum_{i=0}^{n} f\left(\frac{j}{m}, \frac{i}{n}\right) \binom{m}{j} x^j (1-x)^{m-j} \binom{n}{i} y^i (1-y)^{n-i}.$$

b) Show that $\binom{n}{j} x^j (1-x)^{n-j}$ has a unique maximum at $x = j/n$, which approaches $j^j/j! e^j$ as $n \to \infty$ for j fixed. [*Hint:* Differentiate with respect to x. Then use Stirling's formula $n! = \sqrt{2\pi} \, n^{n+1/2} e^{-n+\varepsilon_n}$, $\varepsilon_n \to 0$.]

7.2 In complex function theory, the following is a theorem. Let G be an open set of the complex plane and suppose (f_n) is a sequence of functions holomorphic in G. If $f_n(z) \to f(z)$ uniformly in z for every compact subset of F, then f is holomorphic in G. Explain why this does not apply to the uniform limit of a sequence of polynomials $p_n(x) \to f(x)$, where the convergence is uniform in x on a real closed bounded interval $[a, b]$. (See [7.3], p. 214, for example.)

7.3 Let E be a closed subset in a finite-dimensional normed vector space V. Show that for any f in V there exists a closest point in E. [*Hint:* Choose any g in E and note that the set of points x in E such that $\|x - f\| \le \|g - f\|$ is closed and bounded.]

7.4 Consider the Banach space of convergent sequences $x = (\xi_n)$ such that $\xi_n \to 0$, with $\|x\| = \max |\xi_n|$. Show that the closed infinite-dimensional subspace W of sequences such that $\sum_{n=1}^{\infty} 2^{-n} \xi_n = 0$ does not contain a vector closest to any vector f not in the subspace. [*Hint:* Let $f = (\alpha_1, \alpha_2, \ldots)$ and $l = \sum_{n=1}^{\infty} 2^{-n} \alpha_n \ne 0$. The sequence $u_1 = -2(l, 0, 0, \ldots) + f$, $u_2 = -\frac{4}{3}(l, l, 0, 0, \ldots) + f$, $u_3 = -\frac{8}{7}(l, l, l, 0, 0, \ldots) + f, \ldots$ is such that $\|u_n - f\| \to l$, and $u_n \in W$. For any $x \in W$, if $\|x - f\| \le l$, then

$$\left|\sum 2^{-n} \alpha_n\right| = \left|\sum 2^{-n} (\alpha_n - \xi_n)\right| \le \sum 2^{-n} |\alpha_n - \xi_n|$$

$$\le |l| \sum_{n<k} 2^{-n} + \frac{|l|}{2} \sum_{n \ge k} 2^{-n} < l,$$

a contradiction. Thus, $\|x - f\| > l$ for all $x \in W$.]

7.5 a) Prove that an inner-product space is uniformly convex. [*Hint:* Use the parallelogram law (Exercise 3.29).]

b) Prove that V^n with the l_1-norm is not uniformly convex.
[*Hint:* $\|(1, 0)\| = \|(0, 1)\| = \|(\frac{1}{2}, \frac{1}{2})\| = 1$ and $\|(1, -1)\| = 2.$]
Draw the unit l_1-sphere. The same for l_∞.

c) Prove that L^p, $1 < p < \infty$ is uniformly convex.

7.6 Show that the point $\bar{x}$ of Theorem 7.3.2 is unique. [*Hint:* If $\|\bar{y}\| = 1$, $\bar{y} \in E''$, then $\frac{1}{2}(\bar{x} + \bar{y}) \in E''$ by convexity and $\frac{1}{2}\|\bar{x} + \bar{y}\| \geq 1$. Then use uniform convexity.]

7.7 a) Prove that V is strictly normed if and only if every closed ball is strictly convex. [*Hint:* $\|\alpha x + \beta y\| = \|x\| = \|y\| = 1$ implies $\|\alpha x + \beta y\| = \alpha\|x\| + \beta\|y\|$, since $\alpha + \beta = 1$. By the strict norm, $\alpha x = \lambda \beta y$ and $\|\lambda \beta y + \beta y\| = 1$. This implies $|\lambda|\beta = \alpha$ and $|\lambda + 1|\beta = 1$. Hence, $y = x$. Conversely, if $\|x + y\| = \|x\| + \|y\|$, then $\|\alpha \bar{x} + \beta \bar{y}\| = 1$, where $\alpha = \|x\|/(\|x\| + \|y\|)$, $\beta = \|y\|/(\|x\| + \|y\|)$ and $\bar{x} = x/\|x\|$, $\bar{y} = y/\|y\|$. By strict convexity, this can only happen if $\bar{x} = \bar{y}$.]

b) Show that l_∞ and $C[a, b]$ are not strictly normed.

7.8 Prove that $g = \|f - \sum_{i=1}^{n} \alpha_i f_i\|$ is a convex function of the α_i. [*Hint:*
$$g(\lambda x + \beta y) = \|f - \sum (\lambda \xi_i + \beta \eta_i) f_i\|$$
$$= \|\lambda(f - \sum \xi_i f_i) + \beta(f - \sum \eta_i f_i)\| \leq \lambda g(x) + \beta g(y).]$$

7.9 a) Let $d(s) = \|sx + (1 - s)y\|$, $0 \leq s \leq 1$, be the distance to the origin of any point on the line segment joining x to y. Prove that $d(s)$ is a convex function. [*Hint:* Let $\lambda + \mu = 1$, $\lambda \geq 0$, $\mu \geq 0$. Then
$$d(\lambda a + \mu b) = \|(\lambda a + \mu b)x + (1 - \lambda a - \mu b)y\|$$
$$= \|\lambda(ax + (1 - a)y) + \mu(bx + (1 - b)y)\| \leq \lambda d(a) + \mu d(b).]$$

b) Use (a) to show that a ball is convex.

c) Let $\|y_1\| = \|y_2\| = 1$. Suppose
$$\frac{\|y_1 + y_2\|}{2} = 1 - \delta, \delta > 0.$$
Show that this implies $\|\lambda y_1 + \mu y_2\| < 1$ for all λ, μ with $\lambda + \mu = 1$, $\lambda > 0$, $\mu > 0$. Thus, if the midpoint of the line segment joining two points on the unit sphere is in the interior of the unit ball, then all points on the line segment are interior. Hence, strict convexity could have been defined by "$\|x\| = \|y\| = 1$ implies $\|\frac{1}{2}(x + y)\| < 1$ for $x \neq y$." [*Hint:* In (a), let $x = \frac{1}{2}(y_1 + y_2)$. Then
$$d(0) = 1, d(1) = 1 - \delta \quad \text{and} \quad d(\mu) = d(\lambda \cdot 0 + \mu \cdot 1)$$
$$\leq (1 - \mu)d(0) + \mu d(1) = 1 - \mu + \mu(1 - \delta)$$
$$= 1 - \mu\delta < 1.$$
Similarly, letting $y = \frac{1}{2}(y_1 + y_2)$, we obtain the points on the other half of the line segment.]

d) Show that $\|f\|$ is a convex function of f.

7.10 Prove: **a)** A convex function defined on an open convex set in R^n is continuous;

b) A twice-differentiable real function $f(x)$, $x \in R$, is convex if and only if $f''(x) \geq 0$;

c) If f is a real convex function on R, then f has one-sided derivatives at each $x \in R$. Generalize this to f defined on R^n.

7.11 Prove: **a)** A uniformly convex space V is strictly normed. [*Hint:* $\|x\| = \|y\| = \frac{1}{2}\|x + y\| = 1$ implies $\|x - y\| < \varepsilon$ for all $\varepsilon > 0$. Hence, the unit sphere in V is strictly convex by Exercise 7.9c. Then use Exercise 7.7.]

b) The converse of (a) when V is finite dimensional. [*Hint:* The set S of pairs $(x, y) \in V \times V$ such that $\|x\| = \|y\| = 1$ and $\|x - y\| \geq \varepsilon$ is compact. The function $\varphi(x, y) = 1 - \frac{1}{2}\|x + y\|$ is continuous and positive on S. Hence $\delta_\varepsilon = \inf_{x \in S} \varphi(x) > 0$ and $\|x - y\| \geq \varepsilon$ implies $1 - \frac{1}{2}\|x + y\| \geq \delta_\varepsilon$, i.e., $\frac{1}{2}\|x + y\| > 1 - \delta_\varepsilon$ implies $\|x - y\| < \varepsilon$.]

7.12 Prove that a finite set of vectors $\{f_1, \ldots, f_n\}$ is linearly independent if and only if the Gram determinant is nonzero. [*Hint:* Linear dependence $\Rightarrow \exists \; \alpha_j$ not all zero and $\sum_1^n \alpha_j f_j = 0 \Rightarrow \sum_{j=1}^n \alpha_j \langle f_j, f_i \rangle = 0, \; 1 \leq i \leq n \Rightarrow \det(\langle f_i, f_j \rangle) = 0$. Conversely, $\det(\langle f_i, f_j \rangle) = 0 \Rightarrow \exists \alpha_j$ not all zero: $\sum_{j=1}^n \alpha_j \langle f_j, f_i \rangle = 0, 1 \leq i \leq n \Rightarrow \langle \sum_{j=1}^n \alpha_j f_j, f_i \rangle = 0, 1 \leq i \leq n \Rightarrow \langle \sum \alpha_j f_j, \sum \alpha_j f_j \rangle = 0$.]

7.13 Prove that a Vandermonde determinant $D = \prod_{0 \leq j < i \leq n} (x_i - x_j)$. [*Hint:* Use induction on n. To reduce the order, multiply column k by x_0 and subtract it from column $k + 1$, $k = n - 1, n - 2, \ldots, 1$. Then since $x_j^{k+1} - x_0 x_j^k = x_j^k(x_j - x_0)$, factor $x_j - x_0$ from row j. Expand by minors of row 1.]

7.14 Let

$$f(\theta) = \sum_{k=0}^n (a_k \cos k\theta + b_k \sin k\theta)$$

and

$$g(\theta) = \sum_0^n (c_k \cos k\theta + d_k \sin k\theta).$$

Prove: If $f(\theta_i) = g(\theta_i)$ for $2n + 1$ points θ_i in $[0, 2\pi)$, then $a_k = c_k$ and $b_k = d_k$ for all k. [*Hint:* Express $\cos k\theta, \sin k\theta$ in terms of $e^{\pm ik\theta}$, making $e^{in\theta} f(\theta)$ a polynomial in $e^{i\theta}$.]

7.15 Let $\{p_0, p_1, \ldots, p_n\}$ be a set of polynomials such that each p_i is of degree $i, 0 \leq i \leq n$. Show that the set is linearly independent and that any polynomial q of degree $\leq n$ can be expressed uniquely as a linear combination of the p_i.
[*Hint:*

$$q(x) = a_0 + a_1 x + \cdots + a_n x^n.$$

Let

$$p_i = b_{i0} + b_{i1}x + \cdots + b_{ii}x^i, \qquad b_{ii} \neq 0.$$

Then $r(x) = q(x) - (a_n/b_{nn})p_n(x)$ is a polynomial of degree $\leq n - 1$. Apply induction to obtain $r(x) = \sum_{i=0}^{n-1} \alpha_i p_i(x)$.]

7.16 a) Use Theorem 7.7.8 to obtain an estimate of the degree n required to uniformly approximate e^x with an accuracy of 10^{-8} by a polynomial on $[-1, 1]$. (See parts (b) and (c).)

b) Compute the Chebyshev coefficients A_k of e^x using

$$A_k = (2/\pi) \int_0^\pi e^{\cos \theta} \cos k\theta \; d\theta = \frac{2}{-i^k} J_k(i),$$

where J_k is the Bessel function of order k. Note that

$$\frac{2}{-i^k} J_k(i) = 2I_k(1)$$

and

$$I_k(x) = \sum_{j=0}^{\infty} \frac{1}{j!(j+k)!} \left(\frac{x}{2}\right)^{k+2j}.$$

(See a Table of Bessel Functions, [7.9].)
(Answer: $\tfrac{1}{2}A_0 = 1.2660658778$
$A_1 = 1.1303182080$
$A_2 = 0.2714953395$
$A_3 = 0.0443368498$
$A_4 = 0.0054742404$
$A_5 = 0.0005429263$
$\vdots$

c) Use the A_k to compute the polynomial $\tfrac{1}{2}A_0 + \sum_1^n A_k T_k(x)$ obtaining $T_k(x)$ by formula (7.7.5) or see a table of Chebyshev coefficients [7.9].

d) Program the single-exchange algorithm for e^x in part (a).

7.17 The *Cesaro means* of a sequence (f_n) are defined as the means $F_n = (1/n)\sum_1^n f_i$, $n = 1, 2, \ldots$. The Cesaro means of a series $\sum_{i=1}^{\infty} a_i$ are the Cesaro means of the sequence of partial sums $f_n = \sum_1^n a_i$. Prove that if f is a continuous function of period 2π, then the Cesaro means of its Fourier series converge uniformly to f. (See [7.6], p. 123.)

7.18 Verify that the Chebyshev polynomials $T_n(x)$ satisfy the differential equation

$$(1 - x^2)T_n''(x) - xT_n'(x) + n^2 T(x) = 0.$$

7.19 *Discrete least-squares approximation.* Let $E = \{x_1, \ldots, x_m\}$ be a set of distinct points. Let $f(x_i)$, $1 \leq i \leq m$, be a set of function values.

a) Determine the least-squares polynomial $p(x) = a_0 + a_1 x + \cdots + a_n x^n$ of degree $\leq n$ which minimizes $\sum_{i=0}^m |f(x_i) - p(x_i)|^2$. [*Hint:* Show that the normal equations are of the form $A^T A a = A^T f$, where $a^T = (a_0, \ldots, a_n)$, $f^T = (f(x_1), \ldots, f(x_m))$ and

$$A = \begin{pmatrix} 1 & x_1 & x_1^2 & \cdots & x_1^n \\ \vdots & & & & \\ 1 & x_m & x_m^2 & \cdots & x_m^n \end{pmatrix}$$

Thus $B = A^T A$ is an $(n+1) \times (n+1)$ symmetric matrix with entries $b_{ij} = \sum_{k=1}^m x_k^{i+j}$, $0 \leq i, j \leq n$, and the vector $A^T f$ has components $\sum_{k=1}^m f(x_k) x_k^i$, $0 \leq i \leq n$, which are the *moments* of f.]

b) Show that the normal matrix $A^T A$ becomes ill-conditioned as n becomes large. [*Hint:* Let $x_k = k/m$, $1 \leq k \leq m$.] Compare the Hilbert matrix, Exercise 4.19c.

c) The set V of functions restricted to the set E is an m-dimensional Hilbert space with the inner product

$$\langle f, g \rangle = \sum_{k=1}^m f(x_k) g(x_k).$$

The subset $\{1, x, x^2, \ldots, x^n\}$ regarded as functions restricted to E is linearly independent if $m > n$ and linearly dependent if $m \leq n$. Let $f_i = x^i$, $0 \leq i \leq n$. Show that the normal matrix A^TA is the Gram matrix $(\langle f_i, f_j \rangle)$ given in (7.4.1) and the moments of f are $\langle f, f_i \rangle$, $0 \leq i \leq n$. Hence, show that the normal equations always have a solution and that the solution is unique if and only if $m > n$. Show that for $m \leq n$, the rank of A^TA is m. Hence, there are $n + 2 - m$ linearly independent solutions $(a_0, \ldots, a_n)$, each giving rise to a least-squares polynomial of degree $\leq n$. Show that the uniqueness of the least-squares vector (Theorem 7.4.1) implies that all least-squares polynomials take the same values on the set E.

d) Let $f \in C[a, b]$ and $(Q_0, \ldots, Q_n)$ be an orthogonal sequence of polynomials on $[a, b]$, with Q_i of degree i. Let $\sum_{i=0}^{n} a_i Q_i$ be the least-squares approximation to f. Prove that $g(x) = f(x) - \sum_{i=0}^{n} a_i Q_i(x)$ has at least $n + 1$ changes of sign in $[a, b]$. (Thus, the least-squares polynomial "oscillates about the graph $y = f(x)$".) [*Hint:* See Theorem 8.3.2 and note that $\langle g, Q_j \rangle = 0$, $0 \leq j \leq n$.]

7.20 Show that the Weierstrass theorem implies that the set of polynomials is everywhere dense in the space $C[a, b]$. (See Definition 1.4 and the discussion following it.)

7.21 a) Use formulas (7.7.2), (7.7.3), (7.7.4) to compute the first few Legendre polynomials $P_n(x)$ as defined in Example (c) of Section 7.7. (Use $\langle f, g \rangle = \int_{-1}^{1} f(x)g(x)\,dx$ and obtain $P_0 = 1$, $a_1 = 0$, $P_1 = x$, $a_2 = 0$, $b_2 = \frac{1}{3}$, $P_2 = \frac{3}{2}x^2 - \frac{1}{2}$, etc.)

b) Specialize (7.7.2), (7.7.3), (7.7.4) to the weight function $\rho = 1$ and the interval $[-1, 1]$ to obtain the recurrence relation for the Legendre polynomials (not normalized)

$$P_n(x) = \frac{2n - 1}{n} x P_{n-1}(x) - \frac{n - 1}{n} P_{n-2}(x).$$

c) For a finite set of points $\{x_0, x_1, \ldots, x_M\}$ compare the amount of computation required to obtain the least-squares polynomial p of degree $\leq n$ by solving the normal equations with the method of finding an orthonormal family of polynomials by formula (7.7.2). [*Hint:* Suppose that $Q_{n-1} = \sum_0^{n-1} c_j x^j$ has been determined, $c_{n-1} \neq 0$. Then

$$\langle xQ_{n-1}, Q_{n-1} \rangle = c_{n-1} \langle x^n, Q_{n-1} \rangle + c_{n-2} \langle x^{n-1}, Q_{n-1} \rangle$$

and

$$1 = \langle Q_{n-1}, Q_{n-1} \rangle = \langle c_{n-1} x^{n-1}, Q_{n-1} \rangle.$$

Thus,

$$\langle x^{n-1}, Q_{n-1} \rangle = 1/c_{n-1}$$

and formula (7.7.3) yields

$$a_n = -c_{n-1} \langle x^n, Q_{n-1} \rangle - c_{n-2}/c_{n-1}.$$

We must compute

$$\langle x^n, Q_{n-1} \rangle = \sum_{j=0}^{n-1} c_j \sum_{i=0}^{M} x_i^{n+j} = \sum_{j=0}^{n-1} c_j s_{n+j},$$

where

$$s_k = \sum_{i=0}^{M} x_i^k, \quad k = 0, 1, \ldots, 2n - 1.$$

[We have assumed a weight function $\rho(x) = 1$ identically.]

d) In Exercise 7.21c, let the $x_i = -1 + 2i/M$. Show that $a_n = 0$ for all n. [*Hint:* Induction on n. Note that $\sum_{i=0}^{M} x_i^{2k+1} = 0, k = 0, 1, \ldots$. We have $Q_0 = 1/\sqrt{M+1}$ and

$$a_1 = -\langle xQ_0, Q_0 \rangle = 1/(M+1) \sum_{i=0}^{M} x_i = 0.$$

$Q_1 = (1/\|\bar{Q}_1\| \sqrt{M+1})x$ and $-a_2 = \text{const} \langle x^2, x \rangle = \text{const} \sum x_i^3 = 0$. Hence, $b_2 = -\|\bar{Q}_1\|$ and $Q_2 = c_{22}x^2 + c_{20}$. Thus, Q_2 has only even powers and

$$\langle xQ_2, Q_2 \rangle = \langle c_{22}x^3 + c_{20}x, c_{22}x^2 + c_{20} \rangle$$

involves only sums of odd powers, hence is zero. Therefore, $\bar{Q}_3$ has only odd powers of x. By induction, if Q_{2k-2} has only even powers and Q_{2k-1} only odd powers, then $a_{2k} = 0$ and $Q_{2k} = xQ_{2k-1} + b_{2k}Q_{2k-2}$ has only even powers.]

e) In part (c), examine the effects of rounding errors on the orthogonality relations.

7.22 Prove the following trigonometric relations:

a) $(\sin(nt/2)/\sin(t/2))^2 = n + 2\sum_{j=1}^{n-1}(n-j)\cos jt$.

[*Hint:*

$$\sin^2\left(\frac{nt}{2}\right) = \frac{1 - \cos nt}{2} = \frac{1}{2}\sum_{j=0}^{n-1}(\cos jt - \cos(j+1)t)$$

$$= \sum_{j=1}^{n} \sin\frac{(2j-1)t}{2} \sin\frac{t}{2}.$$

Also,

$$\sin\frac{(2j-1)t}{2} = \sin\frac{t}{2} + \left(\sin\frac{3t}{2} - \sin\frac{t}{2}\right) + \cdots + \left(\sin\frac{(2j-1)t}{2} - \sin\frac{(2j-3)}{2}t\right)$$

$$= \sin\frac{t}{2} + 2\sin\frac{t}{2}\cos t + \cdots + 2\sin\frac{t}{2}\cos(j-1)t. \ ]$$

b) $\int_0^{\pi/2}\left(\frac{\sin nt}{\sin t}\right)^4 dt = \frac{\pi n(2n^2+1)}{6}.$

[*Hint:* Use part (a) and Parseval's identity to get

$$\int_{-\pi}^{\pi}\left(\frac{\sin(nt/2)}{\sin(t/2)}\right)^4 dt = \pi\left(2n^2 + 4\sum_{j=1}^{n-1}(n-j)^2\right).$$

Also,

$$\sum_{j=1}^{n-1}(n-j)^2 = \sum_{j=1}^{n-1} j^2 = n(n-1)(2n-1)/6.$$

c) $\int_0^{\pi/2} t\left(\frac{\sin nt}{\sin t}\right)^4 dt < \frac{\pi^2 n^2}{4}.$

[*Hint:* In $[0, \pi/2n]$ use the inequality $|\sin nt| \leq n|\sin t|$, which is proved by induction on n. In $[\pi/2n, \pi/2]$, use $\sin t \geq 2t/\pi$, which holds for $0 \leq t \leq \pi/2$, since $\sin t/t$ has a negative derivative. Hence,

$$\int_0^{\pi/2} t \left(\frac{\sin nt}{\sin t}\right)^4 dt < n^4 \int_0^{\pi/2n} t\, dt + \frac{\pi^4}{16} \int_{\pi/2n}^{\pi/2} \frac{dt}{t^3}$$

$$< n^4 \frac{\pi^4}{8n^2} + \frac{\pi^4}{16} \int_{\pi/2n}^{\infty} \frac{dt}{t^3}.]$$

7.23 a) Show that the kernel K of a seminormed vector space V is a subspace of V. [*Hint:* $0 \le \|f + g\| \le \|f\| + \|g\| = 0$ for $f, g \in K$.]

b) Verify that V/K is a vector space. [*Hint:* Show that $\hat{f} + \hat{g} = \widehat{f + g}$ is a well-defined operation by showing that $f \equiv f_1$ and $g \equiv g_1$ implies $f + g \equiv f_1 + g_1$, since $\|(f + g) - (f_1 + g_1)\| \le \|f - f_1\| + \|g - g_1\| = 0$. Similarly, for $\alpha \hat{f}$.

c) Show that $\|\hat{f}\|$ is well-defined by $\|\hat{f}\| = \|f\|$ for any $f \in \hat{f}$. [*Hint:* If $f_1 \in \hat{f}$, then $f = f_1 + f_2$, where $\|f_2\| = 0$. Hence, $\|f\| \le \|f_1\|$ and $\|f_1\| \le \|f\|$.]

7.24 a) Let K be the field of characteristic 2; i.e., $K = \{0, 1\}$ with the operation of addition modulo 2. For $x \in V^n$, define $\|x\|_p = (\sum_1^n |\xi_i|^p)^{1/p}$. Show that $\|x\|_p = 0$ need not imply $x = 0$; i.e., $\|x\|_p$ is a seminorm. [*Hint:* Suppose an even number of the ξ_i are 1.]

b) In part (a), show that $x^T y$ is not an inner product.

7.25 Show that the minimax $M_n[a, b]$ for $f \in C[a, b]$ is continuous on the set G of Theorem 7.5.4. [*Hint:* Let $P^* = (x_0^*, x_1^*, \ldots, x_{n+1}^*)$ be a boundary point of G such that $x_i^* = x_{i+1}^* = x_{i+2}^*$. The B.A.P., π_n^*, for f on the set $E = \{x_0^*, \ldots, x_i^*, x_{i+3}^*, \ldots, x_{n+1}^*\}$ is the interpolating polynomial. Hence, $M_n(E) = 0$. By definition, $M_n(P^*) = 0$. For any P sufficiently close to P^*, say $P = (x_0, x_1, \ldots, x_{n+1})$,

$$M_n(P) \le \max_{0 \le i \le n+1} |f(x_i) - \pi_n^*(x_i)| < \varepsilon$$

by the continuity of f and π_n^* at x_i^*.]

7.26 Suppose the single-exchange algorithm (Section 7.5) is modified as follows:
Step 3. Determine y such that

$$M_n[a, b] \le |r_k(y)|.$$

(Thus, y need not maximize $r_k(x)$.) Show that Theorem 7.5.7 is valid for the modified algorithm. (This shows that errors in finding the y which maximizes $r_k(x)$ need not impair convergence of the single-exchange algorithm. Of course, since $M_n[a, b]$ is unknown, the above condition cannot always be verified. However, the bounds in Theorems 7.7.7 and 7.7.9 (Corollary) can often be used instead.)

7.27 a) In Section 7.7, Example (b), the Chebyshev polynomials $T_n(x)$ were shown to be an orthonormal family on $[-1, 1]$. Verify the following orthogonality relations for the T_n on any finite set of $N + 1$ points x_j of the type $x_j = \cos(j\pi/N), j = 0, \ldots, N$:

$$S_N = \tfrac{1}{2}(T_n(x_0)T_m(x_0) + T_n(x_N)T_m(x_N)) + \sum_{j=1}^{N-1} T_n(x_j)T_m(x_j) = \delta \frac{N}{2},$$

where $\delta = \delta(n, m, N)$ is defined by

$$\delta = \begin{cases} 0 & \text{if } n \not\equiv \pm m \pmod{2N} \\ 1 & \text{if either } n \equiv m \pmod{2N} \text{ or } n \equiv -m \pmod{2N} \\ 2 & \text{if both } n \equiv -m \pmod{2N} \text{ and } n \equiv m \pmod{2N} \end{cases}$$

(Note: $n \equiv m \pmod{2N}$ if and only if $2N$ divides $n - m$.)
[*Hint*:

$$x_j = \cos(j\pi/N) = \cos((2N - j)\pi/N) = x_{2N-j}.$$

$$S_N = \tfrac{1}{2} \sum_{j=0}^{2N-1} \cos \frac{n j \pi}{N} \cos \frac{m j \pi}{N}$$

$$= \tfrac{1}{4} \sum_{j=0}^{2N-1} \left(\cos \frac{(n+m)j\pi}{N} + \cos \frac{(n-m)j\pi}{N} \right)$$

$$= \tfrac{1}{4} \operatorname{Re} \left(\sum_{j=0}^{2N-1} [(e^{(n+m)i\pi/N})^j + (e^{(n-m)i\pi/N})^j] \right)$$

$$= \tfrac{1}{4} \operatorname{Re} \left(\frac{1 - r^{2N}}{1 - r} + \frac{1 - s^{2N}}{1 - s} \right),$$

where $r = e^{(n+m)i\pi/N}$ and $s = e^{(n-m)i\pi/N}$. Recall that $e^{2\pi k i} = 1$ for any integer k and $\operatorname{Re}(z)$ means the real part of the complex number z. Note that this establishes that the $T_n(x)$ are the Gram polynomials (up to multiplicative constants) on such finite sets.]

b) Consider the set of $2N + 1$ equally spaced points $x_j = j\pi/N$, $j = 0, \pm 1, \ldots, \pm N$, in the interval $[-\pi, \pi]$. Show that the following orthogonality relations hold for the functions e^{inx}:

$$\delta = \tfrac{1}{2}(e^{in\pi}e^{-im\pi} + e^{-in\pi}e^{im\pi}) + \sum_{j=-N+1}^{N-1} e^{inj\pi/N} e^{-imj\pi/N},$$

where $\delta = 2N$ if $n \equiv m \pmod{2N}$ and $\delta = 0$ otherwise. [*Hint*: Sum the geometric series $\sum s^j$, where s is as in part (a).]

c) Verify the following orthogonality relations for the family $\{\sin nx, \cos mx\}$ on the finite set of $2N$ equally spaced points, $x_j = j\pi/N$, $j = 0, \ldots, 2N - 1$, in the interval $[0, 2\pi]$:

$$\sum_{j=0}^{2N-1} \cos \frac{n j \pi}{N} \cos \frac{m j \pi}{N} = \delta N$$

$$\sum_{j=0}^{2N-1} \sin \frac{n j \pi}{N} \sin \frac{m j \pi}{N} = \delta N$$

$$\sum_{j=0}^{2N-1} \cos \frac{n j \pi}{N} \sin \frac{m j \pi}{N} = 0,$$

where δ is given in part (a). [*Hint*: See the hint in part (a).]

7.28 Let $f(x)$ be periodic with period 2π and suppose $f(x_j)$ is given for the points $x_j = j\pi/N$, $j = 0, \ldots, 2N - 1$ as in 7.27(c). Obtain the least-squares trigonometric sum $\sum_{j=0}^{n}(a_j \cos jx + b_j \sin jx)$ of order $n < 2N$ on the given set of points x_j. [*Hint*: Use the orthogonality relations in Exercise 7.28c and apply the general result (7.4.2).]

7.29 Verify the following orthogonality relations for the Chebyshev polynomials $T_k(x)$, $k < n+1$, with respect to the finite set of points

$$x_j = \cos \theta_j, \quad \theta_j = \frac{2j+1}{n+1}\frac{\pi}{2}, \quad j = 0, 1, \ldots, n:$$

$$S_n = \sum_{j=0}^{n} T_k(x_j) T_m(x_j) = \begin{cases} 0 & \text{for } k \neq m, \\ \frac{n+1}{2} & \text{for } k = m \neq 0, \\ n+1 & \text{for } k = m = 0. \end{cases}$$

Note that the points x_j are the zeros of $T_{n+1}(x)$. Also, the θ_j are equally spaced in $[0, \pi]$ with $\Delta\theta_j = \pi/(n+1)$, but do not include either end point as in Exercise 7.27. [*Hint:* $T_k(x) = \cos k\theta$, where $x = \cos \theta$.

$$S_n = \sum_{j=0}^{n} \cos k\theta_j \cos m\theta_j.$$

Use the hint in Exercise 7.27a.]

7.30 The *Chebyshev polynomials of the second kind* are defined as the polynomials

$$U_n(x) = \frac{\sin((n+1)\arccos x)}{\sqrt{1-x^2}}, \quad n = 0, 1, 2, \ldots.$$

a) Verify that the U_n are polynomials of degree n in x having leading coefficient 2^n. [*Hint:* $\sin(n+2)\theta + \sin n\theta = 2\cos\theta \sin(n+1)\theta$. Set $x = \cos\theta$ and obtain the recursion,

$$U_{n+1}(x) = 2xU_n(x) - U_{n-1}(x),$$

where $U_0(x) = 1$ and $U_1(x) = 2x$. Then use induction.]

b) Verify that

$$\int_{-1}^{1} U_i(x) U_j(x) \sqrt{1-x^2}\, dx = \delta_{ij}\pi/2.$$

[*Hint:* Make the substitution $x = \cos\theta$.] Hence, U_n is a constant multiple of $J_n^{(1/2, 1/2)}$.

c) Prove that the minimum of $\int_{-1}^{1} |p(x)|\, dx$ over the set of monic polynomials p of degree n is attained for $p = 2^{-n}U_n$. [*Hint:* See [9.3], p. 30.]

7.31 Let A be a bounded linear operator from a Hilbert space X into a Hilbert space Y. Let N_A, R_A be the null space and range of A respectively, and similarly for N_{A^*}, R_{A^*}. Prove the following:

a) $\quad \bar{R}_A^\perp = N_{A^*}, \quad \bar{R}_A = N_{A^*}^\perp,$

b) $\quad \bar{R}_{A^*} = N_A^\perp, \quad \bar{R}_{A^*}^\perp = N_A.$

c) $\quad X = N_A \oplus \bar{R}_{A^*}, \quad Y = \bar{R}_A \oplus N_{A^*}.$

[*Hint:* (a) $y \in \bar{R}_A^\perp$ implies $\langle y, Ax \rangle = 0$ all $x \in X$. Hence, $\langle A^*y, x \rangle = 0$ and $A^*y = 0$. Conversely, $A^*y = 0$ implies $0 = \langle A^*y, x \rangle = \langle y, Ax \rangle$ all x. Hence, $\bar{R}_A^\perp = N_{A^*}$. Now, use Exercise 3.29d.]

d) Let $X = V^n$, $Y = V^m$ and A an $m \times n$ matrix. Relate parts (a), (b) to the fact that the rows of A are orthogonal to $x \in N_A$. [*Hint:* The columns of A^* span R_{A^*}.]

e) A is obviously an injective map of $N_A^\perp$ onto R_A. It can be shown ([3.9], p. 488) that R_A is closed if and only if R_{A^*} is closed. (In the finite-dimensional case, all subspaces are closed.) Assume that R_A is closed. By (b), A is an injective map of R_{A^*} onto R_A. This suggests an alternative definition of the *pseudoinverse* (Exercise 2.13), namely, $A^\dagger$ is a linear operator from Y to X satisfying the two conditions,

i) $\qquad\qquad\qquad A^\dagger A x = x \qquad$ for $x \in R_{A^*}$,

ii) $\qquad\qquad\qquad A^\dagger y = 0 \qquad$ for $y \in N_{A^*} = \overline{R}_A^\perp$.

Show that (i) and (ii) do, in fact, define a unique mapping from Y to X and $A^\dagger$ maps the range of A onto the range of A^*. [*Hint:* Use (c) above and the injectiveness of A on R_{A^*}.]

f) Verify that (by the correspondences in (a), (b))

iii) $(A^\dagger)^\dagger = A$

g) Show that

iv) $A^\dagger A = P_{R(A^\dagger)}$

where $P_{R(A^\dagger)} = P_{R(A^*)}$ is the projection of Y onto $R_{A^*} = R_{A^\dagger}$.

h) Prove

v) $A^\dagger A A^\dagger = A^\dagger$ and $(A^\dagger A)^* = A^\dagger A$.

[*Hint:* $y = y_1 + y_2$ with $y_1 \in R_A$, $y_2 \in N_{A^*}$. By (ii), $A^\dagger y = A^\dagger y_1 \in R_{A^*}$ and $A^\dagger A A^\dagger y_1 = A^\dagger y_1$ by (iii). Self-adjointness follows from (iv).]

i) Prove that

vi) $A A^\dagger A = A$ and $(A A^\dagger)^* = A A^\dagger$.

[*Hint:* By (iii) and (v) or use (i) and (c).]

j) Finally, prove (and compare with Exercise 2.13k)

vii) $A A^\dagger = P_{R(A)}$

(Relations (iv) and (vii) were used as the first definition of pseudoinverse by E. H. Moore, *Bulletin Amer. Math. Soc.*, **26** (1920), 394–395.)

Also see Sections 4.6 and 6.11 again for methods of computing the pseudoinverse.

8.2 OTHER FORMS OF THE INTERPOLATING POLYNOMIAL

The Lagrange coefficients $l_i^{(n)}(x)$ (see formula (8.1.2) above) have the property that $F_i l_j = \delta_{ij}$, where $F_i(p(x)) = p(x_i)$, $0 \leq i \leq n$ for any polynomial p of degree $\leq n$. Thus, the functionals $\{F_0, F_1, \ldots, F_n\}$ and the polynomials $\{l_0^{(n)}, l_1^{(n)}, \ldots, l_n^{(n)}\}$ constitute a biorthonormal set. (See Exercise 2.9c.) For such a set, the general interpolation problem is very easy to solve. For suppose we wish to find v such that $F_i v = y_i$, $0 \leq i \leq n$. Writing $v = \sum_0^n a_j l_j^{(n)}$, we find immediately that $a_i = y_i$, since $F_i l_j^{(n)} = 0$ except for $j = i$ and $F_i l_i^{(n)} = 1$. Unfortunately, if we increase the number of points x_i by one, this raises the dimension and the polynomials $l_i^{(n)}$ no longer form a biorthonormal set with the functionals $F_0, \ldots, F_{n+1}$. We must compute new Lagrange coefficients $l_i^{(n+1)}$ of degree $n+1$. This computational problem can be avoided if we choose a different biorthonormal set by the scheme indicated in Exercise 2.9d. (See also Exercises 8.5 and 8.11.)

Let $\{x_0, x_1, \ldots, x_n\}$ be a set of $n+1$ points. Define the polynomials $\varphi_k(x)$ of degree k by the formulas,

$$\varphi_k(x) = (x - x_0)(x - x_1) \cdots (x - x_{k-1}), \quad 1 \leq k \leq n,$$
$$\varphi_0(x) = 1. \quad (8.2.0)$$

Since $\varphi_k(x)$ is a monic polynomial of degree k, $0 \leq k \leq n$, the $n+1$ polynomials $\varphi_k(x)$ are linearly independent. We shall call them the *Newton coefficients* of the set $(x_0, \ldots, x_n)$. Suppose that we wish to find the interpolating polynomial $L_n(x)$ of degree $\leq n$ which assumes the values y_i on the points x_i, $0 \leq i \leq n$. Since the φ_k are linearly independent we may express the interpolating polynomial as a linear combination, $L_n(x) = \sum_0^n a_k \varphi_k(x)$. If a new point x_{n+1} is added, we simply add on the term $a_{n+1} \varphi_{n+1}(x)$ to obtain $L_{n+1}(x)$. To determine the a_k we must solve the following triangular system:

$$\begin{aligned}
y_0 &= a_0 \\
y_1 &= a_0 + a_1(x_1 - x_0), \\
y_2 &= a_0 + a_1(x_2 - x_0) + a_2(x_2 - x_0)(x_2 - x_1), \\
&\vdots \\
y_n &= a_0 + a_1(x_n - x_0) + \cdots + a_n \varphi_n(x_n).
\end{aligned}$$

The solution is easily found by back substitution to be

$$\begin{aligned}
a_0 &= y_0, \\
a_1 &= (y_1 - y_0)/(x_1 - x_0) \\
a_2 &= \left(\frac{y_2 - y_0}{x_2 - x_0} - \frac{y_1 - y_0}{x_1 - x_0}\right) \frac{1}{(x_2 - x_1)},
\end{aligned}$$

and so on. It is clear that a_k involves only the pairs $(x_0, y_0), \ldots, (x_k, y_k)$. Therefore, if a new pair (x_{n+1}, y_{n+1}) is introduced, the new interpolating polynomial of degree $n+1$ is indeed obtained by computing the additional term $a_{n+1} \varphi_{n+1}(x)$, that is

$$L_{n+1}(x) = L_n(x) + a_{n+1} \varphi_{n+1}(x). \quad (*)$$

The coefficient a_k is called the *divided difference of order k* of $(y_0, y_1, \ldots, y_k)$ with respect to $(x_0, x_1, \ldots, x_k)$. If the y_i are regarded as values of a function f at x_i, we

340 Interpolation

usually express a_k by the notation
$$a_k = [f(x_0), \ldots, f(x_k)]$$
or somewhat more simply by $a_k = f[x_0, \ldots, x_k]$. Thus, we have
$$L_n(x) = \sum_{k=0}^{n} f[x_0, \ldots, x_k](x - x_0)(x - x_1) \cdots (x - x_{k-1})$$
$$= \sum_{k=0}^{n} f[x_0, \ldots, x_k]\varphi_k(x) \quad (8.2.1)$$
which is the *Newton interpolating polynomial*.

Observe that a_n is the coefficient of x^n in $L_n(x)$. Recalling the Lagrange form, $L_n(x) = \sum_{j=0}^{n} f(x_j) l_j^{(n)}(x)$, and since
$$l_j^{(n)}(x) = \frac{\varphi_{n+1}(x)}{(x - x_j)\varphi'_{n+1}(x_j)},$$
we obtain as an alternative formula for a_n, $n = 0, 1, 2, \ldots$,
$$a_n = \sum_{j=0}^{n} \frac{f(x_j)}{\varphi'_{n+1}(x_j)}. \quad (8.2.2)$$
Using (8.2.2) it is easy to verify that
$$f[x_0, \ldots, x_n] = \frac{f[x_0, \ldots, x_{n-1}] - f[x_1, \ldots, x_n]}{x_0 - x_n}. \quad (8.2.2a)$$

(See Exercise 8.4.) It also follows immediately from (8.2.2) that $f[x_0, \ldots, x_n]$ is a symmetric function of $x_0, \ldots, x_n$; i.e., any permutation of the x_i yields the same value for the divided difference. The above formula leads to a recursive scheme for computing divided differences. This is usually arranged in the form of a table as follows:

$$
\begin{array}{l}
x_0 \quad f(x_0) \\
\qquad \diagdown \dfrac{f(x_1) - f(x_0)}{x_1 - x_0} = f[x_0, x_1] \\
x_1 \quad f(x_1) \qquad\qquad\qquad\quad \diagdown \dfrac{f[x_0, x_1] - f[x_1, x_2]}{x_0 - x_2} = f[x_0, x_1, x_2] \\
\qquad \diagdown \dfrac{f(x_2) - f(x_1)}{x_2 - x_1} = f[x_1, x_2] \\
x_2 \quad f(x_2) \qquad\qquad\qquad\quad \diagdown \dfrac{f[x_1, x_2] - f[x_2, x_3]}{x_1 - x_3} = f[x_1, x_2, x_3] \\
\qquad \diagdown \dfrac{f(x_3) - f(x_2)}{x_3 - x_2} = f[x_2, x_3] \\
x_3 \quad f(x_3) \qquad\qquad\qquad\quad \diagdown \dfrac{f[x_2, x_3] - f[x_3, x_4]}{x_2 - x_4} = f[x_2, x_3, x_4] \\
\qquad \diagdown \dfrac{f(x_4) - f(x_3)}{x_4 - x_3} = f[x_3, x_4] \\
x_4 \quad f(x_4)
\end{array}
$$

8.2 Other Forms of the Interpolating Polynomial

If we define functionals $\bar{F}_i$, $0 \leq i \leq n$, by

$$\bar{F}_0(f) = [f(x_0)] = a_0,$$
$$\bar{F}_1(f) = [f(x_0), f(x_1)] = \frac{f(x_0)}{x_0 - x_1} + \frac{f(x_1)}{x_1 - x_0} = a_1,$$
$$\vdots$$
$$\bar{F}_n(f) = [f(x_0), \ldots, f(x_n)] = a_n,$$

we see that $L_n(x) = \sum_{i=0}^{n} \bar{F}_i(f) \varphi_i(x)$ for any function f defined on $\{x_0, \ldots, x_n\}$. In particular, for any Newton coefficient $\varphi_j(x)$, $0 \leq j \leq n$,

$$\varphi_j(x) = \sum_{i=0}^{n} \bar{F}_i(\varphi_j) \varphi_i(x).$$

Since the φ_i are linearly independent, we must have $\bar{F}_i(\varphi_j) = \delta_{ij}$. Hence, $\{\bar{F}_i\}$ and $\{\varphi_j\}$ form a biorthonormal set.

A special set of interpolation points is the set $\{x_i\}$ of zero's of $T_n(x)$, the Chebyshev polynomial of degree n. The interpolating polynomial on this set is given by

$$L_n(x) = \frac{1}{n} \sum_{i=1}^{n} f(x_i)(-1)^i \sin \theta_i \frac{T_n(x)}{x_i - x}. \qquad (8.2.3)$$

Since $T_n(x) = \cos(n \arccos x)$, where $x = \cos \theta$, we have $x_i = \cos \theta_i$ and $\theta_i = (2i - 1)\pi/2n$. (See Exercise 8.9.)

8.2a Equally spaced points; difference operators

If the points x_i are equally spaced, the Newton form of the interpolation polynomial becomes especially convenient. Let $x_i = x_0 + ih$, $0 \leq i \leq n$, where h is the *step size*. We define an operator Δ_h on functions f by the relation

$$\Delta_h f(x) = f(x + h) - f(x).$$

Thus, $\Delta_h f$ is a function whose value at x is the difference in its values at $x + h$ and x. (We usually assume that f is defined at $x + h$; otherwise $\Delta_h f$ is not defined at x.) Since h is generally fixed in the discussion, it is customary to suppress the subscript and write simply Δ. Then Δ is called the *forward difference operator* and $\Delta f(x) = f(x + h) - f(x)$ is called the (first) *forward difference*. The Δ operator may be iterated to form $\Delta^2, \Delta^3, \ldots$, etc. Thus, for example, the *second forward difference* (of f at x with step size h) is

$$\Delta^2 f(x) = \Delta(\Delta f(x)) = \Delta(f(x + h) - f(x))$$
$$= f(x + 2h) - f(x + h) - f(x + h) + f(x) = f(x + 2h) - 2f(x + h) + f(x).$$

In general, $\Delta^{n+1} f = \Delta^n(\Delta f(x))$, $n = 0, 1, 2, \ldots$, where $\Delta^0 = I$, the identity operator. If we regard Δ as an operator on the vector space of all real functions defined on the real line, then clearly $\Delta(f + g)(x) = (\Delta f + \Delta g)(x)$ and $\Delta(\lambda f)(x) = (\lambda \Delta f)(x)$.

342 Interpolation

Hence, Δ is a linear operator, and likewise Δ^n. This shows directly that

$$\Delta^{n+1}f(x) = \Delta^n(\Delta f(x)) = \Delta^n f(x+h) - \Delta^n f(x), \qquad n = 0, 1, 2, \ldots,$$

and suggests the arrangement of a forward difference table to compute the forward differences as follows:

$$
\begin{array}{llll}
x_0 & f(x_0) \\
 & f(x_1) - f(x_0) = \Delta f(x_0) \\
x_1 & f(x_1) & \Delta f(x_1) - \Delta f(x_0) = \Delta^2 f(x_0) \\
 & f(x_2) - f(x_1) = \Delta f(x_1) & & \Delta^2 f(x_1) - \Delta^2 f(x_0) = \Delta^3 f(x_0) \\
x_2 & f(x_2) & \Delta f(x_2) - \Delta f(x_1) = \Delta^2 f(x_1) \\
 & f(x_3) - f(x_2) = \Delta f(x_2) \\
x_3 & f(x_3)
\end{array}
$$

The relation between forward and divided differences is easy to derive.

Lemma Let $x_i = x_0 + ih$, $i = 0, 1, 2, \ldots$. Then

$$\frac{\Delta^n f(x_0)}{n!h^n} = f[x_0, x_1, \ldots, x_n]. \tag{8.2.4}$$

Proof. (By induction on n.) For $n = 1$

$$\frac{\Delta f(x_0)}{h} = \frac{f(x_0 + h) - f(x_0)}{h} = f[x_0, x_1].$$

Assume the result is true for n. Then

$$\Delta^{n+1}f(x_0) = \Delta^n(\Delta f(x_0)) = \Delta^n f(x_1) - \Delta^n f(x_0)$$

$$= n!h^n \frac{f[x_1, \ldots, x_{n+1}] - f[x_0, \ldots, x_n]}{x_{n+1} - x_0}(x_{n+1} - x_0)$$

$$= n!h^n f[x_0, x_1, \ldots, x_{n+1}] \cdot (n+1)h,$$

which establishes (8.2.4) for $n + 1$. ∎

Using (8.2.4) and the Newton coefficients $\varphi_k(x)$ given in (8.2.0), we obtain the *forward difference formula* for the interpolating polynomial for equally spaced points,

$$L_n(x) = f(x_0) + \frac{\Delta f(x_0)}{h}\varphi_1(x) + \frac{\Delta^2 f(x_0)}{2!h^2}\varphi_2(x) + \cdots + \frac{\Delta^n f(x_0)}{n!h^n}\varphi_n(x). \tag{8.2.5}$$

Some simplification is achieved by writing $x = x_0 + th$, t real. Since $x_i = x_0 + ih$, $x - x_i = (t - i)h$. Hence, $\varphi_k(x) = h^k \prod_{i=0}^{k-1}(t - i)$. The polynomials $\psi_k(t) = \prod_{i=0}^{k-1}(t - i)$ are called the *factorial* polynomials. We may express the interpolating polynomial in terms of factorial polynomials as follows:

$$L_n(x_0 + th) = f(x_0) + \Delta f(x_0)\psi_1(t) + \frac{\Delta^2 f(x_0)}{2!}\psi_2(t) + \cdots + \frac{\Delta^n f(x_0)}{n!}\psi_n(t).$$

Another form is obtained by writing symbolically (i.e., formally)

$$\psi_k(t) = t!/(t-k)! = k!\binom{t}{k}.$$

Then we may write

$$L_n(x_0 + th) = f(x_0) + \Delta f(x_0)\binom{t}{1} + \Delta^2 f(x_0)\binom{t}{2} + \cdots + \Delta^n f(x_0)\binom{t}{n}.$$

Other formulas may be derived in a formal way by introducing the *shift operator* E_h defined by

$$E_h f(x) = f(x + h).$$

Again, we assume h fixed and suppress the subscript. Observe that $E = I + \Delta$. Furthermore, $E^n f(x) = f(x + nh)$, for every integer $n \geq 0$. We extend the definition of a power of E by defining $E^t f(x) = f(x + th)$ for any real t. If t is an integer, then

$$(I + \Delta)^t = I + \binom{t}{1}\Delta + \binom{t}{2}\Delta^2 + \cdots + \Delta^t.$$

Hence, for integral t,

$$f(x_0 + th) = E^t f(x_0) = (I + \Delta)^t f(x_0) = \left(I + \binom{t}{1}\Delta + \cdots + \Delta^t\right) f(x_0)$$

$$= f(x_0) + \Delta f(x_0)\psi_1(t) + \frac{\Delta^2 f(x_0)}{2!}\psi_2(t) + \cdots + \frac{\Delta^t f(x_0)}{t!}\psi_t(t),$$

which agrees with the previous formula for $L_n(x_0 + th)$. If t is not an integer, we would obtain a formal infinite series for $f(x_0 + th)$.

Writing $\Delta = E - I$ yields $\Delta^n = (E - I)^n$. By the binomial expansion, we find that

$$\Delta^n = \sum_{i=0}^{n} (-1)^i \binom{n}{i} E^{n-i}.$$

Hence,

$$\Delta^n f(x) = \sum_{i=0}^{n} (-1)^i \binom{n}{i} f(x + (n-i)h) = \sum_{j=0}^{n} (-1)^{n-j} \binom{n}{n-j} f(x + jh), \quad (8.2.6)$$

which is an explicit formula for the nth forward difference. This formula explains the propagation pattern of an error through a forward-difference table. (See Exercise 8.8.)

The *backward difference operator* ∇ is defined by

$$\nabla f(x) = f(x) - f(x - h).$$

344 Interpolation

If E^{-1} is the inverse of E, then $E^{-1} = I - \nabla$. Hence, we may write, formally, $E = (I - \nabla)^{-1}$ and $E^t = (I - \nabla)^{-t}$. This suggests that

$$f(x_0 + th) = \left(I - \binom{-t}{1}\nabla + \binom{-t}{2}\nabla^2 - \cdots\right)f(x_0).$$

In fact, there is an interpolation formula of this type. Consider the points $x_{-i} = x_0 - ih$, $0 \le i \le n$ and let $x = x_0 + th$. Then $x - x_{-i} = (t + i)h$ and

$$\varphi_k(x) = h^k \prod_{i=0}^{k-1}(t+i) = (-1)^k h^k (-t)!/(-t-k)!.$$

Further, $\nabla f(x_0) = f(x_0) - f(x_{-1}) = \Delta f(x_{-1})$ and by induction, $\nabla^n f(x_0) = \Delta^n f(x_{-n})$. From Newton's divided difference formula (8.2.1) we obtain, on replacing x_i by x_{-i}, and using (8.2.4),

$$L_n(x_0 + th) = f(x_0) + f[x_{-1}, x_0]\varphi_1(x) + \cdots + f[x_{-n}, x_{-n+1}, \ldots, x_0]\varphi_n(x)$$

$$= f(x_0) + \frac{\Delta f(x_{-1})}{h}\varphi_1(x) + \frac{\Delta^2 f(x_{-2})}{2!h^2}\varphi_2(x) + \cdots + \frac{\Delta^n f(x_{-n})}{n!h^n}\varphi_n(x)$$

$$= f(x_0) + \frac{\nabla f(x_0)}{h}\varphi_1(x) + \frac{\nabla^2 f(x_0)}{2!h^2}\varphi_2(x) + \cdots + \frac{\nabla^n f(x_0)}{n!h^n}\varphi_n(x),$$

$$L_n(x_0 + th) = \sum_{k=0}^{n} \nabla^k f(x_0) \binom{-t}{k}(-1)^k. \quad (8.2.7)$$

Equation (8.2.7) is the *Newton backward difference form* of the interpolating polynomial. For later use, we observe that for the points $x_{-n+1}, \ldots, x_{-1}, x_0, x_1$, where $x_1 = x_0 + h$, the Newton coefficients become

$$\varphi_k(x) = (x - x_1)\prod_{i=0}^{k-2}(x - x_{-i}) = (t-1)h \cdot th \cdots (t+k-2)h$$

$$= \frac{(-t+1)!}{(-t-k+1)!} h^k (-1)^k.$$

For this set, the backward difference formula becomes

$$L_n(x_0 + th) = \sum_{k=0}^{n} \nabla^k f(x_1) \binom{-t+1}{k}(-1)^k. \quad (8.2.8)$$

Another form of the interpolating polynomial is obtained by indexing the equally spaced points as $x_{\pm i} = x_0 \pm ih$, $0 \le i \le m$; i.e., $x_{-m}, x_{-m+1}, \ldots, x_{-1}, x_0, x_1, \ldots, x_m$ and arranging them in the order $x_0, x_1, x_{-1}, x_2, x_{-2}, \ldots, x_m, x_{-m}$. The divided difference form of $L_n(x)$, $n = 2m$, is

$$L_n(x) = f[x_0] + f[x_0, x_1](x - x_0) + f[x_0, x_1, x_{-1}](x - x_0)(x - x_1)$$
$$+ f[x_0, x_1, x_{-1}, x_2](x - x_0)(x - x_1)(x - x_{-1}) + \cdots$$

Since the divided differences are symmetric functions,
$$f[x_0, x_1, x_{-1}] = f[x_{-1}, x_0, x_1], f[x_0, x_1, x_{-1}, x_2] = f[x_{-1}, x_0, x_1, x_2], \ldots,$$
$$f[x_0, x_1, x_{-1}, \ldots, x_m, x_{-m}]$$
$$= f[x_{-m}, \ldots, x_{-1}, x_0, x_1, \ldots, x_m].$$

Using (8.2.4), with appropriate changes in the index, we have
$$f[x_{-1}, x_0, x_1] = \frac{1}{2!h^2} \Delta^2 f(x_{-1}), f[x_{-1}, x_0, x_1, x_2] = \frac{1}{3!h^3} \Delta^3 f(x_{-1}), \ldots,$$
$$f[x_{-m}, \ldots, x_0, \ldots, x_m] = \frac{1}{2m!h^{2m}} \Delta^{2m} f(x_{-m}).$$

This yields the *Gauss forward form* of the interpolating polynomial of degree $n = 2m$:
$$L_{2m} = L_n(x) = L_n(x_0 + th)$$
$$= f(x_0) + \binom{t}{1} \Delta f(x_0) + \binom{t}{2} \Delta^2 f(x_{-1}) + \binom{t+1}{3} \Delta^3 f(x_{-1})$$
$$+ \binom{t+1}{4} \Delta^4 f(x_{-2}) + \cdots + \binom{t+m-1}{2m-1} \Delta^{2m-1} f(x_{-m+1})$$
$$+ \binom{t+m-1}{2m} \Delta^{2m} f(x_{-m}). \tag{8.2.9}$$

(For $n = 2m - 1$ and the set $x_{-m+1}, \ldots, x_0, \ldots, x_m$, delete the last term.)

For $n = 2m + 1$, (8.2.9) can be rearranged by introducing the *central differences*
$$\delta^2 f(x_i) = \Delta f(x_i) - \Delta f(x_{i-1}) = \Delta^2 f(x_{i-1}) = f(x_{i+1}) - 2f(x_i) + f(x_{i-1})$$
and generally $\delta^{2j} f(x_i) = \delta^2(\delta^{2j-2} f(x_i)) = \Delta^{2j} f(x_{i-j})$. It follows that (with $i = 0$) $\Delta^{2j+1} f(x_{-j}) = \delta^{2j} f(x_1) - \delta^{2j} f(x_0), j = 1, 2, \ldots$. Replacing forward differences by central differences, collecting terms and letting $s = 1 - t$, we obtain

$$L_n(x_0 + sh) = [sf(x_0) + tf(x_1)] + \frac{1}{3!} [t(t^2 - 1^2) \delta^2 f(x_1)$$
$$+ s(s^2 - 1^2) \delta^2 f(x_0)] + \cdots + \frac{1}{(2m+1)!} [t(t^2 - 1^2) \cdots (t^2 - m^2) \delta^{2m} f(x_1)$$
$$+ s(s^2 - 1^2) \cdots (s^2 - m^2) \delta^{2m} f(x_0)]. \tag{8.2.10}$$

Formula (8.2.10) is called *Everett's central difference form* of the interpolating polynomial.

8.3 THE ERROR IN POLYNOMIAL INTERPOLATION

Let f be defined on a subset E of the real line. Let $x_0, \ldots, x_n$ be distinct points in E and let $L_n(x)$ be the interpolating polynomial for f on the x_i. Then $f(x_i) = L_n(x_i)$,

$0 \leq i \leq n$. What can be said about the residual $r(x) = f(x) - L_n(x)$ for other x in E? The classical result is the following.

Theorem 8.3.1 Let $x_0, x_1, \ldots, x_n$ be $n + 1$ distinct points. For any point x, let $m = \min \{x_0, \ldots, x_n, x\}$ and $M = \max \{x_0, \ldots, x_n, x\}$. If f has an $(n + 1)$-st derivative in $[m, M]$, then there is a point $\xi = \xi(x)$ in $[m, M]$ such that

$$r_n(x) = f(x) - L_n(x) = \frac{(x - x_0) \cdots (x - x_n)}{(n + 1)!} f^{(n+1)}(\xi) = \frac{\varphi_{n+1}(x)}{(n + 1)!} f^{(n+1)}(\xi). \quad (8.3.1)$$

Proof. Since both $r_n(x_i) = 0$ and $\varphi_{n+1}(x_i) = 0$, $0 \leq i \leq n$, the result holds for any ξ. Hence, we may suppose $x \neq x_i$. Let $A(x) = r_n(x)/\varphi_{n+1}(x)$ and consider x fixed. Define $F(y) = r_n(y) - A(x)\varphi_{n+1}(y)$. Since $\varphi_{n+1}(x_i) = 0$, we have $F(x_i) = 0$, $0 \leq i \leq n$. Also, $F(x) = 0$, by definition of $A(x)$. By Rolle's theorem of differential calculus, $F'(y)$ has at least one zero between two zero's of $F(y)$. Hence, $F'(y)$ has at least $n + 1$ zero's in $[m, M]$. Applying Rolle's theorem successively to $F', F'', \ldots, F^{(n)}$, we find that $F^{(n+1)}$ has at least one zero, ξ, in $[m, M]$. Hence, $0 = F^{(n+1)}(\xi) = r_n^{(n+1)}(\xi) - A(x) \cdot (n + 1)!$. But $r_n^{(n+1)}(\xi) = f^{(n+1)}(\xi) - L_n^{(n+1)}(\xi) = f^{(n+1)}(\xi)$, since the $(n + 1)$-st derivative of an nth-degree polynomial is zero. Thus, $A(x) = f^{(n+1)}(\xi)/(n + 1)!$ and (8.3.1) follows. ∎

Corollary Let $[a, b]$ be an interval containing the points $x_0, \ldots, x_n$. If the function f has a continuous derivative of order $n + 1$ in $[a, b]$, then $f^{(n+1)}(\xi(x))$ can be extended to be continuous in $[a, b]$.

Proof. Define the function g by

$$g(x) = \begin{cases} \dfrac{f(x) - L_n(x)}{\varphi_{n+1}(x)} & \text{for } x \neq x_i, 0 \leq i \leq n, \\ \dfrac{f'(x_i) - L_n'(x_i)}{\varphi_{n+1}'(x_i)} & \text{for } x = x_i. \end{cases}$$

Clearly, g is continuous at $x \neq x_i$. Applying L'Hospital's rule, we find that $\lim_{x \to x_i} g(x) = g(x_i)$. Hence, g is continuous at the points x_i as well. Since $(n + 1)! g(x) = f^{(n+1)}(\xi(x))$ for $x \neq x_i$, g is the required continuous extension. ∎

Formula (8.3.1) is reminiscent of Taylor's formula with the remainder, with $\varphi_{n+1}(x)$ replacing the factor $(x - x_0)^{n+1}$. Comparing the finite Taylor expansion

$$f(x) = f(x_0) + f'(x_0)(x - x_0) + \cdots + \frac{f^{(n)}(x_0)}{n!}(x - x_0)^n + \frac{f^{(n+1)}(\xi)}{(n + 1)!}(x - x_0)^{n+1}$$

with the Newton divided difference formula

$$f(x) = f[x_0] + f[x_0, x_1]\varphi_1(x) + \cdots + f[x_0, \ldots, x_n]\varphi_n(x) + \frac{f^{(n+1)}(\xi)}{(n + 1)!}\varphi_{n+1}(x)$$

suggests that (except for a factorial) the divided differences are approximations to the derivatives. (Indeed, $f[x_0, x_1]$ is a difference quotient.) To see that this is so, recall that $L_{n+1}(x) = L_n(x) + f[x_0, \ldots, x_{n+1}]\varphi_{n+1}(x)$ is the interpolating polynomial of degree $\leq n + 1$ on the points x_i, $0 \leq i \leq n + 1$. Letting $x = x_{n+1}$, we see that

$$r_n(x_{n+1}) = f(x_{n+1}) - L_n(x_{n+1})$$
$$= L_{n+1}(x_{n+1}) - L_n(x_{n+1}) = f[x_0, \ldots, x_{n+1}]\varphi_{n+1}(x_{n+1}).$$

Since x_{n+1} is arbitrary, we have

$$f(x) = L_n(x) + f[x_0, \ldots, x_n, x]\varphi_{n+1}(x). \tag{8.3.2}$$

Comparing (8.3.1) and (8.3.2), we see that

$$f[x_0, \ldots, x_{n+1}] = \frac{f^{(n+1)}(\xi)}{(n+1)!}, \qquad \min_{0 \leq i \leq n+1} \{x_i\} < \xi < \max_{0 \leq i \leq n+1} \{x_i\}. \tag{8.3.3}$$

From (8.2.4) it follows that

$$\frac{\Delta^n f(x_0)}{h^n} = f^{(n)}(\xi), \qquad x_0 < \xi < x_0 + nh. \tag{8.3.4}$$

These formulas show that if f is a polynomial of degree $\leq n$, then all differences of order $> n$ are zero. Also, if $f^{(n)}$ is continuous, then

$$f[x_0, \ldots, x_n] \to f^{(n)}(x)/n! \qquad \text{as } \max_{0 \leq i \leq n} |x_i - x| \to 0.$$

(See Exercise 8.6.) Thus (8.3.3) and (8.3.4) give finite difference approximations to the derivative. (See Section 8.4.)

For the points $x_i = x_0 + ih$, $0 \leq i \leq n$, which are equally spaced, (8.3.1) becomes

$$r_n(x_0 + th) = \psi_{n+1}(t) h^{n+1} \frac{f^{(n+1)}(\xi)}{(n+1)!},$$

where $\psi_{n+1}(t) = \Pi_{i=0}^{n}(t - i)$. It is easy to show that $|\psi_{n+1}(t)|$ is smaller for t near the center of the interval $[0, n]$, than near the ends of it. (See Exercise 8.7.)

For a given interval, $[a, b]$ say, the behavior of $\|r_n(x)\|_\infty$ as $n \to \infty$ depends on that of $\|\psi_{n+1}(t)\|_\infty$ for $0 \leq t \leq n$ and $\|f^{(n+1)}(\xi)\|_\infty$, $a \leq \xi \leq b$. From Exercise 8.7b we see that for $0 \leq t \leq n$, $|\psi_{n+1}(t)| \leq n!$. Hence,

$$|r_n(x_0 + th)| \leq h^{n+1}|f^{(n+1)}(\xi)| = \left(\frac{b-a}{n}\right)^{n+1}|f^{(n+1)}(\xi)|. \tag{8.3.5}$$

If $\lim_{n \to \infty} \|f^{(n+1)}\|_\infty |b - a|^{n+1}/n^{n+1} \to 0$, then the sequence of interpolating polynomials $(L_n(x))$ of $f(x)$ on equally spaced points in the fixed interval $[a, b]$ converges uniformly to $f(x)$ on $[a, b]$. In particular, if there is a constant K such that $\|f^{(n+1)}\|_\infty \leq K^n$ for $n = 1, 2, \ldots$, then

$$\|r_n\|_\infty \leq \left(\frac{K(b-a)}{n}\right)^n \frac{(b-a)}{n} \to 0 \qquad \text{as } n \to \infty.$$

Another result is that if $\|f^{n+1}\|_\infty \leq K^n(n+1)!$, where $0 < K < 1$, then $\|r_n\|_\infty \to 0$.

(Exercise 8.10.) If f is an entire function, this condition holds. (Exercise 8.10.) More general results are possible. However, the following example of Runge demonstrates that the sequence of Lagrange interpolating polynomials does not always converge to the function being interpolated.

Let $f(x) = 1/(1 + x^2)$ on the interval $[-5, 5]$. It is readily verified that f is infinitely differentiable on the interval $[-5, 5]$, so that it is a well-behaved function. However, we duly note that the extension of f to the complex plane has poles at $x = \pm i$. It can be shown that the sequence of interpolating polynomials $L_n(x)$ on the sets $E_n = \{x_j = -5 + jh, 0 \leq j \leq n, h = 10/n\}$ of equally spaced points converges for $|x| \leq 3.63\ldots$ and diverges for $|x| > 3.63\ldots$. Thus, although the sets E_n become dense in $[-5, 5]$ and $L_n(x_j) = f(x_j)$, this is not enough to cause $\|L_n(x) - f(x)\|_\infty$ to approach zero for all x in $[-5, 5]$. (See Exercise 8.16.) To explain the convergence for $|x| \leq 3.63\ldots$, we must refer the reader to [8.2]. It is proved that if $f(z)$ is analytic in the interior of a *lemniscate* $|(z - z_1)(z - z_2) \cdots (z - z_n)| = r^n$, which contains the interval $[a, b]$ and (x_k) is a sequence of points having $z_1, \ldots, z_n$ as limit points, approached cyclically, then the sequence of interpolating polynomials L_n on $(x_0, \ldots, x_n)$ converges to $f(z)$ inside the lemniscate and diverges outside. In the Runge example, it can be shown that there is a lemniscate passing through $\pm 3.63\ldots$ and $\pm i$.

Another example, discovered by Bernstein, is the function $|x|$ on $[-1, 1]$. Interpolating polynomials $L_n(x)$ on $n + 1$ equally spaced points in $[-1, 1]$ converge to $|x|$ only at $-1, 0$ and $+1$. Even when the points are not equally spaced, there are examples of nonconvergence. A general result due to Faber asserts that for any system of interpolation points there is a continuous function f for which the interpolating polynomials fail to converge uniformly to f. (See [8.1] and [8.2].)

The zeros of the Chebyshev polynomials $T_n(x)$ would appear to have better interpolation properties than equally spaced points, i.e., referring to (8.3.1), it seems reasonable to expect a closer approximation if the x_i, $0 \leq i \leq n$, are the zero's of $T_{n+1}(x)$. In that case, $\varphi_{n+1}(x) = T_{n+1}(x)/2^n$. By Theorem 7.6.1, this choice of interpolating points in $[-1, 1]$ yields the smallest value of $\|\varphi_{n+1}\|_\infty$ and would seem to minimize $\|r_n\|_\infty$. (One must also consider ξ in $f^{(n+1)}(\xi)$.) However, there is an example of a continuous function f such that the sequence of interpolating polynomials $(L_n(x))$ of f on the zeros of $T_{n+1}(x)$ diverges at every point in $(-1, 1)$. Indeed, a much more general situation subsists. By applying Theorem 3.7.1, we can prove that for any infinite *triangular matrix of nodes* $x_{ij} \in [0, 1]$,

$$\begin{array}{cccc} x_{00} & & & \\ x_{10} & x_{11} & & \\ x_{20} & x_{21} & x_{22} & \\ x_{30} & x_{31} & x_{32} & x_{33} \\ \vdots & \vdots & \vdots & \vdots & \ddots \end{array} \qquad (8.3.6)$$

there exists an f in $C[a, b]$ such that the sequence of interpolating polynomials $L_n(x)$ on the points in the nth row, $n = 1, 2, 3, \ldots$, does not converge uniformly. We define

$$A_n f = L_n(x) = \sum_{i=0}^{n} l_i^{(n)}(x) f(x_{ni}),$$

8.3 The Error in Polynomial Interpolation

where $l_i^{(n)}$ are the Lagrange coefficients given in (8.1.2). The A_n are bounded linear operators mapping the Banach space $C[0, 1]$ into itself. Let $g_n(x) = \sum_{i=0}^{n} |l_i^{(n)}(x)|$. Then

$$\|A_n f\|_\infty \leq \|f\|_\infty \sum_{i=0}^{n} |l_i^{(n)}(x)|.$$

Hence, $\|A_n\| \leq \|g_n\|_\infty$. In fact, $\|A_n\| = \|g_n\|_\infty$ and $\|g_n\|_\infty > \log n/8\sqrt{\pi}$. (See Exercise 8.14.) Thus, $\{\|A_n\|\}$ is unbounded. By Theorem 3.7.1, $(A_n f) = (L_n(x))$ does not converge uniformly for some $f \in C[0, 1]$.

If one imposes the condition that $\lim_{h \to 0} \omega(h) \log h = 0$, where $\omega(h)$ is the modulus of continuity of f, then interpolation on the zeros of $T_n(x)$ does produce a convergent sequence of interpolating polynomials. This is the case, for example, if f satisfies a Lipschitz condition on $[-1, 1]$, so that $\omega(h) = Lh$. We shall prove this below.

So far we have considered only Lagrange interpolating polynomials. We have seen that their convergence properties are rather poor. Can the convergence question be given any affirmative answers if we impose stronger interpolation conditions as, for example, if we insist that both the values $f(x_i)$ and the values of the derivative $f'(x_i)$ be interpolated? For simple Hermite interpolation some positive results exist.

Let f be continuous on $[-1, 1]$. Let $H_n(x) = \sum_{i=0}^{n} f(x_i) p_i(x)$, where

$$p_i(x) = (1 - 2(x - x_i) l_i'(x_i)) l_i^2(x).$$

By Exercise 8.11, $H_n(x)$ is the Hermite interpolating polynomial of a function g such that $g(x_i) = f(x_i)$ and $g'(x_i) = 0$. If the x_i are chosen to be the zeros of the Chebyshev polynomial $T_n(x)$, then $H_n(x)$ converges to $f(x)$ uniformly on $[-1, 1]$. (See [8.1], Chapter 3, Section 3 for a proof.) However, the H_n are not the simple Hermite interpolating polynomials of f.

Let

$$H_{2n+1}(x) = \sum_{i=0}^{n} (f(x_i) p_i(x) + f'(x_i) q_i(x)), \qquad (8.3.7)$$

where

$$p_i(x) = (1 - 2(x - x_i) l_i'(x_i)) l_i^2(x),$$
$$q_i(x) = (x - x_i) l_i^2(x),$$
$$l_i(x) = l_i^{(n)}(x) = \frac{\varphi(x)}{\varphi'(x_i)(x - x_i)},$$

all as given in Exercise 8.11. Consider a triangular matrix of nodes in $[a, b]$, as in (8.3.6), and define for each row n, $n = 0, 1, 2, \ldots$, the linear functions

$$v_i^{(n)}(x) = 1 - \frac{\varphi''(x_{ni})}{\varphi'(x_{ni})}(x - x_{ni}), \qquad 0 \leq i \leq n. \qquad (8.3.8)$$

Note that $p_i(x) = v_i^{(n)}(x) l_i^2(x)$, by Exercise 8.11. If all the functions $v_i^{(n)}$ are nonnegative on $[a, b]$, i.e., $v_i^{(n)}(x) \geq 0$, $a \leq x \leq b$, then (8.3.6) is called a *normal matrix*. If $v_i^{(n)}(x) \geq \delta > 0$, then (8.3.6) is said to be a *strictly normal* matrix. For example, in the interval $[-1, 1]$, the zeros x_{ni}, $0 \leq i \leq n$ of the Jacobi polynomials $J_{n+1}^{(p, q)}$ (Section 7.7) for $p, q \leq 0$ form a normal matrix and for $p, q < 0$ a strictly normal matrix. To establish this, we first show that $x_{ni} \in [a, b]$ by proving a more general result in the next theorem and corollary.

Theorem 8.3.2 Let $\{Q_0, \ldots, Q_{n-1}\}$ be the orthogonal polynomials of degrees $0, \ldots, n-1$ respectively with respect to the inner product $\langle f, g \rangle = \int_a^b f(x)g(x)\rho(x)\,dx$. If $f \in C[a, b]$ and $\langle f, Q_i \rangle = 0$, $0 \leq i \leq n-1$, then f must have at least n changes of sign in (a, b) or $f = 0$. (Here, we assume that $\rho(x) > 0$.)

Proof. Since $Q_0 = 1$, $\int_a^b f(x)\rho(x)\,dx = 0$. Since $\rho(x) > 0$, if $f \neq 0$, then $f(x)$ must change sign at least once in (a, b). Suppose f has exactly $m < n$ changes of sign and that these occur at the points $x_1 < x_2 < \cdots < x_m$. The polynomial $\varphi_m(x) = \prod_{i=1}^m (x - x_i)$ is orthogonal to f, since φ_m is a linear combination of $\{Q_0, \ldots, Q_m\}$. But $f(x)\varphi_m(x)$ does not change sign in (a, b). This contradicts $\int_a^b f(x)\varphi_m(x)\rho(x)\,dx = 0$. Hence, $m \geq n$. ∎

Corollary Let (Q_n) be an orthogonal sequence of polynomials on $[a, b]$ with weight function $\rho(x)$. The n zeros of Q_n are simple and all are in (a, b).

Proof. Since Q_n is orthogonal to $Q_0, \ldots, Q_{n-1}$, it must have at least n changes of sign in (a, b). Hence, it must have n distinct zeros in (a, b). Since Q_n is of degree n, the result follows. ∎

Now, recall that $J_n^{(p,q)}(x)$ satisfies the differential equation (7.8.4). Since

$$\varphi(x) = \prod_{i=0}^n (x - x_{ni}) = J_{n+1}^{(p,q)}(x) \quad \text{and} \quad \varphi(x_{ni}) = 0,$$

Equation (7.8.4) yields

$$\frac{\varphi''(x_{ni})}{\varphi'(x_{ni})} = \frac{(p + q + 2)x_{ni} - p + q}{1 - x_{ni}^2}.$$

Therefore, for the functions $v_i^{(n)}$ defined above,

$$v_i^{(n)}(1) = 1 - [(p + q + 2)x_{ni} - p + q]\frac{(1 - x_{ni})}{1 - x_{ni}^2} = -q + \frac{(1 + p)(1 - x_{ni})}{(1 + x_{ni})}.$$

Since $-1 < x_{ni} < 1$ and $p > -1$, it follows that $v_i^{(n)}(1) > -q \geq 0$. Similarly, we find that $v_i^{(n)}(-1) > -p \geq 0$. Since $v_i^{(n)}(x)$ is linear in x, $v_i^{(n)}(x) > \min\{-p, -q\}$ for $-1 \leq x \leq 1$. Thus, (8.3.6) is normal for $p, q \leq 0$ and strictly normal for $p, q < 0$. ∎

Lemma 8.3.1 If (8.3.6) is a normal matrix in $[a, b]$ and $0 < h < (b - a)/2$, then for all $x \in [a + h, b - h]$

$$v_i^{(n)}(x) \geq h/(b - a).$$

Proof. $v_i^{(n)}(x_{ni}) = 1$ and $v_i^{(n)}$ is linear in x. If $v_i^{(n)}$ is a constant, it must be identically 1 and therefore $> h/(b - a)$. If it is a decreasing function in $[a, b]$, then for any $x \leq b - h$,

$$v_i^{(n)}(x) \geq v_i^{(n)}(b - h) = v_i^{(n)}(b) + \frac{h}{b - x_{ni}}(1 - v_i^{(n)}(b)) \geq v_i^{(n)}(b) + \frac{h(1 - v_i^{(n)}(b))}{(b - a)} \geq \frac{h}{b - a}.$$

Similar reasoning gives the result when $v_i^{(n)}$ is increasing. ∎

Lemma 8.3.2 If (8.3.6) is a normal matrix and $0 < h < (b - a)/2$, then for all $x \in [a + h, b - h]$

$$\sum_{i=0}^n |l_i(x)| \leq \sqrt{\frac{(b - a)n}{h}}.$$

8.3 The Error in Polynomial Interpolation

If (8.3.6) is strictly normal with constant $\delta > 0$, then

$$\sum_0^n |l_i(x)| \leq \sqrt{\frac{n}{\delta}}.$$

Proof. Using (8.3.7) and (8.3.8) with $f(x) \equiv 1$, we see that $\sum_{i=0}^n v_i^{(n)}(x) l_i^2(x) = 1$. From Lemma 8.3.1, it follows that

$$\sum_0^n l_i^2(x) \leq (b - a)/h \quad \text{for } a + h \leq x \leq b - h.$$

By the Schwarz inequality

$$\sum_0^n |l_i(x)| \leq n^{1/2} \left(\sum_0^n l_i^2(x) \right)^{1/2},$$

and the result follows. ∎

Theorem 8.3.3 Let f satisfy a Lipschitz condition of order α on $[-1, 1]$ with $\alpha > \frac{1}{2}$. Let (x_{ni}) be the nodes of a normal matrix and $L_n(x)$ the Lagrange interpolating polynomial for f on the set $\{x_{n0}, x_{n1}, \ldots, x_{nn}\}$, $n = 0, 1, 2, \ldots$. Then $\lim_{n \to \infty} L_n(x) = f(x)$ for all $x \in [-1, 1]$. The convergence is uniform in every subinterval $[-1 + h, 1 - h]$, $0 < h < 1$. If the matrix (x_{ni}) is strictly normal, the convergence is uniform on $[-1, 1]$.

Proof. Since f is continuous on $[-1, 1]$, there exists a best approximating polynomial, $\pi_n(x)$, of degree $\leq n$ for f on $[-1, 1]$. (See Theorem 7.5.4.) By Theorem 7.7.9, Corollary, the minimax $M_n \leq 12\omega(\pi/n)$, where ω is a modulus of continuity of f. Since f satisfies a Lipschitz condition of order α, we may take $\omega(h) = Lh^\alpha$ and $M_n \leq 12L(\pi/n)^\alpha$. Now,

$$\|L_n - f\|_\infty \leq \|L_n - \pi_n\|_\infty + \|\pi_n - f\|_\infty.$$

But

$$L_n(x) - \pi_n(x) = \sum_{i=0}^n f(x_{ni}) l_i(x) - \sum_{i=0}^n \pi(x_{ni}) l_i(x),$$

since $\pi_n(x)$ is equal to its own interpolating polynomial. Hence,

$$\|L_n - \pi_n\|_\infty \leq \sum_{i=0}^n |f(x_{ni}) - \pi(x_{ni})| \, |l_i(x)| \leq M_n \sum_0^n |l_i(x)|.$$

Applying Lemma 8.3.2, we obtain for any subinterval $[a + h, b - h]$,

$$\|L_n - f\|_\infty \leq \frac{12 L \pi^\alpha}{n^\alpha} \left(n^{1/2} \left(\frac{b-a}{h} \right)^{1/2} + 1 \right),$$

and for $\alpha > \frac{1}{2}$ this bound approaches zero as $n \to \infty$. If (x_{ni}) is strictly normal, Lemma 8.3.2 yields for all $x \in [-1, 1]$, $\|L_n - \pi\|_\infty = (L\pi^\alpha/\sqrt{\delta}) n^{1/2}/n^\alpha$ and the result follows. ∎

A similar result holds for the Hermite interpolation polynomial.

Theorem 8.3.4 Let $f \in C^1[a, b]$. If (x_{ni}) is a normal matrix of nodes and $H_{2n+1}(x)$ is the simple Hermite interpolating polynomial of f, $n = 0, 1, 2, \ldots$, then for all x in (a, b), $\lim_{n \to \infty} H_{2n+1}(x) = f(x)$ and in any subinterval $[a + h, b - h]$, $h > 0$, the convergence is uniform. If the matrix is strictly normal, then the convergence is uniform on $[a, b]$.

Proof. Let $\pi_{2n}(x)$ be the best approximating polynomial of degree $\leq 2n$ for the derivative $f'(x)$ and $M_{2n} = \|f' - \pi_{2n}\|_\infty$. Define the polynomial $g_{2n+1}(x)$ by

$$g_{2n+1}(x) = f(a) + \int_a^x \pi_{2n}(s)\, ds.$$

Then $\|g_{2n+1} - f\|_\infty \leq (b-a)M_{2n}$. Since $g_{2n+1}(x)$ is a polynomial of degree $\leq 2n+1$, it must coincide with its Hermite interpolating polynomial. From (8.3.7) we obtain

$$g_{2n+1}(x) = \sum_{i=0}^n [g_{2n+1}(x_{ni})p_i(x) + g'_{2n+1}(x_{ni})q_i(x)],$$

$$|H_{2n+1}(x) - g_{2n+1}(x)| \leq M_{2n} \sum_{i=0}^n [(b-a)|p_i(x)| + |q_i(x)|].$$

Since (x_{ni}) is normal, $p_i(x) = v_i^{(m)}(x)l_i^2(x) \geq 0$. Hence, $\sum|p_i(x)| = \sum p_i(x) = 1$ and $|H_{2n+1}(x) - g_{2n+1}(x)| \leq M_{2n}(b - a + \sum|q_i(x)|)$. But by Lemma 8.3.2,

$$\sum_{i=0}^n |q_i(x)| = \sum_{i=0}^n |x - x_i|\, l_i^2(x) \leq (b-a) \sum_{i=0}^n l_i^2(x) \leq (b-a)^2/h.$$

Therefore, for $x \in [a - h, b + h]$, $h > 0$,

$$|H_{2n+1}(x) - f(x)| \leq |H_{2n+1}(x) - g_{2n+1}(x)| + |g_{2n+1}(x) - f(x)|$$
$$\leq M_{2n}(b-a)(2 + (b-a)/h),$$

establishing uniform convergence in every closed subinterval $[a - h, b + h]$ and convergence in (a, b). If (x_{ni}) is strictly normal, $\sum |q_i(x)| \leq (b-a)/\delta$ and we obtain uniform convergence on $[a, b]$. ∎

It is a straightforward exercise to prove that if a row of a triangular matrix of nodes (8.3.6) consists of equally spaced points $x_{ni} = a + i(b-a)/n$, then the matrix is not normal ($n \geq 3$). This together with Runge's example, suggests that high-degree interpolating polynomials on equally spaced points are of limited use in approximating continuous functions. In Section 8.5 and Chapter 9, we shall consider methods which employ interpolating polynomials in a different and effective way.

Computational Aspects One of the basic tenets of the numerical analyst's faith is that any problem in continuous mathematics can be solved numerically by *discretizing* it, that is, by selecting a finite subset, E, of the continuum of points in the original problem and solving the corresponding problem restricted to E. If the number of points in E is sufficiently large, then the solution on E is assumed to be a good approximation to the continuous problem. This idea is intuitively appealing and, indeed, leads to an effective computational procedure. After all, the infinite problem must be reduced to a sequence of finite problems if we are to actually carry out computations. However, the negative results of this section warn us that discretization in a naive or uninformed manner, for example by choosing the points of E to be equally spaced, does not always succeed. It fails in what should be a natural application, the approximation of a well-behaved function by its interpolating polynomial on a set of equally spaced points. In fact, there is no clever way of choosing a fixed sequence of finite sets, as for example, is given in (8.3.6), such that interpolation on these sets yields for all continuous functions a convergent sequence of interpolating polynomials. Yet, approximation and interpolation of a continuous function f are closely related. We have seen that the best approximating polynomial of degree $\leq n$ in the L_∞ norm interpolates f on a set of $n + 1$ points. (The residual function alternates in sign on at least $n + 2$ points.) The same is true of the least squares polynomial (Exercise 7.19d). Therefore, if we could predetermine these "best" sets of points, interpolation would succeed as an approximation method. Unfortunately, as we have seen, these sets depend on f and do not exhibit any regular behavior.

Yet, all is not lost. The numerical analyst's basic intuition remains sound, if discretization is applied with circumspection. In Section 8.5, we shall see that interpolation on equally spaced points can succeed if we interpolate with piecewise-polynomial functions of fixed degree. A similar approach yields successful integration methods in Chapters 9 and 10. These results will be established rigorously. Numerical analysis, as we have said in the Preface, is not purely an empirical science—or worse—a bag of computational tricks.

8.4 NUMERICAL DIFFERENTIATION

If $f(x)$ is differentiable in $[a, b]$, an approximation to the derivative $f'(x)$ can be obtained by differentiating the interpolating polynomial $L_n(x)$ for f on a set of points $a \leq x_0 < x_1 < \cdots < x_n \leq b$. The resulting *numerical differentiation formula* expresses the approximate derivative in terms of the function values $f(x_i)$. The principal use of such a formula is in setting up difference equations to approximate differential equations. In practice, the computation of the derivatives of an empirical function f given as a table of values $\bar{f}(x_i)$ subject to errors ε_i should not be carried out by means of a numerical differentiation formula. As shown in Exercise 8.8, the errors ε_i may be magnified in the computation of differences and they are further magnified by division by the small step size, as required in the formulas for differentiation.

The most frequent case of interest is when the points x_i are equally spaced and we wish to approximate $f'(x_i)$. One may then use formula (8.2.5), for example, and differentiate it. For $n = 1$ this gives

$$L'(x_0) = \frac{\Delta f(x_0)}{h} \varphi'_1(x_0) = \frac{\Delta f(x_0)}{h} = \frac{f(x_0 + h) - f(x_0)}{h}.$$

For $n = 2$, we obtain $\varphi'_2(x) = (x - x_0) + (x - x_1)$,

$$L'_2(x_0) = \frac{\Delta f(x_0)}{h} + \frac{\Delta^2 f(x_0)}{2h^2}(x_0 - x_1) = \frac{1}{h}\left(\Delta - \frac{\Delta^2}{2}\right)f(x_0), \qquad (8.4.1)$$

$$L'_2(x_1) = \frac{1}{h}\left(\Delta + \frac{\Delta^2}{2}\right)f(x_0). \qquad (8.4.2)$$

(See Exercise 8.13 for $L'_n(x)$ for arbitrary x.)

To obtain the error $r'_n(x) = f'(x) - L'_n(x)$, we use Rolle's theorem as in Theorem 8.3.1. (See Exercise 8.12 for details.) However, the error at one of the nodes x_i is

354 Interpolation

obtained easily by differentiating (8.3.2). Since $\varphi_{n+1}(x_i) = 0$, this yields

$$r'_n(x_i) = f[x_0, \ldots, x_n, x_i] \prod_{\substack{j=0 \\ j \neq i}}^{n} (x_i - x_j).$$

For equally spaced points $x_j = x_0 + jh$ and assuming that (8.3.3) is applicable, we obtain

$$r'_n(x_0) = (-1)^n n! h^n \frac{f^{(n+1)}(\xi)}{(n+1)!} = (-1)^n h^n \frac{f^{(n+1)}(\xi)}{n+1}.$$

In particular, for $n = 1$ the preceding calculations give

$$f'(x_0) = \frac{f(x_0 + h) - f(x_0)}{h} - \frac{h}{2} f^{(2)}(\xi). \tag{8.4.3}$$

For $n = 2$, we have by (8.4.1) (dropping the subscript),

$$f'(x) = \frac{1}{h}\left(-\frac{1}{2}f(x + 2h) + 2f(x + h) - \frac{3}{2}f(x)\right) + \frac{h^2}{3} f^{(3)}(\xi). \tag{8.4.4}$$

For $n = 2$ we also have

$$r'_n(x_1) = f[x_0, x_1, x_2, x_1] \cdot (x_1 - x_0)(x_1 - x_2) = \frac{-h^2 f^{(3)}(\xi)}{3},$$

and by (8.4.2) (dropping the subscript 1),

$$f'(x) = \frac{1}{2h}(f(x + h) - f(x - h)) - \frac{h^2}{6} f^{(3)}(\xi), \tag{8.4.5}$$

which is the *centered difference* formula for the derivative, somewhat more accurate and much simpler than (8.4.4).

Since differential equations involving second derivatives are of major importance, we derive a numerical formula for $f''(x)$. This can be done conveniently by differentiating (8.2.9) twice with respect to t. We find that for $n = 2m$ (i.e., the points x_{-1}, x_0, x_1)

$$f''(x) = \frac{f(x + h) - 2f(x) + f(x - h)}{h^2} - \frac{h^2}{12} f^{(4)}(\xi). \tag{8.4.6}$$

(See Exercise 8.12b.)

We repeat the warning expressed earlier, that the above formulas for numerical differentiation must be applied with great caution to empirical data values of f or to computed values in which there is a large rounding error. For example, if formula (8.4.5) is used to obtain an approximate numerical value for $f'(x)$, the step size h will have to be small to make the error term, $-h^2 f^{(3)}(\xi)/6$, small. This will force the values $f(x + h)$ and $f(x - h)$ to be close to each other. Hence, the finite-precision subtraction will cause a loss of significant digits and the error in $f(x + h) - f(x - h)$ will usually be relatively large. The error is amplified further when it is multiplied by the large number $1/2h$. (See Exercise 8.15.) For empirical data values it is recommended that a least-squares polynomial $p(x)$ be obtained for the given values $(x_i, f(x_i))$ and then the derivative $p'(x)$ can be used to approximate $f'(x)$.

8.5 SPLINE FUNCTIONS

In the previous sections, we have studied the behavior of the interpolating polynomial $L_n(x)$ as $n \to \infty$. We have seen that, in general, $L_n(x)$ does not converge to $f(x)$, the function being interpolated. In practice, this is not an obstacle, since interpolating polynomials of high degree are seldom used. Rather the degree is kept low (say $n \leq 5$ or 6) and $f(x)$ is approximated by a function $S(x)$ consisting of segments of interpolating polynomials of low degree, each interpolating $f(x)$ on a different subinterval. Since many functions f can be approximated "locally" by a polynomial of low degree, it seems reasonable to expect that $S(x)$ will yield a better approximation than a single polynomial of high degree. In fact, as the number N of interpolation points becomes large, we might expect to obtain arbitrarily close approximations by keeping the degree n of the interpolating polynomial segments constant. In the next chapter, we shall see that this technique does in fact yield quadrature formulas of arbitrary accuracy. For the quadrature problem, there is no need to impose conditions on the derivatives of S. However, in the approximation problem this is desirable. Thus, one requires of a function S made up of polynomials of degree $\leq n$ not only that $S(x_i) = f(x_i)$ at all interpolation points, but also that S have continuous derivatives of all orders $\leq n - 1$. Thus, the individual polynomial segments must have equal derivatives of orders $\leq n - 1$ at the points where they join. This "smoothness" condition on S gives it the name of "spline," which refers to a flexible strip of material used to draw a smooth curve connecting specified points. We shall illustrate these ideas with the important case of *cubic splines*.

Definition 8.5.1 Consider a set E of interpolation nodes $a = x_0 < x_1 < \cdots < x_N = b$ in the finite interval $[a, b]$. We shall call such a set of points a *grid* or *mesh* on $[a, b]$. Let $Y = (y_0, \ldots, y_N)$ be a prescribed set of ordinates. A function $S(x) = S_E(x) = S_E(Y; x)$ is said to be a *cubic spline* on E interpolating Y if $S(x_i) = y_i$ for $0 \leq i \leq N$, $S \in C^2[a, b]$, and S coincides with a polynomial of degree ≤ 3 on each subinterval $[x_{i-1}, x_i]$. If f is a function such that $y_i = f(x_i)$, $0 \leq i \leq N$, then $S = S_E(f; x)$ is said to be a *cubic spline interpolating f* on E.

Note that the condition that S have a continuous second derivative $S''(x)$ for all x in $[a, b]$ implies that S will generally coincide with a cubic polynomial on each subinterval. If we merely required that S be continuous, then a polygonal arc consisting of straight-line segments joining successive points (x_i, y_i) would suffice.

The existence of a cubic spline on any set E is easily established by the following construction.

We seek a vector $M = (M_0, M_1, \ldots, M_N)$ such that on $[x_{i-1}, x_i]$ the second

356   Interpolation

derivative of S is given by the linear function,

$$S''(x) = M_{i-1}\frac{(x_i - x)}{h_i} + M_i\frac{(x - x_{i-1})}{h_i},$$

where $h_i = x_i - x_{i-1}$, $1 \leq i \leq N$. This implies, on integrating each segment twice with respect to x and determining the pairs of constants of integration to make $S(x_i) = y_i$, that on $[x_{i-1}, x_i]$

$$S(x) = M_{i-1}\frac{(x_i - x)^3}{6h_i} + \frac{M_i}{6h_i}(x - x_{i-1})^3$$

$$+ \left(y_i - \frac{M_i h_i^2}{6}\right)\frac{(x - x_{i-1})}{h_i} + \left(y_{i-1} - \frac{M_{i-1} h_i^2}{6}\right)\frac{(x_i - x)}{h_i}. \qquad (8.5.1)$$

For any choice of the values M_i, (8.5.1) defines a piecewise cubic function of x which is continuous on $[a, b]$ and which has a *smooth* second derivative; i.e., $S(x_i) = y_i$ makes S continuous and $S''(x_i+) = S''(x_i-)$. However, for S to be a spline function, we also require that $S'(x)$ be continuous. This is the case if and only if the derivatives of the cubics agree at the points x_i. Then $S'(x)$ will exist for all x and it will follow that $S''(x)$ exists and is continuous. ($S''(x)$ is then a continuous piecewise linear function which is the derivative of $S'(x)$.) Differentiating (8.5.1), we obtain for x in $[x_{i-1}, x_i]$,

$$S'(x) = -M_{i-1}\frac{(x_i - x)^2}{2h_i} + M_i\frac{(x - x_{i-1})^2}{2h_i} + \frac{y_i - y_{i-1}}{h_i} + (M_{i-1} - M_i)\frac{h_i}{6}.$$

We impose continuity on $S'(x)$ at x_i, $1 \leq i \leq N - 1$. The derivative at x_i using the cubic over $[x_{i-1}, x_i]$ is

$$\frac{M_i h_i}{3} + \frac{M_{i-1} h_i}{6} + \frac{y_i - y_{i-1}}{h_i}$$

and the derivative at x_i using the cubic over $[x_i, x_{i+1}]$ is

$$\frac{-M_i h_{i+1}}{3} - \frac{M_{i+1} h_{i+1}}{6} + \frac{y_{i+1} - y_i}{h_{i+1}}.$$

Upon equating these two expressions and simplifying, we obtain

$$a_i M_{i-1} + 2M_i + c_i M_{i+1} = d_i, \qquad 1 \leq i \leq N - 1, \qquad (8.5.2)$$

where

$$a_i = h_i/(h_i + h_{i+1}), \qquad c_i = 1 - a_i,$$

and

$$d_i = 6\frac{[(y_{i+1} - y_i)/h_{i+1} - (y_i - y_{i-1})/h_i]}{h_i + h_{i+1}}. \qquad (8.5.3)$$

remaining two equations for the M_i are obtained by imposing arbitrary "end conditions" on M_0 and M_N. For convenience we write these as

$$2M_0 + c_0 M_1 = d_0 \quad \text{and} \quad a_N M_{N-1} + 2M_N = d_N, \quad (8.5.4)$$

where the constants are at our disposal. The linear system (8.5.2), (8.5.4) can be expressed in matrix form as

$$AM = d,$$

where $d = (d_0, \ldots, d_N)$ and A is the tridiagonal matrix

$$\begin{bmatrix} 2 & c_0 & 0 & 0 & \cdots & 0 \\ a_1 & 2 & c_1 & 0 & & \vdots \\ 0 & a_2 & 2 & c_2 & & 0 \\ \vdots & & \ddots & \ddots & \ddots & \\ & & & a_{N-1} & 2 & c_{N-1} \\ 0 & \cdots & & 0 & a_N & 2 \end{bmatrix}. \quad (8.5.5)$$

Since $|a_i| + |c_i| = 1 < 2$ for $1 \leq i \leq N - 1$, if we choose $|c_0| < 2$ and $|a_N| < 2$, the matrix A will be strictly diagonally dominant. By Corollary 2 of Theorem 4.4.6, A^{-1} exists. Since $d \neq 0$ can be arranged by choosing $d_0 \neq 0$ or $d_N \neq 0$, a solution M always exists and can be obtained by Gaussian elimination or one of the iterative methods in Section 4.4. (See Theorem 4.4.5.) Using these M_i in (8.5.1), we obtain cubics for each subinterval $[x_{i-1}, x_i]$ which form the desired cubic spline function. We have proved

Theorem 8.5.1 For any grid E of $N + 1$ points on $[a, b]$ and any $Y = (y_0, y_1, \ldots, y_N)$ there exists a cubic spline $S_E(Y; x)$.

Now, we shall show that for a sequence of grids (E_k) which become dense in $[a, b]$ the cubic splines $S_k = S_{E_k}$ interpolating a continuous function f converge uniformly to f on $[a, b]$.

We shall use the notation $h_i^{(k)} = x_i^{(k)} - x_{i-1}^{(k)}$ and

$$|E_k| = \max_i \{h_i^{(k)}\},$$

where E_k is the grid $a = x_0^{(k)} < x_1^{(k)} < \cdots < x_{N_k}^{(k)} = b$.

Theorem 8.5.2 Let $f \in C[a, b]$. Let (E_k) be a sequence of grids such that $\lim_{k \to \infty} |E_k| = 0$ and for all k,

$$\max_i \{|E_k|/h_i^{(k)}\} < \mu < \infty.$$

For each k let S_k be a cubic spline interpolating f on E_k with the end conditions (8.5.4) chosen so that $|E_k|^2 (|d_0^{(k)}| + |d_N^{(k)}|) \to 0$ as $k \to \infty$ and $\sup_k \{|c_0^{(k)}|, |a_{N_k}|\} < 2$. Then S_k converges uniformly to f on $C[a, b]$. Further, if f satisfies a Lipschitz condition on $[a, b]$ of order $0 < \alpha \leq 1$ and $|E_k|^{2-\alpha}(|d_0^{(k)}| + |d_N^{(k)}|)$ is bounded, then

$$\|f - S_k\|_\infty = O(|E_k|^\alpha).$$

Proof. By Exercise 4.18 it follows that for each matrix A_k given by (8.5.5) for the grid E_k

$$\|A_k^{-1}\|_\infty \le \frac{1}{\min\{(2 - c_0^{(k)}), (2 - a_{N_k}^{(k)}), 1\}}.$$

Thus, under the hypotheses on $c_0^{(k)}$, $a_{N_k}^{(k)}$, there is a constant C such that $\|A_k^{-1}\|_\infty \le C < \infty$ for all k. Since, $A_k M^{(k)} = d^{(k)}$,

$$\|M^{(k)}\|_\infty \le C \|d^{(k)}\|_\infty.$$

From (8.5.3) with $y_i = f(x_i)$ it follows that

$$|d_i^{(k)}| \le 6\omega(|E_k|)\left(\frac{1/h_{i+1}^{(k)} + 1/h_i^{(k)}}{h_i^{(k)} + h_{i+1}^{(k)}}\right) \le \frac{6}{|E_k|^2} \mu^2 \omega(|E_k|),$$

where $\omega(\delta)$ is the modulus of continuity of f on $[a, b]$. Thus,

$$|E_k|^2 \|d^k\|_\infty \le 6\mu^2 \omega(|E_k|) + |E_k|^2 (|d_0^{(k)}| + |d_1^{(k)}|).$$

From (8.5.1) and omitting the superscript k, we have for the subinterval $[x_{i-1}, x_i]$,

$$|S_k(x) - f(x)| \le \tfrac{2}{3} \|M\|_\infty |E_k|^2 + \left|\frac{f(x_i) + f(x_{i-1})}{2} - f(x)\right|$$
$$+ |f(x_i) - f(x_{i-1})| \frac{|x_i + x_{i-1} - 2x|}{2h_i}$$

(remembering that $h_i = x_i - x_{i-1}$). From the inequalities on $\|M\|_\infty$ we have

$$|S_k(x) - f(x)| \le \tfrac{2}{3} C(6\mu^2 \omega(|E_k|) + |E_k|^2(|d_0| + |d_1|)) + \omega(|E_k|) + \omega(|E_k|),$$

which establishes uniform convergence as $|E_k| \to 0$. Further, if f satisfies a Lipschitz condition of order α, then $\omega(\delta) = O(\delta^\alpha)$ and the rest of the theorem is proved. ∎

It is clear that the sequence of grids $E_k = \{a, a + h, \ldots, a + kh = b\}$, with $h = (b - a)/k$, $k = 1, 2, \ldots$, satisfies the hypotheses of the theorem. Thus, spline interpolation converges when ordinary Lagrange interpolation does not, as in Runge's example.

This suggests the possibility of using a cubic spline as an approximating function for f. If $f \in C[0, 1]$, then the preceding result indicates that an excessive number of grid points might be required; e.g., to obtain an accuracy of $<10^{-8}$, we might need as many as 10^8 points. The storage and computation of a spline consisting of 10^8 polynomial arcs is obviously impractical. Therefore, splines do not replace best approximating polynomials. However, if the problem is to approximate the derivative $f'(x)$, then splines are recommended. If f' satisfies a Lipschitz condition of order α ($0 < \alpha \le 1$) on $[a, b]$, then for $p = 0, 1$,

$$f^{(p)}(x) - S_k^{(p)}(x) = O(|E_k|^{1+\alpha-p}),$$

provided that $S_k(x)$ is required to satisfy the end conditions $S_k'(x_0) = f'(x_0)$, $S_k'(x_N) = f'(x_N)$. (See [8.4], p. 25 for a proof.) Thus, $S_k'(x)$ converges uniformly to

$f'(x)$. By assuming corresponding properties for f'' and $f^{(3)}$, we obtain better results. For example, if $f^{(3)}$ satisfies a Lipschitz condition of order α, then

$$f^{(p)}(x) - S_k^{(p)}(x) = O(|E_k|^{3+\alpha-p})$$

for $p = 0, 1, 2, 3$. Since $S'_k(x)$ is continuous at grid points and, in fact, $S'_k(x_i)$ is an average of divided differences of f, the cubic spline smooths out errors in these differences caused by errors in the $f(x_i)$. Thus, $S'_k(x_i)$ is often a good numerical estimate of $f'(x_i)$.

The extension of the above results to higher-order splines (e.g., piecewise quintic polynomials) leads to computational difficulties. Indeed, even-degree splines may not exist. (See [8.4].)

To define higher-order splines, it is convenient to suppress the ordinates Y. Thus, we consider only the *nodes* (or *knots*) $x_0 < x_2 < \cdots < x_N$. A functions $S = S(x)$ is said to be a *spline function* (or *spline*) *of degree m with nodes* $x_0, \ldots, x_N$ if S is defined for all $x \in R$, $S \in C^{m-1}(R)$ and in each interval (x_{i-1}, x_i), $i = 0, \ldots, N + 1$ (with $x_{-1} = -\infty, x_{N+1} = \infty$), S is equal to a polynomial of degree $\leq m$. A spline of odd degree $m = 2k - 1$ is called a *natural spline* if in $(-\infty, x_0)$ and (x_N, ∞) it coincides with a polynomial of degree at most $k - 1$, rather than $2k - 1$. For the cubic splines considered above, $k = 2$ and $S''(x) = 0$ for x outside $[a, b]$. Formula (8.5.1) defines the cubic section of $S(x)$ in $[x_{i-1}, x_i]$. A single representation of a degree m spline valid for $x \in R$ is given by

$$S(x) = p(x) + \sum_{i=0}^{N} a_i(x - x_i)_+^m, \qquad (8.5.6)$$

where p is a polynomial of degree $\leq m$ and x_+^m is the *truncated power function* defined by $x_+^m = x^m$ for $x > 0$ and $x_+^m = 0$ for $x \leq 0$. The coefficients a_i and p are uniquely determined by S. (Ex. 8.23.) Clearly, S is a natural spline of degree $m = 2k - 1$ if and only if p is of degree $\leq k - 1$ and

$$\sum_{i=0}^{N} a_i x_i^q = 0, \quad q = 0, \ldots, k - 1.$$

Now, an interpolating cubic spline can be computed by solving (8.5.2). However, the end conditions (8.5.4) are not uniquely determined by the nodes and the ordinates Y. Hence, interpolating splines are not uniquely determined unless further constraints are imposed at the end points. However, if we insist that an interpolating spline be a natural spline, then it is uniquely determined by Y, since we must have $0 = S''(x_0) = M_0$ and $0 = S''(x_N) = M_N$. The remaining unknowns, $M_1, \ldots, M_{N-1}$ are determined uniquely by Eqs. (8.5.2). As before, this is a diagonally dominant tridiagonal system and is easily solved. In [8.11], successive overrelaxation is recommended with a relaxation factor of $4(2 - \sqrt{3})$. (See Section 4.5.) Gaussian elimination can also be used in the special form given in Exercise 4.17a, since pivoting is not required. In the special case of equally spaced nodes, we have $h_i = h$. The matrix in (8.5.5) has an especially simple tridiagonal form, since $a_i = c_i = \frac{1}{2}$ for $1 \leq i \leq N - 1$.

Observe that the equations (8.5.2) give only the cubic polynomial segments of S.

In the case of equally spaced nodes, the representation (8.5.6) can be obtained for the natural spline by using precalculated tables. See [8.5]. For arbitrary spacings, the parameters in the single representation (8.5.6) are difficult to compute, since the computations usually involve ill-conditioned linear systems. This is a result of using the truncated powers as a basis. Other bases are considered in [8.12]. Also, see Exercise 8.23.

We have already remarked on the relation between interpolation and approximation. (See Section 8.3, Computational Aspects.) Splines have been used to obtain best approximations in various norms [8.12]. We shall give one such application in Chapter 9.

8.6 INVERSE INTERPOLATION

In the method of false position (5.2.4), linear interpolation is applied to a function f in order to determine an approximate solution of the equation $f(x) = 0$. Suppose that the inverse function f^{-1} exists for f restricted to some neighborhood $N(x^*)$ of a zero, x^*, of $f(x)$. Since the inverse of a linear function is itself linear, we may think of the regula falsi as applying linear interpolation to f^{-1} and then evaluating $L_1^{-1}(0)$, where L_1^{-1} is the linear interpolating polynomial of f^{-1}. In fact, writing $y = f(x)$ and $x = f^{-1}(y)$, we see that the linear function of y which passes through the points (y_m, x_m) and (y_{m-1}, x_{m-1}) is simply

$$L_1^{-1}(y) = x_m + (y - y_m)\left(\frac{x_m - x_{m-1}}{f(x_m) - f(x_{m-1})}\right).$$

If x_m and x_{m-1} are in $N(x^*)$ and $L_1^{-1}(y)$ is a good approximation to $f^{-1}(y)$, then $L_1^{-1}(0)$ should be a good approximation to $x^* = f^{-1}(0)$. Setting $x_{m+1} = L_1^{-1}(0)$, we obtain (5.2.4). Thus, the regula falsi may be regarded as a case of *inverse interpolation*, i.e. interpolation of an inverse function.

More generally, if we are given $n + 1$ values $y_i = f(x_i)$, $0 \le i \le n$, of f, we may seek the interpolating polynomial $L_n^{-1}(y)$ of degree $\le n$ of $f^{-1}(y)$, assuming that f^{-1} exists for f restricted to an interval containing the x_i. Then to approximate a zero x^* we compute $L_n^{-1}(0)$. This computation can be done without actually determining the polynomial $L_n^{-1}(y)$ if we use Aitken's algorithm (Exercise 8.19b). In this algorithm, we first compute

$$p_{i0}(0) = f^{-1}(y_i) = x_i, \quad i = 0, \ldots, n.$$

Then, for $j = 0, \ldots, n - 1$, we calculate recursively,

$$p_{i,j+1}(0) = \frac{f(x_i)p_{jj}(0) - f(x_j)p_{ij}(0)}{f(x_i) - f(x_j)},$$

for $i = j + 1, \ldots, n$. The final result is $p_{nn}(0) = L_n^{-1}(0) = x_{n+1}$. Using the new set $\{x_1, \ldots, x_{n+1}\}$, the process can be iterated. However, this is better done by Neville's algorithm (Exercise 8.19c) in a somewhat modified scheme of *iterated inverse interpolation* as follows.

8.6 Inverse Interpolation

We start with two initial guesses x_0 and x_1 in the neighborhood $N(x^*)$. (As usual for starting values, these guesses are obtained by whatever ad hoc methods are dictated by the particular function f.) We then compute successively higher-degree inverse interpolations to generate an initial set of $n + 1$ points $x_2, \ldots, x_n$. Using the formula of Exercise 8.19c, we first set $p_{00}(0) = x_0$ and $p_{10}(0) = x_1$ and calculate

$$p_{11}(0) = \frac{f(x_1)p_{00}(0) - f(x_0)p_{10}(0)}{f(x_1) - f(x_0)}.$$

We then compute recursively for $i = 2, \ldots, n$, $x_i = p_{i0}(0) = p_{i-1,i-1}(0)$ according to Neville's formula,

$$p_{i,j+1}(0) = \frac{f(x_i)p_{i-1,j}(0) - f(x_{i-j-1})p_{ij}(0)}{f(x_i) - f(x_{i-j-1})}, \tag{8.6.1}$$

where for each i the index j runs from 0 to $i - 1$. Fixing n as the degree of the inverse interpolating polynomials to be determined by iteration, we generate a sequence of points $x_i = p_{i0}(0)$ for $i > n$. Thus, for each $i = n + 1, n + 2, \ldots$, we compute $p_{ij}(0)$ by (8.6.1) for $j = 0, \ldots, n$ and set $p_{i0}(0) = p_{i-1,n}(0)$. It is a simple matter to verify that iterated linear inverse interpolation ($n = 1$) is the same as the regula falsi.

The question of convergence of the sequence (x_i) to x^* can be answered by considering the formula (8.31), which gives the error in polynomial interpolation. Again, we let $y = f(x)$ and $L_n^{-1}(y)$ be the interpolating polynomial for $f^{-1}(y)$ on the points (y_i, x_i), $0 \le i \le n$, where $y_i = f(x_i)$. From Section 8.2, we recall that the Lagrange form of $L_n^{-1}(y)$ is given by

$$L_n^{-1}(y) = \sum_{j=0}^{n} f^{-1}(y_j) l_j^{(n)}(y) = \sum_{j=0}^{n} x_j l_j^{(n)}(y),$$

where

$$l_j^{(n)}(y) = \frac{\varphi_{n+1}(y)}{(y - y_j)\varphi'_{n+1}(y_j)},$$

$$\varphi_{n+1}(y) = (y - y_0) \cdots (y - y_n).$$

Hence

$$l_n^{(j)}(0) = \frac{(-1)^n \prod_{i=0}^{n} y_i}{y_j \varphi'_{n+1}(y_j)}.$$

Now, in iterated inverse interpolation, we take $x_{n+1} = L_n^{-1}(0)$. Therefore

$$x_{n+1} = (-1)^n \prod_{i=0}^{n} y_i \sum_{j=0}^{n} \frac{x_j}{y_j} \frac{1}{\varphi'_{n+1}(y_j)}.$$

By formula (8.3.1), assuming that f^{-1} has a derivative of order $n + 1$ in the appropriate interval,

$$f^{-1}(y) - L_n^{-1}(y) = A(y)\varphi_{n+1}(y),$$

where

$$A(y) = \frac{1}{(n+1)!} \frac{d^{n+1}}{dy^{n+1}} f^{-1}(y)\big|_{y=\eta}.$$

Since $x^* = f^{-1}(0)$, we have $x^* - x_{n+1} = (-1)^{n+1} A \prod_{i=0}^{n} y_i$, where $A = A(0)$. By the mean-value theorem, $y_i = f(x_i) = (x_i - x^*) f'(\xi_i)$. Hence,

$$x^* - x_{n+1} = A \prod_{i=0}^{n} (x^* - x_i) f'(\xi_i). \tag{8.6.2}$$

Assuming that $f'(x)$ is bounded on the interval in question, we have for some constant c,

$$c|x^* - x_{n+1}| \leq \prod_{i=0}^{n} c|x^* - x_i|.$$

Now, suppose all x_i, $0 \leq i \leq n$, are sufficiently close to x^* so that $c|x^* - x_i| < 1$. (This requires that our initial guesses be sufficiently close to x^*.) It follows that $c|x^* - x_{n+1}| < 1$. Let $M = \max(c|x^* - x_i|, 0 \leq i \leq n)$. Then there exist $a_i > 0$ such that $c|x^* - x_i| = M^{a_i}$, and we have

$$M^{a_{n+1}} \leq M^{a_0 + \cdots + a_n},$$

whence

$$a_{n+1} \geq a_0 + \cdots + a_n.$$

In the iterative procedure described above, the next point x_{n+2} is obtained by applying inverse interpolation to the set $\{x_1, \ldots, x_{n+1}\}$ and so on. Thus, x_{n+k} is computed from $\{x_{k-1}, \ldots, x_{n+k-1}\}$ and by induction applied to the previous analysis, $c|x^* - x_i| < 1$ for $k - 1 \leq i \leq n + k - 1$. Therefore, for all $k \geq 1$,

$$a_{n+k} \geq a_{k-1} + \cdots + a_{n+k-1}.$$

Since the $a_i > 0$, it follows that (a_i) is a monotone increasing sequence and $a_i \to \infty$. (If $\lim_{i \to \infty} a_i = L < \infty$, the above inequality would imply $L \geq (n+1)L$.) But then

$$c|x^* - x_i| = M^{a_i} \to 0, \quad \text{since } 0 \leq M < 1.$$

This proves that $x_i \to x^*$.

Some idea of the rate of convergence can be gleaned by considering the case $n = 2$. From (8.6.2), by induction, we have for $k \geq 1$,

$$|x^* - x_{n+k}| = |A_k| \prod_{i=0}^{n} |x^* - x_{i+k-1}|.$$

Let $e_i = |x^* - x_i|$ and $r_i = \log e_i$. Then

$$r_{k+n} = r_{k+n-1} + \cdots + r_{k-1} + \log|A_k|. \tag{8.6.3}$$

Equation (8.6.3) is an example of a linear difference equation of the type (10.1.9) treated in Section 10.1. For $n = 2$, (8.6.3) becomes

$$r_m - r_{m-1} - r_{m-2} - r_{m-3} = \log|A_k|.$$

The general solution of this equation depends on the roots of the polynomial equation,

$$z^3 - z^2 - z - 1 = 0.$$

Analysis of the roots shows that the real root of approximately 1.839 is the root of largest modulus. From the results in Section 10.1 it follows that for large m, r_m is proportional to $(1.839)^m$. Hence,

$$\frac{r_{m+1}}{r_m} \to 1.839,$$

and for m sufficiently large,

$$e_{m+1} = e_m^{1.839}.$$

This shows that the rate of convergence is between linear and quadratic (asymptotically as $m \to \infty$).

Rather than use iterated inverse quadratic interpolation to solve $f(x) = 0$, we could do as suggested in Chapter 5, that is, we could apply quadratic interpolation directly to $f(x)$ in an iterative procedure. Thus, for three points x_{n-2}, x_{n-1}, x_n, we determine the quadratic interpolating polynomial $L_2(x)$ for $f(x)$. We then solve $L_2(x) = 0$ and select the closest root to x_n as the new point x_{n+1}. The procedure is repeated for x_{n-1}, x_n, x_{n+1} and so on. This is known as *Muller's method* [5.12]. An error analysis similar to the above analysis for inverse interpolation proves that the method converges if the three initial guesses x_0, x_1, x_2 are sufficiently close to x^*. Again, the rate of convergence is between linear and quadratic.

8.7 TRIGONOMETRIC INTERPOLATION

In Section 8.1, we remarked that the trigonometric sums of order n are a $(2n + 1)$-parameter linear family of functions. Hence, the general results of that section are applicable. In particular if we have a function $f(x)$ which is continuous and periodic in x of period 2π, we may consider interpolation of f on a set of $2n + 1$ equally spaced points $x_j = j\pi/n$, $j = 0, \ldots, 2n$. Thus, we seek a trigonometric sum,

$$S_n(x) = \sum_{k=0}^{n} (a_{kn} \cos kx + b_{kn} \sin kx), \qquad (8.7.1)$$

such that $S_n(x_j) = f(x_j)$. Since $f(x_0) = f(x_{2n})$, we have $2n$ conditions imposed on $S_n(x)$, which has $2n + 1$ parameters. However, since $S_n(x)$ is periodic also, we shall see that we can take $b_{nn} = 0$ and determine the remaining parameters so that

$$S_n(x_j) = f(x_j), \qquad j = 0, \ldots, 2n - 1. \qquad (1)$$

It will then follow that $S_n(x_{2n}) = f(x_{2n})$. To determine the a_{kn} and b_{kn} we make use of the orthogonality relations in Exercise 7.27c. First, we define an inner product for any two functions φ, ψ restricted to the set $\{x_j\}$ by

$$\langle \varphi, \psi \rangle = \sum_{j=0}^{2n-1} \varphi(x_j)\psi(x_j).$$

In effect, the restrictions of φ and ψ are finite-dimensional vectors. Similarly, the restrictions of $\cos kx$, $\sin kx$ are $2n$-dimensional vectors and they are a linearly independent set, since they are orthogonal; i.e. $\langle \cos mx, \cos kx \rangle = 0$ for $k \neq m$ and $\langle \cos mx, \sin kx \rangle = 0$. Hence,

$$\langle S_n(x), \cos mx \rangle = \sum_k (a_{kn} \langle \cos kx, \cos mx \rangle + b_{kn} \langle \sin kx, \cos mx \rangle)$$

$$= a_{mn} \langle \cos mx, \cos mx \rangle = n a_{mn},$$

and if (1) above is to hold, we must have

$$a_{mn} = (1/n) \sum_{j=0}^{2n-1} S_n(x_j) \cos mx_j = (1/n) \sum_{j=0}^{2n-1} f(x_j) \cos mx_j, \qquad (8.7.2a)$$

and similarly,

$$b_{mn} = (1/n) \sum_{j=0}^{2n-1} f(x_j) \sin mx_j. \qquad (8.7.2b)$$

By the finite-dimensionality, it follows that (8.7.2) is not only necessary but also sufficient for Eqs. (1) above to hold. Since $nx_j = j\pi$, we see that $b_{nn} = 0$, as predicted. This solves the interpolation problem for trigonometric sums insofar as existence is concerned. The necessity of (8.7.2) also establishes uniqueness.

In fact, it is not difficult to show that interpolation by a trigonometric sum of order n is possible on any set of $2n + 1$ points x_j such that

$$0 \leq x_0 < x_1 < \cdots < x_{2n} < 2\pi.$$

(See Exercise 8.20.)

A special type of trigonometric interpolation arises from polynomial interpolation when we consider the interpolation points $x_j = \cos n\theta_j$. Letting $g(\theta) = f(\cos \theta)$ and taking $\theta_j = (2j + 1)\pi/2(n + 1)$, $j = 0, \ldots, n$, we see that the corresponding x_j are the zero's of $T_{n+1}(x)$. In Theorem 8.3.3, we proved that the sequence of interpolating polynomials L_n on such sets $\{x_j\}$ converges uniformly to any f satisfying a Lipschitz condition. Expanding $L_n(x)$ in terms of Chebyshev polynomials, we have

$$L_n(x) = \sum_{k=0}^{n} A_{kn} T_k(x)$$

and making the transformation $x = \cos \theta$, we obtain a trigonometric sum

$$S_n(\theta) = \sum_{k=0}^{n} A_{kn} \cos k\theta \qquad (8.7.3)$$

such that $S_n(\theta_j) = L_n(x_j) = f(x_j) = g(\theta_j)$. Thus, $S_n(\theta)$ interpolates $g(\theta)$ on the set $\{\theta_j\}$ in $[0, \pi]$. In fact, $S_n(\theta)$ is the interpolation trigonometric sum for $g(\theta)$ on the set of $2(n + 1)$ equally spaced points $\theta_j = \pm(2j + 1)\pi/2(n + 1)$, $j = 0, \ldots, n$, on the interval $[-\pi, \pi]$ whenever $g(\theta) = g(-\theta)$. The coefficients A_{kn} in (8.7.3) can be determined directly by using the orthogonality relations for the $T_k(x)$ given in

Exercise 7.29. Thus,

$$A_{kn} = \frac{2}{n+1} \sum_{j=0}^{n} g(\theta_j) \cos k\theta_j = \frac{2}{n+1} \sum_{j=0}^{n} f(x_j) T_k(x_j). \tag{8.7.4}$$

For fixed k, we see that as $n \to \infty$, $A_{kn} \to A_k$, where

$$A_k = \frac{2}{\pi} \int_0^\pi g(\theta) \cos k\theta \, d\theta$$

are the Fourier coefficients of g. This follows easily because the sums in (8.7.4) are Riemann sums with $\Delta\theta_j = \pi/(n+1)$. Similar remarks apply to (8.7.2) above. Thus, if we consider a partial sum of (8.7.3),

$$S_{nm}(\theta) = \sum_{k=0}^{m} A_{kn} \cos k\theta, \tag{8.7.5}$$

$S_{nm}(\theta)$ converges as $n \to \infty$ to the Fourier sum of $g(\theta)$. As we know, the Fourier sum is a least-squares trigonometric sum. The same is true of $S_{nm}(\theta)$, as we can see by applying the orthogonality relations of the $\cos k\theta$ on the finite set $\{\theta_j\}$, i.e. $S_{nm}(\theta)$ is the least-squares cosine sum of order $\leq m$ of $g(\theta)$ on the set $\{\theta_j\}$. A similar result is valid for partial sums $S_{nm}(x)$ of (8.7.1) above on the set $\{x_j\}$. These results are comparable to Theorem 7.7.0, which asserts the convergence of *discrete* (i.e., finite set) least-squares polynomials.

8.8 CURVE FITTING AND SMOOTHING OF DATA

We have considered the problem of interpolation of a function f when given N values $(x_i, f(x_i))$ on the graph of f. Various solutions based on polynomials p such that $p(x_i) = f(x_i)$ have been discussed. Now, suppose that f is unknown and further that $f(x_i)$ cannot be determined exactly. Instead, N approximate values $y_i = f(x_i) + r_i$ are obtained, either by measurement of some physical quantity or by computation subject to errors such as rounding errors. We would like to determine a function f which "best fits" the data points (x_i, y_i). If f has certain smoothness properties, such as continuous derivatives of a certain order, we say that the data has been *smoothed*. When the errors r_i are random variables, it is unlikely that polynomials or other simple functional forms such as trigonometric sums can be found which pass exactly through all the data points. In fact, since $y_i \neq f(x_i)$, we do not attempt to solve such an interpolation problem for the (x_i, y_i). Instead, we select an $(n + 1)$-parameter family of functions, $\varphi(b_0, b_1, \ldots, b_n, x)$, and seek to determine values, b_j^*, of the parameters b_j such that the function $f(x) = \varphi(b_0^*, b_1^*, \ldots, b_n^*, x)$ fits the data best, according to some criterion. The selection of the family φ may be based on the hypothesis that some member, f, of the family will yield exactly the mean (or expected) value of the random quantities y_i. Thus, the y_i are considered to be values of N random variables having some statistical distribution and the expected value, $E(y_i) = f(x_i)$ is presumed to be the "true" value, which would be obtained if the error $r_i = 0$, $i = 1, \ldots, N$.

Let us now further suppose that the family φ is linear in the parameters b_j. For example, φ could be the family of all polynomials in x of degree $\leq n$. In this case, we may consider the system of N equations,

$$y_i = b_0 + x_i b_1 + x_i^2 b_2 + \cdots + x_i^n b_n,$$

called the *conditional* or *observational* equations, and seek to determine the b_j which satisfy them. As we have said, in general there is no exact solution. Therefore, following the suggestion at the end of Section 7.4, we determine the b_j so as to minimize the mean-square error, $\sum_{i=1}^{N} (y_i - \varphi(b_0, \ldots, b_n, x_i))^2 \rho_i$, where the ρ_i are *weights* depending on the accuracy of the y_i. In the polynomial case, $f(x)$ is the least-squares polynomial on the finite set $\{x_i\}$ and may be determined by the methods of Section 7.7 based on orthogonal polynomials. For trigonometric sums we again have orthogonalization techniques for certain sets of points, as we saw in Section 8.7. More generally, we may consider x to be a vector. Then $\varphi(b, x) = b_0 x_0 + b_1 x_1 + \cdots + b_n x_n$, so that the conditional equations have the form,

$$y_i = b_0 x_{i0} + b_1 x_{i1} + \cdots + b_n x_{in}, \quad 1 \leq i \leq N \tag{8.8.1}$$

and a least-squares solution of the b_j is again possible. The determination of the parameters b_j is referred to as a *linear regression analysis* when the y_i are treated as random variables. Since the values b_j^* of the b_j found as a least-squares solution depends on the particular values of the y_i used in the conditional equations, these values can also be viewed as random variables having a statistical distribution determined by the distributions of the y_i. The least-squares solution b_j^* is a best estimate of the b_j in a statistical sense to be explained briefly now.

A *random variable* y is a real variable with which there is associated a *distribution function* $F(y)$ such that $0 \leq F(y) \leq 1$, $F(-\infty) = 0$, $F(+\infty) = 1$, $F(y)$ is non-decreasing (as y increases) and continuous on the right $(F(y+) = F(y))$. The probability that $a < y \leq b$ is given by $F(b) - F(a)$. The *mean* or *expected value*, $E(y)$, of y is defined as the Lebesgue–Stieltjes integral,

$$E(y) = \mu = \int_{-\infty}^{\infty} y \, dF(y).$$

(The function F introduces a measure on the real line (Section 1.6) such that the interval (a, b) has measure $F(b) - F(a)$. See [8.9].) The *variance*, $\sigma^2(y)$, of y is defined by

$$\sigma^2(y) = \int_{-\infty}^{\infty} (y - \mu)^2 \, dF(y),$$

and σ is called the *standard deviation* of y. Since σ^2 is the second moment about the mean, it measures the dispersion or spread of y about the mean; i.e. a small σ would restrict the variation of y about its mean value μ.

Returning to the linear regression analysis (8.8.1), we make the hypothesis that there exist $\bar{b}_0, \ldots, \bar{b}_n$ such that for all $1 \leq i \leq N$,

$$E(y_i) = \bar{b}_0 x_{i0} + \cdots + \bar{b}_n x_{in}. \tag{8.8.2}$$

Here, the x_{ij} are known constants. (For example, in the polynomial case, $x_{ij} = x_i^j$ and x_i is a prescribed value of x.) If we determine the b_i by solving the normal equations corresponding to (8.8.1) (see Section 7.4), then the least-squares solution $b^* = (b_0^*, \ldots, b_n^*)$ is a particular linear function of $y = (y_1, \ldots, y_N)$. Indeed, the normal equations are

$$X^T X b^* = X^T y, \qquad (8.8.3)$$

where X is the $N \times (n+1)$-matrix (x_{ij}). Since the b_j^* are random variables, we may speak of their means and variances. If $E(b_j^*) = \bar{b}_j$, where the $\bar{b}_j$ are the values which satisfy (8.8.2), we call b_j^* an *unbiased estimate* of $\bar{b}_j$. If $\sigma^2(b_j^*)$ is a minimum over the class of all unbiased linear estimates of $\bar{b}_j$ (i.e. linear in y), then b_j^* is called a *minimum variance unbiased estimate* of $\bar{b}_j$ or a *best estimate* of $\bar{b}_j$. In fact, the solution b^* of (8.8.3) is the best estimate of $\bar{b}$ which satisfies (8.8.2) provided that the matrix X has rank $n+1$. This result is known as the *Gauss–Markov theorem*. (See Exercise 8.21.) We have assumed that all $\sigma^2(y_i) = \sigma^2$, that is, all y_i have the same variances. In the general situation the random variables y_i may be correlated statistically. Thus, (y_i, y_j) may have a joint distribution $F(y_i, y_j)$. (See [8.9].) The covariance σ_{ij} of this distribution is defined by

$$\sigma_{ij} = E((y_i - \mu_i)(y_j - \mu_j)) = \int_{-\infty}^{\infty} \int_{-\infty}^{\infty} (y_i - \mu_i)(y_j - \mu_j) \, dF(y_i, y_j).$$

Note that $\sigma_{ii} = \sigma^2(y_i)$. The matrix (σ_{ij}) is called the *covariance matrix* of the vector y. It is obviously symmetric and can be shown to be nonnegative definite. In this case, the best estimate b^* is obtained by solving the modified normal equations,

$$X^T (\sigma_{ij})^{-1} X b^* = X^T (\sigma_{ij})^{-1} y.$$

We close this section with some remarks on nonlinear estimation by least squares. Thus, suppose the family $\varphi(b_0, \ldots, b_n, x)$ is not linear in the parameters b_j. For example, the family $\varphi(b_0, b_1, x) = b_0 e^{b_1 x}$ is nonlinear in b_1. If we know that

$$E(y_i) = \varphi(b_0, b_1, \ldots, b_n, x_i)$$

how can we estimate the b_j? One technique that can often be used successfully is to linearize φ with respect to the b_j. Thus,

$$\varphi(b_0, \ldots, b_n, x) = \varphi(b_0^{(0)}, \ldots, b_n^{(0)}, x) + \sum_{j=0}^{n} \frac{\partial \varphi}{\partial b_j} \Delta b_j + \varepsilon,$$

where $\Delta b_j = b_j - b_j^{(0)}$ and the partials are evaluated at $b^{(0)}$. The conditional equations become

$$y_i - \varphi(b^{(0)}, x_i) = \sum_{j=0}^{n} \frac{\partial \varphi}{\partial b_j} (b^{(0)}, x_i) \Delta b_j \qquad (8.8.4)$$

and we solve (8.8.4) for the Δb_j, again obtaining a least-squares solution. This yields $b^{(1)} = b^{(0)} + \Delta b$ and the procedure is repeated with $b^{(0)}$ replaced by $b^{(1)}$. This is called the *method of differential corrections*. It can be used to do curve fitting even when the

y_i are not regarded as random variables. Convergence of the iterates $b^{(k)}$ depends on having a good initial estimate $b^{(0)}$. Of course, it must be recognized that the nonlinear least squares problems of minimizing

$$g(b_0, \ldots, b_n) = \sum_{i=1}^{n} (y_i - \varphi(b_0, b_1, \ldots, b_n, x_i))^2$$

with respect to the b_j may not have a solution for certain φ. When φ is nonlinear, the results of Section 7.4 do not apply. Even if there exists $(b_0^*, \ldots, b_n^*)$ which minimizes $g(b_0, \ldots, b_n)$, the solution need not be unique. We shall consider the general problem of minimizing nonlinear functions in Chapter 12.

REFERENCES

8.1. E. W. CHENEY, *Introduction to Approximation Theory*, McGraw-Hill, New York (1966)
8.2. P. J. DAVIS, *Interpolation and Approximation*, Blaisdell, New York (1963)
8.3. I. P. NATANSON, *Konstruktive Funktionentheorie*, translated from the Russian by K. Bögel, Akademie-Verlag, Berlin (1955)
8.4. J. H. AHLBERG, E. N. NILSON, and J. L. WALSH, *The Theory of Splines and their Applications*, Academic Press, New York (1967)
8.5. T. N. E. GREVILLE (editor), *Theory and Applications of Spline Functions*, Academic Press, New York (1969)
8.6. J. F. STEFFENSEN, *Interpolation*, Chelsea, New York (1950)
8.7. I. J. SCHOENBERG, "On Hermite-Birkhoff interpolation," *J. Math. Anal. Appl.* **16**, 538–543 (1966)
8.8. D. FERGUSON, "The question of uniqueness for G. D. Birkhoff interpolation problems," *Journal of Approximation Theory* **2**, 1–28 (1969)
8.9. H. CRAMER, *Mathematical Methods of Statistics*, Princeton University Press, Princeton, N.J. (1958)
8.10. I. J. SCHOENBERG and A. WHITNEY, "On Polya frequency functions. III. The positivity of translation determinants with an application to the interpolation problem by spline curves," *Trans. Amer. Math. Soc.* **74**, 246–259 (1953)
8.11. T. N. E. GREVILLE, "Spline functions, interpolation and numerical quadrature," in *Mathematical Methods for Digital Computers* **2**, edited by A. Ralston and H. S. Wilf, Wiley, New York, Chapter 8 (1967)
8.12. L. L. SCHUMAKER, "Some algorithms for the computation of interpolating and approximating spline functions," in *Theory and Application of Spline Functions*, edited by T. N. E. Greville, Academic Press, New York (1969)
8.13. H. B. CURRY and I. J. SCHOENBERG, "On Polya frequency functions. IV. The fundamental spline functions and their limits, "*J. d'Analyse Math.* **17**, 71–107 (1966)

EXERCISES

8.1 Let V be the 2-dimensional space of polynomials $\{a_0 + a_2 x^2\}$. Let $F_i(p) = p(x_i)$, $1 \le i \le 2$, where $-1 \le x_i \le 1$. Show that F_1, F_2 are not linearly independent if $x_1 = -x_2$ or, of course, $x_1 = x_2$. Hence, the interpolation problem has no solution in these cases, in general. (Here, p is in V.)

8.2 a) Consider the 6-parameter family of polynomials in two variables $\varphi(a, x, y) = a_0 + a_1 x + a_2 y + a_3 x^2 + a_4 xy + a_5 y^2$. Show that the interpolation problem does not always have a solution, i.e., for a given set of 6 distinct points (x_i, y_i) there are prescribed values $(z_1, \ldots, z_6)$ such that there is no polynomial with $\varphi(a, x_i, y_i) = z_i$, $1 \le i \le 6$.

b) In fact, show that unisolvence fails for the family $\varphi(a, x, y) = a_0 + a_1 x + a_2 y$. [*Hint:* Choose (x_i, y_i, z_i), $1 \le i \le 3$ on a straight line.]

8.3 a) Let

$$\varphi(a, x, y) = \sum_{j=0}^{m} \sum_{i=0}^{n} a_{ij} x^i y^j$$

be the $(m+1)(n+1)$ parameter family of polynomial functions on some rectangle E. Show that the functions $\{1, x, y, xy, \ldots, x^i y^j, \ldots, x^n y^m\}$ are a linearly independent subset of $\varphi(a, x, y)$. Then show that the functionals $F_{kl}(\varphi(a, x, y)) = \varphi(a, x_k, y_l)$ are linearly independent for any rectangular grid $(x_k, y_l) \in E$, $0 \le k \le n$, $0 \le l \le m$. [*Hint:*

$$\varphi(a, x, y) = \sum_{j=0}^{m} \left(\sum_{i=0}^{n} a_{ij} x^i \right) y^j = \sum_{j=0}^{m} p_j(x) y^j.$$

For each x, $\varphi(a, x, y)$ has at most m zeros. Hence, $\varphi(a, x, y) \equiv 0$ in E implies $p_j(x) \equiv 0$, $c \le x \le d$. Hence, $a_{ij} = 0$. Now $F_{kl}(\varphi(a, x, y)) = \sum a_{ij} x_k^i y_l^j = 0$ for all k, l implies $\sum_{j=0}^{m} p_j(x_k) y_l^j = 0$. Since the y_l, $0 \le l \le m$ are distinct, this implies $p_j(x_k) = 0$ for all $0 \le j \le m$. Since x_k, $0 \le k \le n$ are distinct, this implies p_j is the zero polynomial, i.e., $a_{ij} = 0$. Now use Exercise 2.9b.]

b) Generalize 8.3(a) to three variables x, y, z.

8.4 a) Derive formula (8.2.2a). [*Hint:*

$$f[x_0, \ldots, x_{n-1}] = \sum_{j=0}^{n-1} \frac{f(x_j)}{\varphi'_n(x_j)},$$

$$f[x_1, \ldots, x_n] = \sum_{j=1}^{n} \frac{f(x_j)}{\psi'_n(x_j)}, \qquad \psi_n(x) = (x - x_1) \cdots (x - x_n).$$

$$\psi'_n(x_j) = \varphi'_n(x_j) \frac{(x_j - x_n)}{(x_j - x_0)}, \qquad j < n.$$

Hence,

$$f[x_1, \ldots, x_n] = \sum_{j=1}^{n-1} \frac{f(x_j)}{\varphi'_n(x_j)} \left(\frac{x_j - x_0}{x_j - x_n} \right) + \frac{f(x_n)}{\psi'_n(x_n)}$$

$$f[x_0, \ldots, x_{n-1}] - f[x_1, \ldots, x_n] = \sum_{j=0}^{n-1} \frac{f(x_j)(x_0 - x_n)}{\varphi'_n(x_j)(x_j - x_n)} - \frac{f(x_n)(x_0 - x_n)}{\psi'_n(x_n)(x_0 - x_n)}$$

$$= (x_0 - x_n) \sum_{j=0}^{n} \frac{f(x_j)}{\varphi'_{n+1}(x_j)}.$$

b) Prove that if $f(x)$ is a polynomial of degree $\le n$, then $f[x_0, x]$, $x \ne x_0$, is a polynomial in x of degree $\le n - 1$. By induction show that $f[x_0, \ldots, x_r, x]$ is a polynomial in x of degree $\le n - r - 1$.

8.5 a) Letting $\{1, x, \ldots, x^n\}$ be the basis in V and letting $F_1, \ldots, F_n$ be the linear functionals defined by $F_i(p(x)) = p(x_i)$, $1 \le i \le n$, where the x_i are distinct, show that the scheme in

Exercise 2.9d applied to the x^j and F_i yields the Newton coefficients $\varphi_0, \ldots, \varphi_n$ and the divided differences $f[x_0, \ldots, x_i]$, $0 \leq i \leq n$ as the biorthonormal set.

b) Generalize (a) to two variables x and y by choosing the basis

$$\{1, x, y, x^2, xy, y^2, x^3, x^2y, xy^2, y^3, \ldots\}$$

in the space of polynomials

$$p(x, y) = \sum_{i=0}^{n} \sum_{j=0}^{m} a_{ij} x^i y^j$$

and taking $F_{kl}(p) = p(x_k, y_l)$ where $\{x_k\}$ is a set of $n + 1$ distinct points and $\{y_l\}$ is a set of $m + 1$ distinct points. (See Exercise 8.17.)

8.6 a) Let f have a continuous nth derivative in the interval min $\{x_i\} \leq x \leq$ max $\{x_i\}$, where the x_i, $0 \leq i \leq n$, are distinct. Prove that

$$f[x_0, \ldots, x_n] = \int_0^1 dt_1 \int_0^{t_1} dt_2 \cdots \int_0^{t_{n-1}} dt_n f^{(n)}\bigl(t_n(x_n - x_{n-1}) + \cdots + t_1(x_1 - x_0) + x_0\bigr) \quad \text{where } n \geq 1.$$

[*Hint:* Use induction, proving first that $f[x_0, x_1] = \int_0^1 dt_1 f'(t_1(x_1 - x_0) + x_0)$ by changing the variable of integration to $s = t_1(x_1 - x_0) + x_0$.]

b) For $x_i = x_0 + ih$, $0 \leq i \leq n$, show that

$$\lim_{h \to 0} \frac{\Delta^n f(x_0)}{h^n} = f^{(n)}(x_0).$$

[*Hint:* Use formula (8.3.3) and the continuity of $f^{(n)}$ at x_0.]

c) The integral in part (a) is a continuous function of the x_i. Therefore, it defines the continuous extension of $f[x_0, \ldots, x_n]$ when the x_i are not distinct. Using the first mean-value theorem for integrals prove that $f[x_0, \ldots, x_n] = f^{(n)}(\xi)/n!$, min $\{x_i\} \leq \xi \leq$ max $\{x_i\}$, holds even when the x_i are not distinct. This proves that $f[x_0, \ldots, x_n] \to f^{(n)}(x_0)/n!$ as $\max_{1 \leq i \leq n} |x_i - x_0| \to 0$.

d) Let $f' \in C[a, b]$ and let $f''(x)$ be continuous in some neighborhood of x_1 in $[a, b]$. Prove that $df[x, x_1]/dx$ is continuous in x in $[a, b]$. [*Hint:* Calculate $df[x, x_1]/dx$ for $x \neq x_1$. Write its Taylor expansion about x_1. Allow x to approach x_1 and use the continuity of f''.)

e) Extend the result in (d) to $df[x_0, \ldots, x_n, x]/dx$. (Use part (c).)

8.7 a) Let $\psi_{n+1}(t) = \prod_{i=0}^{n}(t - i)$ be the factorial polynomial of degree $n + 1$. Prove that for n odd $\psi_{n+1}(n/2 - s) = \psi_{n+1}(n/2 + s)$ and for n even $\psi_{n+1}(n/2 - s) = -\psi_{n+1}(n/2 + s)$. [*Hint:* $\psi_{n+1}(n/2 - s)$ and $\psi_{n+1}(n/2 + s)$ are polynomials of degree $n + 1$ in s and both have the roots $n/2, n/2 - 1, n/2 - 2, \ldots, -n/2$. Hence, they differ by a constant which must be $(-1)^{n+1}$.]

b) Let $0 < t + 1 < n/2$, where $t + 1$ is not an integer. Prove that $|\psi_{n+1}(t + 1)| < |\psi_{n+1}(t)|$. Similarly, if $n/2 \leq t \leq n$, then $|\psi_{n+1}(t)| < |\psi_{n+1}(t + 1)|$, t not an integer.

[*Hint*:

$$\frac{\psi_{n+1}(t+1)}{\psi_{n+1}(t)} = \frac{|(t+1)t(t-1)\cdots(t-n+1)|}{|t(t-1)\cdots(t-n+1)(t-n)|} = \frac{|t+1|}{|t-n|}$$

$$= \frac{t+1}{(n+1)-(t+1)} \leq \frac{n/2}{(n+1)-n/2} < 1. \,]$$

Hence, show that the maximum of $|\psi_{n+1}(t)|$ occurs in $[0, 1]$ and therefore that $|\psi_{n+1}(t)| < n!$.

c) Draw the graph of $\psi_n(t)$ for $n = 6$ and verify that its magnitude is generally smaller in the middle of the interval $[0, 6]$ as indicated by part (b).

8.8 Let $f_i = f(x_i)$ and suppose an error ε is made in the computation of f_4. Verify the following propagation of error pattern and explain it by reference to formula (8.2.6).

$$\begin{array}{cccccc}
f_0 & & & & & \\
 & \Delta f_0 & & & & \\
f_1 & & \Delta^2 f_0 & & & \\
 & \Delta f_1 & & \Delta^3 f_0 & & \\
f_2 & & \Delta^2 f_1 & & \Delta^4 f_0 + \varepsilon & \\
 & \Delta f_2 & & \Delta^3 f_1 + \varepsilon & & \\
f_3 & & \Delta^2 f_2 + \varepsilon & & \Delta^4 f_1 - 4\varepsilon & \\
 & \Delta f_3 + \varepsilon & & \Delta^3 f_2 - 3\varepsilon & & \\
f_4 + \varepsilon & & \Delta^2 f_3 - 2\varepsilon & & \Delta^4 f_2 + 6\varepsilon & \\
 & \Delta f_4 - \varepsilon & & \Delta^3 f_3 + 3\varepsilon & & \\
f_5 & & \Delta^2 f_4 + \varepsilon & & \Delta^4 f_3 - 4\varepsilon & \\
 & \Delta f_5 & & \Delta^3 f_4 - \varepsilon & & \\
f_6 & & \Delta^2 f_5 & & \Delta^4 f_4 + \varepsilon & \\
 & \Delta f_6 & & \Delta^3 f_5 & & \\
f_7 & & \Delta^2 f_6 & & & \\
 & \Delta f_7 & & & & \\
f_8 & & & & &
\end{array}$$

8.9 Verify formula (8.2.3). [*Hint:* With $x = \cos\theta$,

$$\lim_{x \to x_i} \frac{T_n(x)}{x - x_i} = \lim_{\theta \to \theta_i} \frac{\cos n\theta}{\cos\theta - \cos\theta_i} = \lim_{\theta \to \theta_i} \frac{n \sin n\theta}{\sin\theta} = \frac{n(-1)^{i+1}}{\sin\theta_i}$$

By continuity of $T_n(x)/(x - x_i)$ the result follows.]

8.10 Let $f(x) = \sum_0^\infty a_i x^i$ be an entire function (i.e., $\lim_n \sup \sqrt[n]{|a_n|} = 0$). Show that on any interval $[a, b]$, $L_n(x) \to f(x)$ as $n \to \infty$, where L_n is the interpolating polynomial on any set of $n+1$ distinct points $x_i \in [a, b]$. [*Hint:* $|r_n(x)| \leq ((b-a)/n)^{n+1} |f^{(n+1)}(\xi)|$ by (8.3.5). For $n > N$, $|f^{(n+1)}(\xi)|/(n+1)! \leq K^{n+1}$, where $0 < K < 1/(b-a)$ since f is entire.

$$|r_n| \leq \frac{(n+1)!}{n^{n+1}} ((b-a)K)^{n+1} \to 0. \,]$$

8.11 a) Let $H = H_{2n+1}(x)$ be the simple Hermite interpolating polynomial of f on the set $x_0, x_1, \ldots, x_n$; i.e., $H(x_i) = f(x_i)$ and $H'(x_i) = f'(x_i)$, $0 \le i \le n$. (See Example 2 of Section 8.1.) Show that $H_{2n+1}(x) = \sum_{i=0}^{n}(f(x_i)p_i(x) + f'(x_i)q_i(x))$, where p_i and q_i are polynomials of degree $\le 2n+1$ such that $p_i(x_j) = \delta_{ij}$, $p_i'(x_j) = 0$, $q_i(x_j) = 0$ and $q_i'(x_j) = \delta_{ij}$. Show that $p_i(x) = [1 - 2(x - x_i)l_i'(x_i)]l_i^2(x)$ and $q_i(x) = (x - x_i)l_i^2(x)$, where $l_i(x)$ are the Lagrange coefficients of formula (8.1.2).
[*Hint:*

$$l_i(x) = \frac{\varphi(x)}{\varphi'(x_i)(x - x_i)},$$

where

$$\varphi(x) = \varphi_{n+1}(x) = \prod_{i=0}^{n}(x - x_i).$$

Hence

$$l_i'(x) = \frac{\varphi'(x)(x - x_i) - \varphi(x)}{\varphi'(x_i)(x - x_i)^2}.$$

By L'Hospital's rule,

$$l_i'(x_i) = \lim_{x \to x_i} \frac{\varphi''(x)(x - x_i) + \varphi'(x) - \varphi'(x)}{2\varphi'(x_i)(x - x_i)} = \frac{\varphi''(x_i)}{2\varphi'(x_i)}.\ ]$$

b) Derive the following formula for the error in simple Hermite interpolation as given in part (a):

$$r_{2n+1}(x) = f(x) - H_{2n+1}(x) = \frac{f^{(2n+2)}(\xi)}{(2n+2)!}\left(\prod_{i=0}^{n}(x - x_i)\right)^2 = \frac{f^{(2n+2)}(\xi)\varphi_{n+1}^2(x)}{(2n+2)!},$$

where $\min\{x_i, x\} < \xi < \max\{x_i, x\}$. [*Hint:* Use Rolle's theorem as in the derivation of the error in Lagrange interpolation, but applying it to the function $F(y) = f(y) - H_{2n+1}(y) - A(x)\varphi_{n+1}^2(y)$, where $A(x) = r_{n+1}(x)/(\varphi_{n+1}(x))^2$. Note that $F'(y)$ has n distinct zeros by Rolle's theorem in addition to the $n+1$ zeros x_i.]

c) Generalize Hermite interpolation by showing that there exists a unique polynomial $p(x)$ such that $p^{(r)}(x_i) = f^{(r)}(x_i)$, $r = 0, 1, \ldots, k_i - 1$, for $n+1$ distinct points $x_0, \ldots, x_n$. The derivatives of f of order $k_i - 1$ are assumed to exist at the x_i. (When all $k_i = 2$, we have simple Hermite interpolation.) [*Hint:* The degree of p is $N = \sum_{i=0}^{n} k_i - 1$. Use a modification of Newton's interpolation polynomial (8.2.1), e.g. for two points x_0, x_1 and $k_0 = 3$, $k_1 = 2$, set up the triangular scheme based on the polynomials

$$\varphi_{00} = 1, \quad \varphi_{10}(x) = (x - x_0), \quad \varphi_{20}(x) = (x - x_0)^2,$$
$$\varphi_{01}(x) = (x - x_0)^3, \quad \varphi_{11}(x) = (x - x_0)^3(x - x_1)$$

and setting $p(x) = \sum a_{ij}\varphi_{ij}(x)$ recursively according to the scheme,

$f(x_0) = a_{00}$
$f'(x_0) = a_{10}$
$f''(x_0) = 2a_{20}$
$f(x_1) = a_{00} + a_{10}(x_1 - x_0) + a_{20}(x_1 - x_0)^2 + a_{01}(x_1 - x_0)^3$
$f'(x_1) = a_{10} + 2a_{20}(x_1 - x_0) + 3a_{01}(x_1 - x_0)^2 + a_{11}\varphi_{11}'(x_1).$

and since

$$f[x; y_0, \ldots, y_m, y] = \sum_{k=0}^{n} f[x_0, \ldots, x_k; y_0, \ldots, y_m, y]\varphi_k(x)$$
$$+ \varphi_{n+1}(x)f[x_0, \ldots, x_n, x; y_0, \ldots, y_m, y],$$

then

$$r(x, y) = \psi_{m+1}(y)f[x; y_0, \ldots, y_m, y] + \varphi_{n+1}(x)f[x_0, \ldots, x_n, x; y]$$
$$- \psi_{m+1}(y)\varphi_{n+1}(x)f[x_0, \ldots, x_n, x; y_0, \ldots, y_m, y].$$

Finally, assuming differentiability as needed,

$$r(x, y) = \frac{\varphi_{n+1}(x)}{(n+1)!} \frac{\partial^{n+1}}{\partial x^{n+1}} f(\xi, y) + \frac{\psi_{m+1}(y)}{(m+1)!} \frac{\partial^{m+1}}{\partial y^{m+1}} f(x, \eta)$$

$$- \frac{\varphi_{n+1}(x)\psi_{m+1}(y)}{(n+1)!(m+1)!} \frac{\partial^{m+n+2}}{\partial x^{n+1} \partial y^{m+1}} f(\xi_1, \eta_1).$$

8.18 a) Let $\{x_0, \ldots, x_n\}$ and $\{y_0, \ldots, y_n\}$ be two sets of $n+1$ distinct points and consider the triangular array T of points, $\{(x_i, y_j) : i + j \leq n\}$. For any $f(x, y)$ defined on T obtain a polynomial $p(x, y)$ of degree at most n in x and y together such that $p(x_i, y_j) = f(x_i, y_j)$. [*Hint:* In Exercise 8.17, take $m = n - k$ to obtain

$$p(x, y) = \sum_{k=0}^{n} \sum_{i=0}^{n-k} \varphi_k(x)\psi_i(y) f[x_0, \ldots, x_k; y_0, \ldots, y_i].\ ]$$

b) Show that $p(x, y)$ in part (a) is unique.

[*Hint:*

$$p(x, y) = \sum_{k=0}^{n} x^k \sum_{i=0}^{n-k} a_{ki} y^i.\ ]$$

8.19 a) Let f be a real-valued function defined on the set $E = \{x_0, \ldots, x_n\} \subset [a, b]$. Let $F \subset E$ be a subset not containing two points of E, x_i and x_j, say. Let p_i be the interpolating polynomial for f on the set $F \cup \{x_i\}$ and similarly, let p_j be the interpolating polynomial on $F \cup \{x_j\}$. Verify that the interpolating polynomial, p, on $F \cup \{x_i, x_j\}$ is given by the recursive formula,

$$p(x) = \frac{(x_i - x)p_j(x) - (x_j - x)p_i(x)}{x_i - x_j}.$$

b) *Aitken's algorithm* for generating the interpolating polynomial $L_n(x)$ on the set E in part (a) is a recursive scheme which generates a triangular array of polynomials p_{rs} as follows:

$$p_{r0}(x) = f(x_r), \qquad r = 0, \ldots, n,$$

$$p_{r,s+1}(x) = \frac{(x_r - x)p_{ss}(x) - (x_s - x)p_{rs}(x)}{x_r - x_s},$$

$$r = s+1, \ldots, n, s = 0, \ldots, n-1.$$

Use induction on s to prove that p_{rs} is the interpolating polynomial on the set $\{x_0, x_1, \ldots, x_{s-1}, x_r\}$, and hence $p_{nn} = L_n$.

c) *Neville's algorithm* for generating L_n by a recursive scheme is as follows:

$$p_{r0}(x) = f(x_r), \quad r = 0, 1, \ldots, n,$$

$$p_{r, s+1}(x) = \frac{(x_r - x)p_{r-1, s}(x) - (x_{r-s-1} - x)p_{rs}(x)}{x_r - x_{r-s-1}},$$

$$r = s + 1, \ldots, n, \, s = 0, \ldots, n - 1.$$

Use induction on s to prove that p_{rs} is the interpolating polynomial on the set $\{x_{r-s}, x_{r-s+1}, \ldots, x_r\}$ and thus $p_{nn} = L_n$.

8.20 Show that the trigonometric functions $\{1, \cos x, \sin x, \ldots, \cos nx, \sin nx\}$ satisfy the Haar condition (Section 7.6) on $[0, 2\pi)$. [*Hint:* Let $\{x_j\}$ be $2n + 1$ distinct points in $[0, 2\pi)$. Form the generalized Vandermonde determinant

$$D = \begin{vmatrix} 1 & \cos x_0 & \sin x_0 & \cdots & \cos nx_0 & \sin nx_0 \\ 1 & \cos x_1 & \sin x_1 & & \cos nx_1 & \sin nx_1 \\ \vdots & & & & & \\ 1 & \cos x_{2n} & \sin x_{2n} & & \cos nx_{2n} & \sin nx_{2n} \end{vmatrix}.$$

Convert to complex exponentials. Rearrange columns to get ascending powers of e^{ix_j}. Then expand as in the standard Vandermonde determinant to get

$$D = (-1)^{n(n-1)/2} 2^{2n^2} \prod_{j=1}^{2n} \prod_{k=0}^{j-1} \sin\left(\frac{x_j - x_k}{2}\right),$$

where $x_k < x_j$ for $k < j$.]

8.21 a) Let y_i, $1 \le i \le N$, be random variables with expectations $\mu_i = E(y_i)$ and covariance matrix $V = (\sigma_{ij})$. (See Section 8.8.) Consider the random vector $y^T = (y_1, \ldots, y_N)$ with $E(y^T) = \mu^T = (\mu_1, \ldots, \mu_N)$. Let $A = (a_{ij})$ be an arbitrary $m \times N$ matrix. Show that the random vector $z = Ay$ has expectation vector given by $E(z) = A\mu$ and covariance matrix AVA^T. [*Hint:*

$$E(z_i) = \sum_{j=1}^{N} a_{ij} E(y_j) = \sum a_{ij} \mu_j.$$

By definition, the covariance matrix of z is

$$E[(z_i - E(z_i))(z_j - E(z_j))] = E((Ay - A\mu)(Ay - A\mu)^T)$$
$$= AE(y - \mu)(y - \mu)^T)A^T. \,]$$

b) Let b^* be the solution of the normal equations (8.8.3). Let $\bar{b}$ be determined by (8.8.2); i.e. $E(y) = X\bar{b}$. Show that $E(b^*) = \bar{b}$. [*Hint:* $E(b^*) = (X^TX)^{-1}X^TE(y)$ by part (a).]

c) Show that b_j^* is a minimum variance unbiased estimate of $\bar{b}_j$, $j = 0, \ldots, n$, where b^* is the solution of (8.8.3) as in part (b). Assume $V = \sigma^2 I$ is the covariance matrix of y. [*Hint:* Let $b_j = c^T y = \sum_{i=1}^{N} c_i y_i$ be any other linear unbiased estimate of $\bar{b}_j$. Then

$$E(b_j) = c^T E(y) = c^T X\bar{b} = e_j^T \bar{b},$$

where $e_j^T = (0, \ldots, 1, 0, \ldots, 0)$ has 1 as its jth component. Thus, $c^T X = e_j^T$. By part (a),
$$\sigma^2(b_j) = c^T c \sigma^2.$$
We must determine c to minimize $c^T c \sigma^2$ subject to the constraint $X^T c = e_j$. This can be done by the Lagrange multiplier rule. (See Chapter 12.)]

8.22 Verify that the right-hand side of Eqs. (8.5.3) is three times the value of the second derivative at x_i of the parabola passing through the points (x_{i-1}, y_{i-1}), (x_i, y_i) and (x_{i+1}, y_{i+1}). For equally spaced points, use formula (8.4.6) to show this. Also, show that $d_i = 6f[x_{i-1}, x_i, x_{i+1}]$, where $f(x_i) = y_i$.

8.23 a) Let $\mathscr{S} = \mathscr{S}_m(x_0, \ldots, x_N)$ be the space of all splines of degree m with nodes $x_0, \ldots, x_N$. Show that any $S \in \mathscr{S}$ can be represented in the form given in Eq. (8.5.6), that is, $\{1, x, \ldots, x^m, (x - x_0)_+^m, \ldots, (x - x_N)_+^m\}$ is a basis for $\mathscr{S}$. [*Hint:* Choose points $t_0 < \cdots < t_{m+N+1}$ and write $S(t_j) = p(t_j) + \sum_{i=0}^{N} a_i (t_j - x_i)_+^m$, $j = 0, \ldots, m + N + 1$. Form the matrix of this linear system (with unknowns $a_0, \ldots, a_N$ and $b_0, \ldots, b_m$, the coefficients of p). Verify that its determinant is positive if and only if $t_j < x_j < t_{m+j+1}$, $j = 0, \ldots, N$.]

b) The matrix in part (a) is ill-conditioned unless N and m are both small. To obtain a band matrix we seek a basis in which the functions are nonzero on the smallest possible number of intervals (x_{i-1}, x_i). One such basis are the *B-splines* [8.13], defined as follows. Choose added nodes $x_{-m-1} < x_{-m} < \cdots < x_{-1} < x_0$ and $x_N < x_{N+1} < \cdots < x_{m+N+1}$. Let $f(x, y) = (m + 1)(x - y)_+^m$. Define
$$M_i(x) = f[x, x_{i-m-1}, \ldots, x_i], \quad 0 \le i \le m + N + 1,$$
as the divided differences of f. Prove that any $S \in \mathscr{S}$ has a unique expansion
$$S(x) = \sum_{i=0}^{m+N+1} a_i M_i(x), \qquad x_{-1} \le x \le x_{N+1}.$$
[*Hint:* Choose points t_j as in part (a) and set
$$S(t_j) = \sum_{i=0}^{m+N+1} a_i M_i(t_j), \qquad 0 \le j \le m + N + 1.$$
Solve for the a_i.]

CHAPTER 9

NUMERICAL INTEGRATION

9.1 INTRODUCTION

Let f be a bounded real-valued function defined on the interval $[a, b] \subset R$. If a and b are finite, the Riemann integral $\int_a^b f(x)\, dx$ is defined as the limit of the sums

$$S_n = \sum_{i=0}^{n} f(x_i')(x_{i+1} - x_i),$$

$$a = x_0 < x_1 < \cdots < x_{n+1} = b, \qquad x_i \le x_i' \le x_{i+1},$$

where $\max_i |x_{i+1} - x_i| \to 0$ as $n \to \infty$. For a half-infinite interval we define

$$\int_a^\infty f(x)\, dx = \lim_{b \to \infty} \int_a^b f(x)\, dx$$

if the limit exists.

(For $(-\infty, \infty)$ we can use the Cauchy principal value given by $\int_{-\infty}^{\infty} f(x)\, dx = \lim_{b \to \infty} \int_{-b}^{b} f(x)\, dx$. Until further notice, we shall take a and b to be finite and f to be bounded. It is known [9.6] that a bounded function f is Riemann integrable on a finite interval $[a, b]$ if and only if the set of discontinuities of f has measure zero. It follows that $|f|$ is Riemann integrable if f is. The indefinite integral $G(x) = \int_a^x f(t)\, dt$ is a differentiable function at any point x at which f is continuous and $G'(x) = f(x)$. If $G(x)$ is known and can be evaluated by a finite procedure, then $\int_a^b f(x)\, dx = G(b) - G(a)$. If f is a combination of elementary functions such as those listed in a table of integrals [9.1], [9.2], [9.8], [9.9] then G can be expressed in "closed form" as a finite combination of elementary functions. However, in general this is not possible and we must seek methods of evaluating $\int_a^b f(x)\, dx$ approximately. In fact, even if a closed formula exists, it may involve transcendental functions such as $\sin x$ which cannot be computed by a finite procedure and must therefore be approximated. One could appeal to the definition of the integral and use an approximating sum S_n. For a sufficiently large n and small *mesh size* ($\max_i |x_{i+1} - x_i|$), S_n gives an arbitrarily close approximation to the integral. However, it is possible to obtain finite sums which give better approximations for the same amount of computation.

Note that the set of (Riemann) integrable functions f is a subspace V_R of the vector space of functions bounded on $[a, b]$. We may use $\|f\|_\infty$ as a norm. With respect to this norm, V_R is a closed subspace of the space of bounded functions. This follows from the fact that the limit, f, of a uniformly convergent sequence of Riemann integrable functions (f_m) is Riemann integrable. Furthermore, $\lim_m \int_a^b f_m(x)\, dx = \int_a^b f(x)\, dx$. We shall confine our discussion in this chapter to real-valued f. Many of the results are valid for vector-valued $f = (f_1, \ldots, f_n)$, although some care must be exercised in the proofs. The integral defines a bounded linear functional S,

$$Sf = \int_a^b f(x)\, dx, \qquad (9.1.1)$$

on the subspace V_R. Indeed,

$$|Sf| \leq \int_a^b |f(x)|\, dx \leq \|f\|_\infty (b - a).$$

For the function $f(x) = 1$ for all x, we have $|Sf| = (b - a)\|f\|_\infty$. Hence, $\|S\| = b - a$. Of course, if a different norm is used in the function space, then $\|S\|$ will be different. For example, if we use the L_1-norm, $\|f\|_1 = \int_a^b |f(x)|\, dx$, then obviously $\|S\| = 1$.

Since $\left|\int_a^b (f(x) - g(x))\, dx\right| \leq \|f - g\|_\infty (b - a)$, this suggests that we try to find a function g which is a good approximation to f in the uniform norm and which is itself integrable in closed form. If f is continuous on $[a, b]$, the choice of a polynomial for g is strongly indicated. For example (see Eq. (8.1.2)), if we approximate f by its Lagrange interpolating polynomial $L_n(x) = \sum_{i=0}^n f(x_i) l_i^{(n)}(x)$ on a set of $n + 1$ points x_i in $[a, b]$, then we obtain

$$\int_a^b f(x)\, dx = \int_a^b L_n(x)\, dx + R_n = \sum_{i=0}^n f(x_i) \int_a^b l_i^{(n)}(x)\, dx + R_n$$

$$= \sum_{i=0}^n a_i f(x_i) + R_n = F(f) + R_n,$$

where the $a_i = \int_a^b l_i^{(n)}(x)\, dx$ are computed in closed form by integrating the polynomials $l_i(x)$, R_n is the error term, and $F(f) = \sum_{i=0}^n a_i f(x_i)$ is a linear functional on the space of functions f defined on $[a, b]$ which is an approximation to the functional S. Thus, numerical integration may be regarded as a problem in the approximation of functionals, namely, the approximation of the nonconstructive functional S by a finitely computable functional.

9.2 LINEAR QUADRATURE FUNCTIONALS

Definition 9.2.1 Let V be a subspace of the vector space of functions integrable on $[a, b]$. Let S be defined by (9.1.1). For any linear functional $F \in V^*$, let $R = S - F$. Then R is called the *remainder functional* of F. If $R(p) = 0$ for all polynomials p of degree $\leq n$ and $R(x^{n+1}) \neq 0$, then F is called a *linear quadrature functional of degree* n.

Suppose F_n is a linear quadrature functional of degree n. We have, for any integrable f and any polynomial p_n of degree $\leq n$, $R_n f = Sf - F_n f = Sf - Sp_n + Sp_n - F_n f = S(f - p_n) + F_n(p_n - f)$, since $Sp_n = F_n p_n$. If F_n is continuous, as is generally the case, then $|R_n f| \leq \|S\|\,\|f - p_n\|_\infty + \|F_n\|\,\|p_n - f\|_\infty \leq (\|S\| + \|F_n\|)\|p_n - f\|_\infty$. If f is continuous, then there exists a sequence of polynomials (p_n) such that $\|f - p_n\|_\infty \to 0$. If the F_n are uniformly bounded, i.e., if there exists $K > 0$ such that $\|F_n\| < K$ for all n, then $R_n f \to 0$. Thus, we have proved the following.

Theorem 9.2.1 Let (F_n) be a uniformly bounded sequence of linear quadrature functionals defined on $C[a, b]$. If F_n is of degree at least n, then $\lim_{n\to\infty} F_n f = \int_a^b f(x)\,dx$ for any $f \in C[a, b]$.

A quite general form of linear quadrature functional is obtained by taking F to be a linear combination of functionals,

$$F(f) = \sum_{i=0}^{m_0} a_{i0} f(x_{i0}) + \sum_{i=0}^{m_1} a_{i1} f'(x_{i1}) + \cdots + \sum_{i=0}^{m_n} a_{in} f^{(n)}(x_{in}), \quad (9.2.1)$$

assuming that the nth derivative $f^{(n)}(x)$ exists in $[a, b]$. Formula (9.2.1) includes most of the quadrature functionals used in practice. Therefore, it is important to obtain a formula for the remainder functional of this type of quadrature functional.

Let us first consider the important special case in which $a_{ij} = 0$ for $j \geq 1$ and the a_{i0} are determined by integrating the interpolating polynomial on the points $x_{i0} = x_i, 0 \leq i \leq n$. Such a functional is called an *interpolatory quadrature functional*. Using the Lagrange form of the interpolating polynomial, we obtain

$$F_n(f) = \int_a^b L_n(x)\,dx = \sum_{i=0}^n f(x_i) \int_a^b \frac{\varphi_{n+1}(x)}{(x - x_i)\varphi_n'(x_i)}\,dx$$

$$= \sum_{i=0}^n f(x_i) \int_a^b l_i^{(n)}(x)\,dx. \quad (9.2.2)$$

Hence, the coefficients $a_{i0}, 0 \leq i \leq n$, are the values of the integrals in this summation.

Remark In fact, any functional of the form $F_n f = \sum_{i=0}^n a_i f(x_i)$ which is of degree at least n, must be interpolatory, that is, its coefficients must be given by (9.2.2). In proof, take $f = l_j^{(n)}, 0 \leq j \leq n$. Since $l_j^{(n)}(x)$ is a polynomial of degree n, $F_n l_j^{(n)} = S l_j^{(n)}$. But $l_j^{(n)}(x_i) = \delta_{ij}$. Hence,

$$F_n l_j^{(n)} = \sum_{i=0}^n a_i l_j^{(n)}(x_i) = a_j = S l_j^{(n)}.$$

Applying formula (8.3.1) for the residual $f(x) - L_n(x)$, we obtain for the remainder

$$R(f) = \int_a^b f(x)\,dx - \int_a^b L_n(x)\,dx = \frac{1}{(n+1)!} \int_a^b f^{(n+1)}(\xi_x) \varphi_{n+1}(x)\,dx. \quad (9.2.3)$$

Alternatively, using (8.3.2) we find that

$$R(f) = \int_a^b f[x_0, \ldots, x_n, x] \varphi_{n+1}(x)\,dx.$$

Since $f^{(n+1)} = 0$ if f is a polynomial of degree $\leq n$, it follows directly that an interpolatory quadrature functional using $n + 1$ distinct points is at least of degree n.

If $\varphi_{n+1}(x)$ does not change sign in $[a, b]$, and $f^{(n+1)}$ is continuous, then by

Theorem 8.3.1, Corollary, we can apply the first mean-value theorem to obtain

$$R(f) = \frac{f^{(n+1)}(\alpha)}{(n+1)!} \int_a^b \varphi_{n+1}(x)\, dx, \qquad \min\{x_i, a\} < \alpha < \max\{x_i, b\}. \quad (9.2.4)$$

A particular case of interest in the solution of differential equations gives rise to the points $x_{-i} = x_0 - ih$, $0 \le i \le n$, when we wish to integrate $f(x)$ from $x_0 = a$ to $b = a + h = x_1$, using the interpolating polynomial on the x_{-i}. (One could say that we are "extrapolating" f over $[a, a+h]$ using its values on $[a - nh, a]$.) Using the Newton backward form (8.2.7), and writing $x = x_0 + th$, we have

$$F_n(f) = \int_a^{a+h} L_n(x)\, dx = h \int_0^1 L_n(t)\, dt = h \sum_{k=0}^n c_k \nabla^k f(x_0), \quad (9.2.5a)$$

where

$$c_k = (-1)^k \int_0^1 \binom{-t}{k} dt.$$

A straightforward calculation yields the following values for the c_k: $c_0 = 1$, $c_1 = \frac{1}{2}$, $c_2 = \frac{5}{12}$, etc. (See Exercise 10.5(c).) Further, since $\varphi_{n+1}(x)$ does not change sign in $[a, a+h]$ and

$$\int_a^{a+h} \frac{\varphi_{n+1}(x)}{(n+1)!}\, dx = \int_a^{a+h} \frac{\prod_{i=0}^n (x - x_{-i})}{(n+1)!}\, dx = (-1)^{n+1} h^{n+2} \int_0^1 \binom{-t}{n+1} dt,$$

the remainder given by (9.2.4) is

$$R_n(f) = S(f) - L_n(f) = h^{n+2} f^{(n+1)}(\alpha) c_{n+1}, \qquad a - nh < \alpha < a + h. \quad (9.2.5b)$$

Another case of interest involves the same set of interpolation points

$$x_{-i} = x_0 - ih, \quad 0 \le i \le n, \qquad \text{and} \qquad a = x_{-1},\ b = x_0.$$

Here, again $\varphi_{n+1}(x)$ does not change sign in $[a, b]$. Letting $x = x_0 + th$, we have

$$F_n(f) = \int_{x_{-1}}^{x_0} L_n(x)\, dx = h \int_{-1}^0 L_n(t)\, dt = h \sum_{k=0}^n d_k \nabla^k f(x_0), \quad (9.2.6a)$$

where

$$d_k = (-1)^k \int_{-1}^0 \binom{-t}{k} dt.$$

Further,

$$R_n(f) = h^{n+2} f^{(n+1)}(\alpha) d_{n+1}. \quad (9.2.6b)$$

The simplest case is obtained with $n = 0$, which yields the *rectangular rule*

$$F_0 f = \int_{x_{-1}}^{x_0} L_0(x)\, dx = h f(x_0).$$

The remainder is

$$f^{(1)}(\alpha) \int_{x_{-1}}^{x_0} \varphi_1(x)\, dx = f^{(1)}(\alpha) \int_{x_{-1}}^{x_0} (x - x_{-1})\, dx = f^{(1)}(\alpha)(-1)h \cdot h \int_{-1}^{0} \binom{-t}{1} dt$$

$$= -f^{(1)}(\alpha) \frac{h^2}{2}, \qquad x_{-1} < \alpha < x_0.$$

Thus,
$$\int_{x_{-1}}^{x_0} f(x)\, dx = hf(x_0) - \frac{h^2}{2} f'(\alpha).$$

If we take the case $n = 1$, we obtain

$$\int_{x_{-1}}^{x_0} L_1(x)\, dx = h(f(x_0) - \tfrac{1}{2} \nabla f(x_0)) = (h/2)(f(x_0) + f(x_{-1})),$$

giving the area under the linear segment joining $(x_0, f(x_0))$ to $(x_{-1}, f(x_{-1}))$. The remainder is

$$\frac{f^{(2)}(\alpha)}{2} \int_{x_{-1}}^{x_0} \varphi_2(x)\, dx = \frac{f^{(2)}(\alpha)}{2} \int_{x_{-1}}^{x_0} (x - x_0)(x - x_{-1})\, dx$$

$$= f^{(2)}(\alpha)(-1)^2 h^2 \cdot h \int_{-1}^{0} \binom{-t}{2} dt = -f^{(2)}(\alpha) \frac{h^3}{12}.$$

Thus, we obtain the familiar *trapezoidal rule*,

$$\int_a^{a+h} f(x)\, dx = \frac{h}{2}(f(a) + f(a+h)) - \frac{h^3}{12} f^{(2)}(\alpha), \qquad a < \alpha < a + h. \quad (9.2.7)$$

The trapezoidal rule uses the interpolating polynomial at the end points a and b of the interval of integration. In general, if $F(f) = \int_a^b L_n(x)\, dx$, where $L_n(x)$ is the interpolating polynomial on the set of equally spaced points

$$a = x_0, x_0 + h, \ldots, x_0 + nh = b$$

then F is called a *closed Newton–Cotes* quadrature functional. Now, we have seen that such a functional must have the form $\sum_{i=0}^n a_i f(x_i)$. The coefficients a_i may be determined in several ways. We can evaluate the integrals in (9.2.2) or we can use any of the many forms of $L_n(x)$ given in Chapter 8. Alternatively, we may make use of the fact that $F(p)$ is exact, that is, $R(p) = 0$ for any polynomial p of degree $\leq n$. In particular, $R(p) = 0$ for $p = 1, x, x^2, \ldots, x^n$. This gives $n + 1$ conditions on the coefficients a_i. The trapezoidal rule could have been derived in this way for $n = 1$. Let us apply this method to the case $n = 2$. For convenience, we make the change of variable $x = x_0 + th$ and take $a = x_0 - h = x_{-1}$, $b = x_0 + h = x_1$ so that $h = (b - a)/2$. Then

$$S(f) = \int_a^b f(x)\, dx = h \int_{-1}^{1} f(x_0 + th)\, dt.$$

Since $R(p) = 0$ for $p = 1, t,$ and t^2, we must have $hS(1) = hF(1)$, $hS(t) = hF(t)$ and $hS(t^2) = hF(t^2)$. Now, $S(1) = \int_{-1}^{1} 1 \cdot dt = 2$, $S(t) = \int_{-1}^{1} t\, dt = 0$, and $S(t^2) = \int_{-1}^{1} t^2\, dt = \frac{2}{3}$. Hence,

$$\left.\begin{array}{r} a_{-1} + a_0 + a_1 = 2 \\ -a_{-1} + 0 \cdot a_0 + a_1 = 0 \\ a_{-1} + 0 \cdot a_0 + a_1 = \frac{2}{3} \end{array}\right\}.$$

Solving this system, we find $a_{-1} = a_1 = \frac{1}{3}$ and $a_0 = \frac{4}{3}$. Thus, we obtain the formula known as *Simpson's rule*,

$$\int_{x_{-1}}^{x_1} f(x)\, dx = (h/3)(f(x_{-1}) + 4f(x_0) + f(x_1)) + R_2(f). \tag{9.2.8}$$

It is easy to see that $R_2(p) = 0$ for p of degree 3 also, since

$$h \int_{-1}^{1} t^3\, dt - h/3(-1 + 4 \cdot 0 + 1) = 0.$$

This increase in degree is true for all closed Newton–Cotes functionals F_n where the number of subintervals n is even. For suppose f is a polynomial of degree $n + 1$. Then $f^{(n+1)} = (n+1)!$ and (9.2.3) becomes

$$R_n(f) = \int_a^b \varphi_{n+1}(x)\, dx.$$

Let $x' = x - (x_0 + \frac{1}{2}nh)$. Since $\varphi_{n+1}(x) = \prod_{i=0}^{n}(x - x_i)$, and $x_i = x_0 + ih$, $0 \leq i \leq n$, $a = x_0$, $b = x_n$, we obtain

$$R_n(f) = \int_{-nh/2}^{nh/2} \prod_{i=0}^{n}(x' - x_i')\, dx',$$

where $x_i' = (i - n/2)h$, $(x - x_i = x' + x_0 + (n/2)h - x_0 - ih = x' - (i - n/2)h$.) Since $x_i' = -x_{n-i}'$,

$$\prod_{i=0}^{n}(x' - x_i') = x' \prod_{i=0}^{(n/2)-1}(x'^2 - x_i'^2)$$

is an odd function of x'. Hence, $R_n(f) = 0$ for f a polynomial of degree $\leq n + 1$. Now if $f(x) = x^{n+2}$, then $f[x_0', \ldots, x_n', x'] = x'$. (See Exercise 8.4b.) Thus, we see that $R_n(x^{n+2}) \neq 0$, proving that the closed Newton–Cotes functional F_n is of degree $n + 1$ when n is even.

To obtain an explicit formula for the remainder $R(f)$ in Simpson's rule, we cannot conveniently use (9.2.3) because $\varphi_3(x) = (x - x_0)(x - x_{-1})(x - x_1)$ changes sign at the point $x_0 = (a + b)/2$. Therefore, we shall develop a general formula for $R(f)$, due to Peano, which can be applied to any quadrature functional F given by (9.2.1).

Theorem 9.2.2 Let $S(f) = \int_a^b f(x)\, dx$. Let F be a quadrature functional given by formula (9.2.1). Suppose that the remainder functional $R = S - F$ is zero for all polynomials of degree $\leq n$. Then for any $f \in C^{n+1}[a, b]$,

$$R(f) = \int_a^b f^{(n+1)}(t)K(t)\, dt, \tag{9.2.9}$$

where, as a function of t, K is given by

$$K(t) = (1/n!)R((x - t)_+^n); \tag{9.2.10}$$

that is, R is applied to the function $(x - t)_+^n$ as a function of x,

$$(x - t)_+^n = \begin{cases} (x - t)^n & \text{for } x > t, \\ 0 & \text{for } x \leq t. \end{cases} \tag{9.2.11}$$

Proof. We start with Taylor's formula with the remainder

$$f(x) = f(a) + f'(a)(x - a) + \cdots + \frac{f^{(n)}(a)}{n!}(x - a)^n + \frac{1}{n!}\int_a^x f^{(n+1)}(t)(x - t)^n\, dt.$$

Since $R((x - a)^j) = 0, 0 \leq j \leq n$, by hypothesis,

$$R(f) = (1/n!)R\int_a^x f^{(n+1)}(t)(x - t)^n\, dt. \tag{1}$$

Let $g(x, t) = f^{(n+1)}(t)(x - t)^n$. Note that the partials $\partial^k g/\partial x^k$ exist for $k \leq n$ and are continuous in x (uniformly with respect to t). Therefore, we may differentiate under the integral sign to get

$$\frac{\partial}{\partial x}\int_a^x g(x, t)\, dt = f^{(n+1)}(x)(x - x)^n + \int_a^x f^{(n+1)}(t)\frac{\partial}{\partial x}(x - t)^n\, dt$$

$$= \int_a^x f^{(n+1)}(t)n(x - t)^{n-1}\, dt$$

$$= \int_a^b f^{(n+1)}(t)n(x - t)_+^{n-1}\, dt = \int_a^b f^{(n+1)}(t)\frac{\partial}{\partial x}(x - t)_+^n\, dt,$$

and similarly for higher derivatives. Now, from (1) above, letting $S_t(g(x, t)) = \int_a^x g(x, t)\, dt$, we have

$$n!R(f) = RS_t(g(x, t)) = (S - F)S_t(g) = SS_t(g) - FS_t(g).$$

But

$$FS_t(g(x, t)) = F\int_a^x g(x, t)\, dt = F\int_a^b f^{(n+1)}(t)(x - t)_+^n\, dt.$$

According to (9.2.1) and the preceding remark on differentiation,

$$F\int_a^x g(x, t)\, dt = \sum_{i=0}^{m_0} a_{i0}\int_a^b f^{(n+1)}(t)(x_{i0} - t)_+^n\, dt + \cdots$$

$$+ \sum_{i=0}^{m_n} a_{in}\int_a^b \frac{\partial^n}{\partial x^n}(x - t)_+^n\bigg|_{x=x_{in}} f^{(n+1)}(t)\, dt$$

$$= \int_a^x dt f^{(n+1)}(t)$$

$$\left(\sum_{i=0}^{m_0} a_{i0}(x_{i0} - t)_+^n + \cdots + \sum_{i=0}^{m_n} a_{in} \frac{\partial^n}{\partial x^n} (x - t)_+^n \bigg|_{x=x_{in}} \right)$$

$$= S_t f^{(n+1)}(t) F(x - t)_+^n.$$

$$SS_t(g(x, t)) = S \int_a^x f^{(n+1)}(t)(x - t)^n \, dt = S \int_a^b f^{(n+1)}(t)(x - t)_+^n \, dt$$

$$= \int_a^b dx \int_a^b f^{(n+1)}(t)(x - t)_+^n \, dt$$

$$= \int_a^b dt f^{(n+1)}(t) \int_a^b (x - t)_+^n \, dx = \int_a^x dt f^{(n+1)}(t) S(x - t)_+^n$$

$$= S_t f^{(n+1)}(t) S(x - t)_+^n.$$

Hence,

$$n! \, R(f) = S_t f^{(n+1)}(t) S[(x - t)_+^n] - S_t f^{(n+1)}(t) F[(x - t)_+^n]$$
$$= S_t f^{(n+1)}(t) R[(x - t)_+^n],$$

as was to be proved. ■

Corollary If the *Peano kernel* $K(t)$ does not change sign in $[a, b]$, then

$$R(f) = \frac{f^{(n+1)}(\xi)}{(n + 1)!} R(x^{n+1}). \quad (9.2.12)$$

Proof. By the first mean-value theorem and (9.2.9),

$$R(f) = f^{(n+1)}(\xi) \int_a^b K(t) \, dt, \quad a \le \xi \le b.$$

Taking $f(x) = x^{n+1}$, we obtain

$$R(x^{n+1}) = (n + 1)! \int_a^b K(t) \, dt$$

and the result follows. ■

Now, let us use this result to evaluate the remainder in (9.2.8). First, however, we transform the variable by letting $x = x_0 + sh$. The remainder operator is

$$R_2(f) = R_2(g) = h \int_{-1}^1 g(s) \, ds - (h/3)(g(-1) + 4g(0) + g(1)),$$

where $g(s) = f(x_0 + sh)$ and the interval $[a, b]$ is transformed to $[-1, 1]$. By (9.2.9), since the Simpson quadrature functional has zero remainder for polynomials of degree ≤ 3, we take $n = 3$ to obtain $R_2(f) = \int_{-1}^1 g^{(4)}(t) K(t) \, dt$.

By (9.2.10), we have

$$K(t) = \tfrac{1}{6} R_2((s-t)_+^3)$$
$$= \tfrac{1}{6} \int_{-1}^{1} (s-t)_+^3 \, ds - \tfrac{1}{18}((-1-t)_+^3 + 4(-t)_+^3 + (1-t)_+^3).$$

But $(-1-t)_+^3 = 0$, since $-1 \le t$ for $-1 \le t \le 1$. Also,

$$(0-t)_+^3 = \begin{cases} (-t)^3 & \text{for } t \le 0, \\ 0 & \text{for } 0 \le t. \end{cases}$$

Finally, $(1-t)_+^3 = (1-t)^3$ for $-1 \le t \le 1$. Since $(s-t)_+^3 = 0$ for $s \le t$,

$$\int_{-1}^{1} (s-t)_+^3 \, ds = \int_{t}^{1} (s-t)^3 \, ds = (1-t)^4/4.$$

Therefore,

$$6K(t) = \begin{cases} (1-t)^4/4 + (4/3)t^3 - (1-t)^3/3, & \text{for } t \le 0, \\ (1-t)^4/4 - (1-t^3)/3 & \text{for } t \ge 0. \end{cases}$$

Expanding and simplifying the two expressions for $K(t)$, one finds that $K(t) = (-1 + 6t^2 - 8t^3 + 3t^4)/72$ for $t \ge 0$ and $K(t) = (-1 + 6t^2 + 8t^3 + 3t^4)/72$ for $t \le 0$. Hence, $K(t) = K(-t)$. Also, $K(t) = -(1-t^3)(1+3t)/72 \le 0$ for $0 \le t \le 1$. Hence, $K(t) \le 0$ for $-1 \le t \le 1$ and we can apply formula (9.2.12) to obtain

$$R_2(f) = \frac{g^{(4)}(\tau)}{4!} R_2(s^4), \qquad -1 \le \tau \le 1$$

$$R_2(s^4) = h \int_{-1}^{1} s^4 \, ds - (h/3)(1+0+1) = -4h/15.$$

Since $g^{(4)}(\tau) = h^4 f^{(4)}(\xi)$, $x_{-1} \le \xi \le x_1$, we get

$$R_2(f) = -h^5 f^{(4)}(\xi)/90. \tag{9.2.13}$$

Rewriting (9.2.8) with $x_{-1} = a$, $x_1 = a + 2h$, we obtain Simpson's rule explicitly as

$$\int_{a}^{a+2h} f(x) \, dx = \frac{h}{3}(f(a) + 4f(a+h) + f(a+2h))$$
$$- \frac{h^5}{90} f^{(4)}(\xi), \qquad a \le \xi \le a+2h. \tag{9.2.14}$$

(For an alternative derivation see Exercise 9.4d.)

Further results on Newton–Cotes quadrature formulas will be found in Exercises 9.3, 9.4 and 9.5.

9.3 COMPOSITE QUADRATURE FUNCTIONALS

Quadrature functionals such as those given by the trapezoidal rule and Simpson's rule can be used to obtain arbitrarily close approximations to $Sf = \int_a^b f(x) \, dx$ by employing them in a composite manner as follows. We divide the interval $[a, b]$ into n equal subintervals by the points $x_i = x_0 + ih$, $0 \leq i \leq n$, with $x_0 = a$ and $x_n = b$. Then we apply the quadrature formula in successive groups of subintervals.

For example, the *composite trapezoidal* rule is defined by applying (9.2.7) to successive subintervals $[x_i, x_i + 1]$. This yields

$$F_h f = \frac{h}{2}(f(x_0) + 2f(x_1) + 2f(x_2) + \cdots + 2f(x_{n-1}) + f(x_n)).$$

The remainder R_{1h} is easily calculated to be

$$R_{1h}f = \sum_{i=0}^{n-1}\left(\int_{x_i}^{x_{i+1}} f(x)\,dx - \frac{h}{2}(f(x_i) + f(x_{i+1}))\right) = \sum_{i=0}^{n-1} R_{1i}f,$$

where by (9.2.7) we have

$$R_{1i}f = \frac{-h^3}{12}f^{(2)}(\xi_i), \qquad x_i \leq \xi_i \leq x_{i+1}.$$

If $|f^{(2)}(x)| \leq M$ for all x in $[a, b]$, then

$$|R_{1h}f| \leq \frac{h^3}{12}Mn = \frac{h^2}{12}M(b-a) = O(h^2).$$

If $f^{(2)}(x)$ is continuous in $[a, b]$, then

$$n \min \{f^{(2)}(\xi_i)\} \leq \sum_{i=0}^{n-1} f^{(2)}(\xi_i) \leq n \max \{f^{(2)}(\xi_i)\}$$

and $\sum_{i=0}^{n-1} f^{(2)}(\xi_i) = nf^{(2)}(\xi)$ for some $a \leq \xi \leq b$. In this case we have

$$\int_a^b f(x)\,dx = \frac{h}{2}(f(x_0) + 2f(x_1) + \cdots$$
$$+ 2f(x_{n-1}) + f(x_n)) - \frac{h^2}{12}(b-a)f^{(2)}(\xi), \qquad a \leq \xi \leq b. \quad (9.3.1)$$

One sees, immediately, that $R_{1h} \to 0$ as $h \to 0$. Thus, for any $f \in C^2[a, b]$, Sf can be approximated with arbitrarily small error by choosing h sufficiently small.

By applying Simpson's rule to the subintervals $[x_0, x_2], [x_2, x_4], \ldots, [x_{n-2}, x_n]$, we obtain the *composite Simpson's rule*,

$$\int_a^b f(x)\,dx = \frac{h}{3}(f(x_0) + 4f(x_1) + 2f(x_2) + 4f(x_3) + \cdots$$
$$+ 4f(x_{n-1}) + f(x_n)) - \frac{(b-a)}{180}h^4 f^{(4)}(\xi), \qquad a \leq \xi \leq b \quad (9.3.2)$$

where n must be even. (See Exercise 9.2.) If n is odd, we obtain a composite formula of the same order of accuracy (in h) by first using the formula in Exercise 9.3 for

$\int_{x_0}^{x_3} f(x)\,dx$ and then using the composite Simpson rule over the interval $[x_3, x_n]$, which has an even number of subintervals as required. For any function $f \in C^4[a, b]$, the remainder $R_{2h} = -(b - a)h^4 f^{(4)}(\xi)/180 \to 0$ as $h \to 0$.

In fact, both the composite trapezoidal rule and the composite Simpson's rule converge for any continuous function f. However, this does not follow from Theorem 9.2.1. To see this for the composite Simpson's rule, let

$$F_m(f) = \frac{(b-a)}{6m}(f(x_0) + 4f(x_1) + 2f(x_2) + \cdots + 4f(x_{2m-1}) + f(x_{2m})), \quad (9.3.3)$$

for $m = 1, 2, 3, \ldots$. Comparing this with (9.3.2), we see that F_m is the composite Simpson functional for the interval $[a, b]$ with $n = 2m$ subintervals. Let $f \in C[a, b]$ with $\|f\|_\infty = \max_{a \le x \le b} \{|f(x)|\}$. Then

$$|F_m(f)| \le \frac{|b-a|(1+6m)\|f\|_\infty}{6m} \le \tfrac{7}{6}|b-a|\,\|f\|_\infty. \quad (9.3.4)$$

Hence, $\|F_m\| \le \tfrac{7}{6}|b - a|$ for all m, so that (F_m) is a uniformly bounded sequence. However, in Theorem 9.2.1 it is also required that F_m be of degree at least m. It follows from (9.3.2) that all F_m in (9.3.3) are of degree 3 on $[a, b]$, since $f^{(4)}(x) \ne 0$, $a \le x \le b$, for some polynomial f of degree 4. However, for any polynomial p, we observe that $F_m p \to Sp$ as $m \to \infty$, since by (9.3.2) the remainder $-(b - a)h^4 p^{(4)}(\xi)/180 \to 0$ as $h = (b - a)/2m \to 0$. Since the set of polynomials is everywhere dense in $C[a, b]$, Theorem 3.7.2 applies and we conclude that $(F_m f)$ converges for every $f \in C[a, b]$, that is, the sequence of linear operators F_m on $C[a, b]$ to R is strongly convergent. Indeed, $F_m f \to Sf$, since for any $\varepsilon > 0$ there exists a polynomial p such that $\|f - p\|_\infty < \varepsilon/4|b - a|$ and an integer N such that for $m > N$, $|Sp - F_m p| < \varepsilon/3$. It follows that

$$\begin{aligned}|Sf - F_m f| &\le |Sf - Sp| + |Sp - F_m p| + |F_m p - F_m f| \\ &\le \|S\|\,\|f - p\|_\infty + |Sp - F_m p| + \tfrac{7}{6}|b - a|\,\|p - f\|_\infty < \varepsilon.\end{aligned} \quad (9.3.5)$$

Similarly, using (9.3.1), we can prove that the composite trapezoidal rule converges. In fact, consider any composite quadrature formula based on a closed Newton–Cotes formula $h \sum_0^n a_i f(x_i)$. Such a functional has the form

$$F_m f = h \sum_{j=0}^{m} \sum_{i=0}^{n} a_i f(x_{ij}), \quad x_{ij} = a + (i + nj)h,\ x_{mn} = b. \quad (9.3.6)$$

The remainder for $f \in C^{n+2}[a, b]$ and n even is

$$R_h f = -K_n h^{n+3} \sum_{j=0}^{m} f^{(n+2)}(\xi_j)/(n + 2)!$$

For n odd and $f \in C^{n+1}[a, b]$,

$$R_h f = -K_n h^{n+2} \sum_{j=0}^{m} f^{(n+1)}(\xi_j)/(n + 1)!$$

9.3 Composite Quadrature Functionals

The positive constants K_n are given by Exercise 9.4c, e. This yields $|R_hf| \leq Mh^{n+2}$ and $|R_hf| \leq Mh^{n+1}$ for even and odd n respectively, where M is a constant depending on $\|f^{(n+2)}\|_\infty$ and $\|f^{(n+1)}\|_\infty$ respectively. Hence, for f a polynomial, $R_hf \to 0$ as $h \to 0$. Also,

Hence,
$$|F_mf| \leq \frac{(b-a)}{n(m+1)}(m+1)\sum_{i=0}^n |a_i|\,\|f\|_\infty.$$

$$\|F_m\| \leq (b-a)\sum_{i=0}^n |a_i|/n,$$

so that the functionals F_m are uniformly bounded. Hence, inequality (9.3.5) applies once more and we have proved the following general result.

Theorem 9.3.1 Let (F_m) be a sequence of composite quadrature functionals based on a closed Newton–Cotes formula as in (9.3.6). For any $f \in C[a, b]$, $F_mf \to \int_a^b f(x)\,dx$ as $m \to \infty$.

Computational Aspects To apply the trapezoidal composite quadrature formula in actual computation, a desired accuracy ε is specified. Using the error estimates in (9.3.1) a value of h is chosen by setting $h^2M(b-a)/12 < \varepsilon$, where $M = \|f^{(2)}\|_\infty$. If M is unknown, choose h arbitrarily. Compute F_nf according to (9.3.1). Then cut h in half and compute $F_{2n}f$. Continue until two successive values are obtained which differ by less than ε. In fact, it is the relative error in Sf which should be the criterion of accuracy. As in most convergent procedures, we compute until two successive values agree to the desired number of digits. Of course, this stopping criterion must be used with caution. (See Exercise 4.23.)

In actual computation, rounding errors are committed. Suppose $Ff = \sum_{i=0}^n a_i f(x_i)$. Let us consider the rounding error ρ_i in the computation of $f(x_i)$. Let $f_i = f(x_i) + \rho_i$ be the finite-precision value actually used. Ignoring other rounding errors (i.e., they arise in computing the sum of products $a_i f_i$ and can be analyzed by formula (4.2.16)), we obtain the rounded value

$$\overline{Ff} = \sum_{i=0}^n a_i f_i = \sum_0^n a_i f(x_i) + \sum_0^n a_i \rho_i.$$

We call $r = \sum_0^n a_i \rho_i$ the *accumulated rounding error*. We assume, as is very often the case, that $|\rho_i| < \rho$. Then

$$|r| \leq \rho \sum_0^n |a_i|. \tag{1}$$

Since Ff is exact for polynomials of degree 0 in any case of interest, taking $f(x) \equiv 1$ we find that $\sum_0^n a_i = b - a$. If all a_i are of the same sign, then $\sum_0^n |a_i| = |\sum_0^n a_i| = b - a$. Hence, in this case, $|r| < \rho(b-a)$. However, if not all a_i are of the same sign, then $\sum_0^n |a_i| > |\sum_0^n a_i| = b - a$, and the bound in (1) above is larger, permitting a larger accumulated rounding error. In fact, if $\rho_i = \rho\,\text{sign}\,(a_i)$, the largest error is

$\rho \sum_0^n |a_i|$, which is greater than $\rho(b - a)$. This suggests that quadrature formulas in which all coefficients have the same sign (as in Simpson's composite rule) produce smaller rounding errors.

A statistical analysis of rounding error (Exercise 9.7) shows that the mean square rounding error is minimized by taking all $a_i = (b - a)/n$. Hence, quadrature functionals in which all coefficients are close in value, behave better with respect to rounding error than others in which the coefficients vary in value and/or sign. Furthermore, formulas in which all $a_i \geq 0$ have desirable convergence properties, as we shall see in the next section. In Exercise 9.5, the closed Newton–Cotes formulas for $3 \leq n \leq 8$ are given. It can be seen that the coefficients of these higher-order formulas become less satisfactory in so far as these rounding error criteria are concerned.

9.4 GAUSSIAN QUADRATURE

Applying Theorem 9.2.1 and Theorem 3.7.1 on the uniform boundedness of strongly convergent sequences, we can obtain a general result on the convergence of certain sequences of linear quadrature functionals.

Theorem 9.4.1 Let

$$F_n f = \sum_{i=1}^{k_n} a_{in} f(x_{in}), \qquad x_{in} \in [a, b], n = 1, 2, \ldots,$$

be a sequence of quadrature functionals such that F_n is at least of degree n on the interval $[a, b]$. Let $Sf = \int_a^b f(x)\,dx$. A necessary and sufficient condition that $F_n f \to Sf$ for all $f \in C[a, b]$ is that there exist a constant M such that $\sum_{i=1}^{k_n} |a_{in}| \leq M$ for all n.

Proof. Sufficiency. F_n and S are continuous linear operators on $C[a, b]$ to the one-dimensional space R. Since

$$|F_n f| \leq \sum |a_{in}| |f(x_{in})| \leq M \|f\|_\infty,$$

the sequence (F_n) is uniformly bounded. By Theorem 9.2.1, $F_n f \to Sf$ for all $f \in C[a, b]$; i.e., (F_n) is strongly convergent on $C[a, b]$. Thus, the condition is sufficient.

Conversely, if $F_n f \to Sf$ for all $f \in C[a, b]$, then by Theorem 3.7.1, (F_n) must be uniformly bounded; i.e., $\|F_n\| < M$. Taking $f_n \in C[a, b]$ to be such that $f_n(x_{in}) = a_{in}/|a_{in}|$, $i = 1, \ldots, k_n$, and $\max_{a \leq x \leq b} |f_n(x)| = 1$ we have

$$|F_n f_n| = |\sum a_{in} f(x_{in})| = \sum |a_{in}| \leq \|F_n\| \|f_n\|_\infty = \|F_n\| < M.$$

This proves the necessity. ∎

Corollary Let $F_n f = \sum_{i=1}^{k_n} a_{in} f(x_{in})$. Suppose further that all $a_{in} \geq 0$. If F_n is of degree at least n, then $F_n f \to Sf$ for all $f \in C[a, b]$.

Proof. Let p be the zero degree polynomial 1. Then $\sum |a_{in}| = \sum a_{in} = F_n p = Sp = b - a$. Taking $M = b - a$, we can apply the theorem to obtain the result. ∎

9.4 Gaussian Quadrature

An important example of a sequence of quadrature functionals which satisfy the hypotheses of the preceding corollary is afforded by the *Gaussian quadrature functionals*. Such a functional is of the form $Ff = \sum_{i=0}^{n} a_i f(x_i)$, previously encountered in our discussion of the interpolatory quadrature functionals in Section 9.2 in which the x_i are arbitrary and in the special case of Newton–Cotes functionals in which the x_i are equally spaced. In the Gaussian formulas, the x_i are no longer arbitrary nor are they equally spaced. Rather, the x_i are strategically placed so that F will have as high a degree as possible. Thus, we seek to determine the $2n + 2$ parameters x_i and a_i, $0 \le i \le n$, so that

$$Rp = Sp - Fp = \int_a^b f(x)\,dx - \sum_{i=0}^{n} a_i p(x_i)$$

will be zero for polynomials p of degree $\le m$, where m is as large as possible.

Suppose that the x_i have been so determined. Since the a_i can always be chosen to make F interpolatory, the degree of F must be at least n. But if F is of degree at least n, then $a_i = \int_a^b l_i^{(n)}(x)\,dx$ and $Rf = \int_a^b \varphi_{n+1}(x) f[x_0, \ldots, x_n, x]\,dx$. (See Remark following Eq. (9.2.2).) Let the maximum degree of F be $n + r$, $r \ge 0$. For any polynomial of degree $n + r$, the $(n + 1)$st divided difference $f[x_0, \ldots, x_n, x]$ is a polynomial in x of degree $\le r - 1$. (Exercise 8.4b.) Hence, a necessary and sufficient condition that $Rp = 0$ for all polynomials p of degree $n + r$ is that

$$\int_a^b \varphi_{n+1}(x) x^j\,dx = 0, \qquad 0 \le j \le r - 1. \tag{9.4.1}$$

But this means that $\varphi_{n+1}(x)$ must be the monic polynomial orthogonal on $[a, b]$ to all polynomials of degree $\le r - 1$. Since $\varphi_{n+1}(x) = \prod_{i=0}^{n}(x - x_i)$ is of degree $n + 1$, we can take $r - 1 = n$. However, $r > n + 1$ would imply that φ_{n+1} is orthogonal to itself, and therefore would be the zero polynomial, which is impossible. Thus, the x_i can be so chosen that F has maximum degree $2n + 1$ and this degree is attained if the x_i are taken to be the zero's of the orthogonal polynomial of degree $n + 1$ on $[a, b]$, which, according to Theorem 8.3.2 (Corollary), are all in (a, b). We have proved

Theorem 9.4.2 Let $Ff = \sum_{i=0}^{n} a_i f(x_i)$. The functional F has maximum degree $2n + 1$ on $[a, b]$ if and only if the x_i are the zeros of the orthogonal polynomial of degree $n + 1$ on $[a, b]$ (with weight function $\rho(x) = 1$) and

$$a_i = \int_a^b l_i^{(n)}(x)\,dx = \int_a^b \frac{\varphi_{n+1}(x)}{(x - x_i)\varphi'_{n+1}(x_i)}\,dx.$$

A functional F is called a *Gaussian quadrature functional* if it satisfies the conditions of Theorem 9.4.2.

Theorem 9.4.3 Let F be a Gaussian quadrature functional. The coefficients a_i are all positive.

394 Numerical Integration

Proof. The functional F has degree $2n + 1$. Let

$$p_j(x) = \frac{\varphi_{n+1}^2(x)}{(x - x_j)^2}, \qquad 0 \le j \le n.$$

The p_j are polynomials of degree $2n$. Hence,

$$Sp_j = \int_a^b p_j(x) \, dx = \sum_{i=0}^n a_i p_j(x_i).$$

But $p_j(x_i) = 0$ for $j \ne i$ and

$$p_j(x_j) = \prod_{\substack{i=0 \\ i \ne j}}^n (x_j - x_i)^2 > 0.$$

Hence, $Sp_j = a_j p_j(x_j)$, or

$$a_j = \frac{1}{p_j(x_j)} \int_a^b p_j(x) \, dx = \frac{1}{p_j(x_j)} \int_a^b \frac{\varphi_{n+1}^2(x)}{(x - x_j)^2} \, dx > 0. \blacksquare$$

Corollary Let (F_n) be the sequence of Gaussian quadrature functionals on $[a, b]$. For any $f \in C[a, b]$, $F_n f \to \int_a^b f(x) \, dx$.

Proof. By Theorems 9.4.2 and 9.4.3, the F_n satisfy the hypotheses of the Corollary to Theorem 9.4.1. $\blacksquare$

To obtain an expression for the error in Gaussian quadrature, we choose points $y_0, \ldots, y_n$ in $[a, b]$ which are distinct from the nodes $x_0, \ldots, x_n$ in the Gaussian quadrature functional $F = \sum_0^n a_i f(x_i)$. Let $L_{2n+1}(x)$ be the interpolating polynomial for f on the set of points $\{x_0, \ldots, x_n, y_0, \ldots, y_n\}$. Then

$$f(x) = L_{2n+1}(x) + r_{2n+1}(x)$$

and

$$r_{2n+1}(x) = \prod_{i=0}^n (x - x_i)(x - y_i) \cdot f[x_0, \ldots, x_n, y_0, \ldots, y_n, x].$$

Now,

$$SF = SL_{2n+1} + Sr_{2n+1} = FL_{2n+1} + Sr_{2n+1} = Ff + Sr_{2n+1},$$

since F is of degree $2n + 1$ and $f(x_i) = L_{2n+1}(x_i)$.

Therefore, $Rf = (S - F)f = Sr_{2n+1}$, that is,

$$Rf = \int_a^b \prod_0^n (x - x_i)(x - y_i) f[x_0, \ldots, x_n, y_0, \ldots, y_n, x] \, dx.$$

If $f^{(2n+2)}(x)$ is continuous, then the integrand is a continuous function of $(y_0, \ldots, y_n)$. Allowing $y_i \to x_i$, we obtain

$$Rf = \int_a^b (\varphi_{n+1}(x))^2 f[x_0, \ldots, x_n, x_0, \ldots, x_n, x] \, dx.$$

By the first mean-value theorem,

$$Rf = f[x_0, \ldots, x_n, x_0, \ldots, x_n, \xi'] \int_a^b (\varphi_{n+1}(x))^2 \, dx, \quad a < \xi' < b,$$

$$Rf = \frac{f^{(2n+2)}(\xi)}{(2n+2)!} \int_a^b (\varphi_{n+1}(x))^2 \, dx. \tag{9.4.2}$$

A simpler derivation of (9.4.2) is obtained by using simple Hermite interpolation. (See Exercise 9.6.)

Computational Aspects The Legendre polynomials $P_n(x)$ (Example (c) in Section 7.7) are orthogonal on $[-1, 1]$ with respect to the weight function $\rho = 1$. The integral over any finite interval $[a, b]$ may be transformed into an integral over $[-1, 1]$ by the substitution $x' = 2x/(b - a) + (a + b)/(a - b)$. Thus

$$\int_a^b f(x) \, dx = \frac{(b-a)}{2} \int_{-1}^1 f((b-a)x'/2 + (a+b)/2) \, dx'$$

and we may restrict our attention to the interval $[-1, 1]$. The Legendre polynomials could be computed as in Exercise 7.21 and the zeros determined. However, the required *nodes* x_i and *weights* a_i for the Gaussian quadrature functional $Ff = \sum_0^n a_i f(x_i)$ have been tabulated. See [9.3] and [9.4]. Thus, Ff can be evaluated for any f defined on $[-1, 1]$.

9.4(a) GAUSSIAN QUADRATURE WITH ASSIGNED NODES

The nodes x_i in a Gaussian quadrature functional are interior points of the interval of integration, $[a, b]$. In some applications, it is useful to have a functional in which one or both of the end points of the interval are among the nodes. For example, if a composite functional is to be constructed, then the end point of one subinterval would be the initial point of the next, thereby reducing the number of function evaluations. In certain problems, such as the Sturm–Liouville problem of Section 7.8, the values of an unknown function $y = f(x)$ are specified at $x = a$ and $x = b$ and the remaining values are to be found by using a quadrature formula of the type

$$\int_a^b f(x) \, dx = Af(a) + Bf(b) + \sum_1^{n-1} a_i f(x_i) + Rf,$$

where the remaining nodes x_i are to be determined by some other conditions. This leads us to the problem of constructing a quadrature functional F of the type

$$Ff = \sum_{i=0}^{n-m} a_i f(x_i) + \sum_{j=1}^{m} b_j f(z_j), \tag{9.4.3}$$

where the m nodes z_j ($1 \leq m \leq n$) are prescribed fixed nodes and the x_i and the coefficients a_i, b_j are to be determined so as to make F have maximum degree.

Theorem 9.4.4 The functional F given by (9.4.3) with prescribed nodes z_j not in the

396   Numerical Integration

interior of $[a, b]$ is of maximum degree $2n - m + 1$ if and only if F is interpolatory and the x_i are the zeros of the orthogonal polynomial of degree $n - m + 1$ with respect to the weight function $|\rho(x)| = |(x - z_1) \cdots (x - z_m)|$.

Proof. As in the proof of Theorem 9.4.2, we note that for any choice of the x_i, the coefficients a_i, b_j can be chosen to make F interpolatory on the set $E = \{x_0, \ldots, x_{n-m}, z_1, \ldots, z_m\}$. Thus, F is at least of degree n. Let the maximum degree of F be $n + r$, $r \geq 0$. Let $l_i^{(n)}(x)$ be the Lagrange coefficients for the set E. Then, as before,

$$a_i = \int_a^b l_i^{(n)}(x)\,dx, \qquad 0 \leq i \leq n - m,$$

and

$$b_j = \int_a^b l_{j+n-m}^{(n)}(x)\,dx, \qquad 1 \leq j \leq m;$$

i.e., F is interpolatory on the set E. A necessary and sufficient condition that the remainder $Rp = 0$ for all polynomials p of degree $n + r$ is (by (9.4.1)) that

$$\int_a^b \varphi_{n-m+1}(x)\rho(x)x^j\,dx = 0, \qquad 0 \leq j \leq r - 1,$$

where $\varphi_{n-m+1}(x) = (x - x_0) \cdots (x - x_{n-m})$. Thus, φ_{n-m+1} must be the monic polynomial orthogonal, with respect to the nonnegative weight function $|\rho(x)|$, on $[a, b]$ to all polynomials of degree $\leq r - 1$. Since φ_{n-m+1} is of degree $n - m + 1$, we can take $r = n - m + 1$. ∎

In Section 7.7, the existence of a sequence of orthogonal polynomials with respect to any positive weight function ρ on a finite interval $[a, b]$ was proved by giving a method of constructing the sequence. The definition of inner product given by (7.7.1) remains valid if ρ has a finite number of zeros in $[a, b]$. It applies to the function $|\rho|$ of Theorem 9.4.3, since the nodes z_j are outside (a, b). (They may include the end points a and b.) Therefore, the orthogonal polynomials φ_n of the theorem do indeed exist and the explicit construction of Section 7.7 could be used to obtain them. If we translate the interval $[a, b]$ to $[-1, 1]$, there is a somewhat easier method which expresses them in terms of the Legendre polynomials (Section 7.7, Example (c)). The latter may be obtained from existing tables ([9.2], [9.9]) or they may be computed as in Exercise 7.21b. Therefore, it is convenient to express φ_n in these terms.

The Legendre polynomials P_i, $0 \leq i \leq n + 1$, are linearly independent. Therefore, the monic polynomial $\rho\varphi_{n-m+1}$, which is of degree $n + 1$, can be expressed as a linear combination,

$$\rho\varphi_{n-m+1} = \frac{1}{c_{n+1}} P_{n+1} + d_n P_n + \cdots + d_0 P_0,$$

where c_{n+1} is the leading coefficient of P_{n+1}. We multiply both sides by P_i and integrate to obtain

$$0 = \int_{-1}^{1} \rho(x)\varphi_{n-m+1}(x)P_i(x)\,dx = d_i \int_{-1}^{1} P_i^2(x)\,dx, \qquad 0 \leq i \leq n - m.$$

We have used the fact that φ_{n-m+1} is orthogonal to all polynomials of degree $\leq n - m$. Hence, the coefficients $d_i = 0$, $0 \le i \le n - m$, and

$$-\rho \varphi_{n-m+1} + \frac{1}{c_{n+1}} P_{n+1} + d_n P_n + \cdots + d_{n-m+1} P_{n-m+1} = 0. \tag{9.4.4}$$

But $\rho(x) = (x - z_1) \cdots (x - z_m)$. Hence, for $1 \le j \le m$.

$$(1/c_{n+1})P_{n+1}(z_j) + d_n P_n(z_j) + \cdots + d_{n-m+1} P_{n-m+1}(z_j) = 0 \tag{9.4.5}$$

We may combine (9.4.4) and (9.4.5) and regard it as a system of $m + 1$ homogeneous linear equations in the "unknowns" $1, d_n, \ldots, d_{n-m+1}$. Since a nontrivial solution exists, the determinant must be zero. Setting the determinant equal to zero, we obtain the desired relationship.

For example, if $m = 1$ and $z_1 = -1$, we get $\rho(x) = x + 1$ and

$$\begin{vmatrix} -\rho(x)\varphi_n(x) + \left(\dfrac{1}{c_{n+1}}\right) P_{n+1}(x) & P_n(x) \\ (1/c_{n+1})P_{n+1}(-1) & P_n(-1) \end{vmatrix} = 0.$$

Hence,

$$\varphi_n(x) = \left(\frac{1}{(x+1)c_{n+1}}\right)\left[P_{n+1}(x) - P_n(x)\frac{P_{n+1}(-1)}{P_n(-1)}\right].$$

Using the recursion formula in Exercise 7.21b, we find that

$$\frac{P_n(-1)}{P_{n-1}(-1)} = \frac{-(2n-1)}{n} - \left(\frac{n-1}{n}\right)\frac{P_{n-2}(-1)}{P_{n-1}(-1)}.$$

If $P_{n-1}(-1)/P_{n-2}(-1) = -1$, it follows that $P_n(-1)/P_{n-1}(-1) = -1$. Together with $P_1(-1)/P_0(-1) = -1$, we have an inductive proof that $P_n(-1)/P_{n-1}(-1) = -1$ for all n. Therefore,

$$\varphi_n(x) = \frac{1}{(x+1)c_{n+1}}[P_{n+1}(x) + P_n(x)], \tag{9.4.6}$$

so that in this case the nodes x_i in (9.4.3) are the zeros of the polynomial,

$$\frac{P_{n+1}(x) + P_n(x)}{x + 1}$$

Formula (9.4.3) becomes

$$Ff = \sum_{i=0}^{n-1} a_i f(x_i) + b_1 f(-1). \tag{9.4.7}$$

Since this formula is interpolatory on the set $E = \{x_0, \ldots, x_{n-1}, -1\}$, we may apply the fact that $a_i = \int_{-1}^{1} l_i^{(n)}(x)\, dx$, where $l_i^{(n)}$ is the Lagrange coefficient with respect to E. (See (8.1.2) and the formula for $l_i^{(n)}$ preceding Eq. (8.2.2).) Hence,

$$a_i = \int_{-1}^{1} \frac{(x+1)\varphi_n(x)}{(x - x_i)\varphi_n'(x_i)(x_i + 1)}\, dx.$$

Let $Q_n = k_n \varphi_n$ be the normalization of φ_n such that $\int_{-1}^{1}(x+1)Q_n^2(x)\,dx = 1$, that is,

$$k_n^2 = 1 \Big/ \int_{-1}^{1}(x+1)\varphi_n^2(x)\,dx.$$

Then, by Exercise 9.8d,

$$a_i = \frac{1}{k_{n-1}^2(x_i + 1)\varphi_n'(x_i)\varphi_{n-1}(x_i)}. \tag{9.4.8}$$

According to Theorem 9.4.4, the φ_n are orthogonal on $[-1, 1]$ with respect to the weight function $\rho(x) = x + 1$. By the results of Section 7.8, we see that $\varphi_n = (1/A_n^{(0,1)})J_n^{(0,1)}$, where $A_n^{(0,1)}$ is the leading coefficient of the Jacobi polynomial $J_n^{(0,1)}$. By the Rodrigues formula (Exercise 9.9),

$$A_n^{(0,1)} = (2n + 1)!/2^n n!(n + 1)!.$$

Now, it can be proved [9.12] that the Jacobi polynomials $J_n^{(p,q)}$ satisfy the relation,

$$(p + q + 2n)(1 - x^2)\frac{d}{dx}J_n^{(p,q)} = -n[(p + q + 2n)x + q - p]J_n^{(p,q)}(x)$$
$$+ 2(p + n)(q + n)J_{n-1}^{(p,q)}(x).$$

In the present instance, $p = 0$, $q = 1$ and $J_n^{(0,1)}(x_i) = 0$ for $i = 0, \ldots, n - 1$. Hence, the relation yields

$$(2n + 1)(1 - x_i^2)A_n^{(0,1)}\varphi_n'(x_i) = 2n(n + 1)A_{n-1}^{(0,1)}\varphi_{n-1}(x_i).$$

Using this in the above expression for a_i, we obtain

$$a_i = \frac{A_n^{(0,1)}}{A_{n-1}^{(0,1)}}\frac{(2n + 1)(1 - x_i)}{k_{n-1}^2 2n(n + 1)\varphi_{n-1}^2(x_i)}. \tag{9.4.9}$$

By the results in Exercise 9.9, it follows that

$$a_i = \frac{(2n + 1)^2(1 - x_i)}{2n(n + 1)^2((\bar{J}_{n-1}^{(0,1)}(x_i))^2},$$

where the bar denotes the normalized polynomial. Indeed, by Exercise 9.9c,

$$\bar{J}_{n-1}^{(0,1)} = (n/2)^{1/2}J_{n-1}^{(0,1)} \quad \text{and} \quad a_i = \left(\frac{(2n + 1)}{n(n + 1)}\right)^2 \frac{1 - x_i}{(J_{n-1}^{(0,1)}(x_i))^2}.$$

We can also express a_i in terms of Legendre polynomials. According to (9.4.6),

$$\varphi_{n-1}(x) = \frac{(P_n(x) + P_{n-1}(x))}{c_n(x + 1)}.$$

Since $\varphi_n(x_i) = 0$, we have $P_{n+1}(x_i) = -P_n(x_i)$. Applying this to the recursion formula for Legendre polynomials (Exercise 7.21b), we get

$$nP_{n-1}(x_i) = [(n + 1) + (2n + 1)x_i]P_n(x_i),$$

hence

$$\varphi_{n-1}(x_i) = (2n + 1)2^n(n - 1)!(n!/2n!)P_n(x_i).$$

Substituting in the previous formula for a_i gives the desired expression. By a similar procedure, we obtain

$$b_1 = \int_{-1}^{1}\frac{\varphi_n(x)}{\varphi_n(-1)}dx = \frac{1}{J_n^{(0,1)}(-1)}\int_{-1}^{1}J_n^{(0,1)}(x)\,dx.$$

From properties of the Jacobi polynomials it can then be shown that

$$b_1 = \frac{2}{(n+1)^2}.$$

The resulting quadrature formula,

$$\int_{-1}^{1} f(x)\, dx = \frac{2}{(n+1)^2} f(-1) + \sum_{i=0}^{n-1} a_i f(x_i) + R(f), \tag{9.4.10}$$

is called the *Radau quadrature* formula. The remainder is derived by the same technique used to derive (9.4.2), except that the formula is of degree $2n$. Taking this into account, we see that

$$R(f) = \frac{f^{(2n+1)}(\xi)}{(2n+1)!} \int_{-1}^{1} (1+x)\varphi_n^2(x)\, dx, \quad -1 < \xi < 1.$$

$$= \left(\frac{2^n n!(n+1)!}{(2n+1)!}\right)^2 \int_{-1}^{1} (1+x)[J_n^{(0,1)}(x)]^2\, dx \; \frac{f^{(2n+1)}(\xi)}{(2n+1)!}$$

$$= \frac{2}{n+1}\left(\frac{2^n n!(n+1)!}{(2n+1)!}\right)^2 \frac{f^{(2n+1)}(\xi)}{(2n+1)!}. \tag{9.4.11}$$

In Exercise 9.10, the nodes and coefficients are given for various Radau formulas.

Another case of interest is where $m = 2$ and $z_1 = -1$, $z_2 = 1$ are the fixed nodes in the interval $[-1, 1]$. In this case, the monic polynomial, $\varphi_{n-1}(x) = (x - x_0) \cdots (x - x_{n-2})$, is orthogonal to all polynomials of degree $< n - 1$ with respect to the weight function $(1-x)(1+x)$. Therefore,

$$\varphi_{n-1}(x) = \frac{1}{A_{n-1}^{(1,1)}} J_{n-1}^{(1,1)}(x),$$

where the leading coefficient, $A_{n-1}^{(1,1)}$, of the Jacobi polynomial $J_{n-1}^{(1,1)}(x)$ is given in Exercise 9.9d. Following a line of reasoning similar to the derivation of the Radau coefficients, one can show that

$$a_i = \frac{8n}{(n+1)[(1-x_i^2)J_{n-1}^{(1,1)\prime}(x_i)]^2}.$$

From Exercise 9.9e it follows that

$$a_i = \frac{2}{n(n+1)(P_n(x_i))^2}.$$

Similarly, $b_1 = b_2 = 2/n(n+1)$ and (9.4.3) becomes the *Lobatto quadrature formula*,

$$\int_{-1}^{1} f(x)\, dx = \frac{2}{n(n+1)}(f(-1) + f(1)) + \frac{2}{n(n+1)} \sum_{i=0}^{n-2} (P_n(x_i))^2 f(x_i) + R(f), \tag{9.4.12}$$

where

$$Rf = -\frac{2^{2n+1}[(n-1)!]^4 n^3 (n+1)}{(2n!)^3 (2n+1)} f^{(2n)}(\xi). \quad -1 < \xi < 1. \tag{9.4.13}$$

Some nodes and coefficients for Lobatto formulas are given in Exercise 9.10.

9.4(b) INTEGRATION WITH WEIGHT FUNCTIONS

Integrands in which a *weight function* $p(x)$ is present as a multiplicative factor occur frequently, for example, in the theory of orthogonal families of functions. The preceding theory of quadrature functionals can be extended to cover this more general situation. For a given weight function, $p(x)$, we seek a linear quadrature functional such that

$$\int_a^b p(x)f(x)\,dx = \sum_{i=0}^n a_i f(x_i) + Rf. \tag{9.4.14}$$

If the nodes x_i are specified (e.g., $x_i = a + ih$), then by the preceding theory it is clear that the coefficients a_i can be determined so that $Rf = 0$ for all f which are polynomials of degree $\leq n$. If the x_i are to be determined together with the a_i, as in Gaussian quadrature, we must restrict $p(x)$ to be nonnegative on $[a, b]$ to obtain results related to orthogonal polynomials. It is a simple matter to prove the following generalization of Theorem 9.4.2.

Theorem 9.4.5 Let $p(x) \geq 0$ on $[a, b]$. Let $\{Q_n(x)\}$ be the orthonormal family of polynomials with respect to p. Then in (9.4.14), $Rf = 0$ for all polynomials f of degree $\leq 2n + 1$ if and only if the x_i are the zeros of Q_{n+1} and the a_i are given by

$$a_i = \frac{-c_{n+2}}{c_{n+1}} \frac{1}{Q_{n+2}(x_i)Q'_{n+1}(x_i)}.$$

Proof. See the proof of Theorem 9.4.2 and use Exercise 9.8. ∎

Examples of quadrature with weight functions will be found in the next section.

When the weight function $p(x)$ changes sign at a finite number of points $z_0, z_1, \ldots, z_m$ in $[a, b]$, quadrature formulas can be developed as follows. Let L_m be the interpolating polynomial for f on the set $\{z_i\}$. Then $f(x) = L_m(x) + r_m(x)$, where r_m is given by formula (8.3.2), and

$$\int_a^b p(x)f(x)\,dx = \int_a^b p(x)L_m(x)\,dx + \int_a^b p(x)r_m(x)\,dx.$$

The first integral on the right is given exactly by an interpolatory quadrature functional, $\sum_{i=0}^m a_i f(z_i)$, as in Section 9.1. The second integral is of the form

$$I = \int_a^b p(x)\varphi_{m+1}(x) f[z_0, \ldots, z_m, x]\,dx,$$

where $\varphi_{m+1}(x) = (x - z_0) \cdots (x - z_m)$. The function $w(x) = p(x)\varphi_{m+1}(x)$ does not change sign on $[a, b]$ and can be regarded as a new weight function to which we can apply Theorem 9.4.5, obtaining a quadrature functional of maximum degree for the integral I. Thus, if $\{Q_n\}$ is the orthonormal family of polynomials with respect to the weight function $w(x)$ and $\{x_j\}$ are the zeros of Q_{n+1}, then the functional

$$Ff = \sum_{i=0}^{m} a_i f(z_i) + \sum_{j=0}^{n} b_j f[z_0, \ldots, z_m, x_j] \qquad (9.4.15)$$

is of degree $2n + m + 2$ with respect to the integral $\int_a^b \rho(x) f(x)\, dx$. Indeed, if f is of degree $< 2n + m + 2$, then the divided difference $f[z_0, \ldots, z_m, x]$ is of degree $\leq 2n + 1$ and the second sum is exactly equal to the integral I in this case. If we take $f = \varphi_{m+1} Q_{n+1}^2$, which is a polynomial of degree $2n + m + 3$, we see that $f(z_i) = 0$ and $f[z_0, \ldots, z_m, x_j] = (Q_{n+1}(x_j))^2 = 0$. Hence, $Ff = 0$, but $\int \rho f\, dx = \int w Q_{n+1}^2\, dx = 1$, showing that F is exactly of degree $2n + m + 2$. From the definition of the divided difference (Section 8.2, formula(*)), it follows that

$$f[z_0, \ldots, z_m, x_j] = \frac{f(x_j) - L_m(x_j)}{\varphi_{m+1}(x_j)},$$

provided that $\varphi_{m+1}(x_j) \neq 0$. In this case, (9.4.15) is of the form (9.4.3). If x_j coincides with one of the z_i, then by l'Hospital's rule,

$$f[z_0, \ldots, z_m, x_j] = \frac{f'(x_j) - L'_m(x_j)}{\varphi'_{m+1}(x_j)},$$

so that (9.4.15) involves values of f and its derivative.

A weight function which we have encountered in Chapter 7 is $\rho(x) = (1 - x^2)^{-1/2}$, which gives rise to the Chebyshev polynomials, T_n, as the orthogonal family. Applying the techniques developed above, we obtain the *Gauss–Chebyshev* quadrature formula,

$$\int_{-1}^{1} \frac{f(x)}{(1 - x^2)^{1/2}}\, dx = \sum_{i=0}^{n} a_i f(x_i) + Rf, \qquad (9.4.16)$$

where x_i is the ith zero of T_{n+1} and

$$a_i = -\frac{\pi}{T'_{n+1}(x_i) T_{n+2}(x_i)}.$$

Since $T_n(x) = \cos(n \arccos x)$, the x_i are given explicitly by

$$x_i = \cos\left(\frac{(2i - 1)\pi}{2n + 2}\right), \qquad i = 1, \ldots, n + 1.$$

Also, by direct calculation we find

$$T'_{n+1}(x_i) = \frac{(-1)^{i+1}(n + 1)}{\sin\left(\dfrac{(2i - 1)\pi}{2n + 2}\right)},$$

$$T_{n+2}(x_i) = (-1)^i \sin \pi \left(\frac{2i - 1}{2n + 2}\right),$$

so that the formula for a_i becomes simply

$$a_i = \frac{\pi}{n+1}.$$

This shows that (9.4.16) is a quadrature formula with equal coefficients. As we observed earlier (Computational Aspects, Section 9.3), this is advantageous in controlling rounding errors. The remainder in (9.4.16) is

$$Rf = \frac{\pi}{2^{2n+1}(2n+2)!} f^{(2n+2)}(\xi), \qquad |\xi| < 1.$$

Other formulas with equal coefficients can be derived for other weight functions. These are called *Chebyshev quadrature* formulas. Suppose such a formula exists for the weight function $\rho(x)$ on $[-1, 1]$. Then

$$\int_{-1}^{1} \rho(x)f(x)\,dx = c_n(f(x_0) + \cdots + f(x_n)) + Rf. \tag{9.4.17}$$

To determine the nodes x_i let us assume that $f^{(n+2)}$ exists on the interval $[-1, 1]$. By Taylor's formula,

$$f(x) = f(0) + xf'(0) + \cdots + \frac{x^{n+1}}{(n+1)!} f^{(n+1)}(0) + \frac{x^{n+2}}{(n+2)!} f^{(n+2)}(\xi).$$

It follows that

$$\int_{-1}^{1} \rho(x)f(x)\,dx = \sum_{j=0}^{n+1} \frac{f^{(j)}(0)}{j!} \int_{-1}^{1} \rho(x)x^j\,dx + \frac{1}{(n+2)!} \int_{-1}^{1} \rho(x)f^{(n+2)}(\xi)x^{n+2}\,dx. \tag{1}$$

Letting $x = x_i$ ($0 \le i \le n$), in Taylor's formula, we have

$$c_n \sum_{i=0}^{n} f(x_i) = c_n \bigg((n+1)f(0) + f'(0)\sum_i x_i + \cdots + \frac{f^{(n+1)}(0)}{(n-1)!}\sum_i x_i^{n+1}$$

$$+ \frac{1}{(n+2)!}\sum_i f^{(n+2)}(\xi_i)x_i^{n+2}\bigg). \tag{2}$$

If (9.4.17) is to be exact for all polynomials f of degree $\le n+1$, we must have

$$(n+1) = c_n^{-1} \int_{-1}^{1} \rho(x)\,dx$$

$$\sum x_i = c_n^{-1} \int_{-1}^{1} x\rho(x)\,dx \tag{3}$$

$$\vdots \qquad \vdots$$

$$\sum x_i^{n+1} = c_n^{-1} \int_{-1}^{1} x^{n+1}\rho(x)\,dx,$$

which is a system of $n+2$ equations to be solved for c_n and the x_i. In the theory of equations, it is shown that the sums $s_k = \sum x_i^k$ are related to the elementary symmetric functions A_k of x_i. The A_k are the coefficients of the monic polynomial $\varphi_{n+1}(x) = (x - x_0) \cdots (x - x_n)$. In fact (see, for example B. van der Waerden, *Moderne Algebra*, Springer, Berlin (1937), p. 86), the A_k can be calculated from the triangular system,

$$A_1 = -s_1$$
$$2A_2 + s_1 A_1 = -s_2$$
$$\vdots \qquad\qquad \vdots$$
$$(n+1)A_{n+1} + \cdots + s_{n-1}A_2 + s_n A_1 = -s_{n+1}.$$

Since $c_n = (1/n+1) \int_{-1}^{1} \rho(x)\,dx$ and $s_k = c_n^{-1} \int_{-1}^{1} x^k \rho(x)\,dx$, we can compute the s_k by evaluating these integrals, determine the A_k and then find the zeros of $\varphi_{n+1}(x)$. If these are all real, the Chebyshev formula (9.4.17) exists. It is not known which weight functions yield real roots. For $\rho(x) = 1$, the s_k are easily computed and real solutions are found for $n = 0, 1, \ldots, 6$. For $n = 7$ (8 nodes), there are complex roots. For $n = 8$, the roots are all real. For all $n \geq 9$, some roots are complex. (For a proof see [9.3], Chapter 10.) Nodes for the cases of real roots are tabulated in Exercise 9.12.

It follows from (1), (2), (3) above that

$$Rf = \frac{1}{(n+2)!}\left(\int_{-1}^{1}\rho(x)x^{n+2}f^{(n+2)}(\xi_x)\,dx - c_n \sum_{i=0}^{n} x_i^{n+2}f^{(n+2)}(\xi_i)\right).$$

An algorithm for constructing Gaussian quadrature formulas for a large class of weight functions is given in [9.35].

9.5 INTEGRATION OVER INFINITE INTERVALS, SINGULAR INTEGRALS

Recall that for finite a,

$$\int_a^\infty f(x)\,dx = \lim_{b\to\infty} \int_a^b f(x)\,dx.$$

Therefore, an integral over $[a, \infty)$ involves two limiting processes and there is no general method for approximating the integral. Some idea of the difficulties may be gleaned from the following elementary result.

Theorem 9.5.1 Let $f(x)$ be positive monotone decreasing for $x \geq 0$ and assume that $\int_0^\infty f(x)\,dx$ exists. Then

$$\lim_{h\to 0} h \sum_{i=i}^{\infty} f(ih) = \int_0^\infty f(x)\,dx$$

and the error for any h is $O(h)$.

Proof. Since $f(x)$ is monotone decreasing, we must have $\lim_{x\to\infty} f(x) = 0$. (Otherwise, the integral could not be finite.) Furthermore,

$$0 \leq \int_h^{(n+1)h} f(x)\,dx \leq h\left(f(h) + \cdots + f(nh)\right) \leq \int_0^{nh} f(x)\,dx.$$

Allowing $n \to \infty$, we obtain

$$\int_h^\infty f(x)\,dx \leq h \sum_{i=1}^{\infty} f(ih) \leq \int_0^\infty f(x)\,dx.$$

As a consequence,

$$0 \le \int_0^\infty f(x)\, dx - h \sum_{i=1}^\infty f(ih) \le \int_0^h f(x)\, dx \le hf(0),$$

and the theorem follows. ∎

Certain integrands occur frequently in applications and for these special Gaussian formulas exist. The weight function $\rho(x) = e^{-x}$ often occurs in integrals over $[0, \infty)$. We seek a quadrature formula such that

$$\int_0^\infty e^{-x} f(x)\, dx = \sum_{i=0}^n a_i f(x_i) + Rf, \qquad (9.5.1)$$

where the coefficients a_i and nodes x_i are to be determined so that the formula is of maximum degree.

Recall (Definition 1.6.1, Example 6 following Definition 2.2.1, and Exercise 3.32) that the set of functions, $L_2^{(\rho)}[0, \infty]$, which are square integrable with respect to the nonnegative weight function $\rho(x) = e^{-x}$ form a Hilbert space with inner product

$$\langle f, g \rangle = \int_0^\infty e^{-x} f(x) g(x)\, dx.$$

From the integral calculus we know that $\int_0^\infty e^{-x} x^k\, dx$ exists for $k = 0, 1, 2, \ldots$. Hence, this $L_2^{(\rho)}[0, \infty]$ contains all polynomials and the results of Section 7.7 remain valid for the above inner product. The corresponding family of orthogonal polynomials are the *Laguerre polynomials*, $\mathscr{L}_n(x)$, given by

$$\mathscr{L}_n(x) = e^x \frac{d^n}{dx^n}(x^n e^{-x}).$$

By applying the methods of the preceding section it can be shown that the coefficients a_i in (9.5.1) are

$$a_i = \frac{[(n+1)!]^2}{\mathscr{L}'_{n+1}(x_i)\mathscr{L}_{n+2}(x_i)},$$

and x_i is the ith zero of $\mathscr{L}_{n+1}$. Also, if $f \in C^{2n+2}[0, \infty)$,

$$Rf = \frac{[(n+1)!]^2}{(2n+2)!} f^{(2n+2)}(\xi), \qquad 0 < \xi < \infty.$$

Using the relation

$$x\mathscr{L}'_n(x) = (x - n - 1)\mathscr{L}_n(x) + \mathscr{L}_{n+1}(x),$$

the formula for a_i can be simplified to

$$a_i = \frac{[(n+1)!]^2 x_i}{[\mathscr{L}_{n+2}(x_i)]^2}.$$

9.5 Integration Over Infinite Intervals, Singular Integrals

Nodes and coefficients for $n = 4$ are tabulated in Exercise 9.11.

Another important case is where $\rho(x) = e^{-x^2}$ and the interval is $(-\infty, \infty)$. As above, it can be verified that the orthogonal polynomials exist. They are called *Hermite polynomials* and are given by

$$H_n(x) = (-1)^n e^{x^2} \frac{d^n}{dx^n}(e^{-x^2}).$$

A *Gauss–Hermite* quadrature formula,

$$\int_{-\infty}^{\infty} e^{-x^2} f(x)\, dx = \sum_{i=0}^{n} a_i f(x_i) + Rf, \qquad (9.5.2)$$

can be derived by the methods of the preceding section. In (9.5.2),

$$a_i = \frac{2^{n+1} n! \pi^{1/2}}{H'_{n+1}(x_i) H_n(x_i)} = \frac{2^{n+2}(n+1)! \pi^{1/2}}{[H_{n+2}(x_i)]^2},$$

where x_i is the ith zero of H_{n+1}. Also, if $f^{(2n+2)}$ is continuous,

$$Rf = \frac{(n+1)! \pi^{1/2}}{2^{n+1}(2n+2)!} f^{(2n+2)}(\xi),$$

for some ξ. Nodes and coefficients for $n = 4$ are given in Exercise 9.11. For other n, see a table of zeros and coefficients for Hermite integration [9.2].

In some cases, integration over an infinite interval can be reduced to integration over finite intervals by a change of variable. Note that

$$\int_a^{\infty} f(x)\, dx = \int_a^b f(x)\, dx + \int_b^{\infty} f(x)\, dx.$$

If $x = 1/t$, the second integral becomes

$$-\int_{1/b}^0 \frac{f(1/t)}{t^2}\, dt = \int_0^{1/b} g(t)\, dt,$$

where $g(t) = f(1/t)/t^2$. If $g(t)$ remains bounded as $t \to 0$, the integral may be evaluated by any of the quadrature formula for finite intervals. If $g(t)$ becomes infinite as $t \to 0$, we have the case of a *singular integral*, which we now consider.

Singular integrals are treated by various methods tailored to the particular singularity. We have already seen that certain integrals of the form $\int_a^b \rho(x) f(x)\, dx$, where ρ has a singularity at an end point of the interval, can be computed by a formula of Gauss type. For example, the Chebyshev weight function $(1 - x^2)^{-1/2}$ becomes infinite at both end points ± 1. The Jacobi weight function $(1 - x)^p (1 + x)^q$, $p > -1$, $q > -1$ can be treated by the Gauss quadrature functionals as discussed previously.

In certain cases, the singularity may be "subtracted out". Suppose $\rho(x) = (x - a)^{-\alpha}$, where $0 < \alpha < 1$, and f has derivatives of order up to $k + 1$ at a. Then by Taylor's formula,

$$f(x) = p(x) + \frac{(x-a)^{k+1}}{(k+1)!} f^{(k+1)}(\xi(x)),$$

where

$$p(x) = f(a) + \cdots + \frac{(x-a)^k}{k!} f^{(k)}(a).$$

Therefore,

$$\int_a^b (x-a)^\alpha f(x)\,dx = \int_a^b \frac{f(x) - p(x)}{(x-a)^\alpha}\,dx + \int_a^b \frac{p(x)}{(x-a)^\alpha}\,dx.$$

The second integral can be evaluated by the methods of calculus. The first integral has no singularity and can be approximated by one of the quadrature functionals in previous sections.

9.6 INTEGRATION OF PERIODIC FUNCTIONS

The composite trapezoidal rule (9.3.1) can be improved in accuracy by the following device. Subdivide $[a, b]$ into subintervals $[x_i, x_{i+1}]$ as in (9.3.1). In each subinterval, approximate $f(x)$ by a different interpolating polynomial, using the Everett central difference form (8.2.10) applied successively at $a = x_0, x_1, \ldots, x_n = b$. Integration of the interpolation polynomial centered at x_i yields

$$\int_{x_i}^{x_{i+1}} f(x)\,dx = \frac{h}{2}(f(x_i) + f(x_{i+1})) + hb_1(\delta^2 f(x_i) + \delta^2 f(x_{i+1})) + \cdots$$
$$+ hb_m(\delta^{2m} f(x_i) + \delta^{2m} f(x_{i+1})) + c_m h^{2m+3} f^{(2m+2)}(\xi_i). \quad (9.6.1)$$

(Here, the interpolating polynomial is of degree $2m + 1$, and n is the number of subintervals; n and m are chosen independently.)

Recall that

$$\delta^{2j} f(x_i) = \Delta^{2j} f(x_{i-j}) = \Delta^{2j-1} f(x_{i-j+1}) - \Delta^{2j-1} f(x_{i-j}).$$

Summing (9.6.1) over i causes all central differences to cancel except those at $i = 0$ and $i = n$. We obtain

$$\int_a^b f(x)\,dx = \frac{h}{2}(f(x_0) + 2f(x_1) + \cdots + 2f(x_{n-1}) + f(x_n))$$
$$+ h \sum_{j=1}^m b_j (\Delta^{2j-1} f(x_{n-j}) - \Delta^{2j-1} f(x_{-j})) + c_m h^{2m+3} \sum_{i=0}^{n-1} f^{(2m+2)}(\xi_i). \quad (9.6.2)$$

The first summation is recognized to be the composite trapezoidal rule. The second sum is sometimes called an *end correction*. If $f^{(2m+2)}$ is continuous, we can find ξ such that the third sum is equal to $nf^{(2m+2)}(\xi)$. Hence, the error in the trapezoidal rule with end corrections is

$$c_m h^{2m+2}(b-a)f^{(2m+2)}(\xi);$$

that is, it is $O(h^{2m+2})$ rather than $O(h^2)$ as in (9.3.1). For an alternative form of the end correction and a general discussion of methods of increasing the accuracy of quadrature functionals see [9.3] (Chapter 11).

Now, suppose $f(x)$ is periodic with period $L = b - a$. Then $f(x) = f(x + L)$ and $f(x_i) = f(x_{i+n})$. It follows that $\Delta^{2j-1}f(x_{n-j}) = \Delta^{2j-1}f(x_{-j})$. In this case, the end correction is zero and the composite trapezoidal formula itself is of order h^{2m+2}, provided $f^{(2m+2)}$ exists and is continuous everywhere.

Integrands which are oscillatory but not periodic present a special problem. Such integrands as $f(x) \cos nx$ occur in Fourier transforms. For large n these functions are poorly approximated by low-degree polynomials so that the quadrature techniques of previous sections are inappropriate. A procedure that is effective in many cases is *Filon's method*. In this method, we must be able to expand $f(x)$ in the form

$$f(x) = a_1 f_1(x) + \cdots + a_k f_k(x) + \varepsilon(x),$$

where $|\varepsilon(x)|$ is small and the functions f_i are such that the integrals $\int_a^b f_i(x) \cos nx\, dx$ can be evaluated by the methods of calculus. More generally, if $K(n, x)$ is an oscillatory function on $[a, b]$,

$$\int_a^b f(x)K(n, x)\, dx = \sum_{i=1}^k a_i F_i(n) + \int_a^b \varepsilon(x)K(n, x)\, dx,$$

where

$$F_i(n) = \int_a^b f_i(x)K(n, x)\, dx$$

are to be evaluated directly in elementary terms. If such f_i can be found and $|\varepsilon(x)|$ is sufficiently small, then $\sum_{i=1}^k a_i F_i(n)$ is a good approximation to the integral. Filon's method of obtaining such f_i for $K(n, x) = \cos nx$ is as follows. Divide $[a, b]$ into $2m$ equal subintervals of length $h = (b - a)/2m$ by points $x_0, \ldots, x_{2m}$. On successive intervals $[x_{2i}, x_{2i+2}], i = 0, \ldots, m - 1$ approximate $f(x)$ by an interpolation polynomial p_i of degree ≤ 2 with nodes $x_{2i}, x_{2i+1}, x_{2i+2}$. Evaluate the integral $\int_{x_{2i}}^{x_{2i+2}} p_i(x) \cos nx\, dx$ by integration by parts. Summing over i yields the result,

$$\int_a^b f(x) \cos nx\, dx = h\alpha(f(b) \sin nb - f(a) \sin na) + \beta h C_2 + \gamma h C_1 + Rf, \quad (9.6.3)$$

where

$$C_2 = f(a) \cos na + 2f(a + 2h) \cos n(a + 2h)$$
$$+ 2f(a + 4h) \cos n(a + 4h) + \cdots + f(b) \cos nb,$$
$$C_1 = f(a + h) \cos n(a + h) + f(a + 3h) \cos n(a + 3h) + \cdots$$
$$+ f(b - h) \cos n(b - h),$$

and with
$$\theta = nh,$$
$$\alpha = \frac{1}{\theta} + \frac{\sin 2\theta}{2\theta^2} - \frac{2\sin^2 \theta}{\theta^3},$$
$$\beta = \frac{1 + \cos^2 \theta}{\theta^2} - \frac{\sin 2\theta}{\theta^3},$$
$$\gamma = \frac{4}{\theta^3}(\sin \theta - \theta \cos \theta).$$

Similarly, one obtains for the sine transform,
$$\int_a^b f(x) \sin nx \, dx = h\alpha(f(a) \cos na - f(b) \cos nb) + \beta h S_2 + \gamma h S_1 + Rf, \quad (9.6.4)$$
where
$$S_2 = f(b) \sin nb + f(a) \sin na + 2 \sum_{j=1}^{m-1} f(a + 2jh) \sin n(a + 2jh),$$
$$S_1 = \sum_{j=1}^{m} f(a + (2j - 1)h) \sin n(a + (2j - 1)h).$$
and
$$Rf = h^3 \frac{(b-a)}{12}\left(1 - \frac{1}{16 \cos(\theta/4)}\right) \sin \frac{\theta}{2} f^{(4)}(\xi), \quad a < \xi < b.$$

Now, if α, β and γ are expanded in a Taylor series, we obtain
$$\alpha = \frac{1}{\theta} + \frac{1}{2\theta^2}\left(2\theta - \frac{(2\theta)^3}{3!} + \cdots\right) - \frac{2}{\theta^3}\left(\theta^2 - \frac{\theta^4}{3} + \cdots\right)$$
$$= \frac{2}{45}\theta^3 - \frac{2}{315}\theta^5 + \cdots$$
and similarly,
$$\beta = \frac{1}{3} + \frac{1}{15}\theta^2 - \cdots,$$
$$\gamma = \frac{4}{3} - \frac{2}{15}\theta^2 + \cdots.$$

If h is sufficiently small so that powers of $\theta = nh$ higher than the first can be neglected, then $\alpha = 0$, $\beta = \frac{1}{3}$ and $\gamma = \frac{4}{3}$ and (9.6.3), (9.6.4) both reduce to the composite Simpson's rule (9.3.2).

Other formulas for oscillatory integrals can be found in [9.14]. See [9.34] for a Filon quadrature algorithm.

9.7 EXTRAPOLATION TO THE LIMIT; ROMBERG INTEGRATION

If we reexamine the trapezoidal rule with end correction in (9.6.2), we see that it can be rewritten in the form,
$$Sf = \int_a^b f(x) \, dx = F_h f + \sum_{j=1}^{m} a_j h^{2j} + O(h^{2m+2}), \quad (9.7.1)$$

9.7 Extrapolation to the Limit; Romberg Integration

where $F_h f$ is the composite trapezoidal formula and the sum of powers of h is derived by replacing the differences of f by derivatives at a and b according to Section 8.4. (The a_j are constants involving the values of these derivatives.) In Section 9.3, we proved that $F_h f \to Sf$ as $h \to 0$. If we halve the step size, (9.7.1) becomes

$$Sf = F_{h/2} f + \sum_{j=1}^{m} a_j \left(\frac{h}{2}\right)^{2j} + O(h^{2m+2})$$

Notice that the functional,

$$F_h^{(1)} = \tfrac{1}{3}(4F_{h/2} - F_h), \tag{9.7.2}$$

is such that $F_h^{(1)} f$ approximates Sf with an error of order h^4 rather than h^2 as in the trapezoidal rule. The functional $F_h^{(1)}$ is called the *first extrapolation* of the trapezoidal rule and is an example of a technique known as the *deferred approach to the limit* or *Richardson's extrapolation*. In effect, it is a linear extrapolation (or interpolation) procedure of the following type. Let

$$F(z) = a_0 + a_1 z + \cdots + a_m z^m + O(z^{m+1}), \tag{9.7.3}$$

where the a_j are unknown coefficients. Suppose we wish to determine $a_0 = \lim_{z \to 0} F(z)$. We can obtain an approximate value by computing two values $F(z_1)$, $F(z_2)$ and using the linear interpolating polynomial for $F(z)$. Thus, $F(z_1) = a_0^{(1)} + a_1^{(1)} z_1$ and $F(z_2) = a_0^{(1)} + a_1^{(1)} z_2$. Solving for $a_0^{(1)}$, we get

$$a_0^{(1)} = \frac{z_2 F(z_1) - z_1 F(z_2)}{z_2 - z_1}.$$

To apply this to (9.7.1) we set $z = h^2$, $a_0 = Sf$ and $F(z) = F_h f$. Taking $z_1 = (h/2)^2$ and $z_2 = h^2$, we obtain (9.7.2).

Now, we may consider the sequence of trapezoidal functionals, $F_{0k} = F_{h/2^k}$, $k = 0, 1, 2, \ldots$. Since $F_{0k} f \to Sf$ for all $f \in C[a, b]$, the sequence (F_{0k}) is strongly convergent (Section 3.7) on $C[a, b]$. For each pair F_{0k}, $F_{0, k+1}$ we form the first extrapolation, F_{1k}, according to (9.7.2), that is,

$$F_{1k} = \tfrac{1}{3}(4F_{0, k+1} - F_{0k}). \tag{9.7.4}$$

Since each value, $F_{1k} f$, $k = 0, 1, 2, \ldots$, approximates Sf with error $O(h^4)$ rather than $O(h^2)$, it would seem reasonable to expect the sequence $(F_{1k} f)$ to converge faster than $(F_{0k} f)$. In fact, this acceleration of convergence takes place for many functions f. It is another example of accelerating the convergence of a sequence by a linear interpolation. It may be compared with Aitken's method (Section 5.7). If we let

$$e_k = Sf - F_{0k} f = \sum_{j=1}^{m} a_j (h/2^k)^{2j} + O(h^{2m+2})$$

be the error in the kth term, then $e_{k+1} = (\tfrac{1}{4} + \delta_k) e_k$, where $\delta_k \to 0$ as $k \to \infty$, as required in Theorem 5.7.1. Since the a_j are linear combinations of the derivatives of f at the end points a and b, we may eliminate f from our considerations and view

(9.7.3) as an operator equation. Thus, we consider

$$F(z) = S + zA_1 + \cdots + z^m A_m + O(z^{m+1}), \qquad (9.7.5)$$

where S and the A_i, $i = 1, \ldots, m$, are bounded linear operators and $O(z^{m+1})$ denotes a linear operator which has norm $\leq cz^{m+1}$ for some constant c. Here, z is a scalar variable and the underlying vector space would be the space of functions sufficiently differentiable on some interval containing $[a, b]$. (See Example 8, Section 3.1.) The operators A_i in this case are functionals consisting of linear combinations of the values of derivatives at $x = a$ and $x = b$. However, in the following we may simply regard (9.7.5) as a relation between arbitrary bounded linear operators on some normed space.

Applying the extrapolation process to the sequence (F_{1k}) and continuing thereafter, we can generate an infinite triangular array of operators (F_{jk}) according to the formula,

$$F_{0k} = F(r^k t),$$
$$F_{j+1,k} = \frac{F_{j,k} - r^{j+1} F_{j,k-1}}{1 - r^{j+1}}, \qquad j \geq 0, \qquad (9.7.6)$$

where r and t are constants, $0 < r < 1$, $t > 0$. (For $j = 0$ and $r = \frac{1}{4}$ we obtain (9.7.4) above, keeping in mind that $z = h^2$.)

Now, (9.7.5) implies that

$$\|F(z) - S\| \to 0 \qquad \text{as } z \to 0.$$

We shall show that (9.7.6) tends to accelerate this convergence.

Theorem 9.7.1 Let $F(z)$ be a one-parameter family of operators satisfying (9.7.5) for every $m \geq 1$. For any j such that $A_j \neq 0$, the sequence (F_{jk}) converges to S faster than the sequence $(F_{j-1,k})$ in the sense that

$$\lim_{k \to \infty} \frac{\|F_{jk} - S\|}{\|F_{j-1,k} - S\|} = 0. \qquad (9.7.7)$$

Proof. By induction, it will be shown that

$$F_{jk} = S + A_{j,j+1} (r^k t)^{j+1} + A_{j,j+2} (r^k t)^{j+2} + \cdots$$
$$+ A_{jq} (r^k t)^q + O((r^k t)^{q+1}), \qquad (9.7.8)$$

where for $i \geq j + 1$,

$$A_{ji} = \frac{(1 - r^{1-i})(1 - r^{2-i}) \cdots (1 - r^{j-i})}{(1 - r)(1 - r^2) \cdots (1 - r^j)} A_i. \qquad (9.7.9)$$

For $j = 0$, $A_{0i} = A_i$ and (9.7.8) reduces to (9.7.5). Now, suppose (9.7.8) and (9.7.9) hold for some $j \geq 0$. Applying the second equation of (9.7.6) and using (9.7.8), we obtain an expression for $F_{j+1,k}$ given by (9.7.8) with j replaced by $j + 1$ and $A_{j+1,i}$ given by (9.7.9) with j replaced by $j + 1$.

For $i = j + 1$, (9.7.9) becomes

$$A_{j,j+1} = \frac{(1-r^{-j}) \cdots (1-r^{-1})}{(1-r) \cdots (1-r^j)} A_{j+1}$$

$$= (-1)^j r^{-j(j+1)/2} A_{j+1}.$$

Now, setting $q = j + 1$ in (9.7.8), we obtain

$$F_{jk} = S + (-1)^j r^{-j(j+1)/2}(r^k t)^{j+1} A_{j+1} + O\big((r^k t)^{j+2}\big),$$

whence,

$$\|F_{jk} - S\| = (r^k t)^{j+1} r^{-j(j+1)/2} \|A_{j+1} + O(r^k t)\|.$$

It follows that

$$\frac{\|F_{jk} - S\|}{\|F_{j-1,k} - S\|} = \frac{r^k t}{r^j} \frac{\|A_{j+1} + O(r^k t)\|}{\|A_j + O(r^{kt})\|} \to 0$$

as $k \to \infty$. ∎

Computational Aspects

Romberg's method [9.16] proceeds in a recursive manner by halving the step size h to compute the next term in the sequence $(F_{0k} f)$ and then using the formula,

$$F_{j+1,k} f = \frac{4^{j+1} F_{jk} f - F_{j,k-1} f}{4^{j+1} - 1}, \tag{9.7.10}$$

obtained from (9.7.6) with $r = \frac{1}{4}$. Now, $F_{0k} f$ is the composite trapezoidal formula with step size $h/2^k$. To obtain $F_{0k} f$ from $F_{0,k-1} f$, we only need to calculate the values of f at the new abscissas and sum these values to form the quantity $S_k f$. Then $F_{0k} f = (F_{0,k-1} f + S_k f)/2$. The terms of the accelerated sequences, $F_{jk} f$, can then be computed for $j = 1, \ldots, k$. In this way, the triangular array is built up. The procedure may be terminated when two or three successive diagonal terms of the array agree within an acceptable accuracy. Of course, it should be borne in mind that the procedure may not actually accelerate convergence of the trapezoidal rule if the integrand does not have derivatives of sufficiently high order. For example, if the function $f(x) = \sqrt{x}$ is integrated from 0 to 1, the convergence is not accelerated. This can be verified in actual computation.

Unfortunately, the number of function evaluations in the halving procedure goes up exponentially. To reduce this effect, there are variations of the Romberg procedure in which the interval $[a, b]$ is divided successively into n_k parts where $(n_k : k = 0, 1, 2, \ldots)$ is a sequence of integers chosen such that $n_{k+1}/n_k > 1 + c$, for some positive c. Such schemes are described in [9.15], [9.17]. A specific algorithm for programming is given in [9.18].

9.8 QUADRATURE FUNCTIONALS WITH MINIMUM REMAINDER

In Section 9.4, we determined quadrature formulas of the form $Ff = \sum_{i=1}^n a_i f(x_i)$ which are of maximum degree. It seems plausible to infer that such formulas give better approximations to the integral of a certain function f than formulas of lower degree. In particular, when f can be approximated by a polynomial of moderately high degree, we would expect that the remainder,

$$Rf = \int_a^b f(x)\,dx - \sum_{i=1}^n a_i f(x_i), \tag{9.8.1}$$

would be smaller in absolute value that the remainder for some other functional. Thus, for example, if f is sufficiently differentiable, such functionals can be expected to yield more accurate results than others. However, for functions f of a low order of differentiability these functionals can actually give worse results than formulas such as the composite trapezoidal rule, as we shall see. This raises the question of whether there exist functionals of the above type which minimize $\sup_f \{|Rf|\}$, where f is restricted to some specified class of functions. Such formulas would have to be considered best for that particular class.

Let us first consider f in the class $C^m[0, 1]$ of functions having continuous derivatives of order $\leq m$ on $[0, 1]$. (We can translate an arbitrary finite interval $[a, b]$ to $[0, 1]$.) From calculus we have Taylor's formula with integral remainder,

$$f(x) = \sum_{i=0}^{m-1} \frac{f^{(i)}(0)}{i!} x^i + \int_0^x f^{(m)}(t) \frac{(x-t)^{m-1}}{(m-1)!}\,dt. \tag{9.8.2}$$

For what is to follow, it is convenient to rewrite this in terms of the truncated power function x_+^m. We recall (Eq. (9.2.11)) that $x_+^m = x^m$ for $x > 0$ and $x_+^m = 0$ for $x \leq 0$. The integral remainder in Taylor's formula is readily seen to be equal to the function, q, where

$$q(x) = \int_0^1 f^{(m)}(t) \frac{(x-t)_+^{m-1}}{(m-1)!}\,dt. \tag{9.8.3}$$

Hence, for an $f \in C^m[0, 1]$, $f(x) = p(x) + q(x)$, where p is the polynomial of degree $\leq m - 1$ which appears in (9.8.2). By the linearity of R in (9.8.1), we have $Rf = Rp + Rq$. Now, let us seek to determine the functional F such that $\sup \{|Rf| : \|f^{(m)}\|_\infty \leq M\}$ is minimized; i.e., we try to minimize the remainder over the set of functions f for which $\sup_{0 \leq x \leq 1} |f^{(m)}(x)| \leq M$, where M is some constant. It is obviously necessary to restrict the size of f in some manner, since $|R(cf)| = |c|\,|Rf|$ could otherwise be made arbitrarily large. In fact, this shows that we must have $Rp = 0$ for any polynomial p of degree $\leq m - 1$. Otherwise, since $p^{(m)} = 0$ for such p, we would have $|R(cp)| \to \infty$ as $c \to \infty$ and F could not possibly minimize R over the set in question. Hence, we must have $Rf = Rq$ and $Rx^j = 0$ for $1 \leq j \leq m - 1$. From (9.8.1) and (9.8.3) it follows that (as in Theorem 9.2.2)

$$Rq = Gf^{(m)} = \int_0^1 f^{(m)}(t) K(t)\,dt, \tag{9.8.4}$$

where

$$K(t) = \frac{(1-t)^m}{m!} - \sum_{i=1}^n a_i \frac{(x_i - t)_+^{m-1}}{(m-1)!}. \tag{9.8.5}$$

9.8 Quadrature Functionals with Minimum Remainder

Thus, $\sup |Rq| = \sup \{|Gf^{(m)}| : \|f^{(m)}\|_\infty \leq M\} = M \|G\|$, where $\|G\|$ is given by Definition 3.5.1 with respect to the uniform norm in $C[0, 1]$. For any $f^{(m)} \in C[0, 1]$,

$$|Gf^{(m)}| \leq \|f^{(m)}\|_\infty \int_0^1 |K(t)|\, dt.$$

Hence, $\|G\| \leq \int_0^1 |K(t)|\, dt$. If we set $f^{(m)}(t) = \text{sgn } K(t)$, we obtain $|Gf^{(m)}| = \int_0^1 |K(t)|\, dt$. Although sgn $K(t)$ is not continuous on $[0, 1]$, its discontinuities are simple jumps. Therefore, it can be approximated by continuous functions $h_\varepsilon(t)$ arbitrarily closely in the L_1-norm, so that

$$\int_0^1 (h_\varepsilon(t) - \text{sgn } K(t))K(t)\, dt < \varepsilon,$$

for arbitrary $\varepsilon > 0$. This proves that

$$\|G\| = \int_0^1 |K(t)|\, dt.$$

Consequently, $\sup |Rf|$ will be minimized over the set $\|f^{(m)}\| \leq M$ if and only if we choose the a_i and x_i to minimize $\int_0^1 |K(t)|\, dt$ subject to the constraints $Rx^j = 0$, namely,

$$\sum_{i=1}^n a_i x_i^j = 1/(j + 1), \qquad j = 0, \ldots, m - 1. \tag{9.8.6}$$

We shall not solve this general minimization problem here. (See [9.19] and [9.3] for some interesting cases.) However, when $m = 1$, the solution is particularly simple. In this case, $Rp = 0$ for any constant function p, which holds if and only if $\sum_1^n a_i = 1$. The function,

$$K(t) = 1 - t - \sum_{i=1}^n a_i(x_i - t)^0_+,$$

has a "sawtooth" graph, consisting of linear segments of slope -1 and having jumps of height a_i at the respective points x_i. It is easy to verify that $\int_0^1 |K(t)|\, dt$ is minimized by making all a_i equal, hence $a_i = 1/n$, and choosing the x_i to be equally spaced and midway between the zeros of $K(t)$, hence $x_i = (2i - 1)/2n$. (It is quite easy to see this in a geometric way by looking at the area under the graph.) The resulting functional,

$$Ff = \frac{1}{n} \sum_{i=1}^n f\left(\frac{2i - 1}{2n}\right), \tag{9.8.7}$$

is known as the *repeated midpoint* formula. One readily computes the area under $|K(t)|$ to be $1/4n$, so that for any function $f \in C^1[0, 1]$, the remainder satisfies

$$|Rf| \leq (1/4n) \sup_{0 \leq t \leq 1} |f^{(1)}(t)|.$$

Also see Exercise 9.14.

Another interesting case arises when the x_i are prescribed points and it is required to determine the a_i. For example, suppose the x_i are equally spaced and we wish to construct a functional $Ff = \sum_{i=0}^{n} a_i f(i/n)$ which has minimum remainder for some class of functions f which have a derivative $f^{(m)}$. If $m \geq n + 1$, then $Rp = 0$ for all polynomials p of degree $\leq n$. But this implies that F must be interpolatory. (See the Remark in Section 9.2.) Hence, F is a closed Newton–Cotes formula and the a_i are given by (9.2.2). However, if we require that $m < n + 1$, we impose the constraints (9.8.6), which do not completely determine the a_i. Assuming that $f \in C^m[0, 1]$, we find that $Rf = Rq$ is given by (9.8.4), where now

$$K(t) = \frac{(1-t)^m}{m!} - \sum_{i=1}^{n} a_i \frac{(i/n - t)_+^{m-1}}{(m-1)!}.$$

Again, for $m = 1$, $K(t)$ is a sawtooth function with jumps of heights a_i at the points i/n, $1 \leq i \leq n - 1$. Also, $K(0) = 1 - \sum_1^n a_i = a_0$ and $K(1) = -a_n$. Again, we see that $\int_0^1 |K(t)| \, dt$ is minimized by choosing the a_i so that the area under the graph of $K(t)$ consists of $2n$ equal right triangles. This is accomplished by taking $a_0 = a_n = 1/2n$ and $a_i = 1/n$ for $1 \leq i \leq n - 1$. (Thus, the height of each triangle is $1/2n$.) The result is recognized to be the composite trapezoidal (9.3.1), except that the estimate of the remainder is now only

$$|Rf| \leq \frac{h}{4} \sup_x |f^{(1)}(x)|,$$

where $h = 1/n$.

The preceding analysis can also be carried through using the L_2 norm. This leads to different quadrature functionals which minimize the remainder $|Rf|$ for f in a wider class of functions. Such functionals are studied in [9.31], [9.32], [9.33]. Again, we consider the class of functions $f \in C^m[0, 1]$, but we now restrict our attention to the subset E such that $\|f^{(m)}\|_2 = (\int_0^1 |f^{(m)}(t)|^2 \, dt)^{1/2} \leq M$. This subset obviously includes the previous subset of functions f satisfying $\|f^{(m)}\|_\infty \leq M$. Reasoning as before, we may confine our search to functionals F for which $Rp = 0$ for any polynomial of degree $\leq (m-1)$. Thus, F must satisfy the constraints (9.8.6).

From Eq. (9.8.4) and the Schwarz inequality, we have

$$|Rf| = |Gf^{(m)}| \leq \|f^{(m)}\|_2 \|K\|_2.$$

Considering G as linear functional on $L_2[0, 1]$, we see that the operator norm $\|G\|_2 \leq \|K\|_2$. Now, if we take f to be the function defined by $f^{(m)}(t) = (M/\|K\|_2)K(t)$, then $\|f^{(m)}\|_2 = M$ and $|Gf^{(m)}| = M\|K\|_2$. Therefore,

$$\|G\|_2 = \|K\|_2 = \left(\int_0^1 |K(t)|^2 \, dt \right)^{1/2},$$

and the minimization of $|Rf|$ over the class E is equivalent to the minimization of $\|K\|_2$. Since K is given by formulas (9.8.5) and (9.8.6), it follows that this minimization is equivalent to finding the closest point (in the L_2 metric) to the function

$(1-t)^m/m!$ in the subspace spanned by the functions $(x_i - t)_+^{m-1}$, $i = 1, \ldots, n$, but subject to the constraints (9.8.6). In other words, we must find the linear combination of these n functions which is the least squares approximation to the function $(1-t)^m/m!$ on $[0, 1]$ subject to the constraints (9.8.6). If the nodes x_i are specified, this is just a constrained linear least squares problem. If the nodes are not fixed in advance, but are to be determined along with the coefficients a_i, then the problem is a nonlinear minimization problem with equality constraints. (See Chapter 12.)

Let us consider the case in which the nodes x_i are given. We assume $0 = x_1 < \cdots < x_n = 1$ and that $m < n$. Since system (9.8.6) has rank m, we can solve for $a_1, \ldots, a_m$ in terms of the remaining a_i. The functions $(x_i - t)_+^{m-1}$ are linearly independent. Therefore, there exists a unique least squares approximation, $\sum_{i=m+1}^{n} a_i^*(x_i - t)_+^{m-1}$, to the function $(1-t)^m/m!$ by Theorem 7.3.3. (Also see Section 7.4.) The functional $F^*f = \sum_{i=1}^{n} a_i^* f(x_i)$ yields the minimum remainder R^* on the set E described above in the sense that $\|K^*\|_2 \leq \|K\|_2$ for all K given by formulas (9.8.5), (9.8.6). We shall now give a method of determining K^* by means of splines (Section 8.5). It is based on the next theorem.

Theorem 9.8.1 Let K^* be the Peano kernel of the quadrature functional $F^*f = \sum_{i=1}^{n} a_i^* f(x_i)$, where $0 = x_1 < \cdots < x_n = 1$ are fixed nodes, for approximating the functional $\int_0^1 f(x)\,dx$, where $f \in C^m[0, 1]$. Let $F^*p = 0$ for any polynomial p of degree $\leq m - 1$. Let $S_{2m-1}(f; x)$ be the natural interpolating spline of degree $2m - 1$ for f on the given nodes. Then $\|K^*\|_2 \leq \|K\|_2$ for all K given by Eqs. (9.8.5) and (9.8.6) if and only if for all such f,

$$F^*f = \int_0^1 S_{2m-1}(f; x)\,dx. \tag{9.8.8}$$

Proof. Any K given by formulas (9.8.5), (9.8.6) is the Peano kernel of a functional F such that $Ff = \sum_{i=1}^{n} a_i f(x_i)$ and the remainder Rf given by formula (9.8.1) is zero for f a polynomial of degree $\leq m - 1$. Let $F_1 = F^* - F$. The corresponding remainder is $R_1 = R^* - R$ and $R_1 p = 0$ for any polynomial p of degree $\leq m - 1$. Since

$$K^*(t) = \frac{(1-t)^m}{m!} - \sum_{i=1}^{n} a_i^* \frac{(x_i - t)_+^{m-1}}{(m-1)!},$$

we see that the kernel K_1 of F_1 must be given by

$$K_1(t) = K^*(t) - K(t) = \sum_{i=1}^{n} (a_i^* - a_i) \frac{(x_i - t)_+^{m-1}}{(m-1)!}.$$

Hence, in $[0, 1]$, $K_1(t)$ is a spline of degree $m - 1$ with nodes x_i. Clearly, for $t \geq 1$, $K_1(t) = 0$. By Eq. (9.2.10), we must have $K_1(t) = (1/(m-1)!)R_1(x - t)_+^{m-1}$. For fixed $t \leq 0$, $(x - t)_+^{m-1} = (x - t)^{m-1}$ is a polynomial p of degree $m - 1$ in x for $0 \leq x \leq 1$. Hence, $R_1 p = 0$, so that $K_1(t) = 0$ for $t \leq 0$. Therefore, K_1 is a natural spline of degree $m - 1$. Let g be any function such that $g^{(m)} = K_1$. Then g is a natural spline of degree $2m - 1$. (By m integrations of K_1 we get a polynomial of degree $\leq m - 1$ outside of $[0, 1]$.)

Now, in the Hilbert space $L_2[0, 1]$, we have

$$\langle K^*, K_1 \rangle = \int_0^1 K^*(t)K_1(t)\, dt = \int_0^1 K^*(t)g^{(m)}(t)\, dt.$$

Furthermore,

$$\|K\|_2^2 = \langle K^* - K_1, K^* - K_1 \rangle = \|K^*\|_2^2 - 2\langle K^*, K_1 \rangle + \|K_1\|_2^2.$$

Now, suppose (9.8.8) holds. Then $R^*g = 0$, since $g(t) = S_{2m-1}(g; t)$ by the uniqueness of natural splines. Hence,

$$0 = R^*g = \int_0^1 K^*(t)g^{(m)}(t)\, dt = \langle K^*, K_1 \rangle.$$

This implies

$$\|K\|_2^2 = \|K^*\|_2^2 + \|K_1\|_2^2 \geq \|K^*\|_2^2.$$

Conversely, suppose (9.8.8) does not hold for some $f \in C^m[0, 1]$. Then for the natural spline $g(t) = S_{2m-1}(f; t)$ we have

$$g^{(m)}(t) = \sum_{i=1}^n c_i(x_i - t)_+^{m-1}/(m-1)!$$

by Eq. (8.5.6). If we define $K_1 = g^{(m)}$, then

$$\langle K^*, K_1 \rangle = \int_0^1 K^*(t)g^{(m)}(t)\, dt = R^*g \neq 0, \quad \text{since } F^*f = F^*g.$$

Take $K = K^* + cK_1$. For a suitable value of the scalar c, we obtain a K such that $\|K\|_2 < \|K^*\|_2$. ∎

Note that we can use Eq. (9.8.8) directly to evaluate F^*f without obtaining F^* itself. We must, of course, determine the spline $S_{2m-1}(f; x)$. (See Section 8.5.) Since a spline is a piecewise polynomial, we can integrate it in closed form piece by piece. If we wish to obtain F^*, we must determine the coefficients a_i^*. One way to do this is to apply the method of Lagrange multipliers to the minimization of $\|K\|_2^2$ subject to the constraints (9.8.6). This method gives rise to the linear system,

$$\sum_{j=i+1}^n (x_j - x_i)^{2m-1} a_j^* + \sum_{j=0}^{m-1} \lambda_j x_i^j = \frac{(1 - x_i)^{2m}}{2m}, \quad i = 1, \ldots, n, \quad (9.8.9)$$

which together with Eqs. (9.8.6) forms a nonsingular system of $n + m$ linear equations for the unknowns $a_1^*, \ldots, a_n^*, \lambda_0, \ldots, \lambda_{m-1}$. (See Exercise 9.15.)

The preceding discussion can be generalized to include more general types of functionals such as, for example, integrals of the derivatives of f. The case of variable nodes is treated in [9.33].

9.9 MULTIPLE INTEGRALS

By a region in R^n we mean a set which is the union of a nonempty open set and some subset of its boundary points. We shall restrict our discussion to bounded regions

9.9 Multiple Integrals

with boundaries which are of R^n-measure zero. For example, in R^2 an open set having a boundary consisting of the union of the graphs $\{(x, f_1(x))\}$ and $\{(x, f_2(x))\}$ of two continuous functions f_1 and f_2 would be such a region. For simplicity we shall confine our attention to connected regions. The simplest such region is a (closed) rectangle, D, consisting of all points $x = (\xi_1, \ldots, \xi_n)$ such that $-\infty < a_i \le \xi_i \le b_i < \infty$, $i = 1, \ldots, n$. Let $f(x)$, defined for all $x \in D$, be a real-valued function. The Riemann integral, Sf, as we know from calculus, can be defined by

$$Sf = \lim_{n \to \infty} \sum_{i=1}^{n} f(x_i) \Delta v_i,$$

where Δv_i is the volume of a subrectangle D_i in a partition of D into rectangular subregions. The limit must be such that max $\Delta v_i \to 0$ in the sequence of partitions. Here, $x_i \in D_i$. As usual, we write $Sf = \int_D f(x) \, dv$. The case of an admissible region G which is not a rectangle can be reduced to the rectangular case as follows. Let D be a rectangle which contains G. Define $g(x) = f(x)$ for $x \in D$ and $g(x) = 0$ for $x \in D - G$. We define $\int_G f(x) \, dx = \int_D g(x) \, dx$.

Perhaps, the simplest way to compute multiple integrals is to treat them as iterated integrals. This and other techniques can be illustrated for the case $n = 2$, since the methods generalize in an obvious way to higher-dimensional cases. Let D be the rectangle $a \le x \le b, c \le y \le d$ and suppose $f(x, y)$ is defined and integrable over D. Further, let the function

$$g(x) = \int_c^d f(x, y) \, dy$$

be integrable over $[a, b]$. Then it is a theorem of calculus that

$$Sf = \iint_D f(x, y) \, dA = \int_a^b dx \int_c^d f(x, y) \, dy.$$

More generally, for a region G bounded by two curves $Y_1(x)$ and $Y_2(x)$,

$$Sf = \iint_G f(x, y) \, dA = \int_a^b g(x) \, dx,$$

where

$$g(x) = \int_{Y_1(x)}^{Y_2(x)} f(x, y) \, dy.$$

We may approximate $\int_a^b g(x) \, dx$ by a quadrature functional of the form

$$F_1 g = \sum_{i=0}^{n} a_i g(x_i).$$

Then

$$g(x_i) = \int_{Y_1(x_i)}^{Y_2(x_i)} f(x_i, y) \, dy$$

may be approximated by using another quadrature functional, $F_2\varphi = \sum_{j=0}^{m} b_j \varphi(y_j)$. Since the interval of integration $[Y_1(x_i), Y_2(x_i)]$ may be different for each i, the nodes y_j will depend on i. Thus, letting $f_i(y) = f(x_i, y)$, we obtain

$$F_2 f_i = \sum_{j=0}^{m} b_j f(x_i, y_{ij}).$$

(In general, the coefficients b_j would also depend on i.) This yields the *cubature* formula,

$$Sf = \sum_{i=0}^{n} a_i \sum_{j=0}^{m} b_j f(x_i, y_{ij}) + Rf, \qquad (9.9.1)$$

where the remainder Rf is given by

$$Rf = R_1 g + \sum_{i=0}^{m} a_i R_2 f_i,$$

R_1 and R_2 being the respective remainders of F_1 and F_2. Thus, when iterated integration is possible, numerical evaluation of multiple integrals reduces to numerical quadratures. Note, however, that this does not necessarily imply that what is true for the quadrature formula carries over to the cubature formula which results. For example, the quadrature functional F_1 may be of degree n and $f(x, y)$ may be of degree n in x and y, but the cubature formula need not be exact because $g(x)$ need not even be a polynomial. This depends on the limits of integration $Y_1(x)$ and $Y_2(x)$, of course.

An important source of quadrature functionals is the integration of interpolating polynomials. These functionals depend on the fact that a function $f(x)$ of a single real variable can be interpolated on an arbitrary set of $n + 1$ points by a polynomial in x of degree $\leq n$. As shown in Theorem 8.1.2ff. and Exercises 8.2, 8.3, 8.5b, there is no such result for functions of two or more variables. Therefore, interpolatory cubature formulas for the numerical integration of a function $f(x, y)$ of two variables must be based on special point sets, such as rectangular or triangular grids (Exercise 8.17, 8.18). For a rectangular grid of points (x_i, y_j), $0 \leq i \leq n$, $0 \leq j \leq m$, the interpolating polynomial, $p = \sum_{i=0}^{n} \sum_{j=0}^{m} a_{ij} x^i y^j$, is given by formula (8.1.3) or the alternative form derived in Exercise 8.17. The degree of p is $\leq m + n$ in x and y. The corresponding cubature formula is obtained by integrating $p(x, y)$. The remainder Rf is given by

$$Rf = \int_G \int (f(x, y) - p(x, y)) \, dx \, dy = \int_G \int r(x, y) \, dx \, dy,$$

where the residual function, $r(x, y)$, is given in Exercise 8.17b. The formula for $r(x, y)$ shows that such an interpolatory cubature functional is of degree at least $\mu = \min \{m, n\}$; i.e., it is exact for polynomials which are of degree $\leq \mu$ in x and y together. However, it follows from Exercise 8.18 that we can derive a cubature functional of the same degree by using a triangular grid. For example, if $\mu = n$ and we take the nodes (x_i, y_j) such that $0 \leq i + j \leq n$, the interpolation polynomial is of degree $\leq n$ and its integral yields a cubature formula of degree at least n. Yet, the number of grid points is $(n + 1)(n + 2)/2$, about half as many as for the rectangular grid. In three dimensions, we choose nodes (x_i, y_j, z_k) such that $0 \leq i + j + k \leq n$. Thus, $i + j \leq n - k$ for $k = 0, \ldots, n$ and there are N nodes, where

$$N = \sum_{1}^{n+1} i(i + 1)/2 = \frac{1}{2}\left(\sum_{1}^{n+1} i^2 + \sum_{1}^{n+1} i\right)$$

$$= \frac{1}{2}\left(\frac{(n + 1)(n + 2)(2n + 3)}{6} + \frac{(n + 1)(n + 2)}{2}\right),$$

that is, $N = (n + 1)(n + 2)(n + 3)/6$ function evaluations.

Thus, interpolatory functionals of this type involve a number of function evaluations proportional to n^d, where n is the degree of the functional and d is the dimension of the domain of f. To achieve reasonable accuracy with a low-degree functional one would have to resort to composite functionals as in the one-dimensional case. For a two-dimensional region D this means subdividing D into subregions. If D is a unit square, for example, then from the formula for the residual (Exercise 8.17b) we see that we must subdivide D into N^2 smaller squares, where $h = 1/N$ is the step size required to obtain an error of order $h^{n+1} < \varepsilon$ with a formula of degree n. Thus, the total number of function evaluations is proportional to $(Nn)^2$, or for dimension d to $(Nn)^d$, which is often prohibitive in computational cost. Therefore, for multiple integrals, it is especially important to have quadrature functionals which use a minimum number of nodes to achieve a given degree. Unfortunately, the technique of Gaussian quadrature does not generalize easily. As we saw in Section 9.4, such quadrature formulas depend on the existence of real zeros of a family of orthogonal polynomials. For orthogonal polynomials in two variables the zeros are sometimes complex numbers and even when they are real may lie outside the region of integration, G. For higher-dimensional regions much less is known. (See [9.25], [9.26].)

Of course, if G is the Cartesian product of two lower-dimensional regions D and E say, then we may construct a *product functional* in a natural way as in (9.9.1). For example, if D and E are intervals and G is the two-dimensional rectangle $E \times F$, we can construct a cubature functional F as the product of two quadrature functionals,

$$F_1 f = \sum_{0}^{n} a_i f(x_i), \quad F_2 f = \sum_{0}^{m} b_j f(y_j).$$

If F_1 and F_2 are Gaussian formulas of degrees $2n + 1$ and $2m + 1$ on D and E respectively, then the functional

$$Ff = \sum_{i=0}^{n} a_i \sum_{j=0}^{m} b_j f(x_i, y_j)$$

will be exact for any monomial $x^r y^s$ where $r \leq n$ and $s \leq m$. Indeed,

$$\iint_G x^r y^s \, dx \, dy = \int_D x^r \int_E y^s \, dy \, dx = \int_D x^r \left(\sum_0^m b_j y_j^s \right) dx$$

$$= \sum_{i=0}^n a_i x_i^r \sum_0^m b_j y_j^s = F(x^r y^s).$$

Thus, the degree of F is min $(2n + 1, 2m + 1)$. As we noted above, for regions bounded by two curves $Y_1(x)$ and $Y_2(x)$, the product approach may fail to give functionals of degree min $(2n + 1, 2m + 1)$, since the integral of $x^i y^j$ with respect to y may not be a polynomial in x. This leads us again to consider functionals of the type $\sum_{i=0}^N a_i f(x_i, y_i)$, where the set of nodes (x_i, y_i) need not be the product of two sets. For functionals which are exact for monomials over various regions, see [9.20] and [9.28], for example. For the most part, a general theory of such functionals seems to be lacking. However, the following theoretical result is known [9.30].

Theorem 9.9.1 Let G be a compact set in R^2 with positive area. There exist points $(x_i, y_i) \in G$, $0 \leq i \leq N = (n + 1)(n + 2)/2$, and positive coefficients a_i such that

$$\iint_G x^r y^s \, dx \, dy = \sum_{i=0}^N a_i x_i^r y_i^s$$

whenever $0 \leq r + s \leq n$.

This result generalizes to higher dimensions and to functions other than monomials $x^r y^s$. In [9.28] a method is given for computing the a_i and the points (x_i, y_i). As we observed for quadrature formulas, those with all positive coefficients behave better with respect to rounding errors.

REFERENCES

9.1. D. Bierens de Haan, *Nouvelles tables d'intégrales définies*, Stechert, New York (1957)
9.2. A. Fletcher, J. C. P. Miller, and L. Rosenhead, *An Index of Mathematical Tables*, second edition, Blackwell, London (1962)
9.3. V. I. Krylov, *Approximate Calculation of Integrals*, translated by A. H. Stroud, Macmillan, New York (1962)
9.4. P. Davis and P. Rabinowitz, "Additional abscissas and weights for Gaussian quadrature of high order: values for $n = 64, 80,$ and 96," *J. Research, National Bureau of Standards* **60**, 613–614 (1958)
9.5. I. P. Natanson, *Konstruktive Funktionentheorie*, Akademie-Verlag, Berlin (1955)
9.6. A. E. Taylor, *General Theory of Functions and Integration*, Blaisdell, New York (1965)
9.7. P. Davis and P. Rabinowitz, *Numerical Integration*, Blaisdell, New York (1967)
9.8. W. Gröbner and N. Hofreiter, *Integraltafel* (2 vols.), Springer-Verlag, Vienna (1961)
9.9. M. Abramowitz and I. A. Stegun (editors), *Handbook of Mathematical Functions*, National Bureau of Standards, Applied Math. Series No. 55, Government Printing Office, Washington, D.C.
9.10. P. Rabinowitz, "Abscissas and weights for Lobatto quadrature of high order," *Math. Comp.* **14**, 47–52 (1960)

9.11. R. Radau, "Étude sur les formules d'approximation qui servent à calculer la valeur numérique d'une intégrale définie," *J. Math. Pures Appl.* **6**, 283–336 (1880)
9.12. G. Szego, *Orthogonal Polynomials*, Amer. Math. Soc. Colloq. Publ. **23**, (1959)
9.13. L. N. G. Filon, "On a quadrative formula for trigonometric integrals," *Proc. Roy. Soc. Edinburgh* **49**, 38–47 (1928)
9.14. E. A. Flinn, "A modification of Filon's method of numerical integration," *J. A.C.M.* **7**, 181–184 (1960)
9.15. F. L. Bauer, H. Rutishauser, and E. Stiefel, "New aspects in numerical quadrature," in *Experimental Arithmetic, High Speed Computing and Mathematics*, American Mathematical Society, Providence, R.I. (1963), 199–218
9.16. W. Romberg, "Vereinfachte numerische Integration," *Norske Vid. Selsk. Forh. Trondheim* **28**, 30–36 (1955)
9.17. R. Bulirsch, "Bemerkungen zur Romberg Integration," *Num. Math.* **6**, 6–16 (1964)
9.18. F. L. Bauer, "Algorithm 60, Romberg integration," *Comm. A.C.M.* **4**, 255 (1961)
9.19. A. Sard and L. F. Meyers, "Best approximate integration formulas," *J. Math. Phys.* **29**, 118–123 (1950)
9.20. A. H. Stroud, "Remarks on the disposition of points in numerical integration formulas," *Math. Comp. (MTAC)* **11**, 257–261 (1957)
9.21. A. H. Stroud, "A bibliography on approximate integration," *Math. Comp.* **15**, 52–80 (1961)
9.22. A. H. Stroud and D. Secrest, *Gaussian Quadrature Formulas*, Prentice-Hall, Englewood, N.J. (1966)
9 23. J. R. Slagle, "A heuristic program that solves symbolic integration problems in freshman calculus," *J.A.C.M.* **10**, 507–520 (1963)
9.24. A. Morris, "Elementary indefinite integration theory for computers," *Journal of Computer and System Sciences* **3**, 387–408 (1969.
9.25. A. H. Stroud, "Integration formulas and othogonal polynomials," *SIAM J. Num. Anal.* **4**, 381–389 (1967)
9.26. A. H. Stroud, "Integration formulas and orthogonal polynomials in two variables," *SIAM J. Num. Anal.* **6**, 222–229 (June 1969)
9.27. P. M. Hirsch, "Evaluation of orthogonal polynomials and relationship to evaluating multiple integrals," *Math. Comp.* **22**, 28–285 (1968)
9.28. P. C. Hammer and A. H. Stroud, "Numerical evaluation of multiple integrals. II," *MTAC* **12**, 272–280 (1958)
9.29. J. E. Dennis and A. A. Goldstein, "Cubature and the Tchakaloff cone," *Journal Computer and System Sciences* **3**, 218–220 (1969)
9.30. V. Tchakaloff, Formules de cubature mécaniques à coefficients non négatifs," *Bull. Sci. Math.* **81**, 123–134 (1957)
9.31. A. Sard, *Linear Approximation*, American Mathematical Society, Providence, R. I. (1963)
9.32. I. J. Schoenberg, "On best approximations of linear operators," *Nederl. Akad. Wetesch. Proc. Ser. A* **67**, 155–163 (1964)
9.33. I. J. Schoenberg, "Monosplines and quadrature formulae," in *Theory and Applications of Spline Functions*, edited by T. N. E. Greville, Academic Press, New York (1969)
9.34. S. M. Chase and L. D. Fosdick, "An algorithm for Filon quadrature," *Comm. ACM*, 453–458 (August 1969)
9.35. W. Gautschi, "Algorithm 331, Gaussian quadrature formulas," *Comm. ACM*, 432–436 (June 1968); "Construction of Gauss-Christoffel quadrature formulas," *Math. Comput.* **22**, 251–270 (1968)

422   Numerical Integration

EXERCISES

9.1 Let $Sf(x) = \int_0^1 f(x+t)\,dt$, $Ff(x) = \sum_{i=1}^n a_i f(x+t_i)$, where $\sum_{i=1}^n a_i = 1$.
   a) Show that S and F can be expressed formally in terms of the differential operator $D = d/dx$ by
$$S = \int_0^1 e^{tD}\,dt = \frac{e^D - I}{D}, \qquad F = \sum_{i=1}^n a_i e^{t_i D}.$$
   [*Hint:* Use Exercise 8.13.]
   b) Show that
$$S - F = S(I - S^{-1}F) = S\left(I - \sum_{i=1}^n a_i \frac{De^{t_i D}}{e^D - I}\right).$$
   The *Bernoulli polynomials* $B_j(t)$ are defined as the coefficients of z in the generating function
$$\frac{ze^{tz}}{e^z - 1} = \sum_{j=0}^\infty \frac{z^j}{j!} B_j(t).$$
   Hence, show that formally
$$S - F = S\left(I - \sum_{i=1}^n a_i \left(\sum_{j=0}^\infty \frac{D^j}{j!} B_j(t_i)\right)\right) = -S \sum_{i=1}^n a_i \left(\sum_{j=1}^\infty \frac{D^j}{j!} B_j(t_i)\right)$$
   since $\sum_{i=1}^n a_i B_0(t_i) = \sum_{i=1}^n a_i = 1$.
   c) Show that $SD^j f(x) = \int_0^1 f^{(j)}(x+t)\,dt = f^{(j-1)}(x+1) - f^{(j-1)}(x)$.
   d) Use (a), (b), (c) to obtain the *generalized Euler–MacLaurin summation formula*,
$$\int_0^1 f(x+t)\,dt = \sum_{i=1}^n a_i f(x+t_i) - \sum_{j=1}^\infty \left(\sum_{i=1}^n a_i \frac{B_j(t_i)}{j!}\right)(f^{(j-1)}(x+1) - f^{(j-1)}(x)).$$
   Taking $x = 0$, obtain
$$\int_0^1 f(t)\,dt = \sum_{i=1}^n a_i f(t_i) - \sum_{j=1}^\infty \left(\sum_{i=1}^n a_i \frac{B_j(t_i)}{j!}\right)(f^{(j-1)}(1) - f^{(j-1)}(0)).$$

9.2 Derive formula (9.3.2). [*Hint:* Take n even.
$$\int_a^b f(x)\,dx = \frac{h}{3} \sum_{i=0}^{n/2} (f_{2i} + 4f_{2i+1} + f_{2i+2}) - \sum_{i=0}^{n/2} f^{(4)}(\xi_i) \frac{h^5}{90}.\ ]$$

9.3 Derive the Newton–Cotes formula for four equally spaced points,
$$\int_{x_0}^{x_3} f(x)\,dx = \frac{3h}{8}(f(x_0) + 3f(x_1) + 3f(x_2) + f(x_3)) - \frac{3h^5}{80} f^{(4)}(\xi), \qquad x_0 < \xi < x_3.$$

9.4 a) Let $a = x_0$, $x_i = x_0 + ih$, $0 \le i \le n$ and $x_n = b$. Let $\varphi_{n+1}(x) = \prod_{i=0}^n (x - x_i)$.
   With $x = x_0 + th$,
$$\varphi_{n+1}(x) = h^{n+1} \prod_{i=0}^n (t - i) = h^{n+1} \psi_{n+1}(t).$$

Using Exercise 8.7, verify that $\varphi_{n+1}(x_{n/2} + \sigma) = (-1)^{n+1}\varphi_{n+1}(x_{n/2} - \sigma)$, where $x_{n/2} = (3a + b)/2 = x_0 + (n/2)h$. Also show for $a < \sigma + h \leq x_{n/2}$ and $\sigma \neq x_i$, $|\varphi_{n+1}(\sigma + h)| < |\varphi_{n+1}(\sigma)|$, and for $x_{n/2} \leq \sigma < b$, $\sigma \neq x_i$, $|\varphi_{n+1}(\sigma)| < |\varphi_{n+1}(\sigma + h)|$.

b) Define $\Phi_{n+1}(x) = \int_a^x \varphi_{n+1}(\sigma)\,d\sigma$. Prove: If n is even, then $\Phi_{n+1}(a) = \Phi_{n+1}(b) = 0$ and $\phi_{n+1}(x) > 0$ for $a < x < b$. [*Hint:* For n even, φ_{n+1} is an odd function with respect to the midpoint $x_{n/2}$ by part (a). Hence, $\Phi_{n+1}(b) = 0$. Also, $\varphi_{n+1}(x) < 0$ for $x < a$ since the degree $n + 1$ is odd. Hence, $\Phi_{n+1}(x) > 0$ for $a < x < x_1$ and $\Phi_{n+1}(x) > 0$ for $a < x \leq x_1$. In $[x_1, x_2]$ the absolute magnitude $|\varphi_{n+1}(x)|$ is less than the corresponding magnitude $|\varphi_{n+1}(x - h)|$ in $[x_0, x_1]$. Changing the variable of integration we see that $|\int_{x_1}^{x_2} \varphi_{n+1}(x)\,dx| < |\int_{x_0}^{x_1} \varphi_{n+1}(x)\,dx|$. Thus, $\Phi_{n+1}(x) > 0$ for $a < x < x_2$ and by the same reasoning for $a < x < x_{n/2}$. Then use the antisymmetry of φ_{n+1} with respect to $x_{n/2}$.]

c) Prove that for any $f \in C^{n+2}[a, b]$, the remainder in a closed Newton–Cotes quadrature formula with an even number n of subintervals of step h is given by

$$R_n f = \frac{I_n}{(n+2)!} f^{(n+2)}(\xi), \quad a < \xi < b$$

where

$$I_n = \int_a^b x\varphi_{n+1}(x)\,dx = O(h^{n+3}).$$

Show that $I_n < 0$.

[*Hint:*

$$R_n f = \int_a^b (f(x) - L_n(x))\,dx = \int_a^b \varphi_{n+1}(x) f[x_0, \ldots, x_n, x]\,dx.$$

Integrate by parts, using Exercise 8.6e, to get

$$R_n f = \int_a^b \frac{d}{dx} \Phi_{n+1}(x) f[x_0, \ldots, x_n, x]\,dx$$

$$= \Phi_{n+1}(x) f[x_0, \ldots, x_n, x]\Big|_a^b - \int_a^b \Phi_{n+1}(x) \frac{d}{dx} f[x_0, \ldots, x_n, x]\,dx$$

$$= -\int_a^b \Phi_{n+1}(x) \frac{d}{dx} f[x_0, \ldots, x_n, x]\,dx = -\int_a^b \Phi_{n+1}(x) \frac{f^{(n+2)}(\zeta_x)}{(n+2)}\,dx$$

by Exercise 8.6c. The continuity of $df[x_0, \ldots, x_n, x]/dx$ allows us to apply the mean-value theorem, since $\Phi_{n+1}(x) > 0$ by part (b). Hence

$$R_n f = \frac{-f^{(n+2)}(\alpha)}{(n+2)!} \int_a^b \Phi_{n+1}(x)\,dx, \quad a < \alpha < b. \tag{1}$$

Integrating by parts again gives

$$\int_a^b \Phi_{n+1}(x)\,dx = -\int_a^b x\varphi_{n+1}(x)\,dx > 0.$$

Letting $x = x_0 + sh$ and using Exercise 9.4b, show that

$$I_n = h^{n+3} \int_0^n s\psi_{n+1}(s)\,ds < 0.$$

Since $f^{(n+2)}(\alpha) = 0$ when f is a polynomial of degree $n + 1$, this shows that the closed Newton–Cotes formula for even n is of degree $n + 1$.

d) Apply formula (1) of part (c) to the case $n = 2$ and obtain an alternative derivation of formula (9.2.13), Simpson's rule remainder.

e) Show that a closed Newton–Cotes formula with odd n has the remainder $R_n f = I_n f^{(n+1)}(\xi)/(n + 1)!$, $a < \xi < b$, where

$$I_n = \int_a^b \varphi_{n+1}(x)\,dx = h^{n+2} \int_0^n \prod_{i=0}^n (s - i)\,ds < 0.$$

[*Hint:*

$$R_n f = \int_a^{b-h} \varphi_{n+1}(x) f[x_0, \ldots, x_n, x]\,dx$$

$$+ \int_{b-h}^b \varphi_{n+1}(x) f[x_0, \ldots, x_n, x]\,dx, \quad \varphi_{n+1}(x) = \varphi_n(x)\cdot(x - x_n).$$

Hence,

$$\int_a^{b-h} \varphi_{n+1}(x) f[x_0, \ldots, x_n, x]\,dx = \int_a^{b-h} \frac{d\Phi_n}{dx}(f[x_0, \ldots, x_{n-1}, x] - f[x_0, \ldots, x_n])\,dx.$$

Since n is odd, Exercise 9.4b gives $\Phi_n(b - h) = 0$. Integration by parts gives

$$\int_a^{b-h} \varphi_{n+1}(x) f[x_0, \ldots, x_n, x]\,dx = -\frac{f^{(n+1)}(\xi')}{(n + 1)!} \int_a^{b-h} \Phi_n(x)\,dx$$

$$= Kf^{(n+1)}(\xi'), \quad a < \xi' < b - h.$$

The integral $\int_{b-h}^b$ can be treated by the first mean-value theorem, since $\varphi_{n+1}(x)$ does not change sign and we obtain a term

$$-\frac{f^{(n+1)}(\xi'')}{(n + 1)!} \int_{b-h}^b \varphi_{n+1}(x)\,dx = Lf^{(n+1)}(\xi'').$$

Thus, $Rf = Kf^{(n+1)}(\xi') + Lf^{(n+1)}(\xi'')$. Since K and L are both negative, $Rf = (K + L)f^{(n+1)}(\xi)$ for some $\xi' < \xi < \xi''$. Since

$$\varphi_{n+1}(x) = \frac{d\Phi_n(x)}{dx}\cdot(x - b),$$

integration by parts gives $K + L = I_n$.

9.5 a) Derive the following closed Newton–Cotes formulas:

$$n = 3. \quad \int_{x_0}^{x_3} f(x)\,dx = \frac{3h}{8}\left(f(x_0) + 3f(x_1) + 3f(x_2) + f(x_3)\right)$$

$$- \frac{3h^5}{80} f^{(4)}(\xi)$$

$$n = 4. \quad \int_{x_0}^{x_4} f(x)\,dx = \frac{2h}{45}\left(7f(x_0) + 32f(x_1) + 12f(x_2) + 32f(x_3) + 7f(x_4)\right)$$

$$- \frac{8h^7}{945} f^{(6)}(\xi)$$

$n = 5.$ $\int_{x_0}^{x_5} f(x)\, dx = \frac{5h}{288} (19f(x_0) + 75f(x_1) + 50f(x_2) + 50f(x_3) + 75f(x_4)$
$+ 19f(x_5)) - \frac{275}{12096} f^{(6)}(\xi) h^7,$

where ξ is a point in the open interval of integration. Show that the coefficients a_i in $(h/c) \sum_0^n a_i f(x_i)$ for $n = 6, 7, 8, 9, 10$ are as shown in Table 9.1.

n	$1/c$	a_0	a_1	a_2	a_3	a_4	a_5	a_6	a_7	a_8
6	1/140	41	216	27	272	27	216	41		
7	7/17,280	751	3,577	1,323	2,989	2,989	1,323	3,577	751	
8	4/14,175	989	5,888	−928	10,496	−4,540	10,496	−928	5,888	989
9	9/89,600	2,857	15,741	1,080	19,344	5,778				
10	10/598,752	16,067	106,300	−48,525	272,400	−260,550	427,368			

b) Verify the symmetry of the coefficients. Prove that $h \sum_0^n a_i/c = b - a$.

9.6 Derive formula (9.4.2) by using Exercise 8.11b.
[*Hint*:
$$f(x) = H_{2n+1}(x) + \frac{f^{(2n+2)}(\xi)}{(2n+2)!} \varphi_{n+1}^2(x)$$

and $S(H_{2n+1}) = F(H_{2n+1})$ when $F = \sum_0^n a_i f(x_i)$ is a Gaussian quadrature formula.]

9.7 In Section 9.3 (Computational Aspects), let $f_i = f(x_i) + \rho_i$, where ρ_i, $0 \le i \le n$ are rounding errors. Assume that each ρ_i is a random variable with a distribution that has zero mean and variance σ^2. (See, for example, H. Cramer, *Mathematical Methods of Statistics*, Princeton University Press (1958).) Also assume that the ρ_i are independent random variables. (This assumption is not always justified in practice.) It is an elementary result of statistics that the accumulated rounding error $r = \sum_0^n a_i \rho_i$ has zero mean and variance

$$\sigma^2(r) = \sigma^2 \sum_0^n a_i^2.$$

Since

$$\sum_0^n a_i = b - a \qquad (1)$$

for any practical quadrature functional, the mean square rounding error is minimized by choosing the a_i to minimize $\sigma^2(r)$ subject to the constraint (1). Prove that this minimum is achieved when all $a_i = (b-a)/(n+1)$. [*Hint*: Use the method of Lagrange multipliers, Chapter 12.]

9.8 Let $(Q_n(x))$ be an orthonormal family of polynomials determined by the recursion formula (7.7.2), with a_n, b_n given by (7.7.3), (7.7.4) and Q_n having leading coefficient c_n.

a) Prove that

$$xQ_{n-1}(x) = \frac{c_{n-1}}{c_n} Q_n(x) - a_n Q_{n-1}(x) + \frac{c_{n-2}}{c_{n-1}} Q_{n-2}(x). \qquad (*)$$

[*Hint:* Let $\bar{Q}_n = d_n Q_n$. By (7.7.2), $xQ_{n-1} = d_n Q_n - a_n Q_{n-1} - b_n Q_{n-2}$. Compare coefficients of x^n. Also note that

$$d_n = \langle xQ_{n-1}, Q_n \rangle = \langle xQ_n, Q_{n-1} \rangle = -b_{n+1}.\;]$$

b) Derive the *Christoffel–Darboux identity*,

$$(x-t) \sum_{i=0}^{n} Q_i(x) Q_i(t) = \frac{c_n}{c_{n+1}} [Q_{n+1}(x) Q_n(t) - Q_n(x) Q_{n+1}(t)].$$

[*Hint:* By (*) in part (a),

$$xQ_i(x)Q_i(t) = \frac{c_i}{c_{i+1}} Q_{i+1}(x) Q_i(t) - a_{i+1} Q_i(x) Q_i(t) + \frac{c_{i-1}}{c_i} Q_{i-1}(x) Q_i(t).$$

Interchange x and t, subtract and sum the result over i from 0 to n.]

c) Let $x_1, \ldots, x_n$ be the zeros of $Q_n(x)$ and define

$$A_j = \int_a^b \rho(x) \frac{Q_n(x)}{(x-x_j) Q_n'(x_j)} dx,$$

where ρ is the weight function for the family (Q_n). Show that

$$A_j = -\frac{c_{n+1}}{c_n} \frac{1}{Q_n'(x_j) Q_{n+1}(x_j)}.$$

[*Hint:* Set $t = x_j$ in the Christoffel–Darboux identity, divide through by $x - x_j$, multiply through by $\rho(x)$ and integrate over $[a, b]$. For $i > 0$, $Q_i(x)$ is orthogonal to any zero-degree polynomial. For $i = 0$, we have $\int_a^b \rho(x) Q_0^2\, dx = 1$.]

d) Show that also

$$A_j = \frac{c_n}{c_{n-1}} \frac{1}{Q_n'(x_j) Q_{n-1}(x_j)}.$$

[*Hint:* Use (*) in part (a) with n replaced by $n+1$.]

9.9 a) Verify that the Jacobi polynomials $J_n^{(p,q)}(x)$, which are orthogonal on the interval $[-1, 1]$ with respect to the weight function $(1-x)^p(1+x)^q$ satisfy *Rodrigues'* formula:

$$J_n^{(p,q)}(x) = \frac{(-1)^n}{2^n n!} (1-x)^{-p} (1+x)^{-q} \frac{d^n}{dx^n} [(1-x)^{p+n}(1+x)^{q+n}].$$

[*Hint:* Let $v(x) = (1-x)^{p+n}(1+x)^{q+n}$. For $r < n$, integration by parts yields

$$\int_{-1}^{1} x^r \frac{d^n v}{dx^n} dx = (-1)^n \int_{-1}^{1} \frac{d^n(x^r)}{dx^n} v\, dx = 0,$$

establishing that the function $(1-x)^{-p}(1+x)^{-q} d^n v/dx^n$ is orthogonal to all polynomials of degree $< n$, with weight function $(1-x)^p(1+x)^q$. Here, normalize by $J_n^{(p,q)} = \Gamma(n+p+1)/\Gamma(n+1)\Gamma(p+1)$.]

b) Use Rodrigues' formula to show that the leading coefficient of $J_n^{(0,1)}$ is

$$A_n^{(0,1)} = \frac{(2n+1)!}{2^n n!(n+1)!}.$$

[*Hint:* Differentiate, using Leibniz's rule, and observe that the same coefficient is obtained by applying this rule to $(1/2^n n!) x^{-\alpha-\beta} D^n(x^{\alpha+\beta+2n})$. Here, $D = d/dx$. Leibniz's rule:

$$D^n(uv) = \sum_{r=0}^{n} \binom{n}{r} D^r u D^{n-r} v. \ ]$$

c) Show that

$$\int_{-1}^{1} (1+x)(J_n^{(0,1)}(x))^2 \, dx = \frac{n!}{2^n n!} A_n^{(0,1)} \int_{-1}^{1} (1-x)^n (1+x)^{n+1} \, dx = \frac{2}{n+1}.$$

d) Use the hint in part (b) to show that the leading coefficient of $J_n^{(1,1)}$ is

$$A_n^{(1,1)} = \frac{(2n+2)!}{2^n n! (n+2)!}.$$

e) Use Rodrigues' formula of part (a) to derive the relation,

$$\frac{d}{dx} J_n^{(1,1)}(x) = -\frac{2(n+1)}{1-x^2} P_{n+1}(x) + \frac{2x}{1-x^2} J_n^{(1,1)}(x).$$

[*Hint:* Differentiate the formula and note that $P_{n+1} = J_{n+1}^{(0,0)}$.]

f) Use the hint in part (b) to prove that the leading coefficient of the Legendre polynomial P_n is $(2n)!/2^n (n!)^2$.

9.10 **a)** Write a computer program to do Radau quadrature for the following sets of nodes and coefficients.

	x_i	a_i	
	-1	$\frac{1}{2}$	(b_1)
	$\frac{1}{3}$	$\frac{3}{2}$	

$n = 2$

x_i	a_i	
-1	0.222 222 22	(b_1)
$-0.289\ 897\ 94$	1.024 971 66	
$0.689\ 897\ 94$	0.752 806 12	

$n = 3$

x_i	a_i	
-1	$\frac{1}{8}$	
$-0.575\ 318\ 9$	0.657 688 6	(b_1)
$0.181\ 066\ 3$	0.776 387 0	
$0.822\ 824\ 1$	0.440 924 4	

Test the program by integrating the functions $x^{1/2}$, $1/(1+x^2)$, $1/(1+e^x)$. Compare its accuracy with the trapezoidal rule. (See formula (9.4.10).)

b) Write a computer program for Lobatto quadrature in the following cases:

x_i	$a_i(b_1, b_2)$
± 1	$\frac{1}{3}$
0	$\frac{4}{3}$

$n = 3$

x_i	a_i
± 1	$\frac{1}{6}$
$\pm .447\ 213\ 60$	$\frac{5}{6}$

$n = 4$

x_i	a_i
± 1	$\frac{1}{10}$
$\pm .654\ 653\ 67$	$\frac{49}{90}$
0	$\frac{32}{45}$

Test it as in part (a). (See formula (9.4.12).)

9.11 a) Write a program to do Gauss–Laguerre quadrature for the case $n = 4$ (five points). The nodes and corresponding coefficients are

x_i	a_i
0.263 560	0.521 756
1.413 403	0.398 667
3.596 426	0.075 942 4
7.085 810	0·003 611 76
12.640 801	0.000 023 370 0

b) Do the same for the Gauss–Hermite formula for $n = 4$.

x_i	a_i
0	0.945 309
$\pm 0.958\ 572$	0.393 619
$\pm 2.020\ 183$	0.019 953 2

c) Test these on e^{-x} over $(0, \infty)$ and e^{-x^2} over $(-\infty, \infty)$ respectively.

9.12 Write a computer program to do Chebyshev quadrature for the following cases ($\rho(x) = 1$):

n	x_i
2	0
	$\pm 0.707\ 106\ 781\ 2$
3	$\pm 0.187\ 592\ 474\ 1$
	$\pm 0.794\ 654\ 472\ 3$
4	0
	$\pm 0.374\ 541\ 409\ 6$
	$\pm 0.832\ 497\ 487\ 0$
5	$\pm 0.266\ 635\ 401\ 5$
	$\pm 0.422\ 518\ 653\ 8$
	$\pm 0.866\ 246\ 818\ 1$
6	0
	$\pm 0.323\ 911\ 810\ 5$
	$\pm 0.529\ 656\ 775\ 3$
	$\pm 0.883\ 861\ 700\ 8$
8	0
	$\pm 0.167\ 906\ 184\ 2$
	$\pm 0.528\ 761\ 783\ 1$
	$\pm 0.601\ 018\ 655\ 4$
	$\pm 0.911\ 589\ 307\ 7$

(See (9.4.17)ff. to determine c_n.)

Test the program by integrating the functions in Exercise 9.10a.

9.13 a) Let R be the rectangle $x_{-1} \le x \le x_1, y_{-1} \le y \le y_1$. Obtain a cubature functional for $\int \int_R f(x, y)\, dx\, dy$ by forming the product functional (Section 9.9) F of two Simpson's rules (formula (9.2.8)).

[*Hint:* Let $h = (x_1 - x_{-1})/2$, $k = (y_1 - y_{-1})/2$, $x_0 = x_{-1} + h$ and $y_0 = y_{-1} + k$. Then

$$Ff = \frac{hk}{9}\left(\sum_{\substack{i=\pm 1 \\ j=\pm 1}} f(x_i, y_j) + 4(f(x_{-1}, y_0) + f(x_1, y_0))\right.$$

$$\left. + 4(f(x_0, y_{-1}) + f(x_0, y_1)) + 16f(x_0, y_0)\right). \ ]$$

b) Show that the 9-point functional F in part (a) is exact for all monomials $x^i y^j$ with $0 \leq i + j \leq 3$.

9.14 a) Consider the midpoint formula (9.8.7). Assume that $f \in C^2[0, 1]$. Show that the remainder in this case satisfies

$$|Rf| \leq \frac{\frac{1}{24}\|f^{(2)}\|_\infty}{n^2}; \tag{1}$$

that is, $Rf = O(h^2)$ rather than $O(h)$ as when $f \in C^1[0, 1]$.
[*Hint:*

$$Rf = \sum_{i=0}^{n} \int_{x_i - h/2}^{x_i + h/2} (f(x_i) - f(x_i))\, dx.$$

$$f(x) = f(x_i) + (x - x_i)f'(x_i) + \frac{(x - x_i)^2}{2} f''(\xi). \ ]$$

b) Show that inequality (1) in part (a) is replaced by

$$|Rf| \leq (L/24)/n^2$$

when f has only a first derivative which satisfies a Lipschitz condition with constant L. [*Hint:* Use the mean-value theorem.]

9.15 Derive equations (9.8.9) by applying the method of Lagrange multipliers (Chapter 12) to the minimization of $\|K(t)\|_2^2$ subject to the constraints (9.8.6), where $K(t)$ is given by formula (9.8.5). [*Hint:* Let $g(a_1, \ldots, a_n) = \|K(t)\|_2^2$ and $g_j(a_1, \ldots, a_n) = \sum_{i=1}^n a_i x_i^j$, $j = 0, \ldots, m - 1$. Introduce multipliers λ_j and form the function $G(a_1, \ldots, a_n) = g - \sum_{j=0}^{m-1} \lambda_j g_j$. Set $\partial G/\partial a_i = 0$, $i = 1, \ldots, n$.)

9.16 Let $Ff = a_1 f(-1) + a_2 f(0) + a_3 f(1)$ be a quadrature functional approximating $Sf = \int_{-1}^{1} f(x)\, dx$, and let $R = S - F$. In [9.31, p. 37ff.], the following minimum-remainder formulas are established:

a) If $Rp = 0$ for all polynomials p of degree 0, then $|Rf|$ is minimized with respect to the L_1, L_2, and L_∞ norms of f by setting $a_1 = a_3 = 1/2$ and $a_2 = 1$, that is, the best quadrature formula is the trapezoidal rule in all three norms.

b) Let $Rp = 0$ for all p of degree ≤ 1. Then relative to the L_2 norm the remainder is minimized by taking $a_1 = a_3 = 3/8$ and $a_2 = 10/8$. Relative to the L_∞ norm, the minimum remainder is given by $a_1 = a_3 = \sqrt{2}/4$ and $a_2 = 2 - \sqrt{2}/2$. Relative to the L_1 norm, $a_1 = a_3 = \sqrt{2} - 1$ and $a_2 = 4 - \sqrt{8}$ yields the best formula.

c) Let $Rp = 0$ for all p of degree ≤ 2. The best formula relative to the L_2 norm is given by Simpson's rule; that is, $a_1 = a_3 = 1/3$ and $a_2 = 4/3$. In fact, Simpson's rule still yields the L_2 minimum remainder when we require $Rp = 0$ for all polynomials p of degree ≤ 3.

Prove (a), (b), and (c). [*Hint:* See Exercise 9.15 for the L_2 case. See Eq. (9.8.7)ff for the L_∞ case.]

9.17 The general linear quadrature functional (9.2.1) may involve derivatives of f. One such class of formulas can be derived by replacing f by its Hermite interpolating polynomial (Exercise 8.11), H_{2n+1}.

a) Obtain the Hermite interpolating polynomial H_3 of f on the set $\{x_0, x_1\}$; that is, $n = 1$.

b) Integrate H_3 on the interval $[x_0, x_1]$ to obtain the *improved trapezoidal rule*,

$$\int_{x_0}^{x_1} f(x)\, dx = \frac{h}{2}(f(x_0) + f(x_1)) - \frac{h^2}{12}(f'(x_1) - f'(x_0)) + Rf.$$

c) Using Exercise 8.11b, show that the remainder in part (b) is given by

$$Rf = \frac{h^5}{24} \int_0^1 s^2(s-1) f^{(4)}(x_0 + sh)\, ds.$$

where $h = x_1 - x_0$.

CHAPTER 10

NUMERICAL SOLUTION OF ORDINARY DIFFERENTIAL EQUATIONS

10.1 INTRODUCTION: DIFFERENTIAL EQUATIONS AND DIFFERENCE EQUATIONS

The theory of ordinary differential equations is so extensive that it would be futile for us to attempt even a brief summary. Instead, the reader is referred to one of the standard reference works such as [10.4], [10.5]. However, we must at least define what is meant by a *differential equation*, the *initial-value problem*, and the *boundary-value problem*, since these are the problems we wish to solve numerically.

In general, a system of differential equations is a finite set of equations of the form

$$f_i(t, x_1, \ldots, x_m, x'_1, \ldots, x'_m, \ldots, x_1^{(r)}, \ldots, x_m^{(r)}) = 0, \qquad 1 \leq i \leq m,$$

where the f_i are functions of the $m(r + 1) + 1$ variables shown. We shall immediately rewrite the system in vector form as $F(t, x, x', \ldots, x^{(r)}) = 0$, where $F = (f_1, \ldots, f_m)$, $x = (x_1, \ldots, x_m), \ldots, x^{(r)} = (x_1^{(r)}, \ldots, x_m^{(r)})$. By a solution of such a system we mean a vector function $u(t) = (u_1(t), \ldots, u_m(t))$ such that $F(t, u(t), u'(t), \ldots, u^{(r)}(t)) = 0$ for all t in some interval $[a, b]$, where $u^{(j)}(t) = (u_1^{(j)}(t), \ldots, u_m^{(j)}(t))$, $0 \leq j \leq r$, that is, $u^{(j)}(t)$ is the jth derivative of $u(t)$ with respect to t. (See Section 5.3.) An rth-order system may be replaced by a larger *first-order system*, that is, by a system in the form $F(t, x, x') = 0$. (See Exercise 10.1.) We shall henceforth assume this has been done and we shall consider only first-order systems. Furthermore, we shall assume that it is always possible to solve for x'. Thus, with almost no loss of generality, we may confine our attention to the differential equation,

$$x' = f(t, x), \qquad a \leq t \leq b, \qquad (10.1.1)$$

where henceforth $x = (x_1, \ldots, x_m)$ and $f = (f_1, \ldots, f_m)$, $m \geq 1$. Indeed, we could allow x to be a variable ranging over a Banach space, but we shall not require such generality. Some of our results remain valid for such x and the reader is encouraged to verify this as he goes through the text. In fact, we shall rarely refer to the components $x_1, \ldots, x_m$ of x. To avoid the confusion that sometimes arises from the customary abuse of notation, we call attention now to the subsequent multiple usage of the variable x. In most instances, x is a variable denoting an element of V^m, the m-dimensional real vector space and $\|x\|$ denotes some vector norm. However, it is convenient to denote a solution of (10.1.1) by $x(t)$. We shall resist the temptation to write $x = x(t)$, since then x would also denote a function; i.e., an element of the space of continuous functions defined on $[a, b]$ with values $x(t) \in V^m$. (We denote this space by $C[a, b]$, as in the case $m = 1$.) On the other hand, strictly speaking, $x(t)$ is a value of a function, that is, $x(t) \in V^m$ for each $t \in [a, b]$. We will use $\|x(t)\|$ to denote a

431

vector norm in V^m. (We shall never specify a particular norm.) For the norm of the function $x(t)$, as an element of $C[a, b]$, we shall use $\|x(t)\|_\infty$, where $\|x(t)\|_\infty = \sup_{a \le t \le b} \{\|x(t)\|\}$. Similarly, $\|f(t, x)\|$ denotes the norm of the m-dimensional vector $f(t, x)$ for $(t, x) \in R \times V^m$. If (t, x) is restricted to some compact subset, D, on which f is continuous, then we define $\|f\|_\infty = \|f(t, x)\|_\infty = \sup_{(t, x) \in D} \{\|f(t, x)\|\}$.

A solution of (10.1.1) is a function $x(t)$, $a \le t \le b$ such that

$$\frac{dx(t)}{dt} - f(t, x(t)) = 0, \qquad a \le t \le b.$$

Suppose f is continuous. If we define the nonlinear operator F mapping $C^1[a, b]$ into $C[a, b]$ by $Fu = du/dt - f(t, u(t))$ for any function $u \in C^1[a, b]$, then a solution of (10.1.1) may be viewed as an element of $C^1[a, b]$ such that $Fx(t) = 0$. Thus, the problem of solving a differential equation is another instance of the problem of solving an equation of the form $F(u) = 0$. We shall soon see how this can be transformed into an equivalent fixed-point problem.

Remark A solution $x(t)$ of (10.1.1) must, by definition, be differentiable in $[a, b]$. If f is continuous in a suitable subset D of (t, x)-space, then $dx/dt = f(t, x(t))$ is continuous in $[a, b]$. Hence $x(t)$ is in $C^1[a, b]$. More generally, if f has continuous partial derivatives of order $n \ge 0$, then a solution $x(t)$ of (10.1.1) has a continuous derivative of order $n + 1$. (e.g., $d^2x/dt^2 = \partial f/\partial t + (\partial f/\partial x)(dx/dt)$.)

It is known that (10.1.1) can have many solutions. For example, $x' = Ax$ has the solution $x(t) = ce^{At}$, where c is an arbitrary constant. (See Exercise 3.38.) A unique solution can be specified if the function $x(t)$ is required to satisfy certain *boundary conditions* at $t = a$ and $t = b$. The simplest such conditions are of the form $x(a) = x_0$, where x_0 is a prescribed *initial value*. This gives rise to the *initial-value problem*,

$$\left.\begin{array}{l} x' = f(t, x), \qquad a \le t \le b, \\ x(a) = x_0. \end{array}\right\} \qquad (10.1.2)$$

The solution of (10.1.2) is the function $x(t)$ which is a solution of (10.1.1) in $[a, b]$ and satisfies the *initial condition* $x(a) = x_0$. It is called the *exact solution*. We shall be concerned, for the most part, with *numerical solutions* of (10.1.2).

Definition 10.1.1 Let $G = \{a = t_0 < t_1 < \cdots < t_q \le b\}$ be a finite subset of points in $[a, b]$, called a *grid* (or *net* or *mesh*). In general, the t_i are equally spaced, that is, $t_i = a + ih$, $(0 \le i \le q)$ and h is called the *step size*. By a *numerical solution* of (10.1.2) *of accuracy* $\varepsilon > 0$ *on a grid* G, we shall mean a finite sequence of values $\bar{x}(i)$ such that $\bar{x}(a) = x_0$ and

$$\max_{0 \le i \le q} |\bar{x}(i) - x(t_i)| < \varepsilon.$$

Here, the $x(t_i)$ are the values of the exact solution $x(t)$ of (10.1.2).

Remark For convenience, we shall henceforth write x_i instead of $\bar{x}(i)$. (There is no danger of confusing x_i with the ith component of the vector x, since the latter will

10.1 Introduction: Differential Equations and Difference Equations

seldom be referred to explicitly.) Thus, a numerical solution $(x_i) = (x_0, x_1, \ldots, x_q)$ on a grid $G = \{t_0, t_1, \ldots, t_q\}$ is an element of the space $V^{(m)}(G)$ of vector-valued functions on G. The norm in $V^{(m)}(G)$ is the sup norm, $\|(x_i)\|_\infty = \max_{0 \leq i \leq q} \|x_i\|$. With each continuous function $u(t) \in C[a, b]$ we can associate the element $(u(t_0), \ldots, u(t_q))$ of $V^{(m)}(G)$. This defines a linear operator T_G mapping $C[a, b]$ onto $V^{(m)}(G)$. The operator T_G is clearly not injective. If $x(t)$ is a solution of (10.1.2), it is differentiable, therefore certainly continuous on $[a, b]$. The condition that (x_i) be a numerical solution of accuracy ε on grid G can be rewritten as

$$\|T_G x(t) - (x_i)\|_\infty < \varepsilon.$$

An alternative notion of numerical solution should also be considered. We consider the subspace $\hat{C}$ of $C[a, b]$ consisting of certain piecewise linear functions; that is, $u(t) \in \hat{C}$ if and only if there exists a grid $\{a = t_0, t_1, \ldots, t_q = b\}$ in $[a, b]$ such that for $0 \leq i \leq q - 1$,

$$u(st_i + (1-s)t_{i+1}) = su(t_i) + (1-s)u(t_{i+1}), \qquad 0 \leq s \leq 1.$$

There is an obvious injective linear map $\hat{T}_G$ of $V^{(m)}(G)$ into $\hat{C}$, namely,

$$\hat{T}_G : (x_i) \longmapsto u(t)$$

where $u(t)$ is the piecewise linear function such that $u(t_i) = x_i, 0 \leq i \leq q$. By a *piecewise linear numerical solution on grid G of accuracy $\varepsilon > 0$*, we mean a function $u(t)$ which is the image under $\hat{T}_G$ of an element of $V^{(m)}(G)$ and such that

$$\sup_{a \leq t \leq b} |x(t) - u(t)| < \varepsilon,$$

where $x(t)$ is the solution of (10.1.2).

Before proceeding further with our discussion of numerical solution, we must review some of the theory of the exact solution $x(t)$ of (10.1.2). The existence of $x(t)$ can be proved under the very weak hypothesis that the function f is continuous. We shall suppose this is true for all f considered here. To prove the uniqueness of a solution of (10.1.2), we shall assume further that f satisfies a Lipschitz condition (Definition 1.3.4) with respect to x, that is,

$$\|f(t, x) - f(t, y)\| \leq L\|x - y\|, \qquad (10.1.3)$$

where L is a constant. We shall prove uniqueness below. Since a unique solution exists under these conditions, one might be tempted to find a closed-form expression for it, or a series expansion for it, as in the elementary example $x' = Ax$. (A compendium of equations and such solutions is given in [10.3].) However, most differential equations do not have such solutions, and often it is unknown whether they do. Furthermore, as with definite integrals, closed-form or series solutions do not always provide the simplest or most efficient algorithm for computing a numerical solution. In current practice, a numerical solution is most often determined by a method based on numerical quadrature. This involves replacing (10.1.2) by an "equivalent" *integral equation*. Such an equation contains an integral of a vector function. For any such function $u(t)$, $a \le t \le b$, we define $\int_a^b u(t)\,dt$ as the vector limit of the sums $\sum_{i=1}^n u(t_i')(t_{i+1} - t_i)$, as $\max_i |t_{i+1} - t_i| \to 0$. Here $a = t_1 < t_2 < \cdots t_{n+1} = b$ is any partition of $[a, b]$ into subintervals and t_i' is any point in $[t_i, t_{i+1}]$. The limit is in the vector norm topology. The integral $\int_a^b u(t)\,dt$ is a straightforward generalization of the Riemann integral for real functions. By Theorem 5.3.4, $\int_a^b u(t)\,dt$ exists for any continuous $u(t)$. In fact, when $u(t) = (u_1(t), \ldots, u_m(t))$, we have simply

$$\int_a^b u(t)\,dt = \left(\int_a^b u_1(t)\,dt, \ldots, \int_a^b u_m(t)\,dt \right).$$

Now, consider the *integral equation*

$$x(t) = x_0 + \int_a^t f(s, x(s))\,ds, \qquad a \le t \le b. \tag{10.1.4}$$

Clearly, if $x(t)$ is a solution of (10.1.2), then integration of (10.1.1) yields (10.1.4). Conversely, if $f(s, x)$ is continuous and $x(t)$ is a continuous function satisfying (10.1.4), then differentiation with respect to t yields (10.1.1) and $x(a)$ clearly equals x_0.

Under certain restrictions on f, the right side of (10.1.4) defines a nonlinear operator g which maps a subset X of the Banach space $C[a, b]$ onto itself. For any function $u \in X$, we define $v = g(u)$ to be the function,

$$g(u) = v(t) = x_0 + \int_a^t f(s, u(s))\,ds, \qquad a \le t \le b. \tag{10.1.5}$$

The subset X arises in a natural way from the condition that f is continuous on a closed "rectangle," D, defined by the inequalities $|t - a| \le c$, $\|x - x_0\| \le d$; i.e., on the cartesian product of these two closed balls. Since x is finite dimensional, D is compact and f is bounded on D. Let

$$M = \max_{(t, x) \in D} \{\|f(t, x)\|\}$$

and let $b = a + \min\{c, d/M\}$. If $x(t)$ is a solution of (10.1.2) for $a \le t \le b$, then $\|x(t) - x(a)\| \le d$ for $t - a$ sufficiently small by the continuity of $x(t)$ at a. Let $t^* = \sup\{t : \|x(t) - x(a)\| \le d\}$. Setting $\Delta t = t^* - a$, and applying the general mean-value theorem (Theorem 5.3.3), we obtain

$$\|x(t^*) - x(a)\| \le \Delta t \sup\{\|x'(a + \theta \cdot \Delta t)\|, \qquad 0 < \theta < 1\}.$$

Since $x'(a + \theta \cdot \Delta t) = f(a + \theta \cdot \Delta t, x(a + \theta \cdot \Delta t))$ and $\|x(a + \theta \cdot \Delta t) - x(a)\| \le d$, it follows that $\|x(t^*) - x(a)\| \le M(t^* - a)$. Now, if $t^* - a < d/M$, then $\|x(t^*) - x(a)\| < d$. But again the continuity of $x(t)$ at t^* would imply that $\|x(t) - x(a)\| \le d$ for some $t > t^*$, contradicting the definition of t^*. Hence $t^* - a \ge d/M$, showing that if a solution exists for t in $[a, b]$, then $(t, x(t)) \in D$ for all t in $[a, b]$. This suggests that we limit our attention to the set $X \subset C[a, b]$ of functions $u = u(t)$ such that $(t, u(t)) \in D$ for $a \le t \le b$. For $u \in X$ it follows by (10.1.5) and

Exercise 10.2b that $v(t) = g(u)$ is also in X. Indeed,

$$\|v(t) - x_0\| \leq \int_a^t \|f(s, u(s))\| \, ds \leq M(t - a) \leq M \cdot d/M = d. \quad (10.1.6)$$

We may write (10.1.4) in the form $x(t) = g(x(t))$. Thus, a solution of (10.1.4) over the interval $[a, b]$ is a fixed point of g in the set X, and conversely. Existence and uniqueness of a solution of (10.1.4) and hence of (10.1.2) can be proved by means of the local fixed-point theorem (Theorem 5.1.2) as follows.

For $h \leq b - a$, $g(u)$ is defined for any $u \in C[a, a + h]$. Let $u_0 = u_0(t) \equiv x_0$ for all t in the interval $[a, a + h]$. Let $\bar{B} = \bar{B}(u_0; d)$ be the closed ball in $C[a, a + h]$ centered at the function u_0 and having radius d defined above. Then $u \in \bar{B}$ if and only if

$$\|u - u_0\|_\infty = \max_{a \leq t \leq a+h} \{\|u(t) - u_0(t)\|\} \leq d,$$

that is, if the graph of $u(t)$ lies in the closed rectangle $|t - a| \leq h$, $\|x - x_0\| \leq d$. From (10.1.6) it follows that $v = g(u)$ is in $\bar{B}$ whenever $u \in \bar{B}$. Thus, g maps $\bar{B}$ into itself. Furthermore, by the Lipschitz condition on f,

$$\|g(u_1) - g(u_2)\|_\infty \leq \int_a^{a+h} \|f(s, u_1(s)) - f(s, u_2(s))\| \, ds$$

$$\leq L \int_a^{a+h} \|u_1(s) - u_2(s)\| \, ds \leq L \|u_1 - u_2\|_\infty h,$$

for any $u_1, u_2 \in \bar{B}$. For $h < \min \{1/L, (b - a)\}$, g is contractive on $\bar{B}$. Now, the image $g(\bar{B})$ is a compact subset of $\bar{B}$. This follows from the boundedness and equicontinuity of $g(\bar{B})$. (See Exercise 1.4.) The equicontinuity is immediate, since for any $v = g(u)$, $u \in \bar{B}$, we have

$$\|v(t_1) - v(t_2)\| \leq \int_{t_2}^{t_1} \|f(s, u(s))\| \, ds \leq M|t_2 - t_1| < \varepsilon$$

whenever $|t_2 - t_1| < \varepsilon/M$. By Theorem 5.1.2, there is a unique $u \in \bar{B}$ such that $u = g(u)$. Let $u(a + h) = x_1$. We now consider the operator

$$g_1(u) = x_1 + \int_{a+h}^t f(s, u(s)) \, ds, \quad a + h \leq t \leq a + h + h_1, h_1 < \min \{1/L, (b - a - h)\}.$$

By the same argument, we obtain a unique fixed point u_1 of g_1. Let $u(t) = u_1(t)$, $a + h \leq t \leq a + h + h_1$. This extends the solution $u(t)$ to the interval $[a, a + h + h_1]$. Continuing in this way, we obtain the unique solution of (10.1.4) in $[a, b]$.

The preceding proof suggests that a numerical solution (x_i) of (10.1.2) can be obtained by replacing the integral in (10.1.4) by a quadrature formula. This leads to a difference equation which approximates the integral equation (10.1.4), or equivalently, the differential equation (10.1.1). For example, if we approximate the derivative dx/dt by the first forward difference $(x(t_i) - x(t_{i-1}))/h$ this suggests replacing

(10.1.1) by the difference equation

$$x_i = x_{i-1} + hf(t_{i-1}, x_{i-1}). \tag{10.1.7}$$

The same equation is suggested by the rectangular rule for quadrature in (10.1.4). If $f(t, x)$ is nonlinear in x, this difference equation is nonlinear. In the above example, this causes no difficulty since x_i is expressed explicitly in terms of x_{i-1}. Hence, if the starting value x_0 is given, x_1, x_2, etc., can be computed as long as f is defined. However, if we consider instead the difference equation

$$x_i = x_{i-1} + \frac{h}{2}(f(t_i, x_i) + f(t_{i-1}, x_{i-1})), \tag{10.1.8}$$

which is suggested by the trapezoidal rule, then x_i occurs on both sides of the equation and we must use some sort of iterative procedure (as in Chapter 5) to solve for x_i. If f is linear in x, we obtain an equation of the form $Ax_i = Bx_{i-1}$, where A and B are known matrices. If A^{-1} exists, we may solve for x_i by one of the methods of Chapter 4. For the purposes of this chapter, it suffices to restrict our attention to linear difference equations of the form

$$\sum_{j=0}^{n} c_j x_{i-j} = b_{i-n}, \quad i > n - 1, \tag{10.1.9}$$

where $c_0 c_n \neq 0$ and (b_i) is a given sequence. Equation (10.1.9) is the general linear difference equation with scalar coefficients c_j. If all $b_i = 0$, the equation is said to be *homogeneous*. If we express x_{i-j} in terms of backward differences the equation can be written in the form

$$\sum_{j=0}^{n} a_j \nabla^j x_i = b_{i-n} \tag{10.1.10}$$

which resembles the linear differential equation $\sum_{j=0}^{n} a_j d^j x/dt^j = b(t)$. Much of the familiar theory of such differential equations carries over to difference equations. In particular, the *characteristic polynomial*

$$p(\xi) = c_0 \xi^n + c_1 \xi^{n-1} + \cdots + c_n \tag{10.1.11}$$

plays a key role in the solution of (10.1.9). (See Exercise 10.12.) We briefly summarize the main results for the case in which x is a scalar. However, since the c_j are scalars, the theory for x of dimension $m > 1$ is essentially the same as for the one-dimensional case. (We simply treat each component of x independently.)

We consider the vector space s of infinite sequences (x_i), $i = 0, 1, \ldots$. (See Section 2.2, Example 1.) We shall denote sequences by capital letters. Thus, $X = (x_i)$. To denote the ith term in a sequence X we shall also write $(X)_i$. Thus, $(X)_i = x_i$. Now, we define the *linear nth-order difference operator* $\mathscr{L}$ on s to s by

$$(\mathscr{L}X)_{i-n} = \sum_{j=0}^{n} c_j x_{i-j}, \quad i \geq n.$$

10.1 Introduction: Differential Equations and Difference Equations

Thus, $Y = \mathscr{L}X$ is the sequence (y_i) having $y_i = \sum_{j=0}^{n} c_j x_{i-j+n}$. If $B = (b_i)$, then we can write (10.1.9) in the form

$$\mathscr{L}X = B.$$

Let us consider the homogeneous equation

$$\mathscr{L}X = 0, \qquad (10.1.12)$$

where 0 is the sequence of zeros. Since $\mathscr{L}$ is obviously a linear operator on s to s, if X and Y are solutions of (10.1.12), then $\alpha X + \beta Y$ is also a solution. More generally, if $X_1, \ldots, X_n$ are solutions of (10.1.12), then any linear combination $\sum_{1}^{n} \alpha_k X_k$ is also a solution. To solve (10.1.9) we must choose n starting values $x_0, x_1, \ldots, x_{n-1}$. These values determine $x_n, x_{n+1}, \ldots$ uniquely. Thus, a general solution involves n arbitrary constants. Suppose

$$X_1 = (x_i^{(1)}), \ldots, X_n = (x_i^{(n)})$$

are sequences in s. The *Wronskian*, $W = W(X_1, \ldots, X_n)$ is the sequence $W = (w_i)$ where the w_i are the determinants,

$$w_i = \begin{vmatrix} x_i^{(1)} & x_i^{(2)} & \cdots & x_i^{(n)} \\ x_{i-1}^{(1)} & x_{i-1}^{(2)} & & x_{i-1}^{(n)} \\ \vdots & & & \\ x_{i-n+1}^{(1)} & x_{i-n+1}^{(2)} & \cdots & x_{i-n+1}^{(n)} \end{vmatrix}.$$

Suppose $X_1, \ldots, X_n$ are n solutions of (10.1.12). If $w_i = 0$ for some i, then there exist $\alpha_1, \ldots, \alpha_n$ not all zero which satisfy the linear system,

$$\sum_{k=1}^{n} \alpha_k x_{i-j}^{(k)} = 0, \qquad j = 0, \ldots, n-1.$$

Hence, from (10.1.9) with $b_i = 0$ it follows that $\sum_{k=0}^{n} \alpha_k x_{i+j}^{(k)} = 0$ for $j = 1, 2, \ldots$, and likewise for $j = -n, -n-1, \ldots$. Therefore $\sum_{k=1}^{n} \alpha_k X_k = 0$ and the X_k are linearly dependent. Conversely, if the X_k are linearly dependent solutions, then $W = 0$; i.e., $w_i = 0$ for all i, since the above linear system has a nontrivial solution for all i.

If $X_1, \ldots, X_n$ are linearly independent solutions, then any solution X is a linear combination of $X_1, \ldots, X_n$; i.e., (10.1.12) has at most n linearly independent solutions. In proof, since $W \neq 0$, there exists $\alpha_1, \ldots, \alpha_n$ such that

$$\sum_{k=1}^{n} \alpha_k x_i^{(k)} = x_i, \qquad 0 \leq i \leq n-1.$$

Since the starting values $x_0, \ldots, x_{n-1}$ determine a solution X uniquely, we must have $X = \sum_{1}^{n} \alpha_k X_k$. Clearly, (10.1.12) has at least n linearly independent solutions. Take the n unit vectors $(1, 0, \ldots, 0), (0, 1, 0, \ldots, 0) \ldots (0, \ldots, 0, 1)$ as starting values, for example. The corresponding solutions are linearly independent. Another linearly independent set is obtained by considering the roots ξ_k of the characteristic polynomial (10.1.11). There are two cases.

First, suppose $\xi_1, \ldots, \xi_n$ are distinct. Then the sequences $X_k = (\xi_k^i)$ are n linearly independent solutions of (10.1.12). In proof, note that X_k is a solution, since

$$\sum_{j=0}^{n} c_j \xi_k^{i-j} = p(\xi_k) \cdot \xi_k^{i-n} = 0.$$

438 Numerical Solution of Ordinary Differential Equations

Forming the Wronskian, we obtain the Vandermonde determinant,

$$\omega_{n-1} = \begin{vmatrix} \xi_1^{n-1} & \xi_2^{n-1} & \cdots & \xi_n^{n-1} \\ \vdots & & & \\ \xi_1 & \xi_2 & & \xi_n \\ 1 & 1 & \cdots & 1 \end{vmatrix} = \prod_{i<j} (\xi_i - \xi_j) \neq 0.$$

Hence, the X_k are linearly independent.

Now, suppose ξ_k has multiplicity $m_k + 1$, where $\xi_1, \ldots, \xi_r$ are the distinct roots of (10.1.11). Then

$$\xi_k^i, \, i\xi_k^{i-1}, \, i(i-1)\xi_k^{i-2}, \ldots, i(i-1)\cdots(i-m_k+1)\xi_k^{i-m_k}, \quad 1 \leq k \leq r \quad (10.1.13)$$

are n linearly independent solutions of (10.1.12). (See Exercise 10.12.)

To solve the nonhomogeneous equation $\mathscr{L}X = B$ with starting values $(x_0, \ldots, x_{n-1})$, say, we first find a particular solution Y with starting values $(0, \ldots, 0)$. Let $X_0, \ldots, X_{n-1}$ be n linearly independent solutions of $\mathscr{L}X = 0$ satisfying the initial conditions $(1, 0, \ldots, 0)$, $(0, 1, 0, \ldots, 0), \ldots, (0, \ldots, 1)$ respectively. Writing $X_{n-1} = (x_i^{(n-1)})$, we see that $x_0^{(n-1)} = \cdots = x_{n-2}^{(n-1)} = 0$ and $x_{n-1}^{(n-1)} = 1$. Now let $Y = (y_i)$ be defined by

$$y_i = \frac{1}{c_0} \sum_{k=0}^{i-n} b_k x_{i-1-k}^{(n-1)}, \quad i \geq n$$

$$y_i = 0, \quad 0 \leq i < n. \quad (10.1.14)$$

For $i \geq n$, we have

$$(\mathscr{L}Y)_{i-n} = \sum_{j=0}^{n} c_j y_{i-j} = \frac{1}{c_0} \sum_{j=0}^{n} c_j \sum_{k=0}^{i-j-n} b_k x_{i-j-k-1}^{(n-1)},$$

$$= \frac{1}{c_0} \sum_{j=0}^{n} c_j \sum_{k=0}^{i-n-1} b_k x_{i-j-k-1}^{(n-1)},$$

since $x_0^{(n-1)} = \cdots = x_{n-2}^{(n-1)} = 0$. Therefore, if we set $x_{-1}^{(n-1)} = 0$,

$$(\mathscr{L}Y)_{i-n} = \frac{1}{c_0} \sum_{k=0}^{i-n} b_k \sum_{j=0}^{n} c_j x_{i-j-k-1}^{(n-1)} = \frac{1}{c_0} \sum_{k=0}^{i-n} b_k (\mathscr{L}X_{n-1})_{i-k-1-n}.$$

But $(\mathscr{L}X_{n-1})_{-1} = c_0 x_{n-1}^{(n-1)} + c_1 x_{n-2}^{(n-1)} + \cdots + c_n x_{-1}^{(n-1)}) = c_0$, as we defined $x_{-1}^{(n-1)} = 0$. Also, $(\mathscr{L}X_{n-1})_{i-k-1-n} = c_0 x_{i-k-1}^{(n-1)} + \cdots + c_n x_{i-k-1-n}^{(n-1)} = 0$,

for $k < i - n$, $i \geq n$. Therefore, $(\mathscr{L}Y)_{i-n} = b_{i-n}$, $i \geq n$, and Y is a solution of the nonhomogeneous equation having 0 as starting values. The sequence

$$X = \sum_{j=0}^{n-1} x_j X_j + Y \quad (10.1.15)$$

is a solution of the nonhomogeneous equation $\mathscr{L}X = B$ having prescribed starting values $x_0, \ldots, x_{n-1}$.

Finally, we discuss the conversion of (10.1.9) to a system of first-order equations, assuming that x is a scalar. Introducing the vector $y = (\eta^{(0)}, \ldots, \eta^{(n-1)})$, we let

$$\eta_i^{(0)} = x_i, \, \eta_i^{(1)} = x_{i-1}, \ldots, \eta_i^{(n-1)} = x_{i-n+1}.$$

Then (10.1.9) can be replaced by the equivalent first-order system of n equations

$$c_0 \eta_i^{(0)} + c_1 \eta_i^{(1)} + \cdots + c_{n-1} \eta_i^{(n-1)} = -c_n \eta_{i-1}^{(n-1)} + b_{i-n},$$
$$\eta_i^{(1)} = \eta_{i-1}^{(0)},$$
$$\eta_i^{(2)} = \eta_{i-1}^{(2)},$$
$$\vdots$$
$$\eta_i^{(n-1)} = \eta_{i-1}^{(n-2)}.$$

In matrix notation this becomes

$$Cy_i = Ay_{i-1} + B_i, \tag{10.1.16}$$

where

$$C = \begin{bmatrix} c_0 & c_1 & \cdots & c_{n-1} \\ 0 & 1 & 0 & \cdots & 0 \\ \vdots & & \ddots & & \vdots \\ 0 & \cdots & & & 1 \end{bmatrix} = \begin{bmatrix} c_0 & c_1 & \cdots & c_{n-1} \\ \hline 0 & & I & \end{bmatrix}$$

$$A = \begin{bmatrix} 0 & \cdots & & & -c_n \\ \hline 1 & \cdots & & 0 & 0 \\ \vdots & \ddots & & & \vdots \\ 0 & \cdots & 1 & & 0 \end{bmatrix} = \begin{bmatrix} 0 & -c_n \\ \hline I & 0 \end{bmatrix}$$

and $B_i = (b_{i-n}, 0, \ldots, 0)$. It is easy to verify that

$$C^{-1} = \begin{bmatrix} 1/c_0 & \dfrac{-c_1}{c_0} & \cdots & \dfrac{-c_{n-1}}{c_0} \\ \hline 0 & & I & \end{bmatrix}$$

which exists, since $c_0 \ne 0$. Similarly, one verifies that

$$C^{-1}A = \begin{bmatrix} \dfrac{-c_1}{c_0} & \dfrac{-c_2}{c_0} & \cdots & \dfrac{-c_{n-1}}{c_0} & \dfrac{-c_n}{c_0} \\ \hline & & I & & 0 \end{bmatrix}$$

Forming $\lambda I - C^{-1}A$, we see that $\det(\lambda I - C^{-1}A) = p(\lambda)$, that is, the characteristic polynomial of $C^{-1}A$ is just the characteristic polynomial (10.1.11). (See Exercise 3.35.) Thus the eigenvalues of $C^{-1}A$ are the roots of the characteristic polynomial of the nth-order difference equation (10.1.9). The solution of the homogeneous linear vector difference equation $Cy_i = Ay_{i-1}$ is immediately seen to be

$$y_i = (C^{-1}A)^i y_0.$$

10.2 NUMERICAL SOLUTION BY NUMERICAL QUADRATURE

A numerical solution of the initial-value problem (10.1.2) is determined by a finite sequence (x_i) of values which approximate the values $x(t_i)$ of the exact solution $x(t)$ on a grid of points t_i in $[a, b]$. For most purposes, it suffices to take $t_i = a + ih$,

$0 \le i \le q$. The step size h can be any real number, but it is often chosen by setting $h = (b - a)/q$, where q is an integer, since then $t_q = b$ is one of the grid points. If the grid points are not equally spaced, we can group them into subintervals of equally spaced points and treat each subinterval separately.

Each of the numerical methods which we shall consider yields a numerical solution $(x_i)_h$ for every step size h. We shall be interested in the accuracy of the numerical solution as a function of h. In particular, we would like the numerical solutions to converge to the exact solution as $h \to 0$. If we use the piecewise linear numerical solution $u_h(t)$ on the grid $G_h = \{a + ih\}$, the convergence is naturally taken to be in the uniform norm on $[a, b]$, that is, $\|u_h(t) - x(t)\|_\infty \to 0$ as $h \to 0$. On the other hand, if we use the numerical solution $(x_i)_h \in V^{(m)}(G_h)$, then by the convergence of $(x_i)_h$ as $h \to 0$ we shall mean that

$$\|T_{G_h} x(t) - (x_i)_h\|_h \to 0,$$

where T_{G_h} is the linear operator which maps $x(t)$ onto the sequence $(x(t_0), \ldots, x(t_q))$ and $\|\cdot\|_h$ denotes the supremum over the grid G_h of $\|x(t_i) - x_i\|$. We shall see that for the methods considered the two types of convergence are equivalent and imply that for any fixed t in $[a, b]$ and any $\varepsilon > 0$, we can choose h sufficiently small so that $\|x(t) - x_i\| < \varepsilon$, where $t = t_i = a + ih$.

If a numerical method yields numerical solutions which converge as $h \to 0$, we shall call the numerical method *convergent*. A large class of convergent methods are based on numerical quadrature, as has already been suggested in Section 10.1. The quadrature methods of Chapter 9 can be used to develop numerical solutions as follows. For the exact solution $x(t)$, we have

$$x(t_i) = x(t_{i-1}) + \int_{t_{i-1}}^{t_i} x'(s)\, ds = x(t_{i-1}) + \int_{t_{i-1}}^{t_i} f(s, x(s))\, ds.$$

Let $L_n(s) = (L_n^{(1)}(s), \ldots, L_n^{(m)}(s))$, where the kth component function $L_n^{(k)}(s)$ is the interpolating polynomial of the kth component of the derivative $x'(s)$ on a set of $n + 1$ distinct points. Then

$$x(t_i) = x(t_{i-1}) + \int_{t_{i-1}}^{t_i} L_n(s)\, ds + R_n(i),$$

where the components of the vector $R_n(i)$ are the remainder functionals described in Chapter 9. Let us suppose that the values of $x(t)$ have somehow been determined on the $n + 1$ preceding points $t_{i-1}, t_{i-2}, \ldots, t_{i-n-1}$, $n \ge 0$. Using the quadrature formula (9.2.5a), we obtain for the value at t_i,

$$x(t_i) = x(t_{i-1}) + h \sum_{j=0}^{n} a_j x'(t_{i-j-1}) + R_n(i), \qquad (10.2.1)$$

where $x'(t_k) = f(t_k, x(t_k))$, $k = i - 1, \ldots, i - n - 1$. This suggests that we may determine a numerical solution (x_i) by means of the recursion or *difference equation*,

$$x_i = x_{i-1} + h \sum_{j=0}^{n} a_j x'_{i-j-1}, \qquad i > n, \qquad (10.2.2)$$

where $x'_k = f(t_k, x_k)$. If $x_{i-1}, \ldots, x_{i-n-1}$ are known, then x_i may be computed directly from (10.2.2). The *starting values* $x_0, \ldots, x_n$ must be obtained by some other method which we shall discuss later. More generally, we have

$$x(t_i) = x(t_{i-r}) + \int_{t_{i-r}}^{t_i} L_n(s)\,ds + R_n(i),$$

$$x(t_i) = x(t_{i-r}) + h \sum_{j=0}^{n} a_j f(t_{i-j-1}, x(t_{i-j-1})) + R_n(i) \qquad (10.2.3)$$

where $L_n(s)$ is the interpolating polynomial of $x'(s)$ on the set $\{t_{i-1}, \ldots, t_{i-n-1}\}$ as before, which gives rise to the difference equation,

$$x_i = x_{i-r} + h \sum_{j=0}^{n} a_j f(t_{i-j-1}, x_{i-j-1}). \qquad (10.2.4)$$

A method given by (10.2.4) is called an *explicit method* (or an *open* formula), since (10.2.4) gives an explicit expression for x_i in terms of preceding values in the sequence. The order of the method is defined to be $n + 1$. If we choose instead to interpolate on the set $\{t_i, \ldots, t_{i-n}\}$, then we obtain a *closed* formula and an *implicit method* of order $n + 1$. As above, using the quadrature formula (9.2.6a) and expressing the differences in terms of the ordinates, we obtain

$$x(t_i) = x(t_{i-r}) + h \sum_{j=0}^{n} b_j f(t_{i-j}, x(t_{i-j})) + R_n(i), \qquad (10.2.5)$$

giving rise to the closed formula,

$$x_i = x_{i-r} + h \sum_{j=0}^{n} b_j f(t_{i-j}, x_{i-j}). \qquad (10.2.6)$$

If $x_{i-1}, \ldots, x_{i-n}$ are known, then x_i can be computed by solving (10.2.6). In contrast to the explicit scheme, x_i now occurs on the right-hand side of the equation as well as the left-hand side. Since $f(t, x)$ is generally nonlinear in x, it would be necessary to solve (10.2.6) by some iterative method. Indeed, for h sufficiently small and f satisfying the Lipschitz condition (10.1.3), (10.2.6) has a unique solution obtainable by successive approximations. (See Exercise 10.13.) The iteration can be avoided by using (10.2.6) in conjunction with (10.2.4) in what is called a *predictor–corrector method* as follows.

$$x_i^{(p)} = x_{i-r_p} + h \sum_{j=0}^{n_p} a_j f(t_{i-j-1}, x_{i-j-1}),$$

$$x_i = x_{i-r} + h \sum_{j=1}^{n} b_j f(t_{i-j}, x_{i-j}) + hb_0 f(t_i, x_i^{(p)}). \qquad (10.2.7)$$

Thus, the open formula is used first to obtain a *predicted value* $x_i^{(p)}$, which then serves on the right-hand side of the closed formula as the initial approximation to compute x_i. The value x_i is called the *corrected value*. By analyzing the error in predictor–corrector methods, one can show that if both formulas have the same order, i.e., if $n = n_p$, it is usually unnecessary to iterate further to solve for x_i in the closed formula. In this case, the first iteration, i.e., producing a corrected value from the predicted value as the initial guess, yields an error in the solution which is asymptotically (i.e., as $h \to 0$) equal to the error obtained by solving the closed formula exactly for x_i. (See [10.1], Section 5.3–7.) Thus, for h sufficiently small, one application of the predictor formula followed by one application of the corrector formula produces the numerical solution x_i for each i. Once the starting values $x_0, \ldots, x_n$ have been obtained, the computation of (10.2.7) can be arranged so that only two evaluations of the derivative are required at each step. First, we compute $f(t_{i-1}, x_{i-1})$ and, using the values $f(t_{i-j-1}, x_{i-j-1})$, $1 \le j \le n$, which were computed in previous steps, we compute $x_i^{(p)}$. We then evaluate $f(t_i, x_i^{(p)})$ and this allows us to compute the corrected value x_i, again using the other previously computed n values of f. In automatic computers, the most recent n values of f are stored in memory as the computation proceeds from step to step. This leads to a convenient computer algorithm which minimizes the number of computations.

We shall mention two specific predictor–corrector methods. First, the *Euler predictor–corrector* method uses a predictor obtained by setting $r_p = 1$ and $n_p = 0$ and a corrector obtained by setting $r = 1, n = 1$.

$$x_i^{(p)} = x_{i-1} + hf(t_{i-1}, x_{i-1}),$$

(10.2.7a) Euler predictor–corrector method

$$x_i = x_{i-1} + \frac{h}{2}\left(f(t_{i-1}, x_{i-1}) + f(t_i, x_i^{(p)})\right).$$

The predictor is simply based on the rectangular rule for numerical quadrature and when used by itself, is called *Euler's method*. The corrector is recognized as the trapezoidal rule. Using quadrature formulas (9.2.5a), (9.2.6a) and taking $r = r_p = 1$ and $n_p = n = 3$ yields the *Adams–Moulton method* of order four. By the calculations in Exercise 10.5c, we obtain

$$\begin{aligned}x_i^{(p)} &= x_{i-1} + \frac{h}{24}\bigl(55f(t_{i-1}, x_{i-1}) \\ &\quad - 59f(t_{i-2}, x_{i-2}) + 37f(t_{i-3}, x_{i-3}) - 9f(t_{i-4}, x_{i-4})\bigr), \\ x_i &= x_{i-1} + \frac{h}{24}\bigl(9f(t_i, x_i^{(p)}) \\ &\quad + 19f(t_{i-1}, x_{i-1}) - 5f(t_{i-2}, x_{i-2}) + f(t_{i-3}, x_{i-3})\bigr).\end{aligned}$$
(10.2.8)

Referring to (10.2.1), (10.2.4) and (10.2.5) we see that if exact values $x_k = x(t_k)$, $i - n - 1 \le k \le i$, are used on the right-hand side of any open or closed formula, then we obtain a quantity $\hat{x}_i$ such that

$$x(t_i) - \hat{x}_i = R_n(i).$$

We shall call $R_n(i)$ the *local* (or one-step) *quadrature error*, since it is the error incurred in one step due to the numerical quadrature and using *exact* values of the derivative, $f(t_k, x(t_k))$. In the Adams–Moulton method, we obtain by evaluating formulas (9.2.5b) and (9.2.6b) with $n = 3$,

$$x(t_i) - \hat{x}_i^{(p)} = \frac{251}{720} h^5 x^{(5)}, \qquad \text{Local quadrature error in} \qquad (10.2.9\text{a})$$
$$\text{Adams–Moulton method}$$

$$x(t_i) - \hat{x}_i = -\frac{19}{720} h^5 x^{(5)}. \qquad (10.2.9\text{b})$$

Thus, a fourth-order method is exact for $x(t)$ which are polynomials of degree ≤ 4. A method of order $n + 1$ is exact for any polynomial $x(t)$ of degree $\leq n + 1$ in t, but not for some polynomial of higher degree. (This could be used as the definition of *order*.) Here, we have assumed the existence of $x^{(5)}$, the fifth derivative of the exact solution $x(t)$. (See the remark following (10.1.1).) We have intentionally omitted the argument α of $x^{(5)}$, since it will have a different value for each component function of the vector $x^{(5)}(t)$, and, of course, will be different in the predictor and corrector. In every case, however, $t_{i-4} < \alpha < t_i$. If h is small, so that $[t_{i-4}, t_i]$ is a small interval we may assume that $x^{(5)}$ does not vary much in this interval. In this case, it is possible to obtain an a posteriori estimate of the local quadrature error which is quite useful in actual computation. Let $\xi(t_i)$ be a single component of the vector $x(t_i)$ and ξ_i the corresponding component of $\hat{x}_i$. Assuming that $\xi^{(5)}$ has the same argument in (10.2.9a) and (10.2.9b), we obtain on subtraction,

$$h^5 \xi^{(5)} = \frac{720}{270} (\xi_i - \xi_i^{(p)}).$$

Substituting back in (10.2.9b), we have

$$\xi(t_i) - \xi_i = -\frac{19}{270} (\xi_i - \xi_i^{(p)}). \qquad (10.2.10)$$

Since $\xi_i^{(p)}$, ξ_i are close to (and may be replaced by) the computed predicted and corrected values, this provides an a posteriori estimate, at each step, of the local quadrature error. In practice, this can be used to choose the step size h. If it is desired to keep the relative local quadrature error smaller than ε, say, then after each step, $\max_\xi \frac{19}{270} |\xi_i - \xi_i^{(p)}|/|\xi_i|$ is compared to ε. If this quantity is greater than ε, then h is halved to increase the accuracy. Halving h introduces a small complication in that new values must be computed at the midpoints of the intervals $[t_{i-k-1}, t_{i-k}]$, $1 \leq k \leq 2$, in order to recompute x_i with the new step size in (10.2.8). These values can be obtained by the same method used to obtain the starting values. (See Section 10.5.)

The methods defined by (10.2.4) and (10.2.6) are sometimes called *multistep* methods when $n > 0$. The Euler method ($r = 1, n = 0$) is an example of a *one-step method* (or *single-step* method) to be discussed in Section 10.5. A more general form of multistep method is given in Section 10.6 by formula (10.6.2). Further generalizations are obtained by evaluating the derivative at intermediate points $t_i + \alpha_i h$, $0 < \alpha_i < 1$. Such methods make use of Runge–Kutta techniques (Section 10.5) and are variously referred to as *hybrid* methods, *generalized Runge–Kutta* methods [10.23] and *generalized predictor–corrector* methods [10.17]. The practical choice of a multistep method must be based on the amount of computation per step, the local accuracy (i.e. the order) and on the convergence and stability properties of the method which govern the propagation of error. We shall discuss the latter two properties in Sections 10.3, 10.4 and 10.6.

10.3 ERROR ANALYSIS AND CONVERGENCE OF QUADRATURE METHODS

Let $x(t)$ be the exact solution of (10.1.2) and assume that $f(t, x)$ satisfies the Lipschitz condition (10.1.3) in the rectangle D described in Section 10.1. If the points (t_{i-j-1}, x_{i-j-1}) are in D for $0 \leq j \leq n$ and x_i is given by (10.2.4), then by Exercise 10.3, $\|x_i - x_{i-r}\| \leq rhM$. If t_i is in $[a, b]$, then the point (t_i, x_i) is in D. The quantity $e_i = x(t_i) - x_i$ will be called the error at t_i. If we subtract (10.2.4) from (10.2.3), we have a difference equation for e_i. Thus,

$$e_i = e_{i-r} + h \sum_{j=0}^{n} a_j [f(t_{i-j-1}, x(t_{i-j-1})) - f(t_{i-j-1}, x_{i-j-1})] + R_n(i).$$

Applying the Lipschitz condition, we find that

$$\|e_i\| \leq \|e_{i-r}\| + hL \sum_{j=0}^{n} |a_j| \|e_{i-j-1}\| + \|R_n\|_h. \qquad (10.3.1)$$

To estimate $\|e_i\|$ we consider the difference equation,

$$m_i = m_{i-1} + (hL \sum_{j=0}^{n} |a_j|) m_{i-1} + \|R_n\|_h, \quad i \geq 1, \qquad (10.3.2)$$

with $\quad m_0 = \max \{\|e_0\|, \ldots, \|e_n\|\} \leq hM.$

Note that $m_i > m_{i-1}$ and m_i is a monotone increasing sequence. It is easy to prove by induction on i that $\|e_i\| \leq m_i$. This certainly holds for $0 \leq i \leq n$ by the definition of m_0 and the monotonicity. Assume it is true for $i < k$. Then

$$\|e_k\| < m_{k-r} + hL \sum_{j=0}^{n} |a_j| m_{k-j-1} + \|R_n\|_h,$$

$$< \left(1 + hL \sum_{0}^{n} |a_j|\right) m_{k-1} + \|R_n\|_h = m_k,$$

10.5 One-Step Methods; Runge–Kutta Methods

where $f^{(3)}(y) = (\partial^3 f_k / \partial \eta_i \partial \eta_j \partial \eta_l)$ is applied to $f(y)$ to obtain matrices A_k, $k = 0, \ldots, m$, having entries

$$\sum_{l=0}^{m} \frac{\partial^3 f_k}{\partial \eta_i \partial \eta_j \partial \eta_l} f_l.$$

Now, assuming that $y_{i-1} = y(t_{i-1}) = y(t)$, we calculate the Taylor series expansion of

$$\frac{\delta y(t)}{h} = \frac{y_i - y(t_{i-1})}{h} = \sum_{0}^{n} a_j k_j \quad \text{for } n = 3.$$

From (10.5.2), we find by expanding in Taylor series

$$\begin{aligned}
k_0 &= f(y), \\
k_1 &= f(y) + \beta_{10} h f'(y) \cdot f(y) \\
&\quad + \tfrac{1}{2} \beta_{10}^2 h^2 f''(y) * (f(y))^2 + \tfrac{1}{6} h^3 \beta_{10}^3 f^{(3)}(y) * (f(y))^3 + O(h^4), \\
k_2 &= f(y) + h \beta_{20} f'(y) \cdot f(y) + h \beta_{21} f'(y) \cdot k_1 \\
&\quad + \frac{h^2}{2} f''(y) * [\beta_{20}^2 (k_0, k_0) + 2 \beta_{20} \beta_{21} (k_0, k_1) + \beta_{21}^2 (k_1, k_1)] \\
&\quad + \frac{h^3}{6} f^{(3)}(y) * (\beta_{20} k_0 + \beta_{21} k_1)^3 + O(h^4), \\
k_3 &= f(y) + h f'(y) \cdot (\beta_{30} k_0 + \beta_{31} k_1 + \beta_{32} k_2) \\
&\quad + \frac{h^2}{2} f''(y) * (\beta_{30} k_0 + \beta_{31} k_1 + \beta_{32} k_2)^2 \\
&\quad + \frac{h^3}{6} f^{(3)}(y) * (\beta_{30} k_0 + \beta_{31} k_1 + \beta_{32} k_2)^3 + O(h^4).
\end{aligned} \quad (10.5.8)$$

Expanding the multilinear forms and replacing k_1 and k_2 by their expressions in terms of $f(y)$, $f'(y)$, $f^{(2)}(y)$ and $f^{(3)}(y)$ and simplifying, we obtain a formula for $\delta y(t)/h$ in terms of powers of h and the derivatives of f.

Setting the coefficients of h^0, h^1, h^2, h^3 equal to the corresponding coefficients in $\Delta y(t)/h$, we obtain a set of nonlinear equations for the a_j and β_{jr}. To simplify these equations it turns out to be convenient to introduce the new coefficients, $\beta_1 = \beta_{10}$, $\beta_2 = \beta_{20} + \beta_{21}$ and $\beta_3 = \beta_{30} + \beta_{31} + \beta_{32}$. The equation for terms of order 0 in $\delta y/h$ and $\Delta y/h$ is easily seen to be

$$\sum_{0}^{n} a_j f(y) = f(y).$$

For terms of order 1 in h we have

$$(a_1 \beta_1 + a_2 \beta_2 + a_3 \beta_3) f'(y) \cdot f(y) = \tfrac{1}{2} f'(y) \cdot f(y).$$

For terms of order 2 in h we have the two equations,

$$\tfrac{1}{2}(a_1\beta_1^2 + a_2\beta_2^2 + a_3\beta_3^2)f''(y)*(f(y))^2 = \tfrac{1}{6}f''(y) + (f(y))^2,$$

$$(a_2\beta_1\beta_{21} + a_3(\beta_{31}\beta_1 + \beta_{32}\beta_2))(f'(y))^2 \cdot f(y) = \tfrac{1}{6}(f'(y))^2 \cdot f(y).$$

Finally, for terms of order 3 in h we have four equations corresponding to the four different terms in $y^{(4)}(t)$ as given in (10.5.7),

$$\tfrac{1}{6}(a_1\beta_1^3 + a_2\beta_2^3 + a_3\beta_3^3) f^{(3)}(y)*(f(y))^3 = \tfrac{1}{24}f^{(3)}(y)*(f(y))^3,$$

$$f''*(f' \cdot f, f)(a_2\beta_1\beta_2\beta_{21} + a_3\beta_3(\beta_1\beta_{31} + \beta_2\beta_{32})) = \tfrac{1}{8}(f''(y))*(f' \cdot f, f),$$

$$(f' \cdot f'' * f^2)(a_2\beta_1^2\beta_{21} + a_3(\beta_1^2\beta_{31} + \beta_2^2\beta_{32})) = \tfrac{1}{12}(f' \cdot f'' * f^2),$$

$$(a_3\beta_1\beta_{21}\beta_{32})(f')^3 \cdot f = \tfrac{1}{24}(f')^3 \cdot f.$$

One solution for these equations yields the following formulas known as the *Runge–Kutta fourth-order method*

$$\left.\begin{aligned}
y_i &= y_{i-1} + \frac{h}{6}(k_0 + 2k_1 + 2k_2 + k_3), \\
k_0 &= f(y_{i-1}), \\
k_1 &= f\left(y_{i-1} + \frac{h}{2}k_0\right), \\
k_2 &= f\left(y_{i-1} + \frac{h}{2}k_1\right), \\
k_3 &= f(y_{i-1} + hk_2).
\end{aligned}\right\} \quad (10.5.9)$$

(Other Runge–Kutta formulas are given in [10.15], [10.6], and in Exercise 10.6.) Since

$$\frac{\delta y(t)}{h} = \frac{y_i - y(t_{i-1})}{h} = \frac{\Delta y(t)}{h} + O(h^4),$$

we see that the one-step error in (10.5.9) is of order h^5, that is, for y_i given by (10.5.9) with $y(t) = y_{i-1} = y(t_{i-1})$, $\delta y(t) - \Delta y(t) = y(t_i) - y_i = O(h^5)$. The quantity $\delta y(t) - \Delta y(t)$ is usually called the *local truncation error*, since it results from truncating the Taylor series expanded about the exact value $y(t)$. (See Exercise 10.9 for estimates of the local truncation error.) Thus, the local truncation error of the Runge–Kutta fourth-order method is of the same order in h as the local quadrature error in the Adams–Moulton method. We may assume that the initial error $y(t_0) - y_0 = 0$, since y_0 is the prescribed initial value (ignoring rounding errors). If the starting values y_1, y_2, y_3 for the Adams–Moulton method are determined by (10.5.9), then they will be subject to local truncation errors of order h^5. However, y_3 will also be subject to an *accumulated error* in y_2. Hence, we should like to estimate the error $e_i = y(t_i) - y_i$ in any value y_i determined from y_0 by successive application of the Runge–Kutta fourth-order method. An upper bound for $\|e\| = \max_i \|e_i\|$ can be obtained by the same kind of analysis used to bound the error in predictor–corrector methods. This is indicated in Exercise 10.7, where it is proved that $\|e_i\|$ is of order h^4, hence is comparable to the Adams–Moulton error. This result also proves that the Runge–Kutta fourth-order method is convergent for any f which satisfies a Lipschitz condition.

Computational Aspects

As in multistep methods, in actual computation with finite-precision numbers round-off is a source of error which can overshadow the accumulated truncation error. Again, the build-up of rounding errors can be reduced considerably by using double-precision incrementing. Thus, the double-precision quantity $\bar{\bar{x}}_{i-1}$ is added in a double-precision addition operation to the single-precision increment

$$(h/6)(\bar{k}_0 + 2\bar{k}_1 + 2\bar{k}_2 + \bar{k}_3)$$

to form the double-precision quantity $\bar{\bar{x}}_i$ corresponding to (10.5.9). Here, the $\bar{k}_i$ are the single-precision numbers representing the k_i in (10.5.9); e.g.,

$$\bar{k}_2 = \bar{f}(t_{i-1} + h/2, \bar{x}_{i-1} + (h/2)\bar{k}_1),$$

where the single bar indicates single-precision computation throughout. Although modern computers can perform double-precision operations quite efficiently, such operations usually do require more computer time and always more computer memory space (to store $2s$-digit numbers rather than s-digit numbers). A method which achieves the same reduction in rounding error without using double-precision is the following [10.13, 10.14].

First, we rearrange the computation in (10.5.9) by introducing intermediate quantities $z_j, q_j, p_j, 0 \le j \le 3$, defined by the equations

$$\left.\begin{aligned} z_0 &= y_{i-1}, \\ q_0 &= 0, \\ p_0 &= hf(z_0), \end{aligned}\right\} \quad (10.5.10)$$

$$\left.\begin{aligned} z_1 &= z_0 + p_0/2, \\ q_1 &= p_0, \\ p_1 &= hf(z_1), \end{aligned}\right\} \quad (10.5.11)$$

$$\left.\begin{aligned} z_2 &= z_1 + p_1/2 - q_1/2, \\ q_2 &= q_1/6, \\ p_2 &= hf(z_2) - p_1/2, \end{aligned}\right\} \quad (10.5.12)$$

$$\left.\begin{aligned} z_3 &= z_2 + p_2, \\ q_3 &= q_2 - p_2, \\ p_3 &= hf(z_3) + 2p_2, \end{aligned}\right\} \quad (10.5.13)$$

$$y_i = z_3 + q_3 + p_3/6. \quad (10.5.14)$$

(Strictly speaking, each of the vectors z_j, q_j, p_j should have a second subscript i to indicate that (10.5.10)–(10.5.14) is carried out for each integration step. We suppress the subscript i for economy of notation.) We shall call (10.5.10)–(10.5.14) the *modified Runge–Kutta method*.

Theorem 10.5.0 The value of y_i in (10.5.14) is equal to the value of y_i given by (10.5.9). Furthermore, (10.5.10)–(10.5.14) can be computed using $3n + B$ intermediate storage registers, whereas (10.5.9) requires $4n + B$, B being some constant and n the order of the differential system.

Proof. Comparing (10.5.9) and (10.5.10), we see that $p_0 = hk_0$. From (10.5.11) it follows easily that $z_1 = y_{i-1} + hk_0/2$, which implies that $p_1 = hk_1$. Since $q_1 = hk_0$, we see that $z_2 = y_{i-1} + hk_0/2 + hk_1/2 - hk_0/2 = y_{i-1} + hk_1/2$. Hence, $p_2 = hk_2 - hk_1/2$. Also, $q_2 = hk_0/6$. From (10.5.13) it then follows that $z_3 = y_{i-1} + hk_1/2 + hk_2 - hk_1/2 = y_{i-1} + hk_2$ and $q_3 = hk_0/6 - hk_2 + hk_1/2$, whence $p_3 = hk_3 + 2hk_2 - hk_1$. Combining these expressions in (10.5.14) we get

$$y_i = (y_{i-1} + hk_2) + (hk_0/6 - hk_2 + hk_1/2) + (hk_3 + 2hk_2 - hk_1)/6$$

$$= y_{i-1} + \frac{h}{6}(k_0 + 2k_1 + 2k_2 + k_3),$$

showing that (10.5.14) is equivalent to (10.5.9).

To save storage in recording the components of the $(n + 1)$-dimensional vectors $z_{j+1}, q_{j+1}, p_{j+1}$, the order of computation should be as follows. First, compute the components of z_{j+1} and q_{j+1} together. Each such component involves only the corresponding component of z_j, q_j and p_j. Hence, as each component of z_{j+1} and q_{j+1} is computed it can be stored in the same register as the corresponding component of z_j and q_j respectively. After all components of z_{j+1} and q_{j+1} have been computed, the components of p_{j+1} can be computed and stored in the same registers as the corresponding components of p_j. Finally, the components of y_i are computed and stored in place of z_3. Thus, $3n + B$ storage registers suffice, B being a constant depending on the amount of temporary storage used for computing the derivatives. In comparison, (10.5.9) requires n storage registers for y_{i-1} which must be preserved throughout each integration step so that the arguments $y_{i-1} + hk_0/2$, $y_{i-1} + hk_1/2$ and $y_{i-1} + hk_2$ can be computed. There are n registers required for these arguments, and n for the k_j. Finally, n registers are required to accumulate y_i, making a total of $4n + B$. This completes the proof. ∎

If all components of the derivative are computed in a separate computation (as when a subroutine is used), then n registers must be provided for these components. In method (10.5.9), these registers can be the same as the ones for the k_j. In (10.5.10)–(10.5.14) an additional n registers are required in this case. Thus, in practice, the saving of n storage registers would usually not be realized. The main advantage of the modified Runge–Kutta method is not in the economy of storage utilization, but rather in the reduction in rounding error which becomes possible by using an idea in [10.13]. To motivate the idea, we recall that the basis of double-precision incrementing is the observation that $hf(y)$ is usually smaller in magnitude than y and causes an appreciable rounding error when single-precision addition is used. Similarly, in the modified method, the p_j and q_j are of order h compared to z_j and y_j. An appreciable rounding error would be incurred if single-precision addition were used in the usual way. By introducing some intermediate quantities r_j, $1 \le j \le 4$, this error can be virtually eliminated, that is, results are of an accuracy comparable to that achieved with double-precision incrementing. Yet, single-precision numbers and operations are used throughout.

For the rest of this section, let us assume that single-precision (decimal) floating-point arithmetic with s-digit precision is used in the actual computation. (See [4.22] for a discussion of floating-point arithmetic.) We shall further assume, unless we say otherwise, that a single-precision number is represented by a numeral $\bar{x}$ in normalized form, that is, by a sequence of $s + k$ digits, where the leftmost k digits represent the exponent and the rightmost s digits the mantissa with a nonzero high-order digit. For m a nonnegative integer, we define the operator R_m ("shift right m places") as follows. For any single-precision numeral $\bar{x}$ the quantity $R_m\bar{x}$ is the numeral having a mantissa obtained by shifting the mantissa of $\bar{x}$ to the right m positions, inserting zeros into positions $1, \ldots, m$ and rounding the result by half-adjusting in the $(s + 1)$-st place (i.e., adding 5 (if $\bar{x} \ge 0$) or subtracting 5 (if $\bar{x} < 0$) in the $(s + 1)$st place and then dropping the digits beyond position s). The exponent of $R_m\bar{x}$ is the same as the exponent of $\bar{x}$. Thus, $R_m\bar{x}$ is not in normalized form. The operator L_m ("shift left m places") is defined similarly; i.e. the mantissa of $L_m\bar{x}$ is obtained by shifting the mantissa of $\bar{x}$ left m places, dropping the digits which are then to the left of the decimal point and inserting zeros in positions $s - m + 1, \ldots, s$. The exponent of $L_m\bar{x}$ is the same as the exponent of $\bar{x}$. For an infinite-precision number x, we define $R_m x = 10^{-m}x$ and $L_m x = 10^m x$. Note that if x is represented by an unnormalized floating-point numeral $\bar{x}$ in which the high-order m digits of the mantissa are zeros, then $L_m\bar{x}$ equals (i.e., represents) $10^m x$. (We shall not henceforth distinguish between numerals and the numbers which they represent.) For a vector $\bar{x}$, $L_m\bar{x}$ applies to each component.

The virtual elimination of the rounding error in the floating-point addition $\bar{z}_0 + \bar{p}_0/2$, where $\bar{p}_0 = hf(\bar{z}_0)$ is assumed to have an exponent which is smaller than the exponent of $\bar{z}_0$, can be accomplished by a rather simple procedure. Observe that the mantissa of $\bar{p}_0/2$ will be shifted to the right m digits (m being the difference in the exponents) before the addition, causing the m (or $m - 1$) low-order digits to be lost. To recapture these digits, we first compute an intermediate quantity by shifting $\bar{p}_0/2$ right m digits and then left m digits, that is, we compute $L_m R_m(\bar{p}_0/2)$. It is this latter quantity which is added to $\bar{z}_0$ rather than $\bar{p}_0/2$. Since the m low-order digits of $L_m R_m(\bar{p}_0/2)$ are zeros, then there will be no rounding error in this addition and the result will be the same as $\bar{z}_0 + \bar{p}_0/2$ with the usual rounding. Next, we recover the lost digits by computing the difference $L_m R_m(\bar{p}_0/2) - \bar{p}_0/2$ and carrying this forward in the computation of $\bar{z}_2$. The same device is used to compute each z_j and finally y_i. At this point, the lost digits are recovered in a quantity $q_{4,i}$ and propagated forward to the next integration step.

The quantities r_j defined below allow us to recapture the digits which are lost during floating-point addition as a result of the shifting of mantissas required by unequal exponents. The lost digits are recovered in the q_j. The exponent m_j of (a component of) z_j is generally greater than the exponents of (corresponding components of) p_{j-1} and q_{j-1}. Let μ_j be the exponents of the quantities in rectangular brackets in the equations below. If $\mu_j < m_j$, we define $R^{(j)} = R_{m_j - \mu_j}$ and $L^{(j)} = L_{m_j - \mu_j}$. If $\mu_j \geq m_j$ we define $R^{(j)} = R^0$ and $L^{(j)} = L_0$. The following equations define the *floating-point modification* of the Runge–Kutta fourth-order method [10.14]:

$$\left.\begin{aligned} z_0 &= y_{i-1}, \\ q_0 &= q_{4,i-1}, \quad (q_{40} = 0), \\ p_0 &= hf(z_0), \end{aligned}\right\} \quad (10.5.10a)$$

$$\left.\begin{aligned} r_1 &= L^{(1)}R^{(1)}[p_0/2 - q_0], \\ z_1 &= z_0 + r_1, \\ q_1 &= 3r_1 - (p_0/2 - q_0), \\ p_1 &= hf(z_1), \end{aligned}\right\} \quad (10.5.11a)$$

$$\left.\begin{aligned} r_2 &= L^{(2)}R^{(2)}[(p_1 - q_1)/2], \\ z_2 &= z_1 + r_2, \\ q_2 &= -r_2 - q_1/3 + p_1/2, \\ p_2 &= hf(z_2) - \tfrac{1}{2}p_1, \end{aligned}\right\} \quad (10.5.12a)$$

$$\left.\begin{aligned} r_3 &= L^{(3)}R^{(3)}[p_2], \\ z_3 &= z_2 + r_3, \\ q_3 &= -r_3 + q_2, \\ p_3 &= hf(z_3) + 2p_2, \end{aligned}\right\} \quad (10.5.13a)$$

$$\left.\begin{aligned} r_4 &= L^{(4)}R^{(4)}[\tfrac{1}{6}p_3 + q_3], \\ y_i &= z_3 + r_4, \\ q_{4,i} &= 3[r_4 - (\tfrac{1}{6}p_3 + q_3)]. \end{aligned}\right\} \quad (10.5.14a)$$

Now, if all quantities are infinite-precision real numbers, then R_m can be replaced by 10^{-m} and L_m by 10^m. If, furthermore, all operations are exact arithmetic operations (i.e., without rounding) on real numbers, then a straightforward calculation shows that (10.5.10a)–(10.5.14a) give precisely the same results as (10.5.10)–(10.5.14). Note that, in this case, $q_{4,i} = 0$ for all i. Also, observe that the r_j require only one additional storage register rather than n. The order of computation should be as follows. First, compute corresponding components of r_j, z_j and q_j for a given j. Then compute all the components of p_j. The computation of a component of z_j and q_j requires only the corresponding component of r_j. Since r_j is not used after the jth stage, one storage register suffices for recording all components of all r_j, $1 \leq j \leq 4$.

In practice, (10.5.10a)–(10.5.14a) would be carried out with finite-precision numerals and rounding errors would accompany the operations. The rounding error in a computed quantity $\bar{u}$ due to the inexact operations in (10.5.10a)–(10.5.14a) will be denoted by $\rho(u)$. Thus, $\rho(u)$ is the local error generated in each integration step i. The total rounding error in $\bar{u}$ is defined as $\varepsilon(u) = u - \bar{u}$, where u is the value which would have been obtained using infinite precision throughout. Thus, $\varepsilon(u)$ arises from the local error ($\rho(u)$ and the rounding errors from preceding integration steps. To estimate the number of correct digits in $\bar{y}_i$, one should estimate $\|\varepsilon(\bar{y}_i)\|/\|\bar{y}_i\|$. This can be done when the values $\bar{y}_i$ are actually computed, if we have a bound for $\|\varepsilon_i(\bar{y}_i)\|$. We shall now show how to estimate such a bound. To do this without obscuring the main ideas, we shall make certain assumptions regarding the rounding errors in (10.5.10a)–(10.5.14a). These assumptions are usually satisfied in practice.

Assumption (i). The step size h is sufficiently small so that in the interval under consideration each component of $hf(y)$ is at least two or three orders of magnitude smaller (in absolute value) than the corresponding component of y. Thus, the exponents of $hf(y)$ are at least m less than the corresponding exponents of y, where $m \geq 2$. (This holds for each component.)

Assumption (ii). In the computation of the z_j, there is no rounding error. This is reasonable, since the low-order digits of the r_j are zeros and the shift during the floating-point addition causes no error. There may be an error in the case of a carry beyond the most significant digit, but this would happen in only a small percentage of the integration steps.

Assumption (iii). There is no rounding error in the computation of the r_j. Actually, the divisions could cause an error. However, this will be small compared to the accumulated errors in p_j and q_j.

Assumption (iv). To simplify the error analysis, we shall take $L^{(j)} = L_m$ and $R^{(j)} = R_m$ for all integration steps i and all j (in all components of the vectors). This is tantamount to limiting our attention to an integration interval in which the exponents of $hf(y)$ remain exactly m less than the corresponding exponents of y. In practice, this condition should hold over fairly long subintervals. Under this assumption it follows easily that

$$L_m R_m(u) - L_m R_m(\bar{u}) = (u - \bar{u})(1 + \alpha),$$

where $|\alpha| \leq 10^{m-s+1}$.

Now, again using the bar to denote actual computed single-precision quantities, we obtain in place of (10.5.11a) under assumptions (ii) and (iii),

$$\begin{aligned}
\bar{r}_1 &= L^{(1)}R^{(1)}[\bar{p}_0/2 - \bar{q}_0], \\
\bar{z}_1 &= \bar{z}_0 + \bar{r}_1, \\
\bar{q}_1 &= 3\bar{r}_1 - \bar{p}_0/2 + \bar{q}_0 - \rho(q_1), \\
\bar{p}_1 &= hf(\bar{z}_1) - \rho(p_1).
\end{aligned} \quad (10.5.11\text{b})$$

Similarly, for the other equations in (10.5.12a)–(10.5.14a), we obtain corresponding "(b)" equations, which we shall not bother to write. Here $\rho(q_1)$ and $\rho(p_1)$ denote the local rounding errors. We make one additional assumption regarding these errors.

Assumption (v). All local rounding errors ρ are such that $\|\rho\| \leq hM \cdot 10^{-s}$, where M is an upper bound for $\|f(y)\|$. Furthermore, we shall take $h = 10^{-m}$, so that $\|\rho\| \leq M \cdot 10^{-s-m}$, where this m is the same as in Assumption (iv).

Subtracting the corresponding "(b)" equations from (10.5.11a)–(10.5.14a), we obtain expressions for the total rounding errors ε as follows:

$$\varepsilon(y_i) = \varepsilon(y_{i-1}) + \sum_{j=1}^{4} \varepsilon(r_j),$$

$$\varepsilon(q_{4i}) = 3[\varepsilon(r_4) - \tfrac{1}{6}\varepsilon(p_3) - \varepsilon(q_3)] + \rho(q_{4i}).$$

(10.5.15)

Applying Assumption (iv), we find by subtracting (10.5.14b) from (10.5.14a) that

$$\varepsilon(r_4) = \left(\tfrac{1}{6}\varepsilon(p_3) + \varepsilon(q_3)\right)(1 + \alpha_3),$$

where $|\alpha_3| \leq 10^{m-s+1}$. Continuing to subtract (b) equations from corresponding (a) equations, we obtain

$$\varepsilon(q_{4i}) = 3\left[\left(\frac{\varepsilon(p_3)}{6} + \varepsilon(q_3)\right)(1 + \alpha_3) - \left(\frac{\varepsilon(p_3)}{6} + \varepsilon(q_3)\right)\right] + \rho(q_{4i})$$

$$= 3\alpha_3\left(\frac{\varepsilon(p_3)}{6} + \varepsilon(q_3)\right) + \rho(q_{4i}).$$

$$\begin{cases} \varepsilon(p_3) = h\,\Delta f(z_3) + 2\varepsilon(p_2) + \rho(p_3), \\ \quad \text{where } \Delta f(z_j) = f(z_j) - f(\bar{z}_j), \quad 0 \leq j \leq 3, \\ \varepsilon(q_3) = -\varepsilon(r_3) + \varepsilon(q_2) + \rho(q_3), \\ \varepsilon(z_3) = \varepsilon(z_2) + \varepsilon(r_3), \\ \varepsilon(r_3) = \varepsilon(p_2)(1 + \alpha_2), \end{cases}$$

$$\begin{cases} \varepsilon(q_2) = -\varepsilon(r_2) - \tfrac{1}{3}\varepsilon(q_1) + \tfrac{1}{2}\varepsilon(p_1) + \rho(q_2), \\ \varepsilon(z_2) = \varepsilon(z_1) + \varepsilon(r_2), \\ \varepsilon(p_2) = h\,\Delta f(z_2) - \tfrac{1}{2}\varepsilon(p_1) + \rho(p_2), \\ \varepsilon(r_2) = [\tfrac{1}{2}\varepsilon(p_1) - \tfrac{1}{2}\varepsilon(q_1)](1 + \alpha_1) \end{cases}$$

$$\begin{cases} \varepsilon(p_1) = h\,\Delta f(z_1) + \rho(p_1), \\ \varepsilon(q_1) = 3\varepsilon(r_1) - \dfrac{\varepsilon(p_0)}{2} + \varepsilon(q_0) + \rho(q_1), \\ \varepsilon(r_1) = (\tfrac{1}{2}\varepsilon(p_0) - \varepsilon(q_0))(1 + \alpha_0), \\ \varepsilon(z_1) = \varepsilon(z_0) + \varepsilon(r_1), \end{cases}$$

$$\varepsilon(p_0) = h\,\Delta f(z_0) + \rho(p_0),$$
$$\varepsilon(q_0) = \varepsilon(q_{4,i-1}),$$
$$\varepsilon(z_0) = \varepsilon(y_{i-1}).$$

10.5 One-Step Methods; Runge–Kutta Methods

Combining these results and denoting by $O(\rho)$ various linear combinations of local rounding errors, we obtain

$\varepsilon(q_1) = (1 + \tfrac{3}{2}\alpha_0)\varepsilon(p_0) - (2 + 3\alpha_0)\varepsilon(q_0) + O(\rho),$

$\varepsilon(p_2) = h\,\Delta f(z_2) - \tfrac{1}{2}h\,\Delta f(z_1) + O(\rho),$

$\varepsilon(q_2) = (\tfrac{1}{6} + \tfrac{1}{2}\alpha_1)(1 + \tfrac{3}{2}\alpha_0)h\,\Delta f(z_0) - (\tfrac{1}{6} + \tfrac{1}{2}\alpha_1)(2 + 3\alpha_0)\varepsilon(q_0) - \tfrac{1}{2}\alpha_1 h\,\Delta f(z_1) + O(\rho),$

$\varepsilon(r_2) = \tfrac{1}{2}h\,\Delta f(z_1)(1 + \alpha_1) - (1 + \tfrac{3}{2}\alpha_0)\tfrac{1}{2}h\,\Delta f(z_0)(1 + \alpha_1) + (\tfrac{1}{2}(2 + 3\alpha_0))(1 + \alpha_1)\varepsilon(q_0) + O(\rho),$

$\varepsilon(r_3) = (h\,\Delta f(z_2) - \tfrac{1}{2}h\,\Delta f(z_1))(1 + \alpha_2) + O(\rho),$

$\varepsilon(q_3) = -(1 + \alpha_2)[h\,\Delta f(z_2) - \tfrac{1}{2}h\,\Delta f(z_1)] + \varepsilon(q_2) + O(\rho),$

$\varepsilon(p_3) = h\,\Delta f(z_3) + 2h\,\Delta f(z_2) - h\,\Delta f(z_1) + O(\rho),$

$\varepsilon(r_4) = \tfrac{1}{6}h(1 + \alpha_3)\,\Delta f(z_3) - (\alpha_2 + \tfrac{2}{3})(1 + \alpha_3)h\,\Delta f(z_2) + [(1 + \alpha_3)h(\tfrac{1}{2}\alpha_2 + \tfrac{1}{3}) - \tfrac{1}{2}\alpha_1]\Delta f(z_1)$
$+ (1 + \alpha_3)(\tfrac{1}{6} + \tfrac{1}{2}\alpha_1)(1 + \tfrac{3}{2}\alpha_0)h\,\Delta f(z_0) - (1 + \alpha_3)(\tfrac{1}{6} + \tfrac{1}{2}\alpha_1)(2 + 3\alpha_0)\varepsilon(q_0) + O(\rho).$

Combining the equations for $\varepsilon(r_j)$, we get

$$\sum_1^4 \varepsilon(r_j) = h\,\Delta f(z_0)(\tfrac{1}{6} + O(\alpha)) + h\,\Delta f(z_1)(\tfrac{1}{3} + O(\alpha)) + h\,\Delta f(z_2)(\tfrac{1}{3} + O(\alpha))$$
$$+ h\,\Delta f(z_3)(\tfrac{1}{6} + O(\alpha)) + \varepsilon(q_{4,i-1})(-\tfrac{1}{3} + O(\alpha)) + O(\rho),$$

where the $O(\alpha)$ terms denote linear combinations of the α_j, $0 \le j \le 3$. Hence,

$$\sum_1^4 \varepsilon(r_j) = \tfrac{1}{6}h(\Delta f(z_0) + 2\,\Delta f(z_1) + 2\,\Delta f(z_2) + \Delta f(z_3)) + hO(\alpha)\sum_0^3 \Delta f(z_j) + (O(\alpha) - \tfrac{1}{3})\varepsilon(q_{4,i-1}).$$

Since f is assumed to satisfy the Lipschitz condition (10.1.3), we have

$\|\Delta f(z_j)\| = \|f(z_j) - f(\bar{z}_j)\| \le L\,\|\varepsilon(z_j)\|,$

$\left\|\sum_1^4 \varepsilon(r_j)\right\| \le \dfrac{hL}{6}(\|\varepsilon(z_0)\| + 2\,\|\varepsilon(z_1)\| + 2\,\|\varepsilon(z_2)\| + \|\varepsilon(z_3)\|) + LhO(\alpha)\sum_0^3 \|\varepsilon(z_j)\|$
$+ \|(O(\alpha) - \tfrac{1}{3})\varepsilon(q_{4,i-1})\|.$

Similarly,

$\|\varepsilon(z_1)\| \le \varepsilon(y_{i-1})\|(1 + \tfrac{1}{2}hL + O(\alpha)) + (1 + O(\alpha))\|\varepsilon(q_{4,i-1})\| + O(\rho),$

$\|\varepsilon(z_2)\| \le \|\varepsilon(y_{i-1})\|(1 + hLO(\alpha)) + hL(1 + O(\alpha))\|\varepsilon(z_1)\| + O(\alpha)\|\varepsilon(q_{4,i-1})\| + O(\rho),$

$\varepsilon(z_3) = \varepsilon(z_1) + \varepsilon(r_2) + \varepsilon(r_3),$

$\|\varepsilon(z_3)\| \le \|\varepsilon(y_{i-1})\|(A_1 + B_1 hL + O(\alpha) + O(hL\alpha))$
$+ (A_2 + B_2 hL + O(\alpha))\|\varepsilon(q_{4,i-1})\| + \|O(\rho)\|,$

where the A_i, B_i are constants. It follows that

$$\left\|\sum_1^4 \varepsilon(r_j)\right\| \le O(hL)\|\varepsilon(y_{i-1})\| + (\tfrac{1}{3} + O(\alpha) + O(hL))\|\varepsilon(q_{4,i-1})\| + hL\,\|O(\rho)\|.$$

From (10.5.15), it follows that

$$\|\varepsilon(y_i)\| \le (1 + O(hL)) \|\varepsilon(y_{i-1})\| + (\tfrac{1}{3} + O(\alpha) + O(hL)) \|\varepsilon(q_{4,i-1})\| + Lh\|O(\rho)\|,$$
$$\|\varepsilon(q_{4i})\| \le hLO(\alpha) \|\varepsilon(y_{i-1})\| + hLO(\alpha) \|\varepsilon(q_{4,i-1})\| + \|O(\rho)\|. \tag{10.5.16}$$

Now, let u_i be the 2-dimensional vector $(\|\varepsilon(y_i)\|, \|\varepsilon(q_{4i})\|)$ and let A be the 2×2 matrix

$$A = \begin{bmatrix} e^{KhL} & \tfrac{1}{2}e^{KhL} \\ \alpha KhL & \alpha KhL \end{bmatrix},$$

where K is a constant such that $O(hL) < KhL$ and $O(\alpha) < K\alpha$. Let $d = (1 + \alpha)KL$. Then

$$\|A\|_1 = e^{KhL} + \alpha KhL < e^{(1+\alpha)KhL} = e^{dh}.$$

Consider the difference equation

$$u_i = Au_{i-1} + c \|\rho\| v, \tag{10.5.17}$$

where c is a constant such that $\|O(\rho)\| < c \|\rho\|$ in (10.5.16), $\|\rho\| < hM \cdot 10^{-s}$ by Assumption (v) and v is the vector $(1,1)$. Since $|\alpha| \le 10^{m-s+1}$ by Assumption (iv) and $m - s + 1 < 3$ in practice, we have $\tfrac{1}{3} + O(\alpha) < \tfrac{1}{2}$. We see that $\|\varepsilon(y_i)\| \le \|u_i\|_1$, since the right-hand side of (10.5.16) is majorized by the right-hand side of (10.5.17). The solution of (10.5.17) is given by

$$u_i = A^i u_0 + c \|\rho\| \sum_0^{i-1} A^j v.$$

Therefore,

$$\|u_i\|_1 \le \|A\|_1^i \|u_0\|_1 + 2c\|\rho\| \sum_0^{i-1} \|A\|_1^j,$$

so that

$$\|u_i\|_1 \le e^{dih} \|u_0\|_1 + 2c\|\rho\| \frac{(1 - e^{dih})}{(1 - e^{dh})}.$$

Since $ih = t_i - t_0 \le b - a$, we have

$$\|\varepsilon(y_i)\| \le e^{d(b-a)} \|u_0\|_1 + C\|\rho\|.$$

Now, $\|u_0\|_1 = \|\varepsilon(y_0)\| + \|\varepsilon(q_{40})\| = \|\varepsilon(y_0)\|$. Thus,

$$\|\varepsilon(y_i)\| \le e^{d(b-a)} \|\varepsilon(y_0)\| + C\|\rho\|. \tag{10.5.18}$$

Since $\varepsilon(y_0)$ is the error (or perturbation) in the initial values and since $\|\rho\|/h < 10^{-s}M$, it follows from (10.5.18) that the floating-point modification of the fourth-order Runge–Kutta method is stable (Definition 10.6.2). Furthermore, for $h = 10^{-m}$ and $\varepsilon(y_0) = 0$, it follows that

$$\|\varepsilon(y_i)\| \le CM10^{-s-m}.$$

Since $C = O(1/h)$, this shows that the accumulated rounding error at the ith step is proportional to the number of steps and to 10^{-s-m}, which is the behavior that would be obtained with a precision of $s + m$ digits, or with double-precision incrementing. We summarize these results in the following theorem.

10.5 One-Step Methods; Runge–Kutta Methods

Theorem 10.5.1 Let $y_i = (t_i, x_i)$ be a numerical solution of (10.1.2) obtained by the floating-point modification (10.5.10a)–(10.5.14a). Let f satisfy the Lipschitz condition (10.1.3). Under Assumptions (i)–(v) stated above, the method is stable and the rounding error in the solution is comparable to the error obtained by using double-precision incrementing in the fourth-order Runge–Kutta method.

As a test of Theorem 10.5.1, the equation $dx/dt = x$ was integrated from $t = 0$ (initial value $x(0) = 1$) to $t = 1$ with a step size $h = 0.1$ and a precision of $s = 6$ digits. After ten steps, the computed value was found to be 2.71828, which differs from the exact value, e, by approximately 1.8×10^{-6}. By comparison, the value obtained by the Runge–Kutta fourth-order method was 2.71831. These and other computational results (not presented here) support the theory.

For higher-order Runge–Kutta formulas the systems of algebraic equations for determining the constants a_i and β_{jr} in formula (10.5.2) are quite complicated and do not have unique solutions. Nevertheless, formulas of order greater than 4 have been available for some time [10.27, 10.28] for a single differential equation. However, it has been proved [10.6, 10.22] that for $n \geq 4$, a Runge–Kutta formula must contain at least $n + 2$ derivative function evaluations if it is to be of order $n + 1$. For example, Nystrom's fifth-order method [10.28] (also see Exercise 10.15) requires six function evaluations per step. In [10.21], the fifth-order case is analyzed with $n = 5$ in formula (10.5.2). Thus again six derivative evaluations are required. (This leads to a system of 17 equations in 21 unknowns. Although the equations are highly nonlinear, the solutions can be expressed as rational functions of certain parameters in all but one case.) In [10.15], a sixth-order formula requiring seven function evaluations is derived. A technique for raising the order without increasing the number of derivative evaluations per step is given in [10.29, 10.31, 10.32]. These *pseudo-Runge–Kutta* methods use a linear combination of values obtained from a standard Runge–Kutta method with the attendant disadvantage that the method is no longer one-step and self-starting. For example, the fifth-order methods in [10.29] replace formula (10.5.2) by a formula of the type

$$y_i = y_{i-1} + h \sum_{j=0}^{n} (a_j k_j(y_{i-1}) + b_j k_j(y_{i-2})),$$

where $k_j(y_{i-1})$ and $k_j(y_{i-2})$ are as in (10.5.2), but evaluated at two successive integration steps. For $n = 3$, this means that only four derivative evaluations are required at each step to achieve fifth-order accuracy. However, to get started, two values, y_0 and y_1, are needed. Of course, y_0 is given but y_1 must be computed by a fifth-order Runge–Kutta method. To obtain values for the parameters a_j, b_j and β_{jr} one expands Δy and $y_i - y_{i-1}$ in a Taylor series at t_{i-1}, as in the derivation of the ordinary Runge–Kutta formulas, and equates coefficients of like powers of h. Convergence is established in [10.32].

Our study of nonconvergence of polynomial interpolation as the degree is increased (Chapter 8) cautions us against raising the order of a method indiscriminately. This approach can actually impair the accuracy of numerical solutions of differential equations in cases where there is a singularity in some higher derivative. In fact, higher-order methods may fail to be convergent and may not be stable. We consider this phenomenon in the next section.

10.6 STABILITY OF NUMERICAL SOLUTIONS BY DIFFERENCE METHODS

In Section 10.4, we analyzed the convergence of numerical solutions obtained by solving the difference equations (10.4.1). Formula (10.4.2), which gives an upper bound for the total error $\|E_i\|$, shows that convergence depends on the quantity $\|R_n\|/h = O(h^{n+1})$ arising from the quadrature error, the quantity ρ/h arising from the rounding errors ρ_i and errors in the starting values. Comparing (10.4.1) with the infinite-precision difference equations (10.2.7), we may regard the ρ_i as a perturbation term, that is, as a sequence which perturbs the infinite-precision solution (x_i) into the actual finite-precision solution $(\bar{x}_i)$. Now, in practice, the value of ρ_i will depend on the precision and on the particular mode of arithmetic rounding that is used. Therefore, it is to be expected that two solutions obtained by two different computers will differ because the values of ρ_i will not be the same. (In fact, this can happen with the same computer if two different orderings of the computational steps are used.) However, if a numerical method is to be of practical value, we must demand that the solution $(\bar{x}_i)$ be relatively insensitive to small changes in the ρ_i, that is, small changes in the ρ_i should produce small changes in $(\bar{x}_i)$. Likewise, we must require that $(\bar{x}_i)$ be relatively insensitive to small differences in the starting values, that is, small errors in the starting values must not be amplified as they propagate through the solution $\bar{x}_i$ for i large. Although we seek the solution $x(t)$ in a bounded interval $[a, b]$, in the numerical method we may encounter large values of the index i as h is made small to secure small quadrature errors. Thus, we must consider the growth of starting errors as $i \to \infty$.

Insensitivity to small changes in the starting values and in the values of ρ_i is referred to as *stability* of the method. A method may be stable for certain functions f and not for others. This is analogous to the notion of the stability of the solutions of differential equations. For example, the solutions of $dx/dt = x$, $x(0) = x_0$, in the infinite interval $[0, \infty)$ are unstable. Since $x_0 e^t \to \infty$ as $t \to \infty$, for an arbitrarily close initial value $x_0 + \delta$, the solution $(x_0 + \delta)e^t$ differs from $x_0 e^t$ by an unbounded amount. On the other hand, $dx/dt = -x$ has solutions $x_0 e^{-t}$ which are stable, since $\delta e^{-t} \to 0$ as $t \to \infty$. The proper mathematical formulation of the concept of stability involves continuity of the solution with respect to certain *perturbations*. In this case, we consider perturbations in the starting values and in the quantities ρ_i of equation (10.4.1). Equation (10.4.1) is a special case of a quite general class of difference equations given by the formula

$$\sum_{j=0}^{n} c_j z_{i-j} = h \sum_{j=0}^{n} a_j f(t_{i-j}, z_{i-j}) + h\omega_i, \qquad i \geq n, \qquad (10.6.1)$$

where the c_j and a_j are prescribed scalar coefficients with $c_0 \neq 0$, $|a_n| + |c_n| \neq 0$. The z_i correspond to the single-precision $\bar{x}_i$ of (10.4.1) and for convenience we have set $\rho_i = h\omega_i$. For a given function f the solution (z_i) of (10.6.1) depends on the starting values $z_0, \ldots, z_{n-1}$ and on the sequence $\omega = (\omega_i)$, $i \geq n$. To emphasize this we shall

10.6 Stability of Numerical Solutions by Difference Methods

sometimes denote the solution by $(z_i)_{z_0, \ldots, z_{n-1}, \omega, h}$. Stability is essentially continuity of (z_i) with respect to $z_0, \ldots, z_{n-1}$ and ω. Note that $(z_i)_{x_0, \ldots, x_{n-1}, 0, h} = (x_i)$, the infinite-precision solution. To be precise in our meaning of continuity, we must specify the topology.

Definition 10.6.1 Let $\sigma = (\sigma_0, \ldots, \sigma_{n-1})$ be a sequence of starting values of (10.6.1). Let $\omega = (\omega_i)$ be a bounded sequence (of vectors). Let

$$\|\sigma\|_\infty = \max_{0 \le i \le n-1} \{\|\sigma_i\|\} \quad \text{and} \quad \|\omega\|_\infty = \sup_i \{\|\omega_i\|\}.$$

The sequence $(\sigma, \omega) = (\sigma_0, \ldots, \sigma_{n-1}, \omega_n, \omega_{n+1}, \ldots)$ is called a *perturbation* of (10.6.1) with norm

$$\|(\sigma, \omega)\|_\infty = \max \{\|\sigma\|_\infty, \|\omega\|_\infty\}.$$

Definition 10.6.2 Let $\mathscr{F}$ be a family of functions $f(t, x)$ defined on the rectangle D as in Section 10.1. The difference equation (10.6.1) is said to be *stable* with respect to $\mathscr{F}$ if for any $f \in \mathscr{F}$ there exists $h_0 = h_0(f)$ such that for any $h < h_0$ the following holds: Given $\varepsilon > 0$ and any perturbation (σ, ω), there exists $\delta = \delta(\varepsilon) > 0$ such that

$$\|(z_i)_{\sigma, \omega, h} - (z_i)_{\sigma + \Delta\sigma, \omega + \Delta\omega, h}\|_\infty < \varepsilon$$

whenever $(\Delta\sigma, \Delta\omega)$ is a perturbation satisfying $\|(\Delta\sigma, \Delta\omega)\|_\infty < \delta$. (Otherwise, the method is said to be *unstable*.)

In other words, (10.6.1) is stable for $f \in \mathscr{F}$, if there exists a step size h_0 such that for all $h < h_0$, the solutions (z_i) of (10.6.1) depend continuously on the starting values and the ω_i. Furthermore, this continuity is uniform with respect to h, that is, the δ which bounds the norm of the perturbation depends only on the given ε and not on h. This is possible because we have represented the rounding error as $\rho_i = h\omega_i$, so that holding $\|\omega_i\|_\infty < \delta$ allows $\|\rho_i\|_\infty$ to approach 0 as $h \to 0$. From our analysis of convergence of the predictor–corrector methods (10.4.1), we know that if we make $\rho/h \to 0$ as $h \to 0$, then we obtain convergence of (x_i) to $x(t)$. Thus, we might expect a stable method to converge, by continuity at $h = 0$, $\omega = 0$. On the other hand, the error σ in the starting values does not include the factor h. Therefore, if (10.6.1) is stable, we can expect that starting errors will not be amplified as they propagate through the sequence (z_i) regardless of how large i becomes as $h \to 0$. Again, this behavior is observed in the error analysis of the convergent predictor–corrector methods where the effect of the starting errors is represented by the quantities m_0 and μ_0 in (10.4.2).

These observations suggest a close relationship between stability and convergence. To make this precise, let us state the definition of convergence for method (10.6.1). (See Theorem 10.3.3.)

Definition 10.6.3 The difference method (10.6.1) is said to be *convergent* for a given initial-value problem if

$$\|(z_i)_{\sigma_0, \ldots, \sigma_{n-1}, \omega, h} - x(t_i)\|_\infty \to 0$$

as $\max \{h, \|(\omega_i)\|_\infty, \|\sigma_0 - x(t_0)\|, \ldots, \|\sigma_{n-1} - x(t_{n-1})\|\} \to 0.$

If in (10.6.1) we take all $\omega_i = 0$ (the case of infinite precision), we obtain the equation

$$\sum_{j=0}^{n} c_j z_{i-j} - h \sum_{j=0}^{n} a_j f(t_{i-j}, z_{i-j}) = 0. \tag{10.6.2}$$

Convergence is then a question of continuous dependence of (z_i) on initial values provided that the exact solution $x(t)$ satisfies (10.6.2) with accuracy $\varepsilon(h)$, where $\varepsilon(h) \to 0$ as $h \to 0$. (See Definition 10.6.4 below.)

It would appear that a stable method given by (10.6.1) must be convergent provided that (10.6.1) represents the differential equation (10.1.1) arbitrarily closely as $h \to 0$. We shall see that this is indeed the case. A little more surprising is the truth of the converse, that a convergent method of type (10.6.1) is stable provided it faithfully represents the differential equation. The definition of convergence almost implies continuity of (z_i) at $\omega_i = 0$, $\sigma_i = 0$, since we must have $(z_i) \to x(t)$ as h, ω_i and σ_i approach zero. However, in the definition of stability we are also dealing with $h \neq 0$. Furthermore, Eq. (10.6.1) is generally nonlinear (i.e., f is nonlinear in x) and we cannot immediately conclude that we have continuity at any (σ, ω) if we have it at $\sigma = 0$, $\omega = 0$. Indeed, in Definition 10.6.2 we are careful to say that the continuity of the solution $(z_i)_{\sigma, \omega, h}$ with respect to σ, ω need hold only for $h < h_0$ sufficiently small. Furthermore, if the value of h_0 required to achieve stability is very small compared to the step size required for an acceptable quadrature error, then the method may be impractical because of the excessive amount of computation and precision required.

The connection between the three sources of error, starting errors, rounding errors and local quadrature (or truncation) errors, for the methods defined by (10.6.1) can be clarified by introducing the concept of *consistency*.

Definition 10.6.4 Let $x(t_i)$ be the values of an exact solution of the differential equation (10.1.1). The quantity

$$\sum_{j=0}^{n} c_j x(t_{i-j}) - h \sum_{j=0}^{n} a_j f(t_{i-j}, x(t_{i-j})) = \tau_i \tag{10.6.3}$$

is called the *local truncation error* of the difference method defined by (10.6.1). Let $\|\tau\|_\infty = \max_i \|\tau_i\|$. If $\|\tau\|_\infty/h \to 0$ as $h \to 0$, then we say that the difference equation (10.6.1) is *consistent* with the differential equation (10.1.1). If the consistency relation holds for all f in a class $\mathscr{F}$, we say that (10.6.1) is *consistent* with respect to $\mathscr{F}$.

For example, the open and closed quadrature methods in (10.2.4) and (10.2.6) are consistent with (10.1.1) for f sufficiently differentiable, since

$$\|\tau_i\|_\infty/h = \|R_n(i)\|_\infty/h = O(h^{n+2})/h \to 0 \qquad \text{as } h \to 0.$$

Examples of inconsistent methods can easily be contrived. (See Exercise 10.11). Since there is no point to considering inconsistent methods, we shall assume that the coefficients in (10.6.1) are chosen so that all methods considered henceforth are consistent. We shall also assume that all functions $f(t, x) \in \mathscr{F}$ are Lipschitz continuous in x.

10.6 Stability of Numerical Solutions by Difference Methods

For consistent methods (10.6.1) the fundamental result [10.8, 10.9] is that convergence implies stability and conversely. Furthermore, a criterion for convergence and stability is the following.

Definition 10.6.5 The difference scheme (10.6.1) is said to satisfy the *root condition* if the polynomial

$$p(\xi) = c_0\xi^n + c_1\xi^{n-1} + \cdots + c_{n-1}\xi + c_n, \qquad (10.6.4)$$

has all of its zeros in or on the unit circle and those on the unit circle have multiplicity 1; i.e., $p(\xi_j) = 0$ implies $|\xi_j| \leq 1$ and if $|\xi_j| = 1$ then ξ_j has multiplicity 1.

We note that $p(\xi)$ is the characteristic polynomial of the homogeneous difference equation which results from setting $h = 0$ in (10.6.1). To make the fundamental result more plausible, we present an example of a consistent method which is neither stable nor convergent. Such unsatisfactory methods arise most frequently from attempts to approximate dx/dt by the derivative of the interpolating polynomial of $x(t)$, that is, by applying numerical differentiation to the differential equation (10.1.1), rather than numerical quadrature. (It is easily verified that the difference schemes (10.2.4) and (10.2.6) satisfy the root condition.) Referring to Exercise 8.13, we may use Gregory's formula to obtain approximations to the derivative of arbitrary order of accuracy. The simplest approximation is the first forward difference, which yields as the difference equation approximating (10.1.1),

$$x_i - x_{i-1} = hf(t_{i-1}, x_{i-1}).$$

According to (8.4.3), this is a consistent difference method for differentiable f. (See Exercise 10.11b.) In fact, it is recognized as Euler's method, which we have shown to be a convergent quadrature method. A higher-order approximation is given by formula (8.4.1), which leads to the consistent difference equation,

$$-\tfrac{1}{2}x_i + 2x_{i-1} - \tfrac{3}{2}x_{i-2} = hf(t_{i-2}, x_{i-2}) + h\omega_i. \qquad (10.6.5)$$

This particular instance of (10.6.1) has associated with it the polynomial

$$p(\xi) = -\tfrac{1}{2}\xi^2 + 2\xi - \tfrac{3}{2},$$

which has roots $\xi_1 = 3$ and $\xi_2 = 1$. Hence, (10.6.5) does not satisfy the root condition. Since (10.6.5) is consistent with (10.1.1) for f twice differentiable, let us consider the initial-value problem

$$\frac{dx}{dt} = x, \qquad x(0) = 1, \qquad (10.6.6)$$

which has the solution $x(t) = e^t$. The approximating difference equation is

$$-\tfrac{1}{2}x_i + 2x_{i-1} - \tfrac{3}{2}x_{i-2} = hx_{i-2}, \qquad i \geq 2, \qquad (10.6.7)$$

with starting value $x_0 = 1$. As the second starting value, let us choose $x_1 = e^h$, which is the value of the exact solution at $t_1 = h$. The roots of the characteristic polynomial $p(\xi) = -\tfrac{1}{2}\xi^2 + 2\xi - (\tfrac{3}{2} + h)$ are $\xi_1 = 2 + \sqrt{1 - 2h}$ and $\xi_2 = 2 - \sqrt{1 - 2h}$.

We see that $\xi_1 = 3 - h + O(h^2)$ and $\xi_2 = 1 + h + h^2/2 + O(h^3)$. We find that for $t = ih$,

$$x_1 - x_0 \xi_2 = e^h - \xi_2 = -\tfrac{11}{6} h^3 + O(h^4),$$
$$x_0 \xi_1 - x_1 = \xi_1 - e^h = 2 + O(h),$$
$$\xi_1^i = 3^i (e^{-t/3} + O(h)),$$
$$\xi_2^i = e^t + O(h).$$

Therefore, by the result in Exercise 10.10e, we have for the solution

$$x_i = \frac{1}{2\sqrt{1-2h}} 3^i (e^{-t/3} + O(h))(-\tfrac{11}{6} h^3 + O(h^4)) + \frac{1}{2\sqrt{1-2h}} (e^t + O(h))(2 + O(h)).$$

Holding $t = ih$ fixed and allowing $h \to 0$, we see that the second term converges to the exact solution e^t. The first term behaves asymptotically like $-\tfrac{11}{12} t^3 e^{-t/3} (3^i/i^3)$ as $i = t/h \to \infty$. Hence $x_i \to -\infty$. Since the starting values $(1, e^h)$ are exact for all h and since we have taken $\omega_i = 0$, we see that (10.6.7) is not convergent for the initial-value problem (10.6.6). The same analysis shows that the solution x_i is not stable with respect to the starting values x_0, x_1. For example, if $x_1 = e^h + h^2$, then we obtain a solution $\tilde{x}_i$ of (10.6.7) which behaves asymptotically like $\tfrac{1}{2} t^2 e^{-t/3} 3^i/i^2$ as $i \to \infty$, $ih = t$ fixed. Clearly, $x_i - \tilde{x}_i \to -\infty$ as $h \to 0$ while the perturbance h^2 in x_1 also approaches zero. Therefore, Definition 10.6.2 cannot be satisfied for any h_0.

The above example suggests that the root condition is necessary for both convergence and stability of a difference scheme. Further insight into the situation is gained by analyzing the solution of a general second-order linear homogeneous difference scheme applied to (10.6.6). This gives rise to the equation,

$$ax_i + bx_{i-1} + cx_{i-2} = hx_{i-2}. \tag{10.6.8}$$

The solution given in Exercise 10.10e is

$$x_i = \frac{1}{\xi_1 - \xi_2} (\xi_1^i (x_1 - x_0 \xi_2) + \xi_2^i (x_0 \xi_1 - x_1)), \tag{10.6.9}$$

where ξ_1 and ξ_2 are the roots of $p_h(\xi) = a\xi^2 + b\xi + c - h$. We shall assume that the scheme under consideration is consistent with $dx/dt = 0$. It follows that 1 must be a root of $p(\xi) = a\xi^2 + b\xi + c = 0$. (See Exercise 10.11a.) (This is a reasonable assumption, since any practical difference scheme should be consistent with all equations $dx/dt = f(t, x)$ such that f is Lipschitz continuous in x.) Let λ be the other root of $p(\xi)$. Then λ must be real. We have, as in the above example,

$$\xi_i = \frac{-b \pm \sqrt{b^2 - 4ac + 4ah}}{2a} = \frac{-b \pm \sqrt{b^2 - 4ac}}{2a} + O(h).$$

Thus, we may set $\xi_1 = \lambda + O(h)$ and $\xi_2 = 1 + O(h)$ obtaining

$$\xi_1^i = \lambda^i (e^{Kt} + O(h)), \qquad \xi_2^i = e^t + O(h),$$

10.6 Stability of Numerical Solutions by Difference Methods

where K is a constant. If the root condition is satisfied, $|\lambda| \leq 1$ and $\lambda \neq 1$. If $|\lambda| < 1$, then $\xi_1^i \to 0$ as $i \to \infty$. Also, $\xi_2^i \to e^t$ and

$$(x_0\xi_1 - x_1)/(\xi_1 - \xi_2) \to (1 \cdot \lambda - e^h)/(\lambda - 1) \to 1$$

provided that the starting values converge to the exact values, as required in the definition of convergence. If $\lambda = -1$, then $\xi_1^i \to (-1)^i e^{Kt}$ asymptotically. But $x_1 - x_0\xi_2 \to 0$, since $\xi_2 \to 1$ and $x_1 \to e^h$. Therefore, by (10.6.9) $x_i \to e^t$ and the root condition is sufficient for convergence. If $|\lambda| > 1$, $|\xi_1^i| \to \infty$ so that x_i diverges from e^t as $i \to \infty$, $ih = t$. If $\lambda = 1$, then $b^2 - 4ac = 0$ and $\xi_i = 1 + O(h^{1/2})$. Again, $e^t - x_i \to \infty$. Thus the root condition is necessary.

We shall now establish the result for the general scheme (10.6.1) and initial-value problems (10.1.2) in which f satisfies the Lipschitz condition (10.1.3) in $[a, b] \times V^m$ (Section 10.1). We denote the class of such f by Lip_x. Note that if (10.6.1) is convergent for all initial-value problems in which $f \in \mathrm{Lip}_x$, then it converges for the initial-value problem

$$dx/dt = 0, \qquad x(a) = 0. \qquad (10.6.10)$$

When we set $\omega_i = 0$ in (10.6.1), the solution x_i must converge to the exact solution $x(t) \equiv 0$ as $h \to 0$ and the starting values approach 0. Suppose $p(\xi) = 0$ has a root $|\xi_1| > 1$. If ξ_1 is real, set $x_i = h\xi_1^i$. If ξ is complex, set $x_i = h(\xi_1^i + \bar{\xi}_1^i)$. If $|\xi_1| = 1$ and the multiplicity is > 1, then $x_i = hi\xi_1^i = t\xi_1^i$, is a solution. In all cases, x_i is a solution of (10.6.1) with $\omega_i = 0$ and $f = 0$. Furthermore, the starting values $x_0, \ldots, x_n$ are of order h and therefore they approach 0 as $h \to 0$. However, for fixed $t = ih$, $x_i \to \infty$ as $i \to \infty$. Hence, the numerical solution does not converge to the exact solution. We have proved

Theorem 10.6.1 If the difference scheme (10.6.1) converges for any initial-value problem in which f satisfies the Lipschitz condition (10.1.3), then it must satisfy the root condition.

A similar argument shows that if the root condition is not satisfied, then (10.6.1) cannot be stable for the class Lip_x. We again use the initial-value problem (10.6.10). If $|\xi_1| > 1$ and ξ_1 is a real root of $p(\xi) = 0$, then $x_i = h\xi_1^i$ and $y_1 = 2h\xi_1^i$ are solutions of (10.6.1) with $\omega_i = 0$ and $f = 0$. But $y_i - x_i = h\xi_1^i \to \infty$ as $h \to 0$ with $i = t/h$. Yet,

$$\max_{0 \leq i \leq n} |x_i - y_i| \to 0 \qquad \text{as } h \to 0.$$

Thus, for any h_0, we can find $h < h_0$ and two sets of starting values which differ by less than any $\delta > 0$ such that the corresponding solutions at $i = t/h$ differ by an arbitrarily large amount. For ξ_1 complex, replace ξ_1^i by $(\xi_1^i + \bar{\xi}_1^i)$. For $|\xi_1| = 1$ and multiplicity > 1, $|x_i - y_i| = |t| |\xi_1|^i$ does not approach zero. This proves the following:

Theorem 10.6.2 If (10.6.1) is stable with respect to the class Lip_x, then it must satisfy the root condition.

We have shown that the root condition is necessary for convergence; now we establish the sufficiency.

Theorem 10.6.3 Let $f \in \text{Lip}_x$. Let (10.6.1) be consistent with the differential equation (10.1.1). If the root condition is satisfied, then (10.6.1) is convergent

Proof. Since the difference scheme is consistent, (10.6.3) holds with $\tau_i/h \to 0$. Subtracting (10.6.1) from (10.6.3), we obtain a difference equation for the error $e_i = x(t_i) - z_i$ as follows:

$$\sum_{j=0}^{n} c_j e_{i-j} = b_{i-n}, \qquad i > n, \tag{1}$$

where

$$b_{i-n} = h \sum_{j=0}^{n} a_j [f(t_{i-j}, x(t_{i-j})) - f(t_{i-j}, z_{i-j})] + \tau_i - h\omega_i. \tag{2}$$

The solution of (1), as given by formulas (10.1.14), (10.1.15), is (for each component $\hat{e}_i$ of e_i)

$$\hat{e}_i = \left(\sum_{j=0}^{n-1} \hat{e}_j X_j\right)_i + \frac{1}{c_0} \sum_{k=0}^{i-n} \hat{b}_k \hat{e}_{i-1-k}^{(n-1)}, \tag{3}$$

where $X_0, \ldots, X_{n-1}$ are the solutions of the homogeneous equation (10.1.12) satisfying the initial conditions $(X_j)_k = \delta_{jk}$, $j, k = 0, \ldots, n-1$. We write $X_j = (e_i^{(j)})$. A set of linearly independent solutions of this homogeneous equation is given by (10.1.13). Since the root condition is satisfied, either $|\xi_k| < 1$ with multiplicity $m_k \geq 1$ or $|\xi_k| = 1$ and $m_k = 1$. Therefore, $|\xi_k^i| \leq 1$ and $|i\xi_k^{i-1}| \to 0$ as $i \to \infty$ for all solutions in (10.1.13). Since $X_0, \ldots, X_{n-1}$ are linear combinations of the solutions in (10.1.13), they must be bounded also, that is, $\|e_i^{(j)}\| < M$ for all i, $0 \leq j \leq n-1$. It is convenient to take the bound $M > 1$.

Since $f \in \text{Lip}_x$, we obtain from (2) above (using the sup norm),

$$\|b_{i-n}\| \leq hL \sum_{j=0}^{n} \|e_{i-j}\| |a_j| + \|\tau_i\| + h\|\omega_i\| \tag{4}$$

$$\leq (n+1)hLA\mu_i + \|\tau\|_\infty + h\|\omega\|_\infty$$

where $\mu_i = \max_{0 \leq k \leq i} \{\|e_k\|\}$, $A = \max_j \{|a_j|\}$, $\|\tau\|_\infty = \max_i \|\tau_i\|$ and $\|\omega\|_\infty = \sup_i \|\omega_i\|$. From (3) it follows that for any $i > n$,

$$\|e_i\| \leq nM\mu_{n-1} + \frac{M}{c_0}(i+1-n) \max_{0 \leq k \leq i-n} \{\|b_k\|\}.$$

Using (4), we obtain

$$\|e_i\| \leq nM\mu_{n-1} + \frac{M}{c_0}(i+1-n)h[(n+1)LA\mu_i + \frac{\|\tau\|_\infty}{h} + \|\omega\|_\infty]$$

$$\leq nM\mu_{n-1} + Kih\mu_i + \frac{Mih}{c_0}\left(\frac{\|\tau\|_\infty}{h} + \|\omega\|_\infty\right),$$

10.6 Stability of Numerical Solutions by Difference Methods

where $K = M(n + 1)LA/c_0$. For any $j \leq i$, $\mu_j \leq \mu_i$. Hence for $j \leq i$,

$$\|e_j\| \leq nM\mu_{n-1} + Kih\mu_i + \frac{Mih}{c_0}\left(\frac{\|\tau\|_\infty}{h} + \|\omega\|_\infty\right). \tag{5}$$

Since $\mu_i = \|e_j\|$ for some $j \leq i$, this implies

$$\mu_i \leq Kih\mu_i + nM\mu_{n-1} + \frac{Mih}{c_0}\left(\frac{\|\tau\|_\infty}{h} + \|\omega\|_\infty\right).$$

For $ihK \leq \frac{1}{2}$ (that is, $n \leq i \leq [1/(2Kh)]$, where [ ] denotes the integer part),

$$\mu_i \leq 2M\left(n\mu_{n-1} + \frac{1}{Kc_0}\left(\frac{\|\tau\|_\infty}{h} + \|\omega\|_\infty\right)\right). \tag{6}$$

As $h \to 0$, $\|\tau\|_\infty/h \to 0$ by consistency. Further, as $z_j - x(t_j) \to 0$, $0 \leq j \leq n - 1$, $\mu_{n-1} \to 0$. Hence, by (6), we see that as

$$\max\{h, \max_{0 \leq j \leq n-1}\{\|z_j - x(t_j)\|, \|\omega\|_\infty\}\}$$

approaches zero, so does μ_i approach zero. Hence the numerical solution converges for all t in $[t_0, t_0 + 1/2K]$.

Now, for h sufficiently small, we may consider $z_{[1/2Kh]-j}$, $0 \leq j \leq n - 1$ as starting values for the interval $[t_0 + [1/2Kh]h, t_0 + 2/2K]$. The preceding proof establishes convergence in this interval, since the starting values converge to the exact values as $h \to 0$. Since K is independent of h, we need only continue this procedure a finite number of times to establish convergence in the entire interval $[a, b]$. ∎

Theorem 10.6.4 If (10.6.1) satisfies the root condition, it is stable with respect to Lip$_x$.

Proof. Let $(z_i) = (z_i)_{z_0,\ldots,z_n,\omega,h}$ and $(\tilde{z}_i) = (z_i)_{z_0+\sigma_0,\ldots,z_n+\sigma_n,\omega+\Delta\omega,h}$ be two solutions of (10.6.1) for two sets of starting values and rounding errors. Let $e_i = z_i - \tilde{z}_i$. As in the preceding theorem, we find that

$$\sum_{j=0}^n c_j e_{i-j} = b_{i-n}, \tag{1}$$

where now

$$b_{i-n} = h\sum_{j=0}^n [f(t_{i-j}, z_{i-j}) - f(t_{i-j}, \tilde{z}_{i-j})] + h\Delta\omega_i.$$

The solution of (1) is again given by (3) of the preceding proof. We have, corresponding to (4) in that proof,

$$\|b_{i-n}\| \leq (n + 1)hLA\mu_i + h\|\Delta\omega\|_\infty,$$

and corresponding to (5)

$$\|e_j\| \leq nM\mu_{n-1} + Kih\mu_i + \frac{Mih}{c_0}\|\Delta\omega\|_\infty.$$

472 Numerical Solution of Ordinary Differential Equations

This yields, corresponding to (6), for $i \leq [1/2Kh]$,

$$\mu_i \leq 2M\left(n\mu_{n-1} + \frac{1}{Kc_0}\|\Delta\omega\|_\infty\right). \tag{10.6.11}$$

But

$$\mu_{n-1} = \max_{0 \leq j \leq n-1}\{\|\sigma_j\|\} = \|\sigma\|.$$

Hence μ_i can be made arbitrarily small by choosing the norm of the perturbation, $\max\{\|\sigma\|, \|\Delta\omega\|_\infty\}$, sufficiently small. We can then apply the same argument to

$$[1/2Kh] \leq i \leq [2/2Kh],$$

using the starting values $z_{[1/2Kh]-j}$ and $\tilde{z}_{[1/2Kh]-j}$, $0 \leq j \leq n-1$, and derive formula (10.6.11) for this range of values of i. For $h < h_0$, h_0 sufficiently small, $[1/2Kh] \geq n$, so that there are enough starting values to continue the argument. The argument must be repeated only a finite number of times independently of h, since $[(b-a)/h]/[1/2Kh] < 4K(b-a)$. This completes the proof. ∎

Remark (10.6.11) actually shows that the solution is Lipschitz continuous with respect to a perturbation $(\sigma, \Delta\omega)$ provided $h < h_0$ and h_0 is sufficiently small.

Computational Aspects

Consider the initial-value problem

$$dx/dt = -x, \quad 0 \leq t \leq 5,$$
$$x(0) = 1. \tag{10.6.12}$$

Replacing $x'(t_{i-1})$ by the centered difference $(x_i - x_{i-2})/2h$ as in (8.4.5), we obtain the second-order stable difference equation,

$$x_i + 2hx_{i-1} - x_{i-2} = 0. \tag{10.6.13}$$

The characteristic equation is

$$\xi^2 + 2h\xi - 1 = 0.$$

The roots are

$$\xi_1 = -h + \sqrt{1+h^2} = 1 - h + h^2/2 + O(h^4) = e^{-h} + O(h^3),$$
$$\xi_2 = -h - \sqrt{1+h^2} = -e^h + O(h^3).$$

As in the previous example of the initial-value problem (10.6.6), we take exact starting values $x_0 = 1$ and $x_1 = e^{-h}$. This gives

$$x_1 - x_0\xi_2 = e^{-h} + e^h + O(h^3) = 2 + h^2 + O(h^3),$$
$$x_0\xi_1 - x_1 = e^{-h} + O(h^3) - e^{-h} = O(h^3) = h^3/6 + O(h^4).$$

10.6 Stability of Numerical Solutions by Difference Methods

For $t = ih$, we obtain

$$\xi_1^i = (e^{-h} + O(h^3))^i = e^{-t}(1 + Ate^h h^2 + O(h^4)),$$
$$\xi_2^i = (-1)^i e^t(1 - Ate^h h^2 + O(h^4)).$$

where A is a constant. By Exercise 10.10e, the solution is

$$x_i = \frac{1}{2\sqrt{1+h^2}}\left[2e^{-t}(1 + Ate^h h^2) + h^2 e^{-t} + (-1)^i \frac{h^3 e^t}{6} + O(h^3)\right].$$

Asymptotically, as $h \to 0$, the solution consists of the exact solution e^{-t}, an error term $O(h^2)$ attributable to the truncation error and an error term $((-1)^i h^3 e^t/6)/2$. As an exercise, the reader should compute the solution of (10.6.13) with $x_0 = 1$, $x_1 = e^{-h}$ and $h = 0.1$. If the computation is carried out with a precision of 10 decimal digits, the value at $t = 5$ will be 0.01803 rounded to five places. The exact value is 0.00674 rounded to five places. The error is -0.01129. The truncation error is of the order $h^2 t e^{-t} = 3.5 \times 10^{-4}$. Thus the predominant error is due to the term

$$-(-1)^i h^3 e^t/12 = (-10^{-3})148.4/12 = -0.012,$$

which arises from what may be called the "extraneous solution," ξ_2^i, of (10.6.13), since it does not correspond to a solution of the differential equation. There will always be extraneous solutions when the order of the difference equation is greater than the order of the differential equation.

For comparison, we consider the first-order difference scheme (derived by replacing dx/dt by the first forward difference or, as we know, by the Euler method)

$$x_i = (1 - h)x_{i-1}. \qquad (10.6.14)$$

The solution of this equation, with starting value $x_0 = 1$, is

$$x_i = (1 - h)^i = e^{-t}(1 - te^h h/2 + O(h^2)).$$

The error is a truncation error and is, asymptotically, $te^{-t}h/2$. Taking $t = 5$ and $h = 0.1$ as before, this quantity is about $5(0.00674)(0.05) = 0.0017$ rounded. In fact $x_{50} = (0.9)^{50} = 0.00513$ rounded. Although (10.6.13) has a smaller local truncation error, it yields a less accurate numerical solution of (10.6.12), since starting errors grow at a faster rate than those in (10.6.14). For the finite interval $[0, 5]$, both solutions can, in principle, be made as accurate as necessary by taking a sufficiently small h. In practice, however, the solution of (10.6.13) may necessitate a value of h so small as to be impractical. In theory, stability is a matter of continuity with respect to initial conditions and rounding errors. Indeed, we have seen that the root condition implies Lipschitz continuity. In the stable difference method (10.6.13), the Lipschitz constant is exponentially large.

The preceding analysis can be easily modified to include the case of perturbed starting values $x_0 = 1 + \Delta x_0$, $x_1 = e^{-h} + \Delta x_1 e^{-h}$. There will then be additional error terms of the form

$$\Delta x_0 e^h e^{-t}, \qquad \Delta x_1 e^{-t}, \qquad (-1)^i(\Delta x_0 - \Delta x_1)e^t e^{-h}$$

and terms of higher order in h. The first two terms introduce relative errors in the numerical solution which are of the same order of magnitude as the relative errors in x_0 and x_1. Therefore, they can be tolerated. However, the third term introduces a magnification of the initial relative errors by a factor $(-1)^i e^{2t}$; i.e., the Lipschitz constant referred to above behaves like e^{2t}. If t is large, this may not be tolerable computationally, since it places stringent limitations on the sizes of Δx_0 and Δx_1. Since these errors include rounding errors, this may necessitate high-precision numbers as well as a small value of h.

Thus, the presence of extraneous solutions causes a phenomenon that in practical computation resembles instability; i.e., small changes in initial values of the order of magnitude of rounding errors can produce changes in the numerical solution orders of magnitude larger. This behavior is sometimes referred to as *numerical instability* and the method is said to be only *weakly stable*. It should be noted that the effect of numerical instability can be overcome, in principle, by using a sufficiently high precision and small h. By contrast, the error in an unstable method (in the sense of Definition 10.6.2) cannot be made small in this way. Nevertheless, from a practical standpoint, a stable method which introduces extraneous solutions is to be avoided, since it may require too high a precision and too small a step size to be feasible.

One might ask whether there is any real advantage in considering the general difference method (10.6.2) as opposed to the special case (10.4.1) where $n = 1$ and $c_0 = 1, c_1 = -1$. Presumably, a practical advantage would be to achieve a method of order $>n + 1$ by choosing the c_j suitably. Since there are $2n + 1$ independent coefficients in (10.6.2) after dividing by c_0, it should be possible to achieve order $2n$. This has been shown to be possible [10.8], but unfortunately it is also proved in [10.8] that for orders $>n + 2$ the method is unstable. Thus, if $p(\xi)$ in (10.6.4) satisfies the root condition, then (10.6.2) is of order $\leq n + 2$. In fact, if n is odd, the order of a stable method cannot exceed $n + 1$. (For a proof of these results, see [10.1, p. 229–233].) Thus the Adams–Moulton formulas (10.2.8) are of maximal order 4.

To circumvent this inherent limitation of multistep difference methods given by formula (10.6.2) these methods have been combined with Runge–Kutta techniques requiring derivative evaluations at intermediate points $t_i + \theta_j h$ to yield methods known as *generalized predictor–corrector methods* [10.24, 10.25, 10.17] or *generalized Runge–Kutta methods* [10.23]. The formulas in these methods are of the type

$$y_i = \sum_{j=1}^{n} c_j y_{i-j} + h \sum_{j=0}^{n} a_j f(y_{i-j}) + h \sum_{1}^{r} b_j y'(\theta_j h),$$

where $n - 1 < \theta_j < n$. Here we have written the differential equation in autonomous form, $y' = f(y)$. For $r = 1$, it seems possible to obtain stable methods of order $2n + 1$ for $n \leq 7$ and for $r = 2$ stable methods of order $2n + 2$ seem to be possible for $n \leq 15$ [10.23].

Of course, as we have seen in the example of the stable second-order method (10.6.13), extraneous solutions are introduced if the polynomial p in (10.6.4) has some $c_i \neq 0$ for $i > 1$. It is not difficult to see that the same numerical instability is frequently present in the general case. Let

$$q(\xi) = \sum_{j=0}^{n} a_j \xi^{n-j}$$

be the polynomial corresponding to the second sum in Eq. (10.6.2). If we apply the general scheme (10.6.2) to the equation $x' = \lambda x$, where λ is a constant, we obtain a linear difference equation having the characteristic equation, $g(\xi) = p(\xi) - \lambda h q(\xi) = 0$. We assume that the method is stable and consistent so that all roots r_v of p have modulus ≤ 1, $p(1) = 0$ and $p'(1) = q(1)$ (Exercise 10.11). For h small, the roots ξ_v of $g(\xi) = 0$ are close to the r_v and are analytic functions of h. Hence $\xi_v = r_v(1 + K_v \lambda h + O(h^2))$. To determine K_v, we note that $0 = g(\xi_v) = g(r_v) + g'(r_v)(\xi_v - r_v) + O(h^2)$. Hence $0 = -\lambda h q(r_v) + p'(r_v) r_v K_v \lambda h + O(h^2)$ and $K_v = q(r_v)/r_v p'(r_v)$. Therefore, with $t = nh$,

$$\xi_v^n = r_v^n \exp(\lambda K_v t)(1 + O(h)).$$

If $|r_v| = 1$ and Re $(\lambda K_v) > 0$, an extraneous root ξ_v close to r_v can introduce the exponentially large Lipschitz constant causing numerical instability (or weak stability). If Re $(\lambda K_v) >$ Re (λ), then the solution component corresponding to ξ_v becomes large relative to the solution ξ_1^n, which corresponds to $r_1 = 1$ and represents the desired approximation to the exact solution $e^{\lambda t}$. (Note that $K_1 = 1$.)

10.7 STIFF EQUATIONS AND A-STABILITY

The term *numerical instability* is to some extent a misnomer, since by making the step size h sufficiently small and the precision large enough, the effect of the extraneous solutions can be made tolerably small. However, if the interval of integration $[a, b]$ is large, this will usually demand an h so small and a precision so high that practical computation is not feasible. It might be more appropriate to characterize this behavior as a form of ill-conditioning. A similar phenomenon occurs for certain types of differential equations even when the difference operator has no extraneous roots. Consider, for example, the linear system of two first-order equations,

$$\begin{aligned} u' &= -11u + (9/2)v, \\ v' &= 18u - 11v, \end{aligned} \quad (10.7.1)$$

with initial values $x(0) = x_0 = (u_0, v_0)$. The exact solution is

$$u(t) = \frac{1}{2}\left(u_0 + \frac{v_0}{2}\right)e^{-2t} + \frac{1}{2}\left(u_0 - \frac{v_0}{2}\right)e^{-20t},$$

$$v(t) = \left(u_0 + \frac{v_0}{2}\right)e^{-2t} - \left(u_0 - \frac{v_0}{2}\right)e^{-20t}.$$

Let $x = (u, v)$. We can rewrite system (10.7.1) in vector form as $x' = Ax$, $x_0 =$

(u_0, v_0), where
$$A = \begin{bmatrix} -11 & 9/2 \\ 18 & -11 \end{bmatrix}.$$

Applying Euler's method to the vector equation, we obtain the vector difference equation,
$$x_{i+1} = x_i + hAx_i = (I + hA)x_i,$$
which has the solution
$$x_i = (I + hA)^i x_0. \tag{10.7.2}$$

The matrix A is diagonalized by the similarity transformation BAB^{-1}, where
$$B = \begin{bmatrix} 1 & \frac{1}{2} \\ -1 & \frac{1}{2} \end{bmatrix}, \qquad B^{-1} = \begin{bmatrix} \frac{1}{2} & -\frac{1}{2} \\ 1 & 1 \end{bmatrix}.$$

The eigenvalues of A are -2 and -20. Hence system (10.7.1) can be diagonalized by the same transformation. We see that its solution is $x_i = B^{-1}D^i Bx_0$, where D is the diagonal matrix having entries $1 - 2h$ and $1 - 20h$. Therefore, the solution (10.7.2) in component form is

$$u_i = \frac{1}{2}\left(u_0 + \frac{v_0}{2}\right)(1 - 2h)^i + \frac{1}{2}\left(u_0 - \frac{v_0}{2}\right)(1 - 20h)^i,$$
$$v_i = \left(u_0 + \frac{v_0}{2}\right)(1 - 2h)^i - \left(u_0 - \frac{v_0}{2}\right)(1 - 20h)^i. \tag{10.7.3}$$

For example, suppose $u_0 = 3$ and $v_0 = 2$. Then
$$u(t) = 2e^{-2t} + e^{-20t}$$
$$v(t) = 4e^{-2t} - 2e^{-20t}.$$

The component e^{-20t} decreases rapidly relative to e^{-2t}. To obtain the same behavior in the numerical solution (10.7.3), the step size h must be chosen to be $<1/20$. After sufficiently many steps, $(1 - 20h)^i$ will be so small relative to $(1 - 2h)^i$ that it will be comparable to the rounding error. For example, for $t = 1$, we would have $e^{-20} \leq 2.5 \times 10^{-9}$, while $e^{-2} = 0.135$ approximately. If we wish to integrate over the interval $[0, 2]$, then we must choose a step size sufficiently small so that $(1 - 20h)^i$ represents e^{-20hi} accurately for $ih \leq 1$. Thereafter we would like to be able to increase h by a factor of 10, thereby retaining the same accuracy in $(1 - 2h)^i$ as an approximation to e^{-2t}. This would be possible if we could either drop the term $(1 - 20h)^i$ completely or keep its magnitude $<10^{-8}$. Unfortunately, as Eqs. (10.7.3) show, starting from $t = 1$ with this larger step size would introduce excessive errors in the approximation $(1 - 20h)^i$. To take an extreme case, if we chose $h \geq 0.1$, this term would oscillate and not decrease to zero. In fact, it could grow in amplitude. Thus we are forced to maintain h small.

Equations $x' = Ax$ in which the eigenvalues λ_j of A have negative real parts which are widely dispersed, that is, $|\max_j \operatorname{Re} \lambda_j|/|\min_j \operatorname{Re} \lambda_j|$ is large, say > 10, are called *stiff equations*. (As vibrational systems, they have highly damped transient modes.) The integration of such equations poses the special difficulty illustrated above with respect to h. To obtain sufficient accuracy, we must choose $h \max |\operatorname{Re} \lambda_j|$ small. For an interval in which the transients are highly damped, we would like to increase h. However, a method such as Euler's would then introduce an oscillating error growing in amplitude if $1 - h \max \operatorname{Re} \lambda_j < -1$. This resembles the *numerical instability* introduced by extraneous roots discussed in the previous section. To obviate such behavior, we would like to have a numerical method in which the numerical solution x_i of any differential equation of the type,

$$x' = \lambda x, \qquad \lambda < 0, \tag{10.7.4}$$

converges to zero for any step size h. This leads to a stronger notion of stability [10.19].

Definition 10.7.1 A numerical method is called *A-stable* if any solution (x_i) of the difference equation obtained by applying the method to an equation of type (10.7.4) with any step size $h > 0$ converges to zero, that is, $\lim_{i \to \infty} x_i = 0$ for any $h > 0$.

Actually, we permit λ to be a complex constant with $\operatorname{Re} \lambda < 0$, since this can happen in the matrix case, $x' = Ax$.

If we limit our search for A-stable methods to consistent linear difference methods of the type (10.6.2), we find that there are no such A-stable methods of order > 2 and of all A-stable methods of order 2, the trapezoidal rule yields the smallest error [10.19]. The proof of this result involves some complex function theory. We shall content ourselves with a sketch of the main ideas.

Consider Eq. (10.6.2) applied to Eq. (10.7.4). The resulting difference equation, as we saw in the last paragraph of the previous section, has the characteristic equation,

$$p(\xi) - \lambda h q(\xi) = 0, \qquad \operatorname{Re} \lambda < 0. \tag{10.7.5}$$

Consider the rational function $r(\xi) = p(\xi)/q(\xi)$. If for some ξ_1 with $|\xi_1| > 1$ the real part of $r(\xi_1)$ is negative, then we can set $r(\xi_1) = \lambda h$ for some λ with $\operatorname{Re} \lambda < 0$. Hence ξ_1 is a root of Eq. (10.7.5), which implies $x_i = \xi_1^i$ is a numerical solution of Eq. (10.7.4) and $\lim_{i \to \infty} x_i \neq 0$. Therefore, the method is not A-stable. Hence a necessary condition for A-stability is that $\operatorname{Re}(p(\xi)/q(\xi)) \geq 0$ for $|\xi| > 1$. This condition is also sufficient, since it implies that Eq. (10.7.5) cannot have a root $|\xi_1| \geq 1$. Such a root would imply that $r(\xi_1) = \lambda h$ has a negative real part. (If $|\xi_1| = 1$, then by continuity $\operatorname{Re}(r(\xi)) < 0$ for some $|\xi| > 1$, near ξ_1.) Note that we cannot have $q(\xi_1) = 0$, since this would imply $p(\xi_1) = 0$ contrary to our assumption that p and q have no common factors. This immediately implies:

Theorem 10.7.1 An explicit n-step method cannot be A-stable.

Proof. In an explicit method, the leading coefficient $a_0 = 0$ in $q(\xi)$. Hence $r(\xi)$ behaves like $\alpha \xi^{n-k}$, $k \geq 1$, as $\xi \to \infty$. Therefore $\operatorname{Re}(r(\xi)) < 0$ for some $|\xi| > 1$. ∎

Corollary Euler's method is not A-stable. The trapezoidal rule is A-stable.

Proof. Euler's method is explicit. For the trapezoidal rule the difference equation is

$$x_i - x_{i-1} = \frac{h\lambda}{2}(x_i + x_{i-1}).$$

Hence $p(\xi) = \xi - 1$ and $q(\xi) = (\xi + 1)/2$. It is well known that $z = (\xi - 1)/(\xi + 1)$ maps the exterior of the unit ξ-circle onto the right half of the complex z-plane. ∎

Theorem 10.7.2 The order k of an A-stable consistent linear difference method (10.6.2) cannot be greater than 2. The trapezoidal rule has the smallest error constant for $k = 2$.

Proof. Let $z = (\xi + 1)/(\xi - 1)$, $\xi = (z + 1)/(z - 1)$. Set

$$p_1(z) = \left(\frac{z-1}{2}\right)^n p\left(\frac{z+1}{z-1}\right) = \sum_{j=0}^{n-1} \alpha_j z^j,$$

$$q_1(z) = \left(\frac{z-1}{2}\right)^n q\left(\frac{z+1}{z-1}\right) = \sum_{j=0}^{n} \beta_j z_j.$$

($\alpha_n = 0$, since $p(1) = 0$ by consistency.) Now suppose that (10.6.2) has order k, so that the local truncation error (10.6.3) is given by $-Ch^{k+1}x^{(k+1)}(\theta_i)$, where $\theta_i \to t_i$ as $h \to 0$. Putting $x(t) = e^t$ in (10.6.3) with $t = t_i$ and setting $\xi = e^h$, we get as $\xi \to 1$,

$$p(\xi) - q(\xi) \log \xi \sim -C(\xi - 1)^{k+1},$$

where $y \sim w$ means that the $\lim y/w = 1$. It follows that

$$\log \xi - \frac{p(\xi)}{q(\xi)} \sim C^*(\xi - 1)^{k+1},$$

where $C^* = C/q(\xi)$. Therefore, as $z \to \infty$,

$$\log\left(\frac{z+1}{z-1}\right) - \frac{p_1(z)}{q_1(z)} \sim C^*\left(\frac{2}{z}\right)^{k+1}.$$

We expand the logarithm in powers of $1/z$, obtaining for $k \geq 2$,

$$\frac{p_1(z)}{q_1(z)} = \frac{2}{z} + (\tfrac{2}{3} - 8C')z^{-3} + O(z^{-4}),$$

where $C' = C^*$ if $k = 2$ and $C' = 0$ if $k > 2$. The A-stability implies that $\operatorname{Re}(p_1(z)/q_1(z)) \geq 0$ for $\operatorname{Re} z > 0$. From complex function theory it follows that $sp_1(s)/q_1(s)$ is a nondecreasing function of real s. (See [10.19].) This implies that $2/3 - 8C' \leq 0$, whence we cannot have $k > 2$ and for $k = 2$ we must have $C^* = C' \geq \frac{1}{12}$. The value $C^* = \frac{1}{12}$ is attained by the trapezoidal rule. ∎

We must seek outside the class of methods (10.6.2) for A-stable methods of order > 2. The fourth-order Runge–Kutta method is not A-stable. If we apply it to Eq.

10.18. L. Fox, *The Numerical Solution of Two-Point Boundary Problems in Ordinary Differential Equations*, Oxford University Press, London (1957)
10.19. G. Dahlquist, "A special stability criterion for linear multistep methods," *BIT* **3**, 22–43 (1963)
10.20. Fred T. Krogh, "Predictor–Corrector Methods of High Order with Improved Stability Characteristics," *J.A.C.M.*, 374–385 (July 1966)
10.21. C. R. Cassidy, "The Complete Solution of the Fifth-Order Runge-Kutta Equations," *SIAM J. Numer. Analysis* **6**, No. 3, 432–431 (September 1969)
10.22. J. C. Butcher, "On the attainable order of Runge-Kutta methods," *Math. Comp.* **19**, 408–417 (1965)
10.23. J. C. Butcher, "A multistep generalization of Runge-Kutta methods with four or five stages," *J.A.C.M.* **14**, no. 1, 84–99 (January 1967)
10.24. W. B. Gragg and H. J. Stetter, "Generalized multistep predictor–corrector methods," *J.A.C.M.* **11**, 188–209 (1964)
10.25. C. W. Gear, "Hybrid methods for initial value problems in ordinary differential equations," *J. SIAM Numer. Anal.* **2**, 69–86 (1965)
10.26. J. H. Verner, "The order of some implicit Runge-Kutta methods," *Numer. Math.* **13**, 14–23 (1969)
10.27. W. Kutta, "Beitrag zur naherungsweisen Integration totaler Differentialgleichungen" *Z. Math. Physik* **46**, 435–452 (1901)
10.28. E. J. Nyström, "Über die numerische Integration von Differentialgleichungen," *Acta Soc. Sci. Fennicae* **50**, 1–55 (1925)
10.29. W. B. Gruttke, "Pseudo-Runge-Kutta methods of the fifth order," *J.A.C.M.* **17**, No. 4, 613–628 (October 1970)
10.30. J. D. Lawson, "An order five Runge-Kutta process with extended region of stability," *J. SIAM Numer. Anal.* **3**, 593–597 (1966)
10.31. G. D. Byrne, "Parameters for pseudo-Runge-Kutta methods," *Comm. ACM* **10**, 102–104 (1967)
10.32. G. D. Byrne and R. J. Lambert, "Pseudo-Runge-Kutta methods involving two points," *J.A.C.M.* **13**, 114–123 (1966)
10.33. C. W. Gear, "The automatic integration of stiff ordinary differential equations," *Proc. IFIP Cong.*, Edinburgh(1968)
10.34. O. B. Widlund, "A note on unconditionally stable linear multistep methods," *BIT* **7**, 65–70 (1967)
10.35. W. Liniger and R. A. Willoughby, "Efficient integration methods for stiff systems of ordinary differential equations," *SIAM J. Numer. Anal.* **7**, No. 1, 47–66 (March 1970)
10.36. C. W. Gear, "The automatic integration of ordinary differential equations," *Comm. ACM* **14**, 176–179 (1971)

EXERCISES

10.1 Given the rth-order differential equation

$$\frac{d^r x}{dt^r} = f(t, x, x', \ldots, x^{(r-1)}),$$

show that there exists an "equivalent" system of r first-order equations. [*Hint:* Intro-

duce new unknowns, $y_1 = x, y_2 = x', \ldots, y_r = x^{(r-1)}$. Then
$$y'_1 = y_2, y'_2 = y_3, \ldots, y'_r = f(t, y_1, y_2, \ldots, y_r)$$
is a system of r first-order equations whose solutions $y_1(t) = x(t)$ is a solution of the given rth-order system.]

10.2 a) Let $x(t)$ be a continuous vector function, with $x(t) \in V$, $a \le t \le b$. Prove that $\int_a^b x(t) \, dt$ exists and is a vector having components $\int_a^b x_i(t) \, dt$, where the $x_i(t)$, $1 \le i \le m$, are the component functions of $x(t)$. (See Theorem 5.3.4.)

b) Prove that $\|\int_a^b x(t) \, dt\| \le \int_a^b \|x(t)\| \, dt \le \|x\|_\infty (b - a)$, where $\|x\|_\infty = \sup_{a \le t \le b}\{\|x(t)\|\}$. [*Hint*:
$$\left\|\sum_1^n x(t'_i)(t_{i+1} - t_i)\right\| \le \sum_1^n \|x(t'_i)\| (t_{i+1} - t_i) \le \|x\|_\infty (b - a),$$
and passage to the limit as $n \to \infty$ with $\max \{\Delta t_i\} \to 0$.]

10.3 a) Prove that the coefficients a_j in the open integration formula (10.2.4) satisfy the relation $\sum_0^n a_j = r$. Similarly $\sum_0^n b_j = r$ for the closed formula (10.2.6). [*Hint:* In (10.2.3), if $x'(t) \equiv 1$, then $R_n = 0$ and $x(t_i) = x_i$.]

b) Prove: If x_i is given by (10.2.2) and $\|f(t, x)\|_\infty = M$ for (t, x) in the rectangle D of Section 10.1, then $\|x_i - x_{i-1}\| \le Mh$. Hence show that $(t_i, x_i) \in D$ for all i. Do the same for (10.2.4).

10.4 Show that $(1 + z)^n \le e^{nz}$ by using Taylor's formula
$$e^z = 1 + z + \frac{z^2}{2} e^{\theta z}.$$

10.5 a) Show that the error in the modified Euler method is of order h^2.

b) Show that the error in the Adams–Moulton method is of order h^4, provided that x_0, x_1, x_2, x_3 are determined with an error of order h^4.

c) Formulas of the type (10.2.4) are sometimes referred to as *Adams–Bashforth* formulas. These are obtained from (9.2.5a) by applying the backward difference analog of formula (8.2.6) to express the backward differences in terms of the ordinates. Show that the coefficients a_j in (10.2.4) are related to the coefficients c_k in (9.2.5a) as follows:
$$a_j = (-1)^j \left[c_j + \binom{j+1}{j} c_{j+1} + \cdots + \binom{n}{j} c_n \right]. \tag{1}$$

Obtain a recurrence relation for the c_k. [*Hint:* Introduce the *generating function*
$$\varphi(t) = \sum_{k=0}^\infty c_k t^k = \sum_{k=0}^\infty (-t)^k \int_0^1 \binom{-s}{j} ds.$$

Then $\varphi(t) = \int_0^1 (1 - t)^{-s} \, ds$, by the binomial expansion. Since $(1 - t)^{-s} = e^{-s \log(1-t)}$, we can evaluate the integral and get
$$\frac{-\log(1 - t)}{t} \varphi(t) = \frac{1}{1 - t}.$$

Expanding in power series, we obtain

$$\left(1 + \frac{t}{2} + \frac{t^2}{3} + \cdots\right)(c_0 + c_1 t + \cdots) = 1 + t + t^2 + \cdots,$$

and equate coefficients to get, for $k = 0, 1, 2, \ldots$,

$$c_k + \frac{c_{k-1}}{2} + \frac{c_{k-2}}{3} + \cdots + \frac{c_0}{k} = 1. \quad]$$

Verify that $c_0 = 1, c_1 = \frac{1}{2}, c_2 = \frac{5}{12}, c_3 = \frac{3}{8}, c_4 = \frac{251}{720}$, and then use (1) above to obtain the predictor in the Adams–Moulton fourth-order formula (10.2.8).

d) Formulas of type (10.2.6) are sometimes called *Adams–Moulton* formulas. Using formula (9.2.6a) and a generating-function technique as in part (c) above, obtain the recurrence relation

$$d_k + \frac{1}{2} d_{k-1} + \frac{1}{3} d_{k-2} + \cdots + \frac{1}{k+1} d_0 = 0, \quad k \geq 1.$$

[*Hint:* $\varphi(t) = \sum_{k=0}^{\infty} d_k t^k = -t/\log(1-t)$.]

Verify that $d_0 = 1, d_1 = -\frac{1}{2}, d_2 = -\frac{1}{12}, d_3 = \frac{1}{24}, d_4 = -\frac{19}{720}$, and use (1) above to obtain the corrector in (10.2.8).

10.6 Derive the general second-order Runge–Kutta method. Show that different choices of the parameters yield the formula known as the *predictor–corrector Euler method* (or *Heun* method) given by

$$x_i = x_{i-1} + \frac{h}{2}(k_0 + k_1),$$

$$k_0 = f(t_{i-1}, x_{i-1}),$$

$$k_1 = f(t_{i-1} + h, x_{i-1} + h k_0)$$

and the *modified Euler method* given by

$$x_i = x_{i-1} + hf\left(t_{i-1} + \frac{h}{2}, x_{i-1} + \frac{h}{2} k_0\right),$$

$$k_0 = f(t_{i-1}, x_{i-1}).$$

[*Hint:* Let $y = (t, x)$ as in Section 10.5 and using (10.5.2) with $n = 1$, we have

$$y_i = y_{i-1} + h(a_0 k_0 + a_1 k_1),$$

$$k_0 = f(y_{i-1}), \qquad k_1 = f(y_{i-1} + h\beta_{10} k_0).$$

By (10.5.3), (10.5.4), and (10.5.5),

$$\frac{\Delta y}{h} = f(y) + \frac{h}{2} f'(y) \cdot f(y) + O(h^2).$$

By (10.5.8),

$$k_1 = f(y) + \beta_{10} h f'(y) \cdot f(y) + O(h^2).$$

Thus

$$\frac{\delta y}{h} = a_0 k_0 + a_1 k_1 = a_0 f(y) + a_1 f(y) + a_1 \beta_{10} h f'(y) \cdot f(y) + O(h^2).$$

Equating coefficients of h^0 and h in $\Delta y/h$ and $\delta y/h$, we get

$$a_0 + a_1 = 1, \qquad a_1 \beta_{10} = \tfrac{1}{2}.$$

Taking $a_0 = a_1 = \tfrac{1}{2}$ forces $\beta_{10} = 1$ and yields the Heun method. Taking $a_0 = 0$, $a_1 = 1$, and $\beta_{10} = \tfrac{1}{2}$ gives us the modified Euler method. In both methods, $\delta y = \Delta y + O(h^3)$. Thus the local truncation error is Kh^3. In [10.7], it is shown that the smallest *upper bound* on K is obtained with $a_0 = \tfrac{2}{3}$.]

10.7 Prove that the error $e_i = y(t_i) - y_i$ in the solution y_i obtained by the fourth-order Runge–Kutta method (10.5.9) is $O(h^4)$ for all i if $f(y)$ satisfies a Lipschitz condition. (Assume that the local truncation error is indeed $O(h^5)$, as asserted in the text.) [*Hint:* Let

$$\tilde{k}_0 = f(y(t_{i-1})), \qquad \tilde{k}_1 = f\left(y(t_{i-1}) + \frac{h}{2}\tilde{k}_0\right),$$

$$\tilde{k}_2 = f\left(y(t_{i-1}) + \frac{h}{2}\tilde{k}_1\right), \qquad \tilde{k}_3 = f(y(t_{i-1}) + h\tilde{k}_2).$$

Then

$$y(t_i) = y(t_{i-1}) + \frac{h}{6}(\tilde{k}_0 + 2\tilde{k}_1 + 2\tilde{k}_2 + \tilde{k}_3) + O(h^5).$$

Subtract (10.5.9) from this equation to obtain

$$e_i = e_{i-1} + \frac{h}{6}(\Delta k_0 + 2\Delta k_1 + 2\Delta k_2 + \Delta k_3) + O(h^5),$$

$$\|\Delta k_0\| = \|f(y(t_{i-1})) - f(y_{i-1})\| \leq L \|e_{i-1}\|,$$

$$\|\Delta k_1\| = \left\| f\left(y(t_{i-1}) + \frac{h}{2}\tilde{k}_0\right) - f\left(y_{i-1} + \frac{h}{2}k_0\right) \right\|$$

$$\leq L\left(\|e_{i-1}\| + \frac{h}{2}\|\Delta k_0\|\right) \leq L(1 + hL/2)\|e_{i-1}\|.$$

Similarly,

$$\|\Delta k_2\| \leq L(1 + hL/2 + h^2 L^2/4)\|e_{i-1}\|,$$

$$\|\Delta k_3\| \leq L(1 + hL + h^2 L^2/2 + h^3 L^3/4)\|e_{i-1}\|.$$

Hence,

$$\|e_i\| \leq (1 + hL + h^2 L^2/2! + h^3 L^3/3! + h^4 L^4/4!)\|e_{i-1}\| + O(h^5),$$

$$\|e_i\| \leq \exp(ihL)\left[\|e_0\| + \frac{O(h^5)}{O(h)}\right] = e^{(t_i - t_0)L} O(h^4) \quad \text{if } \|e_0\| = 0. \;]$$

10.8 a) Show that $\sum_0^n a_j = 1$ for the coefficients in formula (10.5.1) if the method is to be exact for constant functions.

b) Show that when $f(t, x) = f(t)$ is independent of x, the fourth-order Runge–Kutta method is Simpson's rule.

10.9 For the one-dimensional case, $x' = f(t, x)$, show that the local truncation error $\delta y - \Delta y$ in the fourth-order Runge–Kutta method is bounded by

$$MN(3.7 + 5.4M + 1.3M^2 + 0.017M^3)h^5$$

where $|f(t, x)| \leq N$ and $|\partial^k f/\partial t^i \partial x^j| \leq M/N^{j-1}, 0 \leq k \leq 4, 0 \leq j \leq 4$. [*Hint:* See [10.10].]

10.10 Consider the general second-order homogeneous difference equation

$$ax_i + bx_{i-1} + cx_{i-2} = 0, \tag{1}$$

with starting values x_0, x_1.

a) Show that (1) is equivalent to the system of two first-order equations

$$CY_i = AY_{i-1}$$

where $Y_i^T = (y_i, z_i), y_i = x_{i+1}, z_i = x_i, i \geq 0$.

$$C = \begin{pmatrix} a & b \\ 0 & 1 \end{pmatrix} \quad \text{and} \quad A = \begin{pmatrix} 0 & -c \\ 1 & 0 \end{pmatrix}.$$

b) Verify that

$$B = C^{-1}A = \begin{pmatrix} -b/a & -c/a \\ 1 & 0 \end{pmatrix}$$

and $\det(\xi I - C^{-1}A) = p(\xi)/a$, where $p(\xi)$ is the characteristic polynomial of (1). Thus, the eigenvalues of B are the roots ξ_1, ξ_2 of $p(\xi) = 0$.

c) Verify that two eigenvectors of B are $(\xi_1, 1)$ and $(\xi_2, 1)$. These are independent if $\xi_1 \neq \xi_2$. Show that

$$\begin{pmatrix} \xi_1 & \xi_2 \\ 1 & 1 \end{pmatrix}^{-1} = \frac{1}{\xi_1 - \xi_2} \begin{pmatrix} 1 & -\xi_2 \\ -1 & \xi_1 \end{pmatrix}.$$

d) Verify that

$$B^i = \frac{1}{\xi_1 - \xi_2} \begin{pmatrix} \xi_1 & \xi_2 \\ 1 & 1 \end{pmatrix} \begin{pmatrix} \xi_1^i & 0 \\ 0 & \xi_2^i \end{pmatrix} \begin{pmatrix} 1 & -\xi_2 \\ -1 & \xi_1 \end{pmatrix}.$$

e) Compute $Y_i = B^i Y_0$, where $Y_0 = (x_1, x_0)$. Show that

$$x_i = z_i = \frac{1}{\xi_1 - \xi_2}(\xi_1^i(x_1 - x_0\xi_2) + \xi_2^i(x_0\xi_1 - x_1)).$$

10.11 a) Show that a necessary condition that (10.6.1) be consistent with (10.1.1) for $f \equiv 0$ is that $\sum_0^n c_j = 0$, hence that $p(1) = 0$, where $p(\xi)$ is the polynomial in (10.6.4).

b) Show that Euler's method $x_i - x_{i-1} = hf(t_{i-1}, x_{i-1})$ is consistent with (10.1.1) for any f which is Lipschitz continuous in x and continuous in t. [*Hint:*

$$dx/dt = f(t, x(t)) = \lim_{h \to 0} (x(t + h) - x(t))/h. \;]$$

c) Show that $z_{i+1} - z_i = h(f(t_{i+1}, z_{i+1}) + f(t_i, z_i))$ is an inconsistent method for the differential equation $dx/dt = f(t, x)$.

d) Let $q(\xi) = \sum_{j=0}^{n} a_j \xi^{n-j}$. Show that a necessary condition for (10.6.1) to be consistent is that $p'(1) = q(1)$, where p is given by (10.6.4). [*Hint:* Apply the method to the initial-value problem $x' = 1$, $x(0) = 0$, noting that $x_i = ih$ and $\sum c_j = 0$ by part (a).]

10.12 a) Let $(\mathcal{L}X)_{i-n} = \sum_{j=0}^{n} c_j x_{i-j}$ be the general nth-order linear difference operator. If ξ is a root of the characteristic polynomial (10.1.11) of multiplicity $m + 1$, verify that $\mathcal{L}(i(i-1) \cdots (i-m+1)\xi^{i-m}) = 0$. [*Hint:*

$c_0 i(i-1) \cdots (i-m+1)\xi^{i-m} + c_1(i-1)(i-2) \cdots (i-m)\xi^{i-m-1} + \cdots$

$+ c_n(i-n) \cdots (i-m-n+1)\xi^{i-m-n} = \dfrac{d^m}{d\xi^m}(\xi^{i-n} p(\xi))$

$= \binom{m}{0} \xi^{i-n-m} p(\xi) + \binom{m}{1} \xi^{i-n-m+1} p^{(1)}(\xi) + \cdots + \xi^{i-m} p^{(m)}(\xi) = 0.$]

b) Verify that the solutions given in (10.1.13) are linearly independent. [*Hint:* if (ξ^{i-r}) is a solution of (10.1.12), then so is $\xi^r(\xi^{i-r})$ by linearity. Hence

$\xi_k^i, i\xi_k^i, i(i-1)\xi_k^i, \ldots, i(i-1) \cdots (i-m_k+1)\xi_k^i$

are solutions. They are linearly independent because

$\xi_k^i \sum_{j=0}^{m_k+1} \alpha_j i(i-1) \cdots (i-j) = 0$

implies $\alpha_j = 0$, since $i, i(i-1), \ldots$ are linearly independent as polynomials in i.]

10.13 Prove that a corrector formula such as (10.2.6) has a unique solution for h sufficiently small and f satisfying (10.1.3). [*Hint:* Let $x_i^{(q)}$ denote the qth iterant in the successive-approximation procedure applied to the mapping

$$g(x) = hb_0 f(t_i, x) + h \sum_{j=1}^{n} b_j f(t_{i-j}, x_{i-j}) + x_{i-r},$$

where we assume that $x_{i-1}, \ldots, x_{i-n}$ are known fixed values. Note that i is arbitrary but fixed in this analysis. We further assume that the initial guess $x_i^{(0)}$ satisfies condition (ii) of Theorem 5.1.3, where now $\bar{B}$ is in the rectangle D described in Section 10.1. (This will be the case if $x_i^{(0)}$ is the predicted value $x_i^{(p)}$ given by a predictor formula like (10.2.4), since then $\|g(x_i^{(0)}) - x_i^{(0)}\| = O(h)$.) For any $x, y \in D$, we have $\|g(x) - g(y)\| \leq h|b_0| \|f(t_i, x) - f(t_i, y)\| \leq hL|b_0| \|x - y\|$. If $h < 1/L|b_0|$, then g is contractive and Theorem 5.1.3 applies.]

10.14 Write a computer program to solve the initial-value problem (10.1.2) for x a vector of arbitrary dimension $m \geq 1$. (Thus m is part of the input data.)

a) Use the Euler predictor–corrector method (10.2.7a). Why is it not possible to use the automatic step-size control based on formula (10.2.10)?

b) Use the Adams–Moulton method with the Runge–Kutta fourth-order method as a starting procedure. Incorporate automatic step-size control.

c) Test these programs by solving:
   i) $dx/dt = x$, $x(0) = 1$, on the interval $[0, 1]$;
   ii) $dx/dt = y$, $dy/dt = -x$,
   first with initial values $x(0) = 0$, $y(0) = 1$ and then $x(0) = 1$, $y(0) = 0$.

d) Test the programs on the equation
$$dx/dt = x^2, \quad x(0) = 1.$$

[*Hint:* Exact solution $x(t) = 1/(1 - t)$. Note that there is a singularity at $t = 1$. Try to integrate to a point near $t = 1$.] [*Warning:* Computer output listings will become excessive unless controlled as the step size is reduced near $t = 1$.]

10.15 Dynamical systems are frequently governed by special systems of second-order differential equations of the type $x'' = f(x)$, $x(a) = x_0$, $x'(a) = x_0'$. (Thus the first derivative x' does not enter explicitly in the differential equation.) Converting to a first-order system by setting $x' = y$, we have $y' = f(x)$ as the second equation. The following special Runge–Kutta method of Nyström [10.28] is a fifth-order method for solving such systems.

$$x_i = x_{i-1} + \frac{h}{192}(23k_0 + 125k_1 - 81k_2 + 125k_3),$$

$$y_i = y_{i-1} + hy_{i-1} + \frac{h^2}{192}(23k_0 + 75k_1 - 27k_2 + 25k_3),$$

$$k_0 = f(x_{i-1}),$$
$$k_1 = f(x_{i-1} + (2/5)hy_{i-1} + (2/25)h^2 k_0),$$
$$k_2 = f(x_{i-1} + (2/3)hy_{i-1} + (2/9)h^2 k_0),$$
$$k_3 = f(x_{i-1} + (4/5)hy_{i-1} + (4/25)h^2(k_0 + k_1)).$$

Verify that Nyström's method is of order 5. [*Hint:* See [10.1, page 183].] Use this method to solve $x'' = -x$, $x(0) = 1$, $x'(0) = 0$ and compare the solution with Problem 10.14(c), part (ii), in the interval $[0, 2\pi]$.

10.16 The special second-order system of Exercise 10.15 can also be solved by a multistep method based on quadrature. Integrating twice, we obtain

$$x(t + h) - x(t) = hx'(t) + \int_t^{t+h} (t + h - s)x''(s)\,ds.$$

Replacing h by $-h$ and adding, we get

$$x(t + h) - 2x(t) + x(t - h) = \int_t^{t+h} (t + h - s)(x''(s) + x''(2t - s))\,ds.$$

a) Verify this relation.

b) Let $t = t_{i-1}$ and replace $x''(s) = f(x(s))$ by an interpolating polynomial using grid points $t_i, t_{i-1}, \ldots, t_{i-n}$ to derive *Cowell's method*,

$$x_i - 2x_{i-1} + x_{i-2} = h^2 \sum_{m=0}^{n} \sigma_m^* \nabla^m f_i,$$

where $f_i = f(x_i)$ and

$$\sigma_m^* = (-1)^m \int_{-1}^0 (-s)\left(\binom{-s}{m} + \binom{s+2}{m}\right) ds.$$

Show that, for $n = 2$, this yields

$$x_i - 2x_{i-1} + x_{i-2} = \frac{h^2}{12}(f(x_i) + 10f(x_{i-1}) + f(x_{i-2})). \tag{1}$$

[*Hint:* See [10.1, Chapter 6].]

c) Let $t = t_i$ in part (a) and again interpolate as in part (b). The result is called *Stormer's method*, and is given by

$$x_{i+1} - 2x_i + x_{i-1} = h^2 \sum_{m=0}^{n} \sigma_m \nabla^m f_i,$$

where

$$\sigma_m = (-1)^m \int_0^1 (1-s)\left(\binom{-s}{m} + \binom{s}{m}\right) ds.$$

For $n = 2$, verify that the method is given by

$$x_{i+1} - 2x_i + x_{i-1} = \frac{h^2}{12}(13f_i - 2f_{i-1} + f_{i-2}). \tag{2}$$

Use formula (2) as a predictor and formula (1) in part (b) as a corrector to integrate the second-order system in Exercise 10.15.

where now L_1 is a Lipschitz constant for (11.1.9). Indeed,

$$L_1 = \sup_{a \le t \le b} \{\|f_x(t, x(t, x_0))\|\},$$

which is finite by the continuity of f_x on the compact set Γ. Furthermore, by Eq. (1) of Theorem 11.1.1, for all t we have $\|\Delta x(t)\| \le \|h\|e^{L(b-a)}$.

Hence there is a constant K such that

$$\frac{\|\Delta x(t) - Y(t) \cdot h\|}{\|h\|} \le \frac{K\varepsilon \sup \|\Delta x(t)\|}{\|h\|} \le \varepsilon K e^{L(b-a)}.$$

Since $\|\Delta x\| \to 0$ as $\|h\| \to 0$ and $\varepsilon \to 0$ as $\|\Delta x\| \to 0$, this shows that $Y(t) \cdot h$ is the strong differential of $x(t, v)$ with respect to v at the point (t, x_0). Therefore $x_v(t, x_0) = Y(t)$ exists and satisfies the conclusions of the theorem.

The continuity of $x_v(t, x_0)$ with respect to x_0 will be shown to follow from the fact that it satisfies Eq. (11.1.6) and the right side of (11.1.6) is continuous in x_0. The latter is a consequence of the assumed continuity of f_x and the continuity of $x(t, x_0)$ proved in Theorem 11.1.1. Thus

$$f_x(t, x(t, v)) = f_x(t, x(t, x_0)) + \varepsilon_1(t),$$

where $\varepsilon_1(t) \to 0$ uniformly as $v \to x_0$. Hence $x_v(t, v)$ is an approximate solution of (11.1.6) satisfying the same initial conditions (11.1.7). Applying inequality (11.1.3) with $\delta = 0$, we see that

$$\|x_v(t, v) - x_v(t, x_0)\| \le \|\varepsilon_1\|(e^{L_1(t-a)} - 1)/L_1,$$

which establishes continuity. If f_x satisfies a Lipschitz condition in x, then

$$\|\varepsilon_1\| \le L_2\|x(t, v) - x(t, x_0)\| \le L_2 e^{L(b-a)}\|v - x_0\|.$$

This establishes the Lipschitz continuity. If v is any starting value in the interior of D and $x(t, v)$ exists, we can apply the previous arguments at v instead of x_0. This completes the proof. ∎

If D is an infinite strip $\{(t, x) : a \le t \le b, \|x\| < \infty\}$, and f satisfies the conditions of the theorem uniformly in D, then for all $v \in V^m$ the solutions $x(t, v)$ and $x_v(t, v)$ exist for $a \le t \le b$.

Remark 1 As has already been noted, x_v is a linear operator on V^m to V^m. Since it is the derivative of the mapping $(t, v) \to x(t, v)$ with respect to v, it must be the Jacobian $(\partial \xi_i / \partial v_j)$, where the ξ_i are the m components of $x(t, v)$ and the v_j are the m components of v. For this reason, x_v is sometimes denoted by $\partial x / \partial v$. Similarly, f_x is denoted by $\partial f / \partial x$, representing the Jacobian $(\partial f_i / \partial x_j)$.

Remark 2 Boundary-value problems for second-order differential equations arise frequently in practice. Thus, instead of (11.1.1), we are often given the problem

$$x'' = f(t, x, x'),$$
$$\psi(x(a), x'(a), x(b), x'(b)) = 0, \qquad (11.1.10)$$

where $\psi : V^{2m} \times V^{2m} \to V^{2m}$, m being the dimension of x, as before. By the method of Exercise 10.1, the system (11.1.10) can be transformed into a first-order system like (11.1.1), but of order $2m$. We simply introduce a new variable y to replace x' and obtain the system

$$x' = y,$$
$$y' = f(t, x, y), \qquad (11.1.10a)$$
$$\psi(x(a), y(a), x(b), y(b)) = 0.$$

Working with the direct sum $z = (x, y)$, $v = (x(a), y(a))$ and applying the above results with the notation of Remark 1 above, we obtain as the variational equation,

$$\left(\frac{\partial z}{\partial v}\right)' = J\left(\frac{\partial z}{\partial v}\right),$$

where

$$J = \begin{bmatrix} 0 & I \\ \left(\dfrac{\partial f}{\partial x}\right) & \left(\dfrac{\partial f}{\partial y}\right) \end{bmatrix}.$$

Since $(\partial z/\partial v) = (\partial x/\partial v, \partial y/\partial v)$, this becomes

$$(\partial x/\partial v)' = (\partial y/\partial v),$$
$$(\partial y/\partial v)' = (\partial f/\partial x)(\partial x/\partial v) + (\partial f/\partial y)(\partial y/\partial v),$$

which can be reduced to a second-order equation for $(\partial x/\partial v)$ by differentiation,

$$\left(\frac{\partial x}{\partial v}\right)'' = \left(\frac{\partial f}{\partial x}\right)\left(\frac{\partial x}{\partial v}\right) + \left(\frac{\partial f}{\partial x'}\right)\left(\frac{\partial x}{\partial v}\right)'. \qquad (11.1.11)$$

The corresponding initial conditions may be expressed as

$$\begin{bmatrix} (\partial x/\partial v) \\ (\partial x/\partial v)' \end{bmatrix} = I_{2m},$$

where $(\partial x/\partial v)$ is an $m \times 2m$ matrix solution of (11.1.11) and I_{2m} is the $2m \times 2m$ identity.

An important special case of (11.1.10) occurs when f is independent of x' and the boundary conditions are simply of the form

$$x(a) = x_0, \qquad x(b) = x_F, \qquad (11.1.12)$$

where x_0 and x_F are prescribed constant vectors. In this case, the only initial values to be varied are $v = x'(a)$ and we wish to determine $x_v(b, v)$ as before. It is easily seen that Theorems 11.1.1 and 11.1.2 remain valid with this restriction on v. Indeed, it simply means that in considering variations h in the initial values we restrict h to lie in a subspace of the space of initial values. Also, we must take I in (11.1.7) to be the restriction of the identity to this subspace and the other initial values must be zero. In the special case being considered, h (and therefore v) are in the m-dimensional

11.1 Existence and Uniqueness of Exact Solutions

space of initial values $x'(a)$ of the derivative. Now, the boundary condition to be satisfied is just

$$\psi(x(a), x'(a), x(b), x'(b)) = x(b) - x_F = 0. \tag{11.1.13}$$

Hence we must determine $v = x'(a)$ such that

$$\varphi(v) = \psi(v, x(b, v)) = x(b, v) - x_F = 0.$$

Clearly, $\varphi'(v) = \psi_1 + \psi_2 \cdot x_v = x_v(b, v)$, and in this case the variational equation (11.1.11) becomes

$$x_v'' = (\partial f/\partial x)x_v, \tag{11.1.14}$$

obtained as a special case of (11.1.11) with $(\partial f/\partial x') = 0$. The initial values for x_v are given by

$$\begin{bmatrix} x_v(a) \\ x_v'(a) \end{bmatrix} = \begin{bmatrix} 0_m \\ I_m \end{bmatrix}. \tag{11.1.15}$$

Remark 3 Observe that in the general case

$$\frac{\partial x_v}{\partial t} = x_v' = f_x(t, x(t, v))x_v$$

is a continuous function of (t, v) when the hypotheses of the theorem are satisfied. Furthermore, noting that $x'(t, v) = \partial x(t, v)/\partial t$, we have

$$\frac{\partial}{\partial v}\left(\frac{\partial x}{\partial t}\right) = \frac{\partial}{\partial v} f(t, x(t, v)) = f_x \cdot \frac{\partial x}{\partial v}.$$

Thus the second cross-partials are both continuous, so that

$$\frac{\partial}{\partial v} x'(t, v) = x_v'.$$

The special boundary-value problem discussed in Remark 2 above, namely,

$$\begin{aligned} x'' &= f(t, x), \\ x(a) &= x_0, \quad x(b) = x_F. \end{aligned} \tag{11.1.16}$$

can be shown to have a unique solution in the one-dimensional case when $f_x \geq 0$. More precisely, we have the following existence theorem.

Theorem 11.1.3 Let $f(t, x)$ be continuous and satisfy a Lipschitz condition with respect to x on the infinite strip, $E = \{(t, x) : a \leq t \leq b, -\infty < x < \infty\}$. Further, let $f_x(t, x) \geq 0$ for $(t, x) \in E$. Then (11.1.16) has a unique solution.

Proof. As before, let $x(t, v)$ be the solution of the initial-value problem $x'' = f(t, x)$, $x(a) = x_0$, $x'(a) = v$. (By the results in Section 10.1, for all v, a unique solution $x(t, v)$ of this initial-value problem exists for $a \leq t \leq b$.) We must show that there exists a unique value, v_0 say, such that $x(b, v_0) = x_F$. For this, it suffices to prove that $x_v(b, v) \geq b - a$, since this implies that $x(b, v)$, as a function of v, has a positive derivative bounded away from zero for all v, and therefore must assume every value x_F exactly once.

Now, $x_v(t) = x_v(t, v)$ satisfies (11.1.14) with initial conditions (11.1.15), which in the one-dimensional case are simply $x_v(a) = 0$, $x'_v(a) = 1$. We assert that $x_v(t) \geq t - a$ for $a \leq t \leq b$. Otherwise, there exists $t \in [a, b]$ such that $x_v(t) < t - a$. Since $x_v(a) = 0$ and $x'_v(a) = 1$, it follows that $x_v(t) > 0$ in a neighborhood of a. Hence there exists $t_0 > a$ such that $x_v(t_0) < t_0 - a$ and $x_v(t) > 0$ for $0 < t \leq t_0$. By the mean-value theorem, $x_v(t_0) = (t_0 - a)x'_v(t_1)$, where $a < t_1 < t_0$. This implies that $x'_v(t_1) < 1$. Again by the mean-value theorem,

$$x'_v(t_1) - x'_v(a) = (t_1 - a)x''_v(t_2), \qquad a < t_2 < t_1,$$

which implies that $x''_v(t_2) < 0$. But, since $f_x \geq 0$, we also have $x''_v(t_2) = f_x(t_2, x(t_2))x_v(t_2) \geq 0$. This contradiction implies that $x_v(t) \geq t - a$ for all $a \leq t \leq b$, as we wished to prove. ∎

Unfortunately, the condition $f_x \geq 0$ which suffices for the existence of a solution is a condition which tends to make the solution unstable. (It suffices to have $f_x \geq \gamma > -\lambda^*$, the smallest eigenvalue of $x'' + \lambda x = 0$, [11.14].) A *stable* solution [11.1] $x(t, v_0)$ is one such that for $\varepsilon > 0$ there exists δ for which $\|x(t, v) - x(t, v_0)\| < \varepsilon$ for all $t > a$ whenever $\|v - v_0\| < \delta$. An *unstable* solution is one which is not stable. Hence a small change or error in the initial value v_0 can grow as $t \to \infty$. In the one-dimensional case, the variational equation

$$\frac{dx_v}{dt} = f_x x_v, \qquad x_v(a) = 1,$$

has the solution

$$x_v(t) = e^{\int_a^t f_x\, dt}.$$

When $f_x \geq 0$, $x_v(t)$ can grow exponentially with t. Since $\Delta x = x(t, v + \Delta v) - x(t, v) = x_v(t, v)\,\Delta v + \varepsilon(t)$, this suggests that Δx also can grow exponentially with t. Thus a small change in the initial value can produce a large change in the solution for t sufficiently large. This behavior is experienced by any numerical solution of the initial-value problem, so that rounding errors and quadrature errors tend to grow from step to step. This, of course, is troublesome in the solution of the initial-value problem and such unstable situations may demand very high-order integration formulas and high computational precision. As we shall see, one method of solving boundary-value problems involves the numerical solution of a sequence of initial-value problems in an iterative procedure. Here instability can have a perturbing effect which slows the rate of convergence of the sequence.

It is important to note that in the preceding discussion we have given preference to the positive t-direction. This is unjustified mathematically when the interval $[a, b]$ is finite. By a change of independent variable $\tau = a + b - t$, we obtain an equivalent boundary problem for the function $\hat{x}(\tau) = x(a + b - \tau)$ with the roles of end points a, b reversed. Thus (11.1.1) becomes

$$\hat{x}' = d\hat{x}/d\tau = -f(a + b - \tau, \hat{x}) = F(\tau, \hat{x}),$$
$$\psi(\hat{x}(b), \hat{x}(a)) = 0. \qquad (11.1.1)_R$$

11.1 Existence and Uniqueness of Exact Solutions

(See Exercise 11.1.) However, (11.1.10) becomes

$$\frac{d^2\hat{x}}{d\tau^2} = \hat{x}'' = f(a + b - \tau, \hat{x}, -\hat{x}') = G(\tau, \hat{x}, \hat{x}'),$$

$$\psi(\hat{x}(b), \hat{x}'(b), \hat{x}(a), \hat{x}'(a)) = 0. \qquad (11.1.10)_R$$

In the special case where f is independent of x' and the boundary conditions are those of (11.1.12), we have

$$\hat{x}'' = f(a + b - \tau, \hat{x}) = G(\tau, \hat{x})$$
$$\hat{x}(b) = x_0, \qquad \hat{x}(a) = x_F. \qquad (11.1.12)_R$$

Letting $v = \hat{x}(a)$, we obtain as the variational equation corresponding to $(11.1.1)_R$,

$$\frac{d\hat{x}_v}{d\tau} = F_{\hat{x}}(\tau, \hat{x}(\tau, x_F))\hat{x}_v,$$

$$\hat{x}_v(a) = I. \qquad (11.1.6)_R$$

Observe that $F_{\hat{x}}(\tau, \hat{x}) = -f_x(a + b - \tau, \hat{x})$. Hence, in the one-dimensional case, if $f_x \geq 0$, then $F_{\hat{x}} \leq 0$ and the situation for $\hat{x}$ would tend to be stable in $(11.1.1)_R$ and $(11.1.6)_R$; i.e. integrating $(11.1.1)_R$ from $\tau = a$ to $\tau = b$ corresponds to integrating (11.1.1) backward from $t = b$ to $t = a$ and gives rise to a stable situation when $f_x \geq 0$. For systems of order $m \geq 2$, the question of stability is much more complicated. For example, for the linear system $x' = Ax$, the identically zero solution is stable if and only if the eigenvalues of the matrix A all have negative real parts. (See, for example, [11.1], Chapter 13.) If we reverse the direction of integration, we obtain the system $x' = -Ax$, having eigenvalues which are the negatives of those of the given system. If A has two eigenvalues λ_1, λ_2 one with negative and the other with positive real part so that the zero solution is unstable, then $-A$ has $-\lambda_1, -\lambda_2$ as eigenvalues and is therefore unstable also. In fact, this happens with the second-order linear system $x'' = c^2 x$. The equivalent system of two first-order equations has matrix

$$A = \begin{bmatrix} 0 & 1 \\ c^2 & 0 \end{bmatrix},$$

which has eigenvalues $\pm c$. Hence the zero solution is unstable. This is reflected in the fact that for $(11.1.12)_R$ the variational equation is

$$\hat{x}_v'' = G_x \hat{x}_v$$

and $G_{\hat{x}} = f_x$, so that there is no change in sign.

If $f(t, 0) = 0$, then we can write $f(t, x) = f_x(t, 0)x + \varepsilon(t, x)$, where $\varepsilon = o(\|x\|)$. Further, if $f_x(t, 0) = A$ is independent of t, we obtain an equation of the form

$$x' = Ax + \varepsilon(t, x), \qquad \text{where} \quad \varepsilon = o(\|x\|).$$

A theorem of Perron (see [11.1]) asserts that the zero solution is stable if the eigen-

498   Boundary-Value Problems for Differential Equations

values of A all have negative real parts. In fact, the zero solution, $\omega_0(t)$, is *asymptotically stable*, which means that $\|\omega(t) - \omega_0(t)\| \to 0$ as $t \to \infty$ for any solution $\omega(t)$ such that $\|\omega(a) - \omega_0(a)\| < \delta$. Thus not only would errors in initial values not grow, but they actually would decrease to zero as $t \to \infty$.

11.1.1   The Adjoint Equation

Consider (11.1.1) again and let φ_i be a component of the vector function $\varphi(v) = \psi(v, x(b, v))$. At a particular v, the directional derivative in the direction h is given by

$$d\varphi_i(v; h) = \frac{d\varphi_i}{ds} = \left(\frac{\partial \psi_i}{\partial v}\right) + \left(\frac{\partial \psi_i}{\partial x}\right)^T \left(\frac{\partial x}{\partial s}\right),$$

where $\partial x/\partial s = y(b) = Y(b)h$ and Y is the solution of the variational equation (11.1.6) satisfying initial conditions (11.1.7). (We take $x_0 = v$.) An alternative formula for $d\varphi_i/ds$ can be obtained which is sometimes more convenient. Let $z^T = (\partial \psi_i/\partial x)^T$ and y be regarded as functions of t (obtained by varying the end point b). Then

$$\frac{d(z^T y)}{dt} = z^T \frac{dy}{dt} + \frac{dz^T}{dt} y$$

$$= z^T f_x Y h + \frac{dz^T}{dt} y = \left(z^T f_x + \frac{dz^T}{dt}\right) y. \qquad (11.1.17)$$

Now, if z is a solution of the equation

$$\frac{dz}{dt} = -f_x^T(t, x(t, v))z, \qquad (11.1.18)$$

then the right-hand side of (11.1.17) is zero. Further, if z is the solution which satisfies the conditions

$$z(b) = \left(\frac{\partial \psi_i}{\partial x}\right)_{b, v}, \qquad (11.1.19)$$

where $(\partial \psi_i/\partial x)_{b, v}$ is the gradient of ψ_i with respect to x evaluated for $x = x(b, v)$, then integration of (11.1.17) yields

$$\left(\frac{\partial \psi_i}{\partial x}\right)^T_{b, v} \left(\frac{\partial x}{\partial s}\right) = z^T(b)y(b) = z^T(a)y(a) = z^T(a)h.$$

If $\partial \psi_i/\partial v = 0$, then $d\varphi_i(v; h) = z^T(a)h$, showing that $z^T(a) = \varphi'(v)$. If we take the special case $\psi(v, x(b, v)) = x(b, v)$, then $(\partial \psi_i/\partial x)^T_{b, v} = (\partial x_i/\partial x)^T = e_i$, the vector having 1 as the ith component and all other components 0. In this case, $\varphi'(v) = x_v(b, v)$, which shows that $x_v(b, v) = Z(a)$, where $Z(t)$ is the matrix solution of

$$\frac{dZ}{dt} = -f_x^T(t, x(t, v))Z, \qquad (11.1.20)$$

which satisfies

$$Z(b) = I. \qquad (11.1.21)$$

11.1 Existence and Uniqueness of Exact Solutions   499

Equation (11.1.20) (or (11.1.18)) is called the *adjoint equation* of the variational equation (11.1.6) (or (11.1.9)). Note that the "initial" conditions for Z are given at $t = b$, so that the adjoint equation must be integrated "backward" from $t = b$ to $t = a$ to obtain $x_v(b, v) = Z(a)$.

Computational Aspects

We have noted that the condition $f_x(t, x(t, v)) > 0$ indicates an unstable solution of (11.1.1) in the one-dimensional case; i.e., errors in initial or intermediate values tend to grow without bound as $t \to \infty$. Likewise, the variational equation tends to have an unstable solution. In this situation, the adjoint equation tends to behave in a stable manner, but since we must integrate backward in t, the result is again unstable. On the other hand, the condition $f_x < 0$ tends to produce stable solutions of the given differential equation, its variational equation and the adjoint equation when the latter is integrated backward in t.

11.1.2   Linear Differential Equations and Linear Boundary Conditions

In Section 10.1, we obtained the integral equation (10.1.4) equivalent to the initial-value problem (10.1.2). When the differential equation is linear, that is, $f(t, x) = A(t)x$, where $A(t)$ is a continuous matrix function, the integral equation takes the form

$$x(t) = x_0 + \int_a^t A(s)x(s)\, ds \qquad (11.1.22)$$

and defines a bounded linear operator T given by

$$(Tx)(t) = T_t x = \int_a^t A(s)x(s)\, ds.$$

The operator T is defined on some appropriate space of functions x, for example $C[a, b]$. The range of T is a subspace of differentiable functions. If we regard x_0 as a constant function, we may write the integral equation in operator form as

$$(I - T)x = x_0$$

and the solution in the form

$$x = (I - T)^{-1} x_0,$$

assuming that the inverse operator exists on the subspace of constant functions. We shall now show, in fact, that the familiar formula,

$$(I - T)^{-1} = \sum_{i=0}^{\infty} T^i,$$

holds. Convergence is in the operator norm induced by the uniform norm $\|x\|_\infty$.

To prove this, we calculate

$$T_t^2 x = \int_a^t A(\xi) \left(\int_a^\xi A(\eta) x(\eta) \, d\eta \right) d\xi = \int_a^t A(\eta) x(\eta) \left(\int_\eta^t A(\xi) \, d\xi \right) d\eta,$$

by a change in the order of integration. Since (using any matrix norm),

$$\left\| \int_\eta^t A(\xi) \, d\xi \right\| \leq \mu(t - \eta),$$

where $\|A(\xi)\| \leq \mu$, it follows that

$$\|(T^2 x)(t)\| \leq \mu^2 \int_a^t (t - \eta) \|x\|_\infty \, d\eta \leq \frac{\mu^2}{2} (t - a)^2 \|x\|_\infty.$$

$$\|T^2 x\|_\infty \leq \mu^2 (b - a)^2 \|x\|_\infty / 2.$$

In the same way, we can prove by induction on i that

$$\|T^i x\|_\infty \leq \frac{\mu^i}{i!} (b - a)^i \|x\|_\infty, \qquad i = 0, 1, 2, \ldots.$$

Hence $\|T^i\| \leq \mu^i (b - a)^i / i!$ and the infinite series for $(I - T)^{-1}$ converges. Indeed,

$$\|(I - T)^{-1}\| \leq e^{\mu(b - a)}. \tag{11.1.23}$$

Therefore the solution, $x(t)$, of the linear initial-value problem can be expressed in the form

$$x(t) = \sum_{i=0}^\infty T_t^i x_0. \tag{11.1.24}$$

It is easy to show that (11.1.24) can also be derived by applying successive approximation to the mapping $x_0 + T_t x$ (Exercise 11.3).

A linear boundary-value problem is one in which the differential equation is linear,

$$x' = A(t)x,$$

and the boundary conditions are likewise linear, that is, of the form

$$Cx(a) + Dx(b) = u, \tag{11.1.25}$$

where C and D are given matrices and u is a given vector. As before, we set $x(a) = v$. The solution having v as initial value is, by (11.1.24), $x(t, v) = \sum_{i=0}^\infty T_t^i v$. Therefore $x(t, v)$ is a linear function of v and the boundary condition,

$$\varphi(v) = Cv + D \left(\sum_0^\infty T_b^i \right) v - u$$

is linear in v also. Hence we may solve $\varphi(v) = 0$ by solving the linear algebraic system

$$\left(C + D \sum_0^\infty T_b^i \right) v = u.$$

This system has a unique solution for each u if the matrix $B = C + D\sum_0^\infty T_b^i$ is nonsingular. Using (11.1.23), one can give various sufficient conditions for this to be true. For example, if C^{-1} exists, then $B = C(I + C^{-1}D\sum_0^\infty T_b^i)$ and a sufficient condition for B^{-1} to exist is that $\|C^{-1}D\sum_0^\infty T_b^i\| < 1$. By virtue of (11.1.23), it suffices that $e^{\mu(b-a)} < 1/\|C^{-1}D\|_\infty$. If $\|C^{-1}D\|_\infty \geq 1$, this condition cannot be valid and is, therefore, of no use. Very often the boundary conditions are of the form (11.1.25) with $C + D = I$ and C of the type

$$\begin{bmatrix} I_k & 0 \\ 0 & 0 \end{bmatrix}.$$

That is, the boundary conditions consist of k equations $x_i(a) = u_i$, $1 \leq i \leq k$, imposing values on the first k components of $x(a)$ and $m - k$ equations $x_i(b) = u_i$, $k + 1 \leq i \leq m$, imposing values at b on the remaining components. More generally, we might assume that $(C + D)^{-1}$ exists. In such a case, we could write

$$B = (C + D)\left(I + (C + D)^{-1}D \sum_{i=1}^\infty T_b^i\right) \qquad (11.1.26)$$

and B^{-1} would exist if $\|(C + D)^{-1}D\|_\infty \|\sum_1^\infty T_b^i\|_\infty < 1$. Again, this can be translated into a condition on μ (the bound of $\|A(\xi)\|$, $a \leq \xi \leq b$), namely,

$$e^{\mu(b-a)} - 1 < 1/\|(C + D)^{-1}D\|_\infty. \qquad (11.1.27)$$

This is satisfied if $\mu(b - a)$ is sufficiently small.

It frequently happens that the differential equations are nonlinear, while the boundary conditions are linear. In such a case, $\varphi(v) = Cv + Dx(b, v) - u$. A zero of φ is a fixed point of the function $F(v) = v - M\varphi(v)$, where M is a constant matrix. If M can be chosen to make F contractive, then the method of successive approximation will yield a unique fixed point, by Theorem 5.1.1. (We assume that $x(b, v)$ is defined for all v.) Now, differentiating, we have

$$F'(v) = I - M(C + Dx_v(b, v)), \qquad (11.1.28)$$

where x_v is a solution of (11.1.6), (11.1.7), the variational system, with $x_0 = v$. Since (11.1.6) is a linear differential equation, a formula like (11.1.24) applies, where now

$$T_t z = \int_a^t f_x(s, x(s, v))z(s)\,ds.$$

Therefore, since $x_v(a) = I$,

$$x_v(t) = \sum_{i=0}^\infty T_t^i I,$$

and

$$F'(v) = I - M\left(C + D\sum_{i=0}^\infty T_b^i I\right).$$

We see that one choice of M is to take $M = (C + D)^{-1}$, when the inverse exists.

This yields

$$F'(v) = (C + D)^{-1} D \sum_1^\infty T_b^i.$$

Again, (11.1.27) is sufficient for $\|F'(v)\| < 1$, where now μ is a bound for $\|f_x(t, x)\|$. This suggests that successive approximation applied to the function $v - (C + D)^{-1}\varphi(v)$ can be used to solve boundary-value problems with linear boundary conditions. For the general case, we prefer Newton's method as described in the next section. However, the successive approximation method has allowed us to prove the following existence theorem [11.2].

Theorem 11.1.4 Let $x' = f(t, x)$, where f and f_x are continuous in the strip $a \leq t \leq b$, $\|x\| < \infty$. Let $\|f_x\| < \mu$ in this strip. If the boundary conditions are given by (11.1.25), where $(C + D)^{-1}$ exists and satisfies (11.1.27), then there exists a unique solution of the boundary-value problem.

11.2 SOLUTION OF BOUNDARY-VALUE PROBLEMS BY ITERATION ON THE INITIAL VALUES

Let us consider the boundary-value problem (11.1.10) in the special case when f does not involve x' and the boundary conditions are given by (11.1.12). We must determine an initial value $v = x'(a)$ such that $\varphi(v) = 0$, where φ is given by (11.1.13). If Newton's method is invoked, we must compute

$$\begin{aligned} v_{n+1} &= v_n - (\varphi'(v_n))^{-1} \varphi(v_n) \\ &= v_n - (x_v(b, v_n))^{-1}(x(b, v_n) - x_F). \end{aligned}$$

Here $x_v(t, v_n)$ is a solution of the variational equation (11.1.14) with $\partial f/\partial x$ evaluated along the solution, $x(t, v_n)$, of (11.1.10) having initial values $x(a) = x_0$, $x'(a) = v_n$. This leads to an iterative procedure which can be summarized as follows:

Step 1. Choose an initial guess v_0 for $x'(a)$.

Step 2. Solve the initial-value problem

$$x'' = f(t, x), \quad x(a) = x_0, \quad x'(a) = v_0, \qquad (11.2.1)$$

by one of the methods of Chapter 10. Assuming grid points $a = t_0 < t_1 < \cdots < t_q = b$, we denote the numerical solution by $(x_k^{(0)} : 0 \leq k \leq q)$. We take $x_q^{(0)}$ as an approximation to $x(b, v_0)$.

Step 3. The variational equation

$$x_v'' = (\partial f/\partial x)_0 x_v, \qquad (11.2.2)$$

with initial values given by (11.1.15), is solved numerically using the same grid points, the subscript zero indicating that the partial derivatives $\partial f_i/\partial \xi_j$ in the Jacobian $\partial f/\partial x$ are evaluated "along the numerical solution"; i.e., with $x = x_k^{(0)}$ at grid point

t_k, $0 \le k \le q$. Since (11.1.15) is a matrix of initial values, the variational equation must be solved m times. (In actual computation, the m solutions would be computed in parallel with the solution of (11.2.1) to avoid recomputation of $(\partial f/\partial x)$ at the grid points.) The final values $x_{vq}^{(0)}$ at the last grid point t_q form a matrix which approximates $x_v(b, v_0)$.

Step 4. The linear algebraic system

$$x_{vq}^{(0)} \Delta v_0 = x_q^{(0)} - x_F$$

is solved for $\Delta v_0 = v_1 - v_0$. The new initial value $v_1 = v_0 + \Delta v_0$ replaces v_0 and the process is repeated. Iteration is continued until a convergence criterion is satisfied by the v_n or until the boundary condition $x_q^{(n)} - x_F = 0$ is satisfied with acceptable accuracy.

This procedure is a modification of one given in [11.4], where the adjoint equation is used instead of the variational equation to determine $x_v(b, v)$. (See Section 11.1.1.) Iteration on the initial values is also referred to as *shooting*. It is important to remember that there is no reason why this method cannot be applied in the negative t-direction; i.e., we can then iterate on the unspecified values at $t = b$. Thus, in the example just considered, we could have chosen $v = x'(b)$ and integrated backward to $t = a$. In fact, this direction is to be preferred in the "unstable" situation referred to in the preceding section, since errors may decrease (i.e., solutions are stable) in the negative t-direction when they increase in the positive t-direction.

There are two questions which arise concerning the convergence of shooting using Newton's method. These questions arise whenever a general iterative method—such as Newton's or the successive approximation method—is applied to a particular class of problems which must be solved numerically. The first question is: What are sufficient conditions for the method to converge in this class of problems? For example, in the boundary-value problem (11.1.1), what conditions on the derivative f and the boundary function ψ will ensure convergence of Newton's iteration carried out with exact solution of the initial-value problem? The second question is: What is the effect of the numerical approximations which usually must be made in actual computation? For example, the numerical solution of each initial-value problem in the shooting method usually is subject to quadrature errors, as discussed in Chapter 10. Therefore we do not compute the exact value $x(b, v_n)$, nor do we compute $x_v(b, v_n)$ exactly. As described in step 3 of the computational procedure, $x(b, v_n)$ is approximated by $x_q^{(n)}$, the value of the numerical solution at grid point $t_q = b$. Thus, for example, if a fourth-order predictor–corrector integration scheme is used to solve the initial-value problem, we will have $\|x_q^{(n)} - x(b, v_n)\|_\infty = O(h^4)$, where h is the step size. (See formula (10.2.9).) Similarly, $x_v(b, v_n)$ is replaced by the numerical solution of the variational equation, $x_{vq}^{(n)}$. Thus, in the special problem (11.1.10), (11.1.12), we actually solve the systems

$$x_{vq}^{(n)} \overline{\Delta v_n} = x_q^{(n)} - x_F,$$
$$\overline{v_{n+1}} = \overline{v_n} + \overline{\Delta v_n},$$

instead of the exact systems

$$x_v(b, v_n) \Delta v_n = x(b, v_n) - x_F,$$
$$v_{n+1} = v_n + \Delta v_n, \qquad n = 0, 1, 2, \ldots.$$

If the numerical errors are kept small, it is to be expected that the computed sequence $(\bar{v}_n)$ will converge whenever the exact sequence (v_n) converges and that the error $\|(x_i^{(n)}) - T_G x(t, \bar{v}_n)\|$ will be of the same order as the quadrature error. (See the Remark following Definition 10.1.1.)

In general, computationally useful answers to these questions are difficult to find. However, in the case of linear boundary conditions of the form

$$\psi(x(a), x(b)) = Cx(a) + Dx(b) - u,$$

where C and D are matrices and u is a given vector, it is possible to give reasonably satisfactory answers. As in Section 11.1.2, we seek a zero of the function $\varphi(v) = Cv + Dx(b, v) - u$ by searching for a fixed point of $F(v) = v - M\varphi(v)$, where M is a matrix chosen to make F contractive.

Consider first the case where M is a constant matrix; i.e., independent of v. For example, we might take $M = (C + D)^{-1}$, as in Section 11.1.2, which would lead to the successive approximation method

$$v_{n+1} = v_n - (C + D)^{-1} \varphi(v_n).$$

(In actual computation, we would solve the linear system, $(C + D)(v_{n+1} - v_n) = -\varphi(v_n)$.) A sufficient condition for convergence, as we saw, is given by (11.1.27), where $\|f_x(t, x)\| \leq \mu$. Now, in the numerical computation, $x(b, v)$ is replaced by x_q, the value obtained at the grid point $t_q = b$ by one of the methods of Chapter 10. We shall assume that a stable method of order h^r is used; e.g., the Adams–Moulton method. Then $\|x(b, v) - x_q\|_\infty = O(h^r)$, where h is the grid size. Instead of computing $\varphi(v)$, we compute $\bar{\varphi}(v) = Cv + Dx_q - u$ and $\bar{F}(v) = v - M\bar{\varphi}(v)$. Since

$$F(v) - \bar{F}(v) = -MD(x(b, v) - x_q),$$

we have

$$\|F(v) - \bar{F}(v)\| \leq \|MD\| \, \|x(b, v) - x_q\|_\infty = O(h^r).$$

The actual successive approximation computation, ignoring rounding errors, is described by

$$\bar{v}_{n+1} = \bar{F}(\bar{v}_n) = F(\bar{v}_n) + O(h^r),$$

and can be analyzed as we did in Theorem 5.1.5. (There we used m as the iteration index.) Referring to the proof of Theorem 5.1.5 (and taking $\Delta a = 0$, since we assume no error in $\bar{v}_0$), we are able to assert immediately that

$$\|\bar{v}_n - v_n\| = O(h^r)$$

and

$$\|\bar{v}_n - v^*\| = O(h^r) + \frac{K^n}{1 - K} \|\bar{v}_1 - \bar{v}_0\|,$$

(In that case, we may take $\delta = 0$.) For this inequality to hold, it is sufficient that
$$(1 - \|(C + D)^{-1}\| \|D\|(e^{\mu(b-a)} - 1))^2 > 4\|(C + D)^{-1}\|^2 \|D\|^2 \|f\|_\infty (b - a)e^{L'(b-a)},$$
which simplifies to
$$\|(C + D)^{-1}\|_\infty \|D\|_\infty < (2(\|f\|_\infty (b - a))^{1/2} e^{L'(b-a)/2} + e^{\mu(b-a)} - 1)^{-1}. \quad (11.2.7)$$
We summarize these results as follows.

Theorem 11.2.1 Consider the differential equation $x' = f(t, x)$ with linear boundary conditions (11.1.25). Let $(C + D)^{-1}$ exist and choose $v_0 = (C + D)^{-1}u$ as initial value. Suppose that the solution $x(t, v_0)$ of the initial-value problem exists for $a \leq t \leq b$. Let

$$\|f\|_\infty = \sup_{a \leq t \leq b} \|f(t, x(t, v_0))\|_\infty$$

and

$$\mu = \sup_{a \leq t \leq b} \|f_x(t, x(t, v_0))\|_\infty$$

be such that (11.1.27) and (11.2.7) hold, where L' is a Lipschitz constant for $f_x(t, x)$ in a rectangle containing the solution curve $\{(t, x(t, v_0)) : a \leq t \leq b\}$. Finally, let

$$\frac{1}{2\gamma}\left(1 - \left(1 - \frac{4\alpha\gamma}{\beta}\right)^{1/2}\right) < r < \frac{1}{2\gamma}\left(1 + \left(1 - \frac{4\alpha\gamma}{\beta}\right)^{1/2}\right),$$

where α, β, γ are defined above. If the initial-value problem has a solution in $[a, b]$ for any initial value v such that $\|v - v_0\| \leq r$, then the boundary-value problem has a solution $x(t, v^*)$ and $v^* = \lim_{n \to \infty} v_n$, where v_n is given by the modified Newton method.

A similar analysis can be carried out for Newton's method applied to linear boundary conditions. Here we use the conditions (i), (ii) in Theorems 5.5.4 or 5.5.5 (Exercise 5.24), which are similar to those in Theorem 5.5.1.

When solutions of the initial-value problem are unstable, the method of shooting tends to be ill-conditioned. This difficulty can sometimes be overcome by doing *multiple* or *parallel shooting*. In this scheme, the interval $[a, b]$ is subdivided into N subintervals $[a_i, b_i]$ with $a_{i+1} = b_i$ and initial-value problems are solved simultaneously on all subintervals by imposing as a boundary condition on $[a_i, b_i]$ the continuity condition $x^{(i)}(a_i) = x^{(i-1)}(b_{i-1})$, where the $x^{(i)}$, $1 \leq i \leq N$, are the solutions on $[a_i, b_i]$ and are treated as new vector variables. The given boundary condition in (11.1.1) becomes $\psi(x^{(1)}(a), x^{(N)}(b)) = 0$. (See [11.8], [11.9].)

Another method which can sometimes be used to overcome the instability problem is discussed in the next section.

11.3 DIFFERENCE METHODS

We consider the general differential equation $x' = f(t, x)$ with linear boundary conditions $Cx(a) + Dx(b) = u$. Suppose we apply one of the stable difference schemes of Section 10.6 to this problem. The differential equation is replaced by the difference equation (10.6.2) and the boundary conditions become $Cx_0 + Dx_N = u$, where x_i is the numerical approximation to $x(t_i)$ at the grid point $t_i = a + ih$, $0 \le i \le N$, $t_N = b$. For example, the Euler method would give

$$x_i - x_{i-1} - hf(t_{i-1}, x_{i-1}) = 0, \quad 1 \le i \le N,$$

which, together with the equation $Cx_0 + Dx_N = u$, is a system of $N + 1$ equations for the $N + 1$ unknowns $x_0, \ldots, x_N$. Another choice, of higher accuracy, is the trapezoidal rule

$$x_i - x_{i-1} - \frac{h}{2}(f(t_{i-1}, x_{i-1}) + f(t_i, x_i)) = 0.$$

A comparable scheme is obtained by using the centered difference formula (8.4.5) to approximate the derivative. This leads to the system of difference equations,

$$x_i - x_{i-1} - hf\left(a + (i - \tfrac{1}{2})h, \frac{x_i + x_{i-1}}{2}\right) = 0,$$

$$Cx_0 + Dx_N = u. \tag{11.3.1}$$

In [11.2], sufficient conditions are given for the system (11.3.1) to have a unique solution. Furthermore, (11.3.1) can be solved iteratively according to the scheme,

$$x_i^{(k+1)} - x_{i-1}^{(k+1)} = hf\left(t_{i-1/2}, \frac{x_i^{(k)} + x_{i-1}^{(k)}}{2}\right),$$

$$Cx_0^{(k+1)} + Dx_N^{(k+1)} = u, \tag{11.3.2}$$

with arbitrary initial guess $x_i^{(0)}$. Observe that $x_i^{(k+1)}$ is determined by solving a linear algebraic system of order $m(N + 1)$, where m is the dimension of the vector x. In fact, if we let $X^T = (x_0, \ldots, x_N)$ and

$$(F(X))^T = \left(u, hf\left(t_{1/2}, \frac{x_1 + x_0}{2}\right), \ldots, hf\left(t_{N-1/2}, \frac{x_N + x_{N-1}}{2}\right)\right), \tag{11.3.3}$$

then we may write (11.3.2) in the vector (or matrix) form,

$$AX^{(k+1)} = F(X^{(k)}), \tag{11.3.4}$$

where

$$A = \begin{bmatrix} C & 0 & 0 & \cdots & 0 & D \\ -I & I & 0 & \cdots & 0 & 0 \\ 0 & -I & I & 0 & \cdots & 0 \\ \vdots & & & \ddots & & \\ 0 & & \cdots & & -I & I \end{bmatrix} \tag{11.3.5}$$

I being the $m \times m$ identity. It is not difficult to show that A is nonsingular if $C + D$ is nonsingular (Exercise 11.6). In that case, (11.3.4) defines a successive approximation method for finding a fixed point of the mapping $G(X) = A^{-1}F(X)$. It can be shown that, under conditions similar to those in Theorem 11.2.1, G is contractive. (See Exercise 11.7.) Under the same conditions, Newton's method is also applicable, but requires the computation of $G'(X)$. If f has continuous second partials, then the error in the numerical solution is of order h^2. This follows from (8.4.5) and Exercise 8.12, since

$$\frac{x_i + x_{i-1}}{2} - x_{i-1/2} = \frac{x_i - 2x_{i-1/2} + x_{i-1}}{2} = h^2 x''_{i-1/2} + O(h^2).$$

Another difference technique is based on spline interpolation of $x(t)$ on a grid $E = \{a = t_0 < t_1 < \cdots < t_N = b\}$. (See Section 8.5 and [8.4], [8.5], [11.6].) We restrict our attention to the one-dimensional second-order equation $x'' = f(t, x, x')$, with arbitrary boundary conditions $\psi_j(x(a), x'(a), x(b), x'(b)) = 0$, $j = 1, 2$. We shall approximate the exact solution $x(t)$ by a spline function $S(t)$, on the grid E, which satisfies the differential equation at the grid points t_i and the boundary conditions at t_0 and t_N. Rather than use (8.5.1) to express the cubic spline $S(t)$ on the interval $[t_{i-1}, t_i]$, we use instead the following formula involving approximate second derivatives:

$$S(t) = x_{i-1} \frac{(t_i - t)}{h_i} + x_i \frac{(t - t_{i-1})}{h_i}$$
$$- \frac{h_i^2}{6} \left(x''_{i-1} \left[\frac{(t_i - t)}{h_i} - \frac{(t_i - t)^3}{h_i^3} \right] + x''_i \left[\frac{(t - t_{i-1})}{h_i} - \frac{(t - t_{i-1})^3}{h_i^3} \right] \right), \quad (11.3.6)$$

where $h_i = t_i - t_{i-1}$. It is easy to verify that $S(t)$ as given by (11.3.6) is a piecewise cubic polynomial function with $S(t_i) = x_i$ and having a continuous second derivative $S''(t)$ on $[a, b]$ with $S''(t_i) = x''_i$. In order that the first derivative $S'(t)$ be continuous, we differentiate and set derivatives equal at the grid points. This imposes the following conditions on the quantities x_i and x''_i:

$$h_i x_{i-2} - (h_i + h_{i-1}) x_{i-1} + h_{i-1} x_i$$
$$= \frac{h_i h_{i-1}}{6} (h_{i-1} x''_{i-2} + 2(h_i + h_{i-1}) x''_{i-1} + h_i x''_i), \quad i = 2, \ldots, N. \quad (11.3.7)$$

Since $S(t)$ must satisfy the differential equation and boundary conditions, we must also require that

$$x''_i = f(t_i, x_i, x'_i), \quad i = 0, \ldots, N \quad (11.3.8)$$

and

$$\psi_j(x_0, x'_0, x_N, x'_N) = 0, \quad j = 1, 2. \quad (11.3.9)$$

In (11.3.8), the quantity $x'_i = S'(t_i)$ is obtained by differentiating (11.3.6) and allowing $t \to t_i$. We get

$$x'_i = \frac{x_i - x_{i-1}}{h_i} + \frac{h_i}{6} (2x''_i + x''_{i-1}), \quad i = 0, \ldots, N. \quad (11.3.10)$$

Thus, if we solve (by iteration) the $2N + 2$ nonlinear equations in (11.3.7), (11.3.8), (11.3.9) for the $2N + 2$ unknowns x_i, x_i'', we obtain a cubic spline $S(t)$ which satisfies the boundary conditions and the differential equation at the grid points t_i. Note that when a uniform mesh size $h_i = h$ is used, (11.3.7) simplifies to a familiar difference formula.

To improve the accuracy of a difference scheme, we may apply the deferred approach to the limit (Section 9.7). It can be shown [11.2] that $x(t_i) = x_i + h^2 e(t_i) + O(h^4)$ for the scheme in (11.3.2), for example. If we regard $x_i = x_i(h) = F_h x(a + ih)$, then formula (9.7.2) is applicable; i.e., $x_i^{(1)} = \frac{1}{3}(4x_{2i}(h/2) - x_i(h))$ approximates $x(a + ih)$ with an error $O(h^4)$.

11.4 LINEAR BOUNDARY-VALUE PROBLEMS, EIGENVALUE PROBLEMS AND VARIATIONAL METHODS

In Section 7.8, we considered the Sturm–Liouville boundary-value problem in a special case. In this section, we consider the second-order differential operator L defined by

$$Lx = -\frac{d}{dt}\left(p(t)\frac{dx}{dt}\right) + q(t)x$$

and construct the equations

$$Lx = y, \qquad (11.4.1)$$

$$Lx = \lambda r x \qquad (11.4.2)$$

where $y = y(t)$, $r = r(t)$, $p(t)$ and $q(t)$ are given functions. We impose linear boundary conditions of the form

$$ax(0) - bx'(0) = 0, \qquad cx(1) + dx'(1) = 0, \qquad (11.4.3)$$

where a, b, c, d are constants. Equations (11.4.1), (11.4.3) form a linear boundary-value problem which can be solved by variational methods discussed in Chapter 12. Likewise, (11.4.2), (11.4.3) form a linear eigenvalue problem solvable by the methods of Chapter 12. This is a consequence of the fact that the operator L is a symmetric linear operator on the Hilbert space $L^2[0, 1]$. (We note that a linear transformation of the variable t would allow us to consider any other closed bounded interval rather than $[0, 1]$.)

To prove that L is symmetric, we restrict our attention to the subspace D_L of functions which have continuous derivatives of order at least 2 and satisfy the boundary conditions (11.4.3). For u, v in D_L we have

$$\langle Lu, v \rangle = -\int_0^1 v \frac{d}{dt}\left(p(t)\frac{du}{dt}\right) dt + \int_0^1 quv \, dt.$$

Integrating the first integral by parts and applying (11.4.3), we get

$$\langle Lu, v \rangle = \int_0^1 \left(p \frac{du}{dt}\frac{dv}{dt} + quv\right) dt + \frac{a}{b} p(0)u(0)v(0) + \frac{c}{d} p(1)u(1)v(1).$$

(We assume $b \neq 0$, $d \neq 0$. If $b = 0$, then $u(0) = 0$ and we omit that term, and similarly if $d = 0$.) We see immediately that $\langle Lu, v \rangle = \langle u, Lv \rangle$, which establishes the symmetry of L.

We assume that p, p' and q are continuous on $[0, 1]$, so that all indicated operations are well defined. Now, in order to apply the methods of Chapter 12, we must show that $\langle Lu, u \rangle > \alpha^2 \langle u, u \rangle$ for some constant $\alpha \neq 0$. For this we must assume further that $p(t) \geq 0$ and $q(t) \geq 0$ on $[0, 1]$ and the integral

$$\int_0^1 \frac{dt}{p(t)}$$

has a finite value. Finally, we assume that a, b, c, d are nonnegative and not both a and c are zero and similarly for b, d. The existence of the required constant α is proved in Exercise 11.8.

In Section 12.3, it is shown in Example 2 that solving $Lx = y$ is equivalent to minimizing the quadratic functional $J(x) = \frac{1}{2} \langle Lx, x \rangle - \langle y, x \rangle$. Various methods of minimizing J are described in Chapter 12. The eigenvalue problem (11.4.2) can also be solved by variational techniques; i.e., by minimizing a functional. One such method is described in Section 12.7.1.

The operator L is not bounded. However, it is possible to obtain an equivalent eigenvalue problem in which the operator is bounded. This is done by "inverting" the operator L; i.e., converting the differential operator into an integral operator. The standard technique for doing this is to construct the *Green's function*, $g(t, x)$, for the boundary-value problem. The solution of (11.4.2), (11.4.3) is equivalent to the solution of the integral equation

$$x(t) = \lambda \int_0^1 g(t, s) r(s) x(s) \, ds,$$

where g is defined in the square region $0 \leq t \leq 1$, $0 \leq s \leq 1$ and satisfies the conditions:

i) $g(t, s)$ is continuous in t for each s.
ii) $g(0, s) = g(1, s) = 0$.
iii) For $t \neq s$, $\partial g/\partial t$ and $\partial^2 g/\partial t^2$ are continuous, and for $t = s$, $g_t(t+, s) - g_t(t-, s) = -1/p(s)$.
iv) $Lg = 0$ for $t \neq s$.

(For a proof of the equivalence of the two problems, see [11.10], or simply apply L to both sides of the integral equation and use conditions (i)–(iv). Conversely, multiplying (11.4.2) by g and using integration by parts yields the integral equation.) To construct g, we solve $Lu = 0$ with $u(0) = 0$, $u'(0) \neq 0$, and $Lv = 0$ with $v(1) = 0$, $v'(1) \neq 0$. Then

$$g(t, s) = \begin{matrix} \alpha u(t) v(s) & \text{for } t < s, \\ \alpha v(t) u(s) & \text{for } t > s, \end{matrix}$$

where α is a constant chosen to make condition (iii) valid, namely,

$$1/\alpha = -p(s)(v'(s)u(s) - u'(s)v(s)).$$

This shows that $g(t, s) = g(s, t)$. If we let $y(t) = (r(t))^{1/2}x(t)$ we obtain the integral equation

$$y(t) = \lambda \int_0^1 K(t, s)y(s)\, ds, \tag{11.4.4}$$

where the kernel $K(t, s) = (r(t))^{1/2}g(t, s)(r(s))^{1/2}$ is a symmetric function. Thus we may write (11.4.4) in operator form as

$$y = \lambda A y,$$

where A is a bounded self-adjoint operator, and the methods of Chapter 12 can be applied to compute λ and y.

The application of the Rayleigh–Ritz technique (Section 12.7) to the solution of nonlinear boundary-value problems is treated in [11.11].

11.5 THE BUBNOV–GALERKIN METHOD

The Ritz method, referred to in the previous section and considered in Chapter 12, applies to self-adjoint boundary-value or eigenvalue problems such as (11.4.1)–(11.4.3). When the operator L is not symmetric or fails to satisfy the condition $\langle Lu, u \rangle > \alpha^2 \langle u, u \rangle$, the Ritz variational approach does not apply. However, the basic idea of Ritz's method—namely, the reduction of an infinite-dimensional problem to a sequence of increasing finite-dimensional problems—can still be applied. This leads to a method known as the *Galerkin* (or *Bubnov–Galerkin*) method. We describe this briefly and refer to [12.17] for further details.

Consider the equation $Lx = y$, where x and y are regarded as elements of a Hilbert space H and $L : H \to H$ need not be linear. For example, L may be a nonlinear differential operator and y a given function in $L^2[a, b]$. If there are boundary conditions, these restrict x to lie in some subspace D of H. It is required to find $x \in D$ such that $Lx = y$. The Galerkin method requires that we find a sequence of elements $(v_1, v_2, \ldots)$, $v_i \in D$ (called *coordinate functions*) which is complete in some sense. We then solve $Lx = y$ relative to the subspaces V^n spanned by $\{v_1, \ldots, v_n\}$, $n = 1, 2, \ldots$. More precisely, we consider a linear combination $x_n = \sum_{j=1}^n \xi_j v_j$ and determine the coefficients ξ_j so that the component of $Lx_n - y$ in V^n is zero. For this it is necessary and sufficient that

$$\langle Lx_n - y, v_i \rangle = 0, \quad i = 1, \ldots, n.$$

This leads to a system of n equations

$$\left\langle L\left(\sum_{j=1}^n \xi_j v_j\right), v_i \right\rangle = \langle y, v_i \rangle, \quad i = 1, \ldots, n \tag{11.5.1}$$

11.13. J. ORTEGA and W. RHEINBOLDT, "On discretization and differentiation of operators with applications to Newton's method," *SIAM J. Numer. Anal.* **3**, 143–156 (1966)

11.14. M. LEES, "Discrete methods for nonlinear two-point boundary-value problems," in *Numerical Solution of Partial Differential Equations*, J. Bramble (editor), Academic Press, New York (1966)

11.15. L. BERS, F. JOHN, and M. SCHECHTER, *Partial Differential Equations*, Wiley (Interscience), New York (1964)

11.16. G. FORSYTHE and W. WASOW, *Finite Difference Methods for Partial Differential Equations*, Wiley, New York (1960)

11.17. R. S. VARGA, *Matrix Iterative Analysis*, Prentice-Hall, Englewood Cliffs, N.J. (1962)

11.18. J. H. BRAMBLE (editor), *Numerical Solution of Partial Differential Equations*, Academic Press, N.Y. (1965)

EXERCISES

11.1 a) Verify that the function $\hat{x}(\tau) = x(a + b - \tau)$ satisfies $(11.1.1)_R$ if $x(t)$ satisfies (11.1.1). [*Hint*: $x'(a + b - \tau) = f(a + b - \tau, x(a + b - \tau))$ and
$$dx/d\tau \big|_\tau = -dx/dt \big|_{t=a+b-\tau}.]$$

b) Do the same for (11.1.10) and $(11.1.10)_R$. Also show that the equivalent first-order system (11.1.10a) can be treated in the same way as (11.1.1).
[*Hint*: If $\hat{x}(\tau) = x(a + b - \tau)$, then $\hat{x}'(\tau) = -x'(a + b - \tau) = -y(a + b - \tau) = -\hat{y}(\tau)$. Thus the equivalent system would be
$$\hat{x}' = -\hat{y},$$
$$\hat{y}' = -f(a + b - \tau, \hat{x}, \hat{y}),$$
$$\psi(\hat{x}(b), \hat{y}(b), \hat{x}(a), \hat{y}(a)) = 0.$$

11.2 Verify that the differential equation in Example 1 of Section 11.1 does not satisfy the hypotheses of Theorem 11.1.3. (This is consistent with nonexistence and nonuniqueness of the boundary-value problem for certain values of d.)

11.3 Derive (11.1.24) by applying successive approximation to the mapping $x_0 + T_t x$. [*Hint*: Let $x^{(0)} = x_0$. Then $x^{(1)} = x_0 + Tx_0, x^{(2)} = x_0 + Tx^{(1)} = x_0 + Tx_0 + T^2 x_0$, etc.]

11.4 Verify that the boundary-value problem $x' = f(t, x)$, $Cx(a) + Dx(b) = u$, has a solution if in the infinite strip $a \le t \le b$, $\|x\| < \infty$, the following conditions are satisfied: f is continuous, f_x is continuous and $\|f_x(t, x)\|_\infty \le L$, $(C + D)^{-1}$ exists and $e^{L(b-a)} < 1 + 1/\|(C + D)^{-1}D\|_\infty$. [*Hint*:
$$\|T_t z\|_\infty \le \int_a^t \|f_x(s)\| \|z(s)\| \, ds \le L(b - a)\|z\|_\infty$$

and so on for T_t^i to give $\|\sum_1^\infty T_t^i\| < e^{L(b-a)} - 1$. Using (11.1.28), show that

$$F'(v) = I - (C + D)^{-1}\left(C + D + D\sum_1^\infty T_t^i\right) = -(C + D)^{-1} D \sum_1^\infty T_t^i.$$

Also see [11.2].]

11.5 Let x be one-dimensional. Consider the linear second-order boundary-value problem, $-x'' + p(t)x' + q(t)x = r(t)$, $x(a) = \alpha$, $x(b) = \beta$. Replace the derivatives x'' and x' by centered differences on a uniform grid $\{t_i = a + ih, 0 \le i \le N\}$.

a) Show that the resulting difference equations form a linear algebraic system with a tridiagonal matrix A. [*Hint:* Use formula (8.4.5) and Exercise 8.12(b) to replace $x''(t_i)$ by $(x_{i+1} - 2x_i + x_{i-1})/h^2$, and $x'(t_i)$ by $(x_{i+1} - x_{i-1})/2h$, $1 \le i \le N - 1$. The other two equations are $x_0 = \alpha$, $x_N = \beta$.]

b) Assume that $\|p\|_\infty < \infty$ and $0 < \mu \le q(t) \le \|q\|_\infty < \infty$. If $h \le 2/\|p\|_\infty$, show that the matrix A in part (a) is strictly diagonally dominant (Section 4.4), and therefore the difference equations have a unique solution. [*Hint:* See [11.2], Chapter 3.]

c) Program the solution of the problem
$$-x'' + 11x' + 12x = 12t + 11, \quad x(0) = 1, x(1) = 1 + 1/e$$
both by the difference method in part (a) and by shooting (using successive approximation and Newton's method.) [*Hint:* The general solution is $x(t) = t + ce^{-t} + de^{12t}$ and $c = 1$, $d = 0$ satisfy the boundary conditions. However, rounding errors will cause e^{12t} to enter. See [11.7], Chapter 5.]

11.6 Let A be the $m(N + 1) \times m(N + 1)$ matrix in (11.3.5). Verify that if $(C + D)^{-1}$ exists, then A^{-1} exists. [*Hint:* When $m = 1$, we have

$$AE_{N-1} \cdots E_1 = \begin{bmatrix} (C+D) & D & \cdots & D \\ 0 & 1 & \cdots & 0 \\ \vdots & & \ddots & \vdots \\ 0 & \cdots & & 1 \end{bmatrix}$$

where the E_i are elementary column operations. This generalizes to arbitrary m by using block operations.)

11.7 Let $G = A^{-1}F$, where F is given by (11.3.3) and A by (11.3.5). Assume that in the infinite strip $a \le t \le b$, $\|x\| < \infty$, the function $f(t, x)$ and its partial derivative $f_x(t, x)$ are continuous and $\|f_x(t, x)\|_\infty \le \mu$. Assume that (11.1.27) holds and also $e^{\mu(b-a)} - 1 < 1/\|(C + D)^{-1}C\|_\infty$. Prove that $\|G'(X)\|_\infty < 1$, and therefore that G is contractive. [*Hint:* $G'(X) = A^{-1}F'(X)$ and the entries in $F'(X)$ are of the form $hf_x(t_{i-1/2}, (x_i + x_{i-1})/2)$.] Note:

$$A^{-1} = \begin{bmatrix} (C+D)^{-1} & -(C+D)^{-1}D & \cdots & -(C+D)^{-1}D \\ (C+D)^{-1} & (C+D)^{-1}C & & \vdots \\ \vdots & \vdots & \ddots & -(C+D)^{-1}D \\ (C+D)^{-1} & (C+D)^{-1}C & \cdots & (C+D)^{-1}C \end{bmatrix}$$

11.8 Verify that $\langle Lu, u \rangle > \alpha^2 \langle u, u \rangle$ for the Sturm-Liouville operator L in Section 11.4, assuming that $p(0)$ and $p(1)$ are positive and either a or c is nonzero. [*Hint:*

$$\langle Lu, u \rangle = \int_0^1 (pu'^2 + qu^2)\, dt + \frac{a}{b} p(0)(u(0))^2 + \frac{c}{d} p(1)(u(1))^2.$$

$$\ge \int pu'^2\, dt + \frac{a}{b} p_0 u_0^2 \ge \mu \left(\int pu'^2\, dt + u_0^2 \right),$$

12.1 Introduction

according to the system of m differential equations (the *state equations*),

$$\frac{dx}{dt} = f(x, u), \tag{12.1.1}$$

by specifying a *control function*, $u(t)$, for time t in the interval $[a, b]$. The control function is usually required to be in some set of *admissible* controls. For example, it is common to assume that $u(t)$ is piecewise continuous on $[a, b]$ and has values which lie in some open subset $\Omega \subset R^q$. Given an admissible control, $u(t)$, a solution $x(t)$ of the differential equations,

$$\frac{dx}{dt} = f(x, u(t)), \tag{12.1.1a}$$

is called a *trajectory corresponding to $u(t)$* and $(x(t), u(t))$ is a *state-control pair*. Now, we impose general boundary conditions,

$$\psi_j(a, x(a), b, x(b)) = 0, \quad 1 \leq j \leq p \tag{12.1.2}$$

to be satisfied by a trajectory and specify an objective function,

$$J(a, x(a), b, x(b)), \tag{12.1.3}$$

of the end-values of the state and time. The *optimal control problem* is to minimize J, that is, to determine a control $u^*(t)$ which has a corresponding trajectory $x^*(t)$ which satisfies (12.1.2) and is such that

$$J(a, x^*(a), b, x^*(b)) \leq J(a, x(a), b, x(b))$$

for all trajectories $x(t)$ which satisfy (12.1.2). Such a control is called an *optimal control*. The problem just described is said to be a *Mayer* problem. It is not difficult to show [12.2] that the Mayer optimal control problem is equivalent to the classical problem of Mayer in the calculus of variations. The problem of Mayer includes many of the variational problems treated in the classical papers on the calculus of variations. For example, probably the first variational problem ever studied, the brachistochrone (shortest time) problem, is a problem of Mayer type. The brachistochrone problem, originally discussed by Galileo in 1630 and solved by the Bernoulli brothers in 1696, deals with a unit mass (e.g. a bead) sliding along a frictionless arc in a uniform gravitational field. The end points, P_0 and P_1, of the arc are fixed, as is the initial velocity at time $t = 0$. Hence, the time of descent from P_0 to P_1 depends only on the shape of the arc. It is required to find that arc, in a prescribed class of arcs joining P_0 to P_1, which minimizes the descent time. We may assume that the arc lies in the (x, y)-plane, so that the equation of motion is given by

$$\frac{d^2s}{dt^2} = g \cos \theta = g \frac{dy}{ds},$$

where g is the gravitational constant, θ is the angle between the vertical gravity vector (in the y-direction) and the tangent to the arc at the point $P : (x, y)$ and s is arc length from P_0 to P. Integrating the equation of motion with respect to t, we obtain $(ds/dt)^2 = 2gy + \text{const}$. By a suitable choice of coordinate system we may take P_0 to be the point $(0, 1)$ and P_1 to be $(b, 2)$. For illustrative purposes, we take the initial velocity to be $(2g)^{1/2}$. This yields (with $y' = dy/dx$),

$$\frac{dt}{dx} = \frac{dt}{ds}\frac{ds}{dx} = \left(\frac{1 + y'^2}{2gy}\right)^{1/2}.$$

The time of descent, J, is obtained by integrating with respect to x. Thus,

$$J = J(y) = \int_0^b \left(\frac{1 + y'^2}{2gy}\right)^{1/2} dx$$

and we must determine in the class of arcs, $y = y(x)$ with $y(0) = 1$ and $y(b) = 2$, that arc which minimizes J. To restate this as an optimal control problem of the Mayer type, we introduce the control function $u(t) = u_1(t) = y'$, at the same time changing the independent variable from x to t.

Next, set $x_1 = y$ and let

$$x_2(t) = \int_0^t \left(\frac{1 + u_1^2}{2gx_1}\right)^{1/2} dt.$$

Then the equations of state become

$$\left.\begin{aligned} \frac{dx_1}{dt} &= u_1, \\ \frac{dx_2}{dt} &= \left(\frac{1 + u_1^2}{2gx_1}\right)^{1/2}. \end{aligned}\right\} \quad (12.1.4)$$

The boundary conditions are

$$\left.\begin{aligned} \psi_1 &= x_1(0) - 1 = 0, \\ \psi_2 &= x_2(0) = 0, \\ \psi_3 &= x_1(b) - 2 = 0, \end{aligned}\right\} \quad (12.1.5)$$

and the objective function J is simply

$$J = x_2(b). \quad (12.1.6)$$

Clearly, (12.1.4), (12.1.5), (12.1.6) is a Mayer optimal control problem equivalent to the brachistochrone problem. We shall present below general computational methods for solving the optimal control problem of Mayer and will illustrate computations with the brachistochrone problem. As in the mathematical programming problem, there are optimal control problems with inequality constraints. We shall defer consideration of problems with inequality constraints until we have disposed of the case of equality constraints. Although the latter may be regarded as the classical extremum problem, in the nonlinear case there is no standard general method of solution comparable to the simplex method. This is true of both the finite-dimensional mathematical programming problems and the infinite-dimensional extremum problems (usually referred to as *optimization problems*.) The selection of a computational method for a nonlinear extremum problem calls for a thorough understanding and (mathematical) analysis of the particular problem. In the next section, we present theoretical background material necessary for such analyses in problems with equality constraints.

12.2 EXTREMUM PROBLEMS WITH EQUALITY CONSTRAINTS

By using the apparatus of normed spaces developed in Chapters 2, 3 and 5, we can subsume both the finite-dimensional case (mathematical programming) and the infinite-dimensional case in a single theory of extremum problems with equality constraints. Throughout this section, X and Y denote normed (real) vector spaces.

Definition 12.2.1 Let J and $g_1, \ldots, g_p$, $p \geq 1$ be real functionals defined on a set $D \subset X$. The set $C(g_i) = \{x \in D : g_i(x) = 0\}$ is called an *equality constraint*. The intersection $C = \bigcap_{i=1}^{p} C(g_i)$ is also called an *equality constraint*.

Let $u \in C$. If there is a neighborhood, N, of u such that $J(x) \geq J(u)$ ($J(x) \leq J(u)$) for all $x \in C \cap N$, then u is called a *local minimum (maximum) point of J on C*.

(We shall limit our attention to minimization, since to maximize J it suffices to minimize $-J$.)

A necessary condition for u to be a local minimum point of J on the constraint C is given by the Lagrange multiplier rule. We shall derive a version of the multiplier rule which holds when X is a pre-Hilbert space. This version includes the multiplier rules of ordinary calculus and of the calculus of variations [12.6] as special cases. All proofs of multiplier rules require some sort of implicit-function theorem. We prove such a theorem in Appendix IV. When X is a Hilbert space and f is a functional defined on a subset $D \subset X$, the existence of the weak derivative $f'_w(u)$ at $u \in D$ implies the existence of a unique element $\nabla_w f(u) \in X$ such that for all $h \in X$,

$$f'_w(u)h = \langle \nabla_w f(u), h \rangle. \tag{12.2.1}$$

(See Definition 5.3.3 and Theorem 3.5.4.) We call $\nabla_w f(u)$ the *weak gradient* of f at u. Note that the weak gradient in a pre-Hilbert space is also defined by (12.2.1). However, we must postulate its existence. Recall that Definitions 5.3.2 and 5.3.3 require that for each $h \in X$ there must be a segment $\{u + sh : 0 \leq |s| < \delta\}$ on which f is defined. Although this does not imply that u is an interior point of D, it does preclude that u be a boundary point of a certain type which can occur when inequality constraints are admitted.

Definition 12.2.2 Let $C = \bigcap_{i=1}^{p} C(g_i)$ ($p \geq 1$) be an equality constraint as in Definition 12.2.1 with X a pre-Hilbert space. Let $u \in C$ be a point at which the gradients $\nabla_w g_i(u)$, $1 \leq i \leq p$, exist. The subspace of all $h \in X$ such that $\langle \nabla_w g_i(u), h \rangle = 0$ for all $i = 1, \ldots, p$ is called the *tangent subspace* of C at u and is denoted by T_u. The set $u + T_u$ is called the *tangent manifold*.

It is clear that T_u is a subspace. The geometric significance of T_u should be apparent. For example, if $X = R^3$ and $p = 1$, then C is a surface, $g_1(x_1, x_2, x_3) = 0$. If the partials $\partial g_1/\partial x_j$, $1 \leq j \leq 3$, exist and are continuous in a neighborhood of $u = (x_1, x_2, x_3)$, then $\nabla_w g_1(u) = \text{grad } g_1(u) = (\partial g_1/\partial x_1, \partial g_1/\partial x_2, \partial g_1/\partial x_3)$, where the partials are evaluated at u. The tangent manifold $u + T_u$ is the plane passing through u and perpendicular to the vector grad $g_1(u)$; i.e. it is the plane tangent to the g_1-surface. Again, if $p = 2$, then C is the curve formed as the intersection of two surfaces $g_1(x) = 0, g_2(x) = 0$. Then $u + T_u$ is the line tangent to this curve at u. Now, if J is a functional to be minimized on C, we can consider the *level surfaces*, $J(x) = k$, where k is a constant. It is fairly clear, from intuitive geometric considerations, that as we decrease k we reach a local minimum k_0 just when the surface $J(x) = k_0$ is tangent to C. The point of tangency, u, is a minimum point and grad $J(u)$ must be perpendicular to T_u. We prove this in the next theorem.

Theorem 12.2.1 i) Let $J, g_1, \ldots, g_p$ $(p \geq 1)$ be real functionals defined on a subset D of a real pre-Hilbert space. Suppose that $u \in D$ is a local minimum point of J on the constraint $C = \bigcap_{i=1}^{p} C(g_i)$.

ii) For any $h \in T_u$, the tangent subspace, let $\nabla_w J(y)$ and $\nabla_w g_i(y)$ exist for all points $y = u + t_0 h + \sum_{i=1}^{p} t_i \nabla_w g_i(u)$, where $(t_0, t_1, \ldots, t_p)$ is any point in some neighborhood, $N(h)$, of the origin in R^{p+1}. Further, assume that the $\nabla_w g_i(y)$ are continuous in the t_i throughout N and that $\nabla_w J(y)$ is continuous in the t_i at the origin.

iii) Finally, suppose that the vectors $\nabla_w g_i(u)$ $(1 \leq i \leq p)$ are linearly independent. Then for any $h \in T_u$, $\langle \nabla_w J(u), h \rangle = 0$.

Proof. For $t = (t_1, \ldots, t_p)$ a real p-dimensional vector and s a real scalar we define

$$y_{st} = u + sh + \sum_{i=1}^{p} t_i \nabla_w g_i(u),$$

where $h \neq 0$ is an arbitrary vector in T_u. (The result is obviously true for $h = 0$.) Next, we define the functions $F_i(s, t) = F_i(s, t_1, \ldots, t_p) = g_i(y_{st})$, $1 \leq i \leq p$. We shall prove that the F_i have the following three properties:

1. $F_j(0, 0) = g_i(u) = 0$;
2. the partials $F_{ij} = \partial F_i/\partial t_j$ and $F_{is} = \partial F_i/\partial s$ exist as continuous functions of (s, t) in some neighborhood N_0 of $(0, 0)$,
3. the Jacobian $(F_{ij}(s, t))$ has rank p for $(s, t) \in N_0$ and $F_{is}(0, 0) = 0$ for $1 \leq i \leq p$.

Property 1 is immediate. To establish Property 2, we observe that

$$F_{ij}(s, t) = \lim_{\Delta t_j \to 0} \frac{g_i(y_{st} + \Delta t_j \nabla_w g_j(u)) - g_i(y_{st})}{\Delta t_j}$$

$$= dg_i(y_{st}, \nabla_w g_j(u)) = \langle \nabla_w g_i(y_{st}), \nabla_w g_j(u) \rangle.$$

By the assumed continuity of $\nabla_w g_i(y_{st})$ in hypothesis (ii) above, it follows that F_{ij} is continuous in (s, t) in some neighborhood N_0 of $(0, 0)$. Similarly, we obtain existence and continuity of $F_{is} = \langle \nabla_w g_i(y_{st}), h \rangle$. To obtain Property 3, we observe that $F_{ij}(0, 0) = \langle \nabla_w g_i(u), \nabla_w g_j(u) \rangle$. Since the set of vectors $\{\nabla_w g_i(u) : 1 \leq i \leq p\}$ is linearly independent by hypothesis (iii), it follows that the $p \times p$ matrix $(F_{ij}(0, 0))$ has rank p. (See Exercise 7.12.) By continuity of the F_{ij} the matrices $(F_{ij}(s, t))$ also have rank p for (s, t) in N_0. Finally, $F_{is}(0, 0) = \langle \nabla_w g_i(u), h \rangle = 0$, since $h \in T_u$.

With these properties established, we can apply the implicit-function theorem to obtain functions $G_i(s)$, $1 \leq i \leq p$, such that $F_i(s, G_1(s), \ldots, G_p(s)) = 0$ for all s in a neighborhood of $s = 0$. Furthermore, $G_i(0) = 0$ and the G_i are continuously differentiable in a neighborhood of $s = 0$. Since $F_{is}(0, 0) = 0$, it follows that the derivative $G_i'(0) = 0$. Hence $\lim_{s \to 0} G_i(s)/s = G_i'(0) = 0$.

Now, define vectors $y_s = u + sh + \sum_{i=1}^p G_i(s)\nabla_w g_i(u)$. For $|s|$ sufficiently small,

$$g_i(y_s) = F_i(s, G_1(s), \ldots, G_p(s)) = 0,$$

that is, $y_s \in \mathcal{C}$ for $|s|$ sufficiently small. Since $\lim_{s \to 0} G_i(s) = 0$, $\lim_{s \to 0} y_s = u$. Using the same methods as above, we obtain for the partial derivatives of the function

$$H(s, t) = J(y_{st}), \qquad H_{t_i}(0, 0) = \langle \nabla_w J(u), \nabla_w g_i(u) \rangle$$

and $H_s(0, 0) = \langle \nabla_w J(u), h \rangle$. Since H_{t_i} and H_s are continuous at $(0, 0)$ and exist in a neighborhood of $(0, 0)$, we obtain (Exercise 5.2c),

$$J(y_s) - J(u) = s\left(\sum_1^p \langle \nabla_w J(u), \nabla_w g_i(u) \rangle G_i'(0) + \langle \nabla_w J(u), h \rangle + \varepsilon\right),$$

$$J(y_s) - J(u) = s(\langle \nabla_w J(u), h \rangle + \varepsilon), \qquad (12.2.2)$$

where $\varepsilon \to 0$ as $s \to 0$. We see that for $|s|$ sufficiently small the right member has the sign of $\langle \nabla_w J(u), h \rangle s$. This quantity must have the same sign as $J(y_s) - J(u)$. Since u is a local minimum, $J(y_s) - J(u) \geq 0$ for $|s|$ sufficiently small. Hence, $\langle \nabla_w J(u), h \rangle s \geq 0$. For $s > 0$, this implies $\langle \nabla_w J(u), h \rangle \geq 0$ and for $s < 0$, this implies $\langle \nabla_w J(u), h \rangle \leq 0$. Hence, $\langle \nabla_w J(u), h \rangle = 0$, as we wished to prove. ∎

We can now state and prove a multiplier rule as in [12.2].

Theorem 12.2.2 (Lagrange Multiplier Rule). *Let $J, g_1, \ldots, g_p$ be real functionals satisfying hypotheses (i) and (ii) of Theorem 12.2.1. If $\nabla_w J(u) \neq 0$, then there exist real scalars $\lambda_0, \lambda_1, \ldots, \lambda_p$ not all zero such that*

$$\lambda_0 \nabla_w J(u) + \sum_{i=1}^p \lambda_i \nabla_w g_i(u) = 0. \qquad (12.2.3)$$

Furthermore, if the p vectors $\nabla_w g_1(u), \ldots, \nabla_w g_p(u)$ are linearly independent, then we may take $\lambda_0 = 1$ and the remaining scalars are unique and not all zero.

Proof. If the vectors $\nabla_w g_i(u)$, $1 \leq i \leq p$, are linearly dependent, then there certainly exist scalars $\lambda_1, \ldots, \lambda_p$ not all zero such that (12.2.3) is satisfied with $\lambda_0 = 0$. Hence, we may turn at once to the case when the $\nabla_w g_i(u)$ are linearly independent. In this case, since the matrix $\langle \nabla_w g_i(u), \nabla_w g_j(u) \rangle$ is nonsingular, we can find $(\lambda_1, \ldots, \lambda_p)$ as the unique solution of the linear system,

$$\sum_{j=1}^{p} \langle \nabla_w g_i(u), \nabla_w g_j(u) \rangle \lambda_j = -\langle \nabla_w g_i(u), \nabla_w J(u) \rangle, \qquad i = 1, \ldots, p.$$

By the previous theorem, $\langle \nabla_w J(u), h \rangle = 0$ for any $h \in T_u$. This means that $\nabla_w J(u)$ itself cannot be in T_u, since that would imply $\nabla_w J(u) = 0$, contrary to hypothesis. Hence, for some index i we must have $\langle \nabla_w g_i(u), \nabla_w J(u) \rangle \neq 0$. It follows that not all $\lambda_1, \ldots, \lambda_p$ are zero.

Now, let $v = \nabla_w J(u) + \sum_{j=1}^{p} \lambda_j \nabla_w g_j(u)$. Since $\langle \nabla_w g_i(u), v \rangle = 0$ for all $i = 1, \ldots, p$, we must have $v \in T_u$. Again by the previous theorem and the definition of T_u, for any $h \in T_u$,

$$\langle v, h \rangle = \langle \nabla_w J(u), h \rangle + \sum_{j=1}^{p} \lambda_j \langle \nabla_w g_j(u), h \rangle = 0.$$

Taking $h = v$, it follows that $v = 0$, as was to be proved. ∎

Alternative proof. Let M be the subspace spanned by the gradients $\nabla_w g_i(u), 1 \leq i \leq p$. Since M is finite-dimensional, it is complete. Also, $M^\perp = T_u$. By the remark following Theorem 3.3.11, the pre-Hilbert space is the direct sum $M \oplus T_u$. Hence, $M = (T_u)^\perp$ and the preceding theorem implies that $\nabla_w J(u) \in M$; i.e. $\nabla_w J(u)$ is a linear combination of the $\nabla_w g_i(u)$, as we wished to prove. ∎

Remark If $\nabla_w J(u) = 0$, the theorem holds with $\lambda_0 = 1$ and $\lambda_i = 0, 1 \leq i \leq p$.

From ordinary calculus we are familiar with second-order sufficient conditions for a local minimum in terms of the sign of the second derivative. These conditions can be generalized to functionals f on Hilbert spaces if we use strong derivatives $f''(x)$ as given by Definition 5.3.5. First, we obtain a second-order necessary condition applicable in a pre-Hilbert space.

Theorem 12.2.3 Let $J, g_1, \ldots, g_p$ be real functionals defined on a subset of a real pre-Hilbert space.

i) Let the strong gradients $\nabla J(x), \nabla g_1(x), \ldots, \nabla g_p(x)$ exist and let $\nabla g_i(x)$ be continuous for x in a neighborhood of the point u.

ii) Let the strong second derivatives $J''(u), g_1''(u), \ldots, g_p''(u)$ exist.

iii) Let there exist multipliers $\lambda_1^*, \ldots, \lambda_p^*$ not all zero such that

$$\nabla J(u) + \sum_{i=1}^{p} \lambda_i^* \nabla g_i(u) = 0$$

and define $f = J + \sum_{i=1}^{p} \lambda_i^* g_i$.

If u is a local minimum point of J on the constraint $C = \bigcap_{i=1}^{p} C(g_i)$, then for all h in the tangent subspace T_u,

$$f''(u)h^2 \geq 0. \tag{12.2.4}$$

Furthermore,

$$0 = \lim_{q\to\infty} \frac{g_i(u_q) - g_i(u)}{\|h_q\|}$$

$$= \lim_{q\to\infty}\left(\left\langle \nabla g_i(u), \frac{h_q}{\|h_q\|}\right\rangle + \frac{\varepsilon}{\|h_q\|}\right) = \langle \nabla g_i(u), h\rangle,$$

showing that there exists $h \in T_u$ which violates (12.2.5). This completes the proof. ∎

As we have pointed out, the positivity condition (12.2.5) in the finite-dimensional case implies that

$$f''(u)h^2 \geq \mu\|h\|^2, \qquad \mu > 0. \tag{12.2.7}$$

In the infinite-dimensional case, we must assume (12.2.7) to obtain a sufficient condition that u be a local minimum point. Before establishing this, we shall prove a result concerning the tangent subspace T_u in a more general situation.

Definition 12.2.3 Let g be a mapping of a subset of a normed vector space X to a normed vector space Y. The set $C = \{x : g(x) = 0\}$ is called an *equality constraint*. For $u \in C$ let $g'_w(u)$ exist and be nonzero. The set of points $h \in X$ such that $g'_w(u)h = 0$ is called the *tangent subspace* of C at u and is denoted by T_u. If $g'_w(u)$ is a surjective mapping, then u is said to be a *normal* point of C.

We observe that Definition 12.2.2 is a special case in which $g(x) = (g_1(x), \ldots, g_p(x))$ is an element of a p-dimensional space. If X is also finite-dimensional, then $g'_w(u) = (\partial g_i/\partial \xi_j)$ and u is normal if the Jacobian has maximum rank.

If X is a Hilbert space, then the orthogonal complement $T_u^\perp$ exists and $X = T_u \oplus T_u^\perp$. (See Theorem 3.3.11.) Hence, any $x \in X$ can be expressed uniquely as $x = h + k$, where $h \in T_u$ and $k \in T_u^\perp$. If u is a normal point, then $g'_w(u)$ is a linear operator which maps $T_u^\perp$ onto Y, since $g'_w(u)x = g'_w(u)k$. Furthermore, $g'_w(u)$ is injective on $T_u^\perp$, since $g'_w(u)k \neq 0$ for all $k \in T_u^\perp$. We shall denote the restriction of $g'_w(u)$ to $T_u^\perp$ by U. By Theorem 3.5.0, the operator U has a bounded linear inverse U^{-1}. We shall now show that at a normal point u the tangent manifold $u + T_u$ is a good approximation to the constraint C in a neighborhood of u.

Lemma 12.2.1 Let $g : X \longrightarrow Y$, where X is a Banach space. Let the strong derivative $g'(x)$ exist in a neighborhood of $u \in X$. Further, let u be a normal point of the constraint $C = \{x : g(x) = 0\}$ and assume there exists a subspace $T_u^\perp$ such that $X = T_u \oplus T_u^\perp$. (i.e., X is a direct sum of the tangent subspace T_u and some other subspace.) Then there exists a neighborhood N of u such that every point in $N \cap C$ is of the form $u + h + k$ with $h \in T_u$ and $\|k\| = o(\|h\|)$.

Proof. Any vector $x \in X$ can be expressed in the form $x = u + h + k$, with $h \in T_u$ and $k \in T_u^\perp$. We define a mapping $f : T_u \oplus T_u^\perp \longrightarrow Y$ by $f(h, k) = g(u + h + k)$. Thus, $f(h, k) = 0$ if and only if $u + h + k \in C$. Clearly, $f(0, 0) = 0$. Also, the partial derivative of f with respect to k exists for (h, k) in a neighborhood of $(0, 0)$ and $f_k(0, 0)k = g'(u)k = Uk$. Since u is a normal point, $(f_k(0, 0))^{-1} = U^{-1}$ exists. Therefore, the hypotheses of the implicit-function theorem (Appendix IV) are satisfied and there exists a function $\varphi(h)$ such that $f(h, \varphi(h)) = 0$ for all $\|h\| \leq r$. Furthermore, by Corollary 1 of the implicit-function theorem, there is a $\rho > 0$ such that for any $\|h\| \leq r$ and $\|k\| \leq \rho$, if $f(h, k) = 0$, then $k = \varphi(h)$. This determines a neighborhood $N = \{u + h + k : \|h\| \leq r$ and $\|k\| \leq \rho\}$ such that if $u + h + k \in N \cap C$, then $k = \varphi(h)$. By Corollary 2 of the implicit-function theorem, $\varphi'(0)h = -U^{-1}f_h(0,0)h = -U^{-1}g'(u)h$. Since $\varphi(0) = 0$, we have

$$\varphi(h) = \varphi(0) + \varphi'(0)h + \varepsilon = -U^{-1}g'(u)h + \varepsilon = \varepsilon,$$

where $\|\varepsilon\|/\|h\| \to 0$ as $\|h\| \to 0$; i.e. $\|k\| = \|\varphi(h)\| = o(\|h\|)$, as was to be proved. ∎

Theorem 12.2.5 Let $J, g_1, \ldots, g_p$ be real functionals defined on a subset of a real pre-Hilbert space X and satisfying conditions (i), (ii) and (iii) of Theorem 12.2.3. Furthermore, assume that u is a normal point of the constraint C. If inequality (12.2.7) is satisfied at u for all $h \in T_u$, then u is a local strong minimum point of J on C.

Proof. Since X is a pre-Hilbert space, T_u is just the orthogonal complement of $T_u^\perp$, the subspace spanned by $\nabla g_i(u)$, $1 \leq i \leq p$. Further, since u is a normal point, the preceding lemma yields a neighborhood N of u such that every point $x \in N \cap C$ is of the form $u + h + k$ with $h \in T_u$ and $\|k\| = o(\|h\|)$. Hence, again with $f = J + \sum_1^p \lambda_i^* g_i$,

$$f(x) - f(u) = f'(u)(h + k) + \tfrac{1}{2}f''(u)(h + k)^2 + \varepsilon,$$

where $\varepsilon/\|h + k\|^2 \to 0$ as $\|h + k\| \to 0$. Now, $f'(u) = 0$ by hypothesis (iii) and $f''(u)(h + k)^2 = f''(u)h^2 + f''(u)(hk + kh) + f''(u)k^2$. Further,

$$\frac{\|h + k\|^2}{\|h\|^2} \to 1 \quad \text{as } \|h\| \to 0,$$

so that

$$\frac{\varepsilon}{\|h\|^2} = \frac{\varepsilon}{\|h + k\|^2} \frac{\|h + k\|^2}{\|h\|^2} \to 0.$$

Therefore, applying (12.2.7), we have

$$\frac{f(x) - f(u)}{\|h\|^2} \geq \frac{\mu}{2} + \frac{1}{2}\left(f''(u)\frac{(hk + kh)}{\|h\|^2} + f''(u)\frac{k^2}{\|h\|^2} + \frac{2\varepsilon}{\|h\|^2}\right),$$

where the quantity in parentheses can be made arbitrarily small by choosing $\|h\|$ sufficiently small. It follows that $f(x) - f(u) > 0$ for all $x \in B \cap N \cap C$, where B is a sufficiently small ball centered at u. Since $J(x) - J(u) = f(x) - f(u)$ for such x, the point u is a strong local minimum of J on C. ∎

A simpler direct proof of sufficiency can be given if we assume that (12.2.7) holds for all $h \in X$, rather than just for $h \in T_u$.

Theorem 12.2.6 Let f be a real functional defined on a subset D of a normed space X. For $u \in D$, let $f'(u) = 0$ and suppose $f''(u)$ exists and satisfies (12.2.7) for all $h \in X$.

Then u is a local minimum point of f.

Proof. For all h such that $u + h \in D$,
$$f(u+h) - f(u) = f'(u)h + \tfrac{1}{2} f''(u)h^2 + \varepsilon$$
$$\geq \frac{\mu}{2} \|h\|^2 + \varepsilon = \|h\|^2 \left(\frac{\mu}{2} + \frac{\varepsilon}{\|h\|^2} \right).$$

Since $\varepsilon / \|h\|^2 \to 0$ as $\|h\| \to 0$, there exists a ball B centered at u such that for $u + h \in B \cap D$,
$$f(u + h) - f(u) \geq \frac{\mu}{4} \|h\|^2 > 0. \quad \blacksquare$$

Remark The previous result also holds for the weak second derivative. (See Exercise 12.11.) Also see Theorems 12.9.3 and 12.9.4.

12.2a THE LAGRANGIAN

The function $f = \lambda_0 J + \sum_{i=1}^{p} \lambda_i g_i$ of the preceding theorems is usually called the *Lagrangian function* of the minimization problem and the λ_i are called *Lagrange multipliers* or—in the finite-dimensional case—*dual variables*. The multipliers can be given an interpretation which is useful in certain practical problems.

Suppose we consider the equality constraints to be of the form $G_i(x, b_i) = g_i(x) - b_i = 0$, $1 \leq i \leq p$, where the b_i are real parameters. The various values of b_i determine the *level surfaces* of g_i. Let $b = (b_1, \ldots, b_p)$ and $G(x, b) = (G_1(x, b_1), \ldots, G_p(x, b_p))$. Further, let $\Lambda = (\lambda_1, \ldots, \lambda_p)$ be regarded as a vector variable. Setting $\lambda_0 = 1$, the Lagrangian f may be regarded as a function of (x, Λ, b) in the space $V = X \oplus V^p \oplus V^p$, that is,
$$f(x, \Lambda, b) = J(x) + \sum_{i=1}^{p} \lambda_i G_i(x, b_i).$$

The partial derivative with respect to x is
$$f_x(x, \Lambda, b) = J'(x) + \sum_{1}^{p} \lambda_i g_i'(x).$$

If we regard $G(x, \Lambda, b) = G(x, b)$ as being a function of Λ also, then the necessary condition of Theorem 12.2.2 can be rephrased by saying that a minimum point u and the corresponding multipliers λ_i^* must be a zero of the system,
$$G(x, \Lambda, b) = 0,$$
$$f_x(x, \Lambda, b) = 0; \quad (12.2.8)$$

that is, $G(u, \Lambda^*, 0) = 0$ and $f_x(u, \Lambda^*, 0) = 0$, since $b = 0$ in Theorem 12.2.2. Now, if we vary b in a neighborhood of 0 (thereby changing the constraints slightly), it should be possible to find x, Λ which satisfy (12.2.8) for each value of b. Thus, $x = x(b)$ and $\Lambda = \Lambda(b)$ are implicit functions of b, the existence of which is obtained by the implicit-function theorem provided that we assume that $J''(x)$ and $g_i''(x)$ exist and are continuous in a neighborhood of u and that $\{g_i'(u)\}$

532 Extremum Problems

is a linearly independent set. This is best seen by working with the direct sum $X \oplus V^p$ and observing that the partial derivative of the mapping F defined by the pair (G, f') in (12.2.8) can be represented by the matrix

$$F_{(x, \Lambda)} = \begin{bmatrix} G_x & G_\Lambda \\ f_{xx} & f_{x\Lambda} \end{bmatrix}$$

and

$$G_\Lambda = 0,$$

$$f_{xx}(x, \Lambda, b) = J''(x) + \sum_{1}^{p} \lambda_i g_i''(x),$$

$$f_{x\Lambda} = (g_1'(x), \ldots, g_p'(x)),$$

$$G_x = (g_1'(x), \ldots, g_p'(x))^\mathsf{T}.$$

For example, if $X = V^n$, the above matrix has the form

$$\begin{bmatrix} \dfrac{\partial g_1}{\partial x_1} & \cdots & \dfrac{\partial g_1}{\partial x_n} & 0 & \cdots & 0 \\ \vdots & & \vdots & \vdots & & \vdots \\ \dfrac{\partial g_p}{\partial x_1} & \cdots & \dfrac{\partial g_p}{\partial x_n} & 0 & \cdots & 0 \\ \dfrac{\partial^2 f}{\partial x_1^2} & \cdots & \dfrac{\partial^2 f}{\partial x_1 \partial x_n} & \dfrac{\partial g_1}{\partial x_1} & \cdots & \dfrac{\partial g_p}{\partial x_1} \\ \vdots & & \vdots & \vdots & & \vdots \\ \dfrac{\partial^2 f}{\partial x_1 \partial x_n} & \cdots & \dfrac{\partial^2 f}{\partial x_n^2} & \dfrac{\partial g_1}{\partial x_n} & \cdots & \dfrac{\partial g_p}{\partial x_n} \end{bmatrix}$$

Since the $g_i'(u)$ are linearly independent, this operator is nonsingular. Hence, $x(b)$ and $\Lambda(b)$ exist. Furthermore, the derivatives $x'(b)$ and $\Lambda'(b)$ exist. Since $G(x(b), b) = 0$, differentiation yields for $1 \leq i \leq p$,

$$\frac{dG_i}{db} = g_i'(x(b))x'(b) + \frac{\partial G_i}{\partial b} = 0,$$

where $\partial G_i/\partial b = (0, \ldots, 0, -1, 0, \ldots, 0)$ has -1 in the ith position. Similarly,

$$f_x(x(b), \Lambda(b), b) = J'(x(b)) + \sum_{1}^{p} \lambda_i(b) g_i'(x(b)) = 0.$$

Now, by the composite function rule (Lemma 5.3.4), letting $\varphi(b) = J(x(b))$, we have

$$\varphi'(b) = J'(x(b))x'(b).$$

Hence, with $\lambda_i^* = \lambda_i(0)$, $x^* = x(0)$,

$$\varphi'(0) - \sum_{1}^{p} \lambda_i^* \frac{\partial G_i}{\partial b} = \left(J'(x^*) + \sum_{1}^{p} \lambda_i^* g_i'(x^*) \right) x'(0) = 0.$$

This shows that

$$\lambda_i^* = -\left.\frac{\partial \varphi}{\partial b_i}\right|_{b=0} = -\left.\frac{\partial J(x(b))}{\partial b_i}\right|_{b=0}.$$

The Saddle-Point Property of the Lagrangian

When a stationary point x^* is actually a local minimum point (Definition 12.2.1), the Lagrangian f often exhibits a *saddle-point* behavior at (x^*, Λ^*), where Λ^* is the corresponding Lagrange multiplier.

Theorem 12.2.7 Let x^* be a local minimum of J on C; that is, $J(x^*) \leq J(x)$ for all $x \in C \cap N$, where N is a neighborhood of x^*. In N, let J and g_i, $1 \leq i \leq p$, have continuous strong derivatives and let $\{g_i'(x)\}$ be linearly independent. Let Λ^* be the Lagrange multiplier such that $f_x(x^*, \Lambda^*) = 0$. Suppose that for each Λ_0 in a neighborhood U of Λ there exists $x_0 \in N$ which is a local minimum point of $f(x, \Lambda_0)$, and further, that the equation $f_x(x, \Lambda_0) = 0$ has a unique solution in N. Then for $x \in N$ and $\Lambda \in U$,

$$f(x^*, \Lambda) \leq f(x^*, \Lambda^*) \leq f(x, \Lambda^*), \tag{12.2.9}$$

$$J(x^*) = f(x^*, \Lambda^*) = \max_{\Lambda \in U} \{\min_{x \in N} \{f(x, \Lambda)\}\}. \tag{12.2.10}$$

Proof. Let $x_0 \in N$ be a local minimum of $f(x, \Lambda_0)$. Then $f_x(x_0, \Lambda_0) = 0$. Hence x_0 is the only local minimum in N and $f(x_0, \Lambda_0) = \min_{x \in N}\{f(x, \Lambda_0)\}$. Since $f_x(x^*, \Lambda^*) = 0$, this implies that $f(x^*, \Lambda^*) = \min_{x \in N}\{f(x, \Lambda^*)\}$. Hence $f(x^*, \Lambda^*) \leq f(x, \Lambda^*)$. The other inequality in (12.2.9) is immediate, since $f(x^*, \Lambda) = J(x^*)$ for any Λ; that is $f(x^*, \Lambda) = f(x^*, \Lambda^*)$.

To establish (12.2.10) we simply observe that

$$\min_{x \in N}\{f(x, \Lambda)\} \leq \min_{x \in N \cap C}\{f(x, \Lambda)\} = J(x^*).$$

This completes the proof. ∎

Remark The existence of a solution of $f_x(x, \Lambda_0) = 0$ for each Λ_0 in a neighborhood of Λ^* can be secured by using the implicit-function theorem (Appendix IV). This requires that the strong second derivatives J'' and g_i'' exist and be continuous in a neighborhood of x^* and that f_{xx} be nonsingular at (x^*, Λ^*). Indeed, if f_{xx} satisfies the positivity condition (12.2.7) in a neighborhood of (x^*, Λ^*), then f will have a local minimum at each (x_0, Λ_0) in this neighborhood such that $f_x(x_0, \Lambda_0) = 0$. If the neighborhood is properly chosen, x_0 will be the unique local minimum (by Corollary 1 of Appendix IV).

The saddle-point condition (12.2.9) is not only a necessary condition for x^* to be a local minimum of J on C, it is also sufficient provided that it holds for all Λ.

Theorem 12.2.8 Let J and g_i, $1 \leq i \leq p$, be real functionals on a subset D of a real normed vector space. Let (12.2.9) hold for the Lagrangian for all x in a neighborhood N of x^* and for all Λ. Then $J(x^*) \leq J(x)$ for all $x \in N \cap C$ and $g_i(x^*) = 0$, $1 \leq i \leq p$; that is x^* is a local minimum of J on C.

Proof. Since $J(x^*) + \sum \lambda_i g_i(x^*) \leq J(x^*) + \sum \lambda_i^* g_i(x^*)$, it follows that all $g_i(x^*) = 0$. Otherwise, if some $g_j(x^*) \neq 0$, we could vary λ_j until the inequality became reversed. Hence

$$J(x^*) = f(x^*, \Lambda^*) \leq J(x) + \sum \lambda_i^* g_i(x) = J(x)$$

for $x \in N \cap C$. Thus x^* is a local minimum of J on C. ∎

12.3 GRADIENT METHODS FOR EXTREMUM PROBLEMS WITHOUT CONSTRAINTS

Let us first consider the problem of minimizing a functional J defined on a pre-Hilbert space X with no constraints at all. Suppose that we know that J is bounded from below; i.e., $J(x) \geq l > -\infty$ for all x, where $l = \inf_{x \in X} J(x)$. Then there exists a sequence (x_n) such that $J(x_n) \to l$ as $n \to \infty$. Such a sequence is called a *minimizing sequence*. Note that (x_n) need not converge. In fact, $J(x)$ may not assume the infimum l as a value for any x. If the problem is to compute a minimum point, then it may be unsolvable. However, in practical computation, except in the simplest cases, we cannot expect to compute a minimum point x^* even when one exists. Most computational methods produce only a sequence (x_n) converging to x^*, which is then a minimizing sequence. Even when a minimum point x^* does not exist, the problem may require only the determination of the infimum l. In that case, a minimizing sequence yields a sequence of values $J(x_n)$ converging to l and one of these values approximates l with sufficient accuracy for the numerical computation. In this situation, that is, where the extremum value of J is being sought, the convergence or nonconvergence of (x_n) may be irrelevant. However, comparing these remarks with the remark following Theorem 7.3.4 concerning closest points, we soon realize that the existence of x^* and the convergence of (x_n) can materially influence the development of computational methods. In any case, we are always dealing with a minimizing sequence and one natural approach to the construction of such a sequence is the following.

Suppose J has a strong gradient $\nabla J(x_n)$. Letting $h = h(x_n) = \nabla J(x_n)$, we have

$$\Delta J(x_n) = J(x_n - s_n h) - J(x_n)$$
$$= -s_n \langle \nabla J(x_n), h \rangle + \varepsilon_n = -s_n(\|\nabla J(x_n)\|^2 + \varepsilon_n/s_n).$$

For $s_n > 0$ sufficiently small, the quantity on the right is negative if $\nabla J(x_n) \neq 0$, so that $J(x_n - s_n h) < J(x_n)$. This suggests that we take

$$x_{n+1} = x_n - s_n \nabla J(x_n), \qquad n = 0, 1, 2, \ldots \tag{12.3.1}$$

as a procedure for generating a sequence (x_n) such that $(J(x_n))$ is a decreasing sequence. Several questions arise immediately. How shall we choose the scalars s_n to insure that $J(x_n)$ is actually decreasing? Under what conditions will (x_n) be a minimizing sequence (i.e., $J(x_n) \to l$)? When is (x_n) a convergent minimizing sequence?

Methods based on formula (12.3.1) are called *gradient methods*, since at each step the next iteration is obtained by moving in the negative direction of the gradient. This, of course, is the direction in which $J(x)$ decreases at the maximum rate. If s_n is chosen

12.3 Gradient Methods for Extremum Problems without Constraints

to minimize $J(x_n - s\,\nabla J(x_n))$ as a function of s at the nth step, the procedure is referred to as a *method of steepest descent*. (See Exercise 12.12.)

Equation (12.3.1) has the form of the rectangular rule (Euler's method, Section 10.2) for the numerical solution of the differential equation,

$$\frac{dx}{dt} = -\nabla J(x), \qquad (12.3.2)$$

with variable step-size s_n. If it is possible to choose $s_n > s > 0$—and we shall see that it often is—then allowing $n \to \infty$ is the same as allowing $t \to \infty$. Thus, if a solution $x(t)$ of (12.3.2) exists for $a \le t < \infty$ and if $\lim_{n \to \infty} x_n = x^*$ exists (i.e. if (12.3.1) yields a convergent sequence), we would expect that $x^* = \lim_{t \to \infty} x(t) + e$, where e is an error arising from discretization. Under suitable conditions, it happens that $e = 0$ and $x^* = \lim_{n \to \infty} x_n = \lim_{t \to \infty} x(t)$; i.e. both limits exist and are equal to an extremum point x^* of J.

We have observed that for s_n sufficiently small, the sequence $(J(x_n))$ is monotone decreasing provided that $\nabla J(x_n) \ne 0$, for otherwise it is just nonincreasing; i.e. $J(x_{n+1}) \le J(x_n)$. Hence, $\lim_{n \to \infty} J(x_n) = \inf_n J(x_n) \ge l$. If $\lim_{n \to \infty} s_n \ne 0$, then we can expect $\lim_{n \to \infty} \nabla J(x_n) = 0$. Otherwise, there is a subsequence such that $\|\nabla J(x_{n_i})\| > \mu > 0$. For convenience set $y_i = x_{n_i}$ and $t_i = s_{n_i}$. Assume further that $J''(x)$ exists and that $\|J''(x)\| \le 1/s_0$ for all x. By Taylor's theorem,

$$\Delta J(y_i) = -t_i \langle \nabla J(y_i), \nabla J(y_i) \rangle + \frac{t_i^2}{2} J''(y)(\nabla J(y_i))^2,$$

where $y = y_i - \theta t_i \nabla J(y_i)$, $0 < \theta < 1$. Hence

$$\Delta J(y_i) = -t_i \|\nabla J(y_i)\|^2 \left(1 - \frac{t_i}{2} J''(y) h_i^2\right),$$

where $h_i = \nabla J(y_i)/\|\nabla J(y_i)\|$. But $\|J''(y_i)h_i^2\| \le 1/s_0$. Therefore if the s_n are chosen so that $\delta \le s_n \le 2s_0 - \delta$, where δ is any number satisfying $0 < \delta \le s_0$, it follows that

$$|\Delta J(y_i)| \ge \delta \mu^2 (1 - t_i/2s_0) > \delta^2 \mu^2/2s_0 > 0,$$

and $\Delta J(y_i) < 0$. Hence $\lim_{n \to \infty} J(x_n) = -\infty$, contradicting our assumption that $l > -\infty$. We have proved:

Theorem 12.3.1 Let J be a functional on a pre-Hilbert space X. Suppose that $\nabla J(x)$ and $J''(x)$ exist with $\|J''(x)\| \le 1/s_0$. Let (x_n) be a sequence defined by a gradient method (12.3.1) with the s_n chosen so that $\delta \le s_n \le 2s_0 - \delta$, where $0 < \delta \le s_0$. Then $(J(x_n))$ is a monotone nonincreasing sequence. If $\inf J(x_n) > -\infty$, then $\lim_{n \to \infty} \nabla J(x_n) = 0$.

Now, suppose a solution $x(t)$ of the differential equation (12.3.2) exists for $0 \le t < \infty$. Let $g(t) = J(x(t))$. Then $g'(t) = J'(x(t)) \cdot x'(t)$, by the composite function rule (Lemma 5.3.4). Since $x'(t) = -\nabla J(x(t))$, we have $g'(t) = -\|\nabla J(x(t))\|^2 \le 0$ for all t. Hence, $g(t)$ is monotone nonincreasing. Since we are assuming that $l = \inf J(x) > -\infty$, we must have $\lim_{t \to \infty} \nabla J(x(t)) = 0$, provided that $\nabla J(x)$ is uniformly continuous in x and bounded. Otherwise, $\lim_{t \to \infty} \sup$

$\|\nabla J(x(t))\| > \mu > 0$ and $\|\nabla J((x(t))\| > \mu$ for infinitely many values of t and for a fixed-size interval containing each such t. (See Remark below.)

This would imply $J(x(t)) = -\int_0^t \|\nabla J(x(s))\|^2 \, ds \to -\infty$ as $t \to \infty$, contrary to our assumption that $l > -\infty$. Thus, we have established a continuous analog of Theorem 12.3.1 as follows.

Theorem 12.3.2 Let $x(t)$ be a solution of (12.3.2) for $0 \leq t < \infty$, with initial condition $x(0) = x_0$. Let $S = \{x : -\infty < l \leq J(x) \leq J(x_0)\}$, where $l = \inf_{x \in X} J(x)$. Suppose $\|\nabla J(x)\| < M$ for $x \in S$ and $\nabla J(x)$ is uniformly continuous on S. Then $J(x(t))$ is a monotone nonincreasing function of t and $\lim_{t \to \infty} \nabla J(x(t)) = 0$.

Remark Since $\|x'(t)\| = \|\nabla J(x(t))\| < M$, we are able to choose $\delta > 0$ such that $\|\Delta x\| = \|x(t + \Delta t) - x(t)\| < \varepsilon$ whenever $|\Delta t| \leq \delta$. By the uniform continuity of $\nabla J(x)$, we can choose ε such that $\|\nabla J(x + \Delta x) - \nabla J(x)\| < \mu/2$ whenever $\|\Delta x\| < \varepsilon$. Hence, if $\|\nabla J(x(t))\| > \mu$, then $\|\nabla J(x(s))\| > \mu/2$ for $|s - t| \leq \delta$. If this holds for arbitrarily large t, $J(x(t)) \to -\infty$ as $t \to \infty$.

So far, we have seen that a gradient method can produce a decreasing sequence $(J(x_n))$ such that $\nabla J(x_n) \to 0$. Since $\nabla J(x^*) = 0$ is a necessary condition for a minimum, it seems reasonable to expect that (x_n) might indeed converge to a minimum point x^* if the second-order sufficient condition (12.2.7) is satisfied.

Theorem 12.3.3 Let J be a real function defined on a Hilbert space X. Let $\nabla J(x)$ and $J''(x)$ exist and be continuous for all $x \in X$. Suppose further that for all h with $\|h\| = 1$,

$$0 < \mu < \langle J''(x)h, h \rangle < M.$$

In the gradient procedure (12.3.1), choose $s_n = s$, where $\delta < s < (2/M) - \delta$ and δ is any positive number such that $\delta < 1/M$. Then the sequence $x_n \to x^*$ and $J(x^*) < J(x)$ for all $x \neq x^*$.

Proof. Let $g(x) = x - s\nabla J(x)$. Then a fixed point x^* of g is a zero of ∇J and conversely. Furthermore, for any $x = x^* + h$, by Taylor's theorem,

$$J(x) - J(x^*) = \langle \nabla J(x^*), h \rangle + \tfrac{1}{2} J''(y) h^2 > \frac{\mu}{2} \|h\|^2 > 0,$$

where $y = x^* + \theta h$, $0 < \theta < 1$. Thus, x^* is a minimum point of J. It remains to show that g has a fixed point. We shall do this by showing that g is contractive. It will then follow, by Theorem 5.1.1, that a unique fixed point x^* of g exists and $x^* = \lim_{n \to \infty} x_n$, where x_n is the sequence of successive approximations given by $x_{n+1} = g(x_n)$, that is, by (12.3.1) with $s_n = s$.

Now, $g'(x) = I - sJ''(x)$. Clearly,

$$1 - sM < \langle (I - sJ''(x))h, h \rangle < 1 - s\mu, \qquad \text{for } \|h\| = 1.$$

By Theorem 3.6.4,

$$\|I - sJ''(x)\| \leq \max\{|1 - sM|, |1 - s\mu|\}.$$

By our choice of s and δ, $|1 - sM| < 1 - \delta M$ and $|1 - s\mu| < 1 - \delta\mu$. Since $1 - \delta M < 1 - \delta\mu < 1$, we see that $\|I - sJ''(x)\| < 1$. By the mean-value theorem, g is contractive. This completes the proof. ∎

12.3 Gradient Methods for Extremum Problems without Constraints 537

Corollary The minimum point x^* of the theorem is the unique minimum point of J. The sequence (x_n) given by the gradient procedure with the choice of s given in the theorem converges to x^* at a linear rate according to (5.1.5) and (5.1.3).

(See the Remark under Computational Aspects below.)

Example 1 Consider the linear least-squares problem discussed in Section 7.4a. It is required to minimize $J(x) = \|Ax - b\|_2^2$, where A is a bounded linear operator on one real Hilbert space X to another Y. Since

$$J(x) = \langle Ax - b, Ax - b \rangle = \langle Ax, Ax \rangle - 2\langle Ax, b \rangle + \|b\|^2$$
$$= \langle A^*Ax, x \rangle - 2\langle A^*b, x \rangle + \|b\|^2,$$

we can easily calculate the directional differential by evaluating $J(x + sh) - J(x)$. This yields

$$dJ(x; h) = \lim_{s \to 0} \frac{J(x + sh) - J(x)}{s} = 2\langle A^*Ax, h \rangle - 2\langle A^*b, h \rangle,$$

from which it follows (by Theorem 5.3.2, since A is continuous) that

$$J'(x) = \nabla J(x) = 2(A^*Ax - A^*b).$$

We see that a zero, x^*, of $\nabla J(x)$ is just a solution of the normal equations and conversely. (See (7.4.4).) Let us calculate the second derivative. We have

$$\frac{J'(x + sh) - J'(x)}{s} = \frac{2A^*(Ax + sAh - b) - 2A^*(Ax - b)}{s}.$$

Passing to the limit as $s \to 0$, we find

$$dJ'(x; h) = 2A^*Ah,$$

which implies that $J''(x) = 2A^*A$. Now, A^*A is a bounded self-adjoint operator on X to X. (See Remark after Theorem 3.5.5.) Also,

$$\langle A^*Ah, h \rangle = \langle Ah, Ah \rangle = \|Ah\|^2 \geq 0 \qquad \text{for all } h \in X.$$

Hence, by the remark following Theorem 12.2.3, if $\nabla J(x^*) = 0$, then x^* is a weak global minimum point of J. As we know from Section 7.4a, x^* need not be unique. However, if $\|Ah\|^2 > \mu \|h\|^2$ for some constant $\mu > 0$, then the preceding theorem and corollary imply that there exists a unique strong global minimum point x^* which is the limit of a sequence x_n defined by a gradient method,

$$x_{n+1} = x_n - 2s(A^*Ax_n - A^*b),$$

with arbitrary starting vector x_0, and s determined as in Theorem 12.3.3. If A is an $m \times n$ matrix, then A^*A is an $n \times n$ matrix of the same rank. To satisfy the condition $\|Ah\|^2 > \mu \|h\|^2 > 0$, it is necessary and sufficient that A have rank n. For the constant M in Theorem 12.3.3 we may take any operator norm $\|A^*A\|$, or since $\|A^*A\| \leq \|A^*\| \|A\| = \|A\|^2$ we can take $M = \|A\|^2$; (e.g., for $\|A\|$ we calculate maximum row or column sums.) Observe that $M = \|A\|^2$ applies to the infinite-dimensional case as well.

Example 2 The more general problem of minimizing a quadratic functional $J(x) = \frac{1}{2}\langle Ax, x \rangle - \langle b, x \rangle$, where A is a bounded self-adjoint operator on a Hilbert space, can also be solved by the gradient procedure of the preceding theorem. Here, $J''(x) = A$, so that it suffices to have $\langle Ah, h \rangle > \mu \|h\|^2$ for some $\mu > 0$. We can take $M = \|A\|$ and $\delta < s < (2/\|A\|) - \delta$, where

538   Extremum Problems

$0 < \delta < 1/\|A\|$. The gradient procedure becomes

$$x_{n+1} = x_n - s(Ax_n - b) \qquad (12.3.3)$$

and (x_n) converges to a unique global minimum point x^* of J. We see that $\nabla J(x^*) = Ax^* - b = 0$, so that x^* is the unique solution of the linear equation $Ax = b$. This demonstrates that the solution of $Ax = b$, when A is a bounded self-adjoint operator on a Hilbert space X satisfying $\langle Ah, h \rangle > \mu \|h\|^2$, is equivalent to the minimization of $\frac{1}{2}\langle Ax, x \rangle - \langle b, x \rangle$. It implies that $Ax = b$ has a unique solution for arbitrary b, which is consistent with the fact that 0 is in the resolvent set of A. The latter fact is a consequence of 0 not being in the numerical range of A (Theorem 3.6.11). Furthermore, $A = A^*$ is closed (Exercise 3.46), so that A^{-1} is closed. This implies (the space being complete) that the domain of A^{-1} is all of X.

Recall that the matrix case was discussed earlier in Section 4.5, where the Gauss–Seidel method applied to a positive definite matrix was viewed as a minimization procedure. Also, compare (12.3.3) with the relaxation methods given by (4.5.4). (Exercise 12.2.)

Returning to our discussion of the gradient procedure, we observe that the hypothesis of Theorem 12.3.1 can be weakened. As is frequently the case, our assumption regarding the boundedness of the second derivative as a function of x, can be replaced by the assumption of Lipschitz continuity of the first derivative. This is a consequence of the mean-value theorem. Thus, if we assume

$$\|\nabla J(x + h) - \nabla J(x)\| \leq (1/s_0) \|h\|,$$

then at any point of x_n of a gradient procedure we have

$$\Delta J(x_n) = J(x_n - s_n \nabla J(x_n)) - J(x_n)$$
$$= -s_n \langle \nabla J(x_n - \theta s_n \nabla J(x_n)), \nabla J(x_n) \rangle$$
$$= -s_n (\|\nabla J(x_n)\|^2 + \langle u, \nabla J(x_n) \rangle),$$

where $\|u\| \leq (s_n/s_0) \|\nabla J(x_n)\|$. Hence, $\Delta J(x_n) = -s_n \|\nabla J(x_n)\|^2 (1 + \varepsilon)$, where $|\varepsilon| \leq s_n/s_0$. Now, if we choose $0 < \delta \leq s_n \leq s_0 - \delta$, then $\delta/s_0 \leq s_n/s_0 \leq 1 - \delta/s_0$, so that $\Delta J(x_n) < 0$ and $|\Delta J(x_n)| \geq \delta^2/s_0 \|\nabla J(x_n)\|^2$. If $\inf J(x_n) > -\infty$, this implies again that $\nabla J(x_n) \to 0$. Again, $J(x_n)$ converges monotonically downward to a limit. We have proved

Theorem 12.3.4   Let J be a functional on a pre-Hilbert space. Let $\nabla J(x)$ exist and be Lipschitz continuous in x with Lipschitz constant $1/s_0$. Let (x_n) be a sequence defined by a gradient method (12.3.1) with the s_n chosen so that $\delta \leq s_n \leq s_0 - \delta$, where $0 < \delta \leq s_0/2$. Then $(J(x_n))$ is a monotone nonincreasing sequence. If $\inf_n J(x_n) > -\infty$, then $\lim_{n \to \infty} \nabla J(x_n) = 0$.

Remark   In the preceding theorems, we required the hypotheses to hold for all x in the space. This requirement can obviously be relaxed as follows. Let x_0 be an arbitrary point in the domain of J and let $S = \{x : J(x) \leq J(x_0)\}$. It is sufficient for S to be convex and for the hypotheses to hold in S.

It is not difficult to prove that the Lipschitz continuity of the preceding Theorem can be weakened to uniform continuity of $\nabla J(x)$. Let

$$\Delta J(x, s) = J(x - s \nabla J(x)) - J(x).$$

12.3 Gradient Methods for Extremum Problems without Constraints

By the mean-value theorem,

$$\Delta J(x, s) = -s \langle \nabla J(y), \nabla J(x) \rangle$$

$$= -s \|\nabla J(x)\|^2 \left(1 + \frac{\langle \nabla J(y) - \nabla J(x), \nabla J(x) \rangle}{\|\nabla J(x)\|^2}\right),$$

where $y = x - \theta s \, \nabla J(x)$, $0 < \theta < 1$. Denote the quantity in parenthesis by $g(x, s)$. Then $g(x, s) = -\Delta J(x, s)/s \, \|\nabla J(x)\|^2$ is continuous in s for $0 < s < 1$. Furthermore, by the continuity of ∇J, we see that $\nabla J(y) \to \nabla J(x)$ as $s \to 0$, so that $g(x, s) \to g(x, 0) = 1$. Now, let $0 < \delta \leq \frac{1}{2}$. Define s_n as follows:

$$s_n = \begin{cases} 1 \text{ if } g(x_n, 1) \geq \delta, \\ \mu_n \in [0, 1] \text{ otherwise, where } \delta \leq g(x_n, \mu_n) \leq 1 - \delta. \end{cases} \quad (12.3.4)$$

To see that μ_n exists if $g(x_n, 1) < \delta$, simply note that $g(x_n, s)$ is continuous in s and therefore assumes every value between $g(x_n, 0) = 1$ and $g(x_n, 1) < \delta$. With this choice of s_n, $\Delta J(x_n, s_n) \leq -s_n \delta \|\nabla J(x_n)\|^2$.

Now, we assume $\|\nabla J(x)\|$ is bounded and $\inf J(x) > -\infty$. Suppose $\nabla J(x_n)$ did not converge to zero. We claim this would imply $\lim_{n \to \infty} \sup s_n > 0$. Otherwise, $\lim s_n = 0$ and $y_n = x_n - \theta_n s_n \nabla J(x_n) \to x_n$. Hence, by the uniform continuity of ∇J, $\nabla J(y_n) - \nabla J(x_n) \to 0$. This would imply $g(x_n, s_n) \to 1$, contradicting (12.3.4). (If $g(x_n, 1) = 1$ for all n, then $\Delta J(x_n) = -\|\nabla J(x_n)\|^2$, implying $\inf J(x) = -\infty$.) Thus, if $\lim_n \sup \|\nabla J(x_n)\| > c > 0$, then $\lim_n \sup s_n > d > 0$ and $\Delta J(x_n) \leq -d\delta c^2$ for infinitely many n. We summarize these results as

Theorem 12.3.5 Let J be a functional on a pre-Hilbert space. Let $\nabla J(x)$ exist, be bounded and uniformly continuous. Let (x_n) be a sequence given by (12.3.1) with s_n given by (12.3.4). Then $(J(x_n))$ is a nonincreasing sequence. If $\inf J(x) > -\infty$, then $(J(x_n))$ converges monotonically downward and $\nabla J(x_n) \to 0$.

If we choose the s_n according to the steepest descent algorithm, then we can again show that $\nabla J(x_n) \to 0$ provided that (x_n) has an accumulation point. This will be the case, for example, when the set $S = \{x : J(x) \leq J(x_0)\}$ is compact.

Theorem 12.3.6 Let ∇J be continuous on the convex hull of S. If a sequence (x_n) determined by the steepest descent algorithm has an accumulation point x^*, then $\nabla J(x^*) = 0$. If S is compact, then $\nabla J(x_n) \to 0$.

Proof. Since $(J(x_n))$ is nonincreasing and $J(x_{n_i}) \to J(x^*)$ for some subsequence (x_{n_i}) which converges to x^*, we must have $J(x_n) \to J(x^*)$. Suppose $\nabla J(x^*) \neq 0$. We shall deduce a contradiction. If $\nabla J(x^*) \neq 0$, we can determine $\sigma > 0$ which minimizes $J(x^* - s\nabla J(x^*))$. Let $u = x^* - \sigma \nabla J(x^*)$ and $u_i = x_{n_i} - \sigma \nabla J(x_{n_i})$. By continuity of ∇J, $u_i \to u$ as $i \to \infty$. Now, $J(u_i) = J(u) + \langle \nabla J(y_i), u_i - u \rangle$, where $y_i = u + \theta(u_i - u)$. Since $J(x^*) = J(u) + \delta$, where $\delta > 0$, we see that $J(u) < J(x_0)$, so that also $J(u_i) < J(x_0)$ for i sufficiently large. Hence, $u, u_i \in S$ and y_i is in the convex hull of S for all sufficiently large i. Therefore, by continuity, $\nabla J(y_i) \to \nabla J(u)$, which means that for i large enough,

$$J(u_i) \leq J(u) + \delta/2 = J(x^*) - \delta/2.$$

Now, $J(x^*) \leq J(x_{n_i+1}) = J(x_{n_i} - s_{n_i} \nabla J(x_{n_i})) \leq J(u_i)$, by the minimal property of s_{n_i}

in steepest descent. This yields the contradiction $J(x^*) \leq J(x^*) - \delta/2$. Therefore, $\nabla J(x^*) = 0$.

If $\nabla J(x_n)$ did not converge to 0, there would be subsequence (y_n) with $\|\nabla J(y_n)\| > \mu > 0$. By compactness, (y_n) would have a convergent subsequence with accumulation point x^* and $\nabla J(x^*) = 0$, which contradicts $\|\nabla J(y_n)\| > \mu$. ∎

Computational Aspects

To apply (12.3.1) in actual practice, we must be able to determine the scalars s_n. If we have an estimate of the Lipschitz constant $= 1/s_0$, or of the bounds on $J''(x)$, then we may take any s_n, $n \geq 1$, which satisfy the inequalities in Theorems 12.3.1, 12.3.3 or 12.3.4 as the case may be. In particular, we can choose all s_n the same and these choices can be made a priori. On the other hand, if we do not have estimates of the required bounds, the s_n may be determined at each iteration of the computation by using (12.3.4); i.e. we compute

$$g(x_n, 1) = \frac{J(x_n) - J(x_n - \nabla J(x_n))}{\|\nabla J(x_n)\|^2}$$

and if $g(x_n, 1) \geq \delta$, where δ is any number in $(0, \frac{1}{2}]$, we take $s_n = 1$. If $g(x_n, 1) < \delta$, we find an $s_n \in [0, 1]$ such that $\delta \leq g(x_n, s_n) \leq 1 - \delta$. This can be done by (linear) interpolation, for example, after computing several values of $g(x_n, s)$, $s \in [0, 1]$. Once an acceptable value of s_n has been found, it will probably serve for several iterations.

Again, we emphasize that the sequence (x_n) need not converge. If the set $S = \{x : J(x) \leq J(x_0)\}$ is compact, then since $x_n \in S$, there is a convergent subsequence (y_n). The limit x^* of such a subsequence must be a stationary point of J whenever the conclusion $\nabla J(x_n) \to 0$ is valid because $0 = \lim \nabla J(y_n) = \nabla J(x^*)$, by continuity of ∇J.

Furthermore, since every accumulation point must be a zero of $\nabla J(x)$, the number of accumulation points cannot be greater than the number of stationary points. If the latter are finite in number, then the number of accumulation points must be finite. If $\nabla J(x_n) \to 0$, then $\|x_{n+1} - x_n\| = s_n \|\nabla J(x_n)\| \to 0$. If the set S is compact, then a sequence (x_n) with a finite number of accumulation points and $\|x_{n+1} - x_n\| \to 0$ must have precisely one accumulation point, that is, (x_n) must be convergent. (Exercise 12.3.) This is the most common situation in practice, since $\nabla J(x)$ will usually have a finite number of zeros. Hence, the sequence (x_n) produced by a gradient procedure will usually converge to the stationary point nearest to x_0, the initial guess.

To apply the steepest descent procedure, we must minimize

$$g_n(s) = J(x_n - s \nabla J(x_n))$$

at each iteration. In most cases, this must be done approximately by computing $g_n(s)$ for several values of s until $g_n(s)$ increases, and then possibly interpolating. In case J is a quadratic functional, then the minimizing value of s can be determined by solving a quadratic polynomial equation.

12.3 Gradient Methods for Extremum Problems without Constraints

Gradient procedures are inherently slow, converging at a linear rate in general, since they are essentially successive approximation methods for finding zero's of $\nabla J(x)$. Some idea of just how slow the convergence can be is gained from a study of the minimization of the quadratic functional $\frac{1}{2}\langle Ax, x\rangle$, where A is a symmetric positive definite matrix. Consider the simple two-dimensional case where A is diagonal. Thus, $J(x) = \frac{1}{2}(a_{11}\xi_1^2 + a_{22}\xi_2^2)$ and the level surfaces of J are confocal ellipses with center at the origin. The minimum point is $x = 0$ and $J(0) = 0$ is the minimum value. If $a_{11} = a_{22}$, the ellipses are concentric circles and the gradient vector is $\nabla J(x) = Ax = (a_{11}\xi_1, a_{22}\xi_2) = a_{11}(\xi_1, \xi_2)$. Hence, $-\nabla J(x)$ is radial in direction and points directly to the minimum point. If steepest descent is used, then for any x_0, $x_1 = x_0 - s_0 \nabla J(x_0)$ would be the minimum point; i.e. one iteration would suffice (ignoring rounding errors). Any other choice of s_n would also yield convergence in a few steps. However, if a_{22}/a_{11} is large, the ellipses are elongated and $-\nabla J(x)$ points primarily toward the ξ_1-axis. For any initial vector x_0 other than $(\xi_1, 0)$ or $(0, \xi_2)$, a gradient procedure would produce a zig-zag motion along the ξ_1-axis, requiring many iterations. Unfortunately, this behavior is observed frequently in practice. One possible remedy is to make a change of variable by means of a linear transformation $x = Ty$. Then $J(x) = J(Ty) = g(y)$ is a new functional to be minimized. Since $\langle \nabla g(y), h\rangle = J'(x) \cdot Th = \langle \nabla J(x), Th\rangle$ (by Lemma 5.3.4 and the fact that $(Ty)' = T$), we see that $\nabla g(y) = T^*\nabla J(x)$. If $y_{n+1} = y_n - s_n \nabla g(y_n)$, then

$$x_{n+1} = Ty_{n+1} = Ty_n - s_n T\nabla g(y_n) = x_n - s_n TT^*\nabla J(x_n). \quad (12.3.5)$$

This suggests an iteration of the form $x_{n+1} = x_n - s_n B \nabla J(x_n)$, where B is a positive definite self-adjoint operator. In [12.3] it is shown that we can take $B = (J''(x_n))^{-1}$, when J'' is positive definite. Except for the factors s_n, this is essentially Newton's method applied to $\nabla J(x)$. In Section 12.3.1, we discuss other methods based on this idea.

Remark We can make the preceding discussion more precise by referring to Theorem 12.3.3, its Corollary and Example 2 following that theorem. We see that here $J''(x) = A$ is a diagonal matrix, so that

$$\langle J''(x)h, h\rangle = a_{11}h_1^2 + a_{22}h_2^2, \quad \text{where } h = (h_1, h_2).$$

If $0 < a_{11} < a_{22}$, then $\mu = a_{11} < \langle J''(x)h, h\rangle < a_{22} = M$ for any $\|h\| = 1$. Hence, $x_{n+1} = x_n - sAx_n = (I - sA)x_n$ and (Theorem 3.6.5)

$$\|I - sA\| = \max\{|1 - sa_{11}|, |1 - sa_{22}|\}.$$

As Fig. 12.1 quickly shows, $\|I - sA\|$ can be made zero if $a_{11} = a_{22}$ (by choosing $s = 1/a_{11}$), but the minimum of $\|I - sA\|$ over s approaches 1 as $a_{22}/a_{11} \to \infty$. In fact, the minimum occurs at $s^* = 2/(a_{11} + a_{22})$ and its value is

$$\|I - s^*A\| = \frac{a_{22} - a_{11}}{a_{22} + a_{11}}.$$

Even if $a_{22}/a_{11} = 10$, so that $\|I - s^*A\| = 0.9$ (approximately), the gradient method

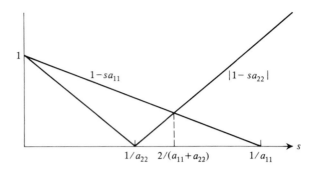

Figure 12.1

may still be feasible. To reduce the initial error by a factor of 10^{-q} would require N iterations, where $N = -q/\log 0.9 = 20q$ approximately. For $q = 8$, $N = 160$. This may not be excessive if the number of computations per iteration is small. For $\|I - s^*A\| = 0.99$ we obtain $N = 220q$ or approximately 1760 iterations for $q = 8$. This is excessive and would usually rule out the use of a gradient procedure.

Note finally that s^* is not the value for steepest descent in this example. Since $J(x - sh) = \frac{1}{2}\langle A(x - sh), x - sh\rangle$ with $h = \nabla J(x) = Ax$, we have $J(x - sh) = \frac{1}{2}(as^2 - 2bs + c)$ with $a = \langle A^3 x, x\rangle$ and $b = \|Ax\|^2$. Hence, $J(x - sh)$ is minimized by taking $s' = \|Ax\|^2/\langle A^3 x, x\rangle$. In the case being considered,

$$s' = \frac{a_{11}^2 \xi_1^2 + a_{22}^2 \xi_2^2}{a_{11}^3 \xi_1^2 + a_{22}^3 \xi_2^2},$$

which approaches $1/a_{22}$ as $a_{22}/a_{11} \to \infty$, whereas $s^* \to 2/a_{22}$. Furthermore, this case illustrates that the steepest descent algorithm need not be the fastest of the gradient procedures. (See Exercise 12.4; also [12.15] and [12.16].)

12.3 (a) Conjugate Direction Methods (Variable Metric Methods)

Let $J(x) = \frac{1}{2}\langle Ax, x\rangle - \langle b, x\rangle + c$ be a quadratic functional. The finite-dimensional case in which $x \in V^m$ and A is a positive definite symmetric matrix of order m has received special attention, since many functions f behave approximately like quadratic functionals in a sufficiently small neighborhood; i.e. in the neighborhood of a minimum point x^* we can set $A = f''(x^*)$ and we have seen (Theorem 12.2.4) that positivity is a sufficient condition for a minimum. In this case, it is possible to construct gradient methods which converge in $\leq m$ iterations (assuming no rounding errors). One of the earliest such methods, the *conjugate gradient* method [12.10] and various related schemes known as *conjugate direction* methods [12.11] are based on the following ideas.

When $J''(x) = A$ is positive definite, a necessary and sufficient condition that x^* be a minimum point of J on a subspace spanned by a set of vectors $\{v_1, \ldots, v_j\}$ is that $\langle \nabla J(x^*), v_i\rangle = 0$, $1 \leq i \leq j$. (Exercise 12.8 and Theorem 12.2.6.) Now let $\{v_1, \ldots, v_m\}$ be a linearly independent set in V^m. For arbitrary $x_0 \in V^m$ consider

12.3 Gradient Methods for Extremum Problems without Constraints 543

$f(t_1, \ldots, t_m) = J(x_0 + \sum_1^m t_i v_i)$. We wish to determine $(s_1, \ldots, s_m)$ which minimizes f. Then $x^* = x_0 + \sum_1^m s_i v_i$. The conjugate direction methods provide vectors v_i which make it possible to determine the s_i one at a time by minimizing in the directions v_i, $1 \leq i \leq m$, successively; i.e. by m one-dimensional minimizations. The vectors v_i are generated successively as each minimization step is completed.

Let $x_j = x_0 + \sum_{i=1}^j s_i v_i$ and set $\Delta x_j = x_j - x_{j-1} = s_j v_j$, $1 \leq j \leq m$. For convenience write $\nabla J_i = \nabla J(x_i) = A x_i - b$. Note that $\nabla J_i - \nabla J_{i-1} = A \Delta x_i = s_i A v_i$. Suppose that each v_{j+1} is determined so that $\langle A v_{j+1}, v_i \rangle = 0$ for $i = 1, \ldots, j$. $\langle x, y \rangle_A = \langle Ax, y \rangle$ is a new inner product (Exercise 3.3d) and the v_j are orthogonal with respect to this A inner product. Such a set of vectors is called a set of *conjugate directions for* J. Suppose further that s_{j+1} is determined so as to minimize $J(x_j + s v_{j+1})$ as a function of s; i.e. $x_{j+1} = x_j + s_{j+1} v_{j+1}$ is the minimum of J on the line $x_j + s v_{j+1}$. Then $\langle \nabla J(x_{j+1}), v_{j+1} \rangle = 0$ (Exercise 12.8). Furthermore, we shall prove by induction that $\langle \nabla J(x_{j+1}), v_i \rangle = 0$ for $1 \leq i \leq j$ as well. This certainly is true for $j = 1$, since $\langle \nabla J(x_1), v_1 \rangle = 0$ because $x_1 = x_0 + s_1 v_1$ is the minimum of J on the line $x_0 + s v_1$. By the induction hypothesis, $\langle \nabla J(x_j), v_i \rangle = 0$ for $1 \leq i \leq j$. Hence, for $1 \leq i \leq j$, $\langle \nabla J_{j+1} - \nabla J_j, v_i \rangle = s_{j+1} \langle A v_{j+1}, v_i \rangle = 0$ by the orthogonality of the v's with respect to the A inner product. Since $\langle \nabla J_j, v_i \rangle = 0$ by hypothesis, it follows that $\langle \nabla J_{j+1}, v_i \rangle = 0$, completing the induction. As a consequence, x_{j+1} must be a minimum point of J on the subspace spanned by $\{v_1, \ldots, v_{j+1}\}$. Since $\{v_1, \ldots, v_m\}$ is a linearly independent set (being orthogonal), the point $x_m = x_0 + \sum_{i=1}^m s_i v_i$ must be the desired minimum point of J on all of V^m. Thus, by constructing the v_j as conjugate directions and minimizing along each v_j in turn we reduce the m-dimensional problem to a succession of m one-dimensional minimizations. The latter are easily solved, since $J(x_j + s v_{j+1})$ is a quadratic polynomial in s and is minimized by taking

$$s = s_{j+1} = -\frac{\langle A x_j - b, v_{j+1} \rangle}{\langle A v_{j+1}, v_{j+1} \rangle} = -\frac{\langle A x_0 - b, v_{j+1} \rangle}{\langle A v_{j+1}, v_{j+1} \rangle}. \tag{12.3.6}$$

The construction of the v_j can be carried out in various ways. One simple method given in [12.12] is defined by the formula,

$$\begin{aligned} v_1 &= \nabla J(x_0), \quad x_0 \text{ arbitrary} \\ v_{j+1} &= \nabla J(x_j) + \alpha_j v_j, \\ \alpha_j &= \|\nabla J(x_j)\|^2 / \|\nabla J(x_{j-1})\|^2, \end{aligned} \tag{12.3.7}$$

$$x_{j+1} = x_j + s_{j+1} v_{j+1}, \quad 1 \leq j \leq m - 1,$$

where again s_{j+1} is given by (12.3.6).

The A-orthogonality is proved by induction. Note that $\langle \nabla J_1, \nabla J_0 \rangle = 0$ because x_1 is a minimum point in the direction $v_1 = \nabla J_0$. Also,

$$\begin{aligned} s_1 \langle v_2, A v_1 \rangle &= \langle \nabla J_1 + \alpha_1 \nabla J_0, \nabla J_1 - \nabla J_0 \rangle \\ &= \|\nabla J_1\|^2 - \alpha_1 \|\nabla J_0\|^2 = 0. \end{aligned}$$

As induction hypothesis assume that $\langle v_i, v_k \rangle_A = 0$ and $\langle \nabla J_i, \nabla J_k \rangle = 0$ for $0 \le i < k \le j$. Then $\langle v_{j+1}, Av_i \rangle = \langle \nabla J_j, Av_i \rangle + \alpha_j \langle v_j, Av_i \rangle$. For $i < j$, $\langle v_j, Av_i \rangle = 0$ by the induction hypothesis. Since $s_i Av_i = \nabla J_i - \nabla J_{i-1}$, $\langle \nabla J_j, Av_i \rangle = 0$ also. For $i = j$, $\langle \nabla J_j, Av_j \rangle = (1/s_j) \|\nabla J_j\|^2$

and

$$\langle v_j, Av_j \rangle = \langle \nabla J_{j-1} + \alpha_{j-1} v_{j-1}, Av_j \rangle = \langle \nabla J_{j-1}, Av_j \rangle = -\|\nabla J_{j-1}\|^2/s_j.$$

By the definition of α_j it follows that $\langle v_{j+1}, Av_j \rangle = 0$, completing the induction on A-orthogonality. Finally,

$$\langle \nabla J_{j+1}, \nabla J_i \rangle = \langle \nabla J_j + s_{j+1} Av_{j+1}, \nabla J_i \rangle = \langle \nabla J_j, \nabla J_i \rangle + s_{j+1} \langle Av_{j+1}, \nabla J_i \rangle. \tag{12.3.8}$$

For $i < j$, $\langle \nabla J_j, \nabla J_i \rangle = 0$ by the induction hypothesis. Also, $\nabla J_i = v_{i+1} - \alpha_i v_i$ by (12.3.7) so that $\langle Av_{j+1}, \nabla J_i \rangle = 0$ by A-orthogonality. For $i = j$, we have $\nabla J_j = v_{j+1} - \alpha_j v_j$ so that $\langle \nabla J_{j+1}, \nabla J_j \rangle = \langle \nabla J_{j+1}, v_{j+1} \rangle - \alpha_j \langle \nabla J_{j+1}, v_j \rangle$. Since x_{j+1} is a minimum in the v_{j+1}-direction and since the v_i are conjugate directions, both inner products on the right must be zero. This completes the inductive proof.

If $J(x)$ is not quadratic in x, a conjugate direction method will usually require an infinite number of iterations. In the method of (12.3.7), it is recommended that the iteration be "restarted" with a steepest descent step by choosing a new x_0 after each cycle of determining the v_j.

There is another class of gradient methods which are derived by combining the idea of conjugate directions with that suggested by formula (12.3.5). The latter may be regarded as an attempt to apply Newton's method to determine zeros of $\nabla J(x)$. When J is quadratic, $J''(x) = A$ and Newton's method becomes

$$x_{n+1} = x_n - A^{-1} \nabla J(x_n),$$

which can be written as

$$A \Delta x_{n+1} = -\nabla J(x_n).$$

When x is finite dimensional, this equation can be solved for Δx_{n+1} by one of the methods of Chapter 4. When J is not quadratic, or if solution of the linear equations is not feasible, we can replace (12.3.5) by an iteration of the form,

$$x_{n+1} = x_n - s_{n+1} H_n \nabla J(x_n), \qquad n \ge 0, \tag{12.3.9}$$

where (H_n) is a sequence of positive definite self-adjoint operators approaching $(J''(x_n))^{-1}$; i.e. $H_n - (J''(x_n))^{-1} \to 0$.

To discover how to construct such H_n we consider the linearization of $\nabla J(x)$, $\nabla J_{n+1} - \nabla J_n = J''(x_n) \Delta x_{n+1}$. This suggests that H_n be determined so that

$$H_n(\nabla J_{n+1} - \nabla J_n) = \Delta x_{n+1}.$$

In fact, since $H_n = A^{-1}$ in the quadratic case, let us require that

$$H_n(\nabla J_i - \nabla J_{i-1}) = \Delta x_i, \qquad 1 \le i \le n. \tag{12.3.10}$$

12.3 Gradient Methods for Extremum Problems without Constraints

Now, (12.3.9) implies that $x_n = x_0 + \sum_{i=1}^{n} s_i v_i$, where $v_i = -H_{i-1} \nabla J_{i-1}$. We have already seen that in the quadratic case if $\langle \nabla J(x_n), v_i \rangle = 0$ for $1 \le i \le n$ and v_{n+1} is chosen so that $\langle \nabla J_i - \nabla J_{i-1}, v_{n+1} \rangle = 0$, $1 \le i \le n$, then $\langle \nabla J_{n+1}, v_i \rangle = 0$ for $1 \le i \le n$. Therefore, if at the nth step the new direction v_{n+1} is chosen orthogonal to all previous gradient changes $\Delta J_i' = \nabla J_i - \nabla J_{i-1}$, then the gradient ∇J_{n+1} at the new point $x_{n+1} = x_n + s_{n+1} v_{n+1}$ must be orthogonal to all previous v_i, $1 \le i \le n$. Note that this is true for any value of s_{n+1}, since

$$\langle \Delta J_{n+1}', v_i \rangle = \langle A(x_{n+1} - x_n), v_i \rangle = s_{n+1} \langle A v_{n+1}, v_i \rangle$$
$$= s_{n+1} \langle v_{n+1}, A v_i \rangle = (s_{n+1}/s_i) \langle v_{n+1}, A(x_i - x_{i-1}) \rangle$$
$$= (s_{n+1}/s_i) \langle v_{n+1}, \Delta J_i' \rangle = 0.$$

If we further choose s_{n+1} to minimize J on the line $x_n + s_{n+1} v_{n+1}$, then $\langle \nabla J_{n+1}, v_{n+1} \rangle = 0$ as well, completing the induction and allowing the iteration to continue with the choice of v_{n+2}. (Since s_1 is chosen to minimize J along v_1, we have $\langle \nabla J_1, v_1 \rangle = 0$ to start the induction.)

If (12.3.9) and (12.3.10) hold, then for $1 \le i \le n$

$$-\langle \Delta J_i', v_{n+1} \rangle = \langle \Delta J_i', H_n \nabla J_n \rangle = \langle H_n \Delta J_i', \nabla J_n \rangle = \langle \Delta x_i, \nabla J_n \rangle = -s_i \langle v_i, \nabla J_n \rangle.$$

Therefore, if $\langle \nabla J_n, v_i \rangle = 0$ for $1 \le i \le n$ and we take $v_{n+1} = -H_n \nabla J_n$, where H_n satisfies (12.3.10), then $\langle \Delta J_i', v_{n+1} \rangle = 0$ for $1 \le i \le n$, fulfilling the conjugate direction condition. This leads to a variety of methods depending on the choice of the H_n. In the positive definite quadratic case, any sequence (H_n) which satisfies (12.3.10) must converge to A^{-1} in m steps, where m is the order of A, because $H_m \Delta J_i' = \Delta x_i = A^{-1} \Delta J_i'$ for $1 \le i \le m$. Since $\Delta J_i' = s_i A v_i$ are m linearly independent vectors, $H_m = A^{-1}$. In the case of a general functional J, the question of convergence is open.

One technique for determining the mth-order matrices H_n, $n = 0, 1, 2, \ldots$, makes use of generalized inverses [12.13]. Let G_n be the $m \times n$ matrix having columns $\Delta J_i' = \nabla J_i - \nabla J_{i-1}$, $i = 1, \ldots, n$. Similarly, let δX_n be the $m \times n$ matrix having columns Δx_i, $1 \le i \le n$. Equations (12.3.10) can be written as the single matrix equation,

$$H_n G_n = \delta X_n. \tag{12.3.11}$$

Referring to Exercise 2.13 (Eqs. (1) and (6)), we see that the general solution of (12.3.11) is

$$H_n = \delta X_n G_n^\dagger + Y(I_m - G_n G_n^\dagger),$$

where $G_n^\dagger$ is the pseudoinverse of G_n. If G_n has rank n, then (Exercise 2.13(i))

$$G_n^\dagger = (G_n^* G_n)^{-1} G_n^*.$$

A generalized inverse $\hat{G}_n$ is given by

$$\hat{G}_n = (G_n^* H G_n)^{-1} G_n^* H, \tag{12.3.12}$$

where H is any mth-order positive definite symmetric matrix. Clearly, $G_n^* H G_n$ is positive definite, so that its inverse exists and obviously $\hat{G}_n G_n = I$. Therefore

$$H_n = \delta X_n \hat{G}_n + Y(I - G_n \hat{G}_n) \tag{12.3.13}$$

also satisfies (12.3.11), Y being an arbitrary matrix. Various choices of H lead to different methods. The method of [12.14] can be derived by setting $H = A^{-1}$ in the first $\hat{G}_n$ and $Y = H$ in the second. Since $A^{-1} G_n = \delta X_n$ in the quadratic case,

$$\hat{G}_n = (G_n^* A^{-1} G_n)^{-1} G_n^* A^{-1} = (\delta X_n^* G_n)^{-1} \delta X_n^*.$$

Using this in (12.3.13), we have

$$H_n = \delta X_n (\delta X_n^* G_n)^{-1} \delta X_n^* + H(I - G_n(G_n^* H G_n)^{-1} G_n^* H). \tag{12.3.14}$$

Thus, at the nth step, having computed x_n, $\nabla J(x_n)$ and the differences $\Delta x_i = x_i - x_{i-1}$, $\Delta J_i' = \nabla J_i - \nabla J_{i-1}$, $1 \leq i \leq n$, we can form the matrices $\delta X_n = (\Delta x_1, \ldots, \Delta x_n)$, $G_n = (\Delta J_1', \ldots, \Delta J_n')$ and compute H_n. The iteration can then continue with the computation of x_{n+1} by (12.3.9), s_{n+1} being determined to minimize J along the $v_{n+1} = -H_n \nabla J(x_n)$ direction.

For efficient computation, it is convenient to derive a recursion formula expressing H_{n+1} in terms of H_n. This is done easily by using the bordering method of obtaining inverses (Exercise 4.20). A straightforward calculation (Exercise 12.9) yields the formula,

$$H_{n+1} = H_n - \frac{(H_n \Delta J'_{n+1})(H_n \Delta J'_{n+1})^*}{\langle \Delta J'_{n+1}, H_n \Delta J'_{n+1} \rangle} + \frac{\Delta x_{n+1} \Delta x_{n+1}^*}{\langle \Delta x_{n+1}, \Delta J'_{n+1} \rangle}, \tag{12.3.15}$$

where $\Delta J'_{n+1} = \nabla J(x_{n+1}) - \nabla J(x_n)$, $\Delta x_{n+1} = x_{n+1} - x_n$ and H_0 is an arbitrary positive definite symmetric matrix (e.g. $H_0 = I$).

When J is not quadratic, the value of s_{n+1} is not given by (12.3.6) of course. Instead, a one-dimensional minimization problem must be solved. This can be done in several ways. One technique is to use cubic interpolation of $J(x_n + s v_{n+1})$ as a function of s. This involves the computation of two values of J and its derivative along the line $x_n + s v_{n+1}$. The derivative is given by

$$\left. \frac{dJ(x_n + s v_{n+1})}{ds} \right|_s = \langle \nabla J(x_n + s v_{n+1}), v_{n+1} \rangle.$$

12.4 GRADIENT METHODS FOR EXTREMUM PROBLEMS WITH EQUALITY CONSTRAINTS

Let J and $g_1, \ldots, g_p$ be real functionals defined on a subset of a pre-Hilbert space X. Consider the equality constraint $C = \bigcap C(g_i)$ of Definition 12.2.1 and let us seek to determine a minimum of J on C. A necessary condition for a minimum point u is given by the multiplier rule, Theorem 12.2.2, if we assume that the weak gradients exist. For computational purposes, we shall require the existence of the strong gradients, $\nabla J(u)$, $\nabla g_i(u)$. If $\lambda_0 \neq 0$ in (12.2.3), then the multiplier rule asserts that

12.4 Gradient Methods for Extremum Problems with Equality Constraints 549

Since $g_i(u) = 0$, we have $g_i(x) = g_i(x) - g_i(u) = \langle \nabla g_i(u), \overline{\Delta x} \rangle + \varepsilon_i(u, \Delta x)$, where $\varepsilon_i/\|\Delta x\| \to 0$ uniformly as $\|\Delta x\| \to 0$, $1 \leq i \leq p$, by (5.3.11) and the uniform continuity of ∇g. (We adopt the notational convention throughout the proof that ε with various marks and subscripts denotes a quantity such that $\varepsilon/\|\Delta x\| \to 0$ uniformly as $\|\Delta x\| \to 0$.) From the uniform continuity of ∇g_i it follows that $g_i(x) = \langle \nabla g_i(x), \overline{\Delta x} \rangle + \tilde{\varepsilon}_i$. Substituting in (12.4.1), we obtain

$$h_G = -\sum_1^p \langle \overline{\nabla g_i(x)}, \overline{\Delta x} \rangle \overline{\nabla g_i(x)} + \varepsilon_G.$$

Hence

$$\langle h_G, \Delta x \rangle = -\|\Delta x\|^2 \sum_1^p \cos^2 \alpha_i + \langle \varepsilon_G, \Delta x \rangle,$$

where the α_i are defined in (v) of Definition 12.4.1. Also

$$\|h_G\|^2 \leq \|\Delta x\|^2 \left(\sum_{i,j=1}^p \cos \alpha_i \cos \alpha_j + w_G \right),$$

where $w_G \to 0$ as $\|\Delta x\| \to 0$. By the Schwarz inequality,

$$\sum_1^p \cos \alpha_i \leq \left(\sum_1^p 1 \right)^{1/2} \left(\sum_1^p \cos^2 \alpha_i \right)^{1/2} = p^{1/2} \left(\sum_1^p \cos^2 \alpha_i \right)^{1/2},$$

$$\sum_1^p \cos \alpha_i \sum_1^p \cos \alpha_j \leq p \sum_{i=1}^p \cos^2 \alpha_i.$$

Hence,

$$\|h_G\|^2 \leq \|\Delta x\|^2 \left(p \sum_1^p \cos^2 \alpha_i + w_G \right).$$

From (12.4.2),

$$\langle h_T, \Delta x \rangle = \frac{-\tan \theta}{\|\nabla \theta\|} \|\Delta x\| \cos \beta. \tag{12.4.6}$$

If $\nabla J_T(x) \neq 0$, consider $x_s = u + s \|\Delta x\| \overline{\Delta x}$. Since $\nabla J_T(x_0) = 0$, $\tan \theta(x_0) = 0$. Applying the ordinary mean-value theorem, we have for some $0 < s < 1$, $\tan \theta(x) = \sec^2 \theta(x_s) \langle \nabla \theta(x_s), \overline{\Delta x} \rangle \|\Delta x\|$.

By condition (iv) of Definition 12.4.1,

$$\langle \nabla \theta(x_s), \overline{\Delta x} \rangle = \langle \nabla \theta(u), \overline{\Delta x} \rangle + w,$$

where $w \to 0$ uniformly as $\|\Delta x\| \to 0$. Similarly, $\sec^2 \theta(x_s) \to 1$ uniformly as $\|\Delta x\| \to 0$. It follows that

$$\tan \theta(x) = \langle \nabla \theta(x), \Delta x \rangle + \hat{\varepsilon},$$

$$\langle h_T, \Delta x \rangle = \begin{cases} -\|\Delta x\|^2 (\cos^2 \beta + w_1) & \text{if } \nabla J_T(x) \neq 0, \\ 0 & \text{if } \nabla J_T(x) = 0 \end{cases}$$

550   Extremum Problems

where $w_1 \to 0$ as $\Delta x \to 0$. Similarly, it can be shown that

$$\|h_T\| = \begin{cases} \|\Delta x\| (|\cos \beta| + w_T) & \text{if } \nabla J_T(x) \neq 0, \\ 0 & \text{if } \nabla J_T(x) = 0, \end{cases}$$

where $w_T \to 0$ as $\Delta x \to 0$. Combining these results, we find

$$\|x + sh - u\|^2 \leq \|\Delta x\|^2 (\varphi(s) + 2sw_1 + s^2 w_2), \tag{12.4.7}$$

where $w_1 \to 0$ and $w_2 \to 0$ as $\Delta x \to 0$ and $\varphi(s) = 1 - 2bs + 2cs^2$, with

$$b = \begin{cases} \sum_1^p \cos^2 \alpha_i + \cos^2 \beta & \text{if } \nabla J_T(x) \neq 0, \\ \sum_1^p \cos^2 \alpha_i & \text{if } \nabla J_T(x) = 0, \end{cases}$$

$$c = \begin{cases} p \sum_1^p \cos^2 \alpha_i + \cos^2 \beta & \text{if } \nabla J_T(x) \neq 0, \\ p \sum_1^p \cos^2 \alpha_i & \text{if } \nabla J_T(x) = 0. \end{cases}$$

Now, $\varphi(s) - (1 - b^2/2c) = 2c(s - b/2c)^2$. We see that $b^2 \leq (p + 2) \sum_{i=1}^p \cos^2 \alpha_i + \cos^2 \beta$, so that for $p \geq 2$ it follows that $b^2/2c \leq 1$. Also, for $p = 1$, $b^2/2c = b/2 \leq 1$. Since $2c \leq 2(p^2 + 1)$ and $b^2 \geq \gamma^2$ by condition (v) of Definition 12.4.1, we have $b^2/2c \geq \gamma^2/2(p^2 + 1) > 0$ for all $x \in N$. Hence, $|\varphi(s)| \leq 1$ whenever $0 \leq s \leq b/2c$. Since $c \leq pb$, it follows that $b/c \geq 1/p$. Therefore, for $d < \frac{1}{2}p$ and $d/2 < s < d$, there exists $\delta > 0$ such that $|\varphi(s)| < 1 - \delta$. Now, let $B' = B(u^*, r')$ be contained in the neighborhood N. For $x \in B'$, we have $\|\Delta x\| = \|x - u\| < r'$. By taking r' small, $\|w_1(\Delta x)\|$ and $\|w_2(\Delta x)\|$ can be made so small that $|\varphi(s) + 2sw_1 + s^2 w_2| < k^2 < 1$ for some constant k and all s with $d/2 < s < d$. It follows that for $x_0 \in B'$ there exists a stationary point $u_0 \in B'$ such that

$$\|x_1 - u_0\| = \|x_0 + sh - u_0\| < k \|x_0 - u_0\|,$$

the constant k being independent of x_0 and u_0. Furthermore, if x_0 is sufficiently close to u^*, we must have $x_1 \in B'$. Therefore, there exists a linearly closest stationary point $u_1 \in B'$ to x_1, and the computation can be repeated with the pair (x_1, u_1). In fact, if $\|x_0 - u^*\| < r = r'(1 - k)/2(1 + k)$, then (as we shall prove by induction) all the points x_n given by (12.4.4) and the corresponding stationary points u_n of condition (v) satisfy $\|x_n - u_n\| \leq k^n \|x_0 - u_0\|$, and all x_n are in B'. It will follow that the sequence (x_n) converges to a stationary point $\tilde{u} \in B'$. We now carry out the inductive step. Suppose $x_i \in B'$ and $\|x_i - u_i\| \leq k^i \|x_0 - u_0\|$ for $0 \leq i \leq n$. Then by the above proof,

$$\|x_{n+1} - u_n\| = \|u_n + sh_n - u_n\| \leq k \|x_n - u_n\|.$$

By condition (v), there exists $u_{n+1} \in N$ such that

$$\|x_{n+1} - u_{n+1}\| \leq \|x_{n+1} - u_n\|$$

and, therefore,
$$\|x_{n+1} - u_{n+1}\| \le k \|x_n - u_n\| \le k^{n+1} \|x_0 - u_0\|.$$
Furthermore, for $0 \le i \le n$,
$$\|x_{i+1} - x_i\| \le \|x_{i+1} - u_i\| + \|u_i - x_i\| \le (k+1)\|x_i - u_i\| \le (k+1)k^i \|x_0 - u_0\|. \tag{12.4.8}$$
Hence,
$$\|x_{n+1} - x_0\| \le \sum_{i=0}^{n} \|x_{i+1} - x_i\| \le (k+1)\|x_0 - u_0\|/(1-k) < r'/2,$$
since $\|x_0 - u_0\| \le \|x_0 - u^*\| < r$. Thus,
$$\|x_{n+1} - u^*\| \le \|x_{n+1} - x_0\| + \|x_0 - u^*\| < r'/2 + r'/2 = r',$$
showing that $x_{n+1} \in B'$ and completing the induction.

It follows that (12.4.8) holds for all i. Hence,
$$\|x_{n+q} - x_n\| \le (k+1)k^n \|x_0 - u_0\|/(1-k), \tag{12.4.9}$$
which implies that (x_n) is a Cauchy sequence. Since the space X is complete, there exists $\tilde{u}$ such that $\lim x_n = \tilde{u}$. Since $\|x_n - u_n\| \to 0$, we also have $\lim u_n = \tilde{u}$. By continuity, $\theta(\tilde{u}) = \lim \theta(u_n) = 0$ (since $\theta(u_n) = 0$) and $g_i(\tilde{u}) = \lim g_i(u_n) = 0$. Thus, $\tilde{u}$ is a stationary point of $\{J, g_i\}$. Allowing $q \to \infty$ in (12.4.9), we obtain
$$\|\tilde{u} - x_n\| \le k^n \|x_0 - u_0\| (1+k)/(1-k) \le k^n \|x_0 - u^*\| (1+k)/(1-k),$$
which completes the proof. ∎

Corollary 1 If all $u_n = u^*$, then for some $k < 1$, $\|x_n - u^*\| \le k^n \|x_0 - u^*\|$.

Corollary 2 If u^* is a quasiregular stationary point such that $\theta(u^* + s\Delta x) \ne 0$ for $0 < s \le 1$ whenever $\theta(u^* + \Delta x) \ne 0$ and condition (v) holds for u^*, then $\|x_n - u^*\| \le k^n \|x_0 - u^*\|$ for some $k < 1$. In this case, u^* is the only stationary point of $\{J, g_i\}$ in the ball B'.

Proof. The proof of the theorem carries through with all $u_n = u^*$. Now let $g(x) = x + sh(x)$ with $d/2 < s < d$. The proof shows that $\|g(x) - g(u^*)\| = \|x + sh - u^*\| \le k\|x - u^*\|$ for any $x \in B'$. By Theorem 5.1.4, u^* is the unique fixed point of g in B', hence is the unique zero of $h(x)$. Since every stationary point of $\{J, g_i\}$ is a zero of h, u^* is the only such stationary point in B'. ∎

Remark Corollary 2 does not imply that there are no other points $x \in B'$ such that $\nabla J_T(x) = 0$. Indeed, it follows from the implicit-function theorem that such stationary points exist on each member of the family of constraints determined by the surfaces $g_i(x) - b_i = 0$ for $|b_i|$ sufficiently small.

Definition 12.4.2 A quasiregular stationary point which satisfies the hypothesis of Corollary 2 is said to be *simple*.

The method defined by (12.4.1)–(12.4.4) will be referred to as the *angular gradient procedure*. We shall now illustrate its use in the computation of eigenvalues and eigenvectors.

12.5 THE EIGENVALUE PROBLEM FOR A SELF-ADJOINT OPERATOR

Let A be a bounded self-adjoint linear operator on a real Hilbert space H. We shall apply the angular gradient procedure to the computation of the eigenvalues and eigenvectors of A [12.23].

Define $J(x) = \langle Ax, x \rangle$ and $g(x) = \|x\|^2 - 1$. Then $g(x) = 0$ is an equality constraint (i.e. the unit sphere). $\nabla J_G(x)$ is the projection of $\nabla J(x)$ on $\nabla g(x)$. These gradients are as follows.

Theorem 12.5.1 Let J and g be as above. For any nonzero $x \in H$,

$$\nabla J(x) = 2Ax,$$
$$\nabla g(x) = 2x,$$
$$\nabla J_G(x) = 2\langle A\bar{x}, \bar{x}\rangle x, \quad (12.5.1)$$
$$\nabla J_T(x) = 2(Ax - \langle A\bar{x}, \bar{x}\rangle x),$$
$$\|\nabla J_T(x)\|^2 = 4\left(\|Ax\|^2 - \frac{\langle Ax, x\rangle^2}{\|x\|^2}\right).$$

Proof. $J(x + sh) - J(x) = s\langle 2Ax, h\rangle + \varepsilon$, where $\varepsilon = s^2\langle Ah, h\rangle$ for any vectors x, h and scalar s. Hence, $\varepsilon/s \to 0$ as $s \to 0$. This shows that the weak gradient $\nabla_W J(x) = 2Ax$. By Theorem 5.3.2, since $2Ax$ is continuous in x, this is also the strong gradient. A similar calculation yields $\nabla g(x) = 2x$. In this example, there is one constraint and the projection $\nabla J_G(x)$ of $\nabla J(x)$ on $\nabla g(x)$ is easily seen to be as in (12.5.1). Since $\nabla J_T = \nabla J - \nabla J_G$, the rest follows easily. ∎

Theorem 12.5.2 A unit vector u is an eigenvector of A if and only if $\nabla J_T(u) = 0$; i.e. every unit eigenvector of a bounded self-adjoint operator A is a stationary point of $\langle Ax, x\rangle$ on the unit sphere and conversely.

Proof. If $Au = \lambda u$, then by (12.5.1)

$$\nabla J_T(u) = 2(\lambda u - \lambda\langle u, u\rangle u) = 0.$$

Conversely, if $\nabla J_T(u) = 0$, then (12.5.1) implies that $Au = \langle Au, u\rangle u$, so that u is an eigenvector. (Compare Exercise 12.13.) ∎

Next, we derive some results concerning the angle θ.

Theorem 12.5.3 For $x \neq 0$, let $\theta(x)$ be the angle in (iv) of Definition 12.4.1. If $\nabla J_T(x) \neq 0$ and $\cos \theta(x) \neq 0$, then $\nabla \theta(x)$ exists and

$$\nabla \theta(x) = \frac{2\langle Ax, x\rangle}{|\langle Ax, x\rangle|\,\|\nabla J_T(x)\|}\left(\frac{\langle Ax, x\rangle}{\|Ax\|^2\,\|x\|}A^2 x + \frac{\langle Ax, x\rangle}{\|x\|^3}x - \frac{2}{\|x\|}Ax\right) \quad (12.5.2)$$

12.5 The Eigenvalue Problem for a Self-Adjoint Operator

Proof. $\theta(x)$ is continuous in x. Let $\Delta\theta = \theta(x + th) - \theta(x)$. Then $\Delta\theta \to 0$ as $t \to 0$. If we can show that $\nabla(\cos \theta)$ exists, then

$$\langle \nabla \cos \theta(x), h \rangle = \lim_{t \to 0} \frac{\cos \theta(x + th) - \cos \theta(x)}{t}$$

$$= \lim_{t \to 0} \frac{\cos(\theta + \Delta\theta) - \cos \theta}{\Delta\theta} \cdot \frac{\Delta\theta}{t}$$

$$= -\sin \theta(x) \langle \nabla\theta(x), h \rangle,$$

showing that $\nabla\theta = -(1/\sin \theta) \nabla \cos \theta$. Now,

$$\cos \theta(x) = \frac{\|\nabla J_G(x)\|}{\|\nabla J(x)\|} = \frac{|\langle Ax, x \rangle|}{\|Ax\| \, \|x\|}.$$

If $\cos \theta(x) \neq 0$, then $J(x) = \langle Ax, x \rangle \neq 0$. By continuity, $J(x + sh) \neq 0$ and has the same sign as $J(x)$ for arbitrary h and $|s|$ sufficiently small. Let $sgJ(x)$ denote the sign of $J(x)$, $F(x) = \|Ax\|$ and $G(x) = \|x\|$. Then

$$\cos \theta(x + sh) = sgJ(x)J(x + sh)/F(x + sh) \cdot G(x + sh).$$

By a simple calculation we find that

$$\left. \frac{dJ(x + sh)}{ds} \right|_{s=0} = 2\langle Ax, h \rangle,$$

$$\left. \frac{dF(x + sh)}{ds} \right|_{s=0} = \frac{\langle A^2 x, h \rangle}{\|Ax\|},$$

$$\left. \frac{dG(x + sh)}{ds} \right|_{s=0} = \frac{\langle x, h \rangle}{\|x\|}.$$

Taking $d \cos \theta(x + sh)/ds$ at $s = 0$, we have

$$sgJ(x)\langle \nabla \cos \theta(x), h \rangle = \frac{\langle 2Ax, h \rangle}{\|Ax\| \, \|x\|} - \frac{\langle Ax, x \rangle}{\|Ax\|^2 \, \|x\|^2} \varphi(x),$$

where

$$\varphi(x) = \frac{\|Ax\|}{\|x\|} \langle x, h \rangle + \frac{\|x\|}{\|Ax\|} \langle A^2 x, h \rangle.$$

Equation (12.5.2) then follows easily. ∎

Corollary 1 Let u be an eigenvector of A belonging to a nonzero eigenvalue λ. There exists a neighborhood N of u in which (12.5.2) holds for all $x \in N$ such that $\nabla J_T(x) \neq 0$.

Proof. $\langle Au, u \rangle = \lambda \langle u, u \rangle \neq 0$. Hence, $\langle Ax, x \rangle \neq 0$ for x in some neighborhood of u. ∎

Corollary 2 $\langle \nabla\theta(x), x \rangle = 0$.

554   Extremum Problems

Proof. Use (12.5.2) to form the inner product, noting that $\langle A^2 x, x \rangle = \langle Ax, Ax \rangle$ by the self-adjointness of A. ∎

Corollary 2 shows that $\langle \nabla \theta(x), \nabla g(x) \rangle = 0$, by (12.5.1). Hence $\nabla \theta(x) \in T_x$, the tangent subspace of the constraint (i.e., the sphere).

Theorem 12.5.4 Let u be an eigenvector of A belonging to a nonzero eigenvalue λ. The one-sided differential $d\theta(u; \bar{h}^+)$ exists for all h and $d(0u + h; \bar{h}^+) \to d\theta(u; \bar{h}^+)$ as $\|h\| \to 0$.

Proof. Recall that $\nabla J_T(u) = 0$, so that $\theta(u) = 0$ and $\lim_{s \to 0} \sin \theta(u + sh)/\theta(u + sh) = 1$. Hence,

$$d\theta(u; h^+) = \lim_{s \to 0+} \frac{\theta(u + sh)}{s} = \lim_{s \to 0+} \frac{\sin \theta(u + sh)}{s}.$$

Referring to (12.5.1), we obtain

$$\sin \theta(u + sh) = \frac{(\|A(u + sh)\|^2 \|u + sh\|^2 - \langle A(u + sh), u + sh \rangle^2)^{1/2}}{\|A(u + sh)\| \|u + sh\|}.$$

We expand the right-hand side, divide by s and pass to the limit. Using $Au = \lambda u$, this yields

$$d\theta(u; h^+) = \frac{\|h\|}{\lambda \|u\|} (\lambda^2 + \|A\bar{h}\|^2 - 2\lambda \langle A\bar{h}, \bar{h} \rangle)^{1/2}.$$

Now, if $x = u + h$ is an eigenvector belonging to λ, then so is h and $d\theta(u; h^+) = 0$. Since $d\theta(u + h; h^+) = d\theta(u; h^+)$, the result is immediate in this case. If h is not an eigenvector belonging to λ and $\|h\| < \|u\|$, then $x = u + h$ cannot be an eigenvector. (Otherwise, $\langle x, u \rangle = 0 = \|u\|^2 + \langle u, h \rangle$, which contradicts $|\langle h, u \rangle| < \|u\|^2$.) Therefore, $\nabla J_T(x) \neq 0$. Furthermore,

$$|\langle Ax, x \rangle - \langle Au, u \rangle| \leq 2 \|A\| (\|u\| + \|h\|) \|h\|,$$

so that $\langle Ax, x \rangle \neq 0$ for $\|h\|$ sufficiently small. Hence, $\nabla \theta(x)$ is given by (12.5.2) and $d\theta(x; \bar{h}^+) = \langle \nabla \theta(x), \bar{h} \rangle$. Let E_λ be the subspace of eigenvectors belonging to λ. Since $H = E_\lambda + E_\lambda^\perp$, we may set $h = h_1 + v$, where $h_1 \in E_\lambda$ and $v \in E_\lambda^\perp$. Then $x = u' + v$, where $u' = u + h_1 \in E_\lambda$. It follows readily that

$$\langle Ax, x \rangle = \langle \lambda u' + Av, u' + v \rangle = \lambda \|u'\|^2 + \langle Av, v \rangle,$$

$$\|Ax\|^2 = \lambda^2 \|u'\|^2 + \|Av\|^2.$$

Therefore

$$\langle \nabla \theta(x), h \rangle = \pm 2(f_1 + f_2 + f_3)/\|\nabla J_T(x)\| \|x\|,$$

where

$$f_1 = \frac{\langle Ax, x \rangle \langle A^2 x, h \rangle}{\|Ax\|^2} = \frac{(\lambda \|u'\|^2 + \langle Av, v \rangle)}{(\lambda^2 \|u'\|^2 + \|Av\|^2)} (\lambda^2 \langle u', h_1 \rangle + \|Av\|^2),$$

$$f_2 = \frac{\langle Ax, x \rangle \langle x, h \rangle}{\|x\|^2} = \frac{(\lambda \|u'\|^2 + \langle Av, v \rangle)}{\|u'\|^2 + \|v\|^2} (\langle u', h_1 \rangle + \|v\|^2),$$

$$f_3 = -2\langle Ax, h \rangle = -2\lambda \langle u', h_1 \rangle - 2\langle Av, v \rangle.$$

12.5 The Eigenvalue Problem for a Self-Adjoint Operator   555

We see that

$$f_1 = \frac{\|v\|^2}{\lambda}\left(\lambda^2 \frac{\langle u', h_1\rangle}{\|v\|^2} + \|A\bar{v}\|^2\right)(1 + O(\|v\|^2)),$$

$$f_2 = \frac{\|v\|^2}{\lambda}\left(\lambda^2 + \lambda^2 \frac{\langle u', h_1\rangle}{\|v\|^2}\right)(1 + O(\|v\|^2)),$$

$$f_3 = \frac{\|v\|^2}{\lambda}\left(-2\lambda\langle A\bar{v}, \bar{v}\rangle - 2\lambda^2 \frac{\langle u', h_1\rangle}{\|v\|^2}\right).$$

Hence

$$f_1 + f_2 + f_3 = \frac{\|v\|^2}{\lambda}(\lambda^2 + \|A\bar{v}\|^2 - 2\lambda\langle A\bar{v}, \bar{v}\rangle)(1 + O(\|v\|^2)).$$

From (12.5.1) we find

$$\|x\| \|\nabla J_T(x)\| = 2((\lambda^2 \|u'\|^2 + \|Av\|^2)(\|u'\|^2 + \|v\|^2) - (\lambda \|u'\|^2 + \langle Av, v\rangle)^2)^{1/2}$$
$$= 2 \|v\| \|u'\| (\lambda^2 + \|A\bar{v}\|^2 - 2\lambda\langle A\bar{v}, \bar{v}\rangle)^{1/2}(1 + O(\|v\|^2)).$$

Therefore

$$\langle \nabla\theta(x), \bar{h}\rangle = \frac{\|v\|}{\lambda \|h\| \|u'\|}(\lambda^2 + \|A\bar{v}\|^2 - 2\lambda\langle A\bar{v}, \bar{v}\rangle)^{1/2}(1 + O(\|v\|^2)).$$

Now $Ah = Ah_1 + Av = \lambda h_1 + Av$. It follows that $\|Av\|^2 = \|Ah\|^2 - \lambda^2\|h_1\|^2$ and

$$\frac{\|v\|^2}{\|h\|^2}\|A\bar{v}\|^2 = \|A\bar{h}\| - \lambda^2\left(1 - \frac{\|v\|^2}{\|h\|^2}\right).$$

Similarly, $\langle Ah, h\rangle = \lambda \|h_1\|^2 + \langle Av, v\rangle$ so that

$$\frac{\|v\|^2}{\|h\|^2}\langle A\bar{v}, \bar{v}\rangle = \langle A\bar{h}, \bar{h}\rangle - \lambda\left(1 - \frac{\|v\|^2}{\|h\|^2}\right).$$

Combining these results, we obtain

$$\frac{\|v\|^2}{\|h\|^2}(\lambda^2 + \|A\bar{v}\|^2 - 2\lambda\langle A\bar{v}, \bar{v}\rangle) = \lambda^2 + \|A\bar{h}\|^2 - 2\lambda\langle A\bar{h}, \bar{h}\rangle.$$

Hence

$$\langle \nabla\theta(x), \bar{h}\rangle = \frac{1}{\lambda \|u'\|}(\lambda^2 + \|A\bar{h}\|^2 - 2\lambda\langle A\bar{h}, \bar{h}\rangle)(1 + O(\|v\|^2)).$$

Since $\|u'\| \to \|u\|$ and $\|v\| \to 0$ as $\|h\| \to 0$, it follows that $\langle \nabla\theta(x), \bar{h}\rangle \to d\theta(u; \bar{h}^+)$ as $\|h\| \to 0$. ∎

Theorem 12.5.5 Let $\lambda \neq 0$ be an eigenvalue of A with unit eigenvector u. Suppose that λ is an isolated point of the spectrum of A; i.e. there is a ball $B(\lambda, r)$, $r > 0$, such that $B(\lambda, r)$ contains no points of the spectrum other than λ. Let $u \in E_\lambda$, be the subspace of eigenvectors belonging to λ. If $E_\lambda \neq H$, then there exists a neighborhood N of u in which $\|\nabla\theta(x)\|$ is bounded away from zero for all x such that $\nabla J_T(x) \neq 0$. Also $\|\nabla\theta(x)\|$ is bounded above for such x.

556   **Extremum Problems**

Proof. For brevity, set $y = Ax$, $z = A\bar{x}$. Then $\cos\theta = \langle \bar{y}, \bar{x}\rangle$, $\|z\| = \|y\|/\|x\|$ and $\langle A\bar{y}, \bar{x}\rangle = \langle \bar{y}, A\bar{x}\rangle = \langle \bar{y}, \bar{z}\rangle$. Also, $\langle z, \bar{x}\rangle = \|z\|\cos\theta$. Using (12.5.2), we obtain for any $x \in H$ such that $\nabla J_T(x) \neq 0$ and $\cos\theta(x) \neq 0$,

$$\|\nabla\theta(x)\|^2 = \frac{4}{\|\nabla J_T(x)\|^2}(4\|z\|^2 - 4\cos\theta\langle A\bar{y}, z\rangle + \cos^2\theta(\|A\bar{y}\|^2 - \|z\|^2)). \quad (12.5.3)$$

Letting x_1 be the projection of x onto E_λ and $x_2 = x - x_1$, we have $Ax_1 = \lambda x_1$ and $\langle x_1, x_2\rangle = 0$. It follows easily that $y = \lambda x_1 + y_2$ and $Ay = \lambda^2 x_1 + Ay_2$, where $y_2 = Ax_2$. Further, $\langle y_2, x_1\rangle = \langle x_2, Ax_1\rangle = \lambda\langle x_2, x_1\rangle = 0$ and similarly, $\langle Ay_2, x_1\rangle = 0$. A simple calculation yields

$$\langle A\bar{y}, z\rangle = \frac{\lambda^3 \|x_1\|^2 + \langle Ay_2, y_2\rangle}{\|y\|\|x\|},$$

$$\|z\|^2 = \frac{\lambda^2 \|x_1\|^2 + \|y_2\|^2}{\|x_1\|^2 + \|x_2\|^2},$$

$$\|A\bar{y}\|^2 = \frac{\lambda^4 \|x_1\|^2 + \|Ay_2\|^2}{\|y\|^2},$$

$$\cos\theta = \frac{\lambda \|x_1\|^2 + \langle Ax_2, x_2\rangle}{\|y\|\|x\|},$$

whence

$$\|A\bar{y}\|^2 - \|z\|^2 = \frac{\|x_1\|^2(\lambda^4 \|x_2\|^2 + \|Ay_2\|^2 - 2\lambda^2 \|y_2\|^2)}{\|y\|^2 \|x\|^2} + w_1,$$

and

$$4(\|z\|^2 - \cos\theta\langle A\bar{y}, z\rangle) = \frac{4\lambda \|x_1\|^2}{\|y\|^2 \|x\|^2}(2\lambda \|y_2\|^2 - \lambda^2\langle Ax_2, x_2\rangle - \langle Ay_2, y_2\rangle) + w_1,$$

where

$$w_1 = \frac{\|Ay_2\|^2 \|x_2\|^2 - \|y_2\|^4}{\|y\|^2 \|x\|^2},$$

$$w_2 = \frac{4(\|y_2\|^4 - \langle Ax_2, x_2\rangle\langle Ay_2, y_2\rangle)}{\|y\|^2 \|x\|^2}.$$

From (12.5.1) we have

$$\|\nabla J_T\|^2 = 4\|y\|^2 (1 - \cos^2\theta)$$

$$= \frac{4\|x_1\|^2}{\|x\|^2}(\lambda^2 \|x_2\|^2 + \|y_2\|^2 - 2\lambda\langle Ax_2, x_2\rangle) + w_3,$$

where

$$w_3 = \frac{4(\|x_2\|^2 \|y_2\|^2 - \langle Ax_2, x_2\rangle^2)}{\|x\|^2}.$$

12.5 The Eigenvalue Problem for a Self-Adjoint Operator

As $x \to u$, we have $\|x_1\| \to 1$, $\|x\| \to 1$, $\|y\| \to |\lambda|$, $\cos^2 \theta \to 1$ and $\|x_2\| \to 0$. Hence, for all x in a sufficiently small neighborhood of u and such that $x_2 \neq 0$, we have

$$\cos^2 \theta \frac{(\|A\bar{y}\|^2 - \|z\|^2)}{\|x_2\|^2} = \frac{1}{\|x_2\|^2}\left(\lambda^2 \|x_2\|^2 + \frac{1}{\lambda^2}\|Ay_2\|^2 - 2\|y_2\|^2\right) + O(\|x_2\|^2),$$

$$\frac{4(\|z\|^2 - \cos\theta\langle A\bar{y}, z\rangle)}{\|x_2\|^2} = \frac{1}{\|x_2\|^2}\left(8\|y_2\|^2 - 4\lambda\langle Ax_2, x_2\rangle - \frac{4}{\lambda}\langle Ay_2, y_2\rangle\right) + O(\|x_2\|^2),$$

$$\frac{\|\nabla J_T\|^2}{\|x_2\|^2} = \frac{4}{\|x_2\|^2}(\lambda^2 \|x_2\|^2 + \|y_2\|^2 - 2\lambda\langle Ax_2, x_2\rangle) + O(\|x_2\|^2).$$

Applying these results to (12.5.3), we get

$$\|\nabla\theta(x)\|^2 = \frac{(1/\lambda^2)\langle(A - \lambda I)^4 \bar{x}_2, \bar{x}_2\rangle + O(\|x_2\|^2)}{\langle(A - \lambda I)^2 \bar{x}_2, \bar{x}_2\rangle + O(\|x_2\|^2)}.$$

The spectrum of $A_\lambda = A - \lambda I$ is the set $\{(\mu - \lambda) : \mu \in \sigma(A)\}$. Since λ is assumed to be an isolated point of $\sigma(A)$, $|\mu - \lambda| > r$ for $\mu \in \sigma(A)$ and $\mu \neq \lambda$. Now, the restriction of A_λ to $E_\lambda^\perp$ has as its spectrum the set $\{\mu - \lambda : \mu \neq \lambda, \mu \in \sigma(A)\}$. Thus, 0 is in the resolvent set of A_λ; i.e. $\|A_\lambda \bar{x}_2\| > m \|\bar{x}_2\|$ for some $m > 0$. Hence, $\langle(A - \lambda I)^2 \bar{x}_2, \bar{x}_2\rangle > m^2 \|\bar{x}_2\|^2$. Similarly, $\langle A_\lambda^4 \bar{x}_2, \bar{x}_2\rangle > m^4 \|\bar{x}_2\|^2$. It follows that $\|\nabla\theta(x)\|$ is bounded away from 0. ∎

Theorem 12.5.6 Let $\lambda \neq 0$ be an eigenvalue of A of multiplicity 1 and u a unit eigenvector belonging to λ. For $x = u + h$, let $\alpha = \arccos\langle\bar{x}, \bar{h}\rangle$ and $\beta = \arccos\langle\overline{\nabla\theta(x)}, \bar{h}\rangle$ as in Definition 12.4.1. There exists a neighborhood N of u and a constant $\gamma > 0$ such that for all $x \in N$ with $\nabla J_T(x) \neq 0$, $\cos^2 \alpha + \cos^2 \beta > \gamma$.

Proof. For any h let $h^* = \langle h, u\rangle u$ and $h' = h - h^*$. Let C be the *translated cone* $\{u + h : \|h^*\|/\|h\| \geq c > 0\}$, where c is an arbitrary constant, $0 < c < 1$. Let B be the ball $\|x - u\| < c/2$. It is easy to see that $\cos^2 \alpha(x)$ is bounded away from zero for $x \in C \cap B$, for suppose $x = u + h \in C \cap B$. Then

$$\cos\alpha = \langle\bar{x}, \bar{h}\rangle = \frac{\|h\| \pm \|h^*\|/\|h\|}{\|x\|}.$$

Hence, either $\cos\alpha \geq 2c/(2 + c) > 0$ or $\cos\alpha \leq -c/(2 - c) < 0$. Since $\langle\nabla\theta(x), x\rangle = 0$, we have $\cos\beta = \langle\overline{\nabla\theta}, \bar{h}\rangle = -\langle\nabla\theta, u\rangle/\|\nabla\theta\| \|h\|$. A rather lengthy calculation yields $\langle Ax, x\rangle = \lambda\mu + \langle Ah', h'\rangle$, $\|Ax\|^2 = \lambda^2\mu + \|Ah'\|^2$, $\|x\|^2 = \mu + \|h'\|^2$, where $\mu = (1 \pm \|h^*\|)^2$. Applying these results to (12.5.2) yields

$$\cos\beta = \frac{-2\langle x, u\rangle}{\|\nabla J_T\| \|x\| \|\nabla\theta\|} \frac{\|h'\|^2}{\|h\|}(\langle(A - \lambda I)^2 \bar{h}', \bar{h}'\rangle + O(\|h\|)).$$

From (12.5.1) we get

$$\frac{\|\nabla J_T\|^2}{\|h'\|^2} = 4\langle(A - \lambda I)^2 \bar{h}', \bar{h}'\rangle + O(\|h\|).$$

In the exterior of cone C, $\|h'\|/\|h\| \geq (1 - c^2)^{1/2} > 0$. Since $\bar{h}'$ is not an eigenvector belonging to λ, $\|(A - \lambda I)\bar{h}'\|^2$ is bounded away from zero for all such h' in the orthogonal complement of the space spanned by u. Therefore, for $\|h\|$ sufficiently small, $\cos^2 \beta$ is bounded away from zero in the exterior of C and the theorem is proved. ∎

Theorem 12.5.7 Let A be a bounded self-adjoint operator on a Hilbert space. If λ is an eigenvalue of A of multiplicity 1 and λ is an isolated point of the spectrum of A, then a unit eigenvector belonging to λ is a simple quasiregular stationary point of $\langle Ax, x \rangle$ on the unit sphere.

Computational Aspects

As a consequence of the preceding theorem, the angular gradient procedure can be used to compute eigenvalues of a bounded self-adjoint operator A. In particular, if A is also completely continuous, its spectrum is at most denumerable and (except for 0) consists entirely of eigenvalues $\{\lambda_1, \lambda_2, \ldots\}$. (See Appendix V.) If $\sigma(A)$ is infinite, then $\lim_{n \to \infty} \lambda_n = 0$. Hence each $\lambda_i \neq 0$ is an isolated point of the spectrum. If λ_i has multiplicity 1, then λ_i and its eigenvectors can be determined by the angular gradient procedure. In fact (Exercise 12.7), the method converges even when λ_i has multiplicity >1. (See [12.8].) Thus, the method can be used to determine eigenvalues of an integral operator,

$$Ax = \int_a^b K(t, s) x(s) \, ds, \tag{12.5.4}$$

where the *kernel* $K(t, s)$ is continuous in the square $a \leq t \leq b, a \leq s \leq b$ and symmetric $(K(t, s) = K(s, t))$. Here, the underlying Hilbert space is $L^2[a, b]$; i.e. $x(t) \in L^2[a, b]$. It can be shown that the integral operator A is completely continuous and self-adjoint (Exercise 12.5).

To apply the angular gradient procedure we must be able to evaluate the integral in (12.5.4) and other integrals such as

$$A^2 x = \int_a^b K(t, s) y(s) \, ds,$$

where $y = Ax$ is given by (12.5.4). In general, a numerical quadrature functional must be used (Chapter 9). This requires that we discretize the functions K, x and y. Choosing a grid $s_j = a + jh$, $t_i = a + ih$, and a suitable quadrature formula, we have

$$y_h(t_i) = h \sum_{j=0}^n a_j K(t_i, s_j) x_h(s_j), \qquad 0 \leq i \leq n,$$

and similar expressions for the other integrals. The vector $x_h = (x_h(s_i) : 0 \leq i \leq n)$ is an approximation to the function $x(s)$. Starting with an initial guess $x_h^{(0)}$, we use the angular gradient procedure to obtain a sequence $(x_h^{(m)} : m = 0, 1, 2, \ldots)$ of vectors. It can be shown that under suitable conditions on the a_j the sequence $(x_h^{(m)})$ converges to a vector x_h and as $h \to 0$ the piecewise linear extensions $\hat{T}_h x_h$ (see Section 10.1) converge to a function $u = u(t)$ which is an eigenfunction belonging to λ (that is $Au = \lambda u$). (See [12.8].)

12.6 COMPUTATIONAL SOLUTION OF THE OPTIMAL CONTROL PROBLEM OF MAYER BY GRADIENT TECHNIQUES

The optimal control problem of Mayer is defined by (12.1.1), (12.1.2), (12.1.3). Solution by the angular gradient procedure is possible, in principle, since we are seeking a stationary point u^* for the functional J subject to equality constraints. However, the gradient $\nabla \theta$ would be difficult to compute. Therefore, we introduce a modification which requires the computation only of ∇J_T instead. However, the modified angular gradient procedure converges only for a special subset of the quasiregular stationary points which we shall call *regular*.

Definition 12.6.1 u is a *regular* stationary point of $\{J, g_i\}$ if it is a stationary point with a neighborhood N satisfying conditions (i)–(iv) of Definition (12.4.1) and, instead of (v), the following condition (v'):
Let
$$\beta = \arccos \langle \overline{\nabla \theta(x)}, \overline{\Delta x} \rangle,$$
$$\alpha_i = \arccos \langle \overline{\nabla g_i(x)}, \overline{\Delta x} \rangle, \quad 1 \leq i \leq p,$$
$$\alpha_0 = \arccos \langle \overline{\nabla J_T(x)}, \overline{\Delta x} \rangle,$$
$$\beta_0 = \arccos \langle \overline{\nabla \theta(x)}, \overline{\nabla J_T(x)} \rangle,$$
where $x \in N$ and $\Delta x = x - u$, u being in the set U_x of closest stationary points to x. Then there exists a constant $\gamma > 0$ such that for all $x \in N$ and some $u \in U_x$, $|\cos \beta_0(x)| > \gamma$ and
$$\sum_{i=1}^{p} \cos^2 \alpha_i(x) + \frac{\cos \alpha_0(x) \cos \beta(x)}{|\cos \beta_0(x)|} > \gamma.$$

We call N a *regular* neighborhood.

The modified angular gradient procedure is defined by the following iteration:
$$\left.\begin{aligned} h_G(x_n) &= -\sum_{i=1}^{p} \frac{g_i(x_n)}{\|\nabla g_i(x_n)\|} \overline{\nabla g_i(x_n)}, \\ h_T(x_n) &= -\frac{\nabla J_T(x_n)}{\|\nabla J_G(x_n)\| |\langle \overline{\nabla \theta(x_n)}, \overline{\nabla J_T(x_n)} \rangle|}, \\ h(x_n) &= h_G(x_n) + h_T(x_n), \\ x_{n+1} &= x_n + s_n h(x_n), \end{aligned}\right\} \quad (12.6.1)$$
where $d/2 < s_n < d$.

The proof of convergence is similar to the proof of Theorem 12.4.1 with the following changes:
$$\langle h_T, \Delta x \rangle = -\frac{\|\nabla J_T\| \|\Delta x\| \cos \alpha_0}{\|\nabla J_G\| \|\nabla \theta\| |\cos \beta_0|} = -\frac{(\tan \theta) \|\Delta x\| \cos \alpha_0}{\|\nabla \theta\| |\cos \beta_0|}.$$

560 Extremum Problems

As before,
$$\tan \theta(x) = \langle \nabla \theta(x), \Delta x \rangle + \hat{\varepsilon}$$
and
$$\langle h_T, \Delta x \rangle = -\|\Delta x\|^2 \left(\frac{\cos \alpha_0 \cos \beta}{|\cos \beta_0|} + w_1 \right),$$

where $w_1 \to 0$ as $\|\Delta x\| \to 0$. Similarly,
$$\|h_T\| = \frac{\|\nabla J_T\|}{\|\nabla J_G\|} \frac{1}{\|\nabla \theta\| |\cos \beta_0|} = \|\Delta x\| \left(\frac{\cos \beta}{|\cos \beta_0|} + w_T \right).$$

We obtain inequality (12.4.7) again, but with
$$b = \sum_{i=1}^{p} \cos^2 \alpha_i + \cos \alpha_0 \cos \beta / |\cos \beta_0|,$$
$$c = p \sum_{i=1}^{p} \cos^2 \alpha_i + \left(\frac{\cos \beta}{|\cos \beta_0|} \right)^2.$$

The conditions on the functions f, ψ_i and J in the Mayer problem which suffice to make a stationary point regular are not known. However, it is possible to calculate h_G and h_T, as given by (12.6.1), if these functions satisfy certain natural differentiability conditions. We now show how these calculations are to be made.

Referring to Section 12.1, we recall that a control $u = u(t)$ is a q-dimensional vector function, $u^T = u^T(t) = (u_1(t), \ldots, u_q(t))$. (The superscript T denotes the transpose.) Let Ω be an open set in R_q. A control is called *admissible* on the closed interval $[t_0, t_1]$ if it is piecewise continuous and the values $\{u(t)\}$ lie in Ω for $t_0 \leq t \leq t_1$. The set, $\bar{C}_q[a, b]$, of all piecewise continuous functions on an interval $[a, b]$ is a subspace of a Hilbert space H which is the direct sum of q copies of $L^2[a, b]$. For $u \in H$ and $v^T = v^T(t) = (v_1(t), \ldots, v_q(t)) \in H$, we define
$$\langle u, v \rangle = \int_a^b (u_1 v_1 + \cdots + u_q v_q) \, dt = \int_a^b u^T v \, dt.$$

For $[t_0, t_1] \subset [a, b]$ the set of admissible controls u on $[t_0, t_1]$ is a subset of $\bar{C}_q$ if we define $u(t) = 0$ for t outside of $[t_0, t_1]$. In what follows, $\bar{C}_q$ is the underlying pre-Hilbert space.

To begin with, let us assume that the initial value $x(t_0) = x_0$ and the initial time t_0 are fixed. (Later, we shall remove this assumption.) Assume that there exists an admissible control $u = u(t)$ such that a trajectory $x(t)$ corresponding to $u(t)$ exists for $t_0 \leq t \leq t_1$ and $x(t_0) = x_0$; i.e. (12.1.1) has a solution and $(x(t), u(t))$ is a state-control pair satisfying (12.1.1a). Now, assume further that the points $(x(t), u(t))$ lie in an open region Γ in which $f(x, u)$ has continuous partials $(\partial f / \partial x)$ and $(\partial f / \partial u)$ with respect to x and u respectively. Consider the one-parameter system of differential equations
$$\frac{dx}{dt} = f(x, u(t) + sh(t)), \tag{12.6.2}$$

where $h = h(t)$ is an arbitrary admissible control on $[t_0, t_1]$ and s is a scalar parameter. Analogous to Theorem 11.1.1, it can be proved that (12.6.2) has a solution $x(t, s)$ on $[t_0, t_1]$ for all sufficiently small $|s|$. (See Exercise 12.6 or [11.1], p. 29, where it is shown that a Lipschitz condition with respect to x suffices, provided f is continuous in u. Furthermore, $x(t, s)$ is continuous in s. Note that with $u(t)$ and $h(t)$ fixed, the right member of (12.6.2) is a function of x, t and the parameter s.) The existence of the continuous partials $\partial f/\partial x$ and $\partial f/\partial u$ implies that $x(t, s)$ has a partial $\partial x/\partial s$ with respect to s which, for any s, satisfies the *variational equation*,

$$\frac{d}{dt}\left(\frac{\partial x}{\partial s}\right) = \left(\frac{\partial f}{\partial x}\right)\left(\frac{\partial x}{\partial s}\right) + \left(\frac{\partial f}{\partial u}\right)h. \quad (12.6.3)$$

This is a direct consequence of Theorem 11.1.2. (See Exercise 12.6.) The m initial values to be used in solving (12.6.3) are all zero.

The final value $x(t_1, s)$ of a solution of (12.6.2) depends on the control $y = u + sh$. We express this by writing $x(t_1, s) = x_1(y)$. Since t_0, t_1 and $x(t_0, s) = x(t_0)$ are being held fixed, we can define the functionals

$$g_i(y) = \psi_i(t_0, x(t_0), t_1, x_1(y)), \quad 1 \leq i \leq p,$$
$$J(y) = J(t_0, x(t_0), t_1, x_1(y)),$$

where the ψ_i specify the boundary conditions (12.1.2) and J is the objective function of (12.1.3). Here, the control y must be admissible and must have corresponding trajectories defined on $[t_0, t_1]$. The latter will be assured if u is such a control and $|s|$ is sufficiently small. Thus, y lies in some subset $D \subset \bar{C}_q$ and g_i and J are real functions defined on D. It is required to find $y \in D$ which minimizes J subject to the equality constraints $g_i(y) = 0$. Such a minimum point must satisfy the Lagrange multiplier rule (Theorem 12.2.2) and, therefore, must be a stationary point of $\{J, g_i\}$ provided the necessary gradients exist. We shall see that ∇J and ∇g_i exist provided that J and ψ_i have continuous first-order partials in some open set, E. It is assumed that the point $P_0 = (t_0, x(t_0), t_1, x(t_1, 0))$ is in E. Since $x(t_1, s)$ is continuous in s, the points $P_s = (t_0, x(t_0), t_1, x(t_1, s))$ are also in E for $|s|$ sufficiently small so that J and ψ_i have continuous partials at such P_s. Further, we assume that the $p \times n$ Jacobian $(\partial \psi/\partial x_1)$ has rank p, where $(\partial \psi/\partial x_1)$ is the matrix of partial derivatives $\partial \psi_i/\partial x_j(t_1)$. Similarly, $(\partial J/\partial x_1)$ is the m-dimensional vector having components $\partial J/\partial x_j(t_1)$. (Here, the $x_j(t_1)$, $1 \leq j \leq m$ are the components of the final value $x(t_1, s)$.) If the partials are evaluated at the point P_0, we indicate this by writing $(\partial J/\partial x_1)_0$ and $(\partial \psi/\partial x_1)_0$. We are now ready to evaluate $\nabla J(u)$ and $\nabla g_i(u)$.

By the definition of the weak differential (Definition 5.3.2) and the usual rules of calculus,

$$dJ(u; h) = \left.\frac{dJ(u + sh)}{ds}\right|_{s=0} = \left(\frac{\partial J}{\partial x_1}\right)_0^T dx_1(u; h).$$

Since

$$dx_1(u; h) = \left.\frac{dx_1(u + sh)}{ds}\right|_{s=0} = \left.\frac{\partial x(t_1, s)}{\partial s}\right|_{s=0},$$

we see that $dx_1(u; h)$ exists and is the final value of a solution of the linear initial-value problem,

$$\frac{dv}{dt} = \left(\frac{\partial f}{\partial x}\right)_0 v + \left(\frac{\partial f}{\partial u}\right)_0 h,$$

$$v(t_0) = 0.$$
(12.6.4)

Equation (12.6.4) is just the variational equation (12.6.3) with all partials evaluated along the points of the state-control pair $(x(t, 0), u(t))$. Now we would like to express $dJ(u; h)$ in the form of an inner product $\langle \nabla J(u), h \rangle$. Recalling the definition of inner product in $\bar{C}_q$, we must find a vector function $w(t)$ such that

$$\left(\frac{\partial J}{\partial x_1}\right)_0^T \frac{\partial x(t_1, s)}{\partial s} = \int_{t_0}^{t_1} w^T(t) h(t)\, dt.$$
(12.6.5)

But $\partial x(t_1, s)/\partial s = v(t_1)$, where $v(t)$ is the solution of (12.6.4). Since $(\partial J/\partial x_1)_0$ is evaluated at the point $(t_0, x(t_0), t_1, x(t_1, 0))$, it is likewise a function of t_1. Call this function $z(t_1)$. Regarding t_1 as variable, we differentiate (12.6.5) with respect to t_1, obtaining

$$\frac{d}{dt_1}(z^T v) = w^T(t_1) h(t_1).$$

Dropping the subscript temporarily, we are led to seek a function $z(t)$ such that

$$z^T \frac{dv}{dt} + \frac{dz^T}{dt} v = w^T(t) h(t).$$

Using (12.6.4), we must have

$$z^T \frac{dv}{dt} = z^T \left(\frac{\partial f}{\partial x}\right)_0 v + z^T \left(\frac{\partial f}{\partial u}\right)_0 h,$$

and we see that if $dz^T/dt = -z^T(\partial f/\partial x)_0$, then $d(z^T v)\, dt$ is of the form $w^T(t) h(t)$ with $w^T = z^T(\partial f/\partial u)_0$. This leads us finally to the equation,

$$\frac{dz}{dt} = -\left(\frac{\partial f}{\partial x}\right)_0^T z,$$
(12.6.6)

which is the adjoint equation of the variational equation (12.6.3). If we solve the adjoint equation by integrating backward in time from t_1 to t_0 with final values (at t_1) taken equal to $(\partial J/\partial x_1)_0$, we obtain a solution, $J_x(t)$, such that

$$J_x^T(t_1) v(t_1) = \int_{t_0}^{t_1} J_x^T(t) \left(\frac{\partial f}{\partial u}\right)_0 h(t)\, dt.$$
(12.6.7)

Since $J_x^T(t_1) v(t_1) = (\partial J/\partial x_1)_0^T \partial x(t_1, s)/\partial x = dJ(u; h)$, it follows that

$$\nabla J(u) = J_x^T(t) \left(\frac{\partial f}{\partial u}\right)_0.$$
(12.6.8)

Hence $\nabla J(u)$ exists and can be calculated by solving the adjoint equations, integrating backward from t_1 to t_0 and using the known values $(\partial J/\partial x_1)_0$ as "initial" values. This assumes that $u(t)$ is known, so that the $x(t_1)$ can be found by integrating the state equations (12.1.1) with given starting value $x(t_0)$. Then $\partial J/\partial x_1)_0$ can be evaluated at the point $(t_0, x(t_0), t_1, x(t_1))$. Note that $(\partial f/\partial x)$ is piecewise continuous in $[t_0, t_1]$, since $u(t)$ is piecewise continuous and f is supposed to have continuous partials. This implies that (12.6.6) has a piecewise differentiable solution in $[t_0, t_1]$ for any initial values.

In similar fashion, one can derive the result that

$$\nabla g_i(u) = \psi_{ix}^T \left(\frac{\partial f}{\partial u}\right)_0, \qquad 1 \le i \le p, \tag{12.6.9}$$

where $\psi_{ix}^T(t)$ is the solution of (12.6.6) having final values $\psi_{ix}^T(t_1) = (\partial \psi_i/\partial x_1)_0$.

Computational Aspects

Equations (12.6.8) and (12.6.9) are valid for any control $u^{(0)} = u^{(0)}(t)$ which satisfies the other conditions imposed on u (e.g. a corresponding trajectory $x^{(0)}(t)$ exists and the points $(x^{(0)}(t), u^{(0)}(t))$ lie in the regions of differentiability of f, J and ψ_i). In applying the modified angular gradient procedure to the Mayer control problem, we first guess a control function $u^{(0)}(t)$ and compute $h_G(u^{(0)})$ and $h_T(u^{(0)})$ by formula (12.6.1). This requires the computation of $\nabla J(u^{(0)})$ by formula (12.6.8) and $\nabla g_i(u^{(0)})$ by (12.6.9). In most cases, the forward integration of the state equation and the backward integration of the adjoint equation must be carried out numerically by one of the methods in Chapter 10. Thus, the problem is discretized and we obtain a finite sequence of values for $\nabla J(u)$ and $\nabla g_i(u)$ at grid points $t_0 + i\Delta t$. We compute ∇J_T by using the Gram matrix $\langle \nabla g_i, \nabla g_j \rangle$ as in system (7.4.1). The inner product integral is replaced by a quadrature formula. Finally, to compute $\langle \nabla \theta(u^{(0)}), \overline{\nabla J_T(u^{(0)})} \rangle$, we observe that this quantity is $d\theta(u^{(0)} + s\overline{\nabla J_T}(u^{(0)}))/ds|_{s=0}$. This derivative is approximated by a difference formula; e.g. by $(\theta(s_1) - \theta(0))/s_1$ where s_1 is small. This requires the computation of $\theta = \arcsin(\|\nabla J_T\|/\|\nabla J\|)$ at the points $u^{(0)}$ and $u^{(0)} + s_1 \overline{\nabla J_T}(u^{(0)})$. Therefore, the computation of ∇J and ∇g_i must be repeated with $u^{(0)} + s_1 \overline{\nabla J_T}(u^{(0)})$ as the control.

We shall illustrate these calculations by applying them to the brachistochrone problem (12.1.4), (12.1.5), (12.1.6). According to (12.6.6), the adjoint equations of (12.1.4) are

$$\left. \begin{aligned} \frac{dz_1}{dt} &= \frac{(1+u^2)^{1/2}}{g^{1/2}(2x_1)^{3/2}} z_2, \\ \frac{dz_2}{dt} &= 0, \end{aligned} \right\} \tag{12.6.10}$$

where we have set $u = u_1$, since the control is one-dimensional. Similarly, we find

$$\frac{\partial f}{\partial u} = \begin{bmatrix} 1 \\ u \\ (2gx_1(1+u^2))^{1/2} \end{bmatrix}.$$

The function $J_x(t) = (J_{x_1}(t), J_{x_2}(t))$ is obtained by integrating (12.6.10) from b to a with final values $J_x(b) = (\partial J/\partial x_1(b), \partial J/\partial x_2(b))$. From (12.1.6) we find that $J_x(b) = (0, 1)$ and the required solution is given by

$$J_{x_1}(t) = -\int_t^b \frac{(1 + u^2(s))^{1/2}}{g^{1/2}(2x_1(s))^{3/2}} ds,$$
$$J_{x_2}(t) = 1.$$
(12.6.11)

Similarly, $\psi_x(t_1) = (\partial \psi/\partial x_1(b), \partial \psi/\partial x_2(b))$ where $\psi = \psi_3 = x_1(b) - 2$ is the only boundary constraint, since ψ_1 and ψ_2 can be regarded as fixed initial conditions. Hence, $\psi_x(t_1) = (1, 0)$. The solution of (12.6.10) satisfying the conditions $(1,0)$ at $t = b$ can be seen to be $\psi_{x_1}(t) = 1, \psi_{x_2}(t) = 0$. Hence, $\nabla \psi = \psi_x^T(\partial f/\partial u) = 1$, whereas

$$\nabla J(u) = J_x^T(\partial f/\partial u) = J_{x_1}(t) + \frac{u(t)}{(2gx_1(t)(1 + u^2(t)))^{1/2}}.$$
(12.6.12)

To compute $\nabla J(u)$, we integrate the state equations (12.1.4) from a to b with initial values $x_1(a) = 1$ and $x_2(a) = 0$ and $u_1 = u(t)$. The solution $x_1(t)$ and the control $u(t)$ are then used in (12.6.10), integrating backward from $t = b$ with values $J_x(b) = (0, 1)$, to obtain $J_x(t)$ as given in (12.6.11). $\nabla J(u)$ as a function of t is then given by (12.6.12). Since there is only one constraint, we have

$$\nabla J_G = \langle \nabla J, \overline{\nabla \psi} \rangle \overline{\nabla \psi},$$

where

$$\langle \nabla J, \nabla \psi \rangle = \int_a^b \left(J_{x_1}(s) + \frac{u(s)}{(2gx_1(s)(1 + u^2(s)))^{1/2}} \right) ds,$$

$$\|\nabla \psi\|^2 = \langle \nabla \psi, \nabla \psi \rangle = \int_a^b 1\, ds = b - a.$$

Formula (12.6.1) with $p = 1$ and $g_1 = \psi_3$ yields

$$h_G(u) = -(x_1(b) - 2)/(b - a).$$

We have immediately that $\nabla J_T = \nabla J - \nabla J_G$ and so we compute $\theta(0) = \arcsin \|\nabla J_T\|/\|\nabla J\|$. Then we repeat the entire process with u replaced by $u + \Delta s\, \overline{\nabla J_T(u)}$, to obtain $\theta(\Delta s)$. We approximate $\langle \nabla \theta(u), \overline{\nabla J_T(u)} \rangle$ by $(\theta(\Delta s) - \theta(0))/\Delta s$ in (12.6.1) and use this to compute $h_T(u)$. The next step, $u^{(1)} = u + s(h_T(u) + h_G(u))$, can then be carried out to obtain a new control $u^{(1)}$ and the process is iterated until successive controls differ by less than a prescribed allowable error.

Returning to the general problem of Mayer, we extend the previous results to the case where $t_0, x(t_0), t_1$ and $x(t_1)$ are all varied simultaneously. We choose the underlying Hilbert space, H^*, to be the direct sum of the spaces H, R^m and R^2. Thus, an arbitrary element $e \in H^*$ is of the form $e = (u(t), x_0, t_{01})$ where $u(t)$ is in the *control space* H, x_0 is in the *initial-value space* R^m and $t_{01} = (t_0, t_1)$ is in the *space of end points*,

Expressing $x_n = \sum_1^n \xi_j v_j$, we obtain n equations

$$\sum_{j=1}^{n} (\langle Av_i, v_j \rangle - \lambda_n \langle v_i, v_j \rangle)\xi_j = 0, \qquad 1 \leq i \leq n, \tag{12.7.6}$$

which together with the equation,

$$\sum_{i,j=1}^{n} \langle v_i, v_j \rangle \xi_i \xi_j = 1 \tag{12.7.7}$$

form a system of $n + 1$ equations for the unknowns $\xi_1, \ldots, \xi_n, \lambda_n$. This system can be solved by Newton's method, for example. Note that we must be able to compute Av_i and the various inner products. Also, $Av_i \notin V^n$ in general. (Otherwise, we would be finding the eigenvalues of a finite-dimensional operator, the restriction of A to V^n.) Newton's method requires an initial guess for x_n and λ_n. For the initial guess for x_n it is reasonable to take x_{n-1}, the minimum point in V^{n-1}. However, a solution of (12.7.6), (12.7.7) need not be a minimum of J on V^n, since these are only necessary conditions. (A solution is a stationary point of J.) To circumvent this difficulty we can proceed as follows.

Suppose $\xi_1, \ldots, \xi_n, \lambda_n$ is a nontrivial solution of (12.7.6). Since (12.7.6) is linear in the ξ_j, the determinant of the nth-order matrix $(\langle Av_i, v_j \rangle - \lambda_n \langle v_i, v_j \rangle)$ must be zero. This determinant is a polynomial $p(\lambda_n)$ of degree n in λ_n. (The leading coefficient is $(-1)^n \langle v_1, v_1 \rangle \cdots \langle v_n, v_n \rangle$.) To see that all the zeros of p are real, simply multiply (12.7.5) by ξ_i and sum over i from 1 to n to obtain

$$\langle Ax_n, x_n \rangle = \lambda_n \langle x_n, x_n \rangle; \tag{12.7.8}$$

that is, if $p(\lambda_n) = 0$, then there exists a nontrivial solution $(\xi_1, \ldots, \xi_n)$ of (12.7.6) and taking $x_n = \sum_{j=1}^n \xi_j v_j$ we recover (12.7.5). Since A is symmetric, (12.7.8) shows that λ_n is real. To satisfy (12.7.7) we normalize x_n, the normalized vector also being a solution of (12.7.6). We denote the normalized vector by x_n again. Therefore $\lambda_n = \langle Ax_n, x_n \rangle = J(x_n)$, so that to minimize J we must take λ_n to be the smallest zero of p. Suppose this is done. We now show that $\lambda_n \to \lambda_{\min}$.

By definition of the infimum, there exists u of norm 1 such that

$$\lambda_{\min} \leq \langle Au, u \rangle \leq \lambda_{\min} + \varepsilon.$$

This is equivalent to $\lambda_{\min} \leq \|u\|_A^2 \leq \lambda_{\min} + \varepsilon$. Now, the completeness of the v_i implies that $u = \sum_{i=1}^{\infty} \alpha_i v_i$, that is, for n sufficiently large, $\|u - \sum_{i=1}^n \alpha_i v_i\|_A < \varepsilon$. Letting $u_n = \sum_1^n \alpha_i v_i$, we have

$$\|u_n\|_A < \|u\|_A + \varepsilon \leq (\lambda_{\min} + \varepsilon)^{1/2} + \varepsilon.$$

Applying (12.7.2), we obtain

$$\|u - u_n\| \leq (1/\alpha)\|u - u_n\|_A < \varepsilon/\alpha,$$

so that

$$\|u_n\| \geq \|u\| - \varepsilon/\alpha = 1 - \varepsilon/\alpha.$$

It follows that

$$\lambda_{\min} \leq \lambda_n \leq \frac{\langle Au_n, u_n \rangle}{\langle u_n, u_n \rangle} \leq \frac{(\lambda_{\min} + \varepsilon)^{1/2} + \varepsilon}{1 - \varepsilon/\alpha}.$$

Since ε is arbitrary, $\lim_{n \to \infty} \lambda_n = \lambda_{\min}$. We summarize the foregoing as

Theorem 12.7.4 Let A be a symmetric operator on a Hilbert space H, with domain dense in H. Suppose $\langle Ax, x \rangle > \alpha^2 \|x\|^2$ for some $\alpha > 0$. Let $\{v_i\}$ be a sequence of vectors which is complete in the norm $\|x\|_A = \langle Ax, x \rangle$ and define V^n to be the subspace spanned by $\{v_1, \ldots, v_n\}$. Let $\lambda_n = \inf_{x \in V^n} \langle Ax, x \rangle / \langle x, x \rangle$. Then

$$\lim_{n \to \infty} \lambda_n = \lambda_{\min} = \inf_{x \in H} \langle Ax, x \rangle / \langle x, x \rangle.$$

Observe that the computation of λ_n as the minimum zero of the polynomial,

$$p(\lambda) = \det (\langle Av_i, v_j \rangle - \lambda \langle v_i, v_j \rangle)$$

may be troublesome. A simplification is effected by orthonormalizing the v_i. With $\langle v_i, v_j \rangle = \delta_{ij}$, the polynomial $p(\lambda)$ becomes the characteristic polynomial of the symmetric matrix $(\langle Av_i, v_j \rangle)$. Hence we can use one of the methods of Sections 6.4–6.6 to obtain its zeros.

What can be said about the convergence of the vectors x_n? We shall give sufficient conditions for the x_n to converge to the invariant subspace, E, of $\lambda_{\min}$. (Note that $Ax_n \neq \lambda_n x_n$, in general.)

Since $H = E + E^\perp$, we can write $x_n = \alpha_n y_n + \beta_n z_n$, where $y_n \in E$ and $z_n \in E^\perp$ are unit vectors. Furthermore, $\alpha_n^2 + \beta_n^2 = \|x_n\|^2 = 1$ and we may take $\alpha_n > 0$. Therefore

$$\|x_n - y_n\|^2 = (\alpha_n - 1)^2 + \beta_n^2 = 2(1 - \alpha_n) \leq 2(1 - \alpha_n^2). \tag{12.7.9}$$

By (12.7.8),

$$\lambda_n = \langle Ax_n, x_n \rangle = \alpha_n^2 \lambda_{\min} + \beta_n^2 \langle Az_n, z_n \rangle$$

so that

$$\lambda_n - \lambda_{\min} = (\alpha_n^2 - 1) \lambda_{\min} + (1 - \alpha_n^2) \langle Az_n, z_n \rangle.$$

Now, assume that

$$\mu = \inf_{\|z\|=1} \{\langle Az, z \rangle : z \in E^\perp\} > \lambda_{\min}.$$

Then

$$\lambda_n - \lambda_{\min} \geq (1 - \alpha_n^2)(\mu - \lambda_{\min}) \tag{12.7.10}$$

and combining this with (12.7.9), we have

$$\|x_n - y_n\|^2 \leq 2(\lambda_n - \lambda_{\min})/(\mu - \lambda_{\min}).$$

Since we have proved that $\lambda_n \to \lambda_{\min}$, this shows that

$$\|x_n - E\| \to 0.$$

The Ritz method can be extended to determine the next eigenvalue greater than $\lambda_{\min}$ when $\lambda_{\min}$ has finite multiplicity, q say. Again, we seek to minimize $\langle Ax, x \rangle$ on V^n, but now subject to the additional constraints $\langle x, u_1 \rangle = 0, \ldots, \langle x, u_q \rangle = 0$, where $u_1, \ldots, u_q$ are the approximate eigenvectors of λ_n. For $q = 1$, this procedure leads to the system (12.7.6) again, except that λ_n must be chosen as the second smallest zero. (For $q > 1$, the computation may not be feasible.)

A generalization of the eigenvalue problem in the form

$$Ax - \lambda Bx = 0, \tag{12.7.11}$$

where both A and B are symmetric and satisfy

$$\langle Ax, x \rangle > \alpha^2 \|x\|^2, \qquad \langle Bx, x \rangle > \alpha^2 \|x\|^2, \alpha > 0,$$

can be solved by the Ritz method. Again, we seek the minimum eigenvalue $\lambda_{\min}$. If we introduce the B inner product $\langle x, y \rangle_B = \langle Bx, y \rangle$, it is easy to prove the following generalizations of the ordinary eigenvalue problem (that is, $B = I$):

1. Any eigenvalue λ of (12.7.11) is real. ($\lambda = \langle Au, u \rangle / \langle Bu, u \rangle$, where u is an eigenvector belonging to λ.)
2. Eigenvectors belonging to distinct eigenvalues are orthogonal in the B inner product.
3. Let $\lambda_{\min} = \inf_x \{\langle Ax, x \rangle / \langle Bx, x \rangle\}$. If there exists u such that $\lambda_{\min} = \langle Au, u \rangle / \langle Bu, u \rangle$, then $\lambda_{\min}$ is the smallest eigenvalue of (12.7.11). Thus $\lambda_{\min}$ can be determined by minimizing $\langle Au, u \rangle$ subject to the constraint $\langle Bu, u \rangle = 1$; that is $\|u\|_B^2 = 1$.

In place of Eqs. (12.7.6), we obtain

$$\sum_{j=1}^n (\langle Av_i, v_j \rangle - \lambda_n \langle Bv_i, v_j \rangle)\xi_j = 0.$$

If the v_i are B-orthogonal, the situation is as before. (A solution by gradient techniques is given in [12.35] for more general B.)

12.8 LINEAR PROGRAMMING

Linear programming deals with the problem of optimizing a real linear objective functional J, defined on a real q-dimensional vector space V and subject to a finite set of linear equality and inequality constraints.

The general linear inequality constraint can be written in the form

$$\sum_{j=1}^q a_{ij} x_j \leq b_i. \tag{12.8.1}$$

By introducing a new variable x_{q+i}, called a *slack variable*, we can replace (12.8.1) by the equation

$$x_{q+i} = b_i - \sum_{j=1}^q a_{ij} x_j,$$

and the special inequality $x_{q+i} \geq 0$. Furthermore, if some x_i, $1 \leq i \leq q$, is not required to be nonnegative, we can introduce two new variables x_i' and x_i'' and define

$$x_i = x_i' - x_i'', \qquad x_i' \geq 0, \qquad x_i'' \geq 0.$$

By these preliminary transformations, we arrive at the following *standard form of the linear programming problem*:

Minimize the linear functional

$$J(x) = \sum_{j=1}^{n} c_j x_j = c^T x \qquad (12.8.2)$$

subject to the linear equality constraints,

$$Ax = b, \qquad (12.8.3)$$

(where A is an $m \times n$ matrix) and the linear inequality constraints,

$$x \geq 0, \qquad (12.8.4)$$

where the vector ordering $x \geq 0$ signifies that $x_i \geq 0$, $1 \leq i \leq n$. Main interest centers on the case where $m < n$ and rank $(A) = m$, since this ensures that there are no redundant or inconsistent equality constraints and that (12.8.3) has infinitely many solutions among which we must search for an optimal one. We shall confine our discussion to this case.

A vector x which satisfies the constraints (12.8.3), (12.8.4) is called a *feasible solution* (or *feasible vector*). A feasible vector which minimizes J is called an *optimal feasible solution*. Let $p = n - m$. Since A has rank m, we can set p of the x_i equal to zero and solve for the remaining m components. This yields a vector solution of (12.8.3) having at least p zero components. Such a vector is called a *basic vector* (or *basic solution*). It is *nondegenerate* if exactly p components are zero. The nonzero variables are called *basic variables* and the others are called *nonbasic variables*.

The standard method for finding an optimal feasible vector when one exists is the simplex method due to G. B. Dantzig [12.26]. This method and its several variations are based on the following fundamental theorem:

Theorem 12.8.1 If an optimal feasible solution exists for the linear programming problem, then there exists a feasible basic vector which is optimal.

Proof. See Appendix VI. The plausibility of the theorem can be demonstrated by considering the two-dimensional case in which $x_1 \geq 0$ and $x_2 \geq 0$ are the given variables. Suppose that there are three linear inequality constraints of type (12.8.1). We introduce slack variables $x_3 \geq 0$, $x_4 \geq 0$, and $x_5 \geq 0$, obtaining constraint (12.8.4) with $n = 5$. In two dimensions, these constraints determine half-planes and the intersection of these half-planes is a convex polygonal region D which we illustrate typically in the figure on page 573.

The lines $x_3 = 0$, $x_4 = 0$, $x_5 = 0$ correspond to the equality in (12.8.1). The dashed lines are the $J = $ constant level "surfaces." The optimal feasible solution in this illustration occurs at the vertex P, and is unique. At this point, two components, x_3 and x_4, are zero. Thus the optimal solution is also a basic solution. (If three of the lines intersected at P, there would be a *degenerate basic solution*.) The simplex method provides a finite efficient algorithm for locating the point P when it exists. Efficiency is

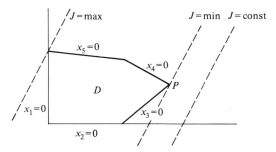

Figure 12.2

of utmost importance in practical computation, since there could be as many as $\binom{n}{m}$ basic vectors. For n and m large, this is a large number. The simplex method is an iterative process which terminates in a finite number of steps. The initial step is to determine a nondegenerate basic solution. To simplify the explanation of the method, we assume that the problem is in standard form, with $m < n$ and rank $(A) = m$. For convenience, we may index the variables so that the first m columns of A are linearly independent. By an elimination process similar to Gauss elimination with pivoting, we can transform the matrix A into the form

$$[I_m, A^{(1)}] = \begin{bmatrix} 1 \cdots 0 & a^{(1)}_{1,m+1}, \ldots, a^{(1)}_{1n} \\ \vdots \ddots & \vdots & \vdots \\ 0 \quad 1 & a^{(1)}_{m,m+1}, \ldots, a^{(1)}_{mn} \end{bmatrix}$$

(This is known as *Gauss–Jordan elimination*. It goes beyond Gauss elimination in that it eliminates the elements above the diagonal as well as those below, making back substitution unnecessary. Note that with maximal pivoting some of the multipliers can be greater than 1. Also, the number of operations is $n^3/2 + O(n^2)$ as compared to $n^3/3 + O(n^2)$ for Gauss elimination (Section 4.2). Therefore the Gauss method is preferred for solving linear systems, whereas the Jordan modification is used in linear programming. The Jordan elimination yields an equivalent system of linear equations in which the variables $x_1, \ldots, x_m$ are completely "uncoupled," so that each may be varied independently.) Actually, Jordan elimination is performed on the augmented matrix,

$$B = \begin{bmatrix} a_{11} \cdots a_{1m} & 0 & a_{1,m+1} \cdots a_{1n} & b_1 \\ \vdots & & & \vdots \\ a_{m1} & a_{mm} & 0 & a_{m,m+1} & a_{mn} & b_m \\ c_1 & c_m & 1 & c_{m+1} & c_n & 0 \end{bmatrix}$$

the result being the matrix,

$$B^{(1)} = \begin{bmatrix} 1 \cdots & 0 & & b^{(1)}_1 \\ \vdots \ddots & \vdots & A^{(1)} & \vdots \\ & 1 & & b^{(1)}_m \\ 0 \quad 0 & 1 & c^{(1)}_{m+1} & c^{(1)}_n & -J_1 \end{bmatrix}.$$

In effect, we have replaced the objective function by an additional equation, $c^T x - J = 0$, corresponding to the level surfaces of J, where J is now just another variable. For any vector $y^T = (x_1^{(1)}, \ldots, x_m^{(1)}, -J, x_{m+1}^{(1)}, \ldots, x_n^{(1)}, -1)$ such that $B^{(1)} y = 0$, we have $J = J_1 + \sum_{i=m+1}^{n} c_i^{(1)} x_i^{(1)}$ as the value of the objective function for $x^{(1)}$ a solution of (12.8.3). If we solve for y by setting $x_{m+1}^{(1)} = \cdots = x_n^{(1)} = 0$, then $x^{(1)}$ is a basic solution and the objective equation becomes $J = J_1$; that is, J_1 is a value of the objective function at a basic vector. If $b_i^{(1)} \geq 0$, $1 \leq i \leq m$, then $x^{(1)}$ is also a feasible vector, since $x_i^{(1)} = b_i^{(1)}$. Let us assume that this is so, since the simplex method will always ensure that it holds initially and it is preserved on iteration, as we shall now see.

First, suppose that all $b_i^{(1)} > 0$. Then $x^{(1)}$ is nondegenerate. If at least one $c_i^{(1)} < 0$, then it is possible to construct another basic feasible solution $x^{(2)}$ with a new value of the objective function $J_2 < J_1$. This is done by increasing the nonbasic variable $x_i^{(1)}$ from 0 to the largest possible positive value which does not make any of the basic variables negative. In this process, the other nonbasic variables are held at 0 and the first basic variable which becomes 0 replaces $x_i^{(1)}$ as a new nonbasic variable. Thus each iteration involves an exchange of a nonbasic variable, $x_s^{(1)}$ say, and a basic variable, $x_r^{(1)}$ say, to obtain a new set of basic variables. The exchange is performed by pivoting; that is, the element $a_{rs}^{(1)}$ is used as a pivot to reduce column s to the form $(0, \ldots, 0, 1, 0, \ldots, 0)$ with 1 in row r, and then columns r and s are interchanged. This gives us a new matrix $B^{(2)}$ having the same form as $B^{(1)}$, but with variables rearranged and $J_2 < J_1$. To ensure that $b_i^{(2)} > 0$, the following simple strategy is used.

First, the nonbasic variable $x_s^{(1)}$ is chosen such that $c_s^{(1)} = \min\{c_i^{(1)}; m+1 \leq i \leq n\}$. (This procedure does not necessarily yield the greatest decrease in J, since we may not be able to increase $x_s^{(1)}$ by much. However, it is easy to do and has been found to be practical.) If all $c_i^{(1)} \geq 0$, then $x^{(1)}$ must be optimal. This follows from Theorem 12.8.1, since if J_1 is not minimum, it must be possible to find a feasible basic vector $x^{(2)}$ with $J(x^{(2)}) < J(x^{(1)})$. However, any such $x^{(2)}$ can be obtained from $x^{(1)}$ only by having at least one of the nonbasic variables $x_i^{(1)}$ increase from 0. This obviously would increase J. Therefore the condition $c_i^{(1)} \geq 0$, $m+1 \leq i \leq n$, terminates the simplex method iteration. If some $c_i^{(1)} = 0$, the solution may not be unique. If all $c_i^{(1)} > 0$, the solution is unique. Assuming $c_s^{(1)} < 0$, we determine r such that

$$\frac{b_r^{(1)}}{a_{rs}^{(1)}} = \min_i \left\{ \frac{b_i^{(1)}}{a_{is}^{(1)}} : a_{is} > 0 \right\}.$$

This has the following effect. To preserve a solution to $Ax = b$, we must set $x_i = b_i^{(1)} - a_{is}^{(1)} x_s$, $1 \leq i \leq n$, as we increase x_s and keep all other nonbasic variables 0. Thus the first basic variable to become 0 is x_r. Conversely, making this basic variable 0 will yield $x_s^{(2)} = b_r^{(1)} / a_{rs}^{(1)}$ and all other basic variables will remain nonnegative; that is, we again have a basic feasible solution $x^{(2)}$ with $J_2 < J_1$. In order to continue the iteration, we must transform $B^{(1)}$ into a matrix $B^{(2)}$ of the same form, but corresponding to the new set of basic variables. This can be done by using $a_{rs}^{(1)}$ as pivot and eliminating all elements in column s other than the element in row r, which is reduced to 1. Finally, columns r and s are interchanged, yielding $B^{(2)}$. Once again we obtain a basic feasible solution with values appearing in the last column as $b_i^{(2)} \geq 0$ and we are ready to continue the iteration.

The choice of r is possible as long as some $a_{is} > 0$. If all $a_{is} \leq 0$, then since $x_i = b_i^{(1)} - a_{is}^{(1)} x_s$, we can increase x_s indefinitely without making any x_i, $1 \leq i \leq n$ negative. In this case, J is unbounded on the feasible set.

If each iteration produces a nondegenerate solution (all $b_i > 0$), then the simplex method will converge to an optimal solution in a finite number of steps. This is clear, since there are a finite number of basic feasible solutions and no basic feasible solution can occur twice (since $J(x^{(k+1)}) < J(x^{(k)})$). If degeneracy occurs at some iteration (that is, some $b_i^{(k)} = 0$, $1 \leq i \leq n$), the pivoting can be done as before (since $a_{rs} > 0$). However, it may not be possible to increase x_s, and the value of J would remain the same. It is possible that a given set of basic variables may ultimately be repeated, resulting in an infinite loop in the iteration. There are procedures to avoid "cycling" of this type, but in practice cycling is rare [12.26]. However, degenerate problems may take longer to solve.

One method of obtaining an initial basic feasible solution is to introduce additional variables $x_{n+1}, \ldots, x_{n+m}$ and solve the new problem of minimizing $W = x_{n+1} + \cdots + x_{n+m}$ with the constraints

$$\begin{aligned}
a_{11}x_1 + \cdots + a_{1n}x_n + x_{n+1} &= b_1 \\
a_{21}x_1 + \cdots + a_{2n}x_n \phantom{{}+ x_{n+1}} + x_{n+2} &= b_2 \\
&\vdots \\
a_{m1}x_1 + \cdots + a_{mn}x_n \phantom{{}+ x_{n+1}} + x_{n+m} &= b_m \\
c_1 x_1 + \cdots + c_n x_n \phantom{{}+ x_{n+1} + x_{n+m}} - J &= 0
\end{aligned}$$

$$x_j \geq 0, \quad 1 \leq j \leq n+m.$$

An initial basic feasible solution for this set is $x_1 = x_2 = \cdots = x_n = J = 0$, $x_{n+i} = b_i$, $1 \leq i \leq m$. A basic feasible solution such that min $W = 0$ is a basic feasible solution for the original problem. If min $W > 0$, the original problem has no solution.

The standard simplex method using Jordan elimination is not always the most efficient computational method (in terms of operations per iteration). When n is much larger than m and there are many zero entries in A, other more efficient procedures exist. Some of these are based on computing matrix inverses. This may seem paradoxical in view of the remarks in Section 4.2b. However, by taking into account the sparseness of the matrix and computing successive inverses efficiently, one can appreciably reduce the overall operation count. Any elimination process, of course, is equivalent to computing an inverse. The standard simplex method may be restated in terms of inverses as follows.

Let us suppose that the first m columns, $A_1, \ldots, A_m$ of A are linearly independent. Form the $(m+1) \times (m+1)$ matrix,

$$C_1 = \begin{bmatrix} A_1 & \cdots & A_m & 0 \\ c_1 & & c_m & 1 \end{bmatrix}$$

and compute C_1^{-1}. The basic solution $x^{(1)}$ having $x_{m+1} = \cdots = x_n = 0$ is given by $x^{(1)} = C_1^{-1}b$, where $b^T = (b_1, \ldots, b_m, 0)$. Row $m + 1$ of C_1^{-1} must be of the form $(-\mu_1, \ldots, -\mu_m, 1) = (-\mu^T, 1)$. The components μ_i of μ are called *simplex multipliers*. Comparing this with the Jordan elimination procedure, we see (multiplying this last row by the jth column of B) that

$$c_j^{(1)} = c_j - \sum_{i=1}^{m} \mu_i a_{ij}, \quad m + 1 \leq j \leq n. \tag{1}$$

The $c_j^{(1)}$ are called *cost factors* for the nonbasic variables. Proceeding as in the standard simplex method, we choose $c_s^{(1)} = \min \{c_j^{(1)}\}$, stopping if $c_s^{(1)} \geq 0$. If $c_s^{(1)} < 0$, we compute the new column vector,

$$\begin{pmatrix} A_s^{(1)} \\ c_s^{(1)} \end{pmatrix} = C^{-1} \begin{pmatrix} A_s \\ c_s \end{pmatrix}. \tag{2}$$

As before, we determine the row r such that

$$b_r^{(1)}/a_{rs}^{(1)} = \min \{b_i^{(1)}/a_{is}^{(1)} : a_{is}^{(1)} > 0\}. \tag{3}$$

Next we wish to interchange columns r and s of the original augmented matrix C_1, forming a matrix C_2, and compute C_2^{-1}. This is done by constructing the augmented matrix

$$\begin{bmatrix} C_1^{-1} & A_s^{(1)} \\ & c_s^{(1)} \end{bmatrix} \tag{4}$$

and pivoting on $a_{rs}^{(1)}$ to reduce the last column to the form $(0, \ldots, 0, 1, 0, \ldots, 0)$, with 1 in position r. If we write the matrix B as $[C_1, B_{m+1}, \ldots, b]$, then

$$C_1^{-1}B = [I_{m+1}, B_{m+1}^{(1)}, \ldots, b^{(1)}] = B^{(1)}.$$

Interchanging columns r and s can be done by postmultiplication by the elementary matrix E_{rs} (the identity with columns r and s interchanged). The Jordan elimination with pivot $a_{rs}^{(1)}$ can also be done by premultiplication by elementary matrices, E_i, as in Section 4.2. Thus

$$E_m \cdots E_1 C_1^{-1} B E_{rs} = B^{(2)} = [I_{m+1}, B_{m+1}^{(2)}, \ldots, b^{(2)}],$$

where $B^{(2)}$ is the transformed augmented matrix at the next step of the standard simplex method. However,

$$BE_{rs} = [C_2, B_{m+1}, \ldots, b] \quad \text{and} \quad C_2^{-1}BE_{rs} = [I_{m+1}, \ldots].$$

This verifies that $C_2^{-1} = E_m \cdots E_1 C_1^{-1}$. In fact, it leads to a method of organizing the computation in which the successive inverses are stored as columns of the elementary matrices E_i. (See [12.27].) The new basic solution is given by

$$x_i^{(2)} = x_i^{(1)} - (b_r^1/a_{rs}^{(1)})a_{is}^{(1)} \quad \text{for } i \neq r \text{ and } x_r^{(2)} = b_r^{(1)} = b_r^{(1)}/a_{rs}^{(1)}. \tag{5}$$

The new inverse C_2^{-1} replaces C_1^{-1} and steps (1) through (5) are repeated until the test for a minimum succeeds (as in the standard simplex method). If the starting method described above is used, then $C_1 = I$.

The procedure just described is called the *revised* (or *inverse*) *simplex method*. It is more efficient than the standard simplex method when A is sparse, for then many of the multiplications by 0 can be eliminated. It is also more economical in its use of internal computer (core) storage, since only each C_k^{-1} needs to be stored in high-speed memory.

12.9 INEQUALITY CONSTRAINTS, MATHEMATICAL NONLINEAR PROGRAMMING

In this section, we consider the minimization of a functional J subject to the inequality constraints $g_i(x) \leq 0$, $1 \leq i \leq p$. We shall find it convenient to introduce the vector-valued function $g(x) = (g_1(x), \ldots, g_p(x))$. Thus $g : D \to V^p$, where D is a subset of a Hilbert space and J is defined on D. The p inequality constraints can then be written in vector form as $g(x) \leq 0$, and we define the constraint set, $C = \{x : g(x) \leq 0\}$. Neither J nor g need be linear.

We shall need to use certain properties of convex functions and we summarize them here. We refer to Exercise 2.16 for the definition of a partially ordered vector space. For a finite-dimensional space, we use the natural ordering given in Exercise 2.16e; that is, $x \geq 0$ if and only if $x_i \geq 0$ for all components x_i.

Definition 12.9.1 Let X be a vector space and let V be a partially ordered vector space. Let $D \subset X$ be a convex set. A function $f : X \longrightarrow V$ is said to be *convex* on D if for all $x, y \in D$ and $0 \leq \alpha \leq 1$,

$$f((1 - \alpha)x + \alpha y) \leq (1 - \alpha)f(x) + \alpha f(y). \tag{12.9.1}$$

We say that f is *convex at* x if there is an open ball $B = B(x; r)$ such that inequality (12.9.1) holds for all $y \in B$. If the inequality is reversed, f is called *concave*.

Remark If V in the above definition has finite dimension p so that $f = (f_1, \ldots, f_p)$ and we use the natural ordering, then f is convex if and only if each f_i is a convex functional.

Theorem 12.9.1 Let $f : X \longrightarrow V$, where X and V are normed vector spaces and V is partially ordered. Let f be convex at x and have a strong derivative $f'(x)$ at x. Then for any y such that (12.9.1) holds,

$$f(y) - f(x) \geq f'(x)(y - x). \tag{12.9.2}$$

Conversely, if (12.9.2) holds for all x and y in a convex set D, then f is convex on D.

Proof. Suppose that (12.9.1) holds. Then

$$f(y) - f(x) \geq (f(x + \alpha(y - x)) - f(x))/\alpha.$$

Allowing $\alpha \to 0$, we obtain (12.9.2).

Conversely, letting $z = \alpha x + \beta y$, where $\alpha + \beta = 1$ and $\alpha \geq 0$, $\beta \geq 0$, and

applying (12.9.2), we obtain
$$f(x) - f(z) \geq f'(z)(x - z), \quad f(y) - f(z) \geq f'(z)(y - z).$$
Multiplying the first inequality by α, the second by β, and adding, we get
$$\alpha f(x) + \beta f(y) - f(z) \geq f'(z)(\alpha x + \beta y - z) = 0,$$
which establishes convexity. ∎

Returning to the problem of minimizing J on the inequality constraint C, we recall the necessary and sufficient conditions given in Theorems 12.2.2, 12.2.3, 12.2.7, 12.2.8 in terms of the Lagrangian in the case of the equality constraint. We shall now develop analogous conditions for the inequality constraint case. To motivate these conditions, let us proceed as in the linear programming problem to introduce slack variables x_{si}, $1 \leq i \leq p$, and replace each inequality $g_i(x) \leq 0$ by the pair $g_i(x) + x_{si} = 0$, $x_{si} \geq 0$. Consider the new space $Y = X \oplus V^p$. Suppose that the minimum of J occurs at an interior point x^* of C. Then $x_{si}^* > 0$, $1 \leq i \leq p$. Forming the Lagrangian for the new equality constraints (and ignoring the constraints $x_{si} \geq 0$), we obtain, with $y = (x, x_{s1}, \ldots, x_{sp})$,
$$f(y, \Lambda) = J(x) + \sum_1^p \lambda_i(g_i(x) + x_{si}).$$
The multiplier rule states that $f_y(y^*, \Lambda^*) = 0$. This implies that
$$\lambda_i^* = \left.\frac{\partial f}{\partial x_{si}}\right|_{y^*} = 0, \quad 1 \leq i \leq p.$$
Similarly, if some subset of the inequality constraints are *inactive*, say $x_{si}^* > 0$ for $i = 1, \ldots, k$ and the remaining constraints are *active* so that $g_i(x^*) = 0$, $i = k + 1, \ldots, p$, we could write the Lagrangian as
$$f(y, \Lambda) = J(x) + \sum_1^k \lambda_i(g_i(x) + x_{si}) + \sum_{k+1}^p \lambda_i g_i(x)$$
and obtain $\lambda_i^* = 0$ for $1 \leq i \leq k$. Since $g_i(x^*) = 0$ for $k + 1 \leq i \leq p$, we obtain $\lambda_i^* g_i(x^*) = 0$, which implies
$$\langle \Lambda^*, g(x^*)\rangle = 0. \tag{12.9.3}$$
Now consider an active constraint, say $g_j(x^*) = 0$. As in Section 12.2a, consider the constraints $g_j(x) = b_j$. We showed that
$$\lambda_j^* = -\left.\frac{\partial J(x(b))}{\partial b_j}\right|_{b=0},$$
where $x^* = x(0)$ and $x(b)$ is the minimum point for the constraint $g_j(x) = b_j$. If we decrease b_j from 0, since $g_j(x(b)) = b_j < 0$, then $x(b)$ is still in the original inequality constraint set C. Since x^* is a minimum point on C, we must have $J(x(b)) \geq J(x^*)$. Hence $\Delta J/\Delta b_j \leq 0$, and allowing $b_j \to 0$, we conclude that $\partial J/\partial b_j \leq 0$. Therefore $\lambda_j^* \geq 0$ for an active constraint. Since $\lambda_i^* = 0$ for an inactive constraint, we have all

12.9 Inequality Constraints, Mathematical Nonlinear Programming

$\lambda_i^* \geq 0$, or in vector form,

$$\Lambda^* \geq 0. \tag{12.9.4}$$

Finally, $f_y(y^*, \Lambda^*) = 0$ implies that

$$J'(x^*) + \sum_{i=1}^{p} \lambda_i^* g_i'(x^*) = 0. \tag{12.9.5}$$

Since $\lambda_i^* \geq 0$, in a Hilbert space (12.9.5) can be interpreted as saying that at a minimum point the negative gradient—$\nabla J(x^*)$ is a linear combination of the gradients $\nabla g_i(x^*)$ with coefficients which are nonnegative; that is, $\nabla J(x^*)$ is in the cone generated by the vectors $\nabla g_i(x^*)$. The effect of this can be seen geometrically in two dimensions in the case of two constraints. Consider the following figure:

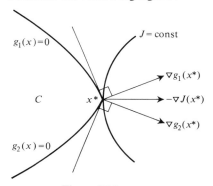

Figure 12.3

A small move from x^* in any direction h which makes an angle less than 90° with $-\nabla J(x^*)$ will decrease J. If $-\nabla J(x^*)$ were not situated, as shown, in the cone generated by $\nabla g_2(x^*)$ and $\nabla g_1(x^*)$, then it would make an angle of less than 90° with one of the directions h pointing into the constraint set C. Hence x^* could not be a minimum point of J on C.

The conditions (12.9.3), (12.9.4), and (12.9.5) are referred to as the *Kuhn–Tucker conditions*. The preceding heuristic argument fails to be a proof that these are necessary conditions, since it assumes that the Lagrange multiplier rule holds in the presence of the inequality constraints $x_{si} \geq 0$. For this we need a special condition to preclude degeneracy at x^*. We state the result here as a theorem and give the proof in Appendix VI.

Theorem 12.9.2 Let J and g_i, $1 \leq i \leq p$, be real functionals defined on an open subset D of a pre-Hilbert space H. Let x^* be a local minimum of J on the constraint $C = \{x : g_i(x) \leq 0, 1 \leq i \leq p, x \in D\}$ and assume that the strong gradients $\nabla J(x^*)$ and $\nabla g_i(x^*)$ exist. Further, assume that for any vector $h \in H$ such that $\langle \nabla g_i(x^*), h \rangle \leq 0$ for all active constraints (that is, $g_i(x^*) = 0$) there exists a "curve" $\varphi(t) \in C$, $0 \leq t \leq 1$, with $\varphi(0) = x^*$ and $\varphi'(0) = \lambda h$ for some $\lambda > 0$. Then there exists a multiplier Λ^* satisfying the Kuhn–Tucker conditions (12.9.3), (12.9.4), (12.9.5).

580   **Extremum Problems**

Proof. See Appendix VI. ∎

Remark The condition that there exists a curve such as $\varphi(t)$ is called the *Kuhn–Tucker constraint qualification*. It rules out such cases as a cusp at x^* and tangency of two active constraints at x^*. In the latter case, the gradients of the active constraints $g_i(x^*) = 0$ would not be linearly independent. Linear independence of these gradients would suffice as a replacement for the Kuhn–Tucker constraint qualification. (See Appendix VI).

To establish sufficient conditions for a local minimum in the case of an equality constraint, we introduced the condition (12.2.5) requiring positive-definiteness of the second derivative, $f''(x^*)$. This is closely related to the convexity of f, as the next two theorems show.

Theorem 12.9.3 Let X be a normed vector space. Let $f: X \longrightarrow V^n$ be defined on a convex set $D \subset X$. If $f''(x)h^2 \geq 0$ for all $x \in D$ and all $h \in X$, then f is a convex function.

Proof. $f = (f_1, \ldots, f_n)$, where the f_i are real functionals. We use the natural ordering in V^n. By the mean-value theorem for functionals,

$$f_i(x + h) - f_i(x) - f_i'(x)h = \tfrac{1}{2} f_i''(x + \theta_i h) h^2 \geq 0.$$

Hence $f(x + h) - f(x) \geq f'(x)h$ and f is convex, by Theorem 12.9.1. ∎

Theorem 12.9.4 Let X and V be normed vector spaces and $D \subset X$ convex. Let V be partially ordered. If $f: X \longrightarrow V$ is convex on D, then $f''(x)h^2 \geq 0$ for all $h \in X$ and $x \in D$.

Proof. By Theorem 12.9.1, for any x and y in X,

$$f(y) - f(x) \geq f'(x)(y - x), \qquad f(x) - f(y) \geq f'(y)(x - y).$$

Adding, we obtain

$$(f'(y) - f'(x))(y - x) \geq 0.$$

Hence

$$f''(x)h^2 = \lim_{t \to 0} \left(\frac{f'(x + th) - f'(x)}{t} \right) h$$

$$= \lim \frac{1}{t^2} (f'(x + th) - f'(x)) th \geq 0,$$

as was to be proved. ∎

With these results, it is not surprising that convexity at x^* can replace the positive-definiteness of $f''(x^*)$ in establishing sufficiency.

Theorem 12.9.5 Let J and g_i, $1 \leq i \leq p$, be real functionals defined on an open set of a normed vector space. Let $J'(x^*)$ and $g_i'(x^*)$ exist and let $g_i(x^*) \leq 0$, $1 \leq i \leq p$. If J and g_i satisfy the Kuhn–Tucker conditions and are convex at x^*, then x^* is a local minimum of J on the constraint $C = \{x : g_i(x) \leq 0, 1 \leq i \leq p\}$.

12.9 Inequality Constraints, Mathematical Nonlinear Programming 581

Proof. Let $B = B(x^*, r)$ be the ball in which convexity holds. Then for any $x \in C \cap B$, by convexity,

$$J(x) - J(x^*) \geq J'(x^*)(x - x^*) = -\sum_1^p \lambda_i^* g_i'(x^*)(x - x^*)$$

$$\geq \sum_1^p \lambda_i^* (g_i(x^*) - g_i(x))$$

$$= -\sum_1^p \lambda_i^* g_i(x) \geq 0,$$

since $\sum_1^p \lambda_i^* g_i(x^*) = 0$ by (12.9.3) and $\lambda_i^* \geq 0$ by (12.9.4). Therefore $J(x) \geq J(x^*)$; that is, x^* is a local minimum of J on C. ∎

As with equality constraints (Section 12.2), we observe that $f''(x^*, \Lambda^*)h^2 \geq 0$ does not suffice for x^* to be a minimum. This does not contradict Theorems 12.9.3 and 12.9.5, since positive semidefiniteness at x^* does not imply that $f(x, \Lambda^*)$ is convex at x^*. Positive definiteness and continuity of $f''(x, \Lambda^*)$ at x^* does imply convexity, and we would have for $x \in C \cap B$, as in Theorem 12.9.5,

$$f(x, \Lambda^*) = J(x) + \sum_{i=1}^p \lambda_i^* g_i(x) \leq J(x),$$

$$f(x^*, \Lambda^*) = J(x^*),$$

$$J(x) - J(x^*) \geq f(x, \Lambda^*) - f(x^*, \Lambda^*) \geq f_x'(x^*, \Lambda^*)(x - x^*) = 0.$$

Hence x^* is a local minimum of J on C. However, one can weaken the positivity requirement. Just as in the equality constraint case (Theorem 12.2.4), it suffices to have $f''(x^*, \Lambda^*)h^2 > 0$ only for h in a cone in the finite-dimensional case.

Theorem 12.9.6 Let $J, g_1, \ldots, g_p$ be real functionals on a subset of a finite-dimensional Hilbert space. Let (x^*, Λ^*) satisfy the Kuhn–Tucker conditions (12.9.3), (12.9.4), and (12.9.5), where x^* is on the constraint $C = \{x : g(x) \leq 0\}$. Let K be the cone of vectors h such that $\langle \nabla g_i(x^*), h \rangle = 0$ for all active constraints $(g_i(x^*) = 0)$ such that $\lambda_i^* > 0$ and $\langle \nabla g_i(x^*), h \rangle \leq 0$ for the active constraints with $\lambda_i^* = 0$. If the Langrangian f is such that $f''(x^*, \Lambda^*)h^2 > 0$ for all $h \in K$, then x^* is a strong local minimum of J on C.

Proof. We follow the proof of Theorem 12.2.4. Thus suppose $u_q = x^* + h_q$ is a sequence such that $g_i(u_q) \leq 0$, $1 \leq i \leq p$, $J(u_q) \leq J(x^*)$ and $h_q \to 0$ as $q \to \infty$. By compactness we may assume $h_q/\|h_q\| \to h$. Since $\lambda_i^* \geq 0$ by condition (12.9.4), we have $f(u_q, \Lambda^*) \leq J(u_q)$. By (12.9.3), $f(x^*, \Lambda^*) = J(x^*)$. Therefore, as in Eq. (12.2.6), we have

$$0 \geq f(u_q, \Lambda^*) - f(x^*, \Lambda^*) = f'(x^*, \Lambda^*)h_q + \tfrac{1}{2}f''(x^*, \Lambda^*)h_q^2 + \varepsilon_q,$$

and it follows that $f''(x^*, \Lambda^*)h^2 \leq 0$. To complete the proof, we show that $h \in K$. First, for any active constraint $g_i(u_q) - g_i(x^*) \leq 0$. This implies that $\langle \nabla g_i(x^*), h \rangle \leq 0$. If $\lambda_i^* > 0$ and $\langle \nabla g_i(x^*), h \rangle < 0$, then $\langle \nabla J(x^*), h \rangle = -\sum_1^p \lambda_i^* < \langle \nabla g_i(x^*), h \rangle > 0$.

This implies that

$$\left\langle \nabla J(x^*), \frac{h_q}{\|h_q\|} \right\rangle > 0$$

for q sufficiently large, and therefore that $J(u_q) > J(x^*)$, contradicting the choice of u_q. Hence $h \in K$ and $f''(x^*, \Lambda^*)h^2 \leq 0$, which contradicts the hypotheses of the theorem. ∎

Computational Aspects

The gradient methods of Sections 12.3 and 12.4 can be modified to solve minimization problems with inequality constraints. These modified gradient procedures are also known as *methods of feasible directions* [12.28]. They are based on the following lemma, which is the contrapositive of Theorem 12.9.2, and the Remark following that theorem.

Lemma 12.9.1 Let J and $g = (g_1, \ldots, g_p)$ be Frechet differentiable on an open set D of a Hilbert space H. Let $C = \{x \in D : g(x) \leq 0\}$ and suppose that $x_0 \in C$ is such that either the Kuhn–Tucker constraint qualification holds at x_0 or that the gradients $\{\nabla g_i(x_0) : i \in I\}$ are linearly independent, where $I = \{i : g_i(x_0) = 0\}$. If the Kuhn–Tucker conditions do not hold at x_0, then there exists a vector Δx_0 such that $x_0 + \Delta x_0 \in C$ and $J(x_0 + \Delta x_0) < J(x_0)$.

The vector Δx_0 is called a *feasible direction* at x_0. There are several ways of determining Δx_0 in the finite-dimensional case where $H = V^n$.

One method given in [12.28] is to determine Δx_0 so as to minimize $F(\Delta x_0) = \langle \nabla J(x_0), \Delta x_0 \rangle$ subject to the constraints $\langle \nabla g_i(x_0), \Delta x_0 \rangle < 0$ and $\|\Delta x_0\|_2 = 1$. The normalization of Δx_0 ensures that $F(\Delta x_0)$ has a finite minimum. An alternative normalization is to require that $\|\Delta x\|_\infty = 1$. Letting $y = \Delta x_0$ and $G_i(y) = \langle \nabla g_i(x_0), y \rangle$, we obtain the linear programming problem,

$$\text{minimize } F(y) = \sum_{i=1}^{n} \left.\frac{\partial J}{\partial x_i}\right|_0 y_i,$$

subject to the linear constraints,

$$G_i(y) + z = 0, \quad z \geq 0, \quad -1 \leq y \leq 1.$$

This can be solved by the simplex method by introducing more slack variables or by special *upper bounding procedures* [12.30]. If a solution (y, z) with $z > 0$ is found, then $y = \Delta x_0$ is a feasible direction and we determine $x_1 = x_0 + s\,\Delta x_0$, choosing s to minimize $J(x_0 + s\,\Delta x_0)$, subject to the constraint $g(x_0 + s\,\Delta x_0) \leq 0$. This is a one-dimensional minimization problem.

12.10 PENALTY FUNCTION TECHNIQUES (SEQUENTIAL UNCONSTRAINED METHODS)

Another frequently used technique for minimizing f subject to the constraints $g_i(x) \leq 0$ is the *penalty function method* [12.33]. In this method, the constrained minimization problem is replaced by a sequence of unconstrained minimization problems. If these unconstrained problems are chosen properly, the sequence of their solutions converges to a solution of the constrained problem. For example, in one approach we form the sequence of functions f_n, $n = 0, 1, 2, \ldots$, given by

$$f_n(x) = f(x) + \tfrac{1}{2} \sum_{i=1}^{p} r_i^{(n)}(\min\{0, -g_i(x)\})^2, \qquad (12.10.1)$$

where the $r_i^{(n)}$ are constants such that $\lim_{n\to\infty} r_i^{(n)} = \infty$, $i = 1, \ldots, p$. Suppose x_n is a minimum point of f_n. It is easy to show that

$$\nabla f(x_n) + \sum_{i \in Q} r_i^{(n)} g_i(x_n) \nabla g_i(x_n) = 0, \qquad (12.10.2)$$

where Q is the set of i such that $g_i(x_n) > 0$. Letting $\lambda_i^{(n)} = r_i^{(n)} g_i(x_n)$ and forming the inner product with $\nabla g_j(x_n)$, $j \in Q$, we obtain the linear system of equations for the "multipliers" $\lambda_i^{(n)}$,

$$\sum_{i \in Q} \langle \nabla g_i(x_n), \nabla g_j(x_n) \rangle \lambda_i^{(n)} = -\langle \nabla f(x_n), \nabla g_j(x_n) \rangle.$$

This is recognized to be the Gram system (7.4.1) used in Theorem 12.2.2 and in the angular gradient method of Eqs. (12.4.1) through (12.4.4). Now $g_i(x_n) = \lambda_i^{(n)}/r_i^{(n)}$ and $r_i^{(n)} \to \infty$. Suppose $x_n \to x^*$. If all gradients are continuous and the Gram matrix is nonsingular at x^*, then $\lim_{n\to\infty} \lambda_i^{(n)} = \lambda_i^*$ exists and $g_i(x_n) \to 0$. Hence $g_i(x^*) = 0$ for $i \in Q$ and $g_i(x^*) \leq 0$ for the remaining i. From (12.10.1) it follows that $\lim_{n\to\infty} f_n(x_n) = f(x^*)$, since for $i \in Q$ we have $\lim_{n\to\infty} r_i^{(n)}(g_i(x_n))^2 = \lambda_i^* g_i(x^*) = 0$. Similarly, from (12.10.2), we see that

$$\lim_{n\to\infty} \nabla f(x_n) = -\sum_{i \in Q} \lambda_i^* \nabla g_i(x^*),$$

$$\lambda_i^* = \lim_{n\to\infty} \lambda_i^{(n)} = \lim_{n\to\infty} r_i^{(n)} g_i(x_n) \geq 0.$$

Taking $\lambda_i^* = 0$ for $i \notin Q$, we see that the Kuhn–Tucker conditions are satisfied at x^*. If the additional conditions of either Theorem 12.9.5 or Theorem 12.9.6 hold as well, then x^* is a local minimum of f on the inequality constraint C.

If we take all $r_i^{(n)} = r^{(n)}$, we obtain a method considered in [12.34]. Since $g_i(x_n) > 0$ for $i \in Q$, the points x_n lie outside the constraint C and this type of penalty function method is called an *exterior point algorithm*. (If Q is empty for some x_n, then $f_n = f$ and $x_n = x^*$ is a minimum of f on C.) If there exists an isolated local minimum point $x^* \in C$, it can be shown that $\lim_{n\to\infty} x_n = x^*$. (See [12.34].) Otherwise the sequence (x_n) need not converge, but under certain compactness conditions on the set of local minima, $\lim_{n\to\infty} f_n(x_n)$ is the local minimum value of f on C. If x^* is on the boundary of C, then the set Q is not empty.

Numerous versions of the penalty-function method have been used in practice. See [12.34] for the finite-dimensional case and [12.36] for the optimal control problem.

REFERENCES

12.1. A. V. Balakrishman and L. W. Neustadt (editors), *Mathematical Theory of Control*, Academic Press, New York (1967)

12.2. E. K. Blum, "The calculus of variations, functional analysis and optimal control problems," in *Topics in Optimization* (edited by G. Leitmann), Academic Press, New York (1967)

12.3. A. A. Goldstein, *Constructive Real Analysis*, Harper and Row, New York (1967)

12.4. M. R. Hestenes, *Calculus of Variations and Optimal Control Theory*, Wiley, N.Y., (1966)

12.5. G. Leitmann, *An Introduction to Optimal Control*, McGraw-Hill, New York (1966)

12.6. L. S. Pontryagin, V. G. Boltyanskii, R. V. Gamkrelidze, and E. F. Mischenko, *The Mathematical Theory of Optimal Processes*, (translated by K. Tirirogoff, edited by L. W. Neustadt), Wiley, New York (1962)

12.7. G. A. Bliss, *Lectures on the Calculus of Variation*, Univ. of Chicago Press, Chicago (1946)

12.8. S. F. McCormick, "A general approach to one-step iterative methods," Ph.D. Thesis, University of Southern California (1971)

12.9. W. Ritz, "Über eine neue Methode zur Losung gewisser Variationsprobleme der mathematischen Physik," *J. Reine Angew. Mathematik* **135**, 1 (1908)

12.10. M. R. Hestenes, "The conjugate gradient method for solving linear systems," *Proceedings of the Symposium on Applied Mathematics*, Vol. 6, McGraw-Hill, New York, 83–102 (1956)

12.11. M. J. D. Powell, "A survey of numerical methods for unconstrained optimization," *SIAM Rev.* **12**, 79–97 (1970)

12.12. R. Fletcher and C. M. Reeves, "Function minimization by conjugate gradients," *Computer J.* **7**, 149–154 (1964)

12.13. J. D. Pearson, "On variable metric methods of minimization," *British Computer J.* **12**, 171–178 (1969)

12.14. R. Fletcher and M. J. D. Powell, "A rapidly convergent descent method for minimization," *Computer J.* **6**, 163–168 (1963)

12.15. L. V. Kantorovich and G. P. Akilov, *Functional Analysis on Normed Spaces* (Chapter 15), Macmillan, New York (1964)

12.16. P. Wolfe, "Convergence conditions for ascent methods," *SIAM Rev.* **11**, 226 (1969)

12.17. S. G. Mikhlin, *Variational Methods in Mathematical Physics*, translated by T. Boddington, Macmillan, New York (1964)

12.18. S. G. Mikhlin, *The Problem of the Minimization of a Quadratic Functional*, translated by A. Feinstein, Holden-Day (1965)

12.19. P. G. Ciarlet, M. H. Schultz, and R. S. Varga, "Numerical methods of high-order accuracy for nonlinear boundary value problems."
 I. "One-dimensional problem," *Numerische Mathematik* **9**, 394–430 (1967)
 II. "Nonlinear boundary conditions," *Numerische Mathematik* **11**, 331–345 (1968)
 III. "Eigenvalue problems," *Numerische Mathematik* **12**, 120–133 (1968)

12.20. M. H. Schultz, "Rayleigh-Ritz-Galerkin methods for multidimensional problems," *SIAM J. Numer. Anal.* **6** (December 1969)

12.21. E. K. Blum, "A convergent gradient procedure in pre-Hilbert spaces," *Pacific J. Math.* **18**, No. 1 (1966)

12.22. E. K. Blum, "Stationary points of functionals in pre-Hilbert spaces," *J. Computer and System Sciences* **1**, No. 1 (April 1967)

12.23. E. K. Blum, "The computation of eigenvalues and eigenvectors of a completely continuous self-adjoint operator, *J. Computer and System Sciences* **1**, No. 4 (Dec. 1967)

12.24. Y. BARD, "Comparison of gradient methods for the solution of nonlinear parameter estimation problems," *SIAM J. Numer. Anal.* **7**, No. 1 (March 1970)

12.25. S. VAJDA, *Mathematical Programming*, Addison-Wesley, Reading, Mass. (1961)

12.26. G. B. DANTZIG, *Linear Programming and Extensions*, Princeton University Press, Princeton, N.J. (1963)

12.27. L. S. LASDON, *Optimization Theory for Large Systems*, Macmillan, New York (1970)

12.28. G. ZOUTENDIJK, *Methods of Feasible Directions*, American Elsevier, New York (1960)

12.29. J. B. ROSEN, "The gradient projection method for nonlinear programming, Part I: Linear constraints," *S.I.A.M. Journal* **8**, 181–217 (1960). "Part II: Nonlinear constraints," *S.I.A.M. Journal* **9**, 514–532 (1961)

12.30. G. B. DANTZIG and R. M. VAN SLYKE, "Generalized upper bounding techniques," *J. Computer and System Sciences* **1**, 213–226 (1967)

12.31. K. C. KAPUR and R. M. VAN SLYKE, "Cutting plane algorithms and state space constrained linear optimal control problems," *J. Computer and System Sciences* **4**, No. 6, 570–605 (December 1970)

12.32. E. S. LEVITIN and B. T. POLYAK, "Constrained optimization methods" (English trans.), *U.S.S.R. Comput. Math. Math. Phys.* (November 1968); *Zh. Vychist. Mat. Mat. Fiz.* **6**, 787–823 (1966)

12.33. R. COURANT, "Calculus of variations and supplementary notes and exercises" (mimeographed). Notes by M. KRUSKAL and H. RUBIN, revised and amended by J. MOSER, New York University (1956)

12.34. A. V. FIACCO and G. P. MCCORMICK, *Nonlinear Programming: Sequential Unconstrained Minimization Techniques*, Wiley, New York (1968)

12.35. G. RODRIGUE, "A variational method for numerical solution of algebraic problems," Ph.D. Thesis, Univ. of Southern California (1971)

12.36. A. V. BALAKRISHNAN, "A computational approach to the maximum principle," *J. Computer and System Sciences* **5**, No. 2, 163–191 (1971)

EXERCISES

12.1 Verify that the function $f(x, y) = x^3 - y^3$ does not have an extremum at $(0, 0)$, while $f'(0, 0) = 0$ and $f''(0, 0)h^2 = 0$ for all $h \in R^2$. Thus the necessary condition (12.2.4) is not sufficient for a minimum.

12.2 Show that the gradient procedure (12.3.3) may be viewed as an iterative procedure of the type (4.4.4), (4.4.5) discussed in Chapter 4. Verify that Q is quasinilpotent for the choice of s given in Theorem 12.3.3 with $M = \|A\|$.

12.3 Show that a sequence of vectors (x_n) which (i) has a finite number of accumulation points (as a set $\{x_n\}$), (ii) is such that $\{x_n\}$ is contained in a compact set S and (iii) $\|x_{n+1} - x_n\| \to 0$, must be convergent. [*Hint:* If $z_1, \ldots, z_k$ are the accumulation points in S and $B(z_i, \varepsilon)$ are disjoint open balls, $S - \bigcup B(z_i, \varepsilon)$ has a finite number of x_n.]

12.4 Verify by example that the steepest descent algorithm does not always yield a gradient procedure which converges at the fastest possible rate. [*Hint:* Use Example 2 following Theorem 12.3.3, taking $b = 0$ and A a 2×2 diagonal matrix. Referring to the Remark under Computational Aspects, Section 12.3, compare the values $x^* = x - s^*Ax$ and $x' = x - s'Ax$. Show that for $x = (\xi_1, \xi_2)$ with $\xi_1 > 0, \xi_2 > 0$ and $\xi_2/\xi_1 < a_{11}/a_{22}$, the point x' is closer to 0, whereas for $\xi_2/\xi_1 > a_{11}/a_{22}$, the point x^* may be closer to 0 (the minimum).]

586   **Extremum Problems**

12.5 Let $x = x(t) \in L^2[a, b]$ and let A be the integral operator

$$Ax = \int_a^b K(t, s)x(s)\,ds,$$

where the kernel $K(t, s)$ is symmetric and continuous on the square $a \le t \le b, a \le s \le b$. Prove that A is a completely continuous self-adjoint operator. [*Hint:* Let $y = Ax$. Then

$$|y(t)| = \left|\int_a^b K(t, s)x(s)\,ds\right| \le \left(\int_a^b |K(t, s)|^2\,ds\right)^{1/2}\left(\int_a^b |x(s)|^2\,ds\right)^{1/2} \le \mu\,\|x\|\,(b-a)^{1/2},$$

where $\mu = \max |K(t, s)|$. Thus, if $\|x\| < r$, the image set $Y = \{Ax : \|x\| < r\}$ is bounded in the uniform norm. Further,

$$|y(t_1) - y(t_2)| \le \left(\int_a^b |K(t_1, s) - K(t_2, s)|^2\,ds\right)^{1/2}\|x\| < \varepsilon$$

when $|t_1 - t_2| < \delta$ is chosen so small that $|K(t_1, s) - K(t_2, s)| < \varepsilon/r$. Since δ is independent of t_1, t_2 and y (by the uniform continuity of K), the family Y is equicontinuous. Hence Y is compact in the uniform norm by Arzela's theorem (Exercise 1.4) and therefore in the L^2-norm.]

12.6 Derive the variational equation (12.6.3). [*Hint:* Introduce the variables $y_i = x_i$, $1 \le i \le m$, $y_{m+1} = s$ and adjoin the extra equation $dy_{m+1}/dt = 0$ to (12.6.2), which assumes the form

$$\frac{dy}{dt} = F(t, y).$$

As initial values take $y_i(t_0) = x_i(t_0)$, $1 \le i \le m$ and $y_{m+1}(t_0) = s$. This converts the parameter s into an initial-value parameter. By Theorem 11.1.2, $y(t, s)$ exists for $|s|$ sufficiently small and

$$\frac{d}{dt}\frac{\partial y}{\partial s} = \frac{\partial F}{\partial y}\frac{\partial y}{\partial s}. \qquad (*)$$

Since $F_i(t, y) = f_i(x_1, \ldots, x_m, u + y_{m+1}h)$, $1 \le i \le m$, and $F_{m+1}(t, y) = 0$, we have

$$\sum_{j=1}^{m+1} \frac{\partial F_i}{\partial y_j}\frac{\partial y_j}{\partial s} = \sum_{j=1}^{m} \frac{\partial f_i}{\partial x_j}\frac{\partial x_j}{\partial s} + \frac{\partial f_i}{\partial u}h.$$

We obtain $\partial x/\partial s$ as the first m components of $(\partial y/\partial s)$, where the latter corresponds to the solution of system (*) with initial values $(0, 0, \ldots, 0, 1)$.]

12.7 Prove that a unit eigenvector u^* belonging to an isolated eigenvalue λ of a bounded self-adjoint operator A on a Hilbert space is a quasiregular stationary point of $\langle Ax, x\rangle$ on the unit sphere. [*Hint:* For $x_n \in B(u^*, r)$, let u_n be the projection of x_n on the invariant subspace E_λ of eigenvectors belonging to λ.]

12.8 Let V^k be a finite-dimensional subspace of V^m spanned by the orthonormal set $\{v_1, \ldots, v_k\}$.

a) Prove that a necessary condition for u to be a minimum point of a functional J on the subspace V^k is that $\langle \nabla J(u), v_i\rangle = 0$, $1 \le i \le k$. [*Hint:* $x \in V^k$ if and only if $\langle x, v_{k+1}\rangle = 0, \ldots, \langle x, v_m\rangle = 0$, where $\{v_1, \ldots, v_m\}$ is an orthonormal set for V^m. Define $g_i(x) = \langle x, v_{k+i}\rangle$, $1 \le i \le m - k$, $\nabla g_i = v_{k+i}$. Use Theorem 12.2.1, after showing that $T_u = V^k$.]

b) Prove that (a) holds when V^k is replaced by $x_0 + V^k$, the translate of V^k by any vector x_0. [*Hint:* Alternative proof, valid if V^k is a subspace of any vector space V. Let $u = x_0 + \sum_{i=1}^{k} s_i v_i$. Then

$$\langle \nabla J(u), v_i \rangle = \frac{dJ}{ds}(u + sv_i)\bigg|_{s=0} = 0$$

is a necessary condition for u to be a minimum point in $x_0 + V^k$.]

12.9 Derive formula (12.3.15) from (12.3.14), using the bordered inverse formula of Exercise 4.20. [*Hint:* Let $B_n = \delta X_n^* G_n$. Then

$$\delta X_{n+1}^* = (\delta X_n, \Delta x_{n+1})^*, \quad G_{n+1} = (G_n, \Delta J_{n+1}'). \tag{1}$$

Since

$$\Delta x_{n+1}^* G_n = (\langle \Delta x_{n+1}, \Delta J_n' \rangle, \ldots, \langle \Delta x_{n+1}, \Delta J_1' \rangle)$$

and

$$\Delta x_{n+1} = s_{n+1} v_{n+1},$$

we have

$$\Delta x_{n+1}^* G_n = 0.$$

Similarly,

$$\delta X_n^* \Delta J_{n+1}' = (\langle \Delta x_1, \Delta J_{n+1}' \rangle, \ldots, \langle \Delta x_n, \Delta J_{n+1}' \rangle)$$

and

$$\Delta J_{n+1}' = A \, \Delta x_{n+1} = s_{n+1} A v_{n+1}, \qquad \Delta x_i = s_i v_i.$$

Hence

$$\delta X_n^* \Delta J_{n+1}' = 0.$$

Since $B_{n+1} = \delta X_{n+1}^* G_{n+1}$, we get

$$B_{n+1} = \begin{bmatrix} B_n & 0 \\ 0 & \Delta x_{n+1}^* \Delta J_{n+1}' \end{bmatrix}$$

By Exercise 4.20,

$$B_{n+1}^{-1} = \begin{bmatrix} B_n^{-1} & 0 \\ 0 & c \end{bmatrix}$$

(where $1/c = \Delta x_{n+1}^* \Delta J_{n+1}'$). Hence

$$\delta X_{n+1} B_{n+1}^{-1} \delta X_{n+1}^* = \delta X_n B_n^{-1} \delta X_n^* + c \, \Delta x_{n+1} \, \Delta x_{n+1}^*.$$

Now let $C_n = G_n^* G_n$:

$$C_{n+1} = \begin{bmatrix} C_n & G_n^* \Delta J_{n+1}' \\ \Delta J_{n+1}'^* G_n & \Delta J_{n+1}'^* \Delta J_{n+1}' \end{bmatrix}$$

Again using Exercise 4.20,

$$C_{n+1}^{-1} = \begin{bmatrix} C_n^{-1} & 0 \\ 0 & 0 \end{bmatrix} + \begin{bmatrix} dC_n^{-1} uv^* C_n^{-1} & w \\ z^* & d \end{bmatrix},$$

588 **Extremum Problems**

$$v^* = \Delta J'^*_{n+1} G_n, \quad u = G_n^* \Delta J'_{n+1}, \quad z^* = -d \Delta J'_{n+1} G_n C_n^{-1}, \quad w = -d C_n^{-1} G_n^* \Delta J'_{n+1},$$

$$1/d = \Delta J'^*_{n+1} \Delta J'_{n+1} - \Delta J'^*_{n+1} G_n C_n^{-1} G_n^* \Delta J'_{n+1}$$

$$= \langle \Delta J'_{n+1}, (I - G_n(G_n^* G_n)^{-1} G_n^*) \Delta J'_{n+1} \rangle.$$

Since H_n is given by (12.3.14) and $\delta X_n^* \Delta J'_{n+1} = 0$, this becomes

$$1/d = \langle \Delta J'_{n+1}, H_n \Delta J'_{n+1} \rangle. \quad]$$

12.10 Weierstrass in 1895 gave the following example to show that a minimizing sequence need not have a limit. Consider the functional

$$J(x) = \int_{-1}^{1} t^2 (dx/dt)^2 \, dt$$

defined on the set X of continuously differentiable functions $x = x(t)$ such that $x(-1) = -1$ and $x(1) = 1$. Clearly, $J(x) \geq 0$. In fact, $\inf_{x \in X} J(x) = 0$. One can see this graphically, but prove that the following sequence is minimizing:

$$x_n(t) = \frac{\arctan nt}{\arctan n}, \quad n = 1, 2, \ldots.$$

[*Hint:*

$$J(x_n) = \int_{-1}^{1} t^2 (x_n')^2 \, dt < \int_{-1}^{1} (t^2 + 1/n^2)(x_n')^2 \, dt.$$

The latter integral is equal to

$$\frac{1/n^2}{(\arctan n)^2} \int_{-1}^{1} \frac{dt}{t^2 + 1/n^2} = \frac{2}{n \arctan n}. \quad]$$

However, there exists no $u \in X$ such that $J(u) = 0$. Prove this. [*Hint:* $J(u) = 0$ implies $u'(t) = 0$ for all t.] Finally, show that J is continuous in the topology

$$\|x\| = \sup_{1 \leq t \leq 1} |x'(t)|.$$

12.11 Prove Theorem 12.2.6 when f' and f'' are replaced by the weak derivatives f'_w and f''_w respectively. [*Hint:* Derive the analog of formula (5.3.25),

$$f(x + sh) = f(x) + sf'_w(x)h + \frac{s^2}{2} f''_w(x)h^2 + \varepsilon,$$

where $\varepsilon/s^2 \to 0$ as $s \to 0$. Use the proof of Theorem 5.3.5, with h replaced by sh.]

12.12 Let f be a real functional on a Hilbert space X with gradient $\nabla f(x)$. Suppose A is a bounded self-adjoint operator on X such that $\langle Ax, x \rangle \geq \mu \|h\|^2$, $\mu > 0$. We consider the A-inner product $\langle x, y \rangle_A = \langle x, A, y \rangle$. The *direction of steepest descent of f with respect to the A-norm*, $\|x\|_A^2 = \langle x, Ax \rangle$, is along a vector u such that $-\langle \nabla f(x), u/\|u\|_A \rangle$ is a minimum of $-\langle \nabla f(x), y \rangle$ on the sphere $\|y\|_A = 1$. Show that $u = -A^{-1} \nabla f(x)$ is the direction of steepest descent in the A-norm. [*Hint:* For any y with $\|y\|_A = 1$,

$$\langle \nabla f(x), y \rangle = \langle \nabla f(x), A^{-1} Ay \rangle = \langle A^{-1} \nabla f(x), Ay \rangle = \langle A^{-1} \nabla f(x), y \rangle_A.$$

By the Schwarz inequality,
$$|\langle \nabla f(x), y\rangle| \leq \|A^{-1}\nabla f(x)\|_A = |\langle \nabla f(x), A^{-1}\nabla f(x)\rangle|. \ ]$$

12.13 Let A be a bounded self-adjoint operator on a real Hilbert space. The function
$$R(x) = \frac{\langle Ax, x\rangle}{\langle x, x\rangle}$$
is called the *Rayleigh quotient*. Prove that u is an eigenvector of A belonging to the eigenvalue λ if and only if $\nabla R(u) = 0$ (that is, u is a stationary point of R) and $\lambda = \langle Au, u\rangle/\langle u, u\rangle$. [*Hint:* Verify that $\nabla R(x) = 2(A - R(x)I)x$.]

12.14 (Duality in Linear Programming) Consider the Standard linear programming problem: Minimize $J(x) = c^T x$ subject to the constraints $Ax = b$ and $x \geq 0$. The *dual problem* is to maximize $K(u) = u^T b$ subject to the constraints $A^T u \leq c$, where the *dual vector* u is unconstrained in sign. (As usual, the superscript T denotes the transpose.) The given standard problem is called the *primal problem*.

a) Let x be a feasible primal solution (that is, $Ax = b$ and $x \geq 0$). Let u be a feasible dual solution ($A^T u \leq c$). Prove that $J(x) \geq K(u)$. [*Hint:* $c^T x \geq u^T Ax$.]

b) Prove that if both the primal and dual problems have feasible solutions, then both have optimal solutions and
$$\min_x J(x) = \max_u K(u).$$

[*Hint:* Both J and K have finite bounds by part (a). Hence there exists a basic optimal solution x^* (Fundamental Theorem). Let $x_B \geq 0$ be the basic part of x^* and A_B the corresponding submatrix; that is, $A_B x_B = b$. Let c_B be the corresponding cost vector, so that $u_B^T = c_B^T A_B^{-1}$ is the multiplier vector. By optimality of x_B, $c_j^{(1)} = c_j - u_B^T A_j \geq$ for all j. Thus u_B is a feasible dual solution and
$$K(u_B) = \min_x J(x).$$
Then use part (a).]

APPENDIX I

PROOF OF CONVERGENCE OF GALERKIN'S METHOD

We shall prove Theorem 11.5.1. Since $L = A + K$, the equation $Lx = y$ can be replaced by $Ax + Kx = y$ and the latter by $x + Tx = z$, where $z = A^{-1}y$ and $T = A^{-1}K$. (There is a fine technical point concerning the fact that $D_A \subset H_A$, which we bypass by considering any solution $x \in H_A$ to be a generalized solution of $Lx = y$.) In Appendix V, it is shown that a completely continuous operator T has isolated eigenvalues, except possibly for 0. Furthermore, any $\lambda \neq 0$ must either be in the point spectrum or in the resolvent set (Theorem V.5). Thus either $x + Tx = 0$ has a nontrivial solution (i.e., an eigenvector belonging to -1) or $(I + T)^{-1}$ exists and is bounded (i.e., -1 is in the resolvent set) and $x + Tx = z$ has a unique solution for all z. This last statement is known as the *Fredholm alternative* and applies to the general case $x - \lambda Tx = z$.

The proof of Theorem 11.5.1 hinges on the following important property of completely continuous linear operators (Definition V.2).

Lemma I.1 Let $T \in L_c(X, X)$ be completely continuous and assume that X is a Banach space which has a basis $\{v_i : i = 1, 2, \ldots\}$. Then for any $\varepsilon > 0$ there exists an n and operators T_1, T_2 such that $T = T_1 + T_2$, $T_1 x \in V^n$ (the finite-dimensional subspace spanned by $\{v_1, \ldots, v_n\}$) and $\|T_2\| \leq \varepsilon$.

Proof. By Definition V.2, if S is the unit sphere, its image $T(S)$ has compact closure, E. By the results on compact subsets of a complete metric space (Section 1.4), E must be totally bounded; i.e., there exists a finite set $\{x_1, \ldots, x_N\} \subset E$ such that any $x \in E$ is within a specified distance ε' of some x_i. Let P_n be the projection onto V^n; that is, for $x = \sum_1^\infty \xi_i v_i$, $P_n x = \sum_1^n \xi_i v_i$. To show that P_n is bounded, consider the linear operator A which maps X onto the space X' of all sequences (ξ_i) such that $\sum_1^\infty \xi_i v_i$ is convergent. In X' we define the norm of a sequence to be the quantity $\sup_n \|\sum_1^n \xi_i v_i\|$. It is not difficult to show that X' is a Banach space. (See [3.3], Section 28.) The operator A is obviously linear and bijective. Further,

$$\|A^{-1}(\xi_i)\| = \|\sum_1^n \xi_i v_i\| \leq \sup_n \|\sum_1^n \xi_i v_i\| = \|(\xi_i)\|.$$

Hence A^{-1} is continuous. By Theorem 3.5.0, A is also continuous. Now,

$$\|P_n x\| \leq \sup_n \|\sum_1^n \xi_i v_i\| = \|Ax\| \leq \|A\| \|x\|.$$

Therefore $\|P_n\| \leq \|A\|$.

Let $R_n = I - P_n$. We assert that there exists n_0 such that $\|R_n y\| < \varepsilon$ for all $y \in E$ and $n \geq n_0$. In proof,

$$\|R_n y\| = \|y - P_n x\| \leq \|y - x_i\| + \|x_i - P_n y\|$$
$$\leq \varepsilon' + \|P_n(x_i - y)\| + \|R_n x_i\| \leq (1 + \|A\|)\varepsilon' + \|R_n x_i\|.$$

590

Choose n_0 so that $\|R_n x_i\| < \varepsilon'$, for $n \geq n_0$ and $i = 1, \ldots, N$. Then $\|R_n y\| \leq \varepsilon$, where $\varepsilon = \varepsilon'/(2 + \|A\|)$. Now $Tx = P_n Tx + R_n Tx$. Define $T_1 = P_n T$ and $T_2 = R_n T$. For $x \in S$, $\|T_2 x\| = \|R_n Tx\| = \|R_n y\| \leq \varepsilon$, which implies $\|T_2\| \leq \varepsilon$. This completes the proof of the lemma. ∎

We return to the equation $x + Tx = z$, where T is completely continuous on the Hilbert space H_A. The Galerkin iteration x_n must satisfy (11.5.1); that is,

$$\langle Lx_n - y, v_i \rangle = 0, \quad 1 \leq i \leq n,$$

where $\{v_i\}$ is a basis for H_A. We suppose that $\{v_i\}$ has been orthonormalized. Since $L = A + K = A(I + T)$, the Galerkin equations become $\langle A[(I + T)x_n - z], v_i \rangle = 0$, $1 \leq i \leq n$, or $\langle (I + T)x_n - z, v_i \rangle_A = 0$. These conditions are satisfied if and only if $P_n(I + T)x_n - P_n z = 0$ or, since $x_n \in V^n$,

$$x_n + T_1^{(n)} x_n = z_n, \tag{I.1}$$

where $z_n = P_n z$ and $T_1^{(n)} = P_n T$. By the preceding lemma, $T - T_1^{(n)} = T_2^{(n)}$ and $\|T_2^{(n)}\| \to 0$ as $n \to \infty$. Since the hypothesis of the theorem implies that $(I + T)^{-1}$ exists, it follows that $(I + T_1^{(n)})^{-1}$ exists for n sufficiently large. Hence, (I.1) has a unique solution $x_n = (I + T_1^{(n)})^{-1} z_n$. Since $z_n \to z$ and $(I + T_1^{(n)})^{-1} \to (I + T)^{-1}$ as $n \to \infty$, it follows that $x_n \to x = (I + T)^{-1} z$, which is equivalent to $Lx = y$. ∎

APPENDIX II

THE HAHN–BANACH THEOREM

Let X be a normed vector space. For any $x_0 \in X$ there exists a bounded linear functional F on X such that $\|F\| = 1$ and $F(x_0) = \|x_0\|$.

The proof will be given in the form of three lemmas.

Lemma II.1 Let X be a real vector space and V a proper subspace of X. Let p be a seminorm on X and f a linear functional on V such that $f(x) \leq p(x)$ for all $x \in V$. Then there exists a linear functional F defined on all of X such that F is an extension of f and $F(x) \leq p(x)$ for all $x \in X$.

Proof. Choose an arbitrary vector $x_0 \in X - V$. Let V_0 be the subspace generated by $V \cup \{x_0\}$; i.e., V_0 consists of all elements of the form $x + \alpha x_0$, where $x \in V$. We extend f to V_0 by defining a linear functional g_0 such that $g_0(x + \alpha x_0) = f(x) + \alpha \beta_0$, where β_0 is a real number chosen so that $f(x) + \alpha \beta_0 \leq p(x + \alpha x_0)$. To show that such a β_0 exists, we observe that β_0 must satisfy the inequality with $\alpha = 1$. Hence, we must have

$$\beta_0 \leq p(x + x_0) - f(x) \tag{1}$$

for all $x \in V$. Similarly, for $-x \in V$ and $\alpha = -1$, we must have

$$-p(-x - x_0) - f(x) \leq \beta_0. \tag{2}$$

Conversely, if β_0 satisfies both of these inequalities for all $x \in V$, then taking $\alpha > 0$ and replacing x by x/α in (1), we get

$$\beta_0 \leq p(x/\alpha + x_0) - f(x/\alpha) = (1/\alpha)p(x + \alpha x_0) - (1/\alpha)f(x)$$

or

$$f(x) + \alpha \beta_0 \leq p(x + \alpha x_0).$$

If $\alpha < 0$, we replace x by x/α in (2) and obtain

$$\beta_0 \geq -p(-x/\alpha - x_0) - f(x/\alpha) = (1/\alpha)p(x + \alpha x_0) - (1/\alpha)f(x).$$

Multiplying by $\alpha < 0$, we get

$$f(x) + \alpha \beta_0 \leq p(x + \alpha x_0).$$

To see that there exists β_0 satisfying (1) and (2), consider any two vectors $y, z \in V$. Then

$$f(y) - f(z) = f(y - z) \leq p(y - z)$$
$$= p(y + x_0 - z - x_0) \leq p(y + x_0) + p(-z - x_0).$$

Hence,

$$-p(-z - x_0) - f(z) \leq p(y + x_0) - f(y)$$

and it follows that

$$a = \sup_{x \in V} \{-p(-x - x_0) - f(x)\} \leq \inf_{x \in V} \{p(x + x_0) - f(x)\} = b.$$

Therefore there exists $a \leq \beta_0 \leq b$ and we have extended f to V_0 in the desired manner.

Now, consider the set E of all linear functionals g which are proper extensions of f to a subspace containing V and satisfy $g(x) \leq p(x)$ for all x in the domain of g. Note that $g_0 \in E$. Let E be partially ordered by defining $g_1 \leq g_2$ to hold if g_2 is an extension of g_1; hence, the domains are such that $D(g_1) \subset D(g_2)$. If $G \subset E$ is a totally ordered subset, let $D = \bigcup_{g \in G} D(g)$. Then D is a subspace. We associate with it the functional g^* defined by $g^*(x) = g(x)$, where $g \in G$ is such that $x \in D(g)$. Clearly, g^* is a well-defined linear functional and $g^* \in G$. Further, $g \leq g^*$ for all $g \in G$; i.e., g^* is an upper bound of G. By Zorn's lemma, E must have a maximal element F. But then the domain $D(F) = X$, since otherwise we can extend it properly as we did, contradicting the maximality of F. This proves the lemma. ∎

Lemma II.2 Let p be a seminorm on the real vector space X. Let $V \subset X$ be a subspace and f a linear functional on V such that $|f(x)| \leq p(x)$ for all $x \in V$. Then there exists a linear functional F on X which is an extension of f and $|F(x)| \leq p(x)$ for all $x \in X$.

Proof. By Lemma 1, there exists F such that $F(x) \leq p(x)$ for all $x \in X$. But then $-F(x) = F(-x) \leq p(-x) = p(x)$, so that $F(x) \geq -p(x)$. Hence $|F(x)| \leq p(x)$. ∎

Lemma II.3 Let V be a proper subspace of a normed vector space X. If $f \neq 0$ is a bounded linear functional on V, then there is a bounded linear functional F on X such that $\|F\| = \|f\|$ and $F(x) = f(x)$ for all $x \in V$; that is, F is an extension of f having the same norm.

Proof. Define $p(x) = \|f\| \|x\|$, $x \in X$. Clearly, p is a norm and $|f(x)| \leq p(x)$ for all $x \in V$. By Lemma 2, there exists an extension F of f such that $|F(x)| \leq \|f\| \|x\|$. Hence $\|F\| \leq \|f\|$. But since F is an extension of f,

$$\|F\| = \sup_{\substack{x \in X \\ x \neq 0}} \frac{|F(x)|}{\|x\|} \geq \sup_{x \in V} \frac{|f(x)|}{\|x\|} = \|f\|.$$

Thus $\|F\| = \|f\|$, as we wished to show. ∎

(Lemma II.3 is sometimes also called the Hahn–Banach theorem.) To prove the Hahn–Banach theorem as stated above we proceed as follows.

Proof of Theorem. Let V be the subspace generated by x_0. For $x = \alpha x_0$ define $f(x) = \alpha \|x_0\|$. Clearly, f is a bounded linear functional on V and $\|f\| = 1$. By Lemma 3, there exists an extension F of f to the whole space X such that $\|F\| = \|f\| = 1$, which proves the theorem. ∎

As an application, we shall prove that $\|T^*\| = \|T\|$ for any $T \in L_c(X; Y)$, X and Y normed vector spaces. We have already shown (Section 3.5, The Adjoint of a Continuous Linear Operator) that $\|T^*\| \leq \|T\|$. We shall now prove that $\|T\| \leq \|T^*\|$.

For any $x^* \in X'$ with $\|x^*\| = 1$ we have $|x^*(x)| \leq \|x^*\| \|x\| = \|x\|$. Hence $\sup_{\|x^*\|=1} |x^*(x)| \leq \|x\|$. By the Hahn–Banach theorem, there exists $x^* \in X'$ such that $\|x^*\| = 1$ and $x^*(x) = \|x\|$. Therefore $\|x\| = \sup_{\|x^*\|=1} |x^*(x)|$. It follows that $\|Tx\| = \sup \{|x^*(Tx)| : \|x^*\| = 1\}$. Now, writing $\langle Tx, x^* \rangle$ for $x^*(Tx)$, we have for $\|x^*\| = 1$,

$$|\langle Tx, x^* \rangle| = |\langle x, T^*x^* \rangle| \leq \|T^*x^*\| \|x\| \leq \|T^*\| \|x\|.$$

Hence $\|Tx\| = \sup |\langle Tx, x^* \rangle| \leq \|T^*\| \|x\|$ and $\|T\| \leq \|T^*\|$.

APPENDIX III

THE STONE–WEIERSTRASS THEOREM

As in Section 7.10, $C[X]$ denotes the algebra of real continuous functions on the set X, and I is the identity element, that is, I is the function $I(x) = 1$ for all $x \in X$. The topology of $C[X]$ is that of the uniform norm, $\|f\|_\infty = \sup_{x \in X}\{|f(x)|\}$. A set of functions A is said to separate points of X if for any two distinct points $x, y \in X$, there exists $f \in A$ such that $f(x) \neq f(y)$.

Theorem Let X be a compact metric space. If A is a subalgebra of $C[X]$ which contains the identity I and separates points of X, then A is dense in $C[X]$; i.e. $C[X] = \bar{A}$.

Proof. For any $f \in C[X]$, let $|f|$ be the function having values $|f(x)|$. Now, we assert that if $f \in \bar{A}$, then $|f| \in \bar{A}$. We must show that $|f|$ is an accumulation point of A. Consider the real function $|\xi|$ of the real variable ξ. Since $|\xi|$ is continuous for $-\|f\|_\infty \leq \xi \leq \|f\|_\infty$, there is a sequence of polynomials $p_n(\xi)$ which converges uniformly to $|\xi|$ on the interval $[-\|f\|_\infty, \|f\|_\infty]$. Since $C[X]$ is an algebra over the reals, the element $p_n(f) = a_0 + a_1 f + \cdots + a_m f^m$ is a function in $C[X]$. In fact, since $\bar{A}$ is a subalgebra and $f \in \bar{A}$, it follows that $p_n(f) \in \bar{A}$. (The verification that $\bar{A}$ is a subalgebra is a simple exercise.)

Since $p_n(\xi) \to |\xi|$, we see that $p_n(f)(\xi) = a_0 + a_1 f(\xi) + \cdots + a_m(f(\xi))^m$ converges uniformly to $|f(\xi)|$ for all $f(\xi)$ in $[-\|f\|_\infty, \|f\|_\infty]$. Hence $p_n(f) \to |f|$ in the uniform norm. For $\varepsilon > 0$ choose n sufficiently large that $\|p_n(f) - |f|\|_\infty < \varepsilon/2$. Fix n and find $g \in A$ such that $\|p_n(f) - g\|_\infty < \varepsilon/2$, since $p_n(f) \in \bar{A}$. Then $\||f| - g\|_\infty < \varepsilon$, showing that every neighborhood of $|f|$ contains a point of A. Therefore $|f| \in \bar{A}$. As a consequence,

$$M(x) = \max\{f(x), g(x)\} = (f(x) + g(x) + |f(x) - g(x)|)/2 \tag{1}$$

is in $\bar{A}$ whenever $f, g \in \bar{A}$ and similarly for

$$\min\{f(x), g(x)\} = (f(x) + g(x) - |f(x) - g(x)|)/2. \tag{2}$$

From the hypothesis that A separates points it follows that for any $x, y \in X$ and real numbers α, β there exists $f_{xy} \in A$ such that $f_{xy}(x) = \alpha$ and $f_{xy}(y) = \beta$. To construct f_{xy}, find $g \in A$ such that $g(x) \neq g(y)$. Then take

$$f_{xy} = \frac{\alpha - \beta}{g(x) - g(y)} g + \frac{\beta g(x) - \alpha g(y)}{g(x) - g(y)} I. \tag{3}$$

Now let $f \in C[X]$ be arbitrary. Consider a ball B of radius $\varepsilon > 0$ centered at f. We shall show that there exists $g \in A \cap B$. For each pair of points $a, b \in X$ there exists $f_{ab} \in A$ such that $f_{ab}(a) = f(a)$ and $f_{ab}(b) = f(b)$, by (3) above. Consider the sets O_{ab} consisting of all points $x \in X$ such that $f_{ab}(x) < f(x) + \varepsilon$. O_{ab} is open and contains the point a. Therefore, the family $(O_{ab})_{a \in X}$ is an open covering of X. Since X is compact, there is a finite subcovering $\{O_{a_1 b}, \ldots, O_{a_n b}\}$. Set $f_b = \min\{f_{a_1 b}, \ldots, f_{a_n b}\}$. By (2) above, $f_b \in A$. For any $x \in X$ there exists a_i such that $x \in O_{a_i b}$, so that

$$f_b(x) \leq f_{a_i b}(x) < f(x) + \varepsilon. \tag{4}$$

Let $O_b = \{x : f_b(x) > f(x) - \varepsilon\}$. Since $f_{ab}(b) = f(b) > f(b) - \varepsilon$ for all $a \in X$, we have $f_b(b) = f_{a_jb}(b) > f(b) - \varepsilon$. Hence $b \in O_b$. Again, the family $(O_b)_{b \in X}$ is an open covering. Let $b_1, \ldots, b_m$ be the points giving a finite subcovering and define $g = \max \{f_{b_i}\} = f_{b_j}$. By (1) above, $g \in A$. By (4) above, $g(x) = f_{b_j}(x) < f(x) + \varepsilon$ for all $x \in X$. But for any x, x is in some O_{b_i} and therefore $g(x) \geq f_{b_i}(x) > f(x) - \varepsilon$. This proves that $\|g - f\|_\infty < \varepsilon$, that is, $g \in A \cap B$. Therefore $f \in \bar{A}$. This completes the proof. ∎

APPENDIX IV

THE IMPLICIT-FUNCTION THEOREM

Consider a system of n equations in $n + m$ variables,

$$\left.\begin{array}{c} f_1(x_1, \ldots, x_m, y_1, \ldots, y_n) = 0, \\ \vdots \\ f_n(x_1, \ldots, x_m, y_1, \ldots, y_n) = 0. \end{array}\right\} \quad (1)$$

If for all $x = (x_1, \ldots, x_m)$ in a certain region, we can solve for $y = (y_1, \ldots, y_n)$, then we may regard the y_i as functions of $x_1, \ldots, x_m$; that is, $y_i = g_i(x_1, \ldots, x_m)$. In vector form, letting $f = (f_1, \ldots, f_n)$ and $g = (g_1, \ldots, g_n)$, we may write (1) as $f(x, y) = 0$. Then $f(x, g(x)) = 0$ and we say that g is defined implicitly by (1). The implicit-function theorem gives sufficient conditions for the existence of g. We shall prove it in the general case where x and y are vectors in arbitrary normed vector spaces X, Y respectively; that is, X, Y need not be finite dimensional. f is a mapping from some subset of the direct sum $X \oplus Y$ (see Exercise 2.11) into a normed space V. We shall require f to have a strong partial derivative f_y (Definition 5.3.6).

Theorem Let $f: X \oplus Y \to V$, where X, Y, V are normed spaces, with Y complete. Suppose that (i) $f(x_0, y_0) = 0$, (ii) $f_y(x, y)$ exists and is a continuous function of (x, y) in a neighborhood of (x_0, y_0) and (iii) $(f_y(x_0, y_0))^{-1}$ exists as a bounded operator. Then there exists a ball $B = B(x_0; r)$, $r > 0$, and a continuous function g defined for all $x \in B$ such that $f(x, g(x)) = 0$. Furthermore, $g(x_0) = y_0$.

Proof. Let $x = x_0 + h$, $y = y_0 + k$. For a given h with $\|h\| < r$ we shall find a unique k such that $f(x_0 + h, y_0 + k) = 0$. This will define $g(x)$. Since h is fixed, we are seeking a zero of the function $f(x_0 + h, y_0 + k)$. We shall apply the theory of Chapter 5, in particular, the fixed-point theorem (5.1.1) based on successive approximation.

We introduce the function φ defined, for $\|h\|$ sufficiently small, by a Newton method scheme as

$$\varphi(h, k) = k - (f_y(x_0 + h, y_0))^{-1} f(x_0 + h, y_0 + k). \quad (2)$$

For h fixed, a fixed point, k, of φ is a zero of $f(x_0 + h, y_0 + k)$ and conversely. We shall show that φ is a contractive mapping when h and k are suitably restricted. Note that $(f_y(x_0 + h, y_0))^{-1}$ exists for $\|h\|$ sufficiently small (by Theorem 3.5.2, Corollary 4).

From (2) above it follows by direct differentiation with respect to k, that for $\|h\| + \|k\|$ sufficiently small,

$$\begin{aligned}\varphi_y(h, k) &= I - (f_y(x_0 + h, y_0))^{-1} f_y(x_0 + h, y_0 + k). \\ &= (f_y(x_0 + h, y_0))^{-1}(f_y(x_0 + h, y_0) - f_y(x_0 + h, y_0 + k)). \end{aligned} \quad (3)$$

By the continuity of f_y, for any $\varepsilon > 0$ there exists $\delta > 0$ such that for $\|h\| < \delta$, $\|k\| < \delta$

$$\|f_y(x_0 + h, y_0) - f_y(x_0 + h, y_0 + k)\| < \varepsilon.$$

Furthermore, δ can be chosen small enough that $(f_y(x_0 + h, y_0))^{-1}$ exists. For such δ and $\|h\| < \delta$, $\|(f_y(x_0 + h, y_0))^{-1}\| < C$, where C is a constant. (Again, see Theorem 3.5.2, Corollary 4.) Thus, C may be fixed in advance and we may choose the above ε so small that $C\varepsilon < \frac{1}{2}$.

With $\|h\| < \delta$ and $\|k\| < \delta$, from (3) it follows that $\|\varphi_y(h, k)\| < \frac{1}{2}$. Therefore, by the mean-value Theorem (5.3.3), for any $\|h\| < \delta$, $\|k_1\| < \delta$, $\|k_2\| = \delta$,

$$\|\varphi(h, k_1) - \varphi(h, k_2)\| \leq \tfrac{1}{2}\|k_1 - k_2\|.$$

Thus, $\varphi(h, k)$ is contractive on the ball $B(0; \delta) \subset Y$ as a function of k for any fixed $\|h\| < \delta$.

Now, observe that $\varphi(h, 0) = -(f_y(x_0 + h, y_0))^{-1} f(x_0 + h, y_0)$. Since f and $(f_y(x_0 + h, y_0))^{-1}$ are continuous in h, we have as $\|h\| \to 0$, $\varphi(h, 0) \to -(f_y(x_0, y_0))^{-1} f(x_0, y_0) = 0$. Hence, for any $\rho \leq \delta$ there exists $r \leq \delta$ such that $\|\varphi(h, 0)\| \leq \rho/2$ for $\|h\| \leq r$. It follows that $\|\varphi(h, k)\| \leq \|\varphi(h, k) - \varphi(h, 0)\| + \|\varphi(h, 0)\| \leq \tfrac{1}{2}\|k\| + \rho/2 \leq \rho$ whenever $\|h\| < r$ and $\|k\| \leq \rho$. We see that for any $\|h^*\| < r$, $\varphi(h^*, k)$ maps the closed ball $\|k\| \leq \rho$ into itself. Since this closed ball is a complete metric space and φ is contractive on it, Theorem 5.1.1 yields a unique fixed point, k^*; i.e. $\varphi(h^*, k^*) = k^*$, which by (2) implies $f(x_0 + h^*, y_0 + k^*) = 0$. We define $g(x_0 + h^*) = y_0 + k^*$ to obtain the function g of the theorem. Since k^* may be obtained by successive approximation with any starting value k_0 in the ball $\|k\| \leq \rho$, we shall take $k_0 = 0$ for all h. Thus, for each $\|h\| < r$, we obtain the sequence $k_n(h) \to k(h)$, where $k(h)$ is the fixed point for a given h. Since $k_n(h) = \varphi(h, k_{n-1}(h))$, we see by induction that all $k_n(h_1) \to k_n(h_2)$ uniformly as $h_1 \to h_2$. Hence, $k(h)$ is continuous in h. This completes the proof. ∎

Corollary 1 For any $\|h^*\| \leq r$ and $\|k^*\| \leq \rho$, if $f(x_0 + h^*, y_0 + k^*) = 0$, then $y_0 + k^* = g(x_0 + h^*)$.

Proof. Considering h^* fixed, k^* is a zero of $f(x_0 + h^*, y_0 + k)$ and, by (2) in the proof of the theorem, k^* must be the unique fixed point of $\varphi(h^*, k)$. ∎

Corollary 2 Let f be as in the statement of the theorem and assume further that the partial derivative $f_x(x, y)$ exists and is continuous for $\|x - x_0\| \leq r$ and $\|y - y_0\| \leq \rho$. Then $g'(x)$ exists for $\|x - x_0\| \leq r$.

Proof. We shall show that $k(h) = g(x_0 + h) - y_0$ is differentiable with respect to h at $h = 0$. Using the notation in the proof of the theorem, we have $k(h) = \lim_{n \to \infty} k_n(h)$, where $k_n(h) = \varphi(h, k_{n-1}(h))$. Since $k(h) = \varphi(h, k(h))$ and $k_0(h) = 0$,

$$\|k(h) - k_1(h)\| = \|\varphi(h, k(h)) - \varphi(h, k_0(h))\|$$
$$\leq C\varepsilon \|k(h) - k_0(h)\| = C\varepsilon \|k(h)\|. \tag{4}$$

Applying formula (5.1.2) to the present case (with $m = 1$ and $L = \tfrac{1}{2}$), we see that $\|k(h)\| \leq 2\|k_1(h)\|$. Now, by the continuity of f_x, we may assume that $\|f_x(x_0 + h, y_0)\| \leq d$ for $\|h\| \leq r$; i.e. we suppose that r was chosen sufficiently small to bound f_x. Since $f(x_0, y_0) = 0$, we have

$$\|f(x_0 + h, y_0)\| = \|f(x_0 + h, y_0) - f(x_0, y_0)\| \leq d\|h\|.$$

Hence

$$\|k_1(h)\| \leq \|(f_y(x_0 + h, y_0))^{-1}\| \|f(x_0 + h, y_0)\| \leq Cd\|h\|,$$

so that

$$\|k(h)\| \leq 2Cd\|h\|. \tag{5}$$

From (2) in the proof of the theorem, we see that

$$k_1(h) = -(f_y(x_0 + h, y_0))^{-1} f(x_0 + h, y_0 + k)$$
$$= (-f_y(x_0, y_0))^{-1} f_x(x_0, y_0)h + \varepsilon_h,$$

where $\|\varepsilon_h\|/\|h\| \to 0$ as $h \to 0$. (We have used the continuity of $f_y(x, y)$ at (x_0, y_0).) Using (4) and (5) above, we obtain

$$\|k(h) - (-f_y(x_0, y_0))^{-1} f_x(x_0, y_0)h\| \leq \|\varepsilon_h\| + 2C^2 d\varepsilon \|h\|. \tag{6}$$

Since $\varepsilon = \|f_y(x_0 + h, y_0) - f_y(x_0 + h, y_0 + k(h))\|$, $\varepsilon \to 0$ as $h \to 0$. Thus (6) shows that $k'(0)$ exists. To complete the proof, we note that for any $x \in B(x_0, r)$ we have the same properties as those used at x_0. Hence $g'(x) = -(f_y(x, y))^{-1} f_x(x, y)$, where $y = g(x)$. ∎

APPENDIX V

RESULTS ON THE SPECTRUM OF LINEAR OPERATORS

Let $T: X \to X$, where X is a normed vector space and T is a linear operator defined on the domain $D \subset X$. The spectrum $\sigma(T)$ is described in Definition 3.6.1 in terms of $T_\lambda = \lambda I - T$. If T_λ has a bounded inverse, then either $\lambda \in \rho(T)$ or $\lambda \in R_\sigma$. However, $\lambda \in R_\sigma$ does not imply that T_λ has a bounded inverse. Thus, if T_λ does not have a bounded inverse, then $\lambda \in \sigma(T)$ but we cannot say in which part of $\sigma(T)$ λ lies. The following condition is useful in this connection.

Definition V.1 We call λ an *approximate eigenvalue* of T if for any $\varepsilon > 0$ there exists $x \in D$ with $\|x\| = 1$ and such that $\|(\lambda I - T)x\| < \varepsilon$. We denote the set of approximate eigenvalues by $\pi(T)$.

Obviously, an eigenvalue is also an approximate eigenvalue.

Theorem V.1 $\lambda \in \pi(T)$ if and only if T_λ does not have a bounded inverse. Hence $\pi(T) \subset \sigma(T)$.

Proof. Suppose T_λ does not have a bounded inverse. By Lemma 3.5.1 (with $X = D$ and Y the range of T), there is no constant $m > 0$ such that $\|T_\lambda x\| \geq m\|x\|$. Hence, for any $\varepsilon > 0$ there exists x such that $\|T_\lambda x\| < \varepsilon \|x\|$; that is, $\lambda \in \pi(T)$.

Conversely, if $\lambda \in \pi(T)$, there exist $x_n \in D$ of unit norm such that $\|T_\lambda x_n\| < 1/n$ for $n = 1, 2, \ldots$. By Lemma 3.5.1, T_λ cannot have a bounded inverse. This completes the proof. ∎

Corollary $C_\sigma(T) \subset \pi(T)$.

Theorem V.2 Let X be a Hilbert space and $T \in L_c(X; X)$ a normal operator. Then $\pi(T) = \sigma(T)$.

Proof. It suffices to prove that $\sigma(T) \subset \pi(T)$, or equivalently, that $\lambda \notin \pi(T)$ implies $\lambda \in \rho(T)$. Thus suppose $\lambda \notin \pi(T)$. Then there exists $\varepsilon > 0$ such that $\|T_\lambda x\| \geq \varepsilon \|x\|$ for all $x \neq 0$. Hence T_λ has a bounded inverse. It remains to prove that $R(T_\lambda)$, the range, is dense in X. Let Y be the closure of $R(T_\lambda)$. If $y \in Y^\perp$, then $0 = \langle y, T_\lambda x \rangle = \langle T_\lambda^* y, x \rangle$ for all $x \in X$. Hence $T_\lambda^* y = 0$. But T_λ is normal and therefore $\|T_\lambda^* y\| = \|T_\lambda y\| \geq \varepsilon \|y\|$ (by the proof of Theorem 3.6.10). Hence $y = 0$; that is, $Y^\perp = \{0\}$ and $Y = X$, as we wished to show. ∎

Corollary 1 The residual spectrum of a normal transformation is empty.

Proof. Suppose $R(T_\lambda)$ is not dense in X. Then $Y^\perp \neq \{0\}$ and there exists $y \neq 0$ such that $0 = \langle y, T_\lambda x \rangle = \langle T_\lambda^* y, x \rangle$ for all x. Hence $T_\lambda^* y = 0$. Since $\|T_\lambda^* y\| = \|T_\lambda y\|$, $T_\lambda y = 0$. Therefore, λ is an eigenvalue of T; i.e., it is not in the residual spectrum. ∎

Corollary 2 If T is a bounded self-adjoint transformation, the residual spectrum is empty.

Definition V.2 Let X, Y be normed vector spaces and $T \in L(X; Y)$. We say that T is *completely continuous* if the image under T of a bounded set has a closure which is compact; that is, $\overline{T(S)}$ is compact for any bounded set $S \subset X$.

Obviously, a completely continuous T is bounded, since the compact set $\overline{T(S)}$ is bounded. Hence T is certainly continuous. The identity I on an infinite-dimensional space X is not completely continuous, since the image of the unit ball is not compact.

Theorem V.3 Let $T : X \to X$ be a completely continuous linear operator. The point spectrum of T is at most denumerable and, if infinite, then 0 is its only accumulation point.

Proof. Let $S_n = \{\lambda \in P_\sigma(T) : |\lambda| \geq 1/n\}$. We shall show that S_n is finite for every $n = 1, 2, \ldots$. Suppose not. Then for some n there are an infinite number of distinct eigenvalues $\lambda_1, \lambda_2, \ldots, \lambda_k, \ldots \in S_n$. Let $u_1, u_2, \ldots$ be corresponding eigenvectors. Since the u_i belong to distinct eigenvalues, they are linearly independent. Let V_k be the subspace spanned by $\{u_1, \ldots, u_k\}$. By the proof of Theorem 3.2.4, there exists $x_k \in V_k$ such that $\|x_k - x\| \geq \frac{1}{2}$ for all $x \in V_{k-1}$ and $\|x_k\| = 1$. The set $\{x_k : k = 1, 2, \ldots\}$ is bounded. We shall show that $\{Tx_k\}$ cannot have a convergent subsequence, hence cannot have a compact closure. This contradicts the complete continuity.

Take any $x = \sum_1^k a_i u_i \in V_k$. Then

$$\lambda_k x - Tx = \sum_1^{k-1} a_i(\lambda_k - \lambda_i)u_i \in V_{k-1}.$$

Hence $v = (\lambda_k I - T)x_k + Tx_j \in V_{k-1}$ for $1 \leq j \leq k - 1$.

But then for $j < k$,

$$\|Tx_k - Tx_j\| = \|\lambda_k x_k - v\| = |\lambda_k| \, \|x_k - (1/\lambda_k)v\| \geq \frac{|\lambda_k|}{2}.$$

This holds for all k and since $|\lambda_k| \geq 1/n$, the vectors Tx_k are separated at least by a distance $1/2n$. Therefore there can be no convergent subsequence. ∎

Theorem V.4 Let T be a completely continuous operator on a Hilbert space. Let $\lambda \neq 0$ be an eigenvalue. The subspace V_λ of eigenvectors belonging to λ is finite-dimensional.

Proof. Suppose V_λ is infinite-dimensional. There is an infinite orthonormal set $\{u_1, u_2, \ldots\}$ in V_λ. Hence, for $i \neq j$,

$$\|Tu_i - Tu_j\|^2 = |\lambda|^2 \, \|u_i - u_j\|^2 = 2|\lambda|^2,$$

and there can be no convergent subsequence. Therefore T is not completely continuous. ∎

Theorem V.5 Let T be a completely continuous linear operator on a Hilbert space. If $\lambda \neq 0$ is an approximate eigenvalue, then $\lambda \in P_\sigma(T)$. Hence 0 is the only possible point in the continuous spectrum.

Proof. There exists a sequence (x_n) of unit vectors such that $\|Tx_n - \lambda x_n\| \to 0$. By the complete continuity, there is a convergent subsequence $Tx_{n_i} \to u$. Hence

$$\|u - \lambda x_{n_i}\| \leq \|u - Tx_{n_i}\| + \|Tx_{n_i} - \lambda x_{n_i}\| \to 0.$$

Therefore $Tu = \lambda \lim_i Tx_{n_i} = \lambda u$. Further, $\|u\| = \lim_i \|\lambda x_{n_i}\| = |\lambda| \neq 0$, so that u is an eigenvector and $\lambda \in P_\sigma(T)$. ∎

We summarize the preceding theorems as follows.

Corollary The spectrum of a completely continuous normal operator consists of the point spectrum, which is at most denumerable, and possibly 0 which may be in $P_\sigma(T)$ or $C_\sigma(T)$. There can be no other value in $C_\sigma(T)$ and the residual spectrum is empty. Further, if $P_\sigma(T) = \{\lambda_n\}$ is infinite, then $\lim_{n\to\infty} \lambda_n = 0$.

Theorem V.6 Let $\mu = \inf_{\|x\|=1} \langle Tx, x\rangle$ and $M = \sup_{\|x\|=1}\langle Tx, x\rangle$, where T is a bounded self-adjoint operator. Then μ and M are in $\sigma(T)$.

Proof. Without loss of generality we may assume $M \geq \mu \geq 0$, since we can always consider $T - \mu I$. Thus $M = \|T\|$ and there exists a sequence of unit vectors (x_n) such that $\langle Tx_n, x_n\rangle = M - \varepsilon_n$, where $\varepsilon_n \to 0$. Now,

$$\|Tx_n - Mx_n\|^2 = \|Tx_n\|^2 - 2M\langle Tx_n, x_n\rangle + M^2$$
$$\leq M^2 - 2M(M - \varepsilon_n) + M^2 = 2M\varepsilon_n.$$

Hence M is in the approximate spectrum of T (Definition V.1). Since a self-adjoint transformation is normal, by Theorem V.2 M is in the spectrum of T. A similar argument applies to μ. ∎

Corollary 1 A bounded self-adjoint operator has a nonempty spectrum.

Corollary 2 A completely continuous self-adjoint operator other than the zero operator has at least one eigenvalue.

Proof. If $T \neq 0$, then $\|T\| \neq 0$ and either μ or M is not zero. Hence μ or $M \in \sigma(T)$. By Theorem V.5, μ or $M \in P_\sigma(T)$. ∎

Theorem V.7 For a completely continuous self-adjoint operator T on a Hilbert space X the spectrum consists of the point spectrum; that is, $\sigma(T) = P_\sigma(T)$. Furthermore, the closed subspace generated by the eigenvectors is X.

Proof. Let E be the closed subspace generated by all eigenvectors of T. Consider the orthogonal complement $E^\perp$. If $y \in E^\perp$, for any eigenvector u, we have

$$\langle Ty, u\rangle = \langle y, Tu\rangle = \lambda\langle y, u\rangle = 0.$$

Hence $T: E^\perp \to E^\perp$. The restriction T_2 of T to $E^\perp$ is completely continuous and self-adjoint. If $E^\perp \neq \{0\}$, by Corollary 2 above, T_2 has an eigenvalue λ and hence a nonzero eigenvector v in $E^\perp$. But $T_2 v = \lambda v$ implies $Tv = \lambda v$ so that $v \in E$, which is impossible. Hence $X = E$ and $T_2 = 0$.

Since $T(E) \subset E$ also, we may always write $T = T_1 + T_2$, where T_1 is the restriction of T to E. It is easy to see that $\sigma(T) = \sigma(T_1) \cup \sigma(T_2)$. Clearly, $\sigma(T_2) \subset C_\sigma(T)$. Conversely, if $\lambda \in C_\sigma(T)$, then $\lambda \in \sigma(T_2)$. ∎

APPENDIX VI

PROOF OF FUNDAMENTAL THEOREM OF LINEAR PROGRAMMING

Proof. (We assume nondegeneracy.) Let x^* be an optimal feasible solution. Index the components so that $x_1, \ldots, x_k$ are all positive and $x_{k+1} = \cdots = x_n = 0$. Consider the submatrix consisting of the first k columns $A_1, \ldots, A_k$ of A and suppose it has rank r. Again reindex so that $A_1, \ldots, A_r$ are linearly independent. Now assume that x^* is not a basic solution. This implies that $k > m \geq r$. We then show how to construct an optimal feasible solution, having at most $k - 1$ nonzero components. This can be continued until $k \leq m$.

$$\sum_{i=1}^{k} x_i^* A_i = \sum_{i=1}^{r} x_i^* A_i + \sum_{r+1}^{k} x_i^* A_i = b.$$

By linear independence,

$$A_k = \sum_{i=1}^{r} \alpha_i A_i,$$

and for any quantity Δx_k,

$$\Delta x_k A_k = \sum_{i=1}^{r} \Delta x_k \alpha_i A_i.$$

Adding, we get

$$\sum_{i=1}^{r} (x_i^* - \Delta x_k \alpha_i) A_i + \sum_{r+1}^{k-1} x_i^* A_i + (x_k^* + \Delta x_k) A_k = b.$$

Let $y = (y_1, \ldots, y_n)$ be the vector given by

$$y_i = \begin{cases} x_i^* - \Delta x_k \alpha_i, & 1 \leq i \leq r, \\ x_i^*, & r < i < k, \\ x_k^* + \Delta x_k, & i = k, \\ 0, & i > k. \end{cases}$$

Then

$$J(y) = \sum_{1}^{k} c_i y_i = \sum_{1}^{k} c_i x_i^* - \sum_{1}^{r} c_i \Delta x_k \alpha_i + c_k \Delta x_k$$

$$= J(x^*) + \Delta x_k \left(c_k - \sum_{1}^{r} c_i \alpha_i \right).$$

If $d = c_k - \sum_{1}^{r} c_i \alpha_i < 0$, we could take

$$\Delta x_k = \max_{1 \leq i \leq r} (\{x_i^*/\alpha_i : \alpha_i < 0\} \cup \{-x_k\}).$$

Since $x_i^* > 0$ for $1 \leq i \leq k$, this would make $\Delta x_k < 0$ and y would be a feasible vector with $J(y) > J(x^*)$. This contradicts the optimality of x^*. Similarly, $d > 0$ leads to a contradiction. Hence we must have $d = 0$. Determining Δx_k in the same way, we obtain a feasible vector y with $J(y) = J(x^*)$ and having at most $k - 1$ nonzero components. This completes the proof. ∎

The Kuhn–Tucker necessary conditions

The functionals J and g_i, $1 \leq i \leq p$, are real functionals defined on an open subset $D \subset H$. $g = (g_1, \ldots, g_p)$ defines the inequality constraint $C = \{x : g(x) \leq 0\}$. x^* is a local minimum of J on C. Let $I = \{i : g_i(x^*) = 0\}$. We assume the *Kuhn–Tucker constraint qualification*, namely, for any $h \in H$ such that $\langle \nabla g_i(x^*), h \rangle \leq 0$ for $i \in I$ there exists a curve $\varphi(t) \in C$, $0 \leq t \leq 1$, with $\varphi(0) = x^*$ and $\varphi'(0) = \lambda h$ for some $\lambda > 0$. We shall prove that (12.9.3), (12.9.4), and (12.9.5) are necessary conditions that must be satisfied by x^* and some $\Lambda^* = (\lambda_1^*, \ldots, \lambda_p^*)$. To prepare the way for the proof, we establish the following basic result concerning linear inequalities.

Farkas' lemma

Let $b, v_1, \ldots, v_r$ be arbitrary vectors in a Hilbert space H and let C be the positive cone generated by the v_i. Then $b \in C$ (that is, $b = \sum_1^r \lambda_i v_i$ with $\lambda_i \geq 0$) if and only if the following condition holds:

$$\langle v_i, x \rangle \geq 0, \quad 1 \leq i \leq r, \quad \text{implies } \langle b, x \rangle \geq 0 \text{ for all } x \in H. \tag{*}$$

Proof. The implication is obvious in one direction, namely, $b \in C$ implies condition (*). That is, $b = \sum_1^r \lambda_i v_i$ with all $\lambda_i \geq 0$ and $\langle v_i, x \rangle \geq 0$ for $1 \leq i \leq r$ implies $\langle b, x \rangle \geq 0$.

To prove the converse, we shall show that $b \notin C$ implies that there exists $x_0 \in H$ such that

$$\langle v_i, x_0 \rangle \geq 0 \quad \text{for} \quad 1 \leq i \leq r \quad \text{and} \quad \langle b, x_0 \rangle < 0. \tag{1}$$

C is a closed convex set. Therefore there exists a unique closest point $y_0 \in C$ to b (Theorems 7.3.2, 7.3.3). We shall prove that the vector $x_0 = y_0 - b$ satisfies (1) above. Note that $b \notin C$ implies $x_0 \neq 0$.

For arbitrary $y \in C$, consider the line $y(t) = ty + (1 - t)y_0$. Again there is a point on this line which is closest to b. It is obtained by minimizing

$$\varphi(t) = \|y(t) - b\|^2 = \|y - y_0\|^2 t^2 + 2t \langle y - y_0, x_0 \rangle + \|y_0 - b\|^2.$$

The minimum of φ occurs at $t_m = -\langle y - y_0, x_0 \rangle / \|y - y_0\|^2$. Suppose that $t_m > 0$. Then $\varphi(t)$ is monotone decreasing in the interval $[0, t_m]$, and for some $t_1 \in (0, 1]$, we would have

$$\|y(t_1) - b\| = \varphi(t_1) < \varphi(0) = \|y_0 - b\|.$$

Since $y(t_1) \in C$, this is a contradiction. Therefore $t_m \leq 0$ and $\langle y - y_0, x_0 \rangle \geq 0$. We assert that $\langle y_0, x_0 \rangle = 0$. This is certainly true when $y_0 = 0$. If $y_0 \neq 0$, choose $y = 0$ in the preceding analysis and again minimize $\varphi(t)$, obtaining $t_m = \langle y_0, x_0 \rangle / \|y_0\|^2 \leq 0$. However, now $t < 0$ yields points in C and the closest, y_0, occurs for $t_m = 0$. Hence again $\langle y_0, x_0 \rangle = 0$. It follows that $\langle y, x_0 \rangle \geq 0$ for all $y \in C$, hence for $y = v_i$. Finally, $\langle b, x_0 \rangle = \langle y_0 - x_0, x_0 \rangle = -\|x_0\|^2 < 0$, establishing (1). ∎

Proof of Theorem 12.9.2. Since x^* is a local minimum, there exists a ball $B = B(x^*, r)$ such that $J(x^*) \leq J(x)$ for all $x \in B \cap C$. There are two cases.

Case (a). $I = \varnothing$ so that $g_i(x^*) < 0$ for all i. For any unit vector $h \in H$,

$$g_i(x^* + sh) = g_i(x^*) + s_i \langle \nabla g_i(x^*), h \rangle + \varepsilon_i,$$

where $\varepsilon_i/s_i \to 0$. Hence, for $s < \delta$ sufficiently small, $g_i(x^* + sh) < 0$ for all i. That is, $x^* + sh \in C$. Therefore

$$0 \leq J(x^* + sh) - J(x^*) = s(\langle \nabla J(x^*), h \rangle + \varepsilon/s).$$

Since $\varepsilon/s \to 0$ as $|s| \to 0$, it follows that $\langle \nabla J(x^*), h \rangle = 0$. Since h was arbitrary, $\nabla J(x^*) = 0$ and we may take $\Lambda^* = 0$ to satisfy the Kuhn–Tucker conditions.

Case (b). $I \neq \emptyset$. Let h be a vector such that $\langle \nabla g_i(x^*), h \rangle \leq 0$ for all $i \in I$. By the constraint qualification, there exists a function $\varphi(t)$ such that $\varphi(0) = x^*$, $\varphi'(0) = \lambda h$ for some $\lambda > 0$ and $\varphi(t) \in C$, $0 \leq t \leq 1$. Hence

$$\varphi(t) = \varphi(0) + t\varphi'(0) + \varepsilon = x^* + t\lambda h + \varepsilon,$$

so that for $t > 0$ sufficiently small, $\varphi(t) \in B \cap C$. It follows by the chain rule (Lemma 5.3.4) that

$$0 \leq J(\varphi(t)) - J(x^*) = t\langle \nabla J(x^*), \varphi'(0) \rangle + \varepsilon',$$

where $\varepsilon'/t \to 0$ as $t \to 0$. Therefore $\langle \nabla J(x^*), h \rangle \geq 0$. By Farkas' lemma applied to $\nabla J(x^*)$ and $-\nabla g_i(x^*)$, $i \in I$, there exist $\lambda_i^* \geq 0$ such that $\nabla J(x^*) = -\sum_{i \in I} \lambda_i^* \nabla g_i(x^*)$. For $i \notin I$, define $\lambda_i^* = 0$. The resulting vector $\Lambda^* \geq 0$ and satisfies the Kuhn–Tucker conditions. ∎

Remark Instead of the Kuhn–Tucker constraint qualification, assume that $\{\nabla g_i(x^*) : i \in I\}$ is a linearly independent set. Now, x^* must be a minimum of J in $B \cap C_0$, where $C_0 = \{x : g_i(x) = 0, i \in I\}$ is an equality constraint and B is a sufficiently small ball. Applying Theorem 12.2.2, we obtain $-\nabla J(x^*) = \sum_{i \in I} \lambda_i \nabla g_i(x^*)$. Again, for $i \notin I$, we set $\lambda_i = 0$. Now, if $\lambda_j < 0$ for some $j \in I$, then $-\nabla J(x^*)$ is not in the positive cone generated by $\{\nabla g_i(x^*) : i \in I\}$. By Farkas' lemma, there exists h such that $\langle \nabla g_i(x^*), h \rangle \geq 0$, $i \in I$, and $\langle -\nabla J(x^*), h \rangle < 0$. Let $K \subset I$ be the set of indices such that $\langle \nabla g_i(x^*), h \rangle = 0$, $i \in K$. Then h is in the tangent subspace of the constraint $C_k = \{x : g_i(x) = 0, i \in K\}$. As in the proof of Theorem 12.2.1, we obtain vectors

$$y_s = x^* - s\left(h + \sum_{i \in K} \frac{G_i(s)}{s} \nabla g_i(x^*)\right)$$

with $y_s \in C_k$ for $|s| < \delta$ sufficiently small and $G_i(s)/s \to 0$ as $s \to 0$. Hence, letting $v_s = \sum_{i \in K} (G_i(s)/s)\nabla g_i(x^*)$, we obtain

$$g_i(y_s) = g_i(x^*) - s(\langle \nabla g_i(x^*), h \rangle + \langle \nabla g_i(x^*), v_s \rangle + \varepsilon_i).$$

Since $g_i(x^*) = 0$, it follows that for $i \in I - K$ and $s > 0$ sufficiently small, we have $g_i(y_s) < 0$. Hence $y_s \in C$ for $s > 0$ sufficiently small. But

$$J(y_s) - J(x^*) = -s(\langle \nabla J(x^*), h \rangle + \langle \nabla J(x^*), v_s \rangle + \varepsilon).$$

Since ε and $v_s \to 0$ as $s \to 0$, this implies $J(y_s) < J(x^*)$, which is a contradiction. Hence all $\lambda_i \geq 0$.

INDEX

A-stability, 475, 477
Accumulation point, 5
Adams-Moulton method, 442, 443, 483
Adjoint equation, 498
Aitken's algorithm, 377
Aitken's Δ^2-method, 213, 214, 228
Angular-gradient method, 551
A-posteriori error estimate in Gauss elimination, 115
Approximation
  best (closest), 269
  best polynomial of degree n, 269
A-priori error estimate in Gauss elimination, 116
Ascoli-Arzela theorem, 19
Autonomous differential equation form, 451

Back substitution, 105, 108
Bairstow's method, 211
Ball, 8
  closed, 8
  open, 8
Banach algebra, 64
Banach space, 84
Basic degenerate solution, 572
Basic nondegenerate vector, 572
Basic solution, 572
Basic variables, 572
Basic vector, 572
Basis
  biorthogonal, 262
  canonical, 27, 29
  Hamel (algebraic), 27
  neighborhood, 6
  Schauder, 55
Bernoulli's method, 210, 224, 227
Bernstein, 266, 348
Bernstein polynomials, 267, 321, 325
Bessel function, 318, 328
Bessel inequality, 58
Best approximating polynomial (BAP), 280

Biorthonormal, 42
Bisection method, 249
Bolzano-Weierstrass theorem, 13
Boundary-conditions, 489
Boundary-value problem, 431, 489
Boundary-values, 489
Brachistochrone problem, 521, 563
Brouwer fixed-point theorem, 166
Bubnov-Galerkin method, 513

Cayley-Hamilton theorem, 90
Cesaro means, 328
Chain rule for differentiation, 183
Characteristic equation, 74, 230
Characteristic polynomial, 74, 436, 467
Chebyshev coefficients, 292
Chebyshev polynomial of the first kind, 291, 294, 348
Chebyshev polynomials of the second kind, 333
Chebyshev system, 290, 335, 337
Cholesky decomposition, 263
Chord method, 176
Closed-graph theorem, 100
Compact support, 95
Completely continuous operator, 600, 513
Component (connected), 18
Condition number of a matrix, 68, 88, 114, 127
  of an eigenvalue, 234
Cone, 46
  convex, 46
  translated, 557
Conjugate, of a normed space, 69
Conjugate direction, 543
  method, 542
Conjugate gradient method, 42
Conjugate space, 38
Consistent difference equation, 466
Constraint
  equality, 520, 523, 529, 546

605

Constraint *continued*
  inequality, 520, 523
Continued fraction, 311
Contractive map, 164
Control
  admissible, 521, 560, 565
  function, 521
  space, 564
  variable, 520
Convergence
  acceleration of, 213
  asymptotic rate of, 130
  average rate of, 130
  Fibonacci, 207
  linear, 170
  of difference method, 465
  of numerical solution, 440
  of operators, 84
  quadratic, 170
  rates of, 215
Convex
  function, 273
  hull, 62
  set, 62
  strictly, 273
Coordinate, 26
  functions, 565, 512
Cost
  factors, 576
  function, 520
Covering, 11
  open, 11
Cowell's method, 487
Critical point, 520

Dantzig, 520, 572
Deflation
  theorem, 239
  Wielandt's, 240
Derivative
  G-, 181   c
  of a product, 221
  strong (Frechet), 182
  strong (Frechet) second derivative, 191
  strong nth derivative, 191
  strong partial, 195
  weak (Gâteaux), 181
  weak partial, 194
  weak total, 194

Determinant, 34
Diagonally dominant, 136
  strictly, 130, 136
Diameter, of a set, 19
Difference
  central, 345, 354
  forward, 341, 342
  method, 507
  Newton backward, 344
  nth forward, 343
  operator, 341
  second forward difference, 341
Difference equation, 172
  homogeneous, 172
  operator, 172
Differential
  second-order, 189
  weak (Gâteaux), 179, 195
  weak partial, 194
Differential equation, 431
Differentiation
  numerical, 353, 373
Direct sum, 37
  external, 43
Directional derivative, 178, 179
  one-sided, 178
Dirichlet boundary-value problem, 514
Dirichlet integral, 305
Divided difference of order k, 339
Dual space, 38
Dual variables, 531

Eigenfunction, 310
Eigenvalues, 73, 74, 230, 310
  approximate, 600
Eigenvector, 74, 230
Elementary divisors, 92
Equations
  linear, 22
  normal, 45, 277
Equicontinuous, 19
Error
  accumulated rounding, 448, 454
  local truncation, 454, 466
  total, 448
Euler predictor-corrector, 442
Euler's method, 442, 450, 467, 477, 535
  modified, 483
Everett's central difference form, 345

Explicit method, 441, 477
Exterior point algorithm, 583
Extremum, 520
  local, 520
  relative, 520

Faber, 348
False position, method of, 176
Feasible direction method, 582
Feasible optimal solution, 572
Feasible solution, 572
Feasible vector, 572
Filter base, 6
First-order difference operator, 174
First-order method, 170
First-order system of differential equation, 431
Fixed point, 164
Floating point, 111
Form
  bilinear, 45
  Hermitian, 87
  homogeneous of degree n, 190
  quadratic, 45
Fourier generalized coefficients, 294
Fourier sum, 294
Fourier-Chebyshev coefficients, 295, 304
Fourier-Chebyshev expansion, 295
Fredholm alternative, 590
Frobenius, 141
Function (mapping), 2
  bijective, 3
  bilinear, 38, 39
  Borel measurable, 17
  bounded, 8
  composite, 3
  concave, 577
  continuous, 7
  convex, 577
  cost, 520
  domain, 29
  essential supremum, 17
  extension, 3
  homeomorphism, 8
  homomorphism, 26
  identity, 3
  image, 3
  injective, 3
  inverse, 3

Function *continued*
  inverse image (pre-image), 3
  isometry, 14
  isomorphism, 26
  Lipschitz, 11
  lower semicontinuous, 16
  measurable, 17
  metric (distance), 8
  modulus of continuity, 10
  multilinear, 39
  objective, 321, 520
  partial, 3
  piecewise continuous, 303
  projection, 3
  quadratic, 190
  range, 29
  rational, 311, 335
  restriction, 3
  simple, 17
  surjective, 3
  symmetric, 40
  truncated power, 359
  uniformly continuous, 10
  upper semicontinuous, 16
Functional
  bounded, 15
  quadratic, 527
  real, 15

Galerkin's method, 590, 513
Galois, 230
Gauss elimination, 104, 35
Gauss forward form, 345
Gauss-Jordan elimination, 573
Gauss-Seidel iteration, 135
Generalized polynomial, 290
Gerschgorin discs, 232
Gerschgorin's theorem, 231
Given's method, 247
Gradient
  method, 534
  weak, 523
Gram matrix, 277
Gram polynomials, 297, 332
Gram-Schmidt process, 57, 295
  modified, 300
Graph
  closed-graph theorem, 74, 100
  of an operator, 73

Green's function, 511
Gregory's formula, 374, 467
Grid, 355, 432
    rectangular, 338

Haar condition, 290
Hadamard's theorem, 137, 231
Hahn-Banach theorem, 592
Heine-Borel theorem, 12
Hessian matrix, 189
Heun method, 483
Hölder inequality, 88
Horner's form, 319
Householder's method, 245
Hybrid method, 444
Hypergeometric equation, 309, 310

Ill-condition
    eigenvalue problem, 233
    matrix, 114
    of degree n, 234
Implicit-function theorem, 596
Implicit method, 441
Initial-value problem, 122, 431, 432
Inner-product, 55
    complex, 86
    norm, 55
    space, 55
Integral
    equation, 433, 434
    Lebesgue, 17
    Riemann, 188, 221
Interior point, 5
Interpolation
    bivariate, 376
    general linear problem, 335
    Hermite-Birkhoff, 373
    inverse, 360
    iterated inverse, 360
    polynomial, 336
    simple Hermite, 336, 349, 372
Interval
    closed, 1
    half-open, 1
    open, 1
    open in R^n, 7
Inverse
    computation, 109
    iteration, 250, 255

Inverse *continued*
    matrix, 30
Isolated point, 5

Jackson's theorem, 306
Jacobi cyclic method, 243
Jacobi iteration, 129
Jacobi matrix, 139
Jacobi method, 240
Jacobi polynomials, 294, 349
Jacobi theorem, 241
Jacobian, 181, 183, 196
Jensen's inequality, 94
Jordan arc, 18
Jordan matrix, 91, 92, 233

Kernel
    integral, 558
    of a seminorm, 269
Knot, 359
Kronecker delta, 30
Kuhn-Tucker conditions, 579, 603
Kuhn-Tucker constraint qualification, 580, 603

Lagrange coefficients, 281, 338, 339
Lagrange interpolation formula, 281
Lagrange multiplier rule, 524
Lagrange multipliers, 531
Lagrangian function, 531
Laplace operator, 514
Least squares
    approximation, 268
    discrete approximation, 328
    polynomial, 269, 297, 355
    problem, 278
    solution, 278
Lebesgue measure, 17
Legendre equation, 311
Legendre polynomials, 294
Linear part of f, 175
Linear programming, 321
Linear operator
    adjoint, 40, 71
    bounded, 52
    continuous, 62
    invariant, 33
    inverse, 30

Linear operator *continued*
  K-homomorphism, 31
  nth-order difference, 436
  nonsingular, 30
  normal, 72
  orthogonal, 72
  rank, 31
  singular, 30
  symmetric, 79
  unitary, 72
Linearization, 175
Lipschitz class, 469
Lipschitz condition, 11
LR algorithm, 251

Mantissa, 110
Matrix, 22
  addition, 24
  algebra, 25
  augmented, 23
  band, 131, 252
  block tridiagonal, 147
  companion, 97
  diagonal, 34
  diagonalizable, 92, 235
  elementary Hermitian unitary, 245
  elementary row (column), 34
  equivalent, 34
  generalized inverse, 44
  Hermitian, 72
  idempotent, 44
  identity, 30
  inverse, 30
  irreducible, 136, 137
  Jordan, 91, 92
  multiplier, 106
  nilpotent, 41
  nonsingular, 30
  normal, 92, 349
  orthogonal, 72
  positive definite, 76, 145
  positive semi-definite, 76
  product, 23
  pseudoinverse, 44, 257, 279, 334
  rank, 30
  reducible, 137
  regular, 44
  similar, 34
  singular, 30

Matrix *continued*
  skew-symmetric, 93
  square, 22
  strictly normal, 349
  symmetric, 72
  transpose, 23
  triangular, 35, 106, 348
  unitary, 72, 87
  upper Hessenberg, 245, 250
Maximum
  global, 520
  local, 523
Mayer problem, 521
Mean-square error, 280, 294
  weighted, 295
Mesh, 290, 355, 432
Metric space, 8
  complete, 13
  discrete, 9
  equivalent, 11
  measurable, 16
  sequentially compact, 13
  totally bounded, 13
Minimax of degree n, 280
Minimum, 520
  global, 273
  local, 523
  polynomial, 98
  strong, 528
  weak, 528
Moments, 328
Monic polynomial, 292
Muller's method, 209, 363
Multiplicity of eigenvalue, 75
Multiplier in Gauss elimination, 104, 106, 108
Multistep method, 443

n-parameter
  family, 335, 338
  linear family, 335, 338
Natural homomorphism, 269
Neighborhood, 5
  base, 6
  filter, 20
Neville formula, 361
Newton coefficients, 339
Newton interpolating polynomial, 340
Newton (-Raphson) iterates, 175

Newton (-Raphson) method, 170, 174, 183, 195
Nonexpansive mapping, 166
  strictly, 166
Nonsingular element, 64
Norm, 50
  equivalent, 50, 52
  l_∞, 67, 112
  l_1, 67, 121
  l_p, 50
  l_2, 67, 68
  of an operator, 63
  spectral, 67
  strict, 55
  uniform, 50
Normal point, 529
Nullity, 30
Numerical solution of differential equations, 432
  of accuracy ε, 432, 433
  piecewise linear, 433
Nystrom, 463, 487

Open formula, 441
Operation count in Gauss elimination, 108
Operator
  algebra, 30
  backward difference, 343
  closed, 73
  completely continuous, 600
  difference, 172, 341
  elliptic, 514
  forward difference, 341
  integral, 64
  linear, 26, 29
  nilpotent, 79
  projection, 62
  quasinilpotent, 78, 124, 134, 136
  scalar product, 29
  shift, 343
  sum, 29
Optimal control problem, 521
Order-of-difference method, 443
Orthogonal polynomials, 295
Orthogonal projection, 62
Orthonormal set, 56
  complete, 59
Overrelaxation
  method, 146
  successive line, 515

Parallelogram law, 95
Parseval identity, 60
Penalty, function method, 582
Permutation matrix, 129, 108, 115, 147
Perron, 141
Perturbation, 133, 464, 465
Pivot
  in Gauss elimination, 105
  partial, 106, 108
Polynomial
  of degree n, 190
  factorial, 342
  interpolating, 335
Power method, 238
Precision, 110
  double, 110, 122
  single, 110
Predictor-corrector method, 441
  generalized, 444, 474
  Euler's method, 483
Principal minor, 36, 43, 76
Principal vector of grade r, 90
Programming
  linear, 520, 571, 602
  mathematical, 520
  nonlinear, 577
Pseudointerior point, 285
Pseudoinverse, 44

Q-factors, 215
Q-linear, 216
Q-order, 216
Q-quadratic, 216
Q-sublinear, 216
Q-superlinear, 216
QR algorithm, 251
Quadrature local error, 443
Quasiregular neighborhood, 547
Quasiregular point, 547
Quasiregular simple point, 551
Quotient convergence factors, 215
Quotient difference algorithm, 227

R-factors, 217
R-linear, 217
R-quadratic, 217
R-superlinear, 217
r-step iterative method, 125
Rayleigh quotient, 589

Rayleigh-Ritz method, 565
Regula falsi, method of, 176, 200
Regular neighborhood, 559
Regular stationary point, 559
Relative error, 114
Relaxation factor, 146
Remes multiple-exchange algorithm, 289
Remes single-exchange algorithm, 287
Reorthogonalization, 300
Residual iteration, 123
   of a function, 280
Residual vector, 115
Resolvent set, 74
   of operator, 77
Riesz representation theorem, 71
Ritz method, 565
Root condition, 465
Root convergence factors, 217
Rounding error
   successive approximation, 171
   sum of products, 110
Runge-Kutta methods, 348, 352, 358, 450, 451, 454, 456, 458, 478
   generalized, 444, 474
   pseudo-, 463

Saddle-point, 533
Scaling, 112
Schwarz inequality, 55, 86
Secant method, 208
Second-order method, 170
Segment line, 62
Self-adjoint, 143
Semigroup, 43
Seminorm, 50
Sequence, 3
   Cauchy, 13
   convergent, 6
   limit, 6
   minimizing, 274, 534
   subsequence, 3
Series
   absolutely convergent, 54
   convergent, 54
Sets
   boundary, 6
   bounded, 2, 8
   cardinality, 3
   Cartesian product, 1

Sets *continued*
   closed, 4
   closure, 5
   compact, 12
   complement, 1
   convex, 62
   countable, 3
   dense, 6
   difference, 1
   disjoint, 1
   distance, 10
   finite, 3
   infimum, 2
   interior, 5
   intersection, 1
   measurable, 16
   nondense, 6
   open, 4, 9
   partially ordered, 2
   supremum, 2
   totally ordered, 2
   union, 1
Shooting
   method, 503
   parallel-method, 507
Simplex method, 322, 520, 572
Simplex multipliers, 576
Single-step method, 444, 450
Singular values, 81, 257
Singular vectors, 257
Slack variable, 571
Space
   Banach, 54
   compact, 12
   complete, 14
   connected, 18
   factor, 269
   Hilbert, 59
   locally compact, 15
   measure, 17
   metric, 8
   normed vector, 50
   null, 30
   pre-Hilbert, 55
   reflexive, 70
   separable, 11
   sequentially compact, 13
   topological, 4
   topological vector, 49

Space *continued*
  unitary, 87
Spectrum, 73, 74
  continuous, 74
  point, 74
  residual, 74
  spectral norm, 67, 81
  spectral radius, 77, 141
Spline
  cubic, 355
  function of degree m, 359
  natural, 359
Stability, 464
  asymptotic, 480
  successive approximations, 172
Stable solution, 132, 133, 496
  asymptotic, 498
  difference equation, 174
  iterative method, 133
State
  equations, 521
  variable, 520
Stationary method, 215
  point, 250, 547
  regular-point, 559
Steepest-descent method, 535
Steffensen's method, 214, 225
Step size, 341
Stiff equations, 475, 477
Stone-Weierstrass theorem, 320, 594
Stormer's method, 488
Sturm sequence, 209, 224
Sturm-Louville problem, 309
Sturm's theorem, 209, 248
Subspace
  closed, 54
  generated, 54
  tangent, 523
Successive approximation, 164
Successive displacements, method of, 146
Surface level, 524, 531

Tangent
  manifold, 523
  subspace, 523, 529
Threshhold (Jacobi) method, 243
Topology, 4
  coarse, 4

Topology *continued*
  discrete, 4
  metriable, 11
  metric space, 9
  order, 4
  product, 7
  relative, 4
  separated (Hausdorff), 6, 11
  strong, 86, 100
  uniform, 100
Trace, of a matrix, 90
Trajectory, 521
Translation invariant metric, 49
Trapezoidal rule, 477
Triangle inequality, 55
Trigonometric sum, 268, 363
  best-sum, 291

Ultra-spherical polynomials, 294
Underrelaxation method, 146
Uniform approximation, 266
Uniform boundedness principle, 85
Uniform convexity, 271, 325
Unisolvent, 337

Vallee-Poussin, 282
Vandermonde's determinant, 282, 291, 327
Variable metric method, 542
Variational equation, 480, 561, 491
Vector
  column, 23
  linearly dependent, 27
  linearly independent, 27
  orthogonal, 56
  row, 23
  scalar multiplication, 24
  space, 26
  sum, 24

Weakly stable method, 474
Weierstrass theorem, 266
Weight function, 293
Well-conditioned matrix, 121
Wielandt-Hoffman theorem, 244, 262
Wilkinson, 247, 250
Wronskian, 437

Young's property, 147